INTERNATIONAL ASTRONOMICAL UNION

UNION ASTRONOMIQUE INTERNATIONALE

STRUCTURE AND DYNAMICS OF ELLIPTICAL GALAXIES

PROCEEDINGS OF THE 127TH SYMPOSIUM OF THE INTERNATIONAL ASTRONOMICAL UNION HELD IN PRINCETON, U.S.A., MAY 27-31, 1986

EDITED BY

TIM DE ZEEUW
The Institute for Advanced Study, Princeton, U.S.A.

D. REIDEL PUBLISHING COMPANY

A MEMBER OF THE KLUWER  ACADEMIC PUBLISHERS GROUP

DORDRECHT / BOSTON / LANCASTER / TOKYO

Library of Congress Cataloging in Publication Data

International Astronomical Union. Symposium (127th: 1986: Princeton, N.J.)
Structure and dynamics of elliptical galaxies: proceedings of the 127th Symposium of the International Astronomical Union, held at the Institute for Advanced Study in Princeton, USA, May 27–31, 1986 / edited by Tim de Zeeuw.
p. cm.
Includes bibliographies and indexes.
ISBN 90–277–2585–3 ISBN 90–277–2586–1 (pbk.)
1. Elliptical galaxies—Congresses. I. Zeeuw, Pieter Timotheus de, 1956–
II. Title.
QB858.4.I58 1986
523.1′12—dc 19

87–18791
CIP

Published on behalf of
the International Astronomical Union
by
D. Reidel Publishing Company, P.O. Box 17, 3300 AA Dordrecht, Holland

Sold and distributed in the U.S.A. and Canada
by Kluwer Academic Publishers,
101 Philip Drive, Assinippi Park, Norwell, MA 02061, U.S.A.

In all other countries, sold and distributed
by Kluwer Academic Publishers Group,
P.O. Box 322, 3300 AH Dordrecht, Holland

Printed in The Netherlands

STRUCTURE AND DYNAMICS OF ELLIPTICAL GALAXIES

TABLE OF CONTENTS

POSTERS

APPENDICES

INDICES

PREFACE

IAU Symposium 127 was held in Princeton on May 28–31, 1986, at the Institute for Advanced Study. There were 150 participants from 19 countries. This was the first IAU Symposium devoted exclusively to elliptical galaxies.

The last decade has been a period of exceptionally rapid progress regarding our understanding of elliptical galaxies, driven on the observational side by a wealth of new photometric and spectroscopic data, and on the theoretical side by the recognition that ellipticals are slowly rotating triaxial systems, instead of rotationally flattened bodies. The 30 invited speakers reviewed all aspects of this progress, its consequences, and the major outstanding problems. Nearly 80 poster contributions were displayed for the duration of the meeting. These contained many of the most recent developments.

The first part of these Proceedings contains the written versions of the invited reviews and the summary, in the order of their presentation. The accompanying discussions are based on the written versions of questions and answers as handed in by the participants. The second part of this book is devoted to the poster contributions. They have been grouped in a thematic order, proceeding from morphology to theory. Indices of objects, subjects, and names are provided.

Unfortunately, none of the invited participants from the USSR were able to attend the Symposium. The Proceedings include the written versions of the planned contributions by Popov and Polyachenko.

The appendices contain English translations of three important Russian theoretical papers. In addition to Antonov's famous paper on the stability of spherical stellar systems, for which several unpublished translations exist, we have included a short paper by Antonov on the radial orbit instability which has attracted much attention recently (see the reviews by Polyachenko and Merritt), and a prescient paper by Kuzmin on triaxial models with three exact isolating integrals of motion. The latter two papers were published in 1973, in the same conference proceedings, and were translated by Alex Sphrintsen at the Canadian Institute for Theoretical Astrophysics in Toronto, with considerable help from David Merritt.

Following recent practice in IAU Symposia Proceedings (see, e.g., IAU 106), the inevitable blank pages that result from the sensible requirement that each paper should start on a right hand page, have been used for photographs. With the exception of the one of comet P/Halley and Centaurus A, they were all taken by Joshua Barnes.

The symposium dinner included a surprise party for Martin Schwarzschild's 74th birthday. All of Martin's former students and postdocs had been invited to participate in "Operation Black–Shield". Many of them attended in person, while others sent their congratulations by letter. S. Chandrasekhar gave a delightful after–dinner speech.

The local organization was in the very able hands of Stefano Casertano and Mary Wisnowsky, with help from Sarah Johns, Page Hartwell and Michelle Sage. Their untiring efforts resulted in a very smoothly run Symposium. The graduate students at Princeton University Observatory helped, for the third IAU Symposium in three years, with the audio visual equipment. Herwig Dejonghe efficiently edited the book of abstracts, and helped organize the poster session. He was assisted by Barbara Pinkham, who typed all the abstracts. She and Margaret Best helped with the editing of the Proceedings.

It is a pleasure to thank John Bahcall and Piet Hut, and the members of the Scientific Organizing Committee, in particular James Binney, for their help with the preparation of the Symposium. Financial support from IAU, in the form of travel grants for about 25 young participants from abroad, is gratefully acknowledged.

Finally, the editor wishes to record his gratitude to the staff of the Institute for Advanced Study, and in particular to the director Harry Woolf, for their hospitality and support of the Symposium and the subsequent preparation of these Proceedings.

It is a general rule that the editor of a Symposium Proceedings inevitably underestimates the amount of work involved. The present editor proved no exception. The publication of the Proceedings was delayed further by illness and a very exciting World Series.

Tim de Zeeuw

Scientific Organizing Committee:
T.S. van Albada (The Netherlands), K.H. Barkhatova (USSR), J.J. Binney (UK), M. Capaccioli (Italy), K.C. Freeman (Australia), G.D. Illingworth (USA), L. Martinet (Switzerland), M. Milgrom (Israel), G. Monnet (France), D.O. Richstone (USA) F. Schweizer (USA), P.T. de Zeeuw (USA, Chair).

Local Organizing Committee:
S. Casertano (Chair), H. Dejonghe, P. Hartwell, S. Johns, B. Pinkham, M. Sage, M. Wisnowksy, P.T. de Zeeuw.

Sponsoring IAU Commissions: 28 & 33.

LIST OF PARTICIPANTS

Aguilar, L.A., Center for Astrophysics, Cambridge, MA, U.S.A.
van *Albada, T.S.*, Kapteyn Laboratorium, Groningen, The Netherlands.
Bacon, R., Observatoire de Lyon, Saint-Genis-Laval, France.
Bahcall, J.N., Institute for Advanced Study, Princeton, NJ, U.S.A.
Bahcall, N.A., Space Telescope Science Institute, Baltimore, MD, U.S.A.
Barnes, J.E., Institute for Advanced Study, Princeton, NJ, U.S.A.
Bertin, G., Scuola Normale Superiore, Pisa, Italy.
Bertola, F., Osservatorio Astronomico, Padova, Italy.
Binks, P.N., Dept. of Theoretical Physics, Oxford University, United Kingdom.
Binney, J.J., Dept. of Theoretical Physics, Oxford University, United Kingdom.
Bishop, J.L., CITA, University of Toronto, Toronto, ON, Canada.
Bland, J., Institute for Astronomy, University of Hawaii, Honolulu, HI, U.S.A.
Burstein, D., Physics Dept., Arizona State University, Tempe, AZ, U.S.A.
Caldwell, N., Fred Lawrence Whipple Obs., Tucson, AZ, U.S.A.
Canizares, C.R., Dept. of Physics, MIT, Cambridge, MA, U.S.A.
Capaccioli, M., Osservatorio Astronomico, Padova, Italy.
Carlberg, R.G., Dept. of Physics, York University, Toronto, ON, Canada.
Casertano, S., Institute for Advanced Study, Princeton, NJ, U.S.A.
Chandrasekhar, S., Lab. for Astroph. and Space Research, Chicago, IL, U.S.A.
Christiansen, W.A., Univ. of North Carolina, Chapel Hill, NC, U.S.A.
Clark, G., Dept. of Physics, MIT, Cambridge, MA, U.S.A.
Cohn, H., Dept. of Astronomy, Indiana University, Bloomington, IN, U.S.A.
Combes, F., Observatoire de Meudon, Meudon, France.
Crane, P., European Southern Observatory, Garching, West Germany
Davies, R.L., Kitt Peak National Observatory, Tucson, AZ, U.S.A.
Dejonghe, H., Institute for Advanced Study, Princeton, NJ, U.S.A.
van *Dishoeck, E.F.*, Center for Astrophysics, Cambridge, MA, U.S.A.
Djorgovski, S., Center for Astrophysics, Cambridge, MA, U.S.A.
Dressel, L., Dept. of Space Phys. and Astronomy, Rice Univ., Houston, TX, U.S.A.
Duncan, M.J., Physical Science, Univ. of Toronto, Scarborough, ON, Canada
Durisen, R.H., Dept. of Astronomy, Indiana University, Bloomington, IN, U.S.A.
Durodie, W.J.L.V., Astronomy Dept., University of Manchester, United Kingdom.
Duschl, W.J., Institute of Astronomy, Cambridge, United Kingdom.
Dyson, F.J., Institute for Advanced Study, Princeton, NJ, U.S.A.
Ebneter, K., Astronomy Dept., University of California, Berkeley, CA, U.S.A.
Efstathiou, G., Institute of Astronomy, Cambridge, United Kingdom.
Evrard, A.E., SUNY, Astronomy Program, Stony Brook, NY, U.S.A.
Fabbiano, G., Center for Astrophysics, Cambridge, MA, U.S.A.
Fabian, A.C., Institute of Astronomy, Cambridge, United Kingdom.
Fall, S.M., Space Telescope Science Institute, Baltimore, MD, U.S.A.
Ford, H.C., Space Telescope Science Institute, Baltimore, MD, U.S.A.
Franx, M., Sterrewacht, Leiden, The Netherlands.
Freedman, W.L., Mt. Wilson and Las Campanas Obs., Pasadena, CA, U.S.A.
Freeman, K.C., Mt. Stromlo Observatory, Canberra, ACT, Australia.
Gerhard, O.E., MPI für Astrophysik, Garching, West Germany
Gilmore, G., Institute of Astronomy, Cambridge, United Kingdom.
van *Gorkom, J.H.*, Princeton University Observatory, Princeton, NJ, U.S.A.
Gregg, M., Mt. Wilson and Las Campanas Obs., Pasadena, CA, U.S.A.
Gunn, J.E., Princeton University Observatory, Princeton, NJ, U.S.A.

Hamabe, M., Dominion Astrophysical Observatory, Victoria, BC, Canada.
Hayes, J.J.E., Dept. of Phys. and Astron., Rutgers University, NJ, U.S.A.
Heggie, D.C., Dept. of Math., University of Edinburgh, United Kingdom.
Hernquist, L., Astronomy Dept., University of California, Berkeley, CA, U.S.A.
Hunter, C., Dept. of Math., Florida State University, Tallahassee, FL, U.S.A.
Hut, P., Institute for Advanced Study, Princeton, NJ, U.S.A.
Illingworth, G.D., Space Telescope Science Institute, Baltimore, MD, U.S.A.
Inagaki, S., Dept. of Astronomy, University of Kyoto, Kyoto 606, Japan.
Jaffe, W., Sterrewacht, Leiden, The Netherlands.
James, R.A., Astronomy Dept., University of Manchester, United Kingdom.
Jarvis, B.J., Observatoire de Genève, Sauverny, Switzerland.
Jedrzejewski, R.I., Mt. Wilson and Las Campanas Obs., Pasadena, CA, U.S.A.
Jesus-Gonzalez, J., Lick Observatory, Santa Cruz, CA, U.S.A.
Kashlinsky, A., NRAO, Charlottesville, VA, U.S.A.
Kent, S.M., Center for Astrophysics, Cambridge, MA, U.S.A.
King, I.R., Astronomy Dept., University of California, Berkeley, CA, U.S.A.
Knapp, G.R., Princeton University Observatory, Princeton, NJ, U.S.A.
Kochhar, R.K., Indian Institute of Astrophysics, Bangalore, India.
Kormendy, J., Dominion Astrophysical Observatory, Victoria, BC, Canada.
Kunze, R., Universitätssternwarte Göttingen, West Germany.
Lake, G., Astronomy Dept., University of Washington, Seattle, WA, U.S.A.
Lauer, T.R., Princeton University Observatory, Princeton, NJ, U.S.A.
Levison, H.F., Dept. of Astronomy, University of Michigan, Ann Arbor, MI, U.S.A.
Longo, G., Osservatorio Astronomico di Capodimonte, Napoli, Italy.
Lugger, P.M., Dept. of Astronomy, Indiana University, Bloomington, U.S.A.
Lupton, R., Space Telescope Science Institute, Baltimore, MD, U.S.A.
Luwel, M., Vrije Universiteit Brussel, Brussels, Belgium.
May, A., Dept. of Theoretical Physics, Oxford University, United Kingdom.
Merritt, D.R., CITA, University of Toronto, Toronto, ON, Canada.
Meylan, G., Astronomy Dept., University of California, Berkeley, CA, U.S.A.
Milgrom, M., Weizmann Institute of Science, Rehovot, Israel.
Möllenhof, C., Landessternw. Heidelberg, Königstuhl, Heidelberg, West Germany.
Mould, J.R., Laboratory of Astrophysics, Caltech, Pasadena, CA, U.S.A.
Nicholson, R.A., Royal Greenwich Observatory, Hailsham, United Kingdom.
Nieto, J.-L., Observatoire du Pic du Midi et de Toulouse, Toulouse, France.
Norman, C.A., Space Telescope Science Institute, Baltimore, MD, U.S.A.
O'Connell, R.W., Leander McCormick Observatory, Charlottesville, VA, U.S.A.
del Olmo Orozco, A., Instituto de Astrofisica de Andalucia, Granada, Spain.
Ostriker, J.P., Princeton University Observatory, Princeton, NJ, U.S.A.
Paczynski, B., Princeton University Observatory, Princeton, NJ, U.S.A.
Palmer, P.L., Queen Mary College, London, United Kingdom.
Panek, M., Copernicus Astronomical Center, Warszawa, Poland.
Peletier, R.F., Kapteyn Laboratorium, Groningen, The Netherlands.
Pence, W.D., Space Telescope Science Institute, Baltimore, MD, U.S.A.
Perea Duarte, J.D., Instituto de Astrofisica de Andalucia, Granada, Spain.
Petrou, M., Dept. of Theoretical Physics, Oxford University, United Kingdom.
Pfenniger, D., Observatoire de Genève, Sauverny, Switzerland.
Pickles, A.J., Obs. del Roque de los Muchachos, Santa Cruz de la Palma, Spain.
Porter, A.C., Caltech, Pasadena, CA, U.S.A.
Prugniel, Ph., Observatoire du Pic du Midi et de Toulouse, Toulouse, France.
Quinn, P.J., Space Telescope Science Institute, Baltimore, MD, U.S.A.

Ratcliff, S.J., Middlebury College, Middlebury, VT, U.S.A.
Recillas-Cruz, E., Instituto de Astronomia, UNAM, Mexico City, Mexico.
Richstone, D.O., Dept. of Astronomy, Univ. of Michigan, Ann Arbor, MI, U.S.A.
Rohlfs, K., Astronom. Inst., Ruhr University, Bochum, West Germany.
Rood, H.J., Institute for Advanced Study, Princeton, NJ, U.S.A.
Rybicki, G.B., Center for Astrophysics, Cambridge, MA, U.S.A.
Sadler, E.M., Kitt Peak National Observatory, Tucson, AZ, U.S.A.
Salucci, P., Scuola Internazionale Superiore di Studi Avanzati, Trieste, Italy.
Sansom, A., Astronomy Centre, University of Sussex, Brighton, United Kingdom.
Sarazin, C.L., JILA, University of Colorado, Boulder, CO, U.S.A.
Schechter, P.L., Mt. Wilson and Las Campanas Obs., Pasadena, CA, U.S.A.
Schneider, D.P., The Institute for Advanced Study, Princeton, NJ, U.S.A.
Schombert, J.M., Caltech, Pasadena, CA, U.S.A.
Schwarzschild, M., Princeton University Observatory, Princeton, NJ, U.S.A.
Schweizer, F., Dept. of Terrestrial Magnetism, Washington, DC, U.S.A.
Seitzer, P., National Optical Astronomy Observatory, Tucson, AZ, U.S.A.
Sellwood, J.A., Astronomy Dept., University of Manchester, United Kingdom.
Sérsic, J.L., Observatorio Astronomico Nacional, Córdoba, Argentina.
Simien, F., Observatoire de Lyon, Saint-Genis-Laval, France.
Sparke, L.S., Kapteyn Laboratorium, Groningen, The Netherlands.
Sparks, W.B., Royal Greenwich Observatory, Hailsham, United Kingdom.
Spergel, D.N., Institute for Advanced Study, Princeton, NJ, U.S.A.
Spitzer, L., Jr., Princeton University Observatory, Princeton, NJ, U.S.A.
Statler, T.S., Princeton University Observatory, Princeton, NJ, U.S.A.
Stauffer, J.R., Dominion Astrophysical Observatory, Victoria, BC, Canada.
Steiman-Cameron, T.Y., JILA, University of Colorado, Boulder, CO, U.S.A.
Stiavelli, M., Scuola Normale Superiore, Pisa, Italy.
Sussman, G.J., MIT, Cambridge, MA, U.S.A.
Teuben, P., Kapteyn Laboratorium, Groningen, The Netherlands.
Tonry, J.L., Dept. of Physics, MIT, Cambridge, MA, U.S.A.
Toomre, A., Dept. of Mathematics, MIT, Cambridge, MA, U.S.A.
Tremaine, S.D., CITA, University of Toronto, Toronto, ON, Canada.
Vader, J.P., Dept. of Astronomy, Yale University, New Haven, CT, U.S.A.
Valentijn, E.A., Kapteyn Laboratorium, Groningen, The Netherlands.
Vandervoort, P.O., Astron. and Astroph. Center, University of Chicago, IL, U.S.A.
Västerberg, A., Stockholm Observatory, Saltsjöbaden, Sweden.
de Vaucouleurs, G., Dept. of Astronomy, Univ. of Texas, Austin, TX, U.S.A.
Vietri, M., Osservatorio Astrofisico di Arcetri, Firenze, Italy.
Villumsen, J.V., Theoretical Astrophysics, Caltech, Pasadena, CA, U.S.A.
Wagner, S.J., Landessternw. Heidelberg, Königstuhl, Heidelberg, West Germany.
White, S.D.M., Steward Observatory, Tucson, AZ, U.S.A.
Whitmore, B.C., Space Telescope Science Institute, Baltimore, MD, U.S.A.
Wilkinson, A., Astronomy Dept., University of Manchester, United Kingdom.
Williams, T.B., Dept. of Phys. and Astron., Rutgers University, NJ, U.S.A.
Wisdom, J., MIT, Cambridge, MA, U.S.A.
van Woerden, H., Kapteyn Laboratorium, Groningen, The Netherlands.
Wyn-Evans, N., Institute of Astronomy, Cambridge, United Kingdom
Wyse, R.F.G., Space Telescope Science Institute, Baltimore, MD, U.S.A.
de Zeeuw, P.T., Institute for Advanced Study, Princeton, NJ, U.S.A.
Zhou, Hong-nan, Dept. of Astronomy, Nanjing Univ., People's Republic of China

INVITED REVIEWS

G. de Vaucouleurs presenting the introductory review.

GENERAL HISTORICAL INTRODUCTION

Gérard de Vaucouleurs
Department of Astronomy
University of Texas
Austin, TX 78712

ABSTRACT. A brief historical review of the discovery and exploration of elliptical galaxies in the past two centuries is presented.

The organizers of this symposium have asked me to review the history of research on the nature and composition of elliptical galaxies as an introduction to and a background for current and future studies. For the reasons of space and perspective this review will not include the research of the past ten years or so, which other communications are bound to cover in greater detail.

1. PRE–HISTORY: FROM 1749 TO 1924

The brighter companion of the Andromeda nebula was discovered by Le Gentil on October 29, 1749. He estimated it to be 1′ in diameter. Charles Messier first saw it in 1757, judged it to be 2′ in diameter in 1764, and included it as No. 32 in his famous catalogue (Messier 1771, 1781). He also published the first drawing of it (Messier 1807), a fine engraving of M31 and its two companions, the second of which had been discovered by Caroline Herschel in 1783 and included by William Herschel as Nr V.18 in his first catalogue (Herschel 1786).

The designation "elliptic" was first applied to non–spiral nebulae by Stephen Alexander (1852). His discussion does not explicitly describe them as stellar systems, but includes a fine description of their various degrees of concentration and quotes a remarkable observation of John Herschel who noticed that their ellipticity decreases toward the center and that, irrespective of the ellipticity of their outline, many have semi–stellar globular nuclei.

The first attempt, that I know of, at describing the apparent brightness distribution in an elliptical nebula was made by Bigourdan, in the course of his micrometric "Observations de Nébuleuses et Amas Stellaires" with the 30–cm refractor of Paris Observatory. In 1890 and 1891 he described the bright semi–stellar central condensation of M32, 3″ to 5″ in diameter, and the smooth fading outward of the surrounding nebulosity, which he illustrated by two small sketches. He noticed that the central nucleus, which he compared to a star of magnitude 10.5 or 11, did not remain as sharp as stars of the same magnitude when the field was illuminated. Observing the center of M31 in 1899 with the same instrument he also noted that

T. de Zeeuw (ed.), Structure and Dynamics of Elliptical Galaxies, 3–16.

its nucleus, about 5″ to 6″ in diameter, "is much less stellar and much fainter than that of M32" (Bigourdan 1900).

The first successful attempt at measuring the luminosity profile of the central part of the Andromeda nebula by photographic photometry (within $\pm 7'$ from the center) was made in 1912–13 by Reynolds, who used his 28–inch reflector and a Hartmann visual microphotometer (Reynolds 1914). Under the, then common, misconception that the nebular might be diffuse material reflecting the light of a central star (the existence of reflection nebulae had been demonstrated by V. M. Slipher in the previous year), Reynolds tried to represent the "light curve"—as he called it—by an inverse square law, $x^2y = const.$, that is, in more familiar notation, $I = I_0/r^2$. It did not work. He then tried a simple modification, $(x+1)^2y = const.$; the modified formula, in which we recognize the familiar form, $I = I_0/(r+a)^2$, worked quite well! Since this formula is asymptotic to the inverse square law at large distances, Reynolds concluded that it supported the basic assumption of light scattering from a central source and thus conflicted with the concept of the nebula as an external galaxy! The relevance of Reynolds' work on M31 to the subject of this conference is, of course, that this same formula was later adopted by Hubble to represent the luminosity profiles of a representative sample of elliptical nebulae which he studied in his early years at Mt. Wilson, but published only in 1930 (see below).

The first spectrographic observation of M32 was made in 1913 by Slipher, with the 24–inch *refractor* of Lowell Observatory (Slipher 1914, 1915). The close agreement between the negative radial velocities of M31 and M32 (-300 km/s) confirmed their physical association.

The great distance and extragalactic nature of M31 (and thus M32) was first demonstrated by Lundmark in 1918, by comparison of the apparent magnitudes of novae and brightest non–variable stars in M31 and in the Galaxy (Lundmark 1919).

In the same year Curtis (1918) discovered the nuclear jet in M87 with the 36–inch Crossley reflector of Lick Observatory.

In 1921 Hubble noticed the presence of 21–st magnitude star–like images around M87 on photographs taken with the 100–inch Hooker reflector of Mt. Wilson Observatory (Hubble 1923).[1] In 1922 Balanowski reported the presence of a "new star" (m = 11.5 and 12.2:) on plates of M87 taken at Pulkowo Observatory on February 24 and March 22, 1919, which was still visible at $m \approx 20$ on a plate taken by Hubble in March 1920 (see Hubble 1923).

2. OPTICAL STUDIES FROM 1925 TO 1945

The classification of "non-galactic" nebulae, introduced by Hubble in 1924–25, included in the "elliptical" class regular, symmetric, unresolved and structureless extragalactic objects, subdivided by ellipticity (*i.e.*, shape) from E0 to E7 (Hubble 1926). In the competing classification scheme of "anagalactic" nebulae presented by Lundmark (1927), the "globular, elliptical, elongated, ovate or lenticular" nebulae

[1] Their interpretation as unresolved images of globular clusters came only three decades later (Baum 1955). Even after the commissioning of the 200–inch reflector, Hubble (1949) was still describing them as "a tenuous atmosphere of supergiant stars".

were subdivided by light "compression" toward the center. Hubble's "ellipticals" fitted among the more concentrated objects.

The first systematic study of the light distribution in elliptical nebulae by photographic photometry was published by Hubble (1930). He used a Koch thermoelectric microphotometer to analyze plates taken in the early 1920's at the Newtonian focus of the Mt. Wilson 100–inch reflector. He found that the luminosity distribution could be represented by a generalized form of Reynolds' formula: $I = I_0/(r + a)^2$, where the fitting parameter was in the range 5″–1″ for 15 nebulae of total magnitudes 8 to 13. Hubble attempted to interpret the observations by a truncated isothermal sphere model, and concluded that elliptical nebulae (i) have no definite edge, the maximum detectable diameter increasing with exposure time (until plate saturation), ii) have all the same standard luminosity profile in normalized units of r/a and I/I_0, and (iii) are relaxed self–gravitating systems in (approximate) statistical equilibrium.[2] Between 1930 and 1934, Stebbins and Whitford made the first photelectric measurements of magnitudes and colors of M32 and other elliptical galaxies with a Potassium hydride cell attached to the Mt. Wilson 100–inch telescope (Stebbins & Whitford 1936, 1937). When the recording microphotometer came into more general use after 1930, tracings of long–exposure photographs at several observatories, including Cambridge, Harvard and Mt. Wilson (see, *e.g.*, Shapley 1934), produced large increases in the maximum detectable diameters of galaxies, especially ellipticals. The first photographic isophotes of NGC 205 were also produced with this instrument by Reynolds (1934), on plates taken with the Helwan Observatory 30–inch reflector.

The first visual, photographic and interferometric high–resolution study of the center of M32 was made in 1934–35 by Sinclair Smith with the Mt. Wilson 100–inch reflector (Smith 1935). From his failure to detect interference fringes in the image of M32 with a makeshift Fizeau–type interferometer with a 55–inch equivalent baseline, he concluded that no star–like central source smaller than 0″08 could be detected in the nucleus. With the 2.5 m mirror at full aperture and various eyepieces (up to x600) he estimated the semi–stellar nucleus to be 1″1 in apparent diameter on two nights of good seeing, when the seeing disks of adjacent stars of comparable magnitude were 0″5, and concluded that the seeing–corrected diameter of the nucleus was 0″8 ± 0″1. He also noted that, with the photographic magnitude (13.4) assigned by Hubble to the "nucleus" (that is, the $1'' - 2''$ image recorded by the shortest exposures), its surface brightness was far in excess of the extrapolated I_0 value derived from the Reynolds–Hubble formula. Hence, there was in the center a spike of light in excess of that predicted by a quasi–isothermal model. Finally, he placed an upper limit of 2% on the amount of linearly polarized light detectable photographically, which refuted the light scattering model resurrected by Ten Bruggencate (1930*a*). Smith concluded that M32 is a very dense system of solar-type *dwarf* stars (Humason had classified the spectrum as G3 "with dwarf characteristics").

---

[2] In retrospect it is rather surprising that neither Hubble, nor anyone else at the time, commented upon the unexpected applicability to E galaxies of an empirical formula introduced by Reynolds to describe the central regions of a *spiral*. Nearly three decades elapsed before the decomposition of the luminosity profiles of M31 into disk and spheroidal components attracted attention to this remarkable similarity between the gross photometric structures of ellipticals and of the bulges of spirals (de Vaucouleurs 1958*b*, 1959).

An important contribution to the surface photometry of elliptical nebulae was published by Redman, assisted by Shirley (1936, 1938), who used the old Common 36–inch reflector of Cambridge Observatory and a newly developed photoelectric microphotometer built by Carroll and Moss (1931) to measure the luminosity profiles of 15 elliptical galaxies. This study includes the first serious discussion of the multiple sources of systematic and accidental errors in photographic photometry of galaxies and the first determination of the point spread function of the instrument.[3] Redman was especially aware that the parameter I_0 of the Reynolds-Hubble formula could not possibly represent the true peak brightness, and correctly concluded: "It is probable that the determination of the true surface luminosity distribution of any elliptical nebula is beyond the present resources of astronomy."

In 1938 Shapley discovered, on small–scale patrol plates taken at Harvard southern station, the Sculptor and Fornax systems, the first examples of a new type of spheroidal system, akin to elliptical galaxies, but of much lower star density (Shapley 1938*a, b*). Resolution into stars with the 60–inch reflector of the Boyden station, and detection of short–period variables, probably of RR Lyrae type, with the 100–inch reflector at Mt. Wilson by Baade and Hubble soon indicated the order of magnitude of the distance and confirmed the dwarf character.

The last notable pre–World War II paper was a dynamical study of NGC 3115 (then the E7 prototype) and NGC 4494 (E0) by Oort (1940), which included luminosity profiles measured on plates taken at the Newtonian focus of the Mt. Wilson 1.5 m telescope.

In 1944 came the well–known successful resolution of the bulge of M31 and its elliptical satellites into red giant stars by Baade, on photographs taken in red light and under exceptional seeing conditions with the Mt. Wilson 100–inch reflector (Baade 1945).

3. SOME HIGHLIGHTS 1945 TO 1975

3.1 The $r^{1/4}$–Law

In 1946 I began a program of photographic photometry of galaxies, the first phase of which was reported in 1948; it included (*a*) a critical review of earlier works, tracing their poor agreement to a variety of faulty observing, calibrating and recording techniques (de Vaucouleurs 1948*a*); (*b*) a detailed discussion of the main sources of systematic and accidental errors in galaxy photometry (de Vaucouleurs 1948*b*), and (*c*) a study of the central regions of two ellipticals (NGC 3379, 4649), one "E7" (NGC 3115) and one bulge–dominated edge–on spiral (NGC 4594), on plates taken at the Cassegrain focus of the 80–cm reflector. The PSF measured out to 20″ had a Gaussian core of dispersion $\sigma = 1''25$. Although the exposures were short (5–60 min.) and only a limited range of intensities could be measured, about 3 dex (7–8 mag), it was sufficient to discover that the luminosity profiles of the spheroids could be well fitted with a very simple formula of the form $\log I = A - B^{1/4}$ (de Vaucouleurs 1948*c, d*).[4] Although it has no free fitting parameter,

---

[3] Unfortunately, as I showed later (de Vaucouleurs 1958*b*), their calibration had a large scale error.

[4] This formula was soon found to apply equally well to the distribution of galaxies in rich, Coma-type clusters (de Vaucouleurs 1948*e*) and, later, to the

this formula fitted the average of Hubble's observations extremely well and, over a range in excess of 4 dex ($> 10\,mag$), the average of the (then) five best observed ellipticals (de Vaucouleurs 1953). Because, unlike the Reynolds–Hubble formula, its surface integral converges, it leads to a rational definition of the total (asymptotic) luminosity, and of effective scale factors (I_e, r_e) which have been widely used in many studies.[5]

The space density distribution in a spherical system obeying the $r^{1/4}$–law in projection was first derived by Poveda, Iturriaga & Orozco (1960), and more precisely by Young (1976) for the range of normalized distances $10^{-6} \leq s = r/r_e \leq 260$ (from the centers of the densest systems, such as M32, to the outermost fringes outside clusters).[6]

3.2 New or Peculiar Types of Elliptical Galaxies

Between 1948 and 1953 the first radio galaxies were discovered and identified with peculiar E galaxies, beginning with Cen A and Vir A located by means of the "sea interferometer," near Sydney, and soon identified with NGC 5128 and 4486 by Bolton, Stanley & Slee (1948, 1949) and confirmed by Mills (1952*a*, *b*, 1953). Then the radio source For A, detected by Stanley and Slee (1950), was identified with NGC 1316 by Shklovskii & Kholopov (1952) and, independently, by Mills and de Vaucouleurs (1953, 1954).

About 1950 Zwicky discovered the first small, high–surface brightness "compact" elliptical galaxies, with apparently sharp boundaries, the prototype of which is NGC 4486 B, near M87 (Zwicky 1963, 1971)

In 1956 another type of low–luminosity elliptical galaxy, having a semi–stellar nucleus, was discovered by Reaves on plates of the Virgo cluster taken with the Lick 20-inch astrograph (Reaves 1956, 1962).

A new type of supergiant elliptical galaxy, distinguished by an extended envelope (and occasionally multiple cores, as in NGC 6166), was identified by Mathews, Morgan & Schmidt (1964). These objects (labeled cD by Morgan) are often near the centers of rich clusters, but some have been found in groups and poor clusters.[7]

3.3 Truncated Isothermal Models and the King Formula

Hubble's attempt to fit the observed luminosity distribution by a density truncated

---

distribution of luminosity in the bulges of early–type spirals, such as M31 (de Vaucouleurs 1958*b*, 1974*b*) and of globular clusters in galaxies (de Vaucouleurs 1977, 1978). Recent computer simulations show that this distribution easily results from the gravitational collapse of a system of collisionless particles under a broad range of initial conditions and assumptions (Binney 1982; van Albada 1982).

[5] Note that the definition of effective parameters requires only that the total luminosity be finite; it is not restricted to E galaxies.

[6] It has been objected that the $r^{1/4}$ law is unsatisfactory, because it leads to a diverging central density at $r = 0$. This is akin to objecting to the law of perfect gases, $pV/T = k$, on the grounds that it implies that $p \to \infty$ when $V \to 0$.

[7] An extensive elliptical envelope or "corona" had already been detected on long–exposure photographs and photoelectric scans of NGC 5128 made at Mt. Stromlo Observatory in the 1950's (de Vaucouleurs & Sheridan 1957).

isothermal sphere was purely ad hoc and had little theoretical basis. A first attempt to provide a self-consistent theoretical model was made by Belzer, Gamow & Keller (1951). Between 1953 and 1956 Woolley and collaborators at Mt. Stromlo developed detailed analytical models of velocity truncated isothermal distributions with a mass function which could reproduce the $r^{1/4}$-law reasonably well (Woolley 1954; Woolley & Robertson 1956). This type of model was developed further in the 1960's by Michie (1963), Fish (1964) and King (1966*a*, *b*), who also introduced a widely used semi-empirical fitting function (King 1962):

$$(f/k)^{\frac{1}{2}} = \left[1 + (r/r_c)^2\right]^{-\frac{1}{2}} - \left[1 + (r_t/r_c)^2\right]^{-\frac{1}{2}},$$

where r_c is the "core" radius, and r_t is the "tidal" radius. This formula gives a very good representation of star counts in tidally-limited globular clusters and low-density spheroidal galaxies of the Sculptor and Fornax types (Hodge 1961*a*, *b*; 1962; de Vaucouleurs & Ables 1968; Hodge and Michie 1969). It does not fit normal E galaxies well (assuming $f/k \approx I/I_0$), particularly the central spike, and the "tidal radius" r_t often reflects more the detection threshold of the photometry than any physical boundary to the galaxy, as illustrated by the table.

Underlying all these models was the concept of statistical equilibrium by two-body relaxation, on a time scale much longer than the Hubble time. The way out of this difficulty came when Lynden-Bell (1967) introduced the concept of fast relaxation during an early collapse phase of galaxy formation and when Wolfe and Burbidge (1970) suggested the possible presence of massive black holes in the centers of E galaxies to account for the central spike.[8]

3.4 Apparent Ellipticities and True Shapes

Until recently, all attempts to find the apparent ellipticity of the two-dimensional image (the "nebula") on the sky, and the three-dimensional shape of the object (the "galaxy") in space, have been statistical in nature, and depended on the long-unquestioned assumption that the isophotal surfaces are homothetic oblate spheroids with axes $a = b > c$. As early as 1926, Hubble had attempted to determine the frequency function of the true axis ratio c/a from the relative frequencies of apparent axis ratios of 85 nebulae of types E0–E7. He concluded that the distribution of c/a is nearly uniform from 0.3 to 1.0.

Hubble's data were reanalyzed by Ten Bruggencate (1930*a*, *b*) who curiously concluded that the true shape of all ellipticals is actually E7! This conclusion was criticized by Machiels (1930*a*) who confirmed Hubble's results. Further discussion by Machiels (1930*b*, 1933) of the axis ratios among the objects classified E in the Shapley-Ames (1930) survey of the Virgo cluster, and in their (1932) survey of the brighter galaxies, reconfirmed Hubble's conclusions.

In 1950 a reanalysis of the more than 200 E objects in the Shapley-Ames catalogue led Wyatt (1950) to a different view: he found a significant excess of

---

[8] Numerous computer simulations have since confirmed that density distributions closely approximating the $r^{1/4}$ distribution can be obtained through the collapse of collisionless systems under a wide range of initial conditions and assumptions, but it is still moot whether a massive black hole is the correct interpretation for the central spike.

A Short History of the "Tidal Radius" of NGC 3379

Source of Photometry	Range $r <$	Threshold μ_B	Source of r_t	r_t
Dennison (1954)	3′4	25.3	King (1962)	5′
Miller and Prendergast (1962)	4.4	27.4	King (1966*a*)	7′8
de Vaucouleurs and Capaccioli (1979)	> 16′	> 30.9	deV.&C.(1979)	20′
Kormendy (1977)	No tidal truncation, but "tidal extension"			

highly flattened systems over nearly spherical ones, and a definite deficiency of systems of intermediate flattening (E3).[9]

These early discussions were confused by inclusion of lenticular galaxies and poorly–resolved S0/a and Sa spirals (E: objects in S–A catalogue).

Fresh material on 48 E, E^+ objects in the Survey of Southern Galaxies with the Mt. Stromlo 30–inch Reynolds reflector (de Vaucouleurs 1956), and using both estimated type and measured axis ratio, led again to agreement with a uniform distribution of true ellipticities, but the sample was too small for definite conclusions.

The (First) "Reference Catalogue" (de Vaucouleurs & de Vaucouleurs 1964, RC1) provided a much larger, more homogeneous all–sky sample which was first anlyzed by Sandage, Freeman & Stokes (1970), and by several others later. The "Second Reference Catalogue" (de Vaucouleurs, de Vaucouleurs & Corwin, 1976, RC2) provided still better and richer material, which was first analyzed by de Vaucouleurs & Pence (1973), and later by Binney and de Vaucouleurs (1981). All these studies indicated that, if E galaxies are homocentric oblate spheroids, the most frequent true ellipticity is E3–E4, and spherical and highly flattened systems are rare.[10] The problem is complicated by the fact that the apparent ellipticity of the isophotes depends on the brightness level, as was noted by Redman and Shirley (1938) who, from their study of 15 ellipticals concluded that "isophotal contours do not always have a constant ratio of minor to major axis...There is no uniform

[9] That is precisely the opposite of what is indicated by modern studies; see e.g., de Vaucouleurs (1974*a*); Binney & de Vaucouleurs (1981).

[10] More recently, it was realized that statistical studies of the ellipticity distribution alone cannot unequivocally define the true shapes of E galaxies, whether oblate, prolate or triaxial, and that photometric and kinematic information is needed to solve the problem.

tendency in the changes of shape of the contours, either to become more circular with diminishing surface brightness, or vice versa."

Two decades later these conclusions were generally confirmed by Hazen (1960), in her study of 15 Virgo cluster ellipticals with the Williams and Hiltner isophotometer, except that in this sample the ellipticity was generally increasing with radius up to a maximum reached near the effective radius, followed by a decrease in the outer parts. Two more decades elapsed before the trend of ellipticity with radius became again, during the past ten years, an active subject of investigation, together with the phenomenon of "isophote twists" or rotation of the major axes, first reported for the central parts of IC 1459, type E3–4, and, possibly NGC 1549, type E0, by Evans (1951),[11] and first measured by Hazen in 1957 on equidensity tracings of two E2 Virgo cluster ellipticals (NGC 4459 and 4472).

3.5 Integrated Colors and Color Distribution

The first two–color photoelectric photometry of galaxies by Stebbins & Whitford, using the Mt. Wilson telescopes, published in 1937, included 30 "ellipticals" (some recognized today as lenticulars). It showed them to be redder then spirals and slightly redder than suggested by their G or G–K spectral classification by Humason. This was confirmed by their work with a longer baseline (Stebbins and Whitford 1952).

Pettit (1954) reported, from two–color photometry of 58 ellipticals measured through two or more field apertures with the Mt. Wilson reflectors, that 52% are redder near the center, 32% bluer and 16% unchanged.[12]

Integrated magnitudes and colors of 20 ellipticals (including Local Group dwarf systems) measured by photographic surface photometry were included in the catalogue of Holmberg (1959).

The first *multi*–color photometry of M32 by Stebbins and Whitford (1948) showed significant departures from stellar spectra of matching P–V color. This effect is a consequence of the compositeness of the stellar population.

In 1957–58 I began a long–term program of photoelectric photometry of galaxies in the UBV system, initially with the 20– and 42–inch reflectors at Lowell Observatory (continued at McDonald Observatory with the 36– and 82–inch reflectors since 1960). The first results included 15 ellipticals (de Vaucouleurs 1958*a*). Simultaneously, Tifft (1961, 1963) was securing four–color photometry with the 60–inch Mt. Wilson reflector. The two studies were combined, together with earlier two–color data, in the first general catalogue of galaxy colors in the UBV system (de Vaucouleurs 1963), including 21 ellipticals. This study demonstrated clearly the presence of a color gradient (redder near center) in normal E galaxies and gave the first application to distance determinations of the color–luminosity relation reported by Baum (1959, 1961).

In 1966 Wood published 12–color photometry of galaxies with applications to population analysis, but Martin & Bingham (1970) soon demonstrated—by prin-

---

[11] The most clear cut example, NGC 1291, since reclassified as a barred S0/a, is no longer relevant.

[12] However, in a rediscussion of Pettit's data restricted to the larger and brighter objects ($m_{pg} \leq 13.0$), I found that actually all are redder near the center by 0.1–0.2 mag, and that the mean color index, $< P - V > = 0.87$, is slightly redder than the mean (0.85) for lenticulars (de Vaucouleurs 1959).

cipal component analysis—that only *three* independent parameters (of which two are dominant) can be defined by multicolor photometry.

3.6 Masses, Luminosities and Velocity Dispersion

The first estimate of the mass of M32, $M \approx 2.5.10^{10} M_\odot$ for an assumed distance of 0.5 Mpc), was made by Martin Schwarzschild (1954), from an asymmetry in the shape and rotation curve of M32 tentatively attributed to dynamical interaction.

The first formulation of the virial theorem for a spherical galaxy having a constant M/L ratio, and a projected luminosity distribution given by the $r^{1/4}$ law, was given by Poveda (1958) and Poveda, Iturriaga & Orozco (1960): $M_T = 3r_e\sigma_v^2/G$, where r_e = effective radius (in projection), σ_v = central velocity dispersion (in space). The first estimates of the virial mass of M32 using this formula was made by Burbidge, Burbidge & Fish (1961). The masses of other ellipticals were calculated by Poveda (1961), using velocity dispersions measured mainly by Minkowski (1961). This was soon followed by the first systematic discussion of masses and mass/luminosity ratios of E galaxies by Fish (1964).

Three-component (core, bulge, halo) dynamical models of E galaxies were first discussed by Einasto (1972, see also de Vaucouleurs 1974*a*). In these models the halo component, having a low M/L ratio (≈ 3), was supposed to be the only one present in low-density dwarf systems of the Scl, For type, with the bulge and core populations increasingly important in galaxies more massive than $10^9\ M_\odot$.

The first hint of "a possible luminosity effect in the spectra of K-systems" was discovered by Morgan & Mayall (1957), who noticed that absorption lines looked more diffuse in giant systems and suggested that "a line width-absolute magnitude relationship ... may be used for the determination of spectroscopic parallaxes of more distant galaxies".

This possibility was examined by Minkowski (1961) who produced the first plot of velocity dispersion (on a linear scale) versus absolute photographic magnitude for 14 galaxies. The large scatter in the diagram led him to conclude that "the velocity dispersion does not seem to be a good criterion for absolute magnitude," but neither his estimates of σ_v, based on an ingenious, but imprecise analog photographic technique (the unsharp slit), nor the relative distances of the test galaxies, were good enough for definite conclusions.

Although the first quantitative measurements of σ_v in M32 by detailed photographic spectrophotometry and numerical convolution of template star profiles had already been made by Burbidge, Burbidge & Fish (1961), it was not until the early 1970's that new and more reliable methods of measurement by digital television or photoelectric techniques were introduced by Morton & Chevalier (1972, 1973), and at McDonald Observatory (mainly unpublished). But as late as 1973 values of σ_v had been obtained for less than a dozen galaxies, of which five only were ellipticals (NGC 221, 3379, 4486, 4486B, 4494), and agreement between independent estimates was poor (de Vaucouleurs 1974*a*).

3.7 Dawn of a New Era: 1975–76

The traditional concept of elliptical galaxies as oblate spheroids flattened by rotation about the polar axis was still dominant 11 years ago (Larson 1975; Gott 1975; Wilson 1975) when Bertola & Capaccioli (1975) reported their observation that the E5 galaxy NGC 4697 is a slow rotator with a maximum rotation velocity of

$V_r \leq 65\, km\, s^{-1}$. As Binney (1976) was quick to point out, comparison of this value with the velocity dispersion ($\sigma_v = 310\, km\, s^{-1}$) measured by Minkowski, implied that there is not enough kinetic energy in the rotation to account for the observed ellipticity if the galaxy is an oblate spheroid.[13]

In the same year the reality of the correlation between σ_v and absolute magnitude for elliptical galaxies was finally demonstrated by Faber & Jackson (1976). The slope of the $M - \log \sigma_v$ relation was consistent with a simple dynamical model implying L/σ_v = const. Soon afterward, Kormendy (1977) discovered a correlation between the effective parameters (radius and surface brightness, r_e and μ_e) of the $r^{1/4}$–law for a sample of 17 E galaxies (measured mainly by King).

These new discoveries opened a new era in elliptical galaxy research which has developed explosively ever since with the entry into the field of many new workers, the growing application of new detectors and the use of ever more powerful methods of digital data processing which make the pre–1975 era look almost like the age of the horse–and–buggy. It is the subject of this conference to review these more recent developments.

REFERENCES

van Albada, T.S., 1982. *Mon. Not. R. astr. Soc.*, **201**, 939.
Alexander, S., 1852. *Astron. J.*, **2**, 109.
Baade, W., 1945. *Astrophys. J.*, **100**, 137.
Balanowski, I., 1922. *Astr. Nachr.*, **215**, 215.
Baum, W.A., 1955. *Publ. Astr. Soc. Pacific*, **67**, 328.
Baum, W.A., 1959. *Publ. Astr. Soc. Pacific*, **71**, 106.
Baum, W.A., 1961. In IAU Symposium No. 15, *Problems of Extragalactic Research*, p. 20, ed. G.C. McVittie (MacMillan, New York).
Belzer, J., Gamow, G., & Keller, G. 1951. *Astrophys. J.*, **113**, 166.
Bertola, F., & Capaccioli, M., 1975. *Astrophys. J.*, **200**, 439.
Bigourdan, G., 1900. *Ann. Obs. de Paris*, **0.95**, 1904.
Binney, J.J., 1976. *Mon. Not. R. astr. Soc.*, **177**, 19.
Binney, J.J., 1982. *Mon. Not. R. astr. Soc.*, **200**, 951.
Binney, J.J., & de Vaucouleurs, G., 1981. *Mon. Not. R. astr. Soc.*, **194**, 679.
Bolton, G., Stanley, G.J., & Slee, O.B., 1948. *Nature*, **162**, 141.
Bolton, G., Stanley, G.J., & Slee, O.B., 1949. *Nature*, **164**, 101.
Burbidge, G.R., Burbidge, E.M., & Fish, R.A., 1961. *Astrophys. J.*, **134**, 251.
Carroll, J.A., & Moss, E.B., 1931. *Mon. Not. R. astr. Soc.*, **91**, 191.
Curtis, H.D., 1918. *Publ. Lick Obs.*, **13**.
Dennison, E.W., 1954. Ph. D. Thesis, Michigan University.
Einasto, J., 1972. In IAU Symposium No. 44, *External Galaxies and Quasi–Stellar Objects*, p. 37, ed. D.S. Evans (Reidel, Dordrecht).
Evans, D.S., 1951. *Mon. Not. R. astr. Soc.*, **111**, 526.
Faber, S.M., & Jackson, R.E., 1976. *Astrophys. J.*, **204**, 668.
Fish, R.A., 1964. *Astrophys. J.*, **139**, 284.
Gott, J.R., 1975. *Astrophys. J.*, **201**, 296.
Hazen, M., 1960. *Astrophys. J.*, **132**, 306.

[13] Ironically, with revised values of σ_v and V_r, today NGC 4697 is regarded as one of the faster rotating ellipticals in terms of the reduced velocity ratio V_r/σ_v.

Herschel, W., 1786. *Phil. Trans. R. Soc. London*, **76**, 457.
Hodge, P.W., 1961*a*. *Astron. J.*, **66**, 249.
Hodge, P.W., 1961*b*. *Astron. J.*, **66**, 384.
Hodge, P.W., 1962. **67**, 125.
Hodge, P.W., & Michie, R., 1969. *Astron. J.*, **74**, 587.
Holmberg, E., 1959. *Medd. Lund Obs.*, **136**.
Hubble, E., 1923. *Publ. Astr. Soc. Pacific*, **35**, 261.
Hubble, E., 1926. *Astrophys. J.*, **64**, 321.
Hubble, E., 1930. *Astrophys. J.*, **71**, 231.
Hubble, E., 1949. *Publ. Astr. Soc. Pacific*, **61**, 121.
King, I.R., 1962. *Astron. J.*, **67**, 471.
King, I.R., 1966*a*. *Astron. J.*, **71**, 64.
King, I.R., 1966*b*. *Astron. J.*, **71**, 276.
Kormendy, J., 1977. *Astrophys. J.*, **218**, 33.
Larson, R.B., 1975. *Mon. Not. R. astr. Soc.*, **173**, 671.
Le Gentil, G.J., 1749. *Mém. Acad. Sciences*, 470.
Lundmark, K. 1919. *Astr. Nachr.*, **209**, 369.
Lundmark, K., 1927. *Medd. Uppsala*, **30**, 22.
Lynden–Bell, D., 1967. *Mon. Not. R. astr. Soc.*, **136**, 101.
Machiels, A., 1930. *Bull. Astron. Paris*, **6**, 317.
Machiels, A., 1930. *Bull. Astron. Paris*, **6**, 405.
Machiels, A., 1933. *Bull. Astron. Paris*, **9**, 471.
Martin, W.L., & Bingham, R.G., 1970. *Observatory*, **90**, 13.
Mathews, T.A., Morgan, W.W., & Schmidt, M., 1964, *Astrophys. J.*, **140**, 35.
Messier, C., 1771. *Mém. Acad. Sciences*, 440.
Messier, C., 1781. *Connaissance de Temps for 1784*, 277 (Paris).
Messier, C., 1807. *Mém. Acad. Sciences*, 210.
Michie, R.W., 1963. *Mon. Not. R. astr. Soc.*, **125**, 127.
Miller, R.H., & Prendergast. K., 1962. *Astrophys. J.*, **136**, 713.
Mills, B.Y., 1952*a*. *Austr. Journ. Sci. Res.*, **A**, **5**, 266.
Mills, B.Y., 1952*b*. *Austr. Journ. Sci. Res.*, **A**, **5**, 456.
Mills, B.Y., 1953. *Austr. Journ. Physics*, **6**, 452.
Mills, B.Y., & de Vaucouleurs, G., 1953. *Observatory*, **73**, 252.
Mills, B.Y., & de Vaucouleurs, G., 1954. *Observatory*, **74**, 248.
Minkowski, R., 1961. In IAU Symposium No. 15, *Problems of Extragalactic Research*, p. 112, ed. G.C. McVittie (MacMillan, New York).
Morgan, W.W., & Mayall, N.U., 1957. *Publ. Astr. Soc. Pacific*, **69**, 291.
Morton, D.C., & Chevalier, R.A., 1972. *Astrophys. J.*, **174**, 489.
Morton, D.C., & Chevalier, R.A., 1973. *Astrophys. J.*, **179**, 55.
Oort, J.H., 1940. *Astrophys. J.*, **91**, 273.
Pettit, E., 1954. *Astrophys. J.*, **120**, 413.
Poveda, A., 1958. *Bol. Obs. Tonantzintla*, **17**.
Poveda, A., Iturriaga, R., & Orozco, I., 1960. *Bol. Obs. Tonantzintla*, **20**, 3.
Poveda, A., 1961. *Astrophys. J.*, **134**, 910.
Reaves, G., 1956. *Astron. J.*, **61**, 69.
Reaves, G., 1962. *Publ. Astr. Soc. Pacific*, **74**, 392.
Redman, R.O., & Shirley, E.G., 1936. *Mon. Not. R. astr. Soc.*, **96**, 588.
Redman, R.O., & Shirley, E.G., 1938. *Mon. Not. R. astr. Soc.*, **98**, 613.
Reynolds, J.H., 1914. *Mon. Not. R. astr. Soc.*, **74**, 132.
Reynolds, J.H., 1934. *Mon. Not. R. astr. Soc.*, **94**, 519.

Sandage, A., Freeman, K.C., & Stokes, N.R., 1970. *Astrophys. J.*, **160**, 831.
Schwarzschild, M., 1954. *Astron. J.*, **59**, 273.
Shapley, H., 1934. *Mon. Not. R. astr. Soc.*, **94**, 806.
Shapley, H., 1938*a*. *Harvard Obs. Bull.*, **908**, 1.
Shapley, H., 1938*b*. *Nature*, **142**, 715.
Shapley, H., & Ames, A., 1930. *Ann. Harvard Coll. Obs.*, **88**, No. 1.
Shapley, H., & Ames, A., 1932. *Ann. Harvard Coll. Obs.*, **88**, No. 2.
Shklovskii, I.S., & Kholopov, P.N., 1952. *Astr. Circ. Ac. Sci. USSR*, **131**.
Slipher, V.M., 1912. *Lick Obs. Bull.*, **2**, 26.
Slipher, V.M., 1914. *Publ. Am. Astr. Soc.*, **3**, 98.
Slipher, V.M., 1915. *Pop. Astron.*, **23**, 21.
Smith, S., 1935. *Astrophys. J.*, **82**, 192.
Stanley, G.J., & Slee, O.B., 1950. *Austr. Journ. Physics*, **3**, 234.
Stebbins, J., & Whitford, A.E., 1936. *Astrophys. J.*, **83**, 424.
Stebbins, J., & Whitford, A.E., 1937. *Astrophys. J.*, **86**, 247.
Stebbins, J., & Whitford, A.E., 1948. *Astrophys. J.*, **108**, 413.
Stebbins, J., & Whitford, A.E., 1952. *Astrophys. J.*, **115**, 384.
Ten Bruggencate, H., 1930a. *Zs. für Astrophys.*, **1**, 275.
Ten Bruggencate, H., 1930*b*. *Zs. für Astrophys.*, **2**, 83.
Tifft, W., 1961. *Astron. J.*, **66**, 390.
Tifft, W., 1963. *Astron. J.*, **68**, 302.
de Vaucouleurs, G., 1948*a*. *J. des Observ.*, **31**, 113.
de Vaucouleurs, G., 1948*b*. *Ann. d' Astrophys.*, **11**, 247.
de Vaucouleurs, G., 1948*c*. *Ann. d' Astrophys.*, **11**, 267.
de Vaucouleurs, G., 1948*d*. *Compte Rendus*, **227**, 548.
de Vaucouleurs, G., 1948*e*. *Compte Rendus*, **227**, 586.
de Vaucouleurs, G., 1953. *Mon. Not. R. astr. Soc.*, **113**, 134.
de Vaucouleurs, G., 1956. *Mem. Commonwealth Obs.* **13**.
de Vaucouleurs, G., 1958*a*. *Lick Obs. Bull.*, **97**.
de Vaucouleurs, G., 1958*b*. *Astrophys. J.*, **128**, 465.
de Vaucouleurs, G., 1959. *Handbuch der Physik*, **53**, 323/328.
de Vaucouleurs, G., 1963. *Astrophys. J. Suppl.*, **8**, 31.
de Vaucouleurs, G., 1974*a*. In IAU Symposium No. 58, *The Formation and Dynamics of Galaxies*, p. 1, ed. J.R. Shakeshaft (Reidel, Dordrecht).
de Vaucouleurs, G., 1974*b*. In IAU Symposium No. 58, *The Formation and Dynamics of Galaxies*, p. 335, ed. J.R. Shakeshaft (Reidel, Dordrecht).
de Vaucouleurs, G., 1977. *Astron. J.*, **82**, 456.
de Vaucouleurs, G., 1978. *Astron. J.*, **83**, 1383.
de Vaucouleurs, G., & Ables, H.D., 1968. *Astrophys. J.*, **151**, 105.
de Vaucouleurs, G., & Capaccioli, M., 1979. *Astrophys. J. Suppl.*, **40**, 699.
de Vaucouleurs, G., & Pence, W., 1973. *Bull. American Astron. Soc.*, **5**, 446.
de Vaucouleurs, G., & Sheridan, K.V., 1957. In IAU Symposium No. 4, *Radio Astronomy*, p. 169, ed. H.C. van de Hulst (Reidel, Dordrecht).
de Vaucouleurs, G., & de Vaucouleurs, A., 1964. *Reference Catalogue of Bright Galaxies*, Univ. of Texas, Austin (RC1).
de Vaucouleurs, G., de Vaucouleurs, A., & Corwin, H.G., 1976. *Second Reference Catalogue of Bright Galaxies*, Univ. of Texas, Austin (RC2).
Wilson, C.P., 1975. *Astron. J.*, **80**, 175.
Wolfe, A.M., & Burbidge, G.R., 1970. *Astrophys. J.*, **161**, 419.
Wood, D.B., 1966. *Astrophys. J.*, **145**, 36.

Woolley, R., 1954. *Mon. Not. R. astr. Soc.*, **114**, 191.
Woolley, R., & Robertson, D.A., 1956. *Mon. Not. R. astr. Soc.*, **116**, 288.
Wyatt, S.P., 1950. *Astron. J.*, **55**, 187.
Young, P., 1976. *Astron. J.*, **81**, 807.
Zwicky, F., 1963. *Astron. J.*, **68**, 301.
Zwicky, F. 1971. *Catalogue of Selected Compact Galaxies*, California Institute of Technology, Pasadena.

DISCUSSION

Rood: A few years ago, Walter Jaffe published a fitting–formula in the *Monthly Notices* (1982, **202**, 995), to describe the luminosity distribution of elliptical galaxies. Jaffe's formula has the same number of fitting parameters as the $r^{1/4}$–formula and implies a stellar distribution function which lends itself nicely to theoretical studies. How well does the Jaffe formula fit the observational data relative to the goodness–of–fit of the $r^{1/4}$–formula?

Jaffe: My suggested form for the light distribution in ellipticals fits all the data for NGC 3379 as well as the $r^{1/4}$ law, and is analytically much more tractable.

de Vaucouleurs: I regret to admit that I was not familiar with your paper, but after checking it seems to be an excellent single formula for the *space* density which projects into a good fit of the observed $I(r)$ of NGC 3379. The analytical expression for the *projection* is tractable but not simple; the $r^{1/4}$ law is simple in projection, but not tractable in space. Your formula implies more mass (light) in the faintest outer parts.

Djorgovski: The $r^{1/4}$ law has no shape parameters. This would imply that *all* surface brightness profiles have the same shape on log–log plots. It is a well–documented fact that they do not: there is a variety of shapes.
Second, I have fitted the $r^{1/4}$ laws to my surface photometry profiles, outside 3 arc sec, in order to avoid seeing effects. The $r^{1/4}$–law is a fair approximation in many cases, but there are highly significant deviations in many cases as well, and there are no evident systematics in these deviations.

de Vaucouleurs: I agree. However, I consider it a *good* feature of the $r^{1/4}$ law that it has no 'fudge' factor. The remarkable fact is that it does fit so well so many systems from E galaxies, to the globular cluster systems of our Galaxy and M31, to the distribution of galaxies in clusters like Coma. This remains as a challenge to theorists.

Lauer: I have compared my observed high–resolution CCD surface photometry to de Vaucouleurs models that are fitted to the envelopes and convolved with the observed point spread function. In no case in my 42 galaxies observed does the de Vaucouleurs law fit the apparent cores. For well resolved galaxies it predicts too much light in the center. For poorly resolved galaxies it predicts too little light.

de Vaucouleurs: I will look at your welcome new data carefully. My impression is that the higher the density of the system the better the fit to the $r^{1/4}$ formula, and the lower the density, the poorer. Of course, the formula was never claimed to fit globular clusters, or low density spheroidals, which are very well fit by the King formula. As you know, I have documented in detail real departures from the $r^{1/4}$ (e.g., excess in the center of N3379, deficiency in M87 outside the central nonthermal source). As Steve Strom remarked once, the $r^{1/4}$ law is a good 'French curve' against which the departures of real galaxies can be studied. Again all the new CCD work is very welcome!

King: I regret that pressure of time has kept you from discussing the techniques to which you have contributed so much. Regarding sky level, you showed how sensitive the outer photometry is to a correct sky determination. You also showed one limit of $\mu = 30.9$. Can you give us precepts for determining a good sky value?

de Vaucouleurs: The 30.9 is a normal value which has a calculated mean error of -1 mag on the bright side and $+\infty$ on the other as shown by the error bar on our Fig. 2 in *Astroph. J. Suppl.*, **40**, 704, (1979). We attach no real significance to it. As we showed in *Astroph. J. Suppl.*, **52**, 465, (1983), real measurements with a m.e.< 0.1 mag can be pushed down to $\mu_B \approx 28\, B/ss$. Below this level—although readings can be secured—their significance is lost in the irreducible cosmic noise (irrespective of detector) due to galactic cirri, fluctuations due to subthreshold stars and galaxies etc. We have discussed these sources of error at length in *Ann. d' Astrophys.*, **11**, 247, (1948) and *Astrophys. J. Suppl.*, (1983), *op. cit.*

Bertola and Kormendy.

CORES OF EARLY-TYPE GALAXIES

John Kormendy[1]
Dominion Astrophysical Observatory
Herzberg Institute of Astrophysics

ABSTRACT

Systematic study of galaxy cores has become possible through improved understanding of seeing and the reduction of photometric errors by the use of CCDs. Many cores are well resolved in a photometry program with the Canada-France-Hawaii Telescope (median stellar FWHM = 0.″80). Core profile shape correlates with galaxy luminosity L: the brightest galaxies have isothermal profiles; fainter ellipticals and bulges have profiles that do not completely flatten inward into a core. This may be due to velocity anisotropies. Core parameters are correlated: more luminous galaxies have larger core radii and fainter central surface brightnesses. Large deviations suggest special events: Fornax A has too small and bright a core for its luminosity; it may be the remnant of a merger with a smaller galaxy. Core mass-to-light ratio $M/L \propto L^{0.2}$, as expected from the metallicity-luminosity relation. A kinematic search shows strong evidence for a central black hole in M31 and weaker evidence in M32 and NGC 3115. The nucleus of M31 rotates very rapidly but has an outer dispersion of only 107 km s^{-1}. This implies that it is a disk; it may have formed from gas falling into the center. Rapid rotation and a central velocity dispersion $> 241 \pm 7$ km s^{-1} imply a central $M/L_V \gtrsim 20 - 35$. Velocity anisotropies are not a major uncertainty because of the rapid rotation. Therefore there is strong evidence for a nuclear point mass of $\sim 2 \times 10^7\ M_\odot$.

1. INTRODUCTION

Cores of elliptical galaxies are of interest for many reasons. The structure and stellar velocity distribution are easiest to understand in cores. Cores also have well-defined physical scales, the central surface brightness μ_0, the core radius r_c at which the surface brightness has fallen by a factor of two, and the central velocity dispersion σ. The systematic properties of these parameters can give insight into galaxy formation and evolution. Perhaps most interesting is the possibility that galaxy nuclei contain massive black holes that are the engines powering nuclear radio activity and quasars. A review of core structure based on photographic photometry is given in Kormendy (1982*a*). Since then, considerable progress has been made through major improvements in resolution and photometric accuracy.

---

[1] Visiting Astronomer, Canada-France-Hawaii Telescope, operated by the National Research Council of Canada, the Centre National de la Recherche Scientifique of France, and the University of Hawaii.

T. de Zeeuw (ed.), Structure and Dynamics of Elliptical Galaxies, 17–36.

2. TWO TECHNICAL BREAKTHROUGHS

Two developments have allowed us to obtain reliable core photometry. First, Schweizer (1979, 1981) convinced us of the importance of seeing corrections, which formerly were made crudely or not at all. He pointed out that the tiny nuclei of M31 and M32 would not be resolved if they were observed as far away as the Virgo Cluster. Moreover, some apparent cores in Virgo were better described by seeing-convolved $r^{1/4}$ laws than by any isothermal or King (1966) model with a physical core (Fig. 1*a*). Schweizer showed further that if the central profile is as "cuspy", say, as an $r^{1/4}$ law, then seeing produces a fake core whose apparent radius $r_{c,app}$ is determined by the effective radius r_e (Schweizer 1979, Fig. 4; Kormendy 1982*a*, Fig. 20). If the apparent core is little or no bigger than the radius predicted from r_e, we cannot be sure that it is resolved. It could be resolved, because seeing effects are relatively small for real cores. But the profile could be an $r^{1/4}$ law right to the center, or the galaxy could be hiding a nucleus like the one in M31. At that time, only M31 and the ellipticals NGC 4649, NGC 4472 and M87 were securely resolved. Schweizer argued that other galaxies might not have cores at all.

Therefore core parameters could only be derived if one *assumed* that galaxies had isothermal cores. Then seeing corrections were calculable and modest. Photometry by King (1978) and others was analyzed in this way by Kormendy (1982*a*, 1984), who found that r_c, μ_0, and σ correlate with each other and with galaxy luminosity L. However, the results were model-dependent: the seeing corrections would have been larger if the profiles were cuspy. Current CCD data are accurate enough for less model-dependent seeing corrections based on deconvolutions (Bendinelli *et al.* 1984, and references therein; Lauer 1985*a, b*). Nevertheless, we need the best possible seeing for accurate work on profile shapes and characteristic parameters, especially since small galaxies have tiny cores.

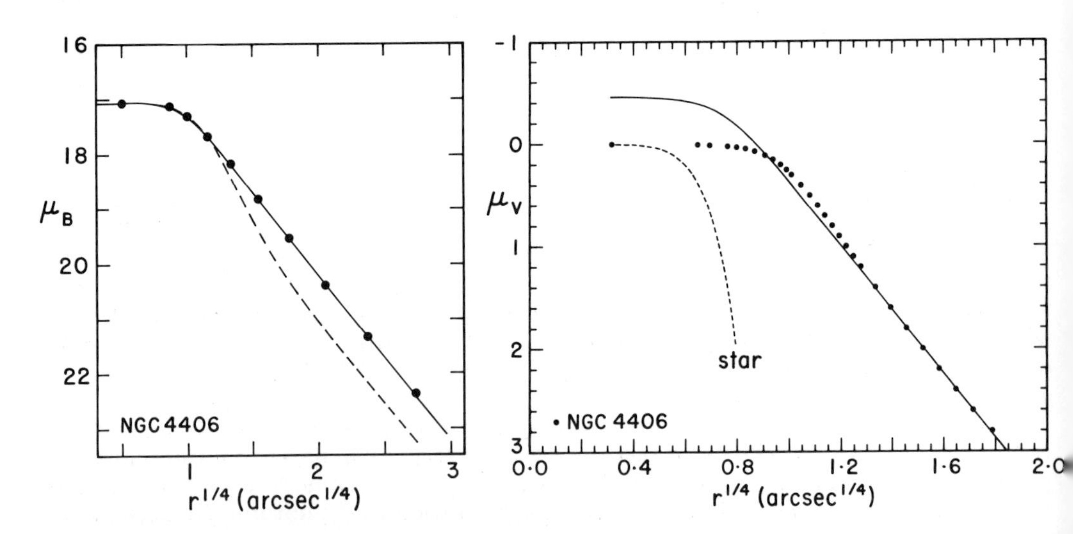

Fig. 1. – (*a*, left) Brightness profile of NGC 4406 (King 1978) compared with a seeing-convolved $r^{1/4}$ law (*solid line*) and a King model fitted at r_c (*dashed line*) (Schweizer 1979). (*b*, right) The profile of NGC 4406 measured in excellent seeing ($\sigma_* = 0\farcs21$) is not consistent with a seeing-convolved $r^{1/4}$ law.

The key to the present paper is the excellent seeing at the Canada-France-Hawaii Telescope. Most of the results discussed here were obtained in a photometry program carried out with an RCA CCD at the Cassegrain focus of the CFHT. The scale was $0\rlap{.}''215$ pixel^{-1}. Figure 2 shows the histogram of Gaussian seeing dispersion radii σ_*. It hardly overlaps with histograms for two well-known studies of cores which were themselves carried out at sites known for good seeing (Mount Wilson Observatory and Lick Observatory). The median $\sigma_* = 0\rlap{.}''34$ is a factor of two better than in previous data. As a result, many cores are well resolved. Figure 1*b* shows the improvement for NGC 4406. The profile is no longer consistent with a seeing-convolved $r^{1/4}$ law. The ratio $r_{c,app}/\sigma_* = 8.4$; this implies that the galaxy is well resolved according to Schweizer's (1979) criterion. Almost all of the bright ellipticals in Virgo and even many that are several times farther away are this well resolved. Lauer (1985*a*, *b*) resolved several additional ellipticals and also found nearly isothermal cores.

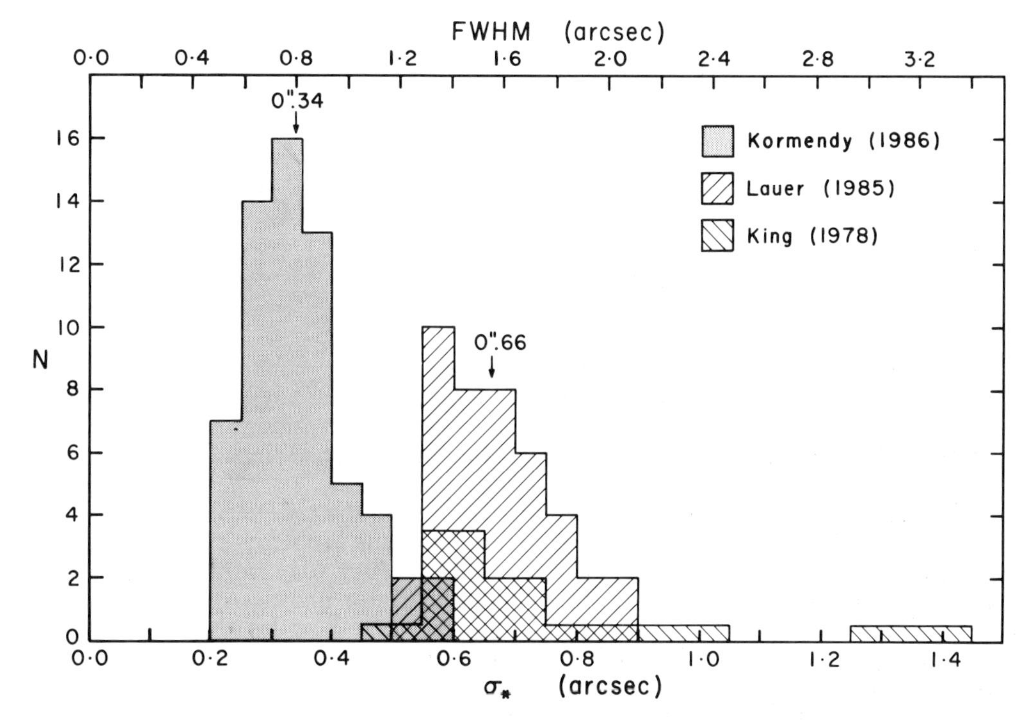

Fig. 2. – Histogram of seeing dispersion radii σ_* for the present CFHT photometry and for two published studies of cores. The arrows indicate median values; they are equal for the King (1978) and Lauer (1985*a*) samples.

It now seems safe to assume that ellipticals generally have cores. Seeing corrections based on deconvolutions will eventually be made, but here I adopt a short-cut (Fig. 3) calculated by convolving a King model with $\log(r_t/r_c) = 2.25$ and the point-spread function of the CFHT. It is convenient to parametrize the corrections in terms of $r_{c,app}/\sigma_*$. Galaxies with $r_{c,app}/\sigma_* \leq 3$ are poorly resolved; their seeing corrections still depend on the assumption that cores are nearly isothermal. In many cases the derived parameters are little better than limits. At the other extreme, galaxies with $r_{c,app}/\sigma_* > 5$ are well resolved; the profile shape and core parameters are largely independent of assumptions.

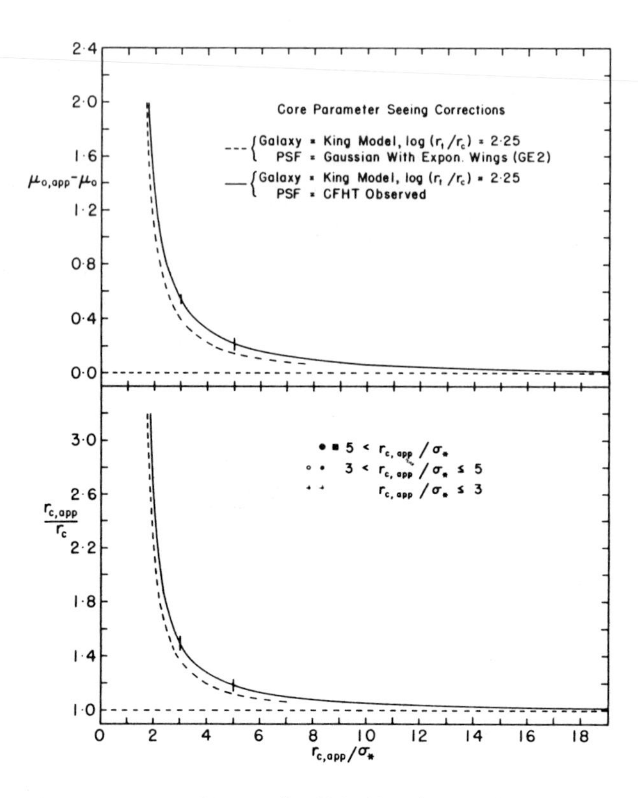

Fig. 3. – Seeing corrections (*solid lines*) used to convert apparent core parameters $r_{c,app}$ and $\mu_{0,app}$ to true values r_c and μ_0. The key shows how symbol sizes are used to indicate small, moderate and large corrections in Figure 6. The corrections derived by Schweizer (1981) are shown as dashed lines.

The second problem with core photometry before the CCD era was systematic errors. Lauer (1985*a*) has shown that there exist serious problems even with careful photographic photometry. Their cause is unclear. They may be due to problems with calibration or developer exhaustion, because fainter photographic photometry is more reliable. In contrast, my CFHT photometry agrees well with Lauer's seeing-corrected profiles but not with his raw data (e. g., Fig. 4). Similarly, our seeing-corrected core parameters agree well provided that we both resolved the galaxy. The mean ratio of core radii determined by Lauer and by me is $<r_{cL}/r_{cK}>= 0.99\pm0.01$ for 4 galaxies with $r_c > 3''$ and Lauer resolution class I ("well resolved"); 1.04 ± 0.03 for all 8 galaxies with resolution class I; 1.24 ± 0.11 for 4 galaxies with resolution class II and 2.3 for 1 galaxy with resolution class III ("unresolved"). This shows (i) that the measurements agree well, (ii) that Lauer's seeing corrections are accurate when they are not too large, and (iii) that when his seeing corrections are large, they are underestimated. The same is probably true of my seeing corrections. In general, CCD photometry by various authors agrees to ~ 0.1 mag arcsec^{-2} over many magnitudes. CCDs should be able to do better. However, this is accurate enough to allow large improvements over photographic work. The following discussion is based entirely on CCD data: those from the CFHT, Lauer's data for 4 galaxies of resolution class I that I have not observed, and Kent's (1983) data on M31.

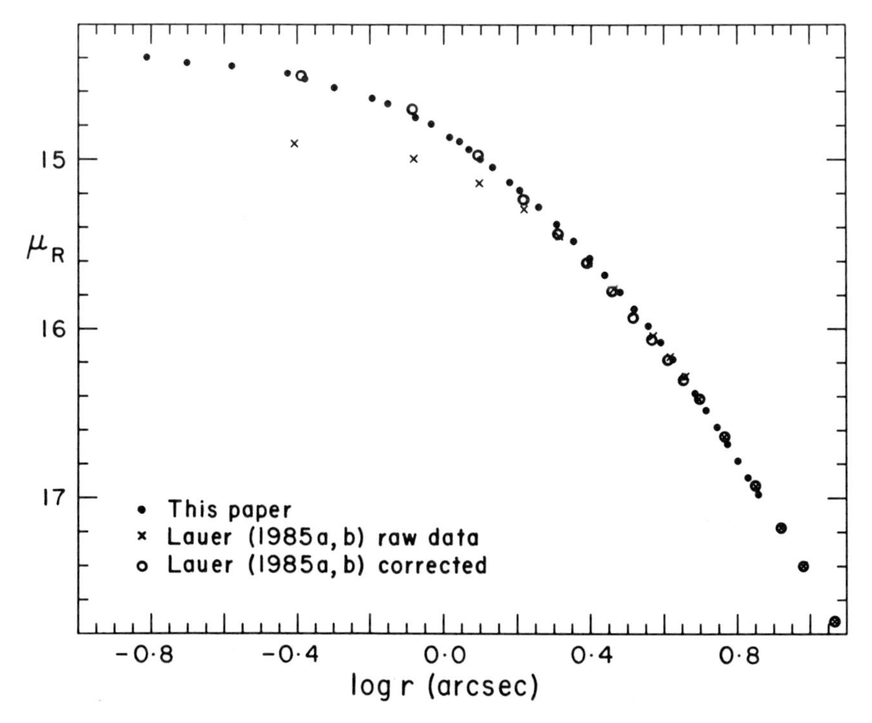

Fig. 4. – Comparison of uncorrected mean profile of NGC 3379 as measured with the CFHT ($\sigma_* = 0\farcs21$) with uncorrected and seeing-corrected profiles from Lauer (1985*a*, *b*).

3. CORE PROFILE SHAPES

The idea that core structure might be simple was suggested in part by early photographic photometry that showed isothermal central profiles. Eventually a number of cores were found that deviated from isothermals (King 1978; Kormendy 1982*a*), but this attracted little attention. Interest in velocity anisotropies was generated about 10 years ago by several nearly simultaneous events. It became clear on theoretical grounds (Binney 1976) that isotropy is not generally expected in ellipticals. In fact, they rotate slowly (Illingworth 1977); their shapes are determined not by rotation but by anisotropic velocity dispersions left over from galaxy formation (Binney 1978). The theoretical arguments applied equally well to cores; there was little more reason to expect them to be isotropic than ellipticals as a whole. The observational situation, however, was unclear. The second event was the claim that a black hole had been detected in the center of M87 based on an inward-rising dispersion profile (Sargent *et al.* 1978) and a brightness profile that did not flatten completely into a core at the center (Young *et al.* 1978). It was soon pointed out that anisotropies could explain these observations as easily as a black hole (Duncan and Wheeler 1980; Binney and Mamon 1982; Richstone and Tremaine 1985). Again it was not clear whether M87 was unusual or whether cores are generally not isothermal.

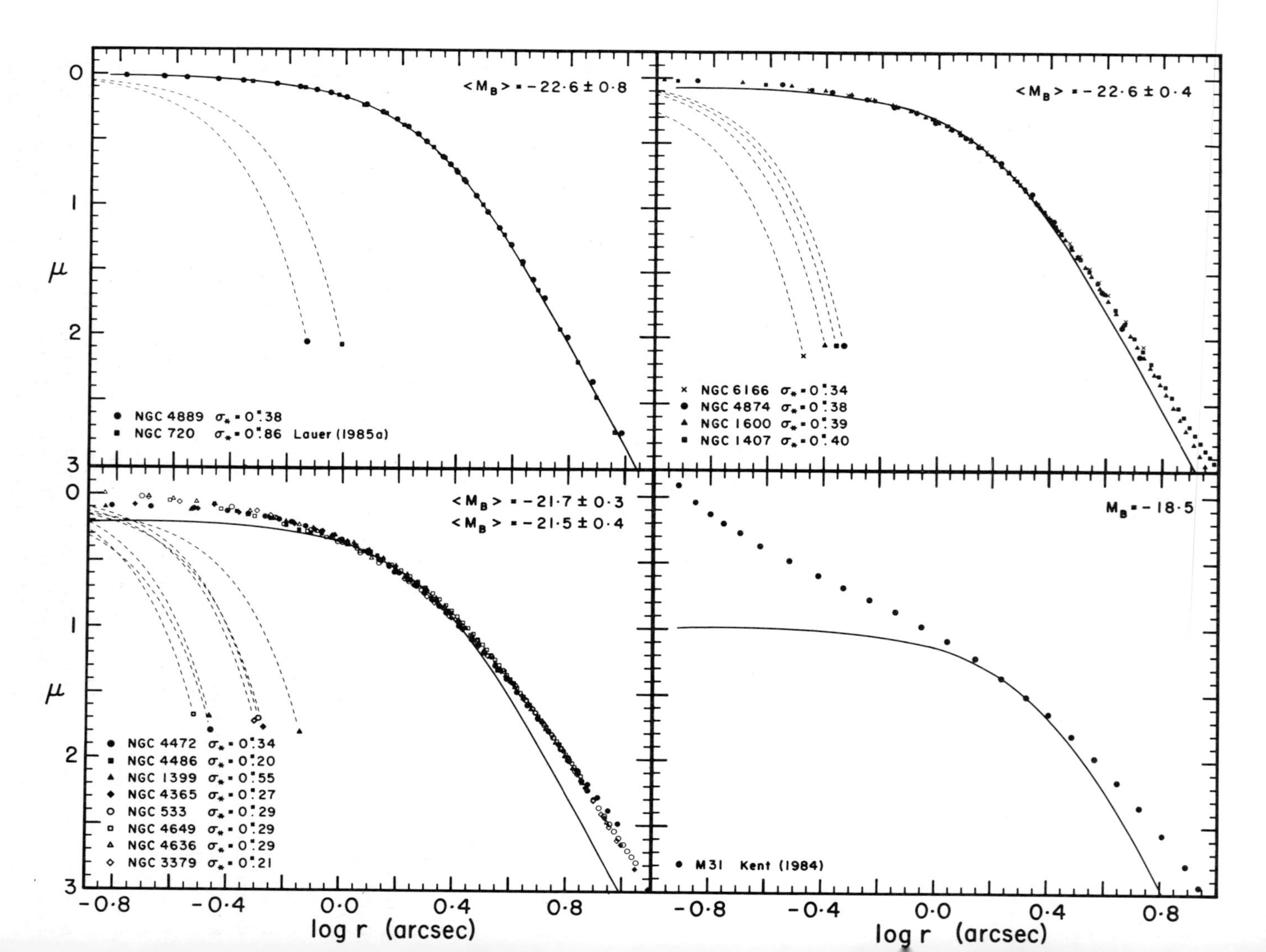

<M_B> = −22·6 ± 0·8
NGC 4889 σ* = 0.″38
NGC 720 σ* = 0.″86 Lauer (1985a)
<M_B> = −22·6 ± 0·4
NGC 6166 σ* = 0.″34
NGC 4874 σ* = 0.″38
NGC 1600 σ* = 0.″39
NGC 1407 σ* = 0.″40
<M_B> = −21·7 ± 0·3
<M_B> = −21·5 ± 0·4
NGC 4472 σ* = 0.″34
NGC 4486 σ* = 0.″20
NGC 1399 σ* = 0.″55
NGC 4365 σ* = 0.″27
NGC 533 σ* = 0.″29
NGC 4649 σ* = 0.″29
NGC 4636 σ* = 0.″29
NGC 3379 σ* = 0.″21
M_B = −18·5
M31 Kent (1984)
μ
log r (arcsec)

Fig. 5. – (Opposite) Composite mean-axis brightness profiles of well-resolved galaxies. The profiles have been shifted by arbitrary amounts in log r and μ (mag arcsec^{-2}) to minimize the scatter. The M31 profile has been shifted to smaller radii by $\delta \log r = -1.2$. Each composite is compared with a projected isothermal model (*solid lines*). The bottom-left panel shows two groups of profiles; those plotted with open symbols are less isothermal near the center. Gaussian star profiles with the observed dispersions σ_* (*dashed lines*) have been scaled with the galaxy profiles to illustrate the relative resolution. Each group of ellipticals is similarly well resolved, so the differences between them are not due to seeing. The mean absolute magnitude $<M_B>$ of each group is also given.

Lauer (1985*b*) was the first to examine core profile shapes in a substantial sample of ellipticals using high-accuracy CCD data. He showed that virtually all cores have nonisothermal brightness profiles; M87 looks just like the rest. These results were extended by Kormendy (1985*a*) using high-resolution CCD photometry from the CFHT. As shown in Figure 5, core profile shape correlates weakly with galaxy luminosity. A few galaxies, usually the brightest first-ranked galaxies in clusters, have isothermal profiles. Fainter galaxies have profiles that do not flatten completely into a core. M31 is the faintest galaxy with a well-resolved core; it is even less isothermal than the ellipticals (Kent 1983). Other bulges are also less isothermal than ellipticals. In addition, many bulges have extra nuclei like that in M31, although the resolution is barely adequate to detect them even with the CFHT (Kormendy 1985*a*).

These data have not yet been modelled. For this we need rotation and velocity dispersion measurements with good resolution at $r \leq r_c$. Currently we have such data only for M31 (§ 5.2). Measurements are in progress with the CFHT. Kinematic evidence for anisotropy exists in several ellipticals; e. g., the core of NGC 1600 falls far below the "oblate line" describing isotropic oblate rotators in the $V_{max}/\sigma - \epsilon$ diagram (Illingworth 1977). Like the overall shapes of ellipticals, the E3 shape of the core of NGC 1600 must be maintained by anisotropy. But despite the hints, we still have little systematic understanding of anisotropies in galaxy cores. Figure 5 suggests that there is considerable regularity from galaxy to galaxy, and that the anisotropies correlate with luminosity.

4. CHARACTERISTIC PARAMETER CORRELATIONS

4.1 Correlations between r_c, μ_0, σ, and M_B

The observed correlations between core parameters of bulges and elliptical galaxies are scaling laws that provide important constraints for theories of galaxy formation. The best known is the Faber-Jackson (1976) observation that more luminous ellipticals have higher central velocity dispersions, $L \propto \sigma^n$, $n \simeq 4 - 6$. More recently, Kormendy (1982*a*, 1984) found that r_c and μ_0 are correlated with each other and with σ and M_B. Similar relations were found by Lauer (1985*b*) from CCD photometry. Kormendy (1985*b*) has extended the L range substantially and derived core parameters from high-resolution CFHT photometry. These results are updated in Figure 6.

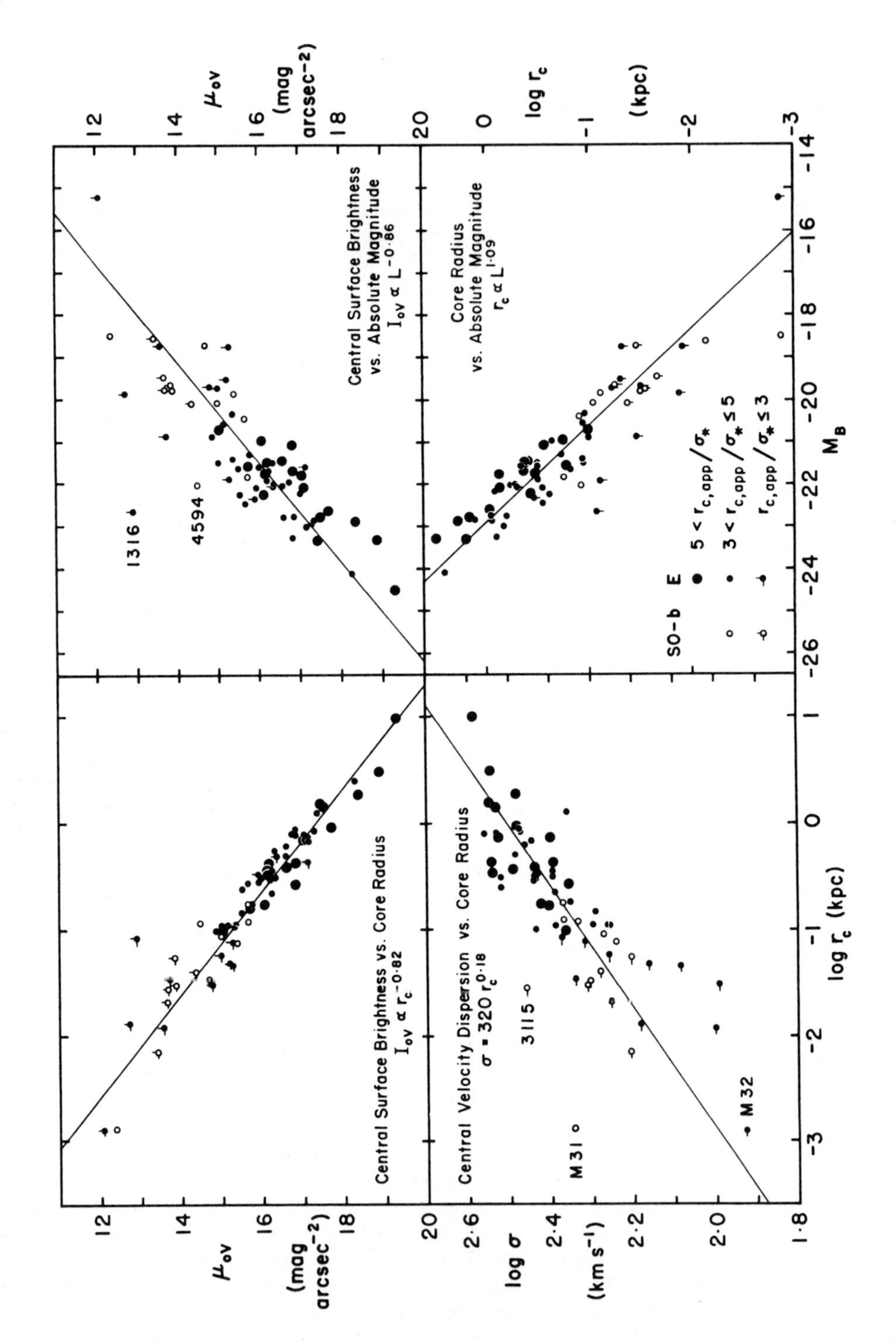
Central Surface Brightness vs. Absolute Magnitude
$I_{ov} \propto L^{-0.86}$
Core Radius vs. Absolute Magnitude
$r_c \propto L^{1.09}$
Central Surface Brightness vs. Core Radius
$I_{ov} \propto r_c^{-0.82}$
Central Velocity Dispersion vs. Core Radius
$\sigma = 320\, r_c^{0.18}$
1316
4594
3115
M31
M32
μ_{ov} (mag arcsec^{-2})
log r_c (kpc)
log σ (km s^{-1})
M_B
log r_c (kpc)
S0-b E
$5 < r_{c,app}/\sigma_*$
$3 < r_{c,app}/\sigma_* \leq 5$
$r_{c,app}/\sigma_* \leq 3$

Fig. 6. – (Opposite) Core parameter correlations for bulges and elliptical galaxies. Distances are based on a Hubble constant of 50 km s^{-1} Mpc^{-1}. Open circles for bulges are derived using all of the observed profile, including any nucleus. Larger points imply better resolution of the core (§ 2). The straight lines are least-squares fits; deviant points (labelled with NGC numbers) are omitted.

More luminous galaxies have larger core radii r_c and fainter central surface brightnesses μ_{0V}. Also, r_c and μ_{0V} correlate with σ, as expected from the Faber-Jackson relation. The least-squares fits are given in Figure 6; they are similar to ones derived in Kormendy (1985*b*). It is interesting to note the variety of galaxies that are consistent with these relations. The high-luminosity end is defined by brightest cluster galaxies, including cDs like NGC 6166. The faintest galaxy plotted is M32; it is very poorly resolved, but its tiny core is normal for its low L. Between these extremes, radio galaxies (e. g., NGC 4874, M87, DA 240, and NGC 6251) and galaxies with X-ray cooling flows (e. g., NGC 4472, NGC 4636, and NGC 4649) are consistent with the correlations. Evidently they are not destroyed either by gas that falls into the center or by the engines that power nuclear activity. Nuclear black holes (if they exist) may be too small to grossly affect core structure: a galaxy with $M_B = -22$ has a core mass of 10^{10} $M_\odot$. Bulges of disk galaxies also essentially satisfy the correlations. Some have small r_c and bright μ_{0V} due to the presence of distinct nuclei (e. g., M31; Light *et al.* 1974). Apart from these nuclei, the cores of bulges and ellipticals are as similar as their global properties.

I mention here only two applications of the core parameter relations to the problem of galaxy formation. Carlberg (1987) shows that phase space densities in the cores of all but the largest ellipticals are much higher than those in galaxy disks. Since phase space density can only decrease during violent relaxation, this implies that ellipticals cannot have formed from stellar disks by dissipationless mergers. A second application is found in Kormendy (1985*b*, 1986). The core parameter relations of dwarf spheroidal galaxies are very different from those for ellipticals: there is a discontinuity between them of >3 orders of magnitude. Dwarf spheroidal galaxies fall on analogous parameter relations for the disks of spiral and irregular galaxies. This confirms a hypothesis that dwarf spheroidals are more closely related to dwarf spiral and irregular galaxies than to ellipticals. They may have evolved from dS + Im galaxies by losing their gas or converting it all into stars. This subject is reviewed in Kormendy (1986).

Much of the scatter in Figure 6 is real. Some is due to distance errors: the adoption of distances based on a Virgocentric flow field will reduce the scatter. Also, some of the scatter is due to a "second parameter" effect; this will be discussed when all of the data are reduced. Here I want to mention one galaxy that departs by large amounts from the correlations. NGC 1316 (Fornax A) has a nucleus that is much too small and bright for the high luminosity of the galaxy. Schweizer (1980) has suggested that NGC 1316 is a merger remnant, based on the existence of ripples in the light distribution, a possible tidal tail, and a central gas disk whose rotation axis is different from that of the stars. The core parameters in Figure 6 could result if NGC 1316 swallowed a disk galaxy with a normal bulge of $M_B \sim -20$ (Kormendy 1984). If star formation were enhanced by the merger, this could also explain the high central surface brightness (Fig. 6, upper left) and the small central mass-to-light ratio (Fig. 7).

4.2 Mass-To-Light Ratios

The existence of a metallicity-luminosity relation for elliptical galaxies suggests that mass-to-light ratios M/L should correlate with L. The parameter relations of § 4.1 predict that $M/L \propto L^{0.18}$ (see also Kormendy 1982*a*, 1984). However, plots of M/L values against M_B have not in the past shown a correlation (Schechter 1980). This situation is unsatisfactory, since the transitive application of several parameter correlations with scatter is dangerous.

Figure 7 shows mass-to-light ratios for the present sample as a function of M_B. Here $M/L = 9\sigma^2/2\pi G I_0 r_c$ is calculated using King's method (King and Minkowski 1972; Richstone and Tremaine 1986). The results are relatively insensitive to seeing corrections, since $I_0 r_c$ varies only slowly with seeing (Schechter 1980). With more accurate core parameters, a correlation between M/L_V and M_B is clearly seen. For ellipticals, $M/L_V \propto L^{0.20\pm0.04}$, similar to the predicted relation and also to the estimate from the metallicity-luminosity relation. From the dependence of giant-branch stellar luminosity on metallicity, Tinsley (1978) derived $M/L_B \propto L_B^{0.13}$.

The mean mass-to-light ratio of old disks provides a welcome check of the present results. In our Galaxy, the disk mass is derived from the Oort-Bahcall analysis of the vertical dynamical equilibrium (Bahcall 1986, and references therein); in other galaxies it is based on a comparison of the vertical σ in face-on galaxies with thicknesses of edge-on galaxies (van der Kruit and Freeman 1986). It is reassuring to see that old disks and ellipticals have similar M/L ratios (Fig. 7).

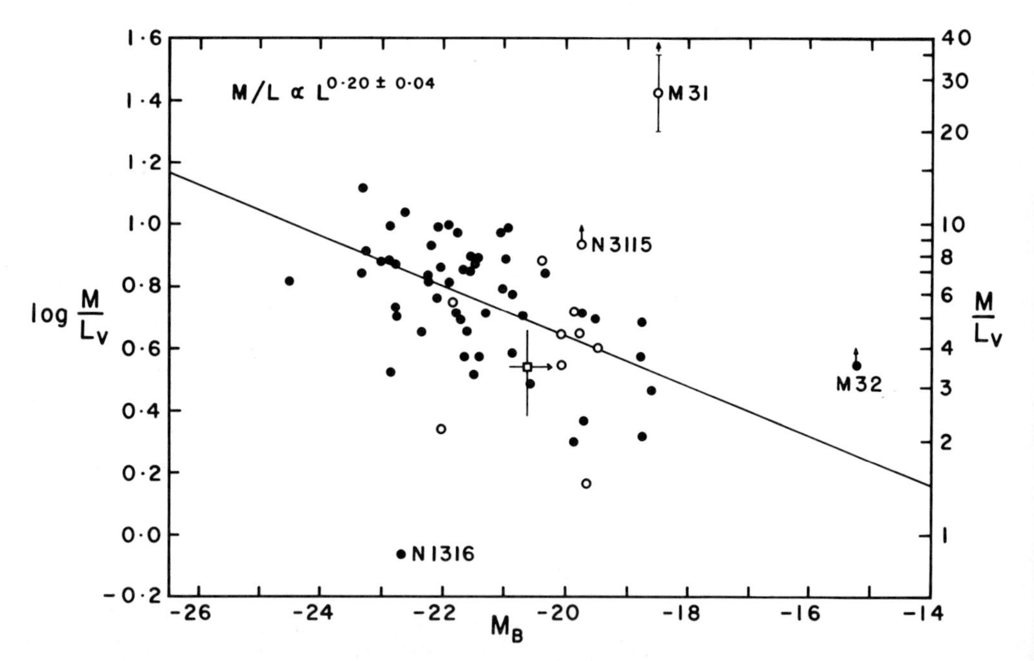

Fig. 7. – Core mass-to-light ratios for bulges and elliptical galaxies. The line is a least-squares fit to the ellipticals. The plus sign is the mean M/L_V for old disks; the error bar is the dispersion of values seen. The point is plotted at the mean luminosity of the galaxies rather than that of the disks (hence the arrow).

An especially interesting implication of Figure 7 follows if the $M/L - L$ correlation holds to galaxies as faint as M32. Richstone (1987) reports that dynamical models of M32 over a large radius range show that $M/L_V = 2.5 \pm 0.1$. This is smaller than the nuclear value, and fits the observed correlation with L. The value is interesting because it is as small as mass-to-light ratios in gobular clusters (Illingworth 1976; Gunn and Griffin 1979; Meylan and Mayor 1986). Dynamical models of globular clusters can account for all of the observed mass with stars and stellar remnants that formed with relatively ordinary mass functions. Richstone's result shows that we need to find little or no dark matter in the inner parts of M32, perhaps not even the kind believed to exist in the galactic disk. The correlations of M/L and metallicity with luminosity then suggest that there is little or no exotic dark matter in the central parts of other ellipticals. It is worth investigating again whether ordinary stellar populations of various metallicities can explain all of the mass in galaxy cores.

5. MASSIVE BLACK HOLES IN GALAXY NUCLEI

5.1 Introduction

The most interesting question we would like to ask of galaxy nuclei is whether they contain massive black holes (hereafter BHs). Compelling motivation is provided by a variety of nuclear activity – modest activity like that in our Galaxy, Seyfert nuclei, radio sources (especially radio continuum jets), and quasars. The activity is believed to originate through the accretion of gas by a nuclear "monster" (e. g., Rees 1984), but direct evidence for BHs is limited. The most conclusive case is the Galaxy, in which the nuclear radio source Sgr A has prodigious power but a size of < 20 AU (Lo *et al.* 1985). For other galaxies I discuss only the evidence based on core dynamics. Published work has been limited by insufficient resolution and a general lack of information about cores. Before we knew that cores are anisotropic, it was easy to misinterpret effects seen in any one galaxy as resulting from a BH. For example, Sargent *et al.* (1978) and Young *et al.* (1978) thought they had detected a BH in M87 because of a velocity dispersion profile and a nonisothermal brightness distribution that turn out to be just like those in other galaxies (Illingworth 1981; Lauer 1985*b*; Kormendy 1985*a*).

Motivated in part by the excellent seeing at the CFHT, I have begun a systematic search for nuclear BHs in early-type galaxies. A wide range of galaxies is being examined, including radio galaxies. However, because of the uncertainty in the velocity distributions of bright ellipticals, I put special emphasis on objects that rotate rapidly, i. e., faint ellipticals and bulges.

5.2 M31

The best evidence for a nuclear BH outside our Galaxy is in M31. Dressler (1984) saw an unresolved central jump in $V(r)$; Walker (1974) observed a maximum rotation velocity of 104 km s^{-1} at $r = 2''$. Dressler also found an increase in σ of 35 % from 145 km s^{-1} to 196 km s^{-1} at the center. He concluded that the mass-to-light ratio increases inward by an order of magnitude; this could be explained by a nuclear BH of mass $\sim 10^7\ M_\odot$. Dressler's observations were obtained with a $2''$ slit, $0\farcs58$ pixels and a seeing of $2''$ FWHM.

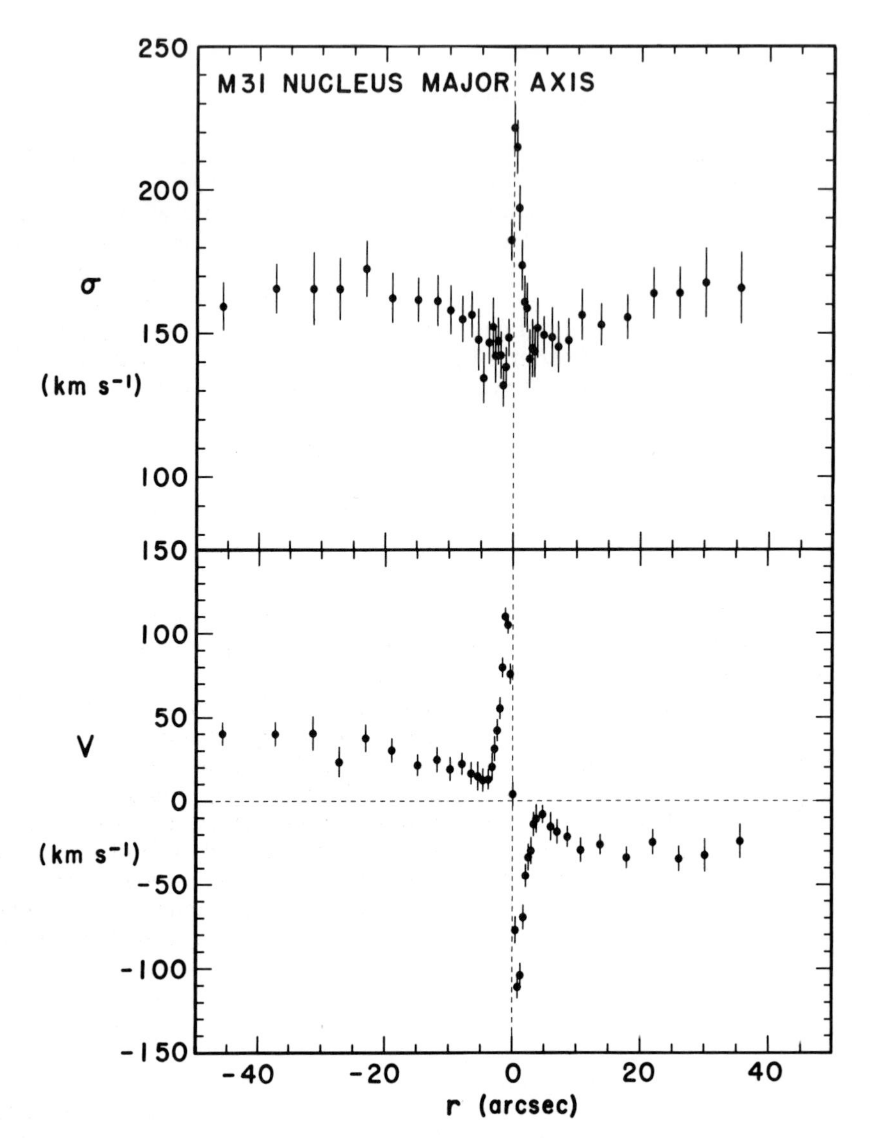

Fig. 8. – Rotation and dispersion profiles along the major axis of the nucleus of M31. Near the center, points are separated by 0.″41.

The present observations were obtained with a slit width of 0.″5, pixels of 0.″41 and a seeing of 1.″2 FWHM. Spectra were taken along the disk major axis, the major axis of the nucleus (26° from the major axis), and the minor axis. Figure 8 shows some of the results. The nucleus spins very rapidly. The central velocity gradient is still unresolved; the apparent peak velocity of 110 km s^{-1} at r = 1.″0 is certainly an underestimate. The dispersion gradient is also unresolved; on one side σ decreases in only 0.″8 from 222 km s^{-1} to the value 145 km s^{-1} in the inner bulge. The asymmetry in the dispersion profile is real (cf. Dressler 1984); it may be due to dust absorption (Nieto *et al.* 1986).

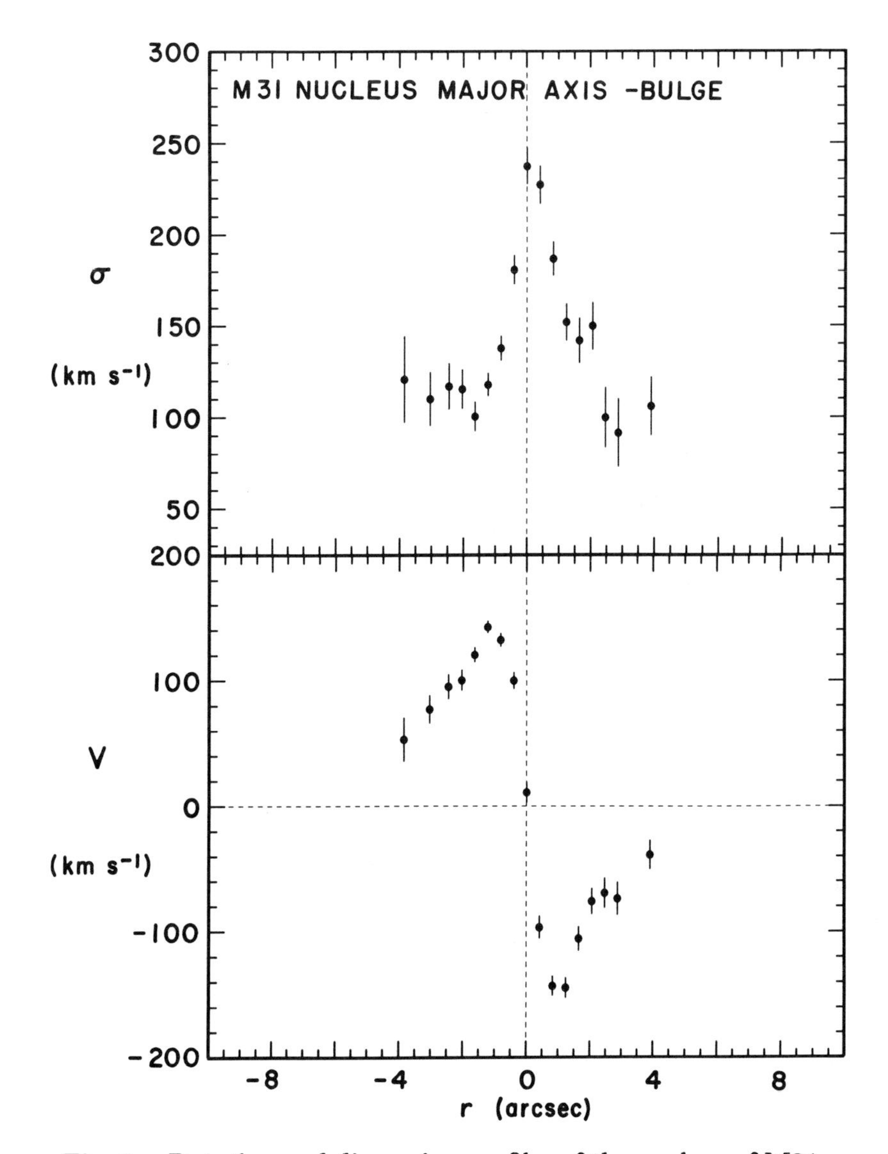

Fig. 9. – Rotation and dispersion profiles of the nucleus of M31 as recalculated after subtraction of the bulge spectrum.

Figure 8 shows that the nucleus is well differentiated from the bulge. In my spectra it outshines the bulge by a factor of 5 at the center; also, $r_{c,app} = 17''$ for the bulge but only 1.″1 for the nucleus. Tremaine and Ostriker (1982) have shown that the bulge and nucleus are, in fact, dynamically independent. I am therefore measuring two superposed but independent stellar populations: stars in the bulge certainly pass in front of and behind the nucleus. Moreover, the bulge has a smaller apparent velocity dispersion than the nucleus. The Fourier Quotient program reacts to such a mixture by weighting the cold component more highly than its relative contribution to the light (Whitmore 1980; McElroy 1983). To get an accurate measurement of the nucleus, I need to subtract the bulge spectrum before running the Fourier program. This is easy to do, since the relative light contributions are easy to determine from the spectrophotometry given the large ratio of bulge and nuclear core radii. The results are shown in Figure 9. As expected, the velocity gradients are larger than in Figure 8. The nucleus has a maximum observed rotation velocity of 143 ± 5 km s^{-1} at $r = 1''.1$ and an apparent central velocity dispersion of 241 ± 7 km s^{-1}.

The surprise in Figure 9 is the dispersion at $r \gtrsim 2''$. Here $\sigma = 107 \pm 5$ km s^{-1}, much smaller than in the bulge. This is clear from the spectra: the absorption lines Mg I λλ 5167 and 5173 Å are partly resolved in the outer nuclear spectrum but not in the total spectrum or in the bulge spectrum that was subtracted. Spectra along all three position angles show this behavior. The mean nuclear dispersion at $r > 2''$ is 107 ± 4 km s^{-1}(internal error). Given the large rotation velocity, this implies that *the nucleus of M31 is a disk.* The "oblate line" describing isotropic rotating spheroids in the $V_{max}/\sigma - \epsilon$ diagram would predict $V_{max}/\sigma = 0.8$ for the E4 shape of the nucleus; an inclined disk would show larger values. *Distinct nuclei like that in M31 may be produced from disk gas that falls into the center in triaxial potentials.* Evidence for the importance of such gas flows is reviewed in Kormendy (1982*a, b*) and in Kormendy and Illingworth (1983). One additional piece of evidence: dust clearly knows how to find the center; it tends to settle in a ring or disk at $r \lesssim r_c$ (Kormendy and Stauffer 1987).

I can now estimate the central mass-to-light ratio. The precise value depends on how much of the central dispersion gradient is unresolved rotation. I consider two extreme cases. If all of the apparent dispersion gradient is unresolved rotation and $\sigma = 107$ km s^{-1} everywhere, then the shallowest possible rotation curve would have to rise to 260 km s^{-1} in the disk plane at $r \simeq 0''.6$ = FWHM/2. The corresponding central M/L_V is ~ 20. Alternatively, if all of the dispersion gradient is real, then King's method corrected for rotation gives $M/L_V = 35$. Since I have not yet made seeing corrections, these are lower limits. Similar values were derived previously (Light *et al.* 1974; Tremaine and Ostriker 1982). However, they are now more secure because of the improved kinematic resolution and because the rapid rotation eliminates the freedom to postulate large velocity anisotropies. Mass-to-light ratios of 20 – 35 are much higher than values in other galaxies (Fig. 7). The unusually high dispersion is emphasized further in Figure 6 (lower left). There is therefore a strong case in M31 for a nuclear BH of mass ~ 2×10^7 $M_\odot$.

5.3 M32

The second reported case of a nuclear BH is M32. Tonry (1984) and Dressler (1984) found an unresolved central jump in $V(r)$ to $\gtrsim 36$ km s^{-1} and a 40 % rise in σ from 60 to 85 km s^{-1} at the center. Both authors concluded that M/L rises toward the center; Tonry estimated a possible BH mass of ~ 5×10^6 $M_\odot$. With the CFHT it is again possible to make a significant improvement in resolution: for Tonry (1984), Dressler (1984) and the present work, seeing FWHM values are, respectively, $2''.5 - 3''$, $2''$, and $1''.00 \pm 0''.05$. Slit width × pixel size values are, respectively, $2'' \times 0''.58$, $2'' \times 0''.58$, and $0''.4 \times 0''.41$.

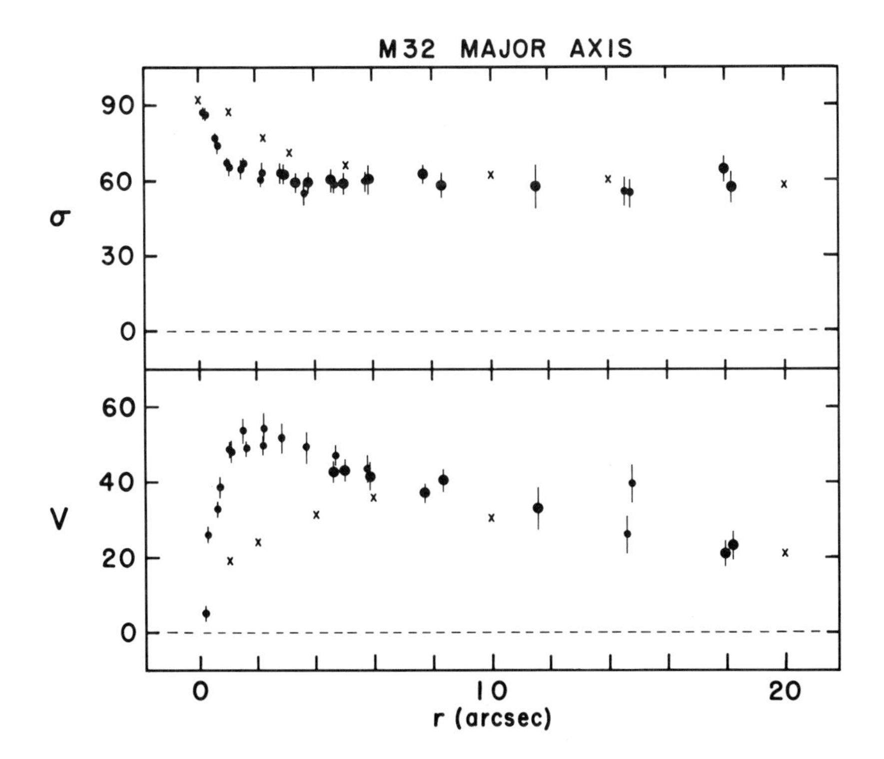

Fig. 10. – Rotation and velocity dispersion curves along the major axis of M32. The crosses show measurements by Tonry (1984); the velocities were taken at position angle 0° (20° from the major axis) and the dispersions at position angle 90°. Tonry's "minor axis" dispersion measurements are shown because they were obtained in better seeing than his PA = 0° data.

The rotation and dispersion curves from the CFHT (Fig. 10) show that the V and σ gradients are much larger at small radii than in published measurements. Also, $V(r)$ reaches > 51 km s^{-1} at $r \simeq 2''$. But the central $\sigma = 87$ km s^{-1} is no larger than before, despite the improvement in resolution. All of the apparent dispersion gradient could be unresolved rotation. The formal core $M/L_V = 4.1$ (corrected for rotation) is plotted in Figure 7. It is not very accurate because of the large seeing corrections, but it is also not especially large for the luminosity of the galaxy. The contrast between M31 and M32 is well illustrated in Figure 6: r_c and μ_{0V} are similar but $\sigma \geq 241$ km s^{-1} in M31 and only ~ 87 km s^{-1} in M32. Since $M/L \propto \sigma^2$, this is a large difference. Based on these results alone, I do not see strong evidence for a nuclear BH.

Richstone (1987) points out that $M/L_V = 2.5 \pm 0.1$ is closely constrained at larger radii. Therefore M/L_V does rise toward the center. Based on his dynamical models, Richstone concludes that M32 remains an interesting BH candidate.

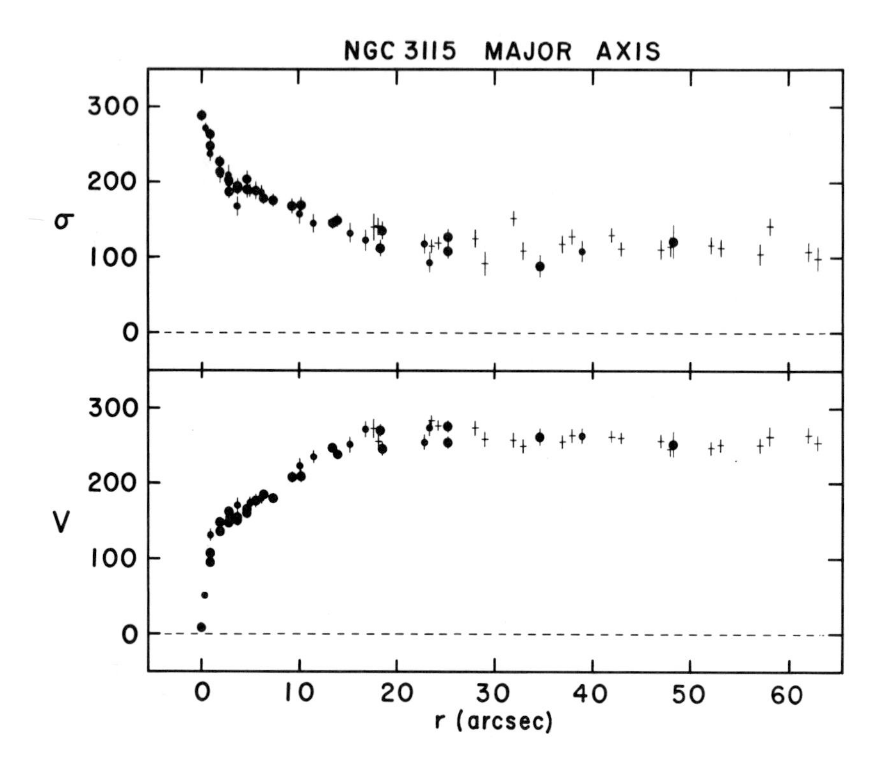

Fig. 11. – Rotation and velocity dispersion curves in NGC 3115. Large points are from an 1800 s exposure and small ones from a 900 s exposure with the CFHT. The plus signs are measurements by Illingworth and Schechter (1982).

5.4 NGC 3115

The present program has produced one new BH candidate. Figure 11 shows rotation and dispersion curves for the edge-on S0 galaxy NGC 3115. The slit width was 0.″8, the pixel size was 0.″92, and the seeing was poor (FWHM = 1.″7). The rotation curve has an unresolved rise to > 120 km s^{-1} at $r = 1''$. The velocity dispersion rises by 50 % from 190 km s^{-1} at $r \simeq 3''$ to 288 ± 8 km s^{-1} at the center. I assume that the velocity distribution is isotropic, because of the rapid rotation (Illingworth and Schechter 1982) and because my photometry shows that there is a nuclear disk at $r \lesssim 2''$. Then the central $M/L_V > 8.6$. This is a lower limit, because I have not corrected for seeing. M/L_V is somewhat large for the luminosity of the bulge (Fig. 7). Also, σ is well outside the distribution of values for other galaxies (Fig. 6, lower left). This suggests the presence of a central mass concentration of $\sim 5 \times 10^8$ M$_\odot$.

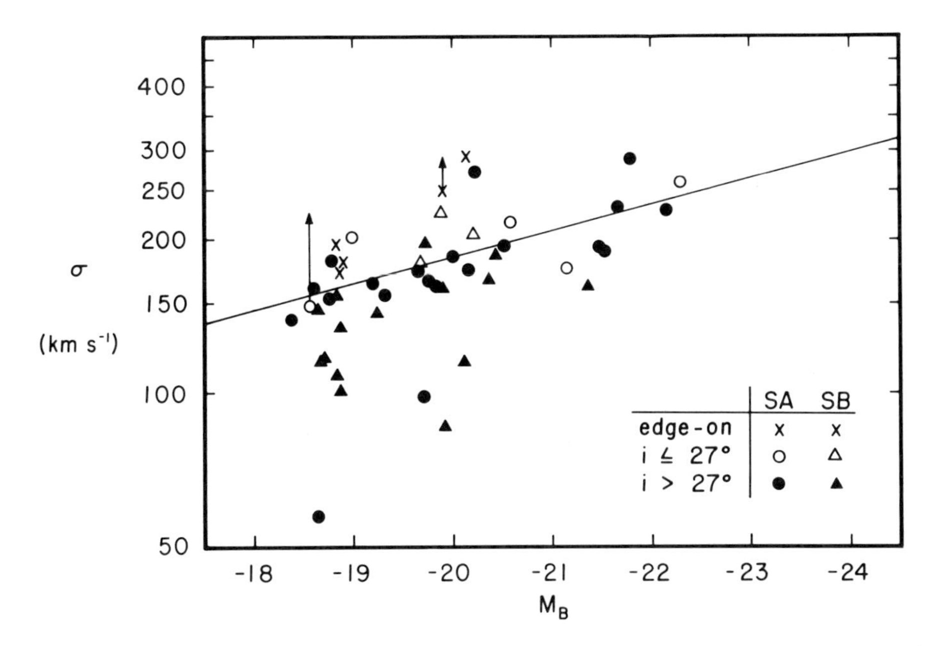

Fig. 12. – Faber-Jackson relation for bulges of various inclinations ($90° - i$). The solid line is the mean relation for SA0–bc galaxies. The arrows show the present measurements of M31 and NGC 3115 (before seeing corrections).

5.5 Ubiquitous Spinning Nuclei and Central Black Holes?

Figure 12 shows the Faber-Jackson relation for bulges of disk galaxies, from Kormendy and Illingworth (1983). We noted that more inclined bulges have larger σ. The zeropoint, say σ at $M_B = -21$, is 208 ± 8 km s^{-1} for SA0-bc galaxies and smaller for SB galaxies but 234 ± 10 km s^{-1} for galaxies within 27° of edge-on. We had no obvious explanation. We noted that the observations could be due to rapid central rotation produced by nuclear mass concentrations, but we didn't really believe it at the time. Given the results of the previous sections, it is interesting to wonder whether rapidly spinning nuclear disks and central mass concentrations are widespread phenomena in galaxy bulges. The search with the CFHT for BHs in galaxy nuclei is continuing.

REFERENCES

Bahcall, J. N. 1986, in *IAU Symposium 117, Dark Matter in the Universe*, ed. J. Kormendy and G. R. Knapp (Dordrecht: Reidel), p. 17.

Bendinelli, O., Parmeggiani, G., and Zavatti, F. 1984, *Astr. Ap.*, **140**, 174.

Binney, J. 1976, *M. N. R. A. S.*, **177**, 19.

Binney, J. 1978, *M. N. R. A. S.*, **183**, 501.

Binney, J., and Mamon, G. A. 1982, *M. N. R. A. S.*, **200**, 361.

Carlberg, R. 1987, in *IAU Symposium 127, Structure and Dynamics of Elliptical Galaxies*, ed. T. de Zeeuw (Dordrecht: Reidel), p. 353.

Dressler, A. 1984, *Ap. J.*, **286**, 97.

Duncan, M. J., and Wheeler, J. C. 1980, *Ap. J. (Letters)*, **237**, L27.
Faber, S. M., and Jackson, R. E. 1976, *Ap. J.*, **204**, 668.
Gunn, J. E., and Griffin, R. F. 1979, *A. J.*, **84**, 752.
Illingworth, G. 1976, *Ap. J.*, **204**, 73.
Illingworth, G. 1977, *Ap. J. (Letters)*, **218**, L43.
Illingworth, G. 1981, in *The Structure and Evolution of Normal Galaxies*, ed. S. M. Fall and D. Lynden-Bell (Cambridge: Cambridge Univ. Press), p. 27.
Illingworth, G., and Schechter, P. L. 1982, *Ap. J.*, **256**, 481.
Kent, S. M. 1983, *Ap. J*, **266**, 562.
King, I. R. 1966, *A. J.*, **71**, 64.
King, I. 1978, *Ap. J.*, **222**, 1.
King, I. R., and Minkowski, R. 1972, in *IAU Symposium 44, External Galaxies and Quasi-Stellar Objects*, ed. D. S. Evans (Dordrecht: Reidel), p. 87.
Kormendy, J. 1982*a*, in *Morphology and Dynamics of Galaxies*, ed. L. Martinet and M. Mayor (Sauverny: Geneva Observatory), p. 113.
Kormendy, J. 1982*b*, *Ap. J.*, **257**, 75.
Kormendy, J. 1984, *Ap. J.*, **287**, 577.
Kormendy, J. 1985a, *Ap. J. (Letters)*, **292**, L9.
Kormendy, J. 1985b, *Ap. J.*, **295**, 73.
Kormendy, J. 1986, in *Nearly Normal Galaxies: From the Planck Time to the Present*, ed. S. M. Faber (New York: Springer-Verlag), in press.
Kormendy, J., and Illingworth, G. 1983, *Ap. J.*, **265**, 632.
Kormendy, J., and Stauffer, J. 1987, in *IAU Symp. 127, Structure and Dynamics of Elliptical Galaxies*, ed. T. de Zeeuw (Dordrecht: Reidel), p. 405.
Lauer, T. R. 1985*a*, *Ap. J. Suppl.*, **57**, 473.
Lauer, T. R. 1985*b*, *Ap. J.*, **292**, 104.
Light, E. S., Danielson, R. E., and Schwarzschild, M. 1974, *Ap. J*, **194**, 257.
Lo, K. Y., Backer, D. C., Ekers, R. D., Kellermann, K. I., Reid, M., and Moran, J. M. 1985, *Nature*, **315**, 124.
McElroy, D. B. 1983, *Ap. J.*, **270**, 485.
Meylan, G., and Mayor, M. 1986, *Astr. Ap.*, **166**, 122.
Nieto, J.-L., Macchetto, F. D., Perryman, M. A. C., di Serego Alighieri, S., and Lelièvre, G. 1986, *Astr. Ap.*, **165**, 189.
Rees, M. J. 1984, *Ann. Rev. Astr. Ap.*, **22**, 471.
Richstone, D. O. 1987, in *IAU Symposium 127, Structure and Dynamics of Elliptical Galaxies*, ed. T. de Zeeuw (Dordrecht: Reidel), p. 261.
Richstone, D. O., and Tremaine, S. 1985, *Ap. J.*, **296**, 370.
Richstone, D. O., and Tremaine, S. 1986, *A. J.*, **92**, 72.
Sargent, W. L. W., Young, P. J., Boksenberg, A., Shortridge, K., Lynds, C. R., and Hartwick, F. D. A. 1978, *Ap. J.*, **221**, 731.
Schechter, P. L. 1980, *A. J.*, **85**, 801.
Schweizer, F. 1979, *Ap. J.*, **233**, 23.
Schweizer, F. 1980, *Ap. J.*, **237**, 303.
Schweizer, F. 1981, *A. J.*, **86**, 662.
Tinsley, B. M. 1978, *Ap. J.*, **222**, 14.
Tonry, J. L. 1984, *Ap. J. (Letters)*, **283**, L27.
Tremaine, S. D., and Ostriker, J. P. 1982, *Ap. J.*, **256**, 435.
van der Kruit, P. C., and Freeman, K. C. 1986, *Ap. J.*, **303**, 556.
Walker, M. F. 1974. *Publ. A. S. P.*, **86**, 861.
Whitmore, B. C. 1980, *Ap. J.*, **242**, 53.
Young, P. J., Westphal, J. A., Kristian, J., Wilson, C. P., and Landauer, F. P. 1978, *Ap. J.*, **221**, 721.

DISCUSSION

Lupton: Using your correlation, at what M_B is the central two–body relaxation time equal to the Hubble time?

Kormendy: Fifty–five galaxies with $-18.6 \geq M_B \geq -24.5$ give $t_{or} \propto L^{2.3\pm0.1}$ with $t_{or} = 10^{10} yr$ at $M_B = -17.4$ (assuming a Hubble constant $H_0 = 50\ km\ s^{-1} Mpc^{-1}$). However, the smallest 14 values of t_{or} belong to galaxies that are very poorly resolved. Since $t_{or} \propto \sigma r_c^2$, and seeing corrections are usually underestimated, I expect that these values of t_{or} are upper limits, that the real relation may be slightly steaper than I have found, and that t_{or} may already equal $10^{10} yr$ at an absolute magnitude brighter than -17.4.

King: You seem to have determined a core radius for M32, yet I remember that the unpublished Stratoscope curves, which I have seen, have a much smaller core radius—if any at all. How did you determine your core radius for M32? This is important, because the derived M/L depends linearly on the r_c assumed (since it is $I_0 r_c$ that goes into the formula, while $I_0 r_c^2$ is conserved).

Kormendy: The value I quoted, $r_c \sim 1.3$ pc, is a lower limit based on a profile that was very poorly resolved ($r_{c,app}/\sigma_* < 3$). I applied the best correction for seeing that I could, using the method discussed in the text, but I would not be surprised if any real core were much smaller than I derived.

Gunn: If you interpret your M31 data as due to motion in a central disk, is the fall of rotation velocity with radius consistent with a Keplerian potential?

Kormendy: Formally the best fit is $v \propto r^{-1}$, but $v \propto r^{-1/2}$ is not excluded by the rather large error bars.

Porter: You said that the different central values of the distribution function in ellipticals and disks argues against formation of ellipticals by dissipationless mergers of disks—but would you expect disks to merge without dissipation? Or do you think the differences in f_0 are large enough to argue against mergers anyway?

Kormendy: This question should really be addressed to Ray Carlberg, whose work I was summarizing. As I understand it, the argument excludes only dissipationless mergers of stellar disks. Any gas in two merging disks could produce high phase—space densities in stars that they form. Also, mergers of galaxies that already contain bulges with high central densities could easily make cores like those we see in ellipticals.

Gerhard: Do you confirm the 2:1 apparent axial ratio of the nucleus of M31, as found in the Stratoscope II observations (Light, Danielson and Schwarzschild, 1974, *Astrophys. J.*, **194**, 257), which would—in the disk interpretation of the nuclear rotation—increase the velocities correspondingly?

Nieto: About the nucleus of M31, I would like to comment on the observations made in the UV at the CFH Telescope with the ESA Photon-Counting-Detector, which is the scientific prototype of the Faint Object Camera of ST. The resolution (FWHM) was 0.65 arcseconds. We found indeed a dust lane in the nucleus oriented along the major axis, i.e., perpendicular to the rotation axis found in 1960 by Lallemand, Duchesne & Walker (*Publ. Astr. Soc. Pacific*, **72**, 76); it is suggested that the nucleus was close to an oblate system. In addition, we found a strong UV excess by comparing our profile with the radial profile obtained by Kent. This result should be however taken with care since it comes from two different instruments, and should be confirmed. (Nieto, Macchetto, Perryman, di Serego Alighieri & Lelièvre, 1986, *Astr. Astrophys.*, **165**, 189). These observations were made in the Time Resolved Imaging Mode of the PCD. We are presently applying image centering and – selection algorithms in order to improve the resolution of these data and to go down to the sub-half arc second level. We may have therefore a few more things to say soon.

Kormendy: These beautiful observations appear to support the hypothesis that the nucleus was made out of infalling disk gas; they show that the interstellar matter reaches all the way in to the nucleus, and suggest that young stars are present.

Sussman: Assuming M31 is 750 kpc distant, 1 arcsec projects to about 4 pc. The nucleus thus has a radius of a few pc, from your measurements. Your dynamical data requires that we stuff a few times 10^7 $M_\odot$ into that radius. This density does not seem terribly inconsistent with a large, purely stellar cluster. Is there strong evidence that I miss here, indicating that the mass is concentrated into an unusual compact object? Is this conclusion supported by the dynamical data alone?

Kormendy: At the luminosity of the bulge of M31, M/L_V values for other bulges and ellipticals lie between ~ 2 and ~ 6. The bulge of M31 itself has a normal M/L_V. But the mass–to–light ratio of the nucleus, calculated in the same way and of course taking into account the very high surface brightness, is large. At $M/L_V = 20 - 40$ (even without seeing corrections) it is far outside the distribution of M/L_V values for other stellar systems which have similar K–star spectra. Of course, it is always possible to explain such a mass–to–light ratio with a sufficiently strange stellar mass function, but the resulting stellar population would be *very* different from any that we have seen elsewhere. In fact, if the blue central colors measured by Nieto *et al.* are confirmed, the indications are, if anything, that the mass–to–light ratio of the stars in the nucleus is lower, rather than higher, than average.

ISOPHOTE SHAPES

R. I. Jedrzejewski
Mount Wilson and Las Campanas Observatories
813 Santa Barbara Street
Pasadena
CA 91101

ABSTRACT. The shapes of the isophotes of elliptical galaxies are discussed. Ellipticity and position angle variations with distance from the centres of galaxies are described, as well as deviations from perfectly elliptical shape. The factors contributing to these deviations, in particular edge-on disks and boxiness, are described and portrayed.

1. INTRODUCTION

As their name suggests, elliptical galaxies are characterised by having isophotes that are very closely elliptical. With this in mind, I will first discuss what we can learn from the ellipse parameters of galaxies, and then describe in what ways the isophotes deviate from this simple picture.

2. ELLIPSE PARAMETERS OF ISOPHOTES

2.1. Ellipticity Variations with Radius

Firstly, it has been known for some time that the ellipticities of the isophotes of elliptical galaxies vary with radius, sometimes strongly so. Many types of behaviour are found (Figure 1), including nearly constant ellipticity (IC 4296), a sharp rise to approximately E4 shape followed by a decline outside roughly 1 effective radius (eg NGC 2865, NGC 3377, NGC 4387), a change from E2 in the central regions to almost round in the outer parts (NGC 3311, NGC 4374) and more complex behaviour (NGC 3923, NGC 4976). Most galaxies seem to fall in the last of these categories.

2.2. Isophote Twists

Recently, the phenomenon of variations of the position angle of the isophotes with radius has been observed, and this has been ascribed to

T. de Zeeuw (ed.), Structure and Dynamics of Elliptical Galaxies, 37–46.

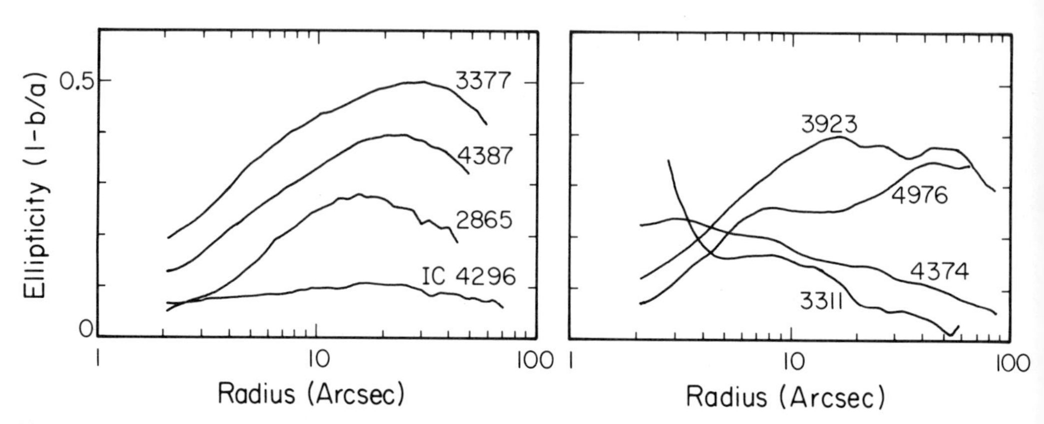

Figure 1. Ellipticity profiles of selected NGC and IC galaxies.

the intrinsic shapes of the galaxies being triaxial. This subject has been nicely described in Mihalas & Binney (1981) and Kormendy (1982).

Note that in the triaxial hypothesis, an isophote twist will only be observed if the axis ratios change with radius, and that no twist will be observed if the galaxy is viewed along any of its principal planes. Also, if the axis ratios change slowly with radius, then we should expect to find few galaxies that are very flattened and have substantial twists (but see Williams, this conference). Observationally, this is found to be the case, which lends support to this combination of hypotheses, but it should be noted that this is not the only model to explain this phenomenon.

Kormendy (1982) argues that some of the observed twists may be tidal in origin, and Gerhard (1983) has produced an N-body model from a merger event with isophote twists that are due to the intrinsic axes of the figure pointing in different directions at different radii, which is stable over many dynamical times. It seems likely that most galaxies are triaxial, but some of the observed twists are possibly due to other effects, such as interaction with a massive neighbour - see, for example, the photograph of M32 in Kormendy (1982).

Note that if ellipticals are triaxial, then these 2 phenomena, ellipticity variations and isophote twists, are caused by the same effect: variations of the intrinsic axial ratios of the triaxial isodensity surfaces with radius. What causes such variations? The models of Wilson (1975) appeared to show convincingly that the variations were linked to rotation, but the discovery that most bright ellipticals do not owe their global shapes to rotation but instead to velocity anisotropies implies that the local shapes are probably anisotropy-driven. For fainter galaxies that are more consistent with isotropic oblate rotators (Davies et al. 1983), Wilson's models are likely to be closer to the truth.

2.3 Examples

Williams (1981) modelled the complex photometric and dynamical behaviour of NGC 596 using a triaxial model, and found three-dimensional models that reproduce the observed behaviour and that would appear relatively normal whichever direction they are viewed from. Since NGC 596 is one of the most twisted galaxies known, other galaxies can probably be fit by even milder triaxial models.

Another galaxy that has a strong twist is NGC 1549, shown in Figure 2. The ellipticity is approximately constant at 0.1, while the position angle twists through about 40 degrees from 5 arcseconds to 60 arcseconds. Note also that the isophotes are not perfectly elliptical, inside 30 arcseconds they are slightly boxy (0.5 to 1% or so) while outside this they have a bar-like perturbation of about the same amplitude. White (1983) describes some structural properties of galaxies formed by merging in N-body simulations; these include box-shaped isophotes (Quinn 1982) for disk-disk mergers and bar-like perturbations if the merger is between spherical progenitors in an oblique collision. Most simulated merger remnants seem to be triaxial and hence may show isophote twists. Malin and Carter (1983) have produced a picture of a shell structure common to NGC 1549 and NGC 1553, only 8 arcminutes distant, so it is clear that some interaction has occurred (or is occurring) between the 2.

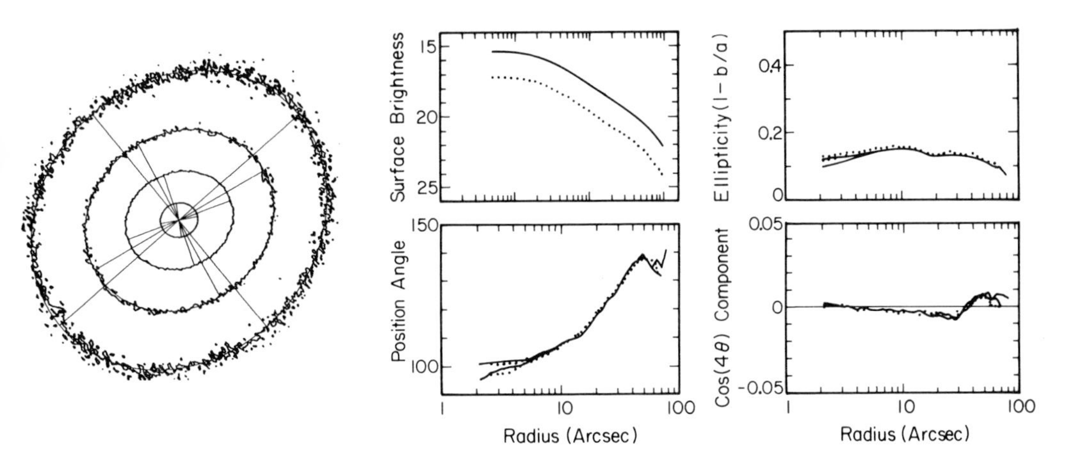

Figure 2. a) Contour map with major and minor axes overlaid and b) surface photometry diagrams for NGC 1549. The solid lines refer to the R band, dotted lines to B.

2.4. Overall Shapes

There are two limiting cases of apparent shape that are seen rarely or never: very flattened galaxies (>E7) and very round objects. My thesis data on 49 ellipticals produced only 2 galaxies that are rounder than E1 at all measured radii (5-100arcsec), and Lauer's (1985a) sample gave only 2 round galaxies out of 42. Similarly, there are no galaxies flatter than E7 that have no disk component. I have also found that very flattened galaxies (>E4.5) almost never have isophotes that are perfectly described by ellipses - either the isophotes are slightly pointed along the major axis, as in NGC 4697, or else they are slightly boxy (eg NGC 6909). These observations indicate that some dynamical processes conspire to inhibit the longevity of very round and very flattened objects, and recent theoretical work on bar instabilities in spherical galaxies with radial anisotropy (Barnes et al. (1985), Merritt and Aguilar (1985)), and on the fire-hose instability in flattened galaxies (Fridman and Polyachenko 1984) is providing fresh insight into the distribution of apparent shapes.

3. DEVIATIONS FROM PERFECT ELLIPSES

The quality of data has now reached the point where we can begin to look for departures from elliptical shape in the isophotes of elliptical galaxies. In particular, CCD's allow investigation of the shapes in the central regions that are not reliably measurable using photographic plates.

3.1. Types of Deviation

What can cause the isophotes of elliptical galaxies to appear non-elliptical? The most obvious agent is dust. Whether in grand dust-lanes (as in Cen A) or more 'blobby' structure (as in NGC 4696) or filaments (Fornax A), dust can certainly alter the shapes of the isophotes of elliptical galaxies. A recent survey by Sadler and Gerhard (1985) indicates that as many as 40% of elliptical galaxies may have dust, although several recent CCD studies (Sparks et al. (1985), Lauer (1985b), Jedrzejewski (1985)) indicate lower numbers of about 20-30%. One notable feature of dust absorption is that it affects blue light more than red, so the ratio of a red image to a blue one will, if the signal-to-noise ratio is high enough, map out the distribution of reddening material. Since dust will be covered by others in these Proceedings, I will say no more about it.

A second cause of non-elliptical isophotes is an edge-on disk component. If the galaxy is mistakenly classified as an elliptical (on the basis of its light profile in the outer parts when the central regions are saturated), but is in fact an edge-on S0 galaxy, the isophotes will show points along the major axis, rather like a lemon. As the galaxy is tilted to become more and more face-on, the points will become softer and softer, and finally disappear from view. The true nature of the galaxy will then be hidden, unless the bulge-to-disk ratio is very small and the disk close to exponential, when we might learn the type from study of the luminosity profile.

Thirdly, we have the phenomenon of 'boxy isophotes'. This term was coined to describe the isophotes of, among others, the bulge of NGC 128. In this case the isophotes look more like an athletics track, consisting of curved ends connected by long 'straights'. It has been found that box-shaped bulges are relatively common in spiral and S0 galaxies (see Jarvis 1986), but not seen up till recently in elliptical galaxies. This has been used as evidence that ellipticals and bulges are different types of animal (Illingworth 1983), but the recent discovery of boxy isophotes in ellipticals negates this.

Lastly, there is the observation that many galaxies show 'shells' or 'ripples'. These appear as sharp edges in the brightness profiles at large distances from the centre of the galaxy, and have been extensively modelled by Quinn. Since they will probably be mentioned at length during this Symposium, I will leave discussion to those more familiar with the subject.

3.2. Displaying the Deviations

The best way to study the small departures from ellipses in the isophotes of elliptical galaxies is to use the fact that the isophotes are, in fact, close to being elliptical. If the ellipse parameters of the isophotes are measured using one of the many fitting programs available (Cawson's PROF, Lauer's VISTA, etc), these numbers can be used to make a noise-free reconstruction of the galaxy that has perfectly elliptical isophotes. The procedure involves making a 'CCD frame', where for each pixel (i,j), one has to determine which isophote that pixel lies upon. The equations are not analytic if the ellipse parameters vary with radius, but a simple iterative scheme almost never runs into problems. However, the ellipse parameters need to be measured very accurately, or else artifacts will be produced at radii where the parameters are wrong.

An example of this technique is shown in Figure 3 for NGC 7144, a 'normal' elliptical galaxy with no obvious distortions of the isophotes. The first picture shows the direct frame of NGC 7144, the second shows the noise-free reconstruction with perfectly elliptical isophotes, and the third shows the difference picture. Note the complete absence of any significant features in the residual map.

3.3. Box-shaped Galaxies

Now the same trick is tried with NGC 4387. The direct frame shows that the isophotes are slightly boxy, and the residual picture reveals a cross-shaped pattern that is characteristic of boxy isophotes (Figure 4a&b). The distortion can best be parametrised by calculating the 4th harmonic of the intensity around the best-fitting ellipse, and a significant $\cos(4\theta)$ term is found in this isophote (Figure 4c).

NGC 4478 shows a similar pattern, and here the galaxy has a significant isophote twist of about 15°. The boxy cross in the residual picture also twists, but interestingly the phase lags behind the position angle twist such that there is a significant $\sin(4\theta)$ term mixed in. NGC 3605 also has boxy isophotes. What do these galaxies all have in common? Note that they are all low-luminosity galaxies with much brighter companions: NGC 4478 is about 9 arcminutes from M87, NGC 4387 is 10 arcminutes from both NGC 4374 and NGC 4406, and NGC 3605 is only 3 arcminutes from NGC 3607. A reasonable guess would be that the influence of the massive neighbour is significant here.

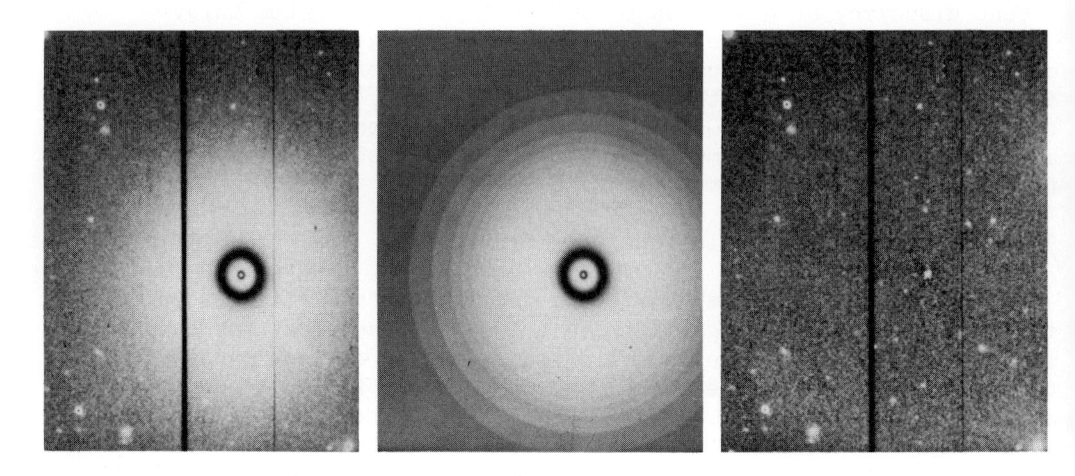

Figure 3. Illustrating the subtraction of a model galaxy with perfectly elliptical isophotes for NGC 7144. a) Direct picture b) Elliptical reconstruction c) Residual picture formed by subtracting b) from a). These are 'pseudo contour' representations, with contours spaced evenly in surface brightness. The dark vertical bars are bad columns on the CCD.

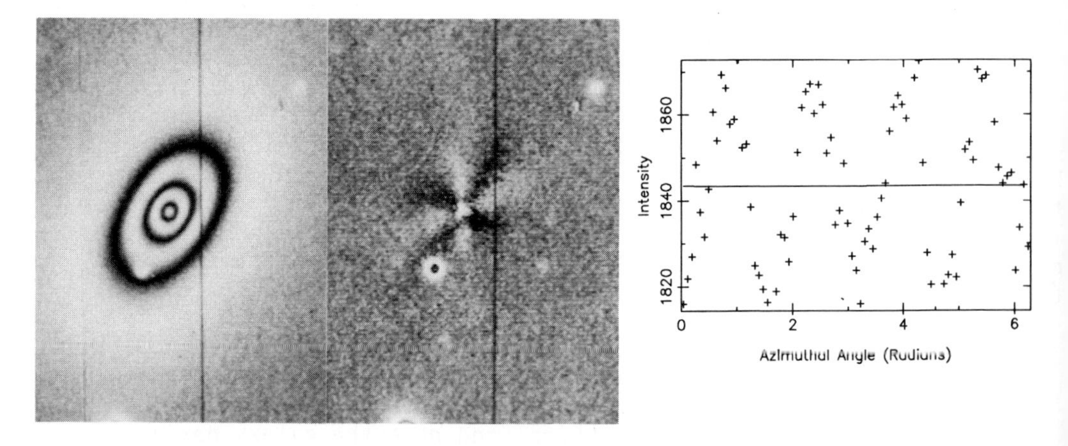

Figure 4. NGC 4387 a) Direct picture b) Residual picture c) Plot of the galaxy intensity as a function of azimuthal angle around the best fitting ellipse for a certain boxy isophote. Note the highly significant cos(4θ) component

These are not the only boxy galaxies. NGC 4374 and NGC 5898 show the same property in their residual maps, as does the extremely flattened NGC 6909. Lauer's study turned up NGC 1600, NGC 4261 and NGC 7785, and these are all bright objects. NGC 4261, another boxy galaxy, has been found by Davies & Birkinshaw (1986) to rotate along its minor axis, and so is probably prolate. My estimate would be that about 20 percent of elliptical galaxies show boxiness at the 0.5%

level or more. It is because we now have CCD's that they are only now beginning to be revealed.

There appears to be a one-to-one correspondence between box-shaped bulges and a cylindrical velocity field (Illingworth 1983), although the sample size is small. We need to know whether the same is true for boxy ellipticals.

3.4. Edge-on disks

Another type of perturbation is that produced by an edge-on disk embedded in the elliptical isophotes. Now the isophotes are slightly pointed along the major axis. Once again, a simple ellipse cannot fit the shape perfectly, and the best that can be done is to increase the flattening slightly - see Figure 5. Now the distortion appears as a $\cos(4\theta)$ term again, but this time the sign is the reverse of the boxy case. This representation is not ideal - some of the intensity from along the major axis has been redistributed onto the minor axis by the reduction in ellipticity such that the average deviation is zero. The residual map of NGC 7029 shows this effect (Figure 6).

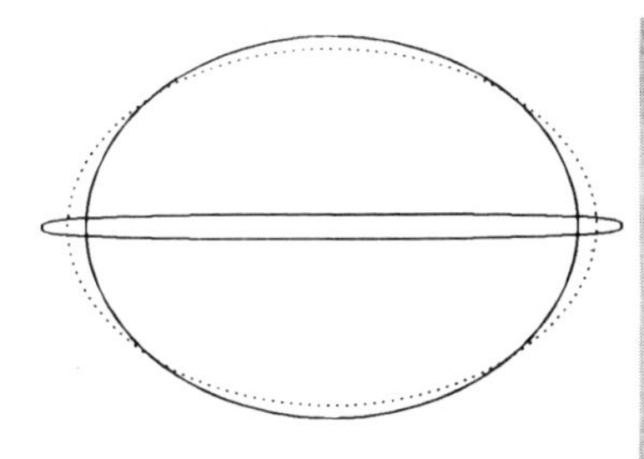

Figure 5. Illustrating the best-fit ellipse (dotted) to an isophote made up of an elliptical bulge and edge-on disk (solid).

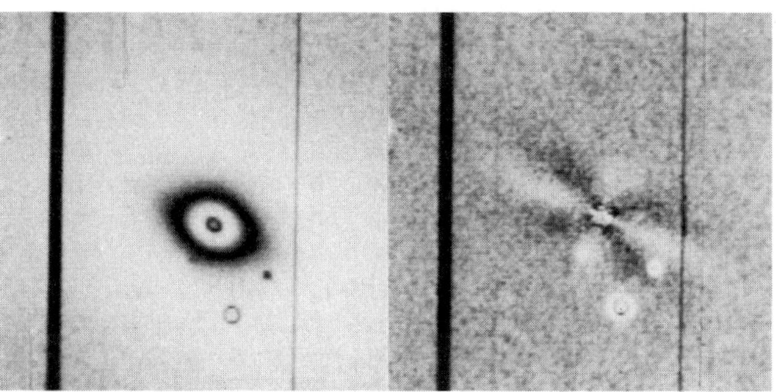

Figure 6. NGC 7029 a) Direct picture b) Residual picture.

The relative numbers of ellipticals that have 'disk-like' perturbations are strongly dependent upon how many S0's creep into the sample. Of the 29 galaxies in my sample that are classified as elliptical in the Revised Shapley-Ames Catalogue (Sandage & Tammann 1981), 7 have significantly boxy isophotes and 4 have significantly pointed isophotes. Of the galaxies classified as E4 or flatter, about half have pointed isophotes. Not all ellipticals are harbouring a faint disk - some very flattened galaxies have isophotes that are boxy rather than pointed.

4. CONCLUSIONS

The isophotes of elliptical galaxies are true ellipses to better than

2% in most cases. The observations of ellipticity gradients and isophote twists are best explained by triaxial models, which have been shown from N-body experiments to be a natural product of many elliptical galaxy formation mechanisms. Boxiness and edge-on disks are the most important contributors to deviations from perfect ellipses, and these deviations are displayed by subtracting a model with perfectly elliptical isophotes. About 20% of galaxies have isophotes that are boxy to 0.5% or more, while about half of all ellipticals may have a weak disk.

ACKNOWLEDGEMENTS

Among the many people that have stimulated my interest and provided helpful insight, I would particularly like to thank Roger Davies, George Efstathiou, Mike Cawson and Paul Schechter.

REFERENCES

Barnes, J., Goodman, J. & Hut, P., (1985). Astrophys. J., 300, 112.
Davies, R.L., & Birkinshaw, M., 1985. Astrophys. J., 302, L45.
Davies, R.L., Efstathiou, G.P., Fall, S.M., Illingworth, G.D. and Schechter, P.L., 1983. Astrophys. J., 266, 41.
Fridman, A.M. & Polyachenko, V.L., 1985. Physics of Gravitating Systems, Volume 2 (New York: Springer-Verlag).
Gerhard, O.E., 1983. Mon. Not. R. astr. Soc., 203, 19P.
Illingworth, G.D., 1983. In Internal Kinematics and Dynamics of Galaxies, ed. E. Athanassoula (Dordrecht: Reidel).
Jarvis, B.J., 1986. Astron. J., 91, 65.
Jedrzejewski, R.I., 1985. Ph.D. Thesis, University of Cambridge.
Kormendy, J., (1982). In Morphology and Dynamics of Galaxies, Twelfth Saas-Fee Advanced Course (Sauverny: Geneva Observatory).
Lauer, T.R., 1985a. Astrophys. J. Suppl. Ser., 57, 473.
Lauer, T.R., 1985b. Mon. Not. R. astr. Soc., 216, 429.
Malin, D.F. & Carter, D., 1983. Astrophys. J., 274, 534.
Merritt, D. & Aguilar, L.A., 1985. Mon. Not. R. astr. Soc., 217, 787.
Mihalas, D. & Binney, J., (1981). Galactic Astronomy (San Francisco: W.H. Freeman).
Quinn, P., 1982. Ph.D. Thesis, Australian National University.
Sadler, E.M. & Gerhard, O.E., 1985. Mon. Not. R. astr. Soc., 214, 177.
Sandage, A. & Tammann, G.A., 1981. A Revised Shapley-Ames Catalogue of Bright Galaxies (Washington: Carnegie Institution of Washington).
Sparks, W.B., Wall, J.V., Thorne, D.J., Jorden, P.R., van Breda, I.G., Rudd, P.J. & Jorgensen, H.E., 1985. Mon. Not. R. astr. Soc., 217, 87.
White, S.D.M., 1983. In Internal Kinematics and Dynamics of Galaxies, ed. E. Athanassoula (Dordrecht: Reidel).
Wilson, C.P., 1975. Astron. J., 80, 175.
Williams, T.B., 1981. Astrophys. J., 244, 458.

DISCUSSION

King: My impression, based on a small sample, was that galaxies with appreciable central ellipticities have steeper rotation curves, while if a galaxy is round in the center but flattened farther out, it has a rotation curve that rises very slowly. An example is NGC 4406. Do you confirm this?

Jedrzejewski: No, in fact, by looking at a few galaxies with ellipticity profiles that rise outwards, reach a peak at about $1r_e$ and then decline slowly, I get the impression that these are more likely to be rotationally supported, and those that you describe are more likely to be non–rotating. I should emphasise that this is only an impression and not based on hard numbers.

White: When you say that a galaxy does not show "pointy" isophotes at the 1% or 0.5% level, what limit does that translate into on the fraction of the light of the system contributed by a nearly edge–on disk?

Jedrzejewski: I haven't tried to model these objects as two–component systems to estimate constraints on the relative brightness of any 'disk' component, but my guess is that for a galaxy with 1% 'pointed' isophotes a disk has a contribution to the total light of 0.1–1%.

Davies: Carter has modeled NGC 4697 and concluded that 2% of the light comes from the disk component. This corresponds to a B_4 of 0.04 - 0.05. Is there a trend in B_4 with luminosity?

Jedrzejewski: There is no real trend, but it does seem that fainter galaxies have a slightly higher fraction of objects with non–elliptical isophotes. These are divided roughly equally between those with 'boxy' and 'pointed' isophotes.

Jarvis: I would like to make two comments. Firstly, there are considerably better examples of elliptical galaxies having box–shaped isophotes than the ones that you have shown. These can be seen to be box–shaped without any image processing at all. Two such examples are IC 3370 and NGC 4125. For the former galaxy, I have obtained surface photometry and kinematics which show that IC 3370 is strongly cylindrically rotating in a similar way that the bulges of disk galaxies having box– or peanut–shaped bulges do.

Jedrzejewski: I agree, but I wasn't particularly looking for boxy galaxies, instead they appeared in a sample of supposedly 'normal' galaxies. By no stretch of the imagination would I call IC 3370 a normal galaxy, but it is interesting to see that the velocity field is similar to that of 'peanut–shaped' bulges.

Schwarzschild: Your beautiful new data seem to me particularly valuable for the following reason. We theoreticians have built, up to now, mostly models with strictly ellipsoidal iso–density surfaces. But I know of no reason why nature should build ellipticals strictly with ellipsoidal shapes. I feel optimistic that models with a larger variety of shapes than you observe can be constructed.

Jedrzejewski: Until the widespread use of CCD's, attempts to measure departures from strictly elliptical isophotes were generally unsuccessful, so it is perhaps not surprising that emphasis was placed on models with perfectly ellipsoidal contours. It must be remembered, however, that in most cases the deviations are *small*, of at most up to a percent or so.

Robert Jedrzejewski and George Djorgovski discuss isophotes.

DISTRIBUTION OF LIGHT: OUTER REGIONS

Massimo Capaccioli
Institute of Astronomy
University of Padova
35122 Padova
Italy

1. INTRODUCTION

The data base of photometric properties of galaxies has grown enormously in the last few years. Surveys in various colors, detailed studies of selected objects, and a swarm of papers with underlying photometric information have invaded the market, fostered by advances in the technology of observations and by improvements and wider availability of reduction facilities. A bibliographical compilation, of great use in the jungle of old and modern references, has been published by Davoust and Pence (1982), and is currently updated (Pence and Davoust 1985).

The preference given by observers to early–type galaxies is in line with the revival of interest for ellipticals, triggered, about 10 years ago, by unexpected kinematical findings (Bertola and Capaccioli 1975, Illingworth 1977). Indeed, photometry has proven a source of original new input to the growing picture on E galaxies; the discoveries of isophotal twisting (whose paternity, although controversial, should be fairly attributed to Williams and Schwarzschild, 1979) and of *shells* (Malin and Carter 1980), are just two examples. The photometric technique has always been a fundamental complement to other lines of investigation; for instance, lacking better information, light profiles are used to describe the projected distribution of matter under the assumption that M/L is constant with radius. Furthermore, after Oemler (1976), surface photometry of early–type galaxies has become a standard tool to investigate tidal interactions, merging and cannibalism. Elliptical and lenticular galaxies are expected to share more homogeneous properties than later types, in absence of external disturbances; they also occur at different levels of clustering. Thus their colors, structure and outer light profiles are all probes of induced modifications and their relation to environmental conditions.

Many modern surveys are based upon CCD data (*e.g.*, Malumuth and Kirshner 1984, Kent 1984, Lauer 1985, Djorgowski 1986). Unfortunately, however, the area covered by presently available detectors (a few *arc–minutes*^{2} at conventional focal–plane scales) is insufficient to encompass large galaxies, which are the most convenient photometric targets (because of better resolution, higher integrated S/N at each isophote, reduced effects of scattered light and local crowding). One consequence is that, so far, our best machinery for surface photometry has contributed primarily to the knowledge of the brighter regions of galaxies. Most, if not all of the modern investigations extending to limiting surface brightness levels ($\sim 10^{-2}$

T. de Zeeuw (ed.), Structure and Dynamics of Elliptical Galaxies, 47–61.

of the night sky luminosity or fainter), are still based *1)* on photographic material collected with large field telescopes, now digitized and reduced with improved techniques, and *2)* on photoelectric observations (aperture photometry and drift scans). A remarkable exception concerns the study of peculiar outer structures with CCD data (*e.g.*, Fort *et al.* 1986).

Before starting this review, it seems worthwhile to comment on the word 'outer' appearing in the title. Cores and outskirts of galaxies pose different problems of both technical and astrophysical nature, justifying the operative separation of light profiles into an inner and an outer region. Still, we should remember that the photometric analysis must be always carried out over the complete luminosity range, since it is often the case that the properties of the outer profile are determined *relative* to those of the inner ones, and vice–versa.

In the following discussion, much space is given to selected technical aspects of the game. Their complexity explains the small enthusiasm for and the skepticism about surface photometry at faint levels, and also accounts for the poor quality of some of the results. In addition to a review of the data and their interpretations, empirical fitting formulae, particularly the $r^{1/4}$ law (de Vaucouleurs 1948), are discussed for the benefit of theorists, commenting on their deviations from observations at large and small scales. Finally, we will speculate about misclassifications and contamination of the elliptical class by lenticulars. Dwarf E galaxies are deliberately ignored, since they appear to form a family which is physically distinct from that of ordinary ellipticals (Kormendy 1985, Ichikawa *et al.* 1986; see the contribution by Nieto in this volume). Previous reviews dealing with this subject were published by de Vaucouleurs (1959, 1979, 1983), Kormendy (1980, 1982), and Capaccioli (1984a).

2. TECHNICAL PROBLEMS AND SOURCES OF ERRORS

The following outline of the technical problems in faint surface photometry is not specific to elliptical galaxies. It is partly based on a paper by Capaccioli and de Vaucouleurs (1983; hereafter CdV), which can be consulted for quantitative information. Similar discussions are scattered in the literature.

2.1. Calibration

Individual pixels of CCD's are linear over a wide range, and sensitivity variations across the detector can be readily corrected. Photographic emulsions, on the contrary, are not linear. This fact does not prevent an accurate intensity calibration, provided that the material is carefully processed and digitized, and that the external sensitometer is properly designed (Fig. 1). Very little can be added to the extensive discussion by de Vaucouleurs (1983). Still, we want to reiterate the complaint about several telescopes whose superb photographic material is degraded by inadequate sensitometry (particularly noteworthy is the light leakage from the densest spots occurring in most of the Kitt Peak type projectors). In any case, calibration errors weight less at faint than bright light levels (the H–D curve is linear at the sky density in well exposed plates). This tolerance justifies the use of an internal calibration whenever strictly needed, *e.g.* for archival plates, although the recovered linearization curve can never extend below the sky density (see Piotto 1986, for an interesting new method of internal calibration, well suited for galaxy photometry). In closing, it is important to emphasize that calibration is not the **limiting factor in photographic surface photometry**.

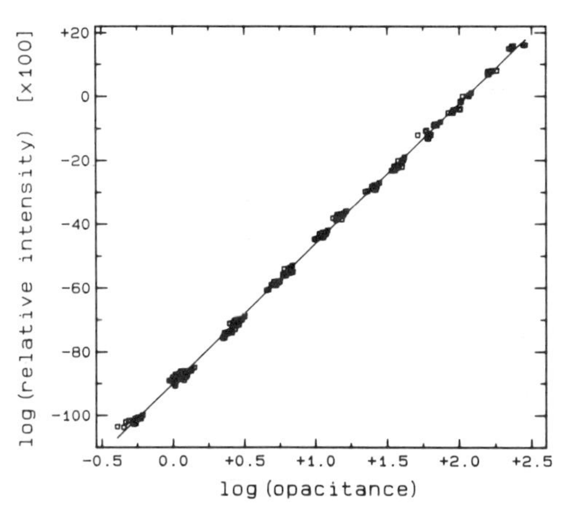

Figure 1. Plot of nominal intensity *vs.* *opacitance* (de Vaucouleurs 1968) for the individual calibration spots of 6 unbaked IIa–O plates exposed at the ESO Schmidt telescope through a GG385 filter. The mean linear relation has a slope of 0.438 with an uncertainty of $\pm 0.3\%$ (Capaccioli and Rampazzo 1986).

2.2. Background removal and related problems

As the distance from center increases, the surface brightness of a galaxy fades smoothly until it merges into the diffuse light of the night sky. Thus, the most critical step in faint surface photometry is the removal of a noisy, low spatial frequency component, which is hundreds of times stronger than the outskirts of the superimposed object. An error in the adopted sky level may have dramatic consequences on the apparent luminosity profiles. For instance, depending on the sign of the systematic error, a pure $r^{1/4}$ profile will mimic a tidal extension or a cutoff (*cf.* Fig. 1 in CdV). The background problem, common to all observational techniques, increases significantly whenever the *blank sky* cannot be recorded simultaneously with the galaxy, since the luminosity of the night glow is not constant (random variations up to 10% may occur in a few minutes; de Vaucouleurs 1958). This is the typical difficulty faced in CCD photometry of galaxies with large angular sizes. *Blank sky* exposures preceding and following the galaxy exposure, hated by observers because they are time consuming, may help to beat the background fluctuations. We will not comment here on other 'tricks' adopted by CCD observers to recover the sky level from a frame completely filled by a galaxy image (for instance, the tautology of seeking it by the assumption that galaxy profile follows a given photometric law).

Since the beginning of galaxy photometry (Hubble 1930), the low frequency component of the local background is estimated by some kind of interpolation of the signal from the *blank sky* areas surrounding the galaxy image. Computer procedures, inspired by the 2–D mapping pioneered by Jones *et al.* (1967), are now currently in use. It would be difficult to list all the various versions and recipes (see, *e.g.*, the median filter approach by Sulentic and Lorre, 1983). In the *Interactive Numerical Mapping Package*, a dedicated procedure developed at the Padova Observatory to reduce PDS scans of plates (Barbon *et al.* 1976a,b), the background contribution is modelled by fitting a canonical 2–D polynomial to the density readings in an 'outer field' encircling the galaxy. Large images, such as bright stars, are masked interactively, while the high frequency components (small stars, emulsion defects, dust, etc.) are taken out by clipping the histogram of the $(O - C)$ residuals. The fitting–rejection procedure is repeated until the standard deviation σ and the skewness of the $(O - C)$ histogram converge. At each cycle, the lowest possible order (from 1 to 5) is chosen for the polynomial on the basis of some graphical tests. The final σ resulting from reduction of limiting Schmidt exposures on unbaked IIa–O plates is of the order of a few 0.001 density units (Capaccioli and Rampazzo 1986); larger values would not permit us to attain the threshold surface brightness $\mu = 28$–29 *B–ss* (abbreviation for *B-mag/arcsec*^{2}). Often a first order polynomial is inadequate to map the complicated pattern of the 'outer field'. Since systematic deviations from linearity may exist (*e.g.*, vignetting in prime focus plates), it is unsafe to prejudicially exclude fitting surfaces other than a plane (*cf.* Schombert 1986).

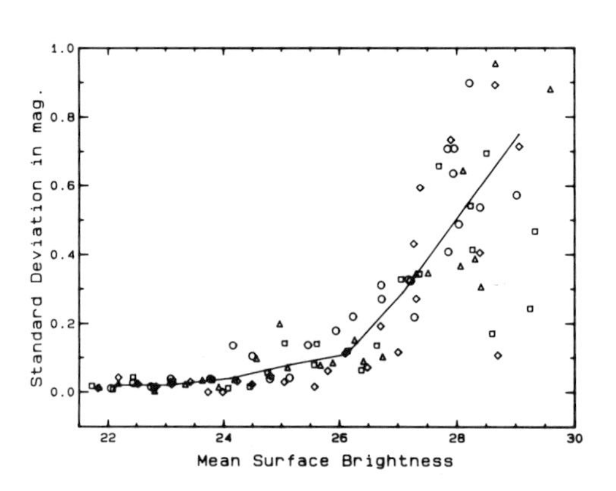

Figure 2. Internal (random) errors *vs.* surface brightness for the main axis light profiles of the S0 galaxy NGC 3115 from two pairs of deep ESO and UK Schmidt plates (Capaccioli *et al.* 1986a). The mean trend (solid line) shows that, at $\mu \simeq 26$ *B–ss*, the uncertainty of the photometric measurements begins to increase rapidly, reaching 100% at $\mu \simeq 28.5$ *B–ss*, in agreement with the threshold estimated by CdV. Consistent values (for the total errors in the V–band) are reported by Schombert (1986).

After several years of use and continuous testing by numerical simulations, the INMP has proven quite effective for removing the background in the absence of complicated situations (see Fig. 2). There are cases, however, in which systematic errors are difficult to control and to evaluate. They occur, for instance, in dense cluster cores, where the outskirts of galaxies overlap and the background becomes a matter of definition, or when a galaxy has close companions, or it is embedded in the corona of a bright star (*e.g.*, NGC 404; Barbon *et al.* 1982) or wrapped within a high–latitude reflection nebulosity (*cf.* CdV). In addition, many methodological issues remain unsolved. They are related to the complexity of the sources contributing to the *blank sky* pattern. For example, low frequency sensitivity variations across a plate and vignetting are different phenomena which, in principle, should be treated separately. Efforts in this direction are rare, and possibly unjustified by the magnitude of the residual sources of systematic and accidental errors. One is the statistical fluctuation in the number density of sub–threshold stars and galaxies, first noted by Miller (1963) and investigated by de Vaucouleurs and Capaccioli (1979; hereafter dVC) in connection with NGC 3379. Unavoidable from both ground and space, this cosmic noise sets at $\mu \simeq 30$ *B–ss* the impassable lower threshold for photometric measurements.

2.3. Scattered light

In addition to classical seeing convolution affecting high frequency structure, the light of a galaxy image is also re–distributed by telescopic, atmospheric and Rayleigh scattering (acting at different scales; see Fig. 4 in CdV). It is mostly the energy of the brighter parts which is shed into the external regions. In their study of NGC

3379, CdV show that the outer ($r > 30"$) radial profile of the scattered light is approximately described by the complete Point Spread Function scaled to the same effective magnitude as the galaxy. With this simple model, it is easily verified that the contamination of the galaxy profile by scattered light increases systematically with radius until the latter becomes the dominant signal; the breakpoint occurs at a surface brightness whose value μ_b depends on the photometric law in the object and its scaling factor. For the model of NGC 3379 ($r_e = 56"$), $\mu_b = 32$ B–ss, and the excess light is less than 0.14 *mag* at all $\mu < 28$ B–ss (CdV). The relation, $\mu_b = \mu_e + 11.44 - 1.52 \log r_e$, holds for an $r^{1/4}$ circular galaxy in the range $30" < r_e < 100"$.

2.4. Light profiles: which and how many ?

A single intensity profile at a given position angle, coupled with the geometrical information (ellipticity and orientation profiles), is often thought sufficient for modelling the 2–D light distribution of an elliptical galaxy. This presumption, together with the need for increasing the significance of measurements at faint levels, has made popular the use of the equivalent profile, in which the surface brightness is related to the radius r^* of the corresponding circularized isophote; for purely elliptical isophotes, $r^* = \sqrt{ab}$. Profiles derived by the ellipse fitting technique fall in this category. Isophotes of real galaxies, however, are often not pure ellipses (Jedrzejewski, this volume), and equivalent profiles (or other *mean* profiles, as in King 1978), may obscure structural details and encourage incorrect conclusions (see Fig. 3 and section 5). In other words, experience suggests that, even for apparently *bona fide* elliptical galaxies, it is always desirable to produce at least two distinct and unsmoothed luminosity profiles along the main axes.

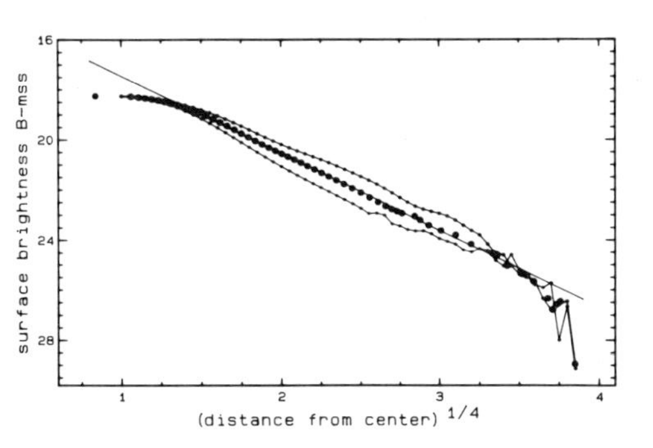

Figure 3. Major and minor axis intensity profiles (small dots) of NGC 4125 from two low resolution plates of the Asiago 1.22–m reflector. The galaxy, classified E4 in RC2, is actually a lenticular with peculiar kinematics (Bertola *et al.* 1982). Large dots trace the equivalent profile, which, contrary to the profiles along the main axes, is well represented by an $r^{1/4}$ law (solid line) down to $\mu \simeq 26$ B–ss.

At very faint levels, especially when statistical fluctuations cause the occurrence of negative intensity values, resolution must be sacrificed in order to reduce the noise. An efficient method of extracting smooth profiles from clean 2–D maps is to sample them with a variable aperture in such a way that S/N remains approximately constant (Capaccioli and Rampazzo 1986). For an $r^{1/4}$ galaxy, a simple rule is to center the aperture along one axis at distances equally spaced in $r^{1/4}$ units, and choose the radius equal to half of the spacing. Anomalous values are easily sorted out by checking the agreement between the geometrical center of the aperture and the light baricenter (no difference should occur at large radial distances). This technique, coupled with an efficient algorithm for removing the stars (*e.g.*, the package 'DAOPHOT' developed for stellar photometry; Stetson 1983), allows a gain of ~ 0.5 *mag* over conventional procedures.

2.5. Photometric zero point calibration

In photographic work the zero point of the photometric scale is usually set by a comparison with photoelectric measurements. In most cases only photoelectric integrated magnitudes within centered apertures are available. The comparison is straightforward, provided that the photographic photometry has no systematic scale error, especially near the galaxy center, where the luminosity gradient is steep. This is not the case for the limiting exposures used to map the faint outskirts of galaxies, due to saturation (*e.g.*, deep Schmidt plates are not reliable for $\mu < 22$ B–ss). Therefore, the outer profiles of galaxies can only be connected by matching to photoelectrically calibrated inner profiles, if available. This procedure, needed in any case to produce the complete light distribution from different exposures, may propagate systematic errors and modify the trend of the final luminosity profile. Often the problem is complicated by the use of heterogeneous material, *e.g.* plates and CCD frames, without proper allowance for the color equations of the various devices. The quality of the result from this pipeline process cannot easily be controlled with photoelectric photometry, *1)* because the data at large apertures is scarce, often old, and unreliable, and *2)* since integration tends to absorb even large errors in the luminosity gradients, transferring them into a shift of the photometric zero point.

2.6. Ultimate limit and photometric standards

Taking into account all sources of errors, CdV estimate that, while it is possible to make *detections*, it is difficult, if not impossible, to obtain *reliable quantitative information* at brightness levels fainter than $\mu \simeq 28.5$ B–ss. Due to the nature of some sources of errors, the limit is unlikly to improve significantly by future technical advances. The solid line in Fig. 2 is indicative of the expected uncertainty in the best possible situation, provided that all systematic errors are removed.

Facing the problem of photometric accuracy, in 1962 the Working Group on Galaxy Photometry of IAU Commission 28 (*IAU Trans.*, XIB, p. 304) stressed the need for standards and proposed a short list of candidates for the various morphological types. However, not much effort has been put into this important project by the community, and so far only three early–type galaxies have been studied in detail, combining a large set of old and new data, namely NGC 3379 (dVC, Nieto 1983a,b, Capaccioli *et al.* 1986b), M87 (de Vaucouleurs and Nieto 1978, 1979) and NGC 3115 (Capaccioli *et al.* 1986a). Several observers have already used this precise information as a reference to evaluate the accuracy of their work. They have simultaneously created a new body of data to be incorporated in a forthcoming revision of the standards themselves.

3. STATE OF THE ART

In a classical paper King (1978) gave light profiles and geometrical parameters for 17 early–type galaxies; the limiting surface brightness was $\mu \simeq 28$ B–ss and, outside cores, the photometric accuracy was usually better than 15% at all $\mu < 26$ B–ss (Capaccioli 1984a). With few exceptions, previously published results are affected by large errors. Figure 5 in Capaccioli (1984b) illustrates the situation for Fraser (1977) and Liller (1960, 1966). The comparison with unpublished data from Benacchio and collaborators, discloses a systematic trend of the magnitude differences *vs.* surface brightness which, as already noted by Burstein (1979), is particularly large for Liller (up to 1 *mag*). The scenario does not improve even for selected well known galaxies (see the case of M87 analysed by Carter and Dixon, 1978). In this situation most of the statements about the outer light profiles of ellipticals made prior to King's (1978) study should be taken with reservations.

Modern studies tend to be more reliable than older ones. Their quality is often clearly established by comparison with a standard galaxy (*e.g.*, Kent 1984, Michard 1985). There are large surveys conceived for statistical purposes, which have a declared limited precision in the individual values (Binggeli *et al.* 1984). Other surveys which are not very deep (*e.g.*, Watanabe 1983, and CCD surveys), and others which claim an accuracy of 0.9 at $\mu = 29$ V–ss (Schombert 1986). The huge project carried out by Lauberts and Valentijn on the ESO Survey plates deserves special mention. It will make available, in a few years, a deep and fairly precise photometric sample for a total of 16,000 galaxies of all types, including 2,500 early–type systems (Lauberts and Valentijn 1984). The flood of data will be filtered and summarized in the forthcoming third edition of the *'Reference Catalogue of Bright Galaxies'* by de Vaucouleurs and collaborators.

However, even for the modern data, errors lie in wait, as can be illustrated by the two following examples.

i. NGC 4406 = M86, a well known early–type giant at the end of the Markarian chain in the Virgo cluster, is a classical target for surface photometry (*cf.* the list of references in Davoust and Pence, 1982, 1985). Its remarkable properties have been compiled by Capaccioli *et al.* (1983). When plotted together, the light profiles from the various sources exhibit a very large scatter at all $\mu > 24$ B–ss. The explanation for this unusually strong disagreement is in the huge halo surrounding M86, which is almost as large as that of M87 (Capaccioli, 1984a). Michard (1985) is the only study where this halo has been taken into account while determining the background level. Uncorrected, the halo results in a spurious increase of the gradient in the outer light profile, with amplitude depending on the individual choice of the fictitious *blank sky.*

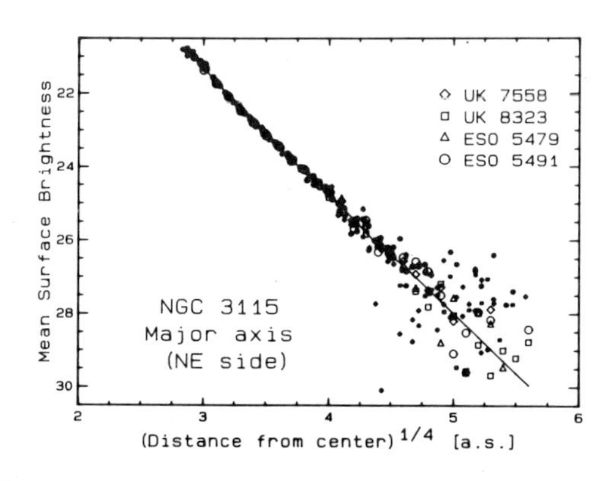

Figure 4. N–E side of the luminosity profile of NGC 3115 from the same four Schmidt plates of Fig. 2, showing the consistency of this material to the limit of detection. Small symbols represent unsmoothed cross–sections, large symbols are measurements with a software defined aperture with variable radius (*cf.* section 2.4). Negative fluctuations were accounted for in computing the mean (solid) line.

ii. NGC 3115 is the prototype of S0 galaxies and one of the standards proposed by the Working Group on Galaxy Photometry. Using a large collection of plate material from various telescopes and a well established methodology, Tsikoudi (1978, 1979) obtained the main axis light profiles extending to $\mu = 28$ B–ss, from which she derived important conclusions about fitting laws and the presence of photometric components. More recently, Capaccioli *et al.* (1986a) have repeated the work, using four deep plates, taken with the ESO and UK Schmidt telescopes, to record the faint outer regions. From the internal agreement (Fig. 4) and the external comparison with the photoelectric measurements by Miller and Prendergast (1968), and from an independent study of NGC 3379 done with similar Schmidt material, these authors estimate an accuracy better than 0.2 *mag* for $\mu < 26$ B–ss (see also Capaccioli *et al.* 1984a). The main axis light profiles of NGC 3115 from the two sources are plotted in Fig. 5; the inner part in Capaccioli *et al.* comes from the reduction of four CFH plates. Besides a zero point difference of about 0.5 *mag*, removed in the figure by a vertical shift of Tsikoudi's data, there is a disagreement between the two sources, starting at $\mu \simeq$ 24 B–ss on both axes and increasing outwards to a maximum of 2 *mag* at the largest radial distance. We face again a problem in the background level, affecting at least one of the two studies.

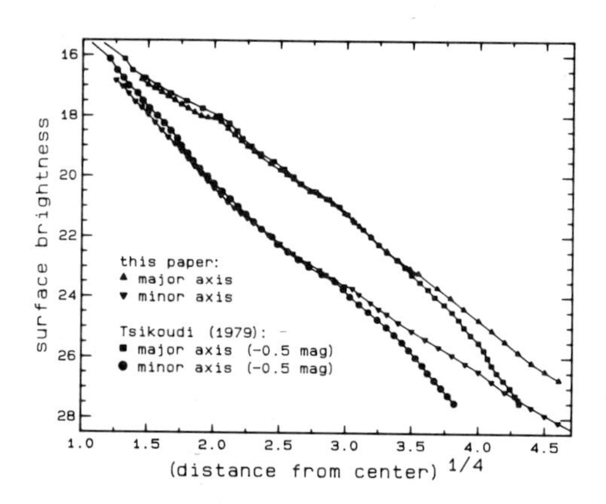

Figure 5. Main axis light profiles of NGC 3115 from Tsikoudi (1978, 1979) and from Capaccioli *et al.* (1986a). The graphical comparison discloses a large systematic discrepancy between the two sources, starting at $\mu \simeq$ 24 B–ss.

If the list of known discrepancies is long, that of unknown mistakes is probably longer. Therefore it seems wise not to accept any photometric result on a single galaxy, which has not been satisfactory tested against a standard, or at least confirmed by independent investigations. Errors on individual cases may still remain, due to the peculiar role played by the background.

4. LIGHT PROFILES AND FITTING LAWS

de Vaucouleurs (1948) found that the light profiles of elliptical galaxies are approximately linear when the surface brightness is given as a function of the 1/4 power of the radial distance from the center: $log(I/I_e) = -3.3307\ [(r/r_e)^{1/4} - 1]$. The so–called $r^{1/4}$ law, applied also to bulges of lenticulars and spirals, is now synonymous with elliptical galaxy. Not only observers are familiar with this law, but more and more theorists use it to build (*e.g.*, Binney 1982b) or to test (*e.g.*, van Albada 1982) their models. With respect to the other empirical formulae reviewed by dVC, the $r^{1/4}$ law has the advantage of giving a reasonable representation of the data, sometimes over a wide surface brightness range, with no need of free parameters. One modest drawback is that deprojection, and related mathematical operations, request cumbersome calculations (Young 1976); for this reason Jaffe (1983) has proposed another fitting law, which results from the projection of a simple analytic formula for the 3–D light distribution. Another problem is that the scale factor r_e is ill defined observationally, since the quantity directly measured is $r_e^{1/4}$. The difficulty is often by–passed replacing r_e with the *half effective aperture*, r_e^*, computed from growth curves. This parameter is not related to the fitting law and, from the methodological point of view, is more suitable to seek correlations with other observational quantities, *e.g.* with effective surface brightness (Kormendy 1980) or with absolute magnitude (Binggeli *et al.* 1984, Romanishin 1986).

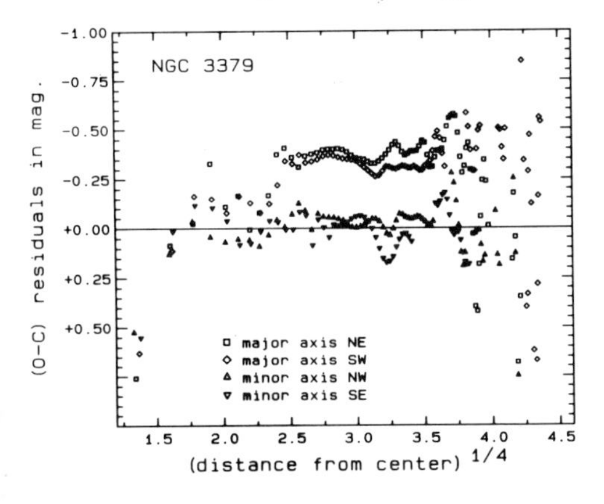

Figure 6. Magnitude differences between the main axis profiles of the standard galaxy NGC 3379 and the $r^{1/4}$ law best fitting the minor axis light distribution (Capaccioli *et al.* 1986b). The waves in the residuals repeat on both sides of each axis.

4.1. Companions

The choice of a parameter–free law to represent the entire light profiles of ellipticals, from the center to the outskirts, conceals the implicit assumption of the existence of a single characteristic photometric component. It is for this reason that we often speak of 'deviations' from the $r^{1/4}$ law as markers of additional components and of modifications induced in the radial distribution of light by the effect of a given environment. Major deviations occur in galaxy centers, where the actual profiles may be either steeper or shallower than the extrapolation of the $r^{1/4}$ law fitting of the outer parts (Schweizer 1979, dVC, Kormendy 1985). Other significant deviations occur in the faint tails of the light profiles, which are tidally truncated in dwarfs (King 1966, Binggeli *et al.* 1984) and exhibit an excess of light ('corona') over the extrapolation of the inner $r^{1/4}$ fit in cD's. This second effect, particularly strong in galaxies dominating dense clusters or groups (Thuan and Romanishin 1981, Sulentic and Lorre 1983, Schombert 1986), is often described by an exponentially truncated Hubble law (Oemler 1976), and is interpreted as the consequence of cannibalism in otherwise $r^{1/4}$ galaxies. Although less pronounced, a similar effect, pointed out by Kormendy (1977), occurs in the outer luminosity profiles of elliptical galaxies having close companions. The tidal stretching bends the profiles in such a way that the residuals from an unbiased $r^{1/4}$ fit over the entire range of measurements (except the galaxy core) have a characteristic parabolic trend. Actually this trend, recalling the behaviour of quasi–isothermal models (de Vaucouleurs 1953, King 1966), seems to be a characteristic of ellipticals, irrespective of the presence of close companion galaxies (*cf.* Fig. 6 in Capaccioli 1984b). In conclusion, even in the most favourable cases (isolated objects), light profiles of ordinary ellipticals show real deviations up to 0.1–0.2 *mag* from the $r^{1/4}$ law, which are possibly systematic and correlated with absolute luminosity (Michard 1985).

4.2. Ripples

The E–W luminosity profile of the standard elliptical NGC 3379 shows a pattern of deviations from the $r^{1/4}$ behaviour which is different from those discussed above. The magnitude residuals from the smooth fitting law, plotted *vs.* distance from center, have a wavy trend with amplitude ~ 0.1 *mag* (Fig. 3 in dVC). The occurrence of such 'waves' has been confirmed by Nieto (1983a,b) and by Kent (1984), who has questioned their amplitude on the basis of CCD photometry in the red (dVC work is in the blue). Further information comes from Capaccioli *et al.* (1986b). From the reduction of two limiting Schmidt exposures they confirm, down to $\mu = 28$ B–ss, the E–W profile given by dVC, and find that the waves in the residuals are symmetrically placed on opposite sides of the major and minor axis profiles (Fig. 6). So far no other *bona fide* elliptical has been discovered to show a similar behaviour, possibly because no other galaxy has been so extensively studied by the photometric technique as NGC 3379. Therefore speculations should wait for additional data. Nevertheless, in the following section we will attempt to explain the waves of the light profile of NGC 3379 by proposing that, as many other E galaxies, this object is a misclassified S0 seen almost face–on. If this is the case, the waves are simply structures in the disk.

5. INTERLOPERS AMONG ELLIPTICAL GALAXIES

Except for the faintest (Ichikawa *et al.* 1986) and possibly the brightest (Schombert 1986) end of the luminosity function of ellipticals, there is no clear evidence of discontinuity in the average photometric properties. This is true at least for those properties which are correlated with integrated parameters (*e.g.*, ellipticity curves have no systematic trend for $e < 0.4$; King 1978, Michard 1985). Still, remarkable differences occur between individual galaxies, which are partly accidental (occasional interactions) and partly intrinsic (Efstathiou and Fall 1984). An external contribution to the scatter may come from contamination of the pure E class by lenticulars. If, as suggested by Burstein (1978) and Simien and de Vaucouleurs (1986), the disk–to–bulge ratio in S0's can assume any value down to zero (a viewpoint which is not in conflict with the dichotomy between E and disk galaxies discussed by Bertola and Capaccioli, 1978), and if light profiles of bulges mimic those of ellipticals (a question which is still open; Kent 1985), then it is not surprising that some low D/B lenticulars have been misclassified.

If it is a generalized phenomenon (*e.g.*, see the very small number of E's in the sample of Corwin *et al.*, 1985), misclassification may have some important astrophysical consequences. First, it alters the relative frequency of disk galaxies, which is a parameter relevant to our understanding of galaxy formation and evolution processes. Second, it may affect the statistics of intrinsic flattening, if the occurrence of interlopers is a function of their apparent flattening, and change the picture of intrinsic shapes, since bulges of lenticulars are oblate spheroids (Binney 1982a). Finally, it may obscure correlations among observational parameters (Capaccioli *et al.* 1984b). In principle, all cases of misclassified lenticulars should be easily identified by accurate photometric and kinematical observations when the faint disk is favourably inclined, *i.e.* when the objects are not far from edge–on. Flat lenticulars have a high rotational velocity (Capaccioli 1979), easy to measure.

Their major axis light and color profiles differ from those along the minor axis; the latter have a shallower and smoother trend in luminosity (Fig. 3), and a steep color gradient (Strom *et al.* 1976). The same is true for structural features such as dust lanes and patches (*cf.* Bertola, this volume), boxiness of the isophotes and their sharpening along the major axis (Capaccioli *et al.* 1986a).

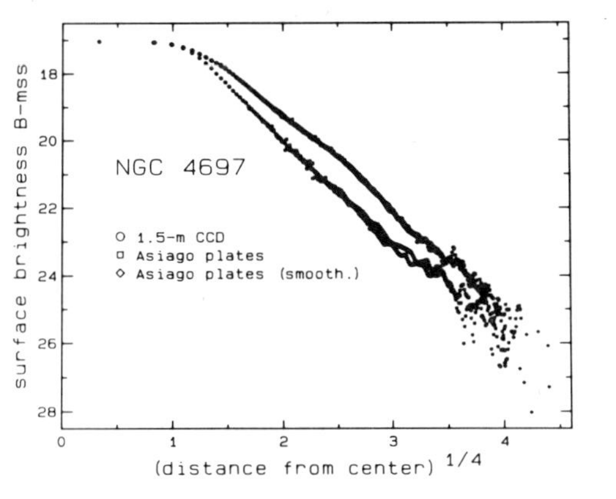

Figure 7. NGC 4697 exhibits the characteristic difference between the trends of major and minor axis light profiles, typical of lenticulars. Again the equivalent profile, not shown in the figure, is close to an $r^{1/4}$ law, as in King (1978).

The situation described above is precisely that of NGC 4125 and NGC 4697, classified ellipticals in the RC2 (de Vaucouleurs *et al.* 1976), and then discovered to be disk galaxies (Fig. 7). It is interesting to note that NGC 4697 is a fast rotator (Davies 1981); its rotation curve has the two relative maxima characteristic of lenticulars (Capaccioli 1979). A list of suspicious cases, such as NGC 4406, 4472 or 4649, may be compiled by comparing RC2 with RSA (Sandage and Tammann 1984) and with the original estimates by de Vaucouleurs (1963).

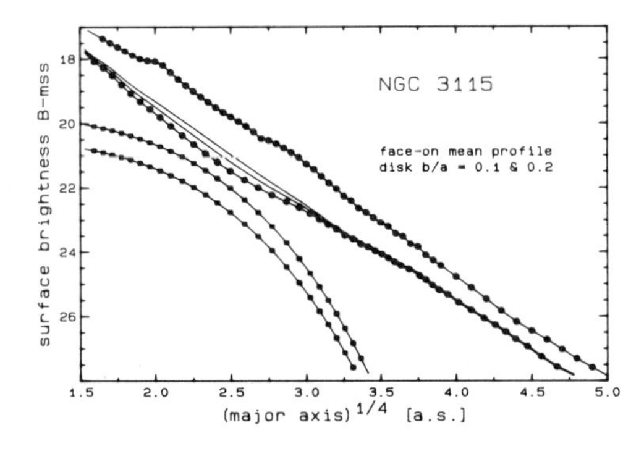

Figure 8. The full lines are face–on light profiles of NGC 3115 computed by adding the deprojected profiles of the bulge (lower dotted curve) and of the disk (triangles) for two values of the intrinsic flattening of the latter (Capaccioli *et al.* 1986a). The observed profile along the major axis (upper dotted curve) is also shown for comparison.

Let us now consider the problem of face–on lenticulars having a faint disk and not very prominent dust/absorption features; strong lanes as in NGC 404 (Sandage 1961) would immediately reveal the nature of the object. The question is: will the light distribution in such a galaxy mimic that of a pure elliptical ? A positive answer has been given by Capaccioli *et al.* (1986a). They were able to compute the face–on light profile of the lenticular NGC 3115, showing that, except for details in the disk, it follows very closely an $r^{1/4}$ law (Fig. 8), while the actual profiles along both axes do not (Fig. 5). In conclusion, unless peculiar features mark the image, or the lens is strong enough to produce the typical 'step' in the light distribution, an almost face–on *bona fide* lenticular can be easily confused with an elliptical galaxy. This is possibly the case for the E1 galaxy NGC 3379, which has a color gradient inconsistent with a spherical object (Strom *et al.* 1976) and a flattening curve similar to that of edge–on S0's (Fig. 9; *cf.* also Michard 1985). As we saw, the fact that, in the mean, its luminosity profiles follow well the $r^{1/4}$ law over a range of 8 *mag*, does not tell us much about its morphological type. Instead, the mysterious wavy residuals with respect to a smooth trend, mentioned above, would be fully understood if *standard elliptical* NGC 3379 were a misclassified lenticular.

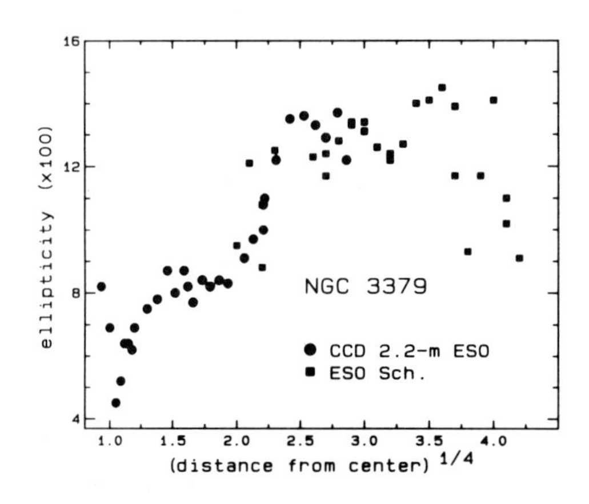

Figure 9. Ellipticity curve of NGC 3379 from Capaccioli *et al.* (1986b), obtained combining CCD and photographic data.

Part of the results presented here come from work done in collaboration with Prof. Gerard de Vaucouleurs and Dr.s Enrico Held, Jean–Luc Nieto and Roberto Rampazzo, to be published in full elsewhere. I would like to thank Dr. Jack W. Sulentic for comments and discussions.

REFERENCES

Barbon, R., Benacchio, L. and Capaccioli, M. 1976a. *Mem. Soc. Astr. It.*, **47**, 263.

Barbon, R., Benacchio, L. and Capaccioli, M. 1976b. *Astron. Astrophys.*, **51**, 25.

Barbon, R., Capaccioli, M. and Rampazzo 1982. *Astron. Astrophys.*, **115**, 388.

Bertola, F. and Capaccioli, M. 1975. *Astrophys. J.*, **200**, 439.

Bertola, F. and Capaccioli, M. 1978. *Astrophys. J. Lett.*, **219**, L95.

Bertola, F., Bettoni, D. and Capaccioli, M. 1982. In: *Internal Kinematics and Dynamics of Galaxies, IAU Symposium No. 100*, (Besançon), ed. E. O. Athanassoula, D. Reidel, p. 311.

Binggeli, B., Sandage, A. and Tarenghi, M. 1984. *Astron. J.*, **89**, 64.

Binney, J. 1982a. *Ann. Rev. Astron. Astrophys.*, **20**, 399.

Binney, J. 1982b. *Mont. Not. R. astr. Soc.*, **200**, 951.

Burstein, D. 1979. *Astrophys. J. Suppl.*, **41**, 435.

Capaccioli, M. 1979. In: *Photometry, Kinematics and Dynamics of Galaxies*, (Austin), ed. D. S. Evans, The Univ. of Texas Press, p. 165.

Capaccioli, M. 1984a. In: *Data Analysis in Astronomy*, ed.s V. Di Gesu, L. Scarsi, P. Crane, J. H. Friedman and S. Levialdi, (Erice), Plenum Press, p. 363.
Capaccioli, M. 1984b. In: *New Aspects of Galaxy Photometry*, (Toulouse), ed. J.–L. Nieto, Springer–Verlag, p. 53.
Capaccioli, M. and de Vaucouleurs, G. 1983. *Astrophys. J. Suppl.*, **52**, 465 (CdV).
Capaccioli, M. and Rampazzo, R. 1986. *preprint.*
Capaccioli, M., Held, E. and Lorenz, H. 1986b. *in preparation.*
Capaccioli, M., Held, E. V. and Nieto, J.–L. 1984a. In: *New Aspects of Galaxy Photometry*, (Toulouse), ed. J.–L. Nieto, Springer-Verlag, p. 265.
Capaccioli, M., Held, E. V. and Nieto, J.–L. 1986a. *Astrophys. J. Suppl.*, submitted.
Capaccioli, M., Held, E. V. and Rampazzo, R. 1984b. *Astron. Astrophys.*, **135**, 89.
Capaccioli,M., Davoust, E., Lelievre, G. and Nieto, J.–L. 1983. In: *Astronomy with Schmidt–type Telescopes, IAU Colloquium No. 78*, (Asiago), ed. M. Capaccioli, D. Reidel, p. 379.
Carter, D. and Dixon, K. L. 1978. *Astron. J.*, **83**, 574.
Corwin, H. G., de Vaucouleurs, A. and de Vaucouleurs, G. 1985. *Southern Galaxy Catalogue*, The Univ. of Texas Monographs in Astronomy, Austin, No. 4.
Davies, R. L. 1981. *Mont. Not. R. astr. Soc.*, **194**, 879.
Davoust, E. and Pence, W. D. 1982. *Astron. Astrophys. Suppl.*, **49**, 631.
de Vaucouleurs, G. 1948. *Ann. Astrophys.*, **11**, 247.
de Vaucouleurs, G. 1953. *Mont. Not. R. astr. Soc.*, **113**, 134.
de Vaucouleurs, G. 1958. *Astrophys. J.*, **128**, 65.
de Vaucouleurs, G. 1959. *Hand. der Phys.*, **53**, 331.
de Vaucouleurs, G. 1963. *Astrophys. J.*, **8**, 31.
de Vaucouleurs, G. 1968. *Appl. Opt.*, **7**, 1513
de Vaucouleurs, G. 1979. In: *Photometry, Kinematics and Dynamics of Galaxies*, (Austin), ed. D. S. Evans, The Univ. of Texas Press, p. 1.
de Vaucouleurs, G. 1983. In: *Astronomy with Schmidt–type Telescopes, IAU Colloquium No. 78*, (Asiago), ed. M. Capaccioli, D. Reidel, p. 367.
de Vaucouleurs, G. and Capaccioli, M. 1979. *Astrophys. J. Suppl.*, **40**, 699 (dVC).
de Vaucouleurs, G. and Nieto, J.–L., 1978. *Astrophys. J.*, **220**, 449.
de Vaucouleurs, G. and Nieto, J.–L., 1979. *Astrophys. J.*, **230**, 697.
de Vaucouleurs, G., de Vaucouleurs, A. and Corwin, H. G. 1976. *Second Reference Catalogue of Bright Galaxies*, Univ. of Texas Press, Austin (RC2).
Djorgowski, S. 1986. *Ph. D. Thesis*, Univ. of California at Berkeley.
Efstathiou, G. and Fall, M. S. 1984. *Mont. Not. R. astr. Soc.*, **206**, 453.
Fort, B. P., Prieur, J.–L., Carter, D., Meatheringam, S. J. and Vigroux, L. 1986. *Astrophys. J.*, **306**, 110.
Fraser, C. 1977. *Astron. Astrophys. Suppl.*, **29**, 161.
Hubble, E. 1930. *Astrophys. J.*, **71**, 231.
Ichikawa, S.–I., Wakamatsu, K.–I. and Okamura, S. 1986. *Astrophys. J. Suppl.*, **60**, 475.
Illingworth, G. 1977. *Astrophys. J. Lett.*, **218**, L43.
Jaffe, W. 1983. *Mont. Not. R. astr. Soc.*, **202**, 995.
Jones, W. B., Obbits, D. L., Gallet, R. M. and de Vaucouleurs, G. 1967. *Publ. Dept. Astron.*, Univ. of Texas, Austin, Ser. 2, **1**, No. 8.
Kent, S. 1984. *Astrophys. J. Suppl.*, **56**, 105.
Kent, S. 1985. *Astrophys. J. Suppl.*, **59**, 115.

King, I. R. 1966. *Astron. J.*, **71**, 64.
King, I. R. 1978. *Astrophys. J.*, **222**, 1.
Kormendy, J. 1977. *Astrophys. J.*, **218**, 333.
Kormendy, J. 1980. In: *Proc. ESO Workshop on Two–dimensional Photometry*, ed.s P. Crane and K. Kjar, (Leiden: Sterrewacht), p. 191.
Kormendy, J. 1982. In: *Morphology and Dynamics of Galaxies, XII Advanced Course of Swiss Soc. Astron. Astrophys.*, ed.s L. Martinet and M. Major, (Sauverny: Geneva Obs.), p. 115.
Kormendy, J. 1985. *Astrophys. J.*, **292**, L9.
Lauberts, A. and Valentijn, E. 1984. In: *New Aspects of Galaxy Photometry*, (Toulouse), ed. J.–L. Nieto, Springer-Verlag, p. 72.
Lauer, T. R. 1985. *Astrophys. J. Suppl.*, **57**, 473.
Liller, M. 1960. *Astrophys. J.*, **132**, 306.
Liller, M. 1966. *Astrophys. J.*, **146**, 28.
Malin, D. F. and Carter, D. 1980. *Nature*, **285**, 643.
Malumuth, E. M. and Kirshner, R. P. 1984. *Astrophys. J.*, **291**, 8.
Michard, R. 1985. *Astron. Astrophys. Suppl.*, **59**, 205.
Miller, R. H. 1963. *Astrophys. J.*, **137**, 733.
Miller, R. H. and Prendergast, K. H. 1968. *Astrophys. J.*, **153**, 35.
Nieto, J.–L. 1983a. *Astron. Astrophys. Suppl.*, **53**, 247.
Nieto, J.–L. 1983b. *Astron. Astrophys. Suppl.*, **53**, 383.
Oemler, A. 1976. *Astrophys. J.*, **209**, 693.
Pence, W. D. and Davoust, E. 1985. *Astron. Astrophys. Suppl.*, **60**, 517.
Piotto, G. 1986. *Ph. D. Thesis*, University of Padova.
Romanishin, W. 1986. *Astron. J.*, **91**, 76.
Sandage, A. 1961. *The Hubble Atlas of Galaxies*, Carnegie Inst. of Washington, Publ. No. 618.
Sandage, A. and Tammann, G. 1984. *A Revised Shapley–Ames Catalogue*, Carnegie Inst. of Washington, Publ. No. 635 (RSA).
Schweizer, F. 1979. *Astrophys. J.*, **233**, 23.
Schombert, J. M. 1986. *Astrophys. J. Suppl.*, **60**, 603.
Simien, F. and de Vaucouleurs, G. 1986. *Astrophys. J.*, **302**, 564.
Strom, S. E., Strom, K. M., Goad, J. W., Vrba, F. J. and Rice, W. 1976. *Astrophys. J.*, **204**, 684.
Stetson, P. B. 1983. *DAOPHOT User's Manual.*
Sulentic, J. W. and Lorre, J. J. 1983. *Astron. Astrophys.*, **120**, 36.
Thuan, T. X. and Romanishin, W. 1981. *Astrophys. J.*, **248**, 439.
Tsikoudi, V. 1978. *Ph.D Thesis*, The Univ. of Texas, Austin.
Tsikoudi, V. 1979. *Astrophys. J.*, **234**, 842.
van Albada, D. 1982. *Mont. Not. R. ast. Soc.*, **201**, 939.
Watanabe, M. 1983. *Ann. Tokyo Astrophys. Obs.*, 2nd Ser., **19**, 121.
Williams, T. B. and Schwarzschild, M. 1979. *Astrophys. J.*, **227**, 56.
Young, P. J. 1976. *Astron. J.*, **81**, 807.

DISCUSSION

Djorgovski: Some ways of plotting the data, e.g., log I vs. $r^{1/4}$, stretch the profiles so that almost anything looks linear. The picture is visually dominated by the strong radial gradient. I think that the only fair and honest way of comparing data to fitting functions and models is to plot the residuals.

Capaccioli: I fully agree.

Burstein: In the on–going investigation of elliptical galaxies by the "Gang of Seven", we compared CCD surface photometry with photoelectric photometry for about 30 galaxies. As far as we can determine, we have as much trouble with different zero points among the photo–electric photometry, as among the CCD data. In particular, we compared CCD four–shooter data on NGC 4697 to $\sim$ 15 photo–electric aperture measures. The mean deviation, (CCD–P.E.) was ± 0.02 mag, *i.e.*, we have not had serious problems in connecting photo–electric and CCD data.

Capaccioli: Very much reassuring! However, I think that your nice results depend also on the fact that you have new photo–electric data. My remark is relative to old published photo–electric data.

Porter: CCD's are potentially more helpful than you imply if they can be used in drift scan mode, being read out continuously as the telescope tracks in the opposite direction. A whole new set of calibration problems arise (sky brightness should be monitored independently, color terms may be different, and the telescope's tracking must be excellent), but you can take strip images long enough to reach "true sky". Those images are very flat, and would allow direct comparison of electronic and photographic photometry over the galaxy's entire range of surface brightness.

Capaccioli: All that is very true. Yet I think it is more economical to combine CCD's with photographic photometry.

Stefano Casertano, chairman of the LOC, has a rare moment of rest with Andrew Pickles and Patricia Vader.

Roger Davies enjoying himself.

THE STELLAR KINEMATICS OF ELLIPTICAL GALAXIES

Roger L. Davies
Kitt Peak National Observatory
National Optical Astronomy Observatories
P. O. Box 26732
Tucson, AZ 85726

ABSTRACT. The kinematic properties of elliptical galaxies are summarized. New developments are discussed in four areas: (i) the Faber-Jackson relation and the role of second parameters (ii) the luminosity-rotation relation (iii) the figures of elliptical galaxies and (iv) the mass-to-light ratio as a function of radius.

1. INTRODUCTION

Earlier reviews of this subject are found in the Proceedings of IAU 100 by Illingworth (1982) and in the Annual Reviews of the same year by Binney. The low rotation velocities of luminous ellipticals (say, $M_B < -20.5$, where I will use $H_o = 50$ km s^{-1} Mpc^{-1} throughout) were discovered by Bertola and Capaccioli (1975) and Illingworth (1977). This result is illustrated in Figure 1 which plots the ratio of rotation velocity over velocity dispersion as a function of ellipticity. The open circles scatter below both the mean expected line for oblate galaxies with isotropic velocity distributions and the median line for prolate isotropic galaxies taken from Binney (1978). Binney concluded that luminous ellipticals are not flattened by rotation, but by anisotropic pressures and therefore do not necessarily have spheroidal figures, but may be triaxial.

In 1982 Kormendy and Illingworth studied the bulges of 8 edge-on spiral galaxies. They discovered that, unlike the ellipticals, these bulges had rotation velocities and flattenings consistent with them being oblate, isotropic rotators. This result was confirmed by Dressler and Sandage (1983). Davies *et al.* 1983 (hereafter DEFIS) studied a sample of 11 low luminosity elliptical galaxies, $-18 < M_B < -21$ and found that they too had kinematics consistent with being oblate figures with isotropic residual velocities that are flattened by rotation. These results are summarized in Figure 1, in which lower luminosity ellipticals are shown as filled circles and the Kormendy and Illingworth bulges as crosses.

T. de Zeeuw (ed.), Structure and Dynamics of Elliptical Galaxies, 63–77.

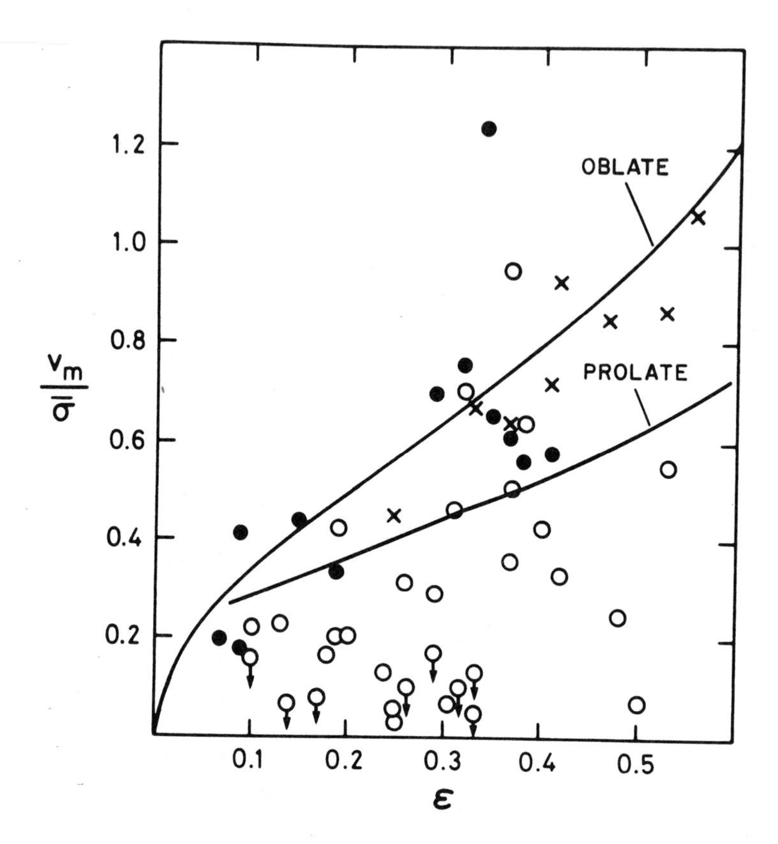

Figure 1: Peak rotation velocity plotted against velocity dispersion; open circles are luminous ellipticals ($M_B < -20.5$), filled circles are lower luminosity ellipticals and crosses are the bulges of spirals [this figure adapted from DEFIS]. The mean line for oblate isotropic galaxies and the median line for prolate isotropic galaxies are shown.

In this review, the observations on stellar kinematics of ellipticals are discussed with regard to four areas within which substantial progress has been made over the last four years:

(i) The Faber-Jackson relation and second parameters.

(ii) Further work on the rotation of early-type galaxies, as a function of luminosity.

(iii) The figures of elliptical galaxies - are they all consistent with being oblate or oblate-triaxial?

(iv) Do ellipticals have constant mass-to-light ratio as a function of radius?

These areas comprise sections 2-5 of this paper, Section 6 presents the conclusions.

2. THE FABER-JACKSON RELATION

The relationship that the luminosity of an elliptical galaxy varies as the fourth power of velocity dispersion, $L \propto \sigma^4$, discovered by Faber and Jackson (1976) has been the subject of much discussion, mostly in the context of generating a distance indicator for ellipticals. In 1981 Terlevich *et al.* used a small sample of well studied galaxies to demonstrate the presence of a 2nd parameter in the FJ relation by noting that, at a given luminosity, galaxies with high values of velocity dispersion also possessed high values of the Mg_2 line index. They suggested that the second parameter was closely related to the intrinsic flattening of galaxies. In the same year, Tonry and Davis used the FJ relation on a large sample of ellipticals to measure local non-uniformities in the Hubble flow, in particular the motion of the Local Group toward the Virgo cluster. Their data confirmed the presence of a second parameter, but did not verify the correlations of Terlevich *et al.* in detail. In 1982 de Vaucouleurs and Olson showed that both color and surface brightness were significant second parameters in a study of 157 galaxies. Efstathiou and Fall (1984) proposed that mass-to-light ratio (M/L) was the second parameter as a result of a principal component analysis of the 67 Tonry and Davis galaxies and an extended sample (35 galaxies) similar to that of Terlevich *et al.*

Recently, the results of the expanded follow-up study to that of Terlevich *et al.* have been presented by Dressler *et al.* (1986). They find that the scatter in the FJ relation can be considerably reduced by using D_Σ, the diameter within which the mean surface brightness is Σ = 20.75 mags arcsec^{-2}, with the relation:

$$D_\Sigma \propto \sigma^{4/3} \qquad (1)$$

being the best fit to the data. They demonstrate that elliptical galaxies populate a plane in the 3-space of absolute magnitude (M_B), $\log\sigma$ and μ_e and demonstrate that the relation (1) produces the minimum scatter in D_Σ. They derive a form of the FJ relation based on (1) explicitly including the effect of surface brightness as a second parameter:

$$L \propto \sigma^{2.65} \langle SB_e \rangle^{-0.65} \qquad (2)$$

where $\langle SB_e \rangle$ is the mean surface brightness within r_e. A similar result has been found by Djorgovski and Davis (1986).

3. ROTATION

Davies et al. (1983) found that low luminosity ellipticals had kinematics consistent with them being oblate isotropic rotators. They concluded that the degree of rotational support decreases as luminosity increases. The mean relation was found to be $v/\sigma \propto L^{-0.4}$, although the rotational properties of luminous ellipticals are much less homogeneous than those of their fainter cousins. This situation is summarized in Figure 2 which shows $\log(v/\sigma)^*$, the ratio of v/σ measured, divided by that expected if the galaxy were an oblate isotopic rotator, against absolute magnitude, ellipticals are shown as filled circles, bulges as crosses. There are no faint galaxies with low values of $(v/\sigma)^*$. Plotted as open circles in Figure 3 are the cD and brightest cluster galaxies from the work of Tonry (1984,1985). These galaxies appear to behave like other luminous ellipticals showing a large scatter in the degree of rotational support, but in the mean having much lower normalized rotational velocities than would be expected for oblate/isotropic models. It appears that central cluster galaxies do not reveal their uniqueness by having exclusively low rotation velocities. The 3 galaxies in Tonry's sample with $\log(v/\sigma)^* > -0.2$ are the central galaxies in Abell 189, Abell 1228 (=IC738) and Abell 2151 (= NGC6041).

Whitmore, Rubin and Ford (1984) measured the stellar kinematics in the bulges of 6 spirals. They noted that the observations of Kormendy and Illingworth (1982) together with their own indicated that the bulges scattered just below the line of oblate isotropic galaxies in a plot of v/σ vs ε. They concluded that the bulges were not fully explained as oblate isotropic rotators. Recently, Jarvis and Freeman (1985) and Fillmore, Boroson and Dressler (1986) have self-consistently modeled their observations of the photometry and kinematics of the disks and bulges of spiral galaxies. They were able to account for the additional flattening of the bulge component due to the presence of the disk, an idea suggested by Monet, Richstone and Schechter (1981). Once this was taken into account, the bulges were found to be consistent with being oblate figures with isotropic velocity residuals. This is not to say that bulges or low luminosity ellipticals may not have some degree of triaxiality, only that it will be small.

Uncertainty over the definition of an elliptical galaxy has arisen because of the detection of dust and weak stellar disks in some ellipticals, e.g. Jedrzejewski (1985), Carter (1986). The surveys of surface photometry of ellipticals have revealed some galaxies with higher order distortions of their isophotes indicative of the presence of a weak disk. The most prominent such disk identified so far is that in NGC 4697 which Carter estimates contributes 2% of the total luminosity and 10-15% of the surface brightness where it is strongest. Such a galaxy with a bulge-to-disk ratio of 50 must surely still be considered an elliptical when discussing its global properties. However, the question arises: could a relation such as that shown in Figure 2 arise because weak disks are more common in low luminosity ellipticals and their presence biases the measurement of rotation velocity? The answer appears to be no; as a result of a surface photometry survey, Peletier et al. (1986) plot the amplitude of the $\cos 4\theta$ terms in the Fourier expansion of the isophote shape, B_4, as a function of absolute magnitude. This coefficient distinguishes disk-like distortions with the disk major axis aligned with the long axis of the ellipse from box-like distortions. They find no obvious excess of low luminosity galaxies

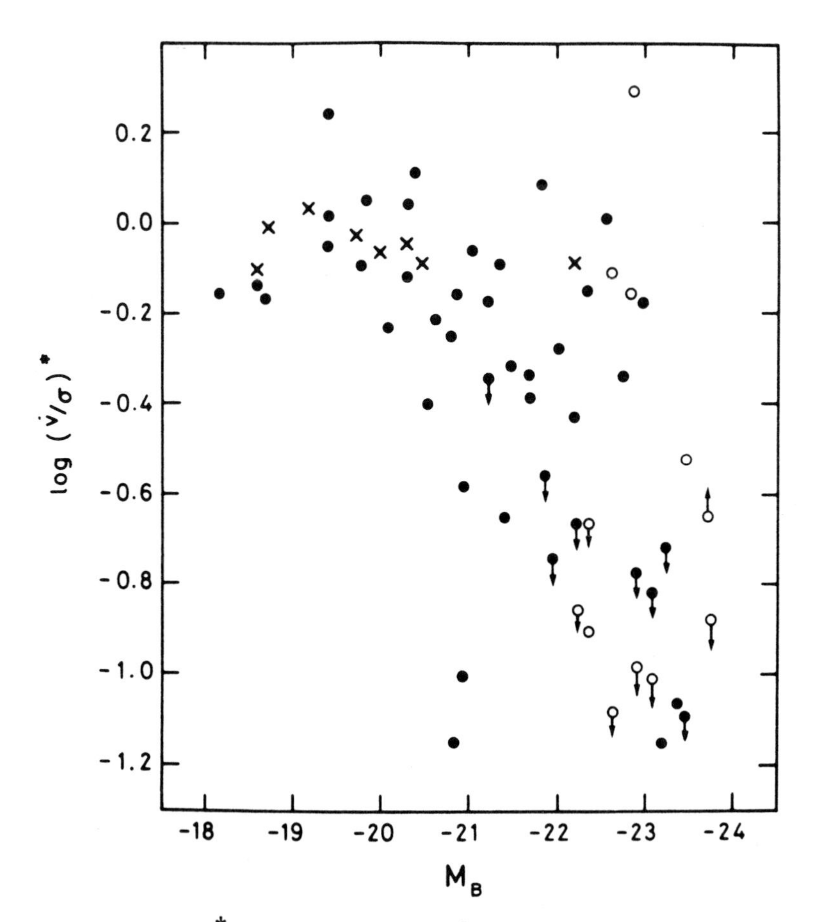

Figure 2: Log $(v/\sigma)^*$, the value of v/σ normalized to the value expected for oblate isotropic galaxies, plotted against absolute magnitude. This plot is taken from DEFIS with the cD and brightest cluster galaxies of Tonry (1984, 1985) added as open circles; filled circles are ellipticals and crosses are bulges.

with disk-like distortions; if anything an excess of boxes is indicated. A similar result is apparent in Djorgovski's thesis (1985) data on ~250 ellipticals. Plots of his higher order parameters as a function of luminosity indicate no trend with absolute magnitude except that the modulus of the amplitude of the higher order terms increases for the lower luminosity systems perhaps indicating that these are more easily tidally distorted than more luminous galaxies.

4. THE FIGURES OF ELLIPTICAL GALAXIES

The low rotation velocities of elliptical galaxies indicate that they are supported by anisotropic pressures and are therefore probably triaxial. Triaxial galaxies have the following properties:

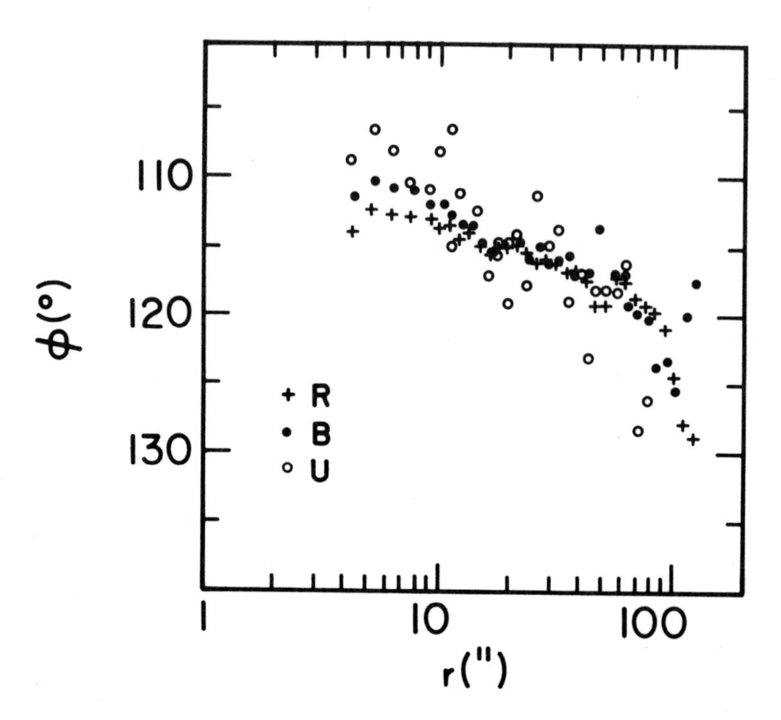

Figure 3a: Position angle as a function of radius taken for NGC 1052 taken from Davis et al. (1985).

(i) The projection onto the sky of concentric, triaxial ellipsoids with a radial gradient in axial ratio produce the appearance of an isophote twist.
(ii) Thus the projected minor axis does not have to be coincident with the projected kinematic minor axis, producing rotation on the apparent minor axis.
(iii) Schwarzschild (1979) showed that a triaxial figure has stable orbits that rotate about both the long and short axes, but that orbits rotating about the intermediate axis are unstable.

4.1. Velocity Mapping

Attempts to determine the figure types of ellipticals have involved mapping the stellar velocity field. In a study of twelve ellipticals hosting powerful ($\log P_{178MHz} > 24$ W.Hz^{-1}) radio sources, Heckman et al. (1985) used 2-3 slit position angles per galaxy and found no evidence for differences between the rotation and minor axes. Davies and Illingworth (1986) did an extensive study of the kinematics of the stars and gas in the active elliptical NGC 1052. The surface photometry of Davis et al. (1985) shows that this galaxy has an isophote twist that continues to radii as low as a few arcseconds (~500 pc). That isophote twist and the stellar kinematic data of Davies and Illingworth are reproduced in Figures 3(a) and (b). At

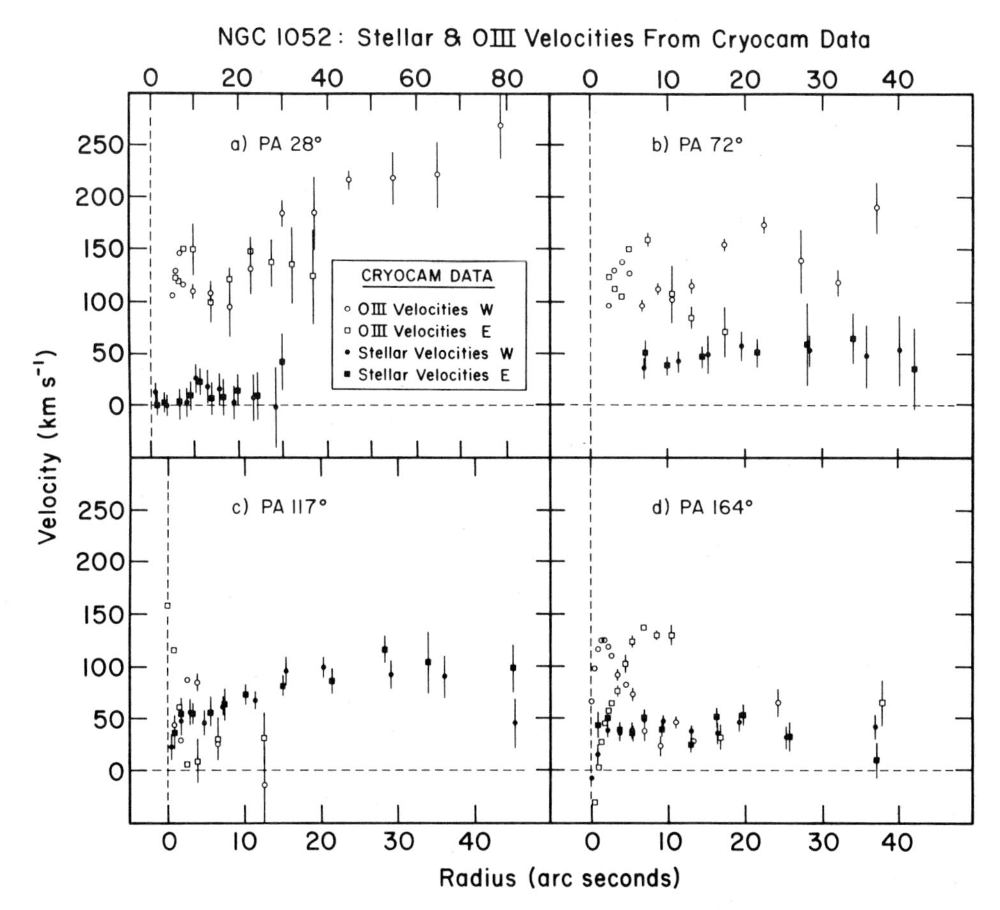

Figure 3b: The stellar velocities (filled symbols) and ionized gas velocities (open symbols) as a function of radius along the major axis (PA 117°), minor axis (PA 28°) and the two intermediate position angles in NGC 1052. (Figure taken from Davies and Illingworth 1986).

position angle 28°, on the photometric minor axis, between radii 10-25 arcsecs there is a shift in velocity seen on both sides of the nucleus (filled squares to the N, and filled circles to the S) indicating a small rotation of 10-15 kms^{-1} along the minor axis. These observations suggest that NGC 1052 is slightly triaxial and probably close to oblate, the difference between the rotation and minor axes being a few degrees at most. The kinematics of the ionized gas confirm this conclusion. In Figure 3(b) the open symbols give the velocities of the ionized gas along the four slit positions. A fit to a thin disk model at a radius of ~ 10 arcsec gives a projected rotation axis for the gas of -41°, that is 69° away from that of the stars. An aperture synthesis map of the HI gas in NGC 1052 has been reported by van Gorkom et al. (1986). This shows that the rotation axis of the gas disk at a radius of 2 arcminutes is -46°, almost

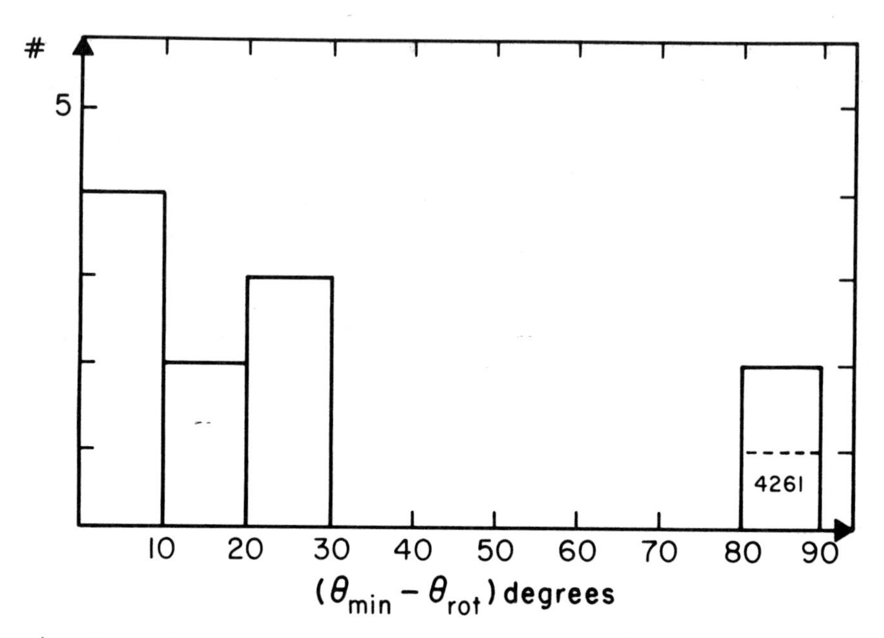

Figure 4: The histogram of the misalignment of the projected minor and rotation axis from the sample of ellipticals studied by Davies and Birkinshaw (1986b).

exactly that of the ionized gas at small radius. Thus the rotation axis of this gas disk remains constant over a factor of 10 in radius and is misaligned with that of the stars. If this gas disk were in a spheroidal galaxy the misalignment would cause the disk to precess, faster at smaller radii, and become warped. Thus the constancy of the rotation axis of the gas disk indicates that indeed NGC 1052 is triaxial and that the gas is rotating about the projected long axis.

In a survey of elliptical galaxies containing weak radio sources (median $\log P_{5GHz} = 21.7$ W.Hz^{-1}Sr^{-1}) Davies and Birkinshaw (1986b) mapped each galaxy with at least four slits and used a kinematic model similar to that of Binney (1985) to determine the rotation axis. Their histogram of the difference between rotation axis and minor axis is shown in Figure 4 (NGC 1052 and NGC 5128 have been added to the sample and 2 galaxies with rotation axis orientations uncertain by more than 30° have been omitted). Most galaxies do not have a significant difference between their rotation and minor axes but at least one of these, NGC 1052, is thought to be triaxial. In addition, three galaxies have ($\theta_{min} - \theta_{rot}$) small but significantly different from zero. These results suggest that most ellipticals are triaxial but closer to oblate than prolate. NGC 4261 was found by Davies and Birkinshaw (1986a) to have a rotation axis 84° ± 4° away from its projected minor axis and to have no isophote twist and a second galaxy with much lower rotation velocity and thus more uncertain rotation axis also falls in this bin in their preliminary analysis. [Williams

(1981) discovered that NGC 596 displays a 60° isophote twist and rotates about an axis that changes with radius, but is nowhere coincident with the projected minor axis. He modeled it with a series of dissimilar nested triaxial ellipsoids and showed that at large radii it rotates about its long axis and is prolate. NGC 4261 shows no isophote twist and no change of rotation axis with radius]. It seems that at least NGC4261 is very likely to be prolate. Levison (1986) has shown that it is possible to populate the orbits of an oblate-triaxial galaxy to generate a streaming velocity about the major axis. This configuration seems less likely than a triaxial-but-close-to-prolate figure for NGC 4261. The data suggests that all ellipticals are triaxial with most being close to oblate spheroidal and a small fraction being close to prolate spheroidal.

4.2. Galaxies with Dust Lanes

For triaxial galaxies with warped minor axis dust lanes van Albada, Schwarzschild and Kotanyi (1982) and Merritt and deZeeuw (1983) showed that, for the dust to be on stable orbits, the action of the Coriolis force and the sense of the warp predicts the sense of figure tumbling. Steiman-Cameron and Durisen (1984) showed that there was a significant chance, 10-20%, that gas would be accreted onto stable orbits rotating about the long axis of a mildly triaxial figure (1:0.98:0.5) as a result of an interaction. Thus galaxies with minor axis dust lanes do not have to be prolate but can be triaxial and close to oblate (as appears to be the case for NGC 1052).

Davies _et al._ 1984, Bertola _et al._ 1985 and Wilkinson _et al._ 1986 applied this idea to NGC5128 (=Centaurus A) and discovered that the stars rotate in the opposite sense to that predicted for the figure. If we assume that models in which the stars counterstream with respect to the figure are unlikely, then this observation indicates that the dust in NGC5128 is not on stable orbits. A similar situation was found for NGC5266 by Caldwell (1984). Perhaps the question to ask now is: while the dust is not on orbits that are formally stable, what is the timescale for decay?

In their comprehensive study Wilkinson _et al._ put limits on the form of a stationary figure for NGC 5128, they showed that if the radio/X-ray jet is constrained to escape perpendicular to the dust lane along the long axis, the galaxy can be oblate triaxial. If that constraint is relaxed, they concluded that the figure could be prolate.

In this section examples of elliptical galaxies that are oblate (the low luminosity ellipticals), oblate/triaxial (NGC 1052, NGC 5128) and prolate/triaxial (NGC 4261, NGC 596) have been identified and it appears that the oblate and triaxial-but-close-oblate galaxies are a significant majority. A small fraction of ellipticals are prolate triaxial. The configurations that are close to spheroidal, either oblate or prolate, may be preferred over more triaxial ($b = 1/2(a+c)$) figures, but many more galaxies need to be mapped to firmly establish this.

5. MASS TO LIGHT RATIO AS A FUNCTION OF RADIUS

5.1. Halos

5.1.1. Stellar Velocity Dispersion as a function of Radius.

Do elliptical galaxies possess high M/L halos analogous to those surrounding spirals? In the specific case of central cluster cD galaxies: does M/L at large radius reach the values inferred from the motions of the galaxies in the cluster, namely $M/L_B \sim 10^2-10^3$. Stellar kinematic data on cD galaxies collected by by Dressler (1979) on IC1101 in the center of Abell 2029, Carter et al. (1981) on IC 2082, Davies and Illingworth (1983) on NGC4889 and NGC4839 and Carter et al. (1985) on 0559-40, PKS 2354-35 and Sersic 40/6 indicate that the velocity dispersion as a function of radius is either constant or rising. Dressler interpreted the rise in velocity dispersion in IC 1101 as evidence for a high M/L component, but Tonry 1983 pointed out that the data could be fitted with constant M/L by increasing the tangential component of the velocity dispersion with radius. This model for IC 1101 required $\sigma_t = 3.7\ \sigma_r$ at the limit of the kinematic data. Subsequent work has recognized the difficulty in assigning a unique interpretation to the dispersion profiles. Several general points are clear:

(i) In models of cDs in clusters with a third component in addition to the stellar galaxy and cluster, that component has an M/L and lengthscale between those of the luminous galaxy and the cluster.

(ii) The local M/L implied at large radius is 2-5 times the central value, so the high cluster values are not directly indicated.

(iii) In all but 2 cases, IC1101 and Sersic 40/6, the dispersion profile can be fitted with $\sigma_t \leqslant 2.5\ \sigma_r$ and constant M/L.

(iv) While only the cD galaxies show increases in velocity dispersion with radius (the best 2 examples being those in (iii)) many normal ellipticals have $\sigma(r)$ approximately constant. There is no reason to believe that those cDs with $\sigma(r) \sim$ constant are particularly unusual.

5.1.2. The Velocities of Multiple Nuclei.

The well known large relative velocities ($\Delta v \sim 1000$ km s^{-1}) between central cluster galaxies and the multiple nuclei embedded in their images led to the idea of very large M/L for these galaxies. This phenomenon is common; Schneider, Gunn and Hoessel (1983) find multiple nuclei in 45% of brightest cluster galaxies. Blandford and Smarr (1983) suggested the idea of deep potential wells in clusters to account for those observations, the so-called "black pit" hypothesis. However, Tonry (1984, 1985) found that, at the radii of the multiple nuclei, the galaxy velocity dispersions were much too low to account for the large relative velocities of the nuclei if they were on circular orbits. He concluded that the nuclei are on highly radial orbits with high line-of-sight velocities and that the M/Ls for cDs and brightest cluster galaxies were the same as those for normal ellipticals.

Merritt (1984) has shown that the effect on cluster evolution of the dynamical friction of galaxies with the background of dark matter is to increase the probability of finding a galaxy very close to the cluster center by a factor of 3 without significantly reducing their velocities. This provides an explanation of Tonry's observations without invoking cannibalism but rather a slower dynamical evolution of clusters.

5.1.3. Other Methods. Dressler, Schechter and Rose (1986) used 6 satellite galaxies around the isolated elliptical NGC720. By comparing the mean projected radius and velocity dispersion of the satellite galaxies with that of the luminous galaxy they concluded that the mass interior to the satellites was 44 times greater than that of the luminous galaxy. Despite questions as to the physical association and orbit distribution of the satellites, if suitable candidate systems can be found this method appears to be a potential means of measuring the mass of ellipticals at large radii. In particular, it could provide independent mass estimates at large radius from those derived from the distribution of X-ray surface brightness.

It appears that the stellar kinematic tracers in elliptical galaxies do not extend far enough in radius to unambiguously detect the presence of a halo. Other methods discussed in this volume such as the presence of shells around ellipticals (Quinn and Hernquist), the velocity dispersion of globular clusters (Mould) and the surface brightness distribution of the X-ray gas (Fabian) provide more compelling evidence for increasing M/L with radius in ellipticals.

5.2 Central Mass Concentrations

The measurements of luminosity and velocity dispersion profiles for M87 by Sargent et al. (1978), Young et al. (1978) and Dressler (1980) have attracted considerable attention. Duncan and Wheeler (1980) pointed out that the inference of a central mass of 5×10^9 M_o was dependent on the assumption of isotropy of the stellar orbits. Binney and Mamon (1982) and Newton and Binney (1984) produced a model that was consistent with the observations and had constant mass-to-light ratio but required the radial component of dispersion to be more than twice the tangential component from 50–200 pc. At maximum the radial component was more than 3 times the tangential. More recently, Richstone and Tremaine (1985) used Schwarzschild's linear programming method to produce models of M87 with M/L_B constant (= 10 ± 2) and radial anisotropies typically less than a factor of 2. The conclusion that emerged was that a central mass in M87 is not required. However, work by Merritt and Aguilar (1985), Barnes, Goodman and Hut (1986), Binney and May (1986) and Palmer and Papaloizou (1986) has demonstrated that even modest radial anisotropies in spherical galaxies can be unstable to bar formation. The question of the stability of spherical galaxies with radial anisotropies of the size required by constant M/L models of ellipticals needs to be resolved.

Finally, other galaxies have now been observed sufficiently well to allow detailed modeling, NGC1052 by Davies and Illingworth (1986) and M31 and M32 by Kormendy, presented in this volume.

6. CONCLUSIONS

Elliptical galaxies populate a plane in the 3-space of absolute magnitude, M_B, log σ and μ_e demonstrating that surface brightness is the second parameter in the Faber-Jackson relation. A new distance indicator for ellipticals, D_Σ, the diameter within which the mean surface brightness is 20.75, has been shown to be optimum. The rms error on relative distances using this indicator are ± 25% for a single galaxy. Studies to utilize this to investigate non-uniformities in the Hubble flow are underway.

In the mean the degree of rotational support in elliptical galaxies decreases with luminosity. Luminous ellipticals ($M_B \leqslant -20.5$) have a range of rotational support from being consistent with oblate/isotropic to having rotation velocities a factor of 10 lower than expected under that hypothesis. The brightest cluster galaxies do not appear to be special in this regard. The question of the rotational properties of luminous bulges such as the Sombrero galaxy (M104) remains unanswered largely because such objects are so rare. Some ellipticals have been shown to possess weak disks but there is no evidence that the presence of such a disk significantly biases the observed kinematics, in particular these disks are not more frequent among low luminosity galaxies and there is no evidence to suggest they are more frequent among faster rotating ellipticals.

It appears that most ellipticals are triaxial and close to oblate with a small fraction being triaxial and close to prolate. The distribution of the differences of rotation and minor axes, ($\theta_{rot} - \theta_{min}$), suggests that forms closer to the spheroids are preferred over perfectly triaxial figures. Only a small number of galaxies have been kinematically mapped; many more are required to firmly establish this conclusion. If one is prepared to make assumptions about the 3D orientation of dust lanes or radio jets this extra information can be used to break the degeneracy of projection onto the sky and further constrain the figure types. As there are well-known cases of the bending of radio jets and warping of dust lanes this approach is not as secure as mapping the stellar velocity field.

The stellar kinematic evidence for an increasing M/L with radius in ellipticals is not decisive. Indications from globular clusters, shells, satellite galaxies and X-rays are more compelling. Evidence for central masses in elliptical galaxies is strengthened by the results on the rotation curve of M32 presented in this volume by Kormendy and by the realization that the radially anisotropic orbital configurations required for constant M/L models of M87 may be unstable.

ACKNOWLEDGMENTS

Stimulating discussions with many individuals have contributed to the discussion presented here. I would particularly like to thank J. Binney, M. Birkinshaw, D. Burstein, A. Dressler, S. Djorgovski, S. Faber, G. Illingworth, R. Jedrzejewski, R. Peletier and S. White.

REFERENCES

Barnes, J., Goodman, J. and Hut, P., 1986, Ap. J., **300**, 112.
Bertola, F. and Capaccioli, M., 1975, Ap. J., **200**, 439.
Bertola, F., Galletta, G. and Zeilinger, W. W., 1985, Ap. J. Letters, **292**, L52.
Binney, J. J., 1978, M.N.R.A.S., **183**, 779.
Binney, J. J., 1982, Annual Reviews of Astr. & Astrophys., **20**, 399.
Binney, J. J., 1985, M.N.R.A.S., **212**, 767.
Binney, J. J. and Mamon, G., 1982, M.N.R.A.S., **200**, 361.
Binney, J. J. and May, A., 1986, M.N.R.A.S., **221**, 13P.
Blandford, R. and Smarr, L. L., 1983, preprint.
Caldwell, N., 1984, Ap. J., **278**, 96.
Carter, D., Efstathiou, G., Ellis, R. S., Inglis, I., and Godwin, J., 1981, M.N.R.A.S., **195**, 15p.
Carter, D., Inglis, I., Ellis, R. S., Efstathiou, G., and Godwin, J. G., 1985, M.N.R.A.S., **212**, 471.
Carter, D., 1986, Ap. J., in press.
Davies, R. L., Efstathiou, G., Fall, S. M., Illingworth, G., and Schechter, P. L., 1983, Ap. J., **266**, 41. (DEFIS)
Davies, R. L. and Illingworth, G. D., 1983, Ap. J., **266**, 516.
Davies, R. L., Danziger, I. J., Fabian, A., Hanes, R., Jones, B. J. T., Jones, J., Morton, D. C., and Pennington, R. L., 1984, B.A.A.S., **16**, 410.
Davies, R. L. and Illingworth, G. D., 1986, Ap. J., **302**, 234.
Davies, R. L. and Birkinshaw, M. 1986(a), Ap. J. Letters, **303**, L45.
Davies, R. L. and Birkinshaw, M. 1986(b), in preparation.
Davis, L., Cawson, M., Davies, R. L. and Illingworth, G., 1985, A. J., **90**, 169.
de Vaucouleurs, G. and Olson, D. W., 1982, Ap. J., **256**, 346.
Djorgovski, S. B., 1985, Ph.D. Thesis, University of California, Berkeley.
Djorgovski, S. B. and Davis, M., 1986, Ap. J., in press.
Dressler, A., 1979, Ap. J., **231**, 659.
Dressler, A., 1980, Ap. J. Letters, **240**, L11.
Dressler, A. and Sandage, A., 1983, Ap. J., **265**, 664.
Dressler, A., Schechter, P. L., and Rose, J. A., 1986, A. J., **91**, 1058.
Dressler, A., Lynden-Bell, D., Burstein, D., Davies, R. L., Faber, S. M., Terlevich, R. and Wegner, G. 1986, Ap. J., in press.
Duncan, M. J. and Wheeler, J. C., 1980, Ap. J. Letters, **237**, L27.
Efstathiou, G. and Fall, S. M., 1984, M.N.R.A.S., **206**, 453.
Faber, S. M., Jackson, R., 1976, Ap. J., **204**, 668.

Fabian, A., this volume p. 155.
Fillmore, J. A., Boroson, T. and Dressler, A., 1986, Ap. J., **302**, 208.
Heckman, T. M., Illingworth, G. D., Miley, G. K., and van Breugal, W. J. M., 1985, Ap. J., **299**, 41.
Illingworth, G. D. 1977, Ap. J. Letters, **218**, L43.
Illingworth, G. D., 1982 in Proc. of IAU 100, 'Internal Kinematics and Dynamics of Galaxies,' p. 257, ed. Athanassoula E.
Jarvis, B. and Freeman, K. 1985, Ap. J., **295**, 324.
Jedrzejewski, R. 1985, Ph.D. Thesis, University of Cambridge.
Kormendy, J. and Illingworth, G. D., 1982, Ap. J., **256**, 460.
Levison, J., 1986, Ph.D. Thesis, University of Michigan.
Merritt, D., 1984, Ap. J. Letters, **280**, L5.
Merritt, D. and Aguilar, L. A., 1985, M.N.R.A.S., **217**, 787.
Merritt, D. and de Zeeuw, T., 1983, Ap. J. Letters, **267**, L19.
Monet, D., Richstone, D. and Schechter, P. L. 1981, Ap. J., **245**, 454.
Mould, J., this volume p. 451.
Newton, A. J. and Binney, J. J., 1984, M.N.R.A.S., **210**, 711.
Palmer, P. and Papaloizou, J. this volume p. 515.
Peletier, R., Davis, L., Davies, R. L., Illingworth, G. D., and Cawson, M. 1986, in preparation.
Quinn, P. and Hernquist, T., this volume, p. 249.
Richstone, D. and Tremaine, S., 1985, Ap. J., **296**, 370.
Sargent, W. L. W., Young, P. J., Boksenberg, A., Shortridge, K., Lynds, C. R. and Hartwick, F. D. A., 1978, Ap. J., **222**, 731.
Schneider, D. P., Gunn, J. E. and Hoessel, J. G., 1983, Ap. J., **268**, 476.
Schwarzschild, M., 1979, Ap. J., **232**, 236.
Steiman-Cameron, T. Y. and Durisen, R. H., 1984, Ap. J., **276**, 101.
Terlevich, R. J., Davies, R. L., Faber, S. M. and Burstein, D., 1981, M.N.R.A.S., **196**, 381.
Tonry, J. and Davis, M., 1981, Ap. J., **246**, 680.
Tonry, J., 1983, Ap. J., **266**, 58.
Tonry, J., 1984, Ap. J., **279**, 13.
Tonry, J., 1985, A. J., **90**, 2431.
Whitmore, B. C., Rubin, V. C., and Ford, W. K., 1984, Ap. J., **287**, 66.
Wilkinson, A., Sharples, R. M., Fosbury, R. A. E., and Wallace, P. T., 1986, M.N.R.A.S., **218**, 297.
Williams, T. B., 1981, Ap. J., **244**, 458.
van Albada, T. S., Kotanyi, C. G., and Schwarzschild, M., 1982, M.N.R.A.S., **198**, 303.
van Gorkom, J. H., Knapp, G. R., Raimond, E., Faber, S. M., and Gallagher, J. S., 1986, A. J., **91**, 791.
Young, P. J., Sargent, W. L. W., Boksenberg, A., Lynds, C. R., and Hartwick, F. D. A., 1978, Ap. J., **222**, 450.

DISCUSSION

Gerhard: I would like to ask you more about NGC 1052, because it might be a galaxy for which the intrinsic axial ratios could be determined by the method described by Mario Vietri and myself in a poster paper outside. Three questions: (i) How detailed a velocity field do you have for this galaxy? (ii) What is its luminosity profile? (iii) What is the radial extent of the gas disk in terms of the scale of the luminosity profile?

Davies: The detailed answer to these questions can be found in *Astrophys. J.*, 1986, **302**, 234, and *Astron. J.*, 1985, **90**, 169. The stellar kinematics will be tabulated in a paper by Binney, Davies and Illingworth that is in preparation.

Capaccioli: NGC 4261 is a boxy galaxy. A few years ago we suggested that it might be an S0. We also know that there is another S0, NGC 4125, showing strong rotation along the photometric minor axis. In the second case there is no major evidence for triaxiality. Could you comment on that?

Davies: There is no reason to suspect that NGC 4261 is an S0. I believe the minor axis rotation in NGC 4125 is in the gas alone.

Kochhar: Why don't low luminosity ellipticals like to have velocity anisotropy?

Davies: Perhaps they have undergone more dissipation. I believe Ray Carlberg will say more about this.

Schechter: I think we must bear in mind that there are S0 galaxies with polar rings which slow significant misalignment with the intrinsic minor axis of the underlying galaxy. Linda Sparke has argued (1986, *Mon. Not. R. astr. Soc.*, **219**, 657) that these rings may be stabilized by their own self-gravity. The circumstances may be similar in some of the dust lane ellipticals, in which case dust lanes might not coincide with the symmetry axes of the underlying galaxy.

Davies: The polar ring galaxies are in general different from the disk we see in NGC 1052 in that they lie generally outside the optical image. In order to have a stable rotation axis over a factor of 10 in radius from the centre of the galaxy outward, the disk in NGC 1052 must have its angular momentum along a principal axis of a triaxial figure.

Jaffe: Does the similarity of form of spiral bulges, which we assume are contained in massive halos, and ellipticals, suggest that ellipticals are surrounded by halos?

Davies: No, in fact there is no evidence for significant dark matter within the luminous part of the galaxies, say within one effective radius.

Binney: At the sort of radii where the bulges can be seen, the work of Kent and others *strongly* suggests that the visible components are solely responsible for any galaxy's force–field.

Martin Schwarzschild and George Djorgovski.

THE MANIFOLD OF ELLIPTICAL GALAXIES

S. Djorgovski
Harvard–Smithsonian Center for Astrophysics
60 Garden St.
Cambridge, MA 02138, USA.

ABSTRACT. Global properties of elliptical galaxies, such as the luminosity, radius, projected velocity dispersion, projected luminosity density, etc., form a two-dimensional family. This "fundamental plane" of elliptical galaxies can be defined by the velocity dispersion and mean surface brightness, and its thickness is presently given by the measurement error-bars only. This is indicative of a strong regularity in the process of galaxy formation. However, all morphological parameters which describe the shape of the distribution of light, and reflect dynamical anisotropies of stars, are completely independent from each other, and independent of the fundamental plane. The M/L ratios show only a small intrinsic scatter in a luminosity range spanning some four orders of magnitude; this suggests a constant fraction of the dark matter contribution in elliptical galaxies.

The problem of the "minimal manifold of galaxies" is, how many and which physical quantities are necessary and sufficient to describe a family of normal galaxies? Such description of important physical variables and relations between them is important for our understanding of the structure, formation, and evolution of galaxies. Brosche (1973) was perhaps the first to clearly state the problem in this way. Many, but not all, properties of elliptical galaxies correlate with luminosity (cf. the review by Kormendy 1982, and references therein), and in almost all cases there is a residual scatter, *not* accountable by the measurement errors, and indicative of a presence of "hidden parameters". The previous studies by Tonry & Davis (1981), Terlevich *et al.* (1981), Efstathiou & Fall (1984), Lauer (1985), and Djorgovski, Davis & Kent (1985) indicated that there is more than one important quantity (the one assumed to be the luminosity), but there was no clear understanding or agreement as what the second parameter may be, or whether there are more than two. In more recent attacks on this problem, Burstein *et al.* (1986) and Dressler *et al.* (1986), and Djorgovski & Davis (1986ab) used new, large, homogeneous data sets, and independently reached essentially the same solutions. I will briefly describe here the results, presented in more detail by Djorgovski & Davis (1986b).

The existence of large and homogeneous data base was necessary for the solution of the minimal manifold problem. We used morphological parameters, radii and magnitudes from the CCD surface photometry survey of $\sim$ 260 early-type

T. de Zeeuw (ed.), Structure and Dynamics of Elliptical Galaxies, 79–88.

galaxies by Djorgovski (1985a), and line strength and central velocity dispersion measurements by Tonry & Davis (1981) and the compilation by Whitmore *et al.* (1985). The details of surface photometry survey, data reductions, error estimates and calibrations are given by Djorgovski (1985a; and in preparation). The selection effects are well understood. The sample excludes diffuse dwarfs, which are probably a completely different family of galaxies (cf. Wirth & Gallagher 1984, Kormendy 1985, or Sandage *et al.* 1985).

In order to parametrize the morphology of galaxies in the sample, a consistent radial scale is needed. This is a non-trivial problem, and we defined and used several different scales, all of them derived from our surface photometry profiles, but independent of any particular isophotal threshold, and thus free of magnitude calibration errors. The results were essentially the same with all radial scales which we used, but slightly better fits are obtained with larger radii: the relations described below are robust, but they are indicative of more global, rather than central, properties of galaxies. In this report, I will use the values of r_e (actually, a semimajor axis, not radius), obtained from the fits of surface brightness to de Vaucouleurs' $r^{1/4}$ formula. This is operationally a parameter from the fit, rather than a half-light radius. All quantities used here were measured at r_e, or within the elliptical isophote whose semi-major axis is r_e. For each galaxy, we thus have measurements of radius, magnitude, mean and local surface brightness, slope of the surface brightness profile, ellipticity, ellipticity gradient, and isophotal twist rate. The magnitudes are defined in appropriate elliptical-isophote apertures; both magnitudes and surface brightness are in the (red) r_G band, defined by Djorgovski (1985b). In addition, there are central velocity dispersion (σ) and line strength measurements from the literature, though not for all galaxies in the survey. There are carefully estimated error-bars for all quantities.

We employed a multi-bivariate statistical analysis in our study, and we intend to repeat the analysis by using different multivariate methods (PCA, MDRACE). We started by investigating the known distance-indicator relations (luminosity or radius *vs.* velocity dispersion, σ, or mean surface brightness), and correlating their residuals with other quantities. It was immediately apparent that the residuals of L – σ and R – σ relations correlate well with the mean surface brightness ($\langle\mu\rangle$, in the usual logarithmic, magnitudes-per-square-arcsec form), and *vice versa*. On the other hand, σ and $\langle\mu\rangle$ do not correlate at all! This, indeed, was the solution: linear combinations of log σ and $\langle\mu\rangle$ with logs of L or R produce excellent fits, with no residual scatter, i.e., not accountable by the measurement errors. A possibility of surface brightness as a "second parameter" in the Faber–Jackson relation was already indicated by de Vaucouleurs & Olson (1982), and Lauer (1985) emphasized possible significance of the three-dimensional luminosity density as a parameter in the core properties of ellipticals. The new distance-indicator relations, which at the same time are the equations of a plane in the L (or R) – σ – $\langle\mu\rangle$ parameter space, are:

$$M(r_e) = -8.62(\log\sigma + 0.10\langle\mu\rangle) + 16.14, \tag{1a}$$

$$\text{or:} \quad L \sim \sigma^{3.45}\langle SB\rangle^{-0.86}. \tag{1b}$$

$$\log r_e = 1.39(\log\sigma + 0.26\langle\mu\rangle) - 6.71, \tag{2a}$$

$$\text{or:} \quad R \sim \sigma^{1.39}\langle SB\rangle^{-0.90}. \tag{2b}$$

Here $\langle SB \rangle$ denotes the mean surface brightness in linear flux units, and r_e is measured in pc ($h = 1$ was assumed). The magnitudes are in the red (r_G) bandpass. The power-law coefficients of σ and $\langle \mu \rangle$ are uncertain by about 10%. These distance-indicator relations represent the fundamental plane of elliptical galaxies viewed edge-on. The Eqs. (1) and (2) were derived independently, and thus do not transform into each other exactly. The Eqs. (2ab) are the better ones, since the magnitudes are afflicted by our zero-point calibration errors, and the radii are not. A good graphical representation in terms of observable quantities is the plane defined by σ and $\langle \mu \rangle$, as shown in Figure 1.

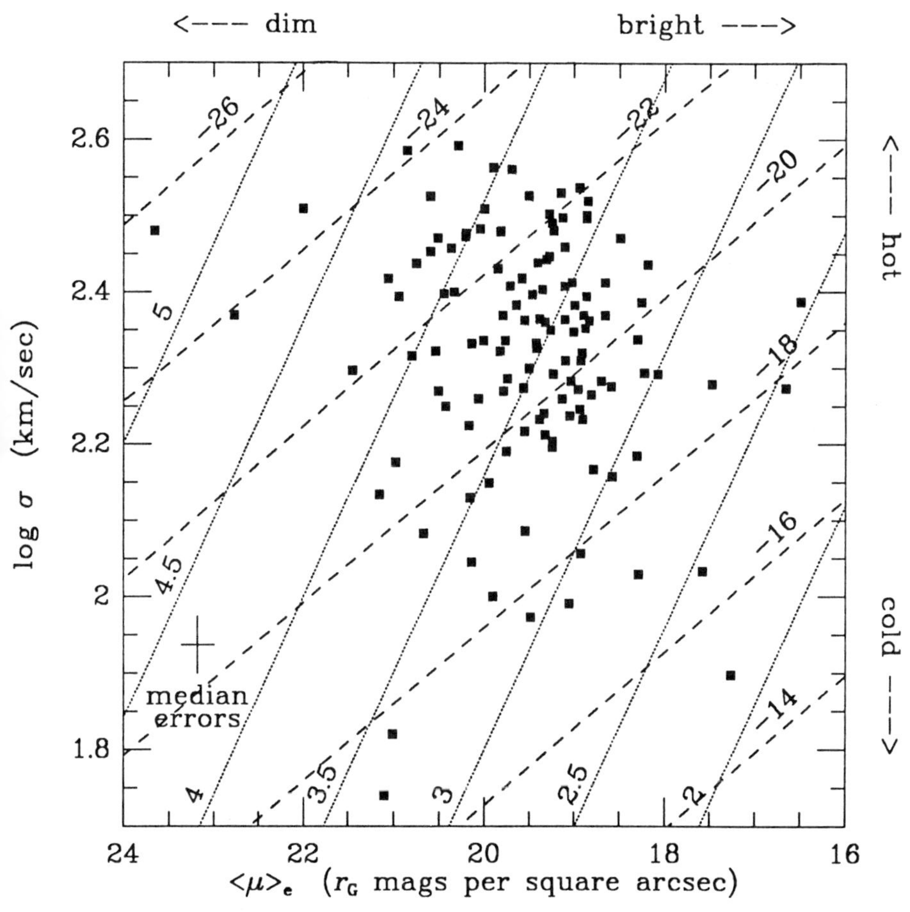

Figure 1: The fundamental plane of elliptical galaxies, shown through the observable quantities, projected central velocity dispersion, σ, measured in km s^{-1}, and mean surface brightness within the r_e isophote, $\langle \mu \rangle$, measured in r_G magnitudes per square arcsec. The dashed lines indicate loci of equal absolute magnitudes (within the r_e isophote, in the r_G band), and the dotted lines indicate loci of equal semimajor axis r_e, measured in h^{-1} pc, and expressed as logarithms (base 10). A Virgocentric infall model with $V_{INF} = 400$ km s^{-1} was assumed in computing the absolute luminosities and radii. The error-bar in the lower left indicates the median errors for the points.

Introduction of line strength as a third parameter does not improve the fits significantly. Line strengths and colors are very well correlated with σ, and can be viewed as additional axes of the parameter space in which this plane is defined. The residuals of new relations do not correlate significantly with any other morphological or spectroscopic quantity, indicating that two dimensions provide an adequate and exhaustive description of the global properties of ellipticals. The plane thickness is completely contained in the present measurement errors, and any cosmic broadening must be very small, on the level of a few percent, or less. The measurement of the intrinsic thickness is is an important target for the future investigations, as it would probe directly the "noise of galaxy formation".

An important question is, how linear are the Eqs. (1 – 2), or, how flat is the "fundamental plane"? First, there is no *a priori* reason why the equations should be pure power-laws, even though the plane appears to be flat within the present measurement errors. A possible physical mechanism which may introduce curvature at the "bright" end, or individual deviations from the plane, could be dissipationless mergers. This can be tested by looking at the positions of apparent cannibals (e.g., galaxies with shells) with respect to the fundamental plane. The very existence of such global regularity, and over several orders of magnitude in luminosity or radius, suggests that elliptical galaxies are formed by a single process, probably dissipative and primordial, but the formation by a large number of mergers is still a possible alternative. Perhaps N-body simulations can be used to test this idea, whether and in which conditions the final locus of multiple mergers is a plane in the L (or R) – σ – μ space, and if so, whether the plane tilt is as observed. It is still possible that there are small differences in the tilts and/or intercepts of the fundamental plane in different large-scale environments, which would reflect, perhaps, some real fluctuations in the process of galaxy formation. Any such deviations, or curvature, would have a considerable importance for the use of Eqs. (1 – 2) as distance indicators: systematic deviations in clusters at different distances (when generally different portions of the luminosity function are sampled) would show up as large peculiar velocities.

It is possible that galaxies in all cluster/field environments lie on the same plane, but populate different portions of it. This may be investigated by looking at the distributions perpendicular to the luminosity axis in the plane, thus factoring out any possible differential selection effects. Galaxian activity (radio power, presence of LINER nuclei, etc.) may also be correlated with the positions in the plane: such correlations, if they exist, could be smeared by projecting them at any of the observable axes, and thus could have passed undetected so far.

Projected velocity dispersions should be influenced by any dynamical anisotropies that may be present. Such anisotropies give rise to the flattening and/or triaxiality of E-galaxies, and thus are reflected in their ellipticities (ϵ), ellipticity gradients, and isophotal twists. Radial-to-tangential anisotropy modifies the radial slope of a surface brightness profile. However, *none* of these shape parameters correlate with velocity dispersion, or any other global property! Moreover, the shape parameters do *not* correlate between themselves, either. There are also no correlations between any of the structural properties and kinematic variables (maximum rotational velocity, V_{max}, its ratio with the mean velocity dispersion, $V_{max}/\langle\sigma\rangle$) from a subset of galaxies of Davies *et al.* (1983), except for the known very weak correlation between V_{max} and ellipticity. In particular, ellipticity does not correlate with either σ or $\langle\mu\rangle$. This renders very difficult or impossible the tests for intrinsic shapes of ellipticals, proposed by several authors in the past: σ and $\langle\mu\rangle$ were expected to correlate with ϵ in the oblate case, and anticorrelate in the prolate case, but a pure scatter is all that is observed.

Since these dynamical quantities are not related to the properties determined by the fundamental plane, which presumably reflect the process of galaxy formation, they must be determined by some other process, or reflect the initial conditions present in the protogalactic (protocluster?) material. There are some indications that the ellipticities and/or galaxy orientations may be related to the properties of the large-scale structure in which they live (Strom & Strom 1978, Djorgovski 1983, and references therein). If I am permitted a crude speculation, and the indications mentioned above are true, then dynamical anisotropies, now frozen in the kinematics of stars, reflect the dynamical anisotropies of the protogalactic material (set, perhaps, by the process of protocluster formation?), whereas the the process of actual galaxy formation, reflected, presumably, in the fundamental plane, determines how is that material going to be cut and assembled in individual galaxies, without any regard for its internal velocity anisotropy. Alternatively, the kinematic anisotropies may be acquired or modified in a stochastic way *after* the galaxies form, by tidal torquing and/or mergers.

Thus, the elliptical galaxies are a "2+n" parameter family, defined by the following observable (projected) quantities: (1) velocity dispersion, (2) surface brightness, and (3,etc.) a variety of shape/anisotropy parameters. A possible interpretation in physical terms is that the two principal parameters are the depth of the potential well (1), and the mean density (2), whereas the multitude of subtle dynamical anisotropies (3,...) determine the details of internal dynamics, and thus the shapes. Note, however, that a full dynamical interpretation of all these observational quantities is afflicted by the projection effects.

In most studies to date, mass-to-light (M/L) ratios have shown a considerable scatter, some of which was (correctly) attributed to the poor and heterogeneous state of photometry available at the time. I have computed the M/L ratios measured within the r_e isophote by using the Poveda's formula (cf. Tonry & Davis 1981); this formula is but a gross approximation: it assumes a spherical galaxy, in which both light and mass follow exactly the $r^{1/4}$ law, and with an isotropic velocity dispersion tensor; all of these assumptions are wrong in differing degrees for all real galaxies, and we do not have the mean, but rather projected central velocity dispersions. Thus, a straight application of this formula will inevitably introduce some scatter in the computed M/L ratios.

Figure 2 shows the histogram of M/L ratios. Note the relatively small spread: the r.m.s. of the histogram is only 0.21 (in log); the median error of measurement is 0.17. This indicates an intrinsic broadening of only about 30%, which should also incorporate any differential shape and anisotropy effects, ignored in my simple computation. Figure 3 plots the computed M/L ratios *vs.* the luminosity. The M/L changes by certainly much less than a factor of ten over some four orders of magnitude in luminosity, and may even be constant. In fact, the only correlation involving M/L which is present in our data is with the velocity dispersion, which is well-known (Faber & Jackson 1976, and references therein). The relative constancy or universality of M/L may imply that the baryons and the dark matter are well mixed on the scales of galaxies, or larger — otherwise, galaxies in different places could have substantially different relative fractions of the dark material.

To summarize, we see a strong regularity in a contrast with a strong cosmic scatter: On one side, there is a set of global properties (luminosity, mass, radius, density, velocity dispersion, metallicity, separately or in various mutual combinations), well described by the thin fundamental plane. On the other hand, there is a set of shape parameters (ellipticities, ellipticity gradients, isophotal twists, surface brightness slopes, rotational properties), reflecting internal dynamical structures of ellipticals, all of which are independent from the first family, and generally independent between themselves as well. The manifold of observable properties of elliptical galaxies is thus split into two distinct sets, one two-dimensional and highly orderded, and the other multi-dimensional, and quite chaotic. The (global) M/L ratios vary very little, and may even be constant. We need a theory of galaxy formation which can account for all this.

Historically, accounting for the shape parameters, all of which were clearly statistically independent, caused some confusion, and prevented some authors (myself included) from seeing forest for the trees: in multivariate analyses there were far too many significant eigen-values and eigen-vectors... Another mistake was to assume that the luminosity is *the* first parameter: we now see that it is a *product* of σ and μ. One may define the luminosity as one of the fundamental plane axes, but then the "second" parameter, perpendicular to that axis, is not anything directly observable or interpretable. Finally, there was always the ideology of *the second* parameter, implicitly assuming or suggesting that there are only two, and some authors even went out to state without a really good substantiation that the ellipticals are a two-parameter family, since that was thought to be the desirable answer, but which we now know is wrong, or rather, an incomplete answer.

The question which remains is, how do galaxies with other morphologies behave? A preliminary investigation with a sample of $\sim$ 50 S0 galaxies, but only for 18 of which we have velocity dispersions, shows that their global properties also form a two-dimensional family, and that the equations (tilt) of their fundamental plane are very similar, and possibly identical, to the plane of ellipticals. Their shape parameters do not correlate with the global properties either. Thus, there seems to be a fundamental continuity between ellipticals and S0's. Whitmore (1984) reached a conclusion that the spirals also form a two-dimensional family, but the observables are quite different there (see also Watanabe, Kodaira & Okamura 1985). There may be some indications of the presence of "second" parameter(s) in the Tully–Fisher relation, perhaps in analogy with our Eqs. (1ab). We may be probing the same fundamental relation for *all* galaxies, but the surface quirks or morphology determine our observables, and make direct comparisons difficult.

Finally, a note on the dynamical models of ellipticals. The distribution of light in elliptical galaxies shows a wide variety of shapes, both in azimuthal and radial sense. The radial surface brightness profiles show a considerable and significant variety of shapes — the ellipticals are *not* well described by the $r^{1/4}$ law, Hubble law, or any other simple formula, or by the King, Binney, or Jaffe models. There is also a variety of ellipticity gradients and isophotal twists. This means that any realistic dynamical model of elliptical galaxies must incorporate several structure parameters, and reproduce this variety; they *have* to be very complex. But the fate of a dynamicist is even worse, if the ellipticals have substantial amounts of dark material, whose radial and azimuthal distribution is completely unknown at the present: any self-consistent dynamical models, in which stars provide both light and mass, are then simply inadequate. All this requires good-quality data, and it may be possible to provide some additional constraints on the true isopotential surfaces from detailed observations of X-ray coronae.

I would like to thank to the staff of Lick Observatory for their help during the surface photometry survey, on which this work is based, and to Marc Davis, who paid the bills. I acknowledge useful conversations with Marc Davis, Ivan King, Sandra Faber, Roger Davies, and many others. This work was supported in part by the NSF grant AST84-19910 to M. Davis, and a Harvard Junior Fellowship to the author.

REFERENCES:

Brosche, P. 1973, *Astron. Astrophys.* **23**, 259.
Burstein, D., Davies, R., Dressler, A., Faber, S., Lynden-Bell, D., Terlevich, R., and Wagner, M. 1986, in *Distances to Galaxies and Deviations from the Universal Expansion*, B. Madore and B. Tully (eds.), p. 123. Dordrecht: D. Reidel.
Davies, R., Efstathiou, G., Fall, M., Illingworth, G., and Schechter, P. 1983, *Astrophys. J.* **266**, 41.
de Vaucouleurs, G., and Olson, D. 1982, *Astrophys. J.* **256**, 346.
Djorgovski, S. 1983, *Astrophys. J. Lett.* **274**, L7.
Djorgovski, S. 1985a, Ph.D. Thesis, University of California, Berkeley.
Djorgovski, S. 1985b, *Publ. Astr. Soc. Pacific* **97**, 1119.
Djorgovski, S., Davis, M., and Kent, S. 1985, in *New Aspects of Galaxy Photometry* J.-L. Nieto (ed.), Lectures in Physics **232**, p. 257, Springer Verlag.
Djorgovski, S., and Davis, M. 1986a, in *Distances to Galaxies and Deviations from the Universal Expansion*, B. Madore and B. Tully (eds.), p. 135. Dordrecht: D. Reidel.
Djorgovski, S., and Davis, M. 1986b, *Astrophys. J.*, in press.
Dressler, A., Lynden-Bell, D., Burstein, D., Davies, R., Faber, S., Wagner, M., and Terlevich, R. 1986, *Astrophys. J.*, in press.
Efstathiou, G., and Fall, M. 1984, *M.N.R.A.S.* **206**, 453.
Faber, S., and Jackson, R. 1976, *Astrophys. J.* **204**, 668.
Kormendy, J. 1982, in *Morphology and Dynamics of Galaxies*, proceedings of the 12th Saas-Fee Advanced Course, Geneva Observatory publication.
Kormendy, J. 1985, *Astrophys. J.* **295**, 73.
Lauer, T. 1985, *Astrophys. J.* **292**, 104.
Sandage, A., Binggeli, B., and Tamman, G. 1985, *Astron. J.* **90**, 1759.
Strom, S., and Strom, K. 1978, *Astron. J.* **83**, 732.
Terlevich, R., Davies, R., Faber, S., and Burstein, D. 1981, *M.N.R.A.S.* **196**, 381.
Tonry, J., and Davis, M. 1981, *Astrophys. J.* **246**, 666.
Watanabe, M., Kodaira, K., and Okamura, S. 1985, *Astrophys. J.* **292**, 72.
Whitmore, B. 1984, *Astrophys. J.* **278**, 61.
Whitmore, B., McElroy, D., and Tonry, J. 1985, *Astrophys. J. Suppl. Ser.* **59**, 1.
Wirth, A., and Gallagher, J. 1984, *Astrophys. J.* **282**, 85.

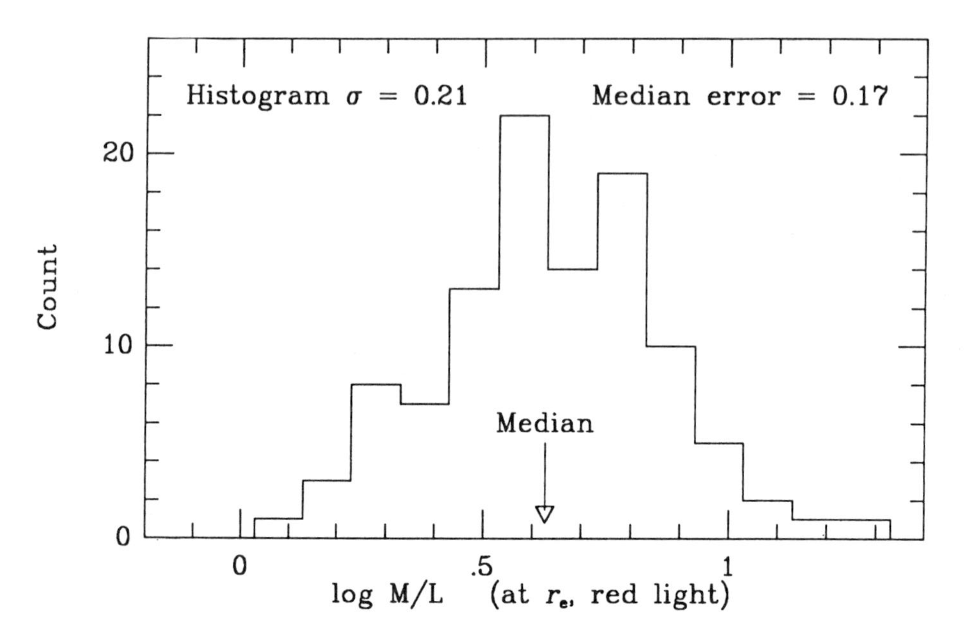

Figure 2: The distribution of (red) M/L ratios, computed with the Poveda's formula, within the r_e isophote. Most of the spread is due to the measurement errors, and the residual scatter is estimated to be $\sim$ 30%, which must also include any possible scatter caused by the application of an approximate formula.

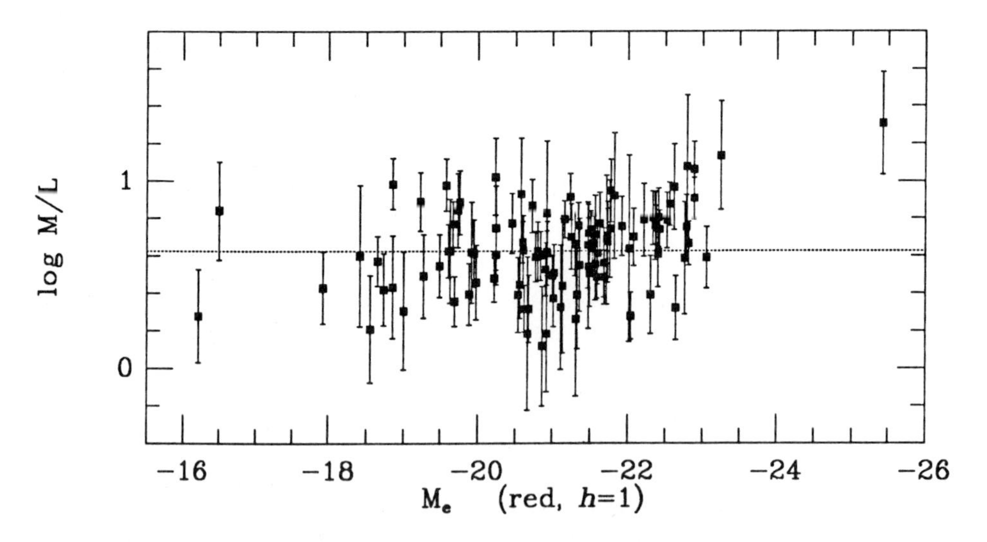

Figure 3: The M/L ratios plotted against the absolute magnitude (M_e), in the r_G band, within the r_e isophote. A Virgocentric infall model with 400 km s^{-1}, and $h = 1$ were used for computation. Dotted line indicates the median M/L. There is no indication that M/L varies with luminosity.

DISCUSSION

Schechter: My understanding of your "fundamental plane" is that you can use any two parameters to predict a third. I would be interested in seeing luminosities and surface brightnesses used to predict velocity dispersions, and then observed minus predicted dispersions plotted versus apparent ellipticity. This might produce a better correlation with ellipticity than either dispersion or surface brightness alone. Have you made such a plot?

Djorgovski: Not directly, but this may be equivalent to introducing the ellipticity in my residual fits. This did not help much. It is a good idea, and I will try it, but I doubt that it will show anything, since there is absolutely no correlation between the ellipticity and surface brightness, velocity dispersion, or any linear combination of the two.

de Zeeuw: There are a few posters outside (Statler, Levison) that show dynamical models with a variety in their velocity dispersion tensors that is possibly larger than you see in real ellipticals.

Djorgovski: Well, good, but the observed variety is such that constraining any models may be very difficult.

White: I would take issue with your statement that the fundamental plane of L or r as a function of $\langle \mu \rangle$ and σ is only a few percent thick. From your data the only safe conclusion is that the plane is thinner than the observed scatter. What is the rms observed scatter of $\log L$ and $\log r$ about the predictions of your relations?

Djorgovski: I cannot really answer that offhand, since there are projection effects, and error correlations. When the new relations are used as distance indicators, the errors of distances expressed in magnitudes are about $0.^m4$, or $\sim 20 - 25\%$. However, please recall that the observed scatter is now given entirely by the error bars, that is, $\chi^2 \sim 1$. This leaves only a little space for intrinsic broadening.

King: First, a comment: you should not say that σ measures the depth of the potential well. It doesn't; in the King models (which, incidentally, don't fit your profiles either) the model parameter W_0 explicitly measures the amount by which σ does *not* measure the depth of the potential well. Second, a question: why didn't you use principal–component analysis, instead of plotting residuals? And a final question, can't you get line strengths from the velocity–dispersion papers?

Djorgovski: About velocity dispersions—yes, you are quite right, but I hope that my intuitive meaning was clear. Then, I *did* use the principal component analysis some three years ago with the fine data by Steve Kent. I got some solutions, but they meant nothing to me, since there were too many significant eigen–values (this is still the case). The overabundance of significant dimensions in the total set of observable parameters prevented me from seeing the fundamental plane which unites a *subset* of fundamental properties, as expanded in my talk. In other words, I did not see the forest for the trees, and a similar thing happened to some other people who used the principal component analysis on this problem. Even if one does apply the principal component analysis correctly (and Efstathiou & Fall addressed some of the important problems), the formal solutions may be quite non-intuitive. My intuition worked much better when I started from the other end, with multi-bivariate statistics. Finally, I did use the line strength measurements from Tonry & Davis, but they are not very good by the present standards, and for a variety of reasons I believe that introduction of line strengths does not help much in this particular problem.

Richstone: If isophotal twists are observed equally often in low and high luminosity ellipticals, and if we believe (as Davies suggests we ought to), that low L ellipticals are oblate isotropic rotators, then what's the logic of using isophotal twists to identify triaxial objects?

Djorgovski: The bulges and the low-luminosity ellipticals are similar in many ways, but not all: For example, the bulges often show boxy isophotes, but the low-luminosity ellipticals do not. Besides, the observed $(V/\sigma)^*$ vs. luminosity correlation is very noisy, and some bulges (and certainly ellipticals) may well be triaxial.

Whitmore: My first comment is a note of caution addressed toward Ivan King's suggestion that the line strength index determined from the Fourier quotient technique be used to estimate metallicity. This will not be useful since the line strength index is on a relative scale, determined by the template star used to make the reductions, rather than on an absolute scale. Since everyone uses different template stars, these are all on a different system.
My second comment is for George. In the interest of trying to compare your two-dimensional parameter space for elliptical galaxies with the two-dimensional space I found for spiral galaxies, do you have any color information for your sample?

Djorgovski: I did not measure colors, and there is no sufficiently large, homogeneous data base in the literature. To the extent that colors measure metallicity in E-galaxies, my remarks to Ivan about line strengths still apply. Faber *et al.* will probably be able to do a better job than what I can.

Davies: The size of the isophote twist as a function of luminosity alone does not indicate the degree of triaxiality. Isophote twists can also be caused by tidal effects, so that if one believes that lower luminosity ellipticals are oblate, then statistically one would expect to find few isophote twists at small radii in them compared to their brighter brothers, *i.e.*, one needs to attach a radius to the isophote twists to make this comparison.

Djorgovski: Quite right. As you know, I have used several different radial scales, and the results are the same at all fiducial radii. Unfortunately, I do not have a sample of bulges.

PROPERTIES OF CD GALAXIES

John L. Tonry
Mass. Inst. of Technology
Room 37-521
Cambridge, MA 02139

ABSTRACT. cD galaxies are the most luminous galaxies in the universe. They are characterized by a surface brightness profile that falls off more slowly with radius than most elliptical galaxies. In most respects D galaxies are a continuous extrapolation from other ellipticals: their M/L and their colors are comparable to other ellipticals, their inner parts are fitted by an $r^{1/4}$ law, and they follow the same relation between L and σ. On the other hand, their luminosity is too bright to be consistent with the luminosity function of other ellipticals and they are always found at the center of a cluster of other galaxies. Being at the center of a cluster of galaxies often endows D galaxies with a very faint, very extended halo of luminosity and multiple nuclei, but these are more properly associated with the cluster than the D galaxy itself. The connection between the formation of cD galaxies and the formation of clusters remains a mystery. It is still unresolved whether cDs are a byproduct of cluster evolution, whether they formed in parallel with clusters, or whether primeval D are galaxies the seed around which clusters accreted.

1. INTRODUCTION

cD galaxies are the largest galaxies in the universe. They are probably the sites of extensive galaxy evolution and they are also intimately linked to clusters of galaxies and the properties of clusters. There are many facts, conjectures, and myths relating to cD galaxies and it is not clear how these fit together. Two recent review articles that have touched on the subject of cD galaxies were written by Dressler (1984) and Sarazin (1986).

I will first discuss what is meant by a cD galaxy and what its defining characteristics are. Next I will describe some of the observations relating to the morphology and dynamics of cD galaxies. Finally I will discuss some theories relating to the origin and evolution of cD galaxies.

2. WHAT IS A CD GALAXY?

Figure 1 shows surface brightness data of a few selected galaxies. N1278 is a typical giant elliptical galaxy. AWM7 is a D galaxy at the center of a poor cluster of galaxies, and A2029 is the brightest cluster galaxy at the center of a rich cluster, as is A1413. The left-hand plots show log surface brightness in magnitudes versus log radius and the right-hand plots show surface brightness versus $r^{1/4}$ (coordinates such that a deVaucouleurs $r^{1/4}$ law fit is a straight line).

cD galaxies were originally defined by Matthews, Morgan, and Schmidt (1964) as being distended superluminous galaxies. "c" is the designation coined by Miss Maury (1897) for stars with unusually narrow H lines, which we now understand to be superluminous, and "D" originated with Morgan

T. de Zeeuw (ed.), Structure and Dynamics of Elliptical Galaxies, 89–98.

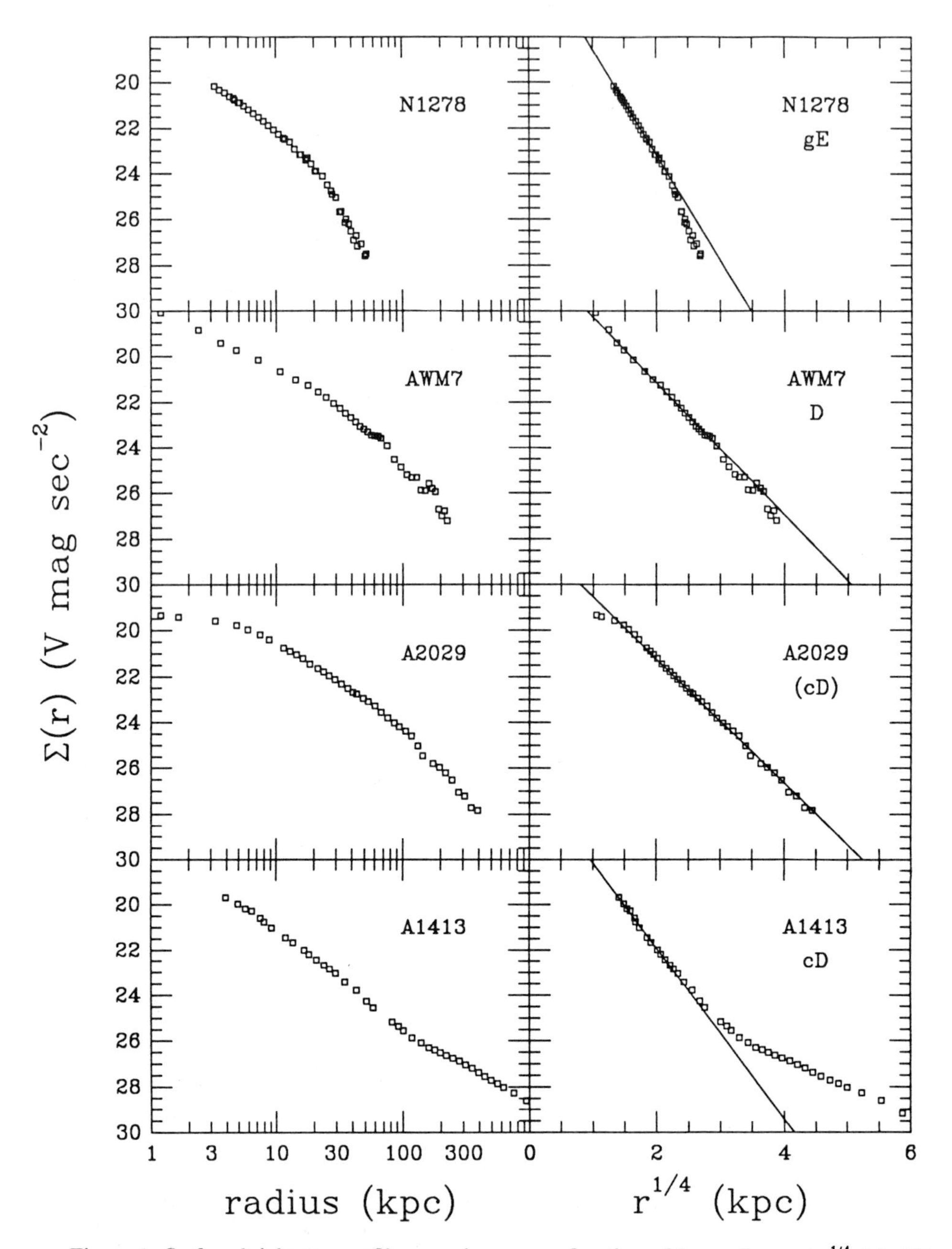

Figure 1. Surface brightness profiles are shown as a function of log radius and $r^{1/4}$ (H_0=50) for four galaxies. $r^{1/4}$ law fits are shown as straight lines. N1278 is a typical giant E galaxy in the Perseus cluster; AWM 7 is a D galaxy in a poor cluster; A2029 is a D galaxy in a very rich cluster; and A1413 is a cD galaxy in a rich cluster. A2029 is often called a cD galaxy although it lacks an extended cD halo. The data are from Schombert (1986).

(1958), meaning "dustless", I believe. AWM7, A2029, and A1413 all satisfy this criterion, but there is a lot of inconsistency in what gets called a D and what gets called a cD. The characteristics that have been traditionally used to identify cD galaxies are luminosity and an apparent halo, very faint and extensive, appearing just at the limits of photographic plates. The primary authority for bona-fide cD galaxies has been Matthews, Morgan, and Schmidt and the extensive catalog of cD galaxies and cluster morphology compiled by Leir and Van den Bergh (1977).

It has been pointed out by Malumuth (1983) and Schombert (1984), among others, that what causes a galaxy to look distended is the logarithmic slope of the surface brightness profile at the plate limit. This governs whether the galaxy looks as though it has a halo. Notice A1413 actually does have a halo beyond the $r^{1/4}$ law fit to the interior, but that this halo doesn't appear until very faint surface brightnesses, fainter than would be visible to the eye on a plate. Oemler (1976) found that a number of classical cD galaxies have these halos, but by no means all.

Figure 2 is a schematic diagram of how surface brightness profiles vary as a function of galaxy luminosity. The four curves show the log surface brightness as a function of log radius for what are labelled an E galaxy, a giant E galaxy, a D galaxy, and a cD galaxy.

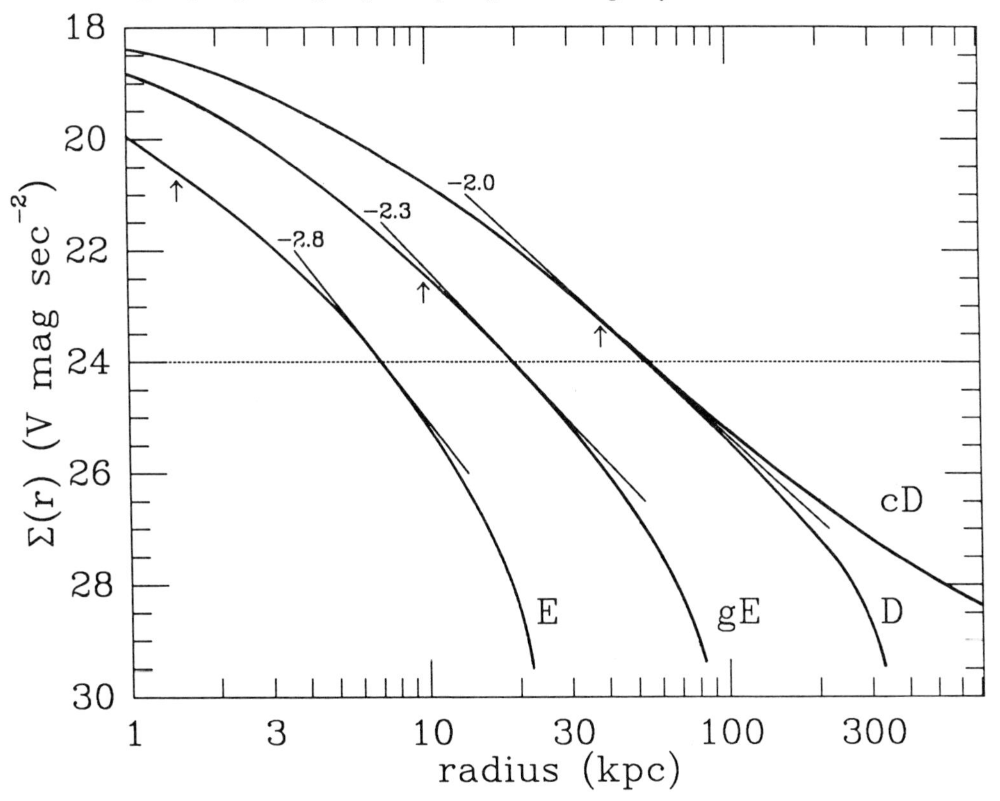

Figure 2. Schematic surface brightness profiles are shown for E, gE, D and cD galaxies as a function of log radius (H_0=50). The tangent line to the profiles at $V = 24$ mag sec^{-2} are drawn and labelled with their slope. The arrows point to the effective radius of each profile, determined by an $r^{1/4}$ law fit to the inner parts of each profile. The cD profile is identical to the D profile brighter than $V \approx 25$. The curves are from Schombert (1986) and are derived from mean galaxy profiles.

Notice how I am modifying the original nomenclature. What I call an E galaxy is an elliptical galaxy whose profile is falling steeply at the plate limit ($V \approx 24$ mag sec^{-2}). A gE galaxy is one whose profile is falling less steeply, a D galaxy even less steeply, and finally a cD galaxy is a D galaxy with an additional halo tacked on. This I think is as faithful as possible to the original definitions while making a quantitative classification possible. The requirement that cD galaxies be superluminous as well as having flat profiles is superfluous because there are very few galaxies that have the flat profiles of D galaxies that are not superluminous. Table 1 shows typical values for a number of parameters derived from $r^{1/4}$ law fits to the inner parts of the profiles shown in figure 2.

Table 1.

Galaxy	dlogΣ/dlog$r$	α	Σ_e	r_e	L/L^*
E	−2.8	0.05	20.5	1.7	0.3
gE	−2.3	0.35	22.5	10.0	1.5
D	−2.0	0.70	23.5	40.0	9.0

Column 1 is the galaxy type, column 2 is the log slope of the profile at V=24 mag sec^{-2}, column 3 is the structure parameter α=dlog$L$/dlog$r$ (see e.g. Hoessel 1980), column 4 is the V surface brightness at the effective radius, column 5 is the effective radius in kpc (H_0=50), and column 6 is the $r^{1/4}$ luminosity in units of L^* (the luminosity of the break in the luminosity function).

E, gE and D galaxies can equally well be characterized by surface brightness at the effective radius as by profile slope at a fixed surface brightness, since the two are equivalent for an $r^{1/4}$ law. This classification is fairly artificial, however, because this is a continuous sequence of galaxy shape and luminosity. The $r^{1/4}$ profiles that Hoessel and Schneider (1986) fitted to their sample of first ranked cluster galaxies show an rms scatter of about 25 percent in r_e at fixed profile slope (or fixed Σ_e). The effective radius of these galaxies increases exponentially with decreasing profile slope, growing by a factor of roughly 30 per unit change in profile slope.

It is important to note that E, gE, D, and cD galaxies are all fitted nicely by $r^{1/4}$ laws for surface brightnesses greater than $V \approx 25$ mag sec^{-2}, and this is the primary means of differentiating Ds and cDs: cDs have an additional halo component above an $r^{1/4}$ law fit to their interiors. This halo starts at about $V \approx 25$ mag sec^{-2} and it has a log slope of about −1.5. According to this scheme A2029 and AWM7 would be D galaxies and only A1413 a cD. This classification will be used for the rest of the article.

Figure 2 also shows the effective radius of these profiles; notice that the surface brightness at the effective radius drops as luminosity increases. This fact, along with a few examples of D galaxies with large core radii and low central surface brightness, has led to the belief that D galaxies have low surface brightness. This is not correct. The surface brightness at r_e drops with increasing L because r_e increases faster than $L^{1/2}$. At any physical radius greater than about 1 kpc, however, D galaxies have higher surface brightness than lower luminosity galaxies. This fact has been stressed by Schombert (1986), and it is also apparent in the data of Hoessel and Schneider (1986). In both data sets there is a clear trend of more luminous galaxies and galaxies with flatter profiles having higher average surface brightness in small central apertures. This does not necessarily imply that the surface brightness at the very center of the galaxy also increases with luminosity; but the luminosity in any aperture of fixed size that includes a significant fraction of the light will increase with luminosity.

3. OBSERVATIONS OF CD GALAXIES

Not suprisingly, all galaxies with cD halos have the shallow profiles of D galaxies; conversely 25 percent of Ds have cD halos (Schombert 1986). D galaxies only occur in the center of galaxy clusters, never in the field, according to Geller and Beers (1983). There are three cases known of cD galaxies that are not at the overall center of a cluster, but they are found centered on subclumps of their own. Beers and Tonry (1986) have found that D galaxies are surrounded by a cusp of cluster galaxies with a projected density distribution which falls off as r^{-1}.

It has long been known that the brightest galaxies in clusters of galaxies are usually exceptionally luminous. A large fraction of these first-ranked cluster galaxies, or brightest cluster galaxies (BCGs), have D galaxy profiles. A classic observation of BCGs is that they vary little in luminosity within a fixed metric aperture (Sandage 1973); this fact has been extensively exploited in attempts to use BCGs as standard candles to measure q_0. This standard luminosity has small positive correlations with galaxy structure parameter α, cluster richness, and cluster morphology (Schneider, Gunn and Hoessel 1983a). BCGs are extremely luminous, too luminous to be consistent with a Schechter function fitted to other elliptical galaxies. It is a non-trivial fact that D galaxies (as a subset of BCGs) are very luminous and span a small range in luminosity (rms of about 0.3 magnitudes).

There have been reports by Hoessel (1980), among others, that D galaxies have larger core radii than gEs. While there are a number of D galaxies with flagrantly large core radii and low central surface brightness, it is not clear whether there is a correlation at fixed luminosity between flatness of surface brightness profile and core radius. There undoubtedly is a correlation between luminosity and core radius, but that does not directly address the question of whether D galaxies have different core structure than gE galaxies of the same luminosity.

D galaxies have the same color (and so presumably stellar population) as gEs and do not appear to have significant color gradients (Lugger 1984). There have been interesting reports by Sastry (1968), Binggeli (1982), and others that D galaxies are slightly more flattened on average than gEs, and that the direction of flattening is weakly aligned with the overall shape of the cluster in which the D galaxy resides.

The galaxy AWM7 in figure 1 is an enormous galaxy in a very sparse cluster. It certainly is a D galaxy but does not have the deviation from an $r^{1/4}$ law characteristic of a cD halo. This appears to be true in general for poor clusters. According to Thuan and Ravanishin (1981), D galaxies in poor clusters never have cD halos although they may be extremely luminous and have extremely flat profiles.

Another property of D galaxies is that they often have "multiple nuclei" which are other, smaller galaxies seen projected on top of the luminosity profile of the D galaxy (within 10 or 20 kpc of the center). Hoessel and Schneider (1986) find that 50% of BCGs have "multiple nuclei." This is not true of the second brightest galaxy, nor does the fraction vary with cluster richness or morphology. These "multiple nuclei" are not deficient in luminosity; they are roughly consistent with being drawn from the overall luminosity function of the cluster.

D galaxies are not only spatially projected at the center of clusters. Quintana and Lawrie (1982) find that they have a much lower rms velocity with respect to the cluster mean than do the rest of the galaxies in the cluster. Thus D galaxies are apparently stationary at the bottom of the cluster potential well.

Observations of the stellar velocities within D galaxies reveal a number of facts. The stellar velocity dispersion never exceeds roughly 400 km/s and is usually between 300 and 350 km/s. D galaxies agree nicely with the $L \propto \sigma^{3.3}$ relation when any excess halo luminosity above an $r^{1/4}$ law fit to the interior is excluded (Tonry 1986). The stellar velocity dispersion never seems to fall with radius, nor do the galaxies ever show any rotation, in contrast with lower luminosity ellipticals which do both. There have been some very intriguing reports by Dressler (1978) and Carter *et al.* (1985) that indicate that in some D galaxies the stellar velocity dispersion actually starts to rise at about 50 kpc (or a surface brightness of $V \approx 23$ mag sec^{-2}). This needs further observation, and more important, needs to be tested as a function of cluster richness, because cluster richness seems to have major effects on the outer parts of D galaxies.

The mass to luminosity ratio of D galaxies is comparable to E galaxies; they are not supermassive. Malumuth and Kirshner (1981) observed a number of Ds and found that their M/Ls were 50% larger than Es. They were limited to a small sample of galaxies, and they used only central velocity dispersions and uncertain core radii, so this result is not beyond question. Tonry (1984) observed the velocity dispersion of two D galaxies as a function of radius and found a mass to luminosity ratio consistent with that of other E galaxies.

A complete sample of multiple nuclei were observed by Tonry (1986), and the rms velocity of these galaxies was found to be 800 km/s. This is so much in excess of the 300 km/s velocity dispersion of the stars in the D galaxies that few if any of these multiple nuclei are bound to the D galaxy. The term "multiple nucleus" is a misnomer: these multiple nuclei are merely sharing the bottom of the cluster potential with the D galaxy. Roughly a quarter of multiple nuclei are moving at less than 300 km/s projected velocity and could be interacting with the D galaxy, but it is surprising that such slow multiple nuclei are only found around low luminosity BCGs. When a BCG is brighter than 2 L^* the multiple nuclei around it all have projected velocities of at least 500 km/s, hinting that it is indeed the slow multiple nuclei that are consumed to make a bright D galaxy.

D galaxies occasionally have significant X-ray fluxes (Forman and Jones 1982). This emission is apparently from a halo of hot gas, but it is not clear whether this gas is bound to the D galaxy or to the cluster in which it resides. The X-ray spectral observations suggest that some of the gas is cool enough to cool further and accrete onto the central D galaxy (Canizares 1981). There have been observations of patchy dust and filaments of gas emission in D galaxies that may corroborate this. Alternatively this gas and dust may be the product of stellar evolution within the D galaxy itself. There has been speculation that if such cooling flows exist, the gas could be converted into low mass stars which could be a significant, high M/L contribution to D galaxies (Fabian *et al.* 1982).

4. ORIGIN AND EVOLUTION OF CD GALAXIES

Despite the fact that there is continuity of many properties between E and D galaxies, it appears that D galaxies have undergone an extra measure of evolution beyond ordinary elliptical galaxies. D galaxies are too bright to be a statistical fluctuation in accord with a continuous extrapolation of the luminosity function of other galaxies and they are relatively constant in luminosity. They are found only at the center of an r^{-1} cluster of other galaxies. These adjacent galaxies frequently appear as multiple nucleus companions, and D galaxies often have an extended cD halo. They are more flattened than elliptical galaxies, and appear to align with their clusters. It is a puzzle whether the process that makes D galaxies surrounds them with a cluster, or whether any galaxy that sits at the center of a cluster would acquire the same properties.

There has been a extensive work by Villumsen (1982), Farouki *et al.* (1983), Duncan *et al.* (1983), May and Van Albada (1984), and Aguilar and White (1986) on N-body simulations of galaxy mergers and accretion which indicate that $r^{1/4}$ laws are a natural formation product. Further, it appears that profiles become more and more flattened as they evolve and one can even get extensive halos. These halos are not very luminous and they are transient, but perhaps the environment of a rich cluster is sufficiently hospitable for them to survive and grow.

The most natural explanation of the origin of D galaxies is that they have formed from galaxy interactions: merging, accretion, and stripping. The question is when? One possibility is that mergers and accretion form D galaxies throughout the history of the cluster right up until the present. This scenario, discussed by Gunn and Tinsley (1976), Ostriker and Hausman (1977) and Hausman and Ostriker (1978), has the largest galaxies in the center of a cluster merge to form a D galaxy and then this galaxy continuing to accrete cluster galaxies, until it eventually grows to its present large size. cD halos are expected as the tidal debris from the galaxy interactions and the high velocity stars that are sprayed off in mergers. This is the "galactic cannibalism" model of cD formation. Richstone and Malumuth (1983) and Malumuth and Richstone (1984) have made detailed calculations indicating that this may be the case.

Several observational points seemed to be explained by this theory. The constancy of BCG luminosity was thought to be the result of homologous growth of the central galaxy, i.e., the galaxy maintains the same shape with rescaled radius and mass. Within a fixed aperture the luminosity could stay constant with growth if the surface brighness of the galaxy decreased as its scale size grew. The multiple nuclei were viewed as the stripped cores of cluster galaxies, drifting about within the halo of the cD. Sandage and Hardy (1973) found an anticorrelation between the luminosity of the second brightest galaxy in a cluster and the luminosity of the brightest. This was taken as evidence that there had been brighter galaxies in the cluster which had merged to become the cD, and present second brightest galaxy was originally much less highly ranked. The current observational evidence shows no indication that surface brightness decreases with luminosity, however. Multiple nuclei are not in the process of merging with the BCG. Schneider, Gunn and Hoessel (1983b) and Schombert (1984) find no anticorrelation between the first and second brightest galaxies in clusters. Thus much of the support for the cannibalism and homologous growth theory has disappeared.

Recent calculations by Miller (1983) and especially Merritt (1985) indicate that there is no present evolution because galaxies have become stripped of their halos and have high velocities, making them relatively immune to dynamical friction. These authors find that the accretion rates of Hausman and Ostriker (1978) and Malumuth and Richstone (1984) to be much too high. They argue that D galaxies formed in the early stages of formation of the cluster before it had virialized, and that the cluster is presently quiescent. There does not seem to be a simple reconciliation of the calculations by Malumuth and Richstone and Merritt because the differences arise mainly as a result of the values assumed for a number of poorly known quantities, such as the mass and extent of halos around cluster galaxies, the density and core radius of the dark matter that gives clusters such high velocity dispersions, and the extent to which clusters are subclustered.

Multiple nuclei and the cusps in galaxy density around D galaxies give ambiguous evidence. On the one hand, the majority of multiple nuclei are not merging with their D galaxies and will not in the near future. Lauer (1986) has extracted the luminosity profiles of some multiple nuclei and finds that they look like ordinary cluster galaxies showing little evidence of strong interaction. On the other hand, low luminosity BCGs apparently do have low velocity companions whereas high luminosity BCGs do not, suggesting that D galaxies evolve until they have consumed all their slow moving neighbors and then stop, and that low luminosity BCGs are currently in the process of growing.

There has also been some work on the possibilty that D galaxies have grown from a rain of gas falling into the center of the cluster and forming stars. This seems implausible to me because of the normal colors of D galaxies and the enormous inflows required by this theory (hundreds of solar masses per year averaged over a Hubble time).

5. SUMMARY

To a large extent D galaxies are a continuous extension from giant elliptical galaxies, characterized by flat surface brightness profiles, high luminosity, and high surface brightness. They have similar colors and *M/L* to elliptical galaxies. They do not vary much in metric luminosity: the rms scatter is about 30 percent. They do not rotate and they have flat or rising stellar velocity dispersion profiles.

In contrast to giant elliptical galaxies, D galaxies are inextricably linked with cluster centers. They only appear at rest at the center of a cluster, surrounded by an r^{-1} entourage of companion galaxies. D galaxies appear to be slightly more flattened than ellipticals and their flattening aligns weakly with the cluster. D galaxies often have cD halos and are surrounded by multiple nuclei. These are probably more intrinsic to the cluster in which the galaxy resides than the D galaxy itself; most of the multiple nuclei would not be bound by the D galaxy alone. Curiously, although cD halos only appear in rich clusters, the size and structure of D galaxies and the frequency of multiple nuclei does not appear to depend on cluster richness.

A number of questions remain unanswered. The constancy of the luminosity of BCGs is real, it is not an artifact of a particular aperture and compensating scale size and surface brightness. It therefore represents a maximum size that galaxies can attain (neglecting cD halos), and the discrepancy between this size and the luminosity function of other galaxies suggests that this cutoff is not caused by statistics, but rather by physics. The causality of the relation between D galaxies and clusters is obscure. Much of the evidence that D galaxies are cannibalism byproducts is not consistent with new observations. It is an open question whether D galaxies were formed at early epochs when the cluster itself was coalescing, and either acted as seeds for the nascent cluster, or grew out of the merging clumps of matter.

REFERENCES.

Aguilar, L. A. and White, S. D. M. 1986, preprint.
Beers, T. C. and Geller, M. J. 1983, *Ap. J.*, 274, 491.
Beers, T. C. and Tonry, J. L. 1986, *Ap. J.*, 300, 557.
Binggeli, B. 1982, *Astron. and Astro.*, 107, 338.
Canizares, C. R. 1981, in *Proceedings of the HEAD Meeting on X-ray Astronomy*, ed. R. Giacconi (Reidel Dordrecht), p. 215.
Carter, D., Inglis, I., Ellis, R. S., Efstathiou, G., and Godwin, J. G. 1985, *MNRAS*, 212, 471.
Dressler, A. 1984, *ARAA*, 22, 185.
Dressler, A. 1979, *Ap. J.*, 231, 659.
Duncan, M. J., Farouki, R. T., and Shapiro, S. L. 1983, *Ap. J.*, 271, 22.
Fabian, A. C., Atherton, P. D., Taylor, K., Nulsen, P. E., 1982, *MNRAS*, 201, L17.
Farouki, R. T., Shapiro, S. L., and Duncan, M. J. 1983, *Ap. J.*, 265, 597.
Forman, W. and Jones, C. 1982, *ARAA*, 20, 547.
Gunn, J. E. and Tinsley, B. M. 1976, *Ap. J.*, 210, 1.
Hausman, M. A. and Ostriker, J. P. 1978, *Ap. J.*, 224, 320.
Hoessel, J. G. 1980, *Ap. J.*, 241, 493.
Hoessel, J. G. and Schneider, D. P. 1986, *A. J.*
Lauer, T. 1986, preprint.
Leir, A. A. and van den Bergh, S. 1982, *Ap. J. Suppl.*, 34, 381.
Lugger, P. M. 1984, *Ap. J.*, 278, 51.
Malumuth, E. M. 1983, Ph.D. thesis, University of Michigan.
Malumuth, E. M. and Kirshner, R. P. 1981, *Ap. J.*, 251, 508.
Malumuth, E. M. and Richstone, D. O. 1984, *Ap. J.*, 276, 413.
Matthews, T. A., Morgan, W. M., and Schmidt, M. 1964, *Ap. J.*, 140, 35.
Maury, A. C. 1897, *Harvard Annals*, 28, *part I*, 5.
May, A. and Van Albada, T. S. 1984, *MNRAS*, 209, 15.
Merritt, D. 1985, *Ap. J.*, 289, 18.
Miller, G. E. 1983, *Ap. J.*, 268, 495.
Morgan, W. M. 1958, *PASP*, 70, 364.
Oemler, A. 1976, *Ap. J.*, 209, 693.
Ostriker, J. P. and Hausman, M. A. 1977, *Ap. J. Lett.*, 217, L113.
Quintana, H. and Lawrie, D. G. 1982, *A. J.*, 87, 1.
Richstone, D. O. and Malumuth, E. M. 1983, *Ap. J.*, 268, 30.
Sandage, A. 1973, *Ap. J.*, 183, 711.
Sandage, A. and Hardy, E. 1973, *Ap. J.*, 183, 743.
Sarazin, C. L. 1986, *Rev. Mod. Phys.*, 58, 1.
Sastry, G. N. 1968, *P.A.S.P.*,80, 252.
Schombert, J. M. 1984, Ph.D. thesis, Yale University
Schombert, J. M. 1986, preprint.
Schneider, D. P., Gunn, J. E., and Hoessel, J. G. 1983a, *Ap. J.*, 264, 337.
Schneider, D. P., Gunn, J. E., and Hoessel, J. G. 1983b, *Ap. J.*, 268, 476.
Thuan, T. X. and Romanishin, W. 1981, *Ap. J.*, 248, 439.
Tonry, J. L. 1984, *Ap. J.*, 279, 13.
Tonry, J. L. 1986, *A. J.*, 90, 2431.
Villumsen, J. 1982, Ph.D. thesis, Yale University.

DISCUSSION

White: I was a little confused by your conclusions. My impression is that the continuity from gE's to D's in profile shape and in relations between metallicity, and velocity dispersion and luminosity, would argue strongly for similar formation mechanisms. One argument against Brightest Cluster Members being drawn from a universal luminosity function has traditionally been based on the small scatter in their absolute magnitudes. How does this argument hold up for the D and cD galaxies you discussed?

Tonry: As I tried to stress, there does seem to be some contradiction in these observations, although I suspect that there is actually something faulty in our understanding of the formation of elliptical galaxies. D and cD galaxies do appear to form a continuous sequence with other elliptical galaxies, but enjoy some special properties such as flat profiles, constant metric luminosities, location of the center of clusters of galaxies and occasional cD halos. Whatever scenario we cook up to explain these properties of D galaxies must also extend in a continuous fashion to other elliptical galaxies—at least to around $1/2\ L^*$ where there may be a discontinuity in the properties of elliptical galaxies.

Spergel: I would like to offer a speculative theory for the formation of cDs. One of the currently fashionable theories of galaxy formation is accretion around cosmic strings. These strings provide the seeds for galaxy formation. Clusters form around the more massive string loops. The more massive loops would thus accrete both other galaxies and stars to form both the cluster and an anomolous galaxy in its center. The large loops could be the source of cD galaxies. One of the remnants of these large loops might be a nucleus with high velocity dispersion and unusual velocity field. My questions are: In what fraction of the cD's do the high velocity nuclei occur? What are the velocity fields of these nuclei?

Tonry: High velocity nuclei ($|v| > 300$ km/s) occur in 2/3 of BCG's with multiple nuclei. There is no hint, however, that there is anything anomalous in the potential there. The stellar velocity dispersion is low (~ 300 km/s) and shows no disturbance in the BCG. The high velocity nuclei appear to be independent galaxies seen in projection.

Gunn: Is the histogram of velocities of the "multiple nuclei" gaussian, or can the large dispersion be due to a bound population and a few interlopers which are accidentally superposed?

Tonry: To within the meager statistics (20 or so velocities) the distribution is a continuous gaussian. There is a hint of bimodality in the distribution, but all velocities are well represented in the distribution from 0 to 1000 km/s.

Kormendy: The classification of D galaxies has been disturbingly ambiguous for many years. I'd like to call attention to this, and to suggest that we return to a "purer" definition of cD galaxies.
When you look at Morgan's original form classification catalogue you find that most D galaxies are S0s. This is consistent with Morgan's description of the D type, which refers to "extra light" in some sort of halo around the galaxy. Morgan

was therefore consistent in his identification of cD galaxies, although their cluster–sized halos are physically very different from S0 disks. Today the term “D galaxy” is not very useful; “S0” or “lenticular”’ more clearly isolate the physically distinct kind of structure that we want to name, i.e., essentially an elliptical plus a disk. Morgan’s term cD galaxy remains useful, because it names an important physical phenomenon, namely an elliptical surrounded by a cluster–sized halo.
Now you are changing the definition of D galaxies to mean “very large elliptical”. This also makes me uncomfortable. Elliptical galaxies have a range of luminosities, and many properties that correlate with luminosity. These include the radial surface brightness gradient, which is shallower in brighter galaxies. You are attaching a special name to some portion of the bright end of the sequence of normal galaxies. This seems to me to be unnecessary, like calling people over 6 feet tall not “people” but something else. I would prefer to attach new names only to new phenomena. So I would advocate not using the term “D galaxy” at all, and going back to Morgan’s original, “pure” definition of a cD as a supergiant elliptical–like galaxy with an extra, cluster–sized halo.

Tonry: I agree with your distaste for the sloppiness with which people are applying the terms “D” and “cD”, but I feel that this is partly the result of the vagueness of the original definition. I feel that the terms “D” and “cD halo” can be usefully applied as descriptions of galaxies as long as one is relatively precise about what one means.

Kochhar: M87 has many properties of D galaxies without being an E galaxy.

Tonry: According to the classification I am using here, M87 *is* a D galaxy: it has a flat profile (and even a bit of cD halo).

COMPACT ELLIPTICAL GALAXIES

Jean-Luc Nieto and Philippe Prugniel
Observatoire du Pic-du-Midi et de Toulouse
14 Avenue Edouard Belin
31400 Toulouse
France

ABSTRACT. We summarize our present knowledge on low-mass high-surface brightness elliptical objects near massive galaxies that are often called M32-type or compact elliptical galaxies. The origin of the low mass of these objects is a controversial matter: is it intrinsic to their formation or produced, as classically believed, by tidal stripping from the massive neighbor? We present new observational data allowing to define better the characteristics of these objects and a simple theoretical model whose consequences support the idea that the precursors of compact ellipticals are related to the low-mass end of the luminosity function of elliptical galaxies.

1. A CLASSICAL VIEW ON COMPACT ELLIPTICALS

Compact elliptical galaxies are those objects limited in size, probably by the tidal field of their large neighbor, as first suggested by King (1962) for M32. Two other galaxies, beside M32, have been classically incorporated into this class: NGC 4486B, at 7'3 from the center of M87, and NGC 5846A, at 0'6 from the center of NGC 5846. Often called "M32-type", this class of elliptical galaxies suffers from several aspects: it is depopulated, numbering only as yet three objects, and the common characteristics of its galaxies are not defined with certainty.

From King's (1962) first discussion and following the classical model proposed by Faber (1973), these (gas-depleted) galaxies, once normal ellipticals (e.g. of the NGC 3379 type) lost their external mass by tidal interaction with the massive neighbor so that their three caracteristics are, classically :

i) they are close to a bright galaxy;
ii) they have a high-surface brightness;
iii) they are tidally-truncated.

The high surface-brightness of these objects is illustrated in Figure 1 showing the blue effective magnitude (in magarcmin^{-2}) versus the absolute magnitude for different types of elliptical-like galaxies, all taken from the Second Reference Catalogue of Bright Galaxies (de

T. de Zeeuw (ed.), Structure and Dynamics of Elliptical Galaxies, 99–107.

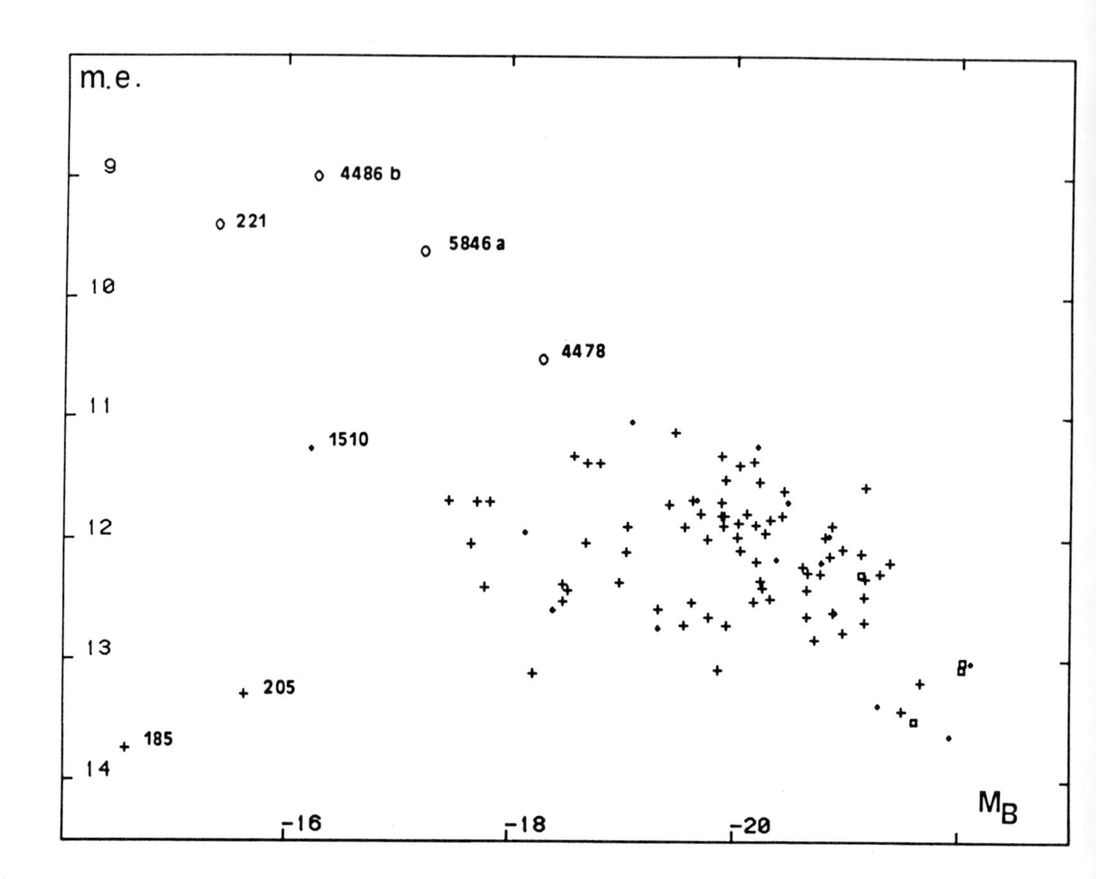

Figure 1: Blue effective magnitude (in mag arcmin^{-2}) versus absolute magnitude for different types of elliptical-like galaxies, taken from the RC2. The absolute magnitudes were calculated after de Vaucouleurs and Olson (1983) or with H_0 = 100 kms^{-1}Mpc^{-1}. Details are in Nieto and Prugniel (1907).

Vaucouleurs et al., 1976; RC2). The three compact ellipticals (and NGC 4478) are represented as well as NGC 185 and NGC 205, prototypes of spheroidal galaxies, normal elliptical galaxies and cDs.

Several papers were addressed to defining the specificity of these three objects : Rood (1965), King and Kiser (1973), Kormendy (1977), Tonry (1981). Almost all other candidates for this class, taken either from the RC2 or from the list of the Virgo cluster cE candidates provided by Binggeli et al. (1985), have not been photometrically investigated for tidal truncation. One of them, NGC 1510, was rejected on the basis of its young star emission spectrum and its HII regions (Eichendorf and Nieto, 1984).

2. "STONES IN THE POND"

Wirth and Gallagher (1984) suggested the possibility for M32-like dwarf galaxies to be the natural low-mass continuation of normal ellipticals, essentially for two reasons :
i) the main body of M32 presents an $r^{1/4}$ profile as normal ellipticals do,
ii) there seems to exist (at least one) "free-flying" M32-like isolated elliptical.

Reason i) means that M32 is not structurally basically different, except for its low mass, from a normal elliptical galaxy. Both reasons i) and ii) suggest that the low mass of the so-called compact objects should not be attributed to the tidal interaction with the neighbor.

In addition, the expected signature of galaxies of this class, e.g. tidal truncation, is by no means observable in the outer regions of M32 because of the strong contamination by the spiral arm of M31 so that, after all, M32 itself could not belong to the M32-type class (see the discussion following this paper)! We shall go back to this point later.

3. WHY STUDY THE SO-CALLED COMPACT ELLIPTICALS?

a) The above considerations raise a fundamental question concerning the formation and evolution of ellipticals as a whole : what is the low-mass end of the luminosity function of elliptical galaxies ? Because of the strong discontinuity of the physical parameters between normal ellipticals and dwarf spheroidals (see Saito, 1978; Kormendy, 1985 and this symposium), other objects but dwarf spheroidals should be the low-mass continuation of normal ellipticals. It is therefore quite crucial to investigate in great detail whether compact ellipticals could be those. In other words, the relevant question is : "are their characteristics (low mass, compactness) intrinsic or due to tidal interaction ?"

If tidal effects are observable, compact ellipticals provide a unique ground to study the mechanisms of galaxy interactions involving two galaxies of quite different relative masses.

4. OUR APPROACH

We have started a long-term project to investigate these questions, based on two-dimensional photometric or spectroscopic observations of known or possible candidate cEs, whether isolated or not. These observations are completed by theoretical considerations. The photometric observations were made with photographic (prime focus CFH plates + Schmidt plates of CERGA) and CCD (Pic du Midi 2m telescope) material. We report here on high-resolution (FWHM ∽ 0.7-1.0 arcsec) CFH photographic observations in the UBV bands of three small-sized early-type galaxies around M87 (NGC 4476, 4478, and 4486B) and of NGC 5846A. Two of them, NGC 4486B and NGC 5846A, were already known as cE. NGC 4478 was suggested to us as candidate for compact elliptical by de Vaucouleurs (1983). For more details, see Prugniel et al.(1987).

5. OBSERVATIONAL RESULTS

Fig. 2a,b,c,d shows the photometric equivalent and geometric profiles of the four galaxies considered. We note the following :

i) NGC 4476 follows a perfect $r^{1/4}$ law. Since it bears a ring of dust, it should be considered as another dusty elliptical galaxy.

ii) NGC 4478 and NGC 5846A appear truncated and do not follow an $r^{1/4}$ law. This is also true to a lesser extent for NGC 4486B. In that sense, these three galaxies seem to behave differently from NGC 4476.....taken as an $r^{1/4}$ prototype obtained with the same data. They fit reasonably well a King model (King, 1966) in their main body. NGC 4478 is then a fourth (but intermediate) case of tidally-truncated object.

iii) There is no apparent color (mass?) gradient in these objects within the uncertainties of our data.

iv) There is a trend in <u>all</u> cases for the outer isophotes to be more circular than the inner isophotes. This is also the case in M32 from our unpublished data (and from Tonry (1984) for the inner regions).

6. TREND FOR CIRCULARITY OF THE EXTERNAL ISOPHOTES

This trend should not pertain specifically to compact ellipticals since most of isolated ellipticals (certainly for other reasons) seem to show this property (di Tullio, 1979). However, there is certainly an idea of non-elongation linked to the intuitive idea of "compactness". If this trend for circularity is related to tidal effects, this means that the effect of tidal stripping would be to attenuate the intrinsic flattening of the isophotes: this is at first sight surprising since evidences for tidal interactions between galaxies reported in the literature often correspond to distorsions and elongations in the direction of the nearby massive galaxy. It seems as if the (intrinsically elongated) shape of the compact object is affected not by the instant potential but instead by the average of this potential over several orbits about the massive galaxy. Numerical simulations should quantify this suggestion.

7. THEORETICAL INVESTIGATIONS: A FRAMEWORK

Following Faber (1973), we shall assume that the compact object was once a normal elliptical galaxy that was captured by the massive neighbor. Dynamical friction (given by the local Chandrasekhar formula) is responsible for slowing down the galaxy on its orbit. The increasing tidal effects modify the characteristics of the companion.

We shall assume also that the present stellar distribution and dynamics follow an energy-truncated isothermal model (King, 1966).

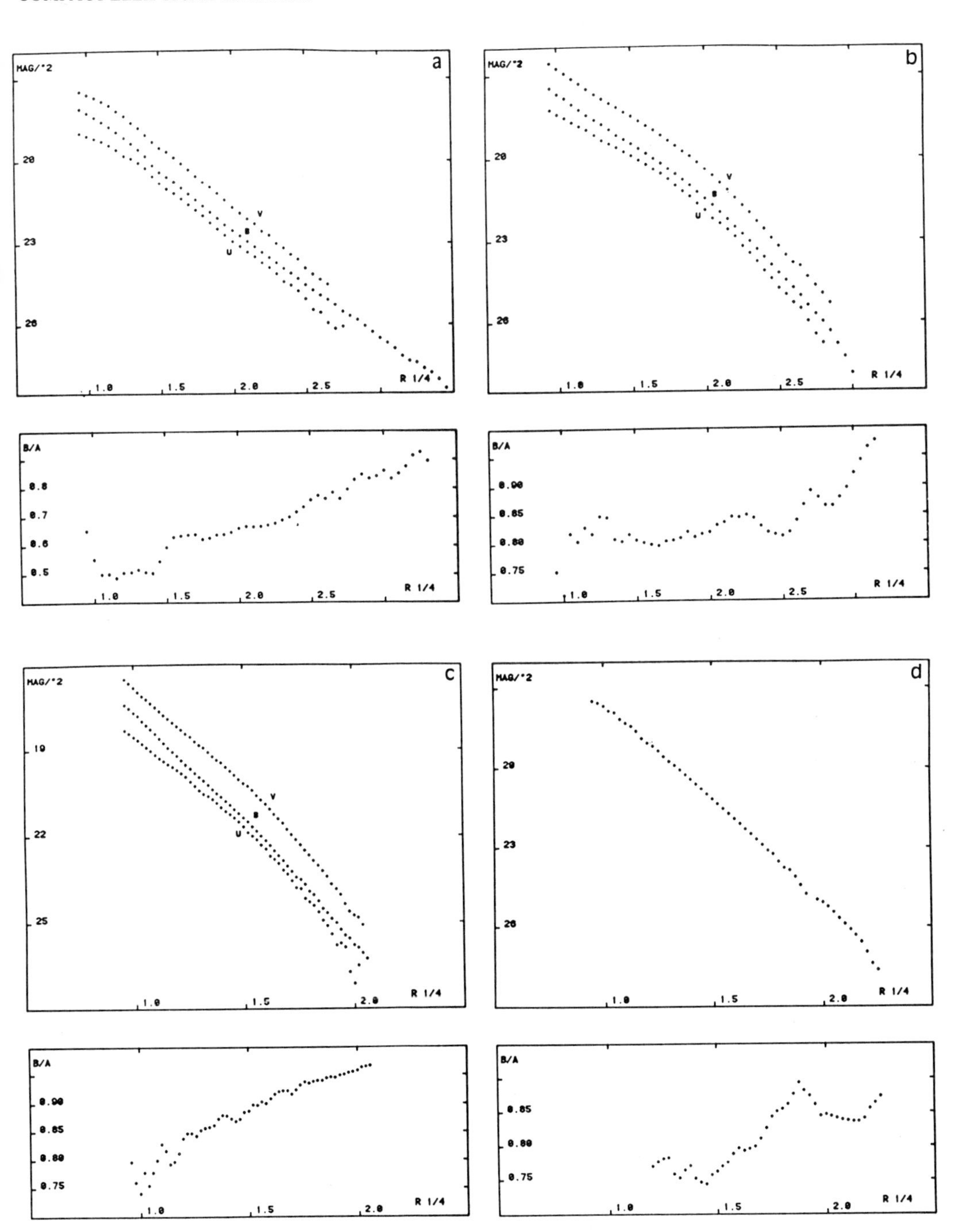

Figure 2: Equivalent photometric and ellipticity profiles of the three companions of M87, NGC 4476(a), NGC 4478(b), NGC 4486B(c) and of NGC 5846A (d).

Since, as currently believed (e.g. Faber, 1973), the stripping should not affect the center of the captured system, we shall assume that the core radius, r_c and the central density are constant during the evolution: only the concentration, $c = r_t/r_c$, where r_t is the tidal radius, varies. (We shall assume that the capture has not affected the structure of the system).

Consequently, in a first approximation, the mass of the object is proportional to its tidal radius. This is at variance with previous considerations (e.g. Byrd (1979) where the mass of the compact was taken constant. For more details, see Nieto and Prugniel (1987).

8. TYPICAL PHOTOMETRIC EVOLUTION OF A COMPACT ELLIPTICAL

In such a framework, a typical galaxy such as NGC 4486B should have evolved from log c = 2.25 (assumed to be typical of normal ellipticals) to log c $\sim$ 1.7 (derived from our data). Assuming a constant M/L during its evolution, the mass of NGC 4486B should have been about three to four times its present mass, which is about 2×10^9 $M_\odot$ after Nieto and Monnet (1984). We conclude that the precursor of NGC 4486B was originally a low-mass object. What could it be? An effect of the capture would be to decrease its binding energy per unit mass and yield a lower surface-brightness object. Since the present object has a high surface brightness, it would be very difficult to conceive that it could come from a low surface-brightness object. It is more logical to assume that the precursor is instead a low-mass elliptical.

9. DECAY TIME SCALE

The orbital decay of a typical compact galaxy (NGC 4486B) can be followed from the time of the capture to present. The decay time scale R/(dR/dt) is much longer for a variable-mass King model than for a constant mass toward the end of the decay. A first estimate seems to indicate that it is even longer than the Hubble time by a large factor. We assume that the decay stops at a disruption radius, r_d, estimated to be reached when the central density of the compact is equal to the mean density of the tidally-limited system orbiting at a distance r_d from the center of the massive neighbor. For compact ellipticals, the disruption radius is of the same order as the core radius, a very small value indeed (3 pc).

This very long time scale for the orbital decay of the compact object means that we should see many such compact ellipticals in the core regions of giant (or normal) galaxies, a prediction that is not confirmed by the observations (Sandage et al., 1985). We see for this three non-exclusive reasons:

i) the constraints on this type of capture are very severe and/or
ii) the destruction rate of such compact objects about massive galaxies is very high and/or
iii) the precursors of such compact objects are very rare.

10. THE PRECURSORS OF COMPACT ELLIPTICALS

We are tempted to conclude from Sect. 8 that the precursors of such compact ellipticals are already low-mass elliptical objects that started to become tidally-truncated with the capture of the massive neighbor. Interestingly enough, one of the possible consequences of the slow decay (Sect 9), e.g. the precursor being rare, is quite consistent with this conclusion, since low-mass -e.g. faint- ellipticals are rare (Binggeli et al., 1985). Therefore, tidal stripping should not be responsible for the low mass of these objects. Even if compact ellipticals look more circular outwards than inwards, their precursors could as well have been elongated throughout.

11. THE LOW-MASS END OF THE LUMINOSITY FUNCTION OF ELLIPTICALS

We summarize the arguments favoring the idea that compact ellipticals (and not dwarf spheroidals) should be associated with the low-mass end of the luminosity function of ellipticals :
i) The luminosity profile of compact ellipticals is closer to an $r^{1/4}$ law than an exponential law (Wirth and Gallagher, 1984).
ii) Their metallicity indices are similar to those of ellipticals (Faber, 1973)
iii) Low-mass isolated ("free-flying") galaxies may exist (e.g. Wirth and Gallagher, 1984). This has to be confirmed (see the discussion following this paper).
iv) There is a well-marked discontinuity between the parametric and physical parameters of normal ellipticals and dwarf spheroidals (e.g. Saito, 1978, Kormendy, 1985).
v) A simple theoretical model calls for a low-mass elliptical precursor (this paper).

12. CONCLUSION

We are attempting a long-term study of compact elliptical galaxies. We derive from the observations presented here that NGC 4478 is a fourth object of this class, and that, in all cases considered, external isophotes are more circular outwards than inwards. A simple theoretical model as well as a few other arguments found in the literature supports the idea that compactness is intrinsic to the way these galaxies were formed. It also suggests that we should find isolated low-mass ellipticals morphologically similar - except for truncation - to these tidally-truncated ellipticals. How rare could they be?

REFERENCES

Binggeli, B., Sandage, A.R. and Tammann, 1985, Astron. J., **90**, 1681
Byrd, G.G., 1979, Astrophys. J., **231**, 32
Eichendorf, W., and Nieto, J.-L., 1984, Astron. Astrophys., **132**,342
Faber, S.M., 1973, Astrophys. J.,**179**, 423
King, I.R., 1962, Astron. J., **67**, 471
King, I.R., 1966, Astron. J., **71**, 64
King, I.R., and Kiser, J., 1973, Astrophys. J., **181**,27
Kormendy, J., 1977, Astrophys. J., **214**, 359
Kormendy, J., 1985, Astrophys. J., **295**, 73
Nieto, J.-L., Monnet, G., 1984, Astron. Astrophys., **139**, 464
Nieto, J.-L., Prugniel, Ph., 1987, in preparation.
Prugniel, Ph., Nieto, J.-L., Simien, F., 1987, in preparation.
Rood, H.J., 1965, Astron. J., **70**, 689.
Saito, M., 1978, Pub. Astron. Soc. Japan, **31**, 181.
Sandage, A.R., Binggeli, B. and Tammann, G.A., 1985, ESO Workshop on the Virgo Cluster, O.-G. Richter and B. Binggeli Ed., p. 239
Tonry, J.L., 1981, Astrophys. J., **261**, L1.
Tonry, J.L., 1984, Astrophys. J., **283**, L27.
Tremaine, S.D., 1976, Astrophys. J., **203**, 72.
di Tullio, G.A., 1979, Astron. Astrophys. Suppl. ts., **37**, 591.
de Vaucouleurs, G., 1983, private communication
de Vaucouleurs G., Olson, D.W., 1983, Astrophys. J., **256**, 346
de Vaucouleurs, G., de Vaucouleurs A., and Corwin, H.G., 1976, Second Reference Catalogue of Bright Galaxies, University of Texas Press, Austin (RC2)
Wirth, A. and Gallagher, J.S., 1984, Astrophys. J., **282**, 85.

DISCUSSION

Richstone: I was under the (perhaps mistaken) impression that M32 is better fit by an $I(R) \propto R^{-1}$ law than by an $r^{1/4}$ law.

Nieto: I think that M32 is the most difficult case to analyse if we want to find a tidal truncation, that, in principle, should occur at rather faint brightness levels. This requires a perfect subtraction of the background light of M31 throughout M32. This background light of M31 is clumpy, and undesirable effects can occur. This is why we left M32 aside for the time being and concentrated on the other cases where the background removal is easier.

Kent: I have recently obtained photometry of M32 with a CCD and an 8 inch telescope. I was able to subtract the background light from M31. The luminosity profile shows a power law with a slope of -1.2 out to 18″ and a de Vaucouleurs profile past that radius. I find no obvious tidal truncation of the profile. Because compact galaxies have such a high characteristic surface brightness, the logarithmic slope of the luminosity profile is quite steep at the limiting radius seen on a photographic plate, even though the shape of the profile is like that of more normal ellipticals.

Kormendy: Tonry found, and I confirm, that the projected brightness profile of M32 is $I(r) \propto r^{-1.2}$ out to $r \sim 20''$.

Djorgovski: The reason that we do not have more "free-flying" compact ellipticals is a very strong selection effect: such galaxies nearby are indistinguishable from L^* galaxies in the background. The reason that the three classical cases are found near other, larger, galaxies is that the big ones draw the attention, so that the small ones are noticed. Tidal interaction may not be important at all.

Nieto: I fully agree.

Sadler: We know from the luminosity function of elliptical galaxies that low-luminosity ellipticals are very common in terms of their space density. Any galaxy sample that is magnitude-limited strongly selects against such objects, giving the false impression that they are scarce. There is certainly no shortage of the low-luminosity galaxies which you suggest as progenitors of tidally-truncated systems.

Nieto: Possibly. The point is that they are difficult to find because of observational effects that makes us focus instead on those near (larger) galaxies.

White: I don't think we have a good idea of what a tidally truncated galaxy *should* look like, so it may be dangerous to identify features in luminosity profiles as evidence of truncation. Perhaps the fact that the outer part of M32 is seen to follow an $r^{1/4}$ law implies that even tidal truncation cannot beat the theorem that de Vaucouleurs is always right!

Jean-Luc Nieto.

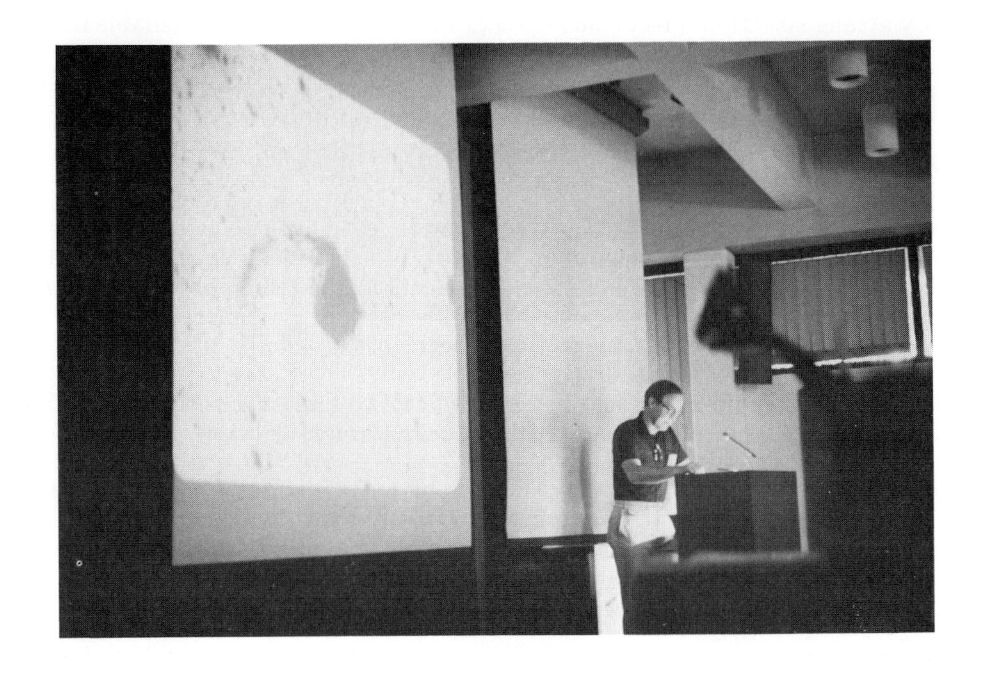

François Schweizer with one of his pet galaxies.

DUST AND GAS: OVERVIEW

François Schweizer
Department of Terrestrial Magnetism
Carnegie Institution of Washington
5241 Broad Branch Road, N.W.
Washington, DC 20015

ABSTRACT. Progress in the 50 years since the discovery of ionized gas in NGC 1052 is reviewed. As discovery has proceeded from H II to H I and recently to 10^7 K gas, the known amount of gas in ellipticals has increased dramatically: from $10^{3-6} M_\odot$ for H II, to $10^{5-9} M_\odot$ for H I in some 15% of E's, to $10^{9-10} M_\odot$ for the X-ray emitting gas. Although a few ellipticals with dust have long been known, recent CCD surveys have revealed dust lanes in nearly half of all E's. The cooler gas — as traced by H I, warm H II, and dust — is generally distributed in the form of a disk that often shows an outer warp. To deduce the true shapes of ellipticals from the geometry and kinematics of such disks has proven a challenging task. There is mounting evidence that many disks consist of gas accreted by mass transfers and mergers, and that some of them may not yet have reached dynamical equilibrium. Some of the major unsolved problems concern the atmospheric balance between sources and sinks of gas, and the importance of cooling flows and galactic winds.

1. OPTICAL OBSERVATIONS 1936 – 1976

If we disregard the discovery of the plasma jet of M87 by Curtis in 1918, which found its explanation only in the mid-1950's, then the discovery of dust and gas in ellipticals dates back 50 years. In describing the morphological class E, Hubble (1936) noted that "small patches of obscuring material are occasionally silhouetted against the luminous background, but otherwise these nebulae present no structural details." And while Mayall (1936) was testing a new spectrograph with good UV transmission for the Crossley telescope, he discovered the [O II] λ3727 emission line in NGC 1052, an E known today as a prototypical LINER galaxy. Three years later, Mayall (1939) had first rough statistics about ionized gas in ellipticals, with one E out of 14 observed with the Crossley showing the [O II] line, and a second E with [O II] emission known from the Mt. Wilson work. By 1956, when the survey by Humason, Mayall, and Sandage was published, the detection rate for the [O II] line in ellipticals was 14% for Lick and 18% for Mt. Wilson.

With the HMS survey completed, the ground work was laid for the three most fundamental papers on gas in ellipticals to appear in the first 40 years. Sandage (1957), in studying the luminosity function and evolution of stars in globular clusters, noted that in M3 about $10^5 M_\odot$ of gas must have been shed

T. de Zeeuw (ed.), Structure and Dynamics of Elliptical Galaxies, 109–124.

by stars evolving from the main sequence to white dwarfs. Drawing parallels to ellipticals with their deeper potential wells and better ability to retain gas, Sandage remarked: "... this gas has probably not escaped from these systems and makes up about one two-hundredth of the total mass. Could this be the origin of the [O II] emission at λ3727 observed in E nebulae?" As we all know, much of the following 25 years has been spent searching for this mass of gas, and only recently have X-ray observations revealed what seems to be the bulk of the gas.

The other two fundamental papers of this period are by Minkowski & Osterbrock (1959) and Osterbrock (1960) and deal with the physics of interstellar matter in ellipticals. Since these two papers are often overlooked in modern work, let me point out how surprisingly much was already known about the ionized gas from these early observations of two E's with the brightest emission lines: NGC 1052 and 4278. Minkowski and Osterbrock found that each galaxy has an extended region of ionized gas at its center, with a spectrum not unlike that of Orion. From the relative intensities of the two components of the [O II] doublet, it followed that $N_e \approx 10 - 300$ cm^{-3} and that the total mass of ionized gas is $10^4 - 10^6\ M_\odot$ ($H_0 = 75$). In both galaxies, the ionized gas seemed to be arranged in a flattened disk and to rotate rapidly. Toward the nucleus, the turbulent motions in the gas were seen to increase and reach values comparable to the velocity dispersion of the stars. Minkowski and Osterbrock pondered about the energy input and suggested that there may be enough UV radiation from blue horizontal-branch stars to explain the observed line emission. They also estimated that collisional excitation due to the turbulent gas motions could marginally supply the energy, the most likely continuous source being mass shed by stars on strongly radial orbits. They did worry, as we still do today, that the mass inflow into the nucleus would be excessive because of rapid cooling. Finally, they guessed that all ellipticals may possess gas, but that ionization conditions may differ.

2. FINDING DUST AND GAS: 4 COMPONENTS

In reviewing the developments of the past 10 years, it seems useful to distinguish four components of the interstellar medium: the dust, cool gas ($T < 10^2$ K), warm gas ($\sim 10^4$ K), and hot gas ($> 10^6$ K).

2.1. Dust

There has been a recent renaissance of interest in dust in ellipticals for two reasons: 1) Bertola and Galletta (1978) suggested that one can find the true shapes of some ellipticals from the orientation of their dust lanes; and 2) the advent of CCD detectors has made it easy to search for weaker dust features. Three techniques are now generally used to find dust from CCD images: digital unsharp masking (Schweizer & Ford 1985), model reconstruction and masking (Lauer 1985), and color mapping (Carter *et al.* 1983; Sparks *et al.* 1985). Figure 1 shows dust and ripples in NGC 5018 revealed by unsharp masking. Several papers in this volume illustrate the other two techniques. All this recent activity has led to several major advances. First, a classification system for dust features introduced by Hawarden *et al.* (1981) has found widespread acceptance. Second, there now exist two main catalogs of dusty E's with about 150 entries combined (Ebneter & Balick 1985; Sparks *et al.* 1985). And third, we now know that at least 25% – 40% of field ellipticals have dust lanes detectable on modern photographs (Sadler & Gerhard

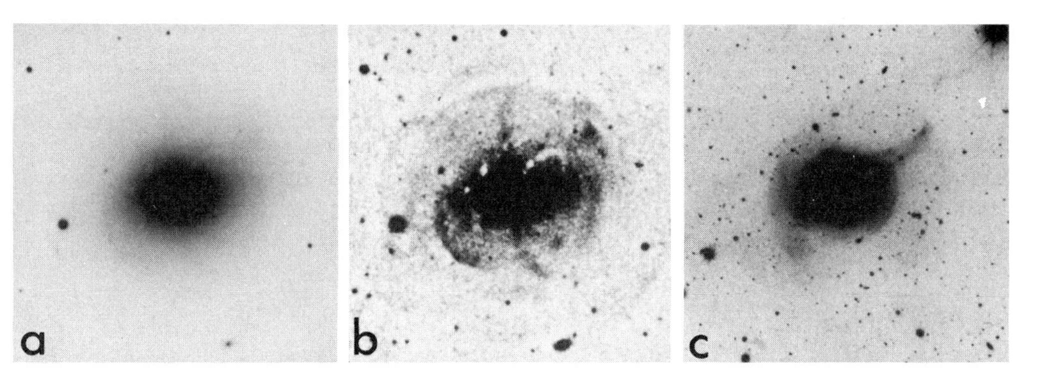

Figure 1. Fine structure in the E3 galaxy NGC 5018. (*a*) $3' \times 3'$ scan of a CTIO 4-m IIIa-J plate; (*b*) unsharply masked image showing dust lanes, inner ripples, and plumes; (*c*) $9' \times 9'$ scan at extremely high contrast showing outer ripples and tails (from Schweizer & Ford 1985).

1985) and that dust masses in these galaxies are of order $10^4 - 10^5\ M_\odot$.

In the past, astronomers have worried that there may be much dust distributed uniformly throughout ellipticals. The main argument against this possibility has been, and still is, the small variation of color index among E's. Infrared observations from IRAS now confirm that there is very little warm dust in them (Rieke & Lebofski 1986). It seems therefore likely that what little dust we see in obscuring patches is what dust there is. Yet major unsolved problems remain: Where does the dust come from, and why has it not yet been destroyed by the hot gas? A possible answer is that the dust is accreted together with cool gas and then survives in the gaseous disk.

2.2. Cool Gas ($<10^2$ K): H I

In the early 1970's, the search for neutral hydrogen in E's turned out to be surprisingly difficult. The apparent lack of cool gas led to the hypothesis and wide acceptance of galactic winds that sweep E's clean of gas. But starting about 1976, H I was finally found in several ellipticals. Among the first to be detected were again NGC 4278 (Bottinelli & Gouguenheim 1977; Gallagher *et al.* 1977) and NGC 1052 (Knapp *et al.* 1978; Fosbury *et al.* 1978). Figure 2 shows a modern VLA map of NGC 1052 made by van Gorkom *et al.* (1986). This map illustrates three points about H I distributions in E's: 1) H I can often be detected beyond the Holmberg radius and is roughly arranged in a disk or ring; 2) such disks often look perturbed and have occasionally appendices reminiscent of tidal tails; and 3) a potential donor galaxy can occasionally be seen nearby in the form of some gas-rich spiral.

From observations with sensitive modern detectors it appears that about 10% – 15% of all E's have measurable amounts of H I (Knapp *et al.* 1985). Total H I masses range from a detection-limited $10^5\ M_\odot$ up to $\sim 5 \times 10^9\ M_\odot$. There also seems to be a clear environmental effect: E's in the Virgo and Pegasus clusters have very little H I, with upper limits of $5 \times 10^7\ M_\odot$ for Virgo (Kumar & Thonnard 1983). Finally, molecules have not yet been discovered except in NGC 5128 (see review by Knapp in this volume).

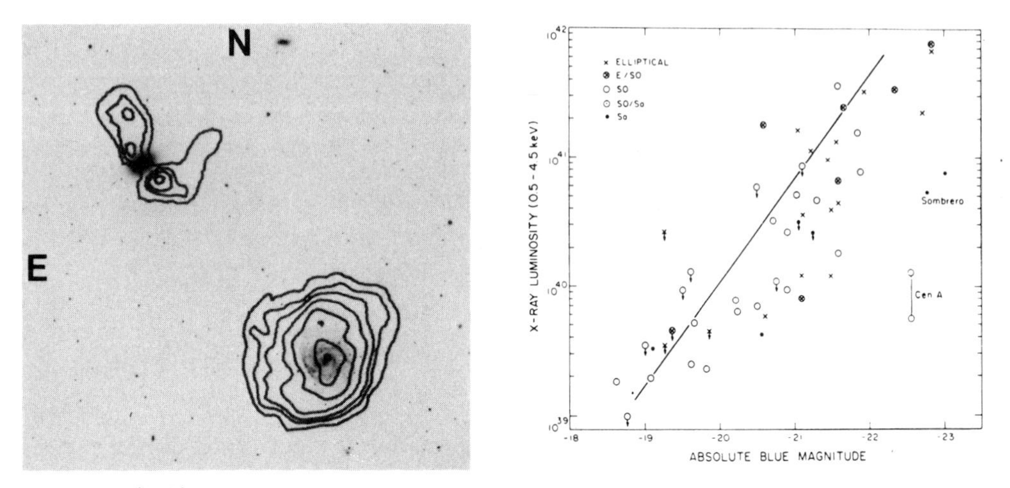

Figure 2. (*left*) VLA map of the H I distribution in NGC 1052 and in spiral companion NGC 1042 (from van Gorkom *et al.* 1986).

Figure 3. (*right*) X-ray luminosity of 55 early-type galaxies plotted against M_B. The line is not a best fit, but is the $L_x - M_B$ relation predicted if all X-ray emission arises from thermal bremsstrahlung of hot gas shed by evolving stars in 10^{10} years (from Forman *et al.* 1985).

2.3. Warm Gas ($\sim 10^4$ K): H II

Virtually all our knowledge about the warm, ionized gas in E's stems from optical spectroscopy. Some nice sample spectra of NGC 4278 and 5077 are shown by Demoulin-Ulrich *et al.* (1984); they are typical LINER spectra and are representative of most ionized regions observed in giant ellipticals. There exist now good spectra obtained with linear detectors for several hundred E's and S0's. Between one and two dozen of these galaxies have been observed in sufficient detail to determine the kinematics of their gas. As a result of improved techniques, the fraction of E's known to possess ionized gas has risen from the canonical 15% of HMS to 40% with [O II] $\lambda 3727$ detected on image-tube plates in the blue (Caldwell 1984b) and even to 55% – 60% with the [N II] $\lambda 6584$ line detected on CCD frames in the red (Phillips *et al.* 1986). A significant part of the success of recent studies stems from a careful, digital subtraction of the galaxy continuum with its bumps and absorption lines.

These and other studies confirm many of Minkowski and Osterbrock's findings of a quarter century ago. Typical central electron densities are $N_e \approx 10^3$ cm^{-3} for $T \approx 10^4$ K. Most ionized regions are small, of order 1 kpc in diameter or less. Some, though, are considerably larger such as in NGC 5128 ($D = 8$ kpc; Graham 1979) and NGC 1316 ($D \approx 16$ kpc; Schweizer 1980). The mass of ionized gas is of order $10^3 - 10^6\ M_\odot$, with Phillips *et al.* finding many E's near the lower limit. The ionized gas always seems to rotate much faster than the stars and is often arranged in an inclined disk. Near the nucleus, the turbulent velocity of the gas increases and reaches values between one-half and one time the velocity dispersion of the stars. The ionized gas, too, shows a strong environmental effect in the sense that nuclear emission is much reduced in dense clusters (Gisler 1978). Finally, a new result not known to Minkowski and Osterbrock is that the presence and strength of emission lines from ionized gas correlates well with the presence of dust and the X-ray luminosity of the galaxy.

2.4. Hot Gas (>10^6 K): X-ray Coronae

The most exciting news of recent years has been the discovery that there does seem to be, after all, a lot of gas in E's and that it is in the form of hot coronae emitting X-ray radiation. As is well known, the EINSTEIN satellite with its high-resolution imaging capability made it possible to identify extended X-ray emission sources with individual, massive E galaxies and to separate them from the general background emission in clusters. One of the first papers to report the detection of individual coronae pointed out already that there seemed to be nearly $10^{10}\ M_\odot$ of 10^7 K gas in NGC 4406 (Forman *et al.* 1979).

Many EINSTEIN observations have been analyzed since. Forman, Jones, and Tucker (1985) and Trinchieri & Fabbiano (1985) summarize observations of 55 early-type galaxies. They find that X-ray coronae around E's seem to be common if not ubiquitous, regardless of whether the E's are isolated or in clusters. In massive ellipticals, the coronae extend out to 100 kpc radius and more, often well beyond the optical isophotal radius R_{25}. Particle energies are of order 1 keV, corresponding to $T \approx 10^7$ K and to r.m.s. velocities for protons of about 300 km s^{-1}, or the same as the stellar velocity dispersions. The masses of hot gas seem to range between $5 \times 10^8\ M_\odot$ and $5 \times 10^{10}\ M_\odot$. The X-ray luminosity depends on nearly the square of the blue luminosity, suggesting that more luminous galaxies have more hot gas per unit blue luminosity (see Fig. 3). On the other hand, there are indications that in the least luminous galaxies a major fraction of the X-ray emission may stem not from hot gas, but from low-mass binaries. Finally, if one assumes that the hot gas is isothermal (which is still under debate), one finds central particle densities of $\sim 0.01 - 0.1$ cm^{-3}, core radii for the gas distribution of 1 – 5 kpc, and short cooling times of order 10^8 yr near the center.

3. CHEMICAL ABUNDANCES

Any discussion of interstellar matter in ellipticals should include chemical abundances. Yet very little work has been done in this direction. In his 1960 paper, Osterbrock pointed out the similarity between the spectrum of the gas in NGC 4278 and that of the Orion nebula. He also pointed out the main difficulty in determining abundances, which is the lack of reliable temperatures and, therefore, the lack of information about the degree of ionization. The difficulty is that weak, temperature-sensitive lines, such as [O III] λ4363, are buried in the continuum radiation of old stars, which itself is riddled with absorption lines. Osterbrock could only conclude that the logarithmic oxygen abundance is $[O/H] \approx 8.3$ or about 1/3 solar if $T = 10^4$ K, but would be $[O/H] \approx 7.4$ or about 1/25 solar if $T = 2 \times 10^4$ K. This basic difficulty remains today, as does the difficulty of choosing between photoionization and collisional ionization models.

On the empirical side, Phillips *et al.* (1986) have found a correlation between the line ratio [N II]$\lambda6584/H\alpha$ and the blue luminosity of E and S0 galaxies, which may reflect abundance variations or changing ionization conditions. Ford & Butcher (1979) have carried out a detailed analysis of the luminous filaments in M87 and find that their nitrogen and sulphur abundances exceed those in planetary nebulae of the Milky Way by factors 2 – 4. Obviously, much more abundance work remains to be done.

4. SPATIAL DISTRIBUTION

The four components of the interstellar matter in ellipticals appear in only two forms of distribution.

The hot gas of the X-ray coronae seems to have a space-filling, approximately spherical distribution. This has increasingly led to the use of the word "atmosphere" to describe it. The radial distribution can be represented by an isothermal distribution remarkably well over a range of $\sim 10^3$ in surface brightness (Forman *et al.* 1985).

The other three components — the dust, H I, and warm ionized gas — all seem to be arranged more or less in the form of a single disk. In different E's these disks are oriented at all conceivable angles with respect to the apparent axes of the projected light distribution. Hawarden *et al.* (1981) found that roughly one third of E's with dust lanes have them oriented along the apparent major axis, another third along the apparent minor axis, and the remaining third have them either at a skewed angle, or strongly distorted, or too irregular to classify. Warps in these disks are frequent and show up either in the dust lanes themselves, or as a progressive change in the rotation axis measured from the inner part of the disk to the outer part.

Young stars are occasionally associated with such disks, as in the well-known case of NGC 5128. Recently, the polar-ring galaxy AM 2020-5050 has been shown to be an elliptical surrounded by a blue ring of stars and a gaseous disk that stretches from the center to beyond this ring (Fig. 4a). Also, I have discovered a very faint, knotty ring of stars around IC 2006, a bona fide southern elliptical shown in Figure 4b, but have not yet been able to detect the ionized gas that presumably goes along with it. This is a good object to observe with a CCD spectrograph at Hα and might yield a reliable M/L ratio in an elliptical far from the center.

5. GAS KINEMATICS AND DYNAMICS

The discovery of inclined and warped gas disks in ellipticals has opened up the possibility to learn about the shape of the potential, *i.e.*, whether it is generally oblate, prolate, or triaxial. As a result, many theoretical and observational papers have been devoted to this subject during the past eight years. Some of them are discussed in more detail in the reviews by Bertola and Davies in this volume. Here, let me try to paint the broad picture, and point out where gains have been made and where more work is needed.

Most papers on the gas kinematics and dynamics can be assigned to one of four subject categories.

The earliest papers attempted to use existing observations to distinguish between oblate and prolate potentials that were assumed to be fixed in space. Bertola & Galletta's (1978) discovery of a class of ellipticals with minor-axis dust lanes suggested that these galaxies would be good candidates for being prolate. Building on this and on Kahn & Woltjer's (1959) notion of an inclined gaseous disk settling due to differential precession, Tohline, Simonson, & Caldwell (1982) studied the existence of preferred planes into which the gas would settle in oblate

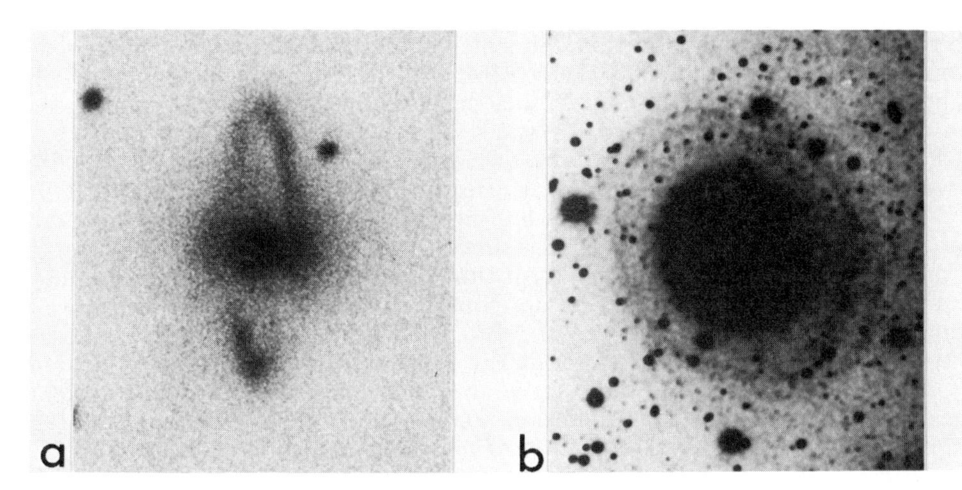

Figure 4. Young stars in outer rings. (*a*) Polar-ring elliptical AM 2020-5050 from Las Campanas 2.5-m plate (see Whitmore *et al.* in this volume); (*b*) IC 2006 (E1) and faint, knotty ring from CTIO 4-m plate.

and prolate potentials; they proposed that of the 12 best observed E's, eight were prolate and only four were oblate. These were the golden days when the structure of ellipticals still looked seductively simple!

Yet, of course, Binney (1978) had already proposed that E's might be triaxial. Papers in the second group then took this refinement into consideration and studied gas orbits in triaxial potentials of fixed orientation. Much marvellous work was done here in Princeton by Martin Schwarzschild (1979) and his collaborators (*e.g.*, Heiligman & Schwarzschild 1979) who showed that in a fixed triaxial potential closed gas orbits *do exist*, that they are nearly elliptical, and that in any given potential these orbits may populate two preferred planes. Steiman-Cameron & Durisen (1982) then demonstrated that polar orbits can be stable and can have unexpectedly large capture areas for gaseous infall. Yet the discovery of more types of possible gas orbits by Merritt & de Zeeuw (1983) raised doubts that a one-to-one correspondance between the apparent orientation of dust lanes and the shape of the potential could be established. These authors emphasized that kinematical information would be needed to test the triaxiality of E's. A recent observational paper that finds evidence for triaxiality based on fixed-orientation models is the study of NGC 1052 by Davies & Illingworth (1986).

In recent years, two major complications have arisen to make the analysis of observations of the gas kinematics more complex: For one, it was realized in the early 1980's that triaxial galaxies could be tumbling; and second, evidence started growing that the observed gas configurations may be transient.

Various people realized that a tumbling triaxial potential could give rise to inclined and warped gas disks due to the Coriolis force. Papers in the third group then analyzed the various possible configurations (Heisler, Merritt, and Schwarzschild 1982; Tohline & Durisen 1982; Merritt & de Zeeuw 1983, see their illustrated Table 1). Especially the beautiful paper by van Albada, Kotanyi, & Schwarzschild (1982) made some precise predictions as to which way the stellar body should tumble if the gaseous disk is observed to rotate in a given direction and has a given warp. Namely, the tilt of the inclined gas disk should be such that, if you look down along the figure rotation axis, the gas orbits in a *retrograde* sense relative to the figure rotation. Sharples *et al.* (1983) applied this theory to the remarkable galaxy NGC 5363, which features an inner dust lane aligned with the apparent minor axis and an outer dust lane aligned with the major axis; they found a tumble period of close to 10^9 years.

Finally, the fourth group of papers have emphasized evidence that inclined and warped gas disks may be transient. Beginning with Graham's (1979) study of NGC 5128, it has been found time and again that (1) the gas rotation axis is decoupled from the stellar rotation axis, and (2) the net angular momentum per unit mass of the gas is much larger than that of the stars. This has forced us to accept the notion that these disks do not consist of gas shed by stars, but have an external origin instead. In the case of NGC 5128 itself, where van Albada *et al.* (1982) made a specific prediction about the sense of rotation of the stellar body, three studies have now found, and agree among themselves, that the rotation is in the sense opposite from the prediction (Davies *et al.* 1984; Bertola *et al.* 1985; Wilkinson *et al.* 1986). Therefore, the warped gaseous disk of at least NGC 5128 is almost certainly a transient phenomenon, making it more difficult to establish the exact shape of the galaxy.

In studying the various observational papers on early-type galaxies with gas disks, I picked eight galaxies which were considered to be the best understood by the authors (with apologies to authors whose pet object I may have overlooked): NGC 1052 (E4), 1316 (D4), 4125 (E6), 5128 (E0p), 5266 (S0), 5363 (E5), 7070A (I0), and 7097 (E5). I find it impressive that four of these eight galaxies show *direct* evidence of having accreted disk material in the recent past: NGC 1316 has ripples or "shells" (Fig. 5a), as do NGC 5128 (Fig. 5b) and NGC 7070A (Sharples *et al.* 1983; Malin & Carter 1983); and NGC 4125 has peculiar crossed streamers (Bertola *et al.* 1984). A fifth galaxy, NGC 7097, has stars and gas *counter-rotating* around the same axis (Caldwell *et al.* 1986), a configuration that can hardly be due to anything else but gas accretion from an external source. The remaining three objects each have a potential donor of gas nearby: NGC 5363 (Sharples *et al.* 1983) is only about three diameters away from the classical grand-design spiral NGC 5364 (Fig. 5c); NGC 1052 is close to the spiral NGC 1042 (Fig. 2); and even NGC 5266 (Caldwell 1984a) is within about seven diameters from the barred Sc galaxy NGC 5266A, though the latter's redshift is not yet known. The reason for suspecting relatively frequent mass transfers in close pairs is, of course, that the cross section for a mass transfer is of order $10 - 10^2$ times greater than the cross section for a merger. An example of a mass transfer taking place is shown in Figure 5d.

In summary, it seems to me that the case for transient gas configurations as a general phenomenon is strong (for detailed theory, see Gunn 1979 and Steiman-Cameron & Durisen 1987). We do seem to have found some galaxies where an oblate potential can nearly be excluded (*e.g.*, Caldwell 1984a), and many recent papers find evidence for triaxiality (*e.g.*, Bertola *et al.* 1985; Davies & Illingworth 1986). Yet we should remain honest and admit that we are still unable to reliably determine the ratios between the three axes of any given galaxy. Only two theoretical papers have so far attempted to model a specific galaxy, NGC 5128, in terms of a recent infall and a transient gas disk (Tubbs 1980; Simonson 1982). I suspect that until such evolutionary models are computed for many more galaxies, we may not reach our goal of determining the intrinsic shapes of ellipticals with confidence.

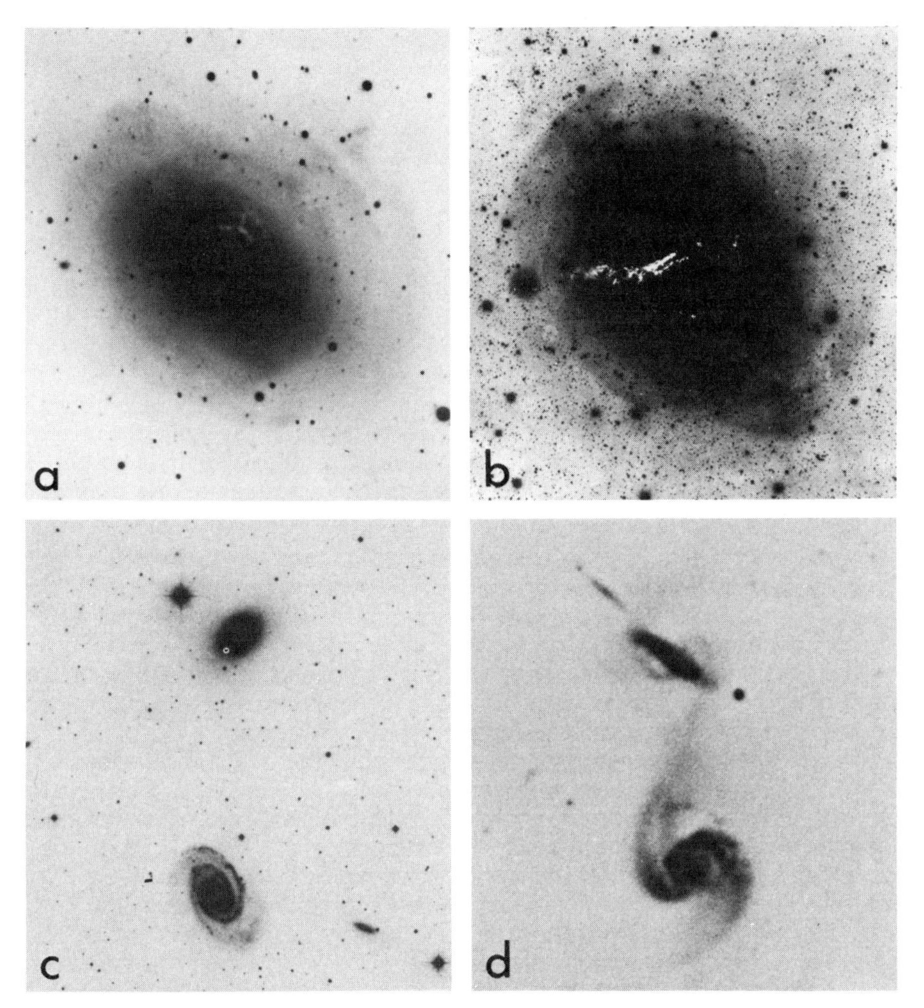

Figure 5. Accretion of disk material in ellipticals. (*a*) Ripples and dust in NGC 1316 (from Schweizer 1980); (*b*) ripples or "shells" in NGC 5128 (from Malin *et al.* 1983); (*c*) NGC 5363 and spiral neighbor NGC 5364; and (*d*) mass transfer between NGC 3808 and edge-on (S0?) companion (from Arp 1966).

6. ATMOSPHERIC BALANCE

There are two main questions concerning the atmospheric balance in ellipticals, and each question consists of various parts: 1) What are the *sources* and *sinks* of gas, and what is the net balance? 2) What are the main *heating* and *cooling* mechanisms, and why is there gas at three different temperatures? The answers to these questions are intertwined and, with one exception, difficult and not yet fully understood. In the brief space available here, I first address the one easy question and then mention a few points that seem important to me.

The relatively easy question concerns the sources of gas in ellipticals. There seem to be two of them. One source is the matter shed by evolving stars, including

Miras, planetaries, novae, and supernovae. The total present mass-loss rate may be estimated from the mass difference between a turnoff star and an average white dwarf ($\Delta M \approx 0.3\,M_\odot$) and the theoretical evolution rate of K giants. The canonical number remains the mass-loss rate of $0.015\,M_\odot\ \mathrm{yr}^{-1}\ (10^9\,L_\odot)^{-1}$ derived by Faber & Gallagher (1976). For average giant ellipticals of $10^{10} - 10^{11}\ L_\odot$, this rate amounts to $\sim 10^9 - 10^{10}\ M_\odot/10^{10}\,\mathrm{yr}$. The second source of gas is infall by accretion or merging. I have estimated earlier from ripple statistics that an average giant E may have experienced 4 – 10 accretion events over a Hubble time (Schweizer 1983). Even if the average accretion event is only the proverbial "small gas-rich galaxy" that falls in, *e.g.*, a Magellanic Cloud, the associated gas mass may be not much less than $10^9\ M_\odot$. The main point is, then, that 4 – 10 such events can supply between a few times $10^9\,M_\odot$ and $10^{10}\,M_\odot$ of gas, or an amount quite comparable with all the gas shed by evolving stars.

What is the *gas inflow*, and is there a sink near the center? Cooling times for the gas are short near the center, of order 10^6 yr for the warm, ionized gas (Spitzer 1942) and $10^8 - 10^9$ yr for the hot X-ray gas (*e.g.*, Forman *et al.* 1985; Canizares *et al.* 1986), though cooling times in the outer envelope may exceed a Hubble time. As Fabian discusses elsewhere in this volume, the short central cooling times supposedly lead to large mass inflows of order $1\,M_\odot\ \mathrm{yr}^{-1}$ in field ellipticals and up to several $10^2\,M_\odot\ \mathrm{yr}^{-1}$ in galaxies located at the centers of clusters. Also in this volume, O'Connell reviews the evidence that star formation in such cooling flows may provide a sink for the gas. We should keep in mind, however, that direct observations of inflow motions do not exist. There is reason to be cautious in relying on calculated gas inflow rates: Recent work by various groups (*e.g.*, Bertschinger & Meiksin 1986) suggests that heat conduction may be significantly more efficient than previously thought and may reduce inflow rates by an order of magnitude.

Personally, I see some serious problems with the concept of cooling flows in ellipticals with gas disks. On the one hand, we believe that the hot X-ray coronae consist of gas shed by the stars of the galaxy. On the other hand, we know that the cooler gas in disks cannot have been shed by these stars because its angular momentum is generally decoupled from that of the stellar body. I conclude that *the observed H I and H II gas disks cannot be manifestations of cooling flows.* This conclusion is strengthened by the presence of significant amounts of dust in the same disks. Dust grains get destroyed rapidly in 10^7 K gas and have lifetimes against sputtering of only $10^6 - 10^7$ yr (Draine & Salpeter 1979). Therefore, the gas disks containing the dust cannot have condensed from the hot coronae. Yet if cooling flows of $\sim 1\,M_\odot\ \mathrm{yr}^{-1}$ exist in these same coronae, funneling masses comparable to those of the disks in $10^8 - 10^9$ yr toward the center, how come cloud-cloud collisions do not destroy the gas disks in similarly short times? To escape the problem of angular-momentum misalignment we could postulate that the hot coronae themselves consist largely of accreted gas, but this does still not solve the problem of the dust. Therefore, I remain skeptical of $\sim 1\,M_\odot\ \mathrm{yr}^{-1}$ cooling flows as a general phenomenon in ellipticals.

What is the *gas outflow*, and is there a sink at infinity? In response to the apparent lack of gas in ellipticals observed in the late 1960's and early 1970's, Johnson & Axford (1971) and Mathews & Baker (1971) introduced the hypothesis of galactic winds powered by supernovae. In such a wind, fresh gas is continuously supplied by evolving stars. The energy released by cloud-cloud collisions occurring with velocities comparable to the random motions of the stars heats this gas to million-degree temperatures, but not enough to drive a wind. It is the combined energy input of many supernovae through their shockfronts that supposedly deposits enough extra heat to drive a hot wind and flush the gas out of the galaxy. The question before us today is whether this hypothesis of galactic winds is still necessary now that we observe about as much gas in X-ray coronae as is expected.

Already in 1979, Norman and Silk concluded that heating by supernovae should be important only during the relatively early phases of elliptical evolution with their presumedly violent star formation. They predicted that supernovae would *not* deliver sufficient energy to keep galactic winds blowing through the later, more quiescent phases of E evolution, especially not in massive galaxies (see also Larson 1974). Recent observations seem to have proven Norman and Silk right on two important counts: 1) Hot coronae have indeed been found to be common and especially pronounced in massive ellipticals; and 2) IRAS observations have opened our eyes to the fact that in colliding and merging galaxies the energy input into the gas can be prodigious. An interesting example is provided by NGC 3256, a pair of colliding spirals believed to be in the process of merging. Graham *et al.* (1984) point out that alone the supernovae associated with the starburst will dump one order of magnitude more energy into the gas than is required to blow it out of the combined potential well of the galaxies.

In conclusion, the answer to the question about gas outflow may be that galactic winds play an important role in ridding merging galaxies of their gas, and perhaps even in helping create some gas-free ellipticals. But these winds may then die down and leave the E's to rebuild their own atmospheres from both internal and external sources.

7. CLOSING REMARKS

How does our knowledge of dust and gas in ellipticals fit in with theories of elliptical formation and evolution? Instead of trying to give a general answer, let me return once more to one of my favorite ellipticals, NGC 5018 (Fig. 1), and comment on a few related issues, including some raised earlier at this Symposium.

First, note that NGC 5018 used to be a bona fide E. De Vaucouleurs classified it as an E3 in the RC2, and Sandage as an E4 in the Revised Shapley-Ames Catalog. Now, based on a new plate taken at Las Campanas, Sandage calls it an S0 because it has too much dust to be an E (private communication). Yet CCD photometry shows no trace of any disk. I believe it does not matter whether we call such galaxies E's or S0's; the story which they tell remains the same.

Second, it is easy to see from Figure 1 that brightness profiles extracted at different position angles across the nucleus will not show a perfect $r^{1/4}$ law. There will be deviations from such a law which cannot be blamed on faulty sky subtraction, but which reflect the presence of fine structure in the galaxy. These deviations will diminish if, as is now usual, mean brightness profiles are derived by some method of azimuthal averaging. Yet Figure 1 cautions us that we should not put too much emphasis on intercomparing such mean profiles. If we do, we may miss the most telling signatures such as ripples and tails.

Third, if we believe in Quinn's (1984) shell models and in tidal tails, NGC 5018 must have accreted some disk material — including stars, dust, and gas — not so long ago. The amount is surprising: From CCD photometry Fort *et al.* (1986) estimate that the extra luminous matter is 9% – 18% of the total. Seitzer and I are finding similar amounts for several northern ellipticals with ripples. Even if perhaps above average, these amounts, together with the estimate of 4 – 10 accretion events per Hubble time, suggest that a significant fraction of the visible matter in ellipticals may stem from a variety of foreign sources.

Dust and gas then seem to fit comfortably into a picture of elliptical formation and evolution based on mergers: Ellipticals may occasionally form from mergers of galaxies of nearly equal mass (Toomre & Toomre 1972). Sporadically they continue accreting companions and matter from galaxies passing by. The gas and dust get stored temporarily in inclined disks, while the luminous matter is incorporated into the main bodies. Although there is evidence that these main bodies may be roughly triaxial, projection effects are unlikely to explain all isophotal twists. Some of the observed twists obviously reflect major recent perturbations, such as in NGC 5018, 1316, and 5128, and must correspond to intrinsic twists of the main axes. I believe the combined data suggest strongly that even in present-day ellipticals some growth and reshaping are still occurring.

I thank Halton C. Arp, William C. Forman, John A. Graham, and Jacqueline H. van Gorkom for their kind permission to reproduce some figures, and gratefully acknowledge partial support from the National Science Foundation through grant AST 83-18845.

REFERENCES

Arp, H.C.: 1966, *Atlas of Peculiar Galaxies* (Pasadena: California Institute of Technology).
Bertola, F., and Galletta, G.: 1978, *Astrophys. J.* **226**, L115.
Bertola, F., Bettoni, D., Rusconi, L., and Sedmak, G.: 1984, *Astron. J.* **89**, 356.
Bertola, F., Galletta, G., Zeilinger, W.W.: 1985, *Astrophys. J. Lett.* **292**, L51.
Bertschinger, E., and Meiksin, A.: 1986, *Astrophys. J. Lett.* **306**, L1.
Binney, J.: 1978, *M.N.R.A.S.* **183**, 779.
Bottinelli, L., and Gouguenheim, L.: 1977, *Astron. Astrophys.* **54**, 641.
Caldwell, N.: 1984a, *Astrophys. J.* **278**, 96.
Caldwell, N.: 1984b, *Publ. A.S.P.* **96**, 287.
Caldwell, N., Kirshner, R.P., and Richstone, D.O.: 1986, *Astrophys. J.* **305**, 136.
Canizares, C.R., Donahue, M.E., Trinchieri, G., Stewart, G.C., and McGlynn, T.A.: 1986, *Astrophys. J.* **304**, 312.
Carter, D., Jorden, P.R., Thorne, D.J., Wall, J.V., and Straede, J.C.: 1983, *M.N.R.A.S.* **205**, 377.
Davies, R.L., *et al.*: 1984, *Bull. A.A.S.* **16**, 410.
Davies, R.L., and Illingworth, G.D.: 1986, *Astrophys. J.* **302**, 234.
Demoulin-Ulrich, M.-H., Butcher, H.R., and Boksenberg, A.: 1984, *Astrophys. J.* **285**, 527.
Draine, B.T., and Salpeter, E.E.: 1979, *Astrophys. J.* **231**, 77.
Ebneter, K., and Balick, B.: 1985, *Astron. J.* **90**, 183.
Faber, S.M., and Gallagher, J.S.: 1976, *Astrophys. J.* **204**, 365.
Ford, H.C., and Butcher, H.: 1979, *Astrophys. J. Suppl.* **41**, 147.
Forman, W., Schwarz, J., Jones, C., Liller, W., and Fabian, A.C.: 1979, *Astrophys. J. Lett.* **234**, L27.
Forman, W., Jones, C., and Tucker, W.: 1985, *Astrophys. J.* **293**, 102.
Fort, B.P., Prieur, J.-L., Carter, D., Meatheringham, S.J., and Vigroux, L.: 1986, *Astrophys. J.* **306**, 110.

Fosbury, R.A.E., Mebold, U., Goss, W.M., and Dopita, M.A.: 1978, *M.N.R.A.S.* **183**, 549.
Gallagher, J.S., Knapp, G.R., Faber, S.M., and Balick, B.: 1977, *Astrophys. J.* **215**, 463.
Gisler, G.R.: 1978, *M.N.R.A.S.* **183**, 633.
Graham, J.A.: 1979, *Astrophys. J.* **232**, 60.
Graham, J.R., Wright, G.S., Meikle, W.P.S., Joseph, R.D., and Bode, M.F.: 1984, *Nature* **310**, 213.
Gunn, J.E.: 1979, in *Active Galactic Nuclei*, ed. C. Hazard and S. Mitton (Cambridge: Cambridge University Press), p. 213.
Hawarden, T.G., Elson, R.A.W., Longmore, A.J., Tritton, S.B., and Corwin, H.G., Jr.: 1981, *M.N.R.A.S.* **196**, 747.
Heiligman, G., and Schwarzschild, M.: 1979, *Astrophys. J.* **233**, 872.
Heisler, J., Merritt, D., and Schwarzschild, M.: 1982, *Astrophys. J.* **258**, 490.
Hubble, E.: 1936, *The Realm of the Nebulae* (New York: Dover), p. 39.
Humason, M.L., Mayall, N.U., and Sandage, A.R.: 1956, *Astron. J.* **61**, 97.
Johnson, H.E., and Axford, W.I.: 1971, *Astrophys. J.* **165**, 381.
Kahn, F.D., and Woltjer, L.: 1959, *Astrophys. J.* **130**, 705.
Knapp, G.R., Gallagher, J.S., and Faber, S.M.: 1978, *Astron. J.* **83**, 139.
Knapp, G.R., Turner, E.L., and Cunniffe, P.E.: 1985, *Astron. J.* **90**, 454.
Kumar, C.K., and Thonnard, N.: 1983, *Astron. J.* **88**, 260.
Larson, R.B.: 1974, *M.N.R.A.S.* **169**, 229.
Lauer, T.: 1985, *M.N.R.A.S.* **216**, 429.
Malin, D.F., and Carter, D.: 1983, *Astrophys. J.* **274**, 534.
Malin, D.F., Quinn, P.J., and Graham, J.A.: 1983, *Astrophys. J. Lett.* **272**, L5.
Mathews, W.G., and Baker, J.C.: 1971, *Astrophys. J.* **170**, 241.
Mayall, N.U.: 1936, *Publ. A.S.P.* **48**, 14.
Mayall, N.U.: 1939, *Publ. A.S.P.* **51**, 282.
Merritt, D., and de Zeeuw, T.: 1983, *Astrophys. J. Lett.* **267**, L19.
Minkowski, R., and Osterbrock, D.: 1959, *Astrophys. J.* **129**, 583.
Norman, C., and Silk, J.: 1979, *Astrophys. J. Lett.* **233**, L1.
Osterbrock, D.E.: 1960, *Astrophys. J.* **132**, 325.
Phillips, M.M., Jenkins, C.R., Dopita, M.A., Sadler, E.M., and Binette, L.: 1986, *Astron. J.* **91**, 1062.
Quinn, P.J.: 1984, *Astrophys. J.* **279**, 596.
Rieke, G.H., and Lebofsky, M.J.: 1986, *Astrophys. J.* **304**, 326.
Sadler, E.M., and Gerhard, O.E.: 1985, *M.N.R.A.S.* **214**, 177.
Sandage, A.: 1957, *Astrophys. J.* **125**, 422.
Schwarzschild, M.: 1979, *Astrophys. J.* **232**, 236.
Schweizer, F.: 1980, *Astrophys. J.* **237**, 303.
Schweizer, F.: 1983, in *Internal Kinematics and Dynamics of Galaxies, IAU Symposium No. 100*, ed. E. Athanassoula (Dordrecht: Reidel), p. 319.
Schweizer, F., and Ford, W.K., Jr.: 1985, in *New Aspects of Galaxy Photometry*, ed. J.-L. Nieto (Berlin: Springer), p. 145.
Sharples, R.M., Carter, D., Hawarden, T.G., and Longmore, A.J.: 1983, *M.N. R.A.S.* **202**, 37.
Simonson, G.F.: 1982, Ph. D. thesis, Yale University.
Sparks, W.B., *et al.*: 1985, *M.N.R.A.S.* **217**, 87.
Spitzer, L., Jr.: 1942, *Astrophys. J.* **95**, 329.
Steiman-Cameron, T.Y., and Durisen, R.H.: 1982, *Astrophys. J. Lett.* **263**, L51.
Steiman-Cameron, T.Y., and Durisen, R.H.: 1987, *Astrophys. J.*, submitted.

Tohline, J.E., and Durisen, R.H.: 1982, *Astrophys. J.* **257**, 94.
Tohline, J.E., Simonson, G.F., and Caldwell, N.: 1982, *Astrophys. J.* **252**, 92.
Toomre, A., and Toomre, J.: 1972, *Astrophys. J.* **178**, 623.
Trinchieri, G., and Fabbiano, G.: 1985, *Astrophys. J.* **296**, 447.
Tubbs, A.D.: 1980, *Astrophys. J.* **241**, 969.
van Albada, T.S., Kotanyi, C.G., Schwarzschild, M.: 1982, *M.N.R.A.S.* **198**, 303.
van Gorkom, J.H., Knapp, G.R., Raimond, E., Faber, S.M., and Gallagher, J.S.: 1986, *Astron. J.* **91**, 791.
Wilkinson, A., Sharples, R.M., Fosbury, R.A.E., and Wallace, P.T.: 1986, *M.N.R.A.S.* **218**, 297.

DISCUSSION

King: In the picture of ellipticals formed by merging of spirals, there is the problem that ellipticals have more globular clusters than spirals. How do you reconcile this?

Schweizer: This problem is not really serious in my view, at least not yet. It is often cited as an objection to the formation of ellipticals through mergers of spirals, yet it seems to be based on two unproven, and possibly wrong, promises: (i) All globular clusters in any galaxy are as old as those in the Milky Way seem to be; and (ii) the number of globular clusters in a merger remnant is the sum of the globular clusters in the pre–merger galaxies.
Concerning the first premise, we know that young and intermediate–age globular clusters exist in the Magellanic Clouds, M33, and M31. It just happens that our Milky Way doesn't seem to possess any. So at least three or four out of five nearby galaxies do know how to make globular clusters long after the Big Bang.
Concerning the second premise, we still know little about the details of star formation in mergers of gas–rich galaxies. Yet we do know that such "super starburst" galaxies as Arp 220 and NGC 6240 produce vast anounts of molecular gas (Joseph, Wright & Wade, 1984, *Nature*, **311**, 132; Rieke, Cutri, Black, Kailey, McAlary, Lebofsky & Elston, 1985, *Astrophys. J.*, **290**, 116) and achieve very high efficiencies of star formation. Both objects seem to be ongoing mergers of about equal–sized disk galaxies. It seems unreasonable to me that both should have dense molecular clouds and form stars at prodigious rates, yet they would not also produce globular clusters. What better environment is there to produce massive clusters than the highly crunched gas in such systems? If this view is correct and ellipticals arise from disk mergers, one would certainly expect the specific globular cluster frequency to be higher in ellipticals than in the pre–merger galaxies.

Norman: You need $\sim$ 10 gas–rich dwarfs per Hubble time to give the gas in ellipticals. Where are these dwarfs or, more precisely, where have they been? A current gas–rich dwarf density of 10 per L^* galaxy is not observed. Were they eaten, are they found in interactions, or have they lost their gas and become very low surface brightness objects?

Schweizer: My statement was only that there seem to have been 4–10 accretion events per average giant elliptical over the age of the Universe. This accretion rate is estimated from the number of ellipticals with ripples (about 50%) that I detected in a CTIO 4-meter survey some years ago, and from estimated ripple lifetimes ($\sim$ 2 Gyr; see IAU Symp. **100**, p. 319, 1983). I believe that the majority of these events were not mergers of small, gas–rich galaxies, but rather less glamorous transfers of mass from the outskirts of neighboring disk galaxies. The reason is that the cross section for mass transfer must be of order $10 - 10^2$ times larger than the cross section for mergers. It takes only a near collision to induce mass transfer, but a very nearly central hit to start a merger. And as we well know, the outskirts of disk galaxies are, on average, quite gas rich. I would, therefore, conclude that foreign gas and stars in ellipticals probably stem from a variety of sources, and that some of the neighbor galaxies that donated material in the past will do it again in the future. Another way to say this is that galaxies seem to be notoriously promiscuous!

Toomre: [Question addressed to Tjeerd van Albada.] After hearing Schweizer mention with some skepticism the test proposed by yourself and Schwarzschild and Kotanyi about the sense of warp in a postulated tumbling triaxial Cen A potential, do *you* yourself agree now that your test has been flunked in that famous example?

van Albada: I will let Martin Schwarzschild answer this.

Schwarzschild: Yes, Alar Toomre, we have flunked the test. But I would only retreat half way. I now think that the truth in this matter—as in so many cases—lies between the extreme pictures. The outer parts of the dust ring in Cen A and other systems, with their twists and rugged appearance, seem not yet settled on stable orbits but the inner parts, for which the viscous settling time is shorter, have it seems managed to settle on stable orbits. Hence, triaxiality is still needed for Cen A to provide a stable orbit for the inner part of the dust ring but fast figure rotation—in a prescribed sense—is not anymore needed to explain the twists of the outer parts which we now assume are relics of the merger geometry.

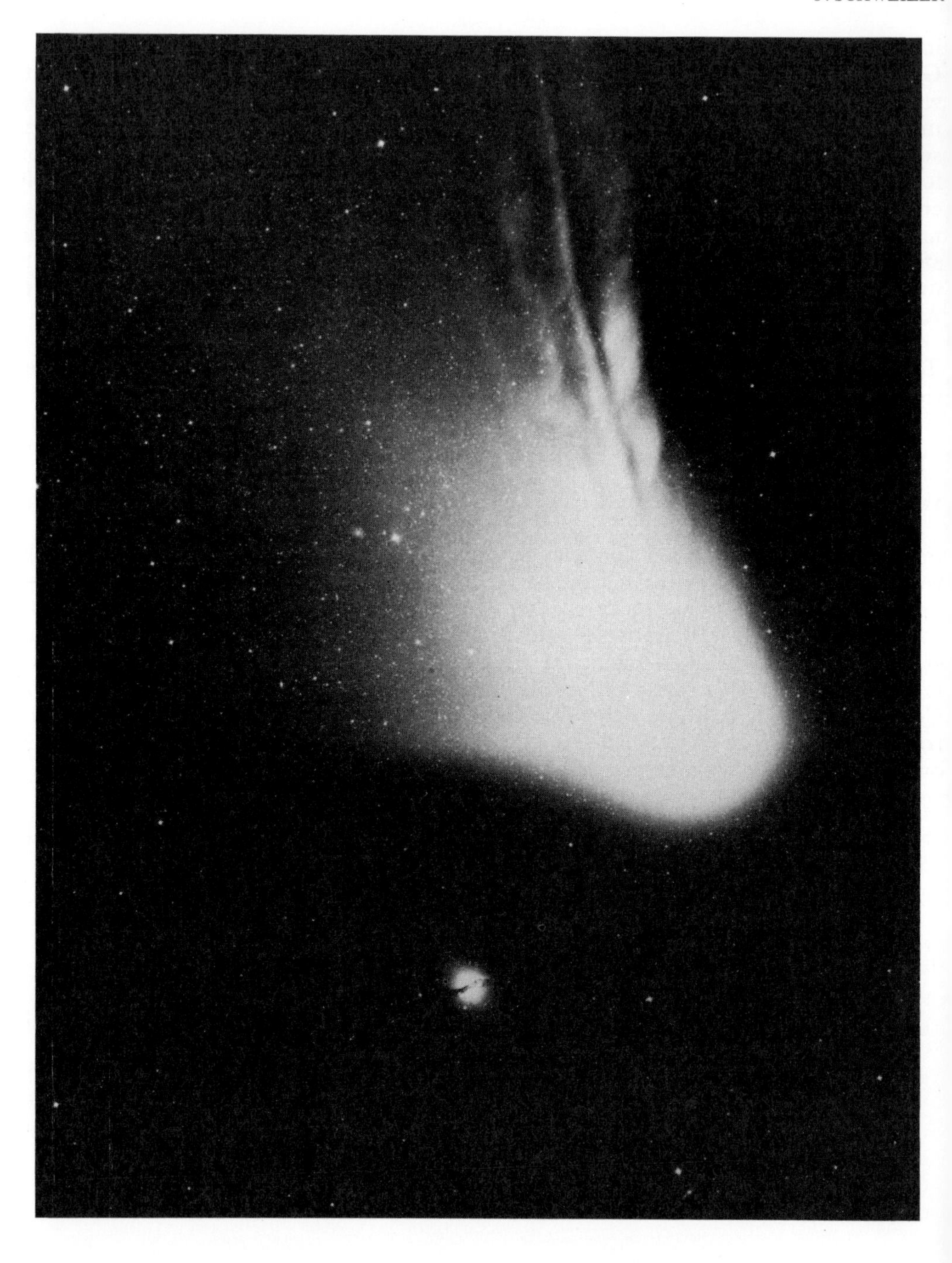

Centaurus A, with comet P/Halley, as presented by Schweizer. The 8^m exposure on a 103aO plate was taken on April 15, 1986, by A. Gomez with the Curtis Schmidt telescope at Cerro Tololo (National Optical Astronomy Observatories).

IONIZED GAS IN ELLIPTICAL GALAXIES

Elaine M. Sadler
Kitt Peak National Observatory
National Optical Astronomy Observatories
P.O. Box 26732
Tucson AZ 85726

ABSTRACT. More than half of all nearby elliptical galaxies contain modest amounts (10^3 - 10^5 $M_\odot$) of ionized gas. In bright elliptical galaxies this gas appears to lie in a rotating, kiloparsec-scale central disk, with a spectrum characteristic of non-thermal ionization. Low-luminosity ellipticals have a clumpy gas distribution and the gas in these galaxies is photoionized by young stars.

1. INTRODUCTION

Francois Schweizer has given an overview of the problem of gas and dust in elliptical galaxies. In this paper, I would like to focus in more detail on the ionized gas component. To put this into perspective, it now seems likely that a 'typical' bright elliptical galaxy has a three-phase interstellar medium, with perhaps 10^9 - 10^{10} $M_\odot$ of hot X-ray gas (Biermann and Kronberg 1983, Nulsen *et al.* 1984, Forman *et al.* 1985) and 10^6 - 10^7 $M_\odot$ of cold (HI or molecular) gas (Jura 1986, Knapp *et al.* 1985). In contrast, the mass of 'warm' (10^4K) gas responsible for the optical emission lines is only 10^3 - 10^5 $M_\odot$ (Phillips *et al.* 1986), and it therefore represents a very small fraction of the total gas content. However, this component gives us much of the information we presently have on the distribution and kinematics of gas in ellipticals, since X-ray observations are still sparse and HI detections rare.

Here, I will discuss the structure and kinematics of ionized gas in 'normal' elliptical galaxies. In doing so, I will draw heavily on the results of a recent, extensive survey of Hα/[NII] emission in early-type galaxies (Phillips *et al.* 1986). This work is based on high-resolution (3 Å FWHM) spectra of 203 members of a complete, magnitude-limited galaxy sample (Sadler 1984) and achieved a detection rate of about 55% to a limiting equivalent width of 0.8 Å in [NII].

2. THE EMISSION LINE SPECTRA OF ELLIPTICAL GALAXIES

The galaxies observed by Phillips *et al.* fall into the three categories shown in Figure 1, with their properties being roughly segregated by

T. de Zeeuw (ed.), Structure and Dynamics of Elliptical Galaxies, 125–133.

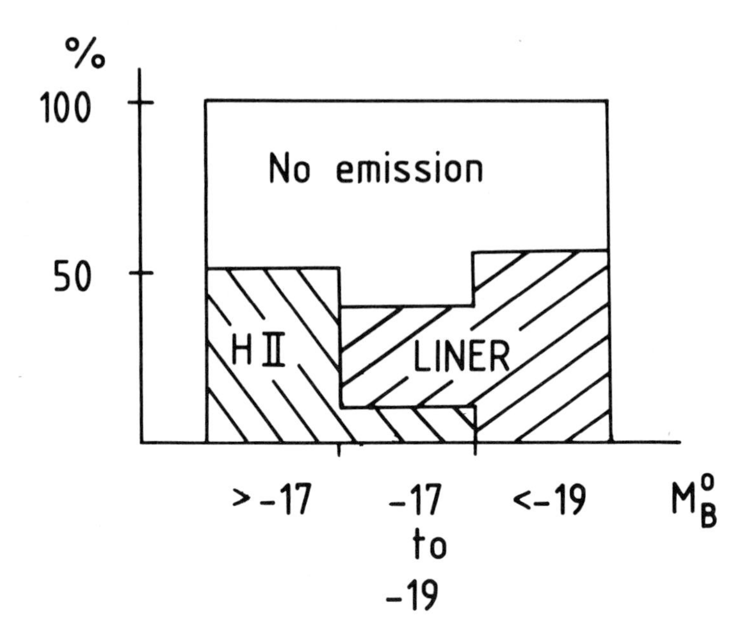

Figure 1: Fraction of E and S0 galaxies with emission spectra in the Phillips et al. survey, divided into three bins in absolute magnitude. Note that in giant ellipticals, the emission is always characteristic of a LINER (i.e. non-thermal ionization), while in the small galaxies the gas has an HII-region-like spectrum, and is ionized by young stars.

luminosity. About 40% showed no evidence for emission lines.

The emission-line spectra of bright elliptical galaxies show line ratios characteristic of LINERs (Low Ionization Nuclear Emission Regions; Heckman 1980), with Hα stronger than [NII] (see Figure 2). Such an emission spectrum can plausibly arise from either shock ionization or ionization by a central non-thermal source, but not from photoionization by young stars. NGC 1052 is the prototype of this class of objects, but all of them show similar spectra. The emission lines are weak and relatively narrow (EW 0.6 - 10 Å, FWHM 200-600 km/s), though a very weak broad-line component may also be present (Filippenko and Sargent 1985). Emission lines are more frequent, and more luminous, in brighter galaxies, suggesting that the ionized gas content increases with galaxy luminosity. The gas content does not, however, appear to be related to other galaxy properties such as color, axial ratio or stellar kinematics.

The emission lines seen in low-luminosity ellipticals resemble those in galactic HII regions, with strong, narrow lines (EW 1-20 Å, FWHM < 200 km/s) and Hα >> [NII]. Since the galaxy sample used by Phillips et al. was apparent magnitude-limited, the actual number of low-luminosity galaxies observed was small, and J.S Gallagher and I have begun a program to identify and study more members of this class.

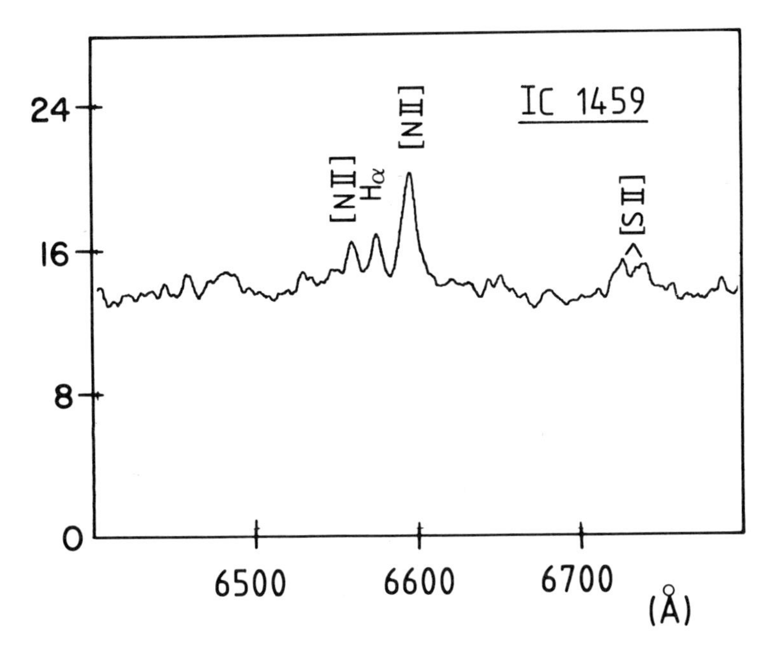

Figure 2: The spectrum of IC 1459, a typical elliptical LINER. Note that the lines are weak relative to the continuum. The vertical scale is in units of 10^{-15} ergs/cm^2/s/Å.

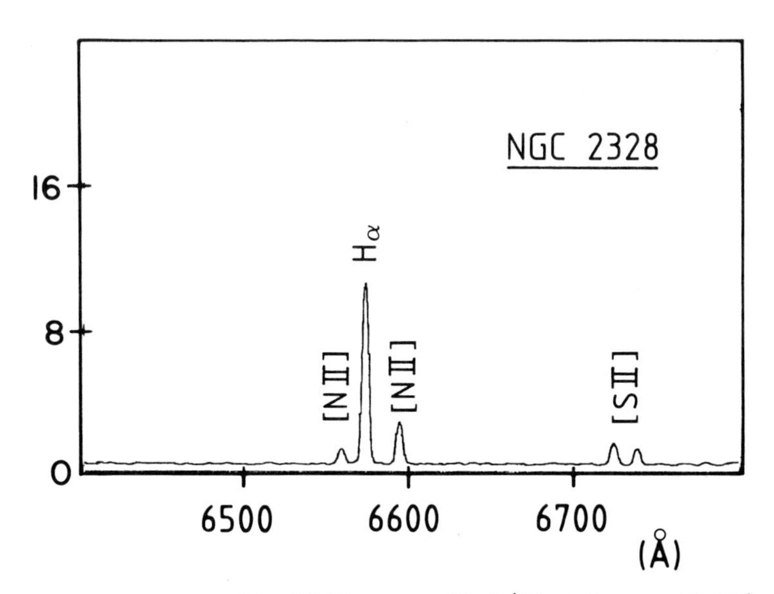

Figure 3: The spectrum of NGC 2328, a small E/SO galaxy. Vertical scale is in units of 10^{-14} ergs/cm^2/s/Å).

Preliminary results from a survey of about 30 such galaxies confirm that somewhere between 30% and 60% of 'normal' small ellipticals contain gas which is photoionized by OB stars. Many of the galaxies which show strong emission lines are isolated, and it seems unlikely that star formation has been triggered by an external mechanism such as interaction with another galaxy. Further study of these galaxies is important in order to understand why star formation with an apparently normal IMF is proceeding in many small ellipticals, even though it is absent in giant Es.

3. STRUCTURE AND KINEMATICS OF THE LINE-EMISSION REGION

3.1 HII-region galaxies

From the Phillips *et al.* spectra, we can state some general properties of the HII-region galaxies. These are low-luminosity galaxies (M_B = -15 to -18 with H_o = 100 km/s/Mpc) with normal elliptical-like appearance on Schmidt plates; the gas is extended over a region 0.5 - 1 kpc in diameter and is not rotating.

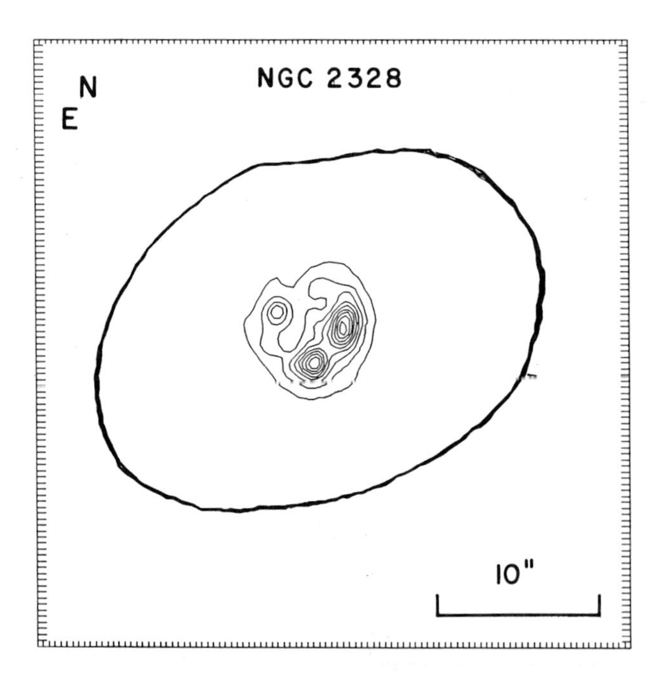

Figure 4: Distribution of emission-line gas in NGC 2328, from narrow-band CCD images obtained at the ESO 2.2m telescope. Contours indicate the strength of the Hα/[NII] emission, which is confined to the central 8-10 arcsec. The galaxy itself has a diameter of about 1.6 arcmin, and the outermost (thicker) line shows a typical continuum isophote.

Differential B-R photometry and Hα imaging of NGC 2328 show that the emission comes from a blue, ring-like structure about 0.5 kpc in diameter and centered on the nucleus of the galaxy. The ring itself is resolved into several condensations. Outside the emission region, the underlying B-V color (0.75 - 0.85) from is typical of an early-type galaxy of this luminosity (M_B = -17.0).

Although it may be unwise to speculate on the results from a single object, the observations of NGC 2328 suggest that we are seeing regions of star formation embedded in a normal (i.e. red) underlying elliptical galaxy. Most of the HII-region galaxies found by Phillips et al. are also IRAS detections, and these objects are clearly worthy of more study.

3.2 LINER galaxies

Determining the structure of the ionized gas in LINER galaxies is more difficult, since we have weak line emission against a strong galaxy continuum. About 25% of the galaxies detected by Phillips et al. had the emission region spatially resolved (at least 7 - 10 arcsec in diameter), and in these the gas lies in a region 200 pc - 2 kpc across, centered on the nucleus of the galaxy. In galaxies with extended emission, the observed gas kinematics are consistent with rapid rotation (V ~ 100-300 km/s). The total mass of ionized gas is typically 10^3 - 10^5 $M_\odot$ (see Figure 5).

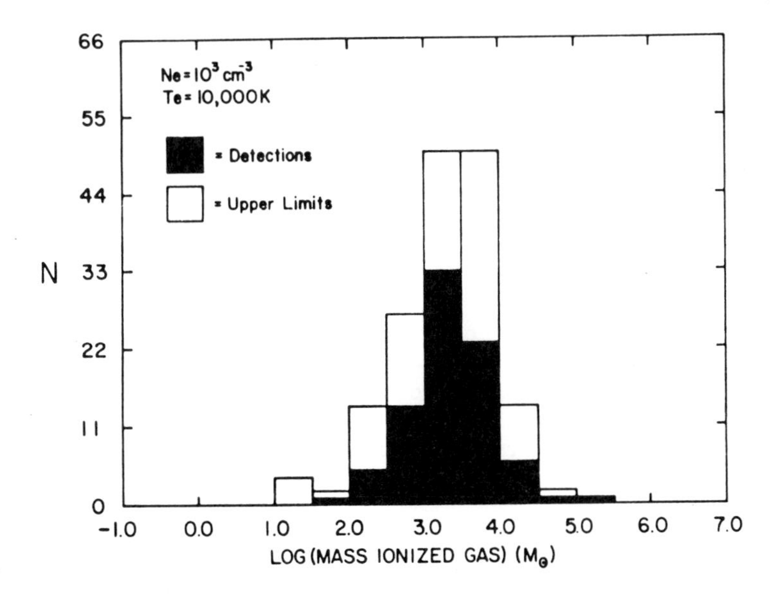

Figure 5: Distribution of ionized gas masses, from Phillips et al. (1986).

For galaxies with strong emission lines (equivalent width 4–5 Å in [NII]), Fabry-Perot methods or CCD imaging in narrow-band filters can be used to determine the structure of the gas directly. These techniques reveal a wide range of features including filaments, warped disks and even spiral structure. F. Bertola, J. Danziger and I have recently observed several bright ellipticals in Hα/[NII] using a CCD at the ESO 2.2m telescope, and Figure 6 shows the gas morphology in two of these galaxies. NGC 4696 is the central galaxy of the Centaurus cluster, and has an X-ray cooling flow (Fabian *et al.* 1982). The ionized gas in this galaxy shows a complex filamentary structure, while that in NGC 5077 resembles a warped disk or bar. Demoulin-Ulrich *et al.* (1985) observed that the gas in NGC 5077 was extended along the minor axis of the galaxy but the higher-resolution data in Figure 6 reveal a strong twist in the gas distribution towards the center. This may be an example of gas settling into a preferred plane (Tohline *et al.* 1982).

The origin of the ionized gas remains uncertain. A cooling flow from the X-ray halo (Nulsen *et al.*1984) appears the most natural source, especially since Phillips *et al.* found a strong correlation between the presence of X-ray gas and emission lines. However, the rotation axes of gas and stars often differ (Bertola *et al.* 1984, Demoulin-Ulrich *et al.* 1985, Caldwell *et al.* 1986), and this has been used to argue that the gas has an external origin. Further work is necessary to clarify this question, and the variety of gas morphologies observed may represent gas of both internal and external origin.

4. CONCLUSION

Most elliptical galaxies have weak optical emission lines and contain modest amounts (10^3 - 10^5 $M_\odot$) of ionized gas in their central regions. Although the amount of gas present is small, it is often possible to determine both the structure and kinematics of the line-emitting region. The results show that high- and low-luminosity ellipticals form two distinct classes in terms of their emission-line properties. Low-luminosity ellipticals show spectra characteristic of HII regions, and the emission can often be resolved into individual blue clumps around the nucleus of an otherwise 'normal' (i.e. red) underlying galaxy. The emission lines are narrow, and the gas does not appear to be in ordered rotation. In contrast, the gas in bright ellipticals generally appears to lie in a rotating, kiloparsec-scale disk and the emission is LINER-like, indicating that the gas is not photoionized by young stars. There is a tendency for larger galaxies to contain more gas, but the gas content is not apparently related to other galaxy properties such as axial ratio or color. The origin of the central gas disk, and the means by which it is ionized, remain open questions.

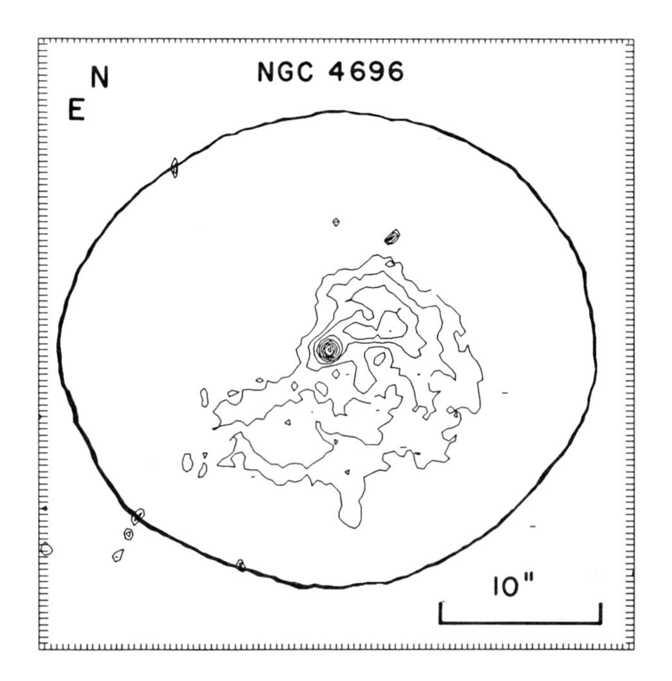

Figure 6: Distribution of emission-line gas in two bright elliptical galaxies, (a) NGC 4696 and (b) NGC 5077. As in Figure 4, contours indicate the strength of Hα/[NII] emission and the thicker line shows a continuum isophote.

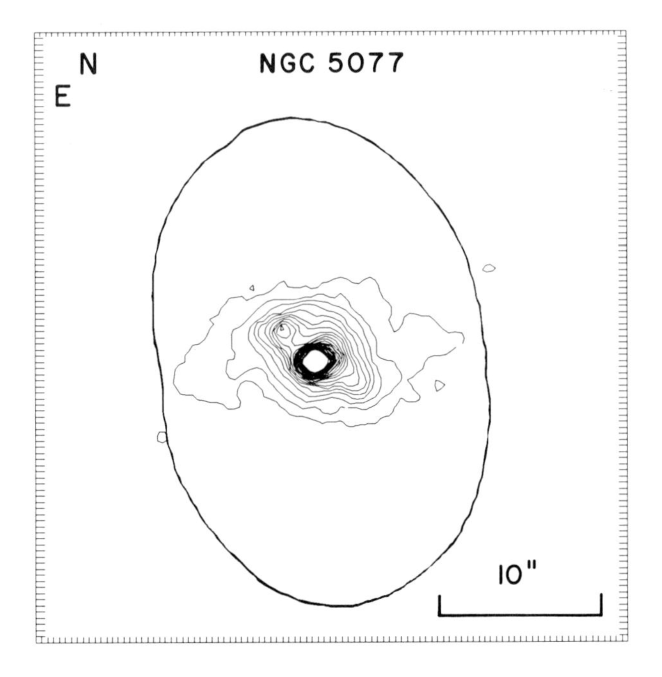

REFERENCES

Bertola, F., Bettoni, D., Rusconi, L. and Sedmak, G. 1984, Astron. J., **89**, 356.
Biermann, P. and Kronberg, P. P. 1983, Astrophys. J., **268**, L69.
Caldwell, N., Kirshner, R. P. and Richstone, D. 1986, Astrophys. J., in press.
Demoulin-Ulrich, M.-H., Butcher, H. and Boksenberg, A. 1985, Astrophys. J. **285**, 527.
Fabian, A. C., Atherton, P. D., Taylor, K. and Nulsen, P. E. J. 1982, Mon. Not. R. Astr. Soc., **201**, 17P.
Filippenko, A. V. and Sargent, W. L. W. 1985, Astrophys. J., Suppl. Ser., **57**, 503.
Forman, W., Jones, C. and Tucker, W. 1985, Astrophys. J., **293**, 102.
Heckman, T. M. 1980, Astron. Astrophys., **87**, 152.
Jura, M. 1986, Astrophys. J., in press.
Knapp, G. R., Turner, E. L. and Cuniffe, P. E. 1985, Astron. J., **90**, 454.
Nulsen, P. E. J., Stewart, G. C. and Fabian, A. C. 1984, Mon. Not. R. Astr. Soc., **208**, 185.
Phillips, M. M., Jenkins, C. R, Dopita, M. A., Sadler, E. M. and Binette, L. 1986, Astron. J., **91**, 1062.
Sadler, E. M. 1984, Astron. J., **89**, 23.
Tohline, J. E., Simonson, G. F. and Caldwell, N. 1982, Astrophys. J., **252**, 92.

DISCUSSION

Djorgovski: It is probably not surprising that the line luminosity scales with the optical luminosity and/or X–ray luminosity. What may be much more interesting to check is whether the ratios of line to optical and X–ray to optical correlate.

Carlberg: Can you estimate the total gas mass added to the stellar core in a Hubble time, using Schweizer's estimate of 10 "accretion events" per lifetime?

Sadler: This is difficult to do just from observations of the ionized gas, since it is such a small fraction of the total gas content. We currently see $10^3 - 10^4 M_\odot$ of ionized gas within 1/2 kpc radius in a typical elliptical, while the dust suggests there may be $\sim 10^6 M_\odot$ of HI in the same region. I don't know how much of this gas comes from "accretion events" and how much from stellar mass loss; but the amount seems to be roughly similar (within an order of magnitude or so) in most of the galaxies we observed.

Richstone: Could the "Liner" Spectra be produced by photoionization by photons from the X–ray emitting gas or is some other mechanism mandated?

Sadler: Ionization by X–ray photons is certainly plausible, and perhaps even the most promising mechanism. I believe Luc Binette and Mike Dopita are currently working on models of this kind.

Porter: One problem with cooling gas inflow models with large $\dot{M}$ is that the central radio sources in the galaxies involved are rather weak, so that most of the mass cannot be reaching the central engine. Do the masses of the gas disks you observe correlate well with the radio source luminosities, and if so, can you conclude from that that the inflow is not even reaching those disks?

Sadler: There is no correlation between gas mass and radio power. We have little or no idea what happens to the gas within about 1 kpc, or how much of it reaches the nucleus.

van Gorkom: You mention a discontinuity in properties of ellipticals at the switch–over point from the $H\alpha$ regime to the Liner regime. Is this discontinuity also reflected in the radio properties?

Sadler: Yes, I think so, in the sense that radio sources are rarely seen in low–density ellipticals. We did see weak radio emission from NGC 2328; but the ratio of radio flux to $H\alpha$ flux is consistent with the radio emission coming from HII regions rather than from a central non–thermal source of the kind seen in giant ellipticals.

Schwarzschild: Are there any observable ionized disks in radio galaxies with clear axes in the center? If so, it would seem to be most instructive to see whether the innermost gas disks are aligned with the innermost radio axis.

Sadler: This is also a question which interests me very much. Unfortunately, most of the galaxies in our sample with a well–defined central gas/dust disk have compact, rather than extended radio sources. Where we do see extended radio emission in such a galaxy, as for example in NGC 4696, both the radio and emission–line components show very complex structure within the central 1 – 2 kpc.

Ford: We are finding galaxies with Liner spectra in which there is a strong correlation between the resolved structure of nonthermal radio sources and the morphology of the ionized gas. This suggests that the gas in these galaxies may be ionized in shocks created when jets or bubbles emanating from the nucleus dissipate mechanical energy in the surrounding ISM. It also should be noted that as we see more and more symmetry in the distribution of ionized gas, it becomes increasingly difficult to account for the ionization by postulating a central, nonthermal source of ionizing photons.

Whitmore: I noticed your spectra included the SII lines. In a paper about 2 years ago, Rubin, Ford and Whitmore (1984, *Astroph. J.*, **281**, L21) found that the NII/SII ratio increases with luminosity in spiral galaxies. Do you know if this relationship also occurs in your ellipticals?

Sadler: The S/N ratio is quite poor at SII, so we have not looked in detail. However, we do note that SII increases in the low luminosity ellipticals.

Whitmore: This is the same as found in spirals. It would be interesting to determine whether the relations join together smoothly as a function of luminosity, and/or mass.

Sadler and Bertola in discussion with the audience.

PROPERTIES OF ELLIPTICAL GALAXIES WITH DUST LANES

Francesco Bertola
Institute of Astronomy
University of Padova
35122 Padova
Italy

1. INTRODUCTION

NGC 5128 (Cen A) has been known for many years as a peculiar elliptical, but only recently has it been recognized as the prototype of a new class of object (Bertola and Galletta 1978). This class of galaxy is characterized by an elliptical–like stellar body crossed along the minor axis by a dust lane, and was comprised initially of five objects. The class was extended in a remarkable way by Hawarden *et al.* (1981), who drew attention also to the existence of ellipticals with the dust lane along the major, and along intermediate axes. Ninety objects are now listed as dust lane ellipticals in the compilation by Ebneter and Balick (1985) and its updated version (Ebneter 1986).

Kotanyi and Ekers (1979) pointed out the tendency of the dust lane to align perpendicularly to the radio source, when present.

Since 1978 it has been realized that the study of this type of object could lead to important information on the intrinsic shape of elliptical galaxies. In fact there is no a priori reason for these galaxies to be oblate spheroids (Binney 1978) since in many cases their rotational velocity is too low to generate a flattened system (Bertola 1972, Bertola and Capaccioli 1975, Illingworth 1977).

The dust lane silhouetted against the luminous body of the galaxy, which otherwise would be classified as a pure elliptical, is interpreted as a disk or a ring seen almost edge–on, the result of a past capture by an early type galaxy of a gas rich system. It is therefore clear that the observed configuration depends on the intrinsic shape of the elliptical galaxy and on the impact parameters.

In this paper we shall describe the morphological, statistical, kinematical, and photometric properties, of the elliptical galaxies with dust lanes.

2. MORPHOLOGY AND DEFINITION OF THE SAMPLE

The appearence of the dust lane, regardless of its orientation with respect to the stellar body, can be described in the following way:

straight: the dust lane lies exactly along an axis of the stellar body and presum-

T. de Zeeuw (ed.), Structure and Dynamics of Elliptical Galaxies, 135–144.

ably represents a disk seen exactly edge-on. A good example is NGC 6702.

curved: a slight curvature is present, suggesting a disk seen almost edge-on. NGC 5485 is an example.

multiple: in a few cases and notably in NGC 1947, a set of parallel dust lanes is present on one side of the galaxy. The lanes could be the result of a system of coplanar rings seen at an angle.

ring-like: a full ring is visible; from its flattening, and on the hypothesis of intrinsic circularity, one can deduce the tilt angle. The best example is NGC 5266 and possible cases are Anon 0151-498 and Anon 0418-583.

warped: the prototype is NGC 5128 where the dust lane, aligned along the minor axis, bends toward the major axis in the outer regions. Additional cases are NGC 5363, NGC 5266, Anon 0151-498, and Anon 1029-459.

In Fig 1 six representative cases are illustrated with high resolution photographs obtained at the prime focus of the 3.9m AAT.

A straightforward question arises from the above morphological description: is the dust distributed to form a disk or is it confined to a ring? While in a few cases the ring is clearly visible, in the case of NGC 5128 there is evidence that the gas associated with the dust forms a disk both in ionized (Bland 1985) and neutral (van Gorkom 1986) hydrogen.

Concerning the extent of the dust lane with respect to the galaxy, a variety of situations are encountered. There are galaxies where the dust lane is confined to the innermost regions of the galaxy (e.g. NGC 4374), and galaxies like NGC 5128 where the dust lane can be followed up to the limit of detection of the luminous stellar material. In Table I we have indicated for each galaxy whether the dust extends to the inner, intermediate, and outer regions. One sees that inner and outer dust lanes are present in approximately equal number. Elliptical galaxies with inner dust lanes are easily found with modern detectors and certainly in the future they will outnumber those with extended dust lanes.

Before proceeding to analyze the properties of ellipticals with dust lanes, it is necessary to define the sample according to precise morphological criteria. In order to do this the catalog of dusty elliptical galaxies and its updated version (Ebneter and Balick 1985, Ebneter 1986) have been used as basic references, carefully selecting from them the ellipticals with well defined dust lanes. Patchy dust structures were excluded, even those regularly distributed along one axis, as in the case of NGC 1316 (Fornax A).

The first part of Table I lists the elliptical galaxies with dust lanes along the minor axis. Due to the special character of this morphology, the identification of these objects is rather unambiguous. In fact we do not know of any disk galaxy, the alternative to an elliptical, where the bulge is elongated perpendicularly to the plane defined by the disk. However, one should be very careful on this point, since there is a tendency in the usual classification schemes to call S0 any galaxy where a stellar body is crossed by a dust lane, regardless of orientation. Thirty objects are listed in Table I as minor axis dust lane ellipticals. In Ebneter's catalog there are 34 objects listed as minor axis dust lane ellipticals, and 23 of these are in common with those of Table I. Also in Table I there are 6 objects which are classified in different way by Ebneter, while 9 cases given by the latter are discarded.

It is worth mentioning why NGC 3108 has been omitted from Table I; in addition to the dust lane crossing the stellar body along the major axis, there is a trace of a luminous ring associated with the dust. This galaxy is probably the

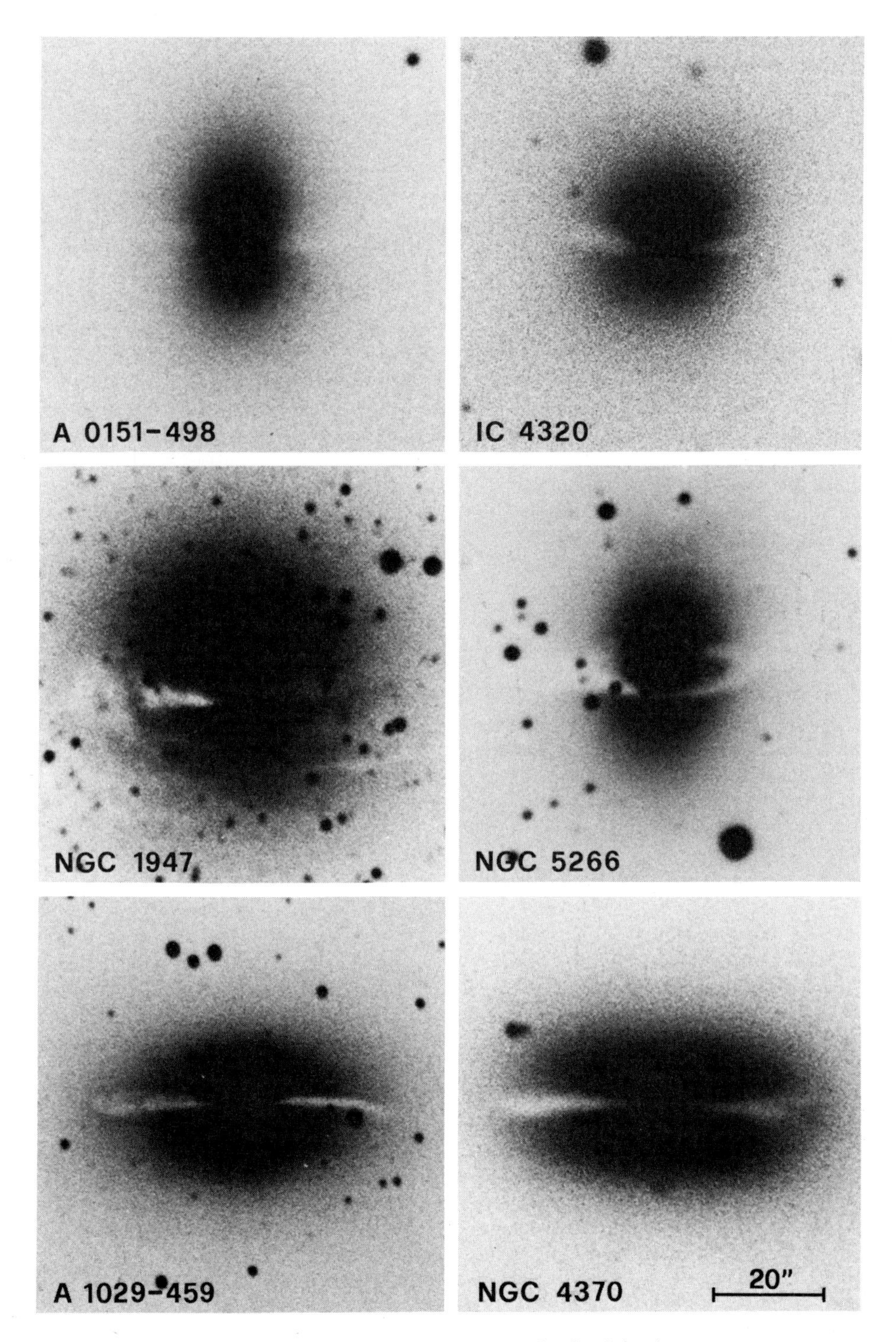

Figure 1. Dust lane ellipticals photographed with the 3.9m AAT.

Table I. Elliptical galaxies with dust lanes

ident.	*Bmag*	V_r	M_B	b/a	V_{max}	σ_0	V_{max}/σ_0	$(V/\sigma)_*$	ext	$\Delta\phi$	Ref.
N 185	11.00	-227	-14.59	.86					N	90	
I 1575				.75					I	90	
N 404	11.30	-39	-17.37	1.00					N	90	
N 662	13.60	5660	-21.67	.63					N	90	
0151-498	13.37	6170	-22.09	.67	50	257	0.19	0.27	E	90	1
0206+355	14.69	11600	-20.60	.87					E	90	
N 1052	11.53	1471	-20.91	.69	96	245	0.39	0.59	N	90	2
N 1297	12.61	1599	-19.90	.91					N	90	
0418-583				1.00					E	90	
N 1947	11.83	1160	-19.88	.90	47	150	0.38	1.12	E	90	3
0609-331		11300		.50	160	265	0.60	0.61	E	90	4
0632-629		8550		.57	145	215	0.67	0.77	E	90	4
0641-412		10900		.86	100	160	0.63	0.41	E	90	4
N 2534	13.80	3517	-20.44	.86					I	90	
1029+544	13.20			.90					I	90	
N 3656	13.40			1.00					I	90	
N 4583	14.00								I	90	
N 5128	7.96	497	-22.60	.77	40	140	0.29	1.04	E	90	5,7
N 5266	12.27	2880	-21.92	.71	140	200	0.70	1.09	E	90	6
I 4320				1.00					E	90	
1352-336				.80					E	90	
N 5363	11.40	1121	-20.50	.56	135	199	0.67	0.75	N	90	1
N 5485	12.40	1985	-20.69	.74					N	90	
1459-724				.70					E	90	
N 5898	12.60	2267	-20.70	1.00					N	90	
N 6251	14.00	6900	-21.70	.83					N	90	
N 6702	13.80	4706	-21.07	.79					N	90	
2105-365				.54					N	90	
N 7432	15.10			.67					I	90	
N 7625	12.80	1622	-20.38	1.00					I	90	
N 1199	12.42	2666	-21.23	.60					N	0	
0557-524				.50					I	0	
1029-459	1352	2760	-20.10	.47	210	260	0.81	1.27	E	0	3
N 3302				.77					E	0	
1040+776				.60					E	0	
N 3665	11.60	2002	-21.27	.86					N	0	
N 4370	14.10	468	-15.76	.53					E	0	
N 5745				.53					E	0	
N 7052	14.00	46.40	-20.84	.56					N	0	
0147-269				.57					E	75	
N 708	14.80	5047	-20.22	.83					I	60	
0219-345				.62					E	40	
N 4374	10.80	952	-21.47	.80	68	290	0.23	0.46	N	60	3
1307-467		1853		.67					I	30	
N 5799		313		0.09					E	40	
N 7070A	13.40	2382		.40	<30	100	<0.3	<0.40	E	35	1

Table I: continued *The information by column, is as follows:*

(1) ... Identification.
(2) ... B magnitude from Nilson (1973) and Lauberts (1982).
(3) ... Radial velocity from Palumbo *et al.* (1983) and from references in 12.
(4) ... Absolute magnitude from B and V_r (H = 50 km sec^{-1} Mpc^{-1}).
(5) ... Axial ratio from Hawarden *et al.* (1981), Nilson (1973) and Lauberts (1982).
(6) ... Rotational velocity.
(7) ... Central velocity dispersion.
(8) ... Ratio of V_{max} and central velocity dispersion.
(9) ... Ratio between V_{max}/σ_0 and the corresponding value for an oblate isotropic rotator.
10)... Extension of the dust lane with respect to the stellar body. E extended; N confined to nuclear region; I intermediate
11)... Angle in degrees of the dust lane to the major axis of the galaxy.
12)... References for kinematical data. For additional references see Ebneter and Balick (1985).
1. Sharples et al 1983; 2. Davies & Illingworth 1986; 3. Bertola *et al.* 1986 4. Möllenhoff & Marenbach 1986; 5. Bertola *et al.* 1985; 6. Varnas *et al.* 1986; 7. Wilkinson *et al.* 1986.

second instance of an elliptical galaxy with an outer luminous ring, the other being AM 2020-5050, recently discovered by Whitmore at al. (1987).

The second part of the Table I lists nine elliptical galaxies with a dust lane along the major axis. They are all in the catalog by Ebneter (1986), who, however, lists 37 objects belonging to this type. The reason for the drastic reduction lies in the fact that an edge–on S0 galaxy can be easily confused with an elliptical galaxy with a dust lane along the major axis. If the disk–to–bulge ratio in an S0 galaxy reaches a value such that the luminous disk is at the limit of detection, then it is very difficult to make a distinction. The size of the disk with respect to the bulge plays also a role in the misclassification of major axis dust lane ellipticals. There are cases in which the presence of a disk is missed if the exposure is too low; and vice–versa inner disks in S0 are cancelled out by a deep exposure. Given such difficulties we tried to select only those cases in which the presence of an even tenuous luminous disk can be ruled out. In connection with the relationship between S0 and major axis dust lane ellipticals it is interesting to remark that the morphology of Anon 1029-459 (Fig.1), where the dust lane along the major axis is strongly warped on both sides, closely resembles that of the S0 NGC 5866 (Sandage 1961) with the only difference that the latter possesses an additional luminous disk. Is the origin of the dust lane in these two galaxies the same?

The last part of Table I lists the elliptical galaxies with a dust lane along an axis intermediate between the major and minor axis. Seven cases were selected out of the 18 given by Ebneter (1986). This configuration can be explained by supposing that we are looking at triaxial galaxies at an angle to the planes where the gas and dust are permitted to settle, contrary to what happens in the cases of major and minor axis dust lanes, where the line of sight lies approximately in these planes. An alternative explanation is that we are looking at gas and dust not yet settled.

It is worth mentioning a couple of elliptical galaxies whose morphology is related to that of dust lane ellipticals, *viz.* NGC 5077 (Bertola *et al.*1986) and NGC 7097 (Caldwell *et al.*1986). These are characterized by a narrow distribution of ionized gas along the minor and major axis respectively. No dust has been detected. Very probably the same mechanism that has produced the dust lane ellipticals has given rise to such peculiar objects.

NGC 1052, which is listed in Table I as a minor axis dust lane elliptical according to images by Sparks *et al.* (1986), is characterized by an HI distribution (van Gorkom *et al.* 1986) much more extended than the dust, but aligned also with the minor axis.

Spindle galaxies with the gaseous ring along the minor axis of the stellar body seem not to be related to minor axis dust lane ellipticals for the following reasons: in several cases the stellar body is clearly not an elliptical galaxy but an S0 with a well defined disk and bulge. In addition there is a tendency for the warps of the gaseous ring (e.g. NGC 4650A) to align with the minor axis of the galaxy in the outer regions, contrary to the behaviour of the minor axis ellipticals. This suggests a different dynamical origin for the two phenomena.

3. THE DISTRIBUTION OF THE INTRINSIC AXIAL RATIOS

If elliptical galaxies are triaxial systems, it is possible to derive the distribution $f(b/a, c/a)$ of the intrinsic axial ratios from that of the projected ones, only if an assumption is made on the length of the middle axis (Benacchio and Galletta 1980; Binney and de Vaucouleurs 1981). Considering the sample of major and minor axis dust lane ellipticals it should be possible to put constraints on f. In fact, since the permitted planes for the dust are those defined by the major and minor axes each with the intermediate axis, the apparent major axis coincides with the true major axis for the minor axis dust lane ellipticals, and the apparent minor axis coincides with the true minor axis for major axis dust lane ellipticals. In this way the apparent distribution is a function of f and of only one projection angle, which is the angle between the intermediate axis of the galaxy and the line of sight.

The distribution of the intrinsic axial ratios of dust lane ellipticals can be constrained further by considering the relative number of minor to major axis dust lane ellipticals. The number of ellipticals with the dust lane along the minor axis listed in Table I is 30, while major axis ones are 9, so that the latter galaxies represent 22% of the total. On the other hand this percentage in the compilation of Ebneter (1986) is 57%, owing to the inclusion among major axis dust lane ellipticals of a large number of galaxies possessing a luminous disk. It should be noted that the survey of Hawarden *et al.* (1981), which has the advantage of being statistically meaningful since it is based on a survey on the Sky Survey plates, lists 6 major axis dust lane ellipticals (excluding doubtful cases) as against 12 minor axis ones. The fact that in our listing, minor axis dust lane ellipticals outnumber major axis ones leads us to belive that there is an intrinsic paucity of the second ones with respect to the first. As has been mentioned already, the potential of a triaxial galaxy which acquires gas (and dust) from outside will force the gas to orbit in the planes defined either by the minor and intermediate axes, or by the major and intermediate axes. The settling toward one of the two planes depends on the orientation of the initial rotation axis of the gas. The shape of the galaxy defines the fraction of the orientation sphere for which the gas will settle onto one of the two planes (Steiman-Cameron 1984). It is therefore clear that the total fraction of objects with dust rings or disks perpendicular to the major axis that we observe as minor axis dust lane ellipticals, is a function of the distribution of the intrinsic flattening, which is again constrained by an observational quantity. In other words the result on the relative number of minor to major axis dust lane ellipticals would indicate a tendency toward prolateness than toward oblateness for our sample of galaxies. A statistical analysis along these lines is in progress by Bertola,Galletta,Vietri and Zeilinger.

4. KINEMATICAL AND PHOTOMETRIC PROPERTIES

There are two different kinematical components in the ellipticals with dust lanes. The first deals with the motions of the stars while the second deals with the gas associated with the dust. As listed in Table I, we possess information on the rotation and velocity dispersion for nine minor axis, one major axis, and two intermediate axis dust lane ellipticals. Data reduction is in progres for NGC 3302, NGC 4370, and NGC 5745 (Bertola *et al.* 1986).

The ionized gas has been detected in all but two minor axis dust lane ellipticals (Anon 0632-629 and Anon 0641-412). As a general result the rotation axis of the gas is perpendicular to that of the stars. This fact has deep implications on the origin of the gas and dust and gives strong support to the idea that they were acquired from outside. A recent and detailed study (Bland 1985) of the ionized gas in NGC 5128 leads to the conclusion that the mass of the gas is insignificant compared with the mass of the stellar component. At this point it should be mentioned that sixteen ellipticals in our sample were detected by the IRAS satellite. Dust mass estimates (Zeilinger 1986) from these data are in the range of $10^5 \div 10^6 M_\odot$, so that the mass of the gas could be at least a couple of order of magnitudes larger. In this connection, an analysis on the possible self–gravitation of the dust rings in ellipticals similar to that for polar rings in S0 (Sparke 1986) is desirable.

Unfortunately no emission lines were detected in the only case of a major axis dust lane elliptical that has been studied (Anon 1029-459). If the gas is acquired from outside one would expect to find cases in which the gas is counter–rotating with respect to stars, as in NGC 7097 (Caldwell *et al.* 1986) and in NGC 4546 (Galletta 1986). From the data in Table I it appears immediately that no slow rotators relative to the isotropic oblate rotator are present among the dust lane ellipticals, although the brightest object reaches $M_B = -22.6$. Whether this fact is due to the small sample at our disposal or is an intrinsic property of dust lane ellipticals, is a point to be checked with additional observations, in order to establish whether the nature of dust lane and normal ellipticals is similar or distinct.

It is worth noting that the rotational velocity of the major axis dust lane elliptical Anon 1029-459 is as high as 210 km sec^{-1}, which is not reached in any other elliptical galaxy. This casts some doubt on whether ellipticals with dust lanes along the major axis, whose numbers have been so drastically reduced in our sample, really exist. This again is a point which needs future observations.

Considerable information on the intrinsic shape of the minor axis dust lane ellipticals can be deduced from kinematical observations. For a minor axis dust lane elliptical the pure oblate configuration is excluded. In fact in this case, if the gas is captured from outside it tends to align with the major axis in the inner regions (Lake and Norman 1983), causing a central discontinuity which is not observed. On the other hand the pure prolate configuration is also excluded in those cases with high values of V_{max}/σ_0 due to the inability of the models to reproduce the case (Richstone and Potter 1982). Therefore it follows that the shape of a minor axis dust lane elliptical is the triaxial one. The rotational velocity which is observed along the major axis of the galaxies is a combination of the streaming velocity of the stars and of the tumbling around an axis in the plane defined by the minor and intermediate axis. In order to give rise to a well defined dust lane, the latter has to see an almost stationary potential. Therefore the conclusion is that the tumbling motion, if any, has to be very small with respect to streaming. Some ideas as to whether we are looking closer to the intermediate axis or to the minor axis, can be derived by the amount of streaming which is higher in the first case than in the second. Then by means of the tensor virial theorem, a lower limit of the axial ratio b/a or c/a can be deduced. If warps are present in the dust lane, an upper limit to these axial ratios can be set by assuming that the warps show the direction of the original infall, on the hypothesis that the warps are transient phenomena (Caldwell 1984; Bertola *et al.* 1985).

Recent photometric studies of dust lane ellipticals concern NGC 5266 (Caldwell 1984, Varnas *et al.* 1986) and about twenty other cases whose reduction is in progress (Bertola *et al.* 1986). All cases are characterized by the fact that the $r^{1/4}$ law is more or less followed. It seems that the photometry does not help so much in discriminating S0's from major axis dust lane ellipticals.

5. THE WARPS AS TRANSIENT OR STATIONARY PHENOMENA

Tubbs (1980) and later Simonson (1982) have produced a model of NGC 5128 where the dust lane represents material captured from the exterior and the warps are comprised of material not yet settled either in the principal plane of a prolate elliptical, or in the plane defined by the minor and intermediate axis of a triaxial one. In this way the warps are transient phenomena and they are destined to disappear and give rise to an elliptical with a straight dust lane, if tumbling is not present. An alternative interpretation has been given by van Albada, Kotanyi and Schwarzschild (1982), who were able to reproduce the phenomenon as a stationary one by assuming a slow tumbling motion around the minor axis of the triaxial figure of the elliptical body. However, the model requires that the tumbling motion has to be retrograde with the respect to the motions of the warps.

Four minor axis dust lane ellipticals with warps have so far studied, *viz.* Anon 0151-498, NGC 5128, and NGC 5266 (where the observed motions along the major axis are prograde), and NGC 5363 (with retrograde motion) (Bertola *et al.* 1985). If tumbling is present, the streaming motions could be either in the same or in opposite way (Freeman 1966, Vietri 1986) so it appears difficult to discriminate between the transient and stationary case. While a satisfactory model has been proposed for NGC 5128 on the first hypothesis, an equally satisfactory alternative model for NGC 5266 assumes counterstreaming with a velocity at 7 kpc twenty times larger than that of tumbling (Varnas et al 1986). As a consequence NGC 5266 would be an almost oblate (10% of triaxiality) galaxy with stable warps. Stability is also suggested by the ordered structure of the dust ring in this galaxy. Of course this does not mean that the gas and dust were not acquired. It is just a matter of how much time has elapsed since the encounter.

There are some other factors that could help in deciding the mechanisms responsible for the warps, as in the case of NGC 5363. In this galaxy the dust is so highly warped in the outer regions that the dust lane runs parallel to the major axis. To explain this in terms of transient warps would require the galaxy to be almost prolate. On the other hand the high rotational velocity suggests rather an almost oblate object. This contradiction is not present in the stationary model.

It is a pleasure to thank the Anglo-Australian Observatory for the unique opportunity which has been given to the author in obtaining the high resolution photographs of Fig.1 through their service photography. In preparing the manuscript I benefited discussions with M.Capaccioli, G. Galletta, M.Vietri and W.Zeilinger. Thanks are due to Peter Usher for reading the manuscript.

REFERENCES

Benacchio, L., and Galletta, G. 1980, *M.N.R.A.S.*, **193**, 885
Bertola, F. 1972 *Proc. 15th Meeting of It. Astr. Soc.*, p. 199.
Bertola,F., Bettoni, D., Danziger, J., and Sadler, E. 1986, *in preparation.*
Bertola, F., and Capaccioli,M. 1975, *Ap. J.*, **200**, 439.
Bertola, F., and Galletta,G. 1978, *Ap. J. (Letters)*, **226**, L115.
Bertola, F., Galletta, G., and Zeilinger, W.W. 1985, *Ap. J. (Letters)*, **292**, L51.
Bertola, F., Galletta, G., and Zeilinger, W.W. 1986, *in preparation.*
Binney, J.J. 1978, *M.N.R.A.S.*, **183**, 501.
Binney, J.J., and de Vaucouleurs, G. 1981 *M.N.R.A.S.*, **194**, 679.
Bland, J. 1985, *PhD Thesis*, University of Sussex.
Caldwell, N. 1984, *Ap. J.*, **278**, 96.
Caldwell, N., Kirshner, R.P., and Richstone D.O. 1986 *Ap. J.*, **305**, 136.
Davies, R.L., and Illingworth, G.D. 1986, *Ap. J.*, **302**, 234.
Ebneter, K. 1986, *private communication.*
Ebneter, K., and Balick,B. 1985, *A. J.*,**90**, 183.
Freeman, K.C. 1966, *M.N.R.A.S.*, **134**, 1.
Galletta, G. 1986, *The Messenger* , in press.
Hawarden, T.G., Elson, R.A.W., Longmore, A.J., Tritton, S.B., and Corwin H.G., Jr 1981, *M.N.R.A.S.*, **196**, 747.
Illingworth,G. 1977, *Ap. J. (Letters)*, **218**, L43.
Kotanyi, C.G., and Ekers, R.D. 1979, *A. and A.*, **73**, L1.
Lake, G., and Norman, C. 1983, *Ap. J.*, **270**, 51.
Lauberts, A. 1982, *ESO/Uppsala Catalogue*, München.
Möllenhoff, C., and Marenbach, G. 1986, *A. and A.*, **154**, 219
Nilson, P. 1973, *Uppsala General Catalogue of Galaxies*, Uppsala.
Palumbo, G.G.C., Tanzella–Nitti, G., and Vettolani, G. 1983 *Catalogue of Radial Velocities of Galaxies*, New York.
Richstone, D.O., and Potter, M.D. 1982, *Nature*, **298**, 728.
Sandage, A.R. 1961, *The Hubble Atlas of Galaxies*, Carnegie Inst. of Washington.
Sharples, L.M., Carter, D., Hawarden, T.G., and Longmore, A.J. 1983, *M.N.R.A.S.*, **202**, 37.
Simonson, G.F. 1982, *PhD Thesis*, Yale University.
Sparke, L.S. 1986, *M.N.R.A.S.*, **219**, 657.
Sparks, W.B., Wall, J.V., Thorne, D.J., Jorden, P.R., van Breda, I.G., Rudd, P.J., and Jorgensen, H.E. 1985, *M.N.R.A.S.*, **217**, 87.
Steiman–Cameron, T.Y. 1984, *PhD Thesis*, Indiana University.
Tubbs, A.D. 1980, *Ap. J.*, **241**, 969.
van Albada, T.S., Kotanyi, C.G., and Schwarzschild, M. 1982, *M.N.R.A.S.*, **198**, 303.

van Gorkom, J.H. 1986, *private communication.*
van Gorkom, J.H., Knapp, G.R., Raimond, E., Faber, S.M., and Gallagher, J.S. 1986, *A. J.*, **91**, 791.
Varnas, S.R., Bertola, F., Galletta, G., Freeman, K., and Carter, D. 1986, *Ap. J.*, in press.
Vietri, M. 1986, *Ap. J.*, **306**, 48.
Whitmore, B., Schweizer, F., and McElroy, D. 1987, *This Meeting*
Wilkinson, A., Sharples, R.M., Fosbury, R.A.E., and Wallace, .P.T. 1986, *M.N.R.A.S.*, **218**, 297.
Zeilinger, W.W. 1986, *private communication.*

DISCUSSION

Nieto: You made a distinction between ring and disk. Is it so physically, or does the distinction come from a different optical depth of the dust?

Bertola: Ring–like dust lanes are so called because of their appearance.

Schechter: Do you know of any systematic search for HI in dust lane ellipticals?

Bertola: No. The only cases I know where HI has been detected are NGC 1052 and NGC 5128.

Ebneter: The morphological classifications of dust lanes in my catalog and in the catalog of Hawarden *et al.* (1981), were based on Sky Survey material. I think it is extremely important for someone to do a deep CCD survey of a large, magnitude—limited sample of E and "S0" galaxies in order to determine the real distribution of major axis vs. minor axis dust lanes in galaxies which lack a *significant* stellar disk.

Whitmore: On the question of whether the material is in a disk or in a ring, I would like to comment that in the cases of S0 galaxies with polar rings, and our one case of an elliptical with an outer ring (AM2020-5050; shown in my poster paper, this volume, p. 413), the extent of the gaseous emission is generally much greater than the optical broad band image of the ring. For example, in AM2020-5050, the ring appears to be about 2″ wide in our V image, and yet $H\alpha$ is observed all the way to the center over a range of about 30″. This suggests that while their may be a gaseous disk, star formation may only occur in a narrow region, thus producing a narrow ring.

Kochhar: Why is it that in some cases the dust lanes are confined to the inner parts of the galaxies, whereas in some other cases, the lanes extend to the outer parts?

Bertola: The data I presented are mainly based on a sample of dust lane ellipticals discovered on the Sky Survey prints. They show extended dust lanes as well as inner ones. Recent CCD surveys tend to reveal mainly inner dust lanes.

THE COLD INTERSTELLAR MEDIUM IN ELLIPTICAL GALAXIES - OBSERVATIONS OF HI AND RADIO CONTINUUM EMISSION

G. R. KNAPP
Department of Astrophysical Sciences
Princeton University
Princeton, NJ 08544, U.S.A.

ABSTRACT. About 10% of nearby elliptical galaxies contain HI, with typical values of M(HI) ~ 5×10^8 $M_\odot$ and M(HI)/L_B ~ 0.03 $M_\odot/L_\odot$. The HI content is unrelated to the stellar content, (unlike the situation in spiral galaxies) suggesting that the HI in early-type galaxies has an external origin and is not produced by mass loss. This conclusion is strengthened by the distribution and kinematics of the HI structures, which lie outside the main optical body of the galaxies, have much larger specific angular momentum than do the stars, and are often highly inclined to the kinematic and distribution axes of the optical bodies.

The HI and stellar kinematics show that the rotation curves of E and SO galaxies are approximately flat from small (a few hundred pc) to large (10-20 kpc) radii, as is the case for spirals. Likewise, large mass-to-light ratios are found for some systems. Comparison with mass models derived from X-ray emission suggests that these may in some cases overestimate the mass.

The presence of HI is shown to enhance the likelihood that an E/SO galaxy has a nuclear radio continuum source, in agreement with models which suggest that the central engine is fuelled by cold gas. Current data suggest that the gas-to-dust ratio for the cold interstellar medium in ellipticals has a value similar to that found in the solar neighborhood.

1. INTRODUCTION

Sensitive data accumulated in the last ten years or so have shown that, after all, elliptical galaxies have an interstellar medium of sorts, observable in almost all of the familiar signatures, i.e. optical line emission from ionized gas, optical line absorption, diffuse X-ray emission, HI emission and absorption, dust absorption at visual wavelengths and diffuse infrared emission from cool dust. This contribution discusses some of the issues raised by these observations which relate to the origin of the cool interstellar medium, i.e. whether its origin is internal or external to the galaxy, how it is

T. de Zeeuw (ed.), Structure and Dynamics of Elliptical Galaxies, 145–154.

affected by the environment of the galaxy, how it is related to the nuclear radio sources seen in some ellipticals, and what dynamical information can be found from the observations.

2. THE GLOBAL HI CONTENT OF ELLIPTICAL GALAXIES

About 10% of nearby ellipticals are currently detected in the HI 21 cm line. The observed HI content ranges from a few $\times$ 10^5 $M_\odot$ for the Local Group dwarfs NGC185 and NGC205 (Johnson and Gottesman 1983) to a few $\times$ 10^9 $M_\odot$ for NGC807 (Dressel, this conference) [H_o = 100 km/sec/Mpc assumed]. Typical values of $M(HI)/L_B$ are ~ 0.03 $M_\odot/L_\odot$, although this quantity shows a very wide variation.

The distribution of $\phi[M(HI)/L_B]$, where $\phi(x)$ is the fraction of galaxies in a sample with a given value of x, can be compared for spiral and elliptical galaxies. For Sc galaxies, for example, ϕ has a well-defined average value and small dispersion (c.f. Figure 1d), reflecting the interrelationships between gas and stars in disk galaxies. The evaluation of ϕ for elliptical galaxies has to take into account the facts that most ellipticals are undetected in the HI line and that the upper limits have similar values to the detections. Methods for estimating ϕ for such data sets are described by Knapp, Turner and Cunniffe (1985), Feigelson and Nelson (1985), and Wardle and Knapp (1986). The resulting distribution, shown in Figure 1a, shows a much wider range of relative HI content than is seen for Scs, suggesting that in Es, the gas and stars are not evolving together. Thus much of the HI seen in Es may have an external origin. The values of $M(HI)/L_B$ in ellipticals range from $\sim 3\times10^{-3}$ to 0.3; the predicted fraction of ellipticals containing HI in this range is $\sim 45\%$, in reasonable agreement with the fraction of ellipticals expected to contain dust (Sadler and Gerhard 1985).

In Figure 1 the distributions of $M(HI)/L_B$ in S0s and S0/as are also given. These distributions show that the global HI properties of galaxies have a steady progression from E$\rightarrow$S0$\rightarrow$S0/a$\rightarrow$spiral. The distribution for S0/as is like that for spirals, with a reasonably well-defined mean value (~ 0.04 $M_\odot/L_\odot$). The distribution for S0s, however, resembles that for ellipticals and thus these two types of galaxies are similar to each other and different from spirals. The two types are, indeed, often difficult to tell apart morphologically (e.g. Kent 1985, and much discussion at this conference).

Despite the skimpy statistics, environmental effects on the occurrence of HI can be observed. For example, Wardle and Knapp (1986) show the distance distribution for their sample of E, S0 and S0/a galaxies for both HI detections and non-detections. The HI detection rate, and the occurrence of S0/a galaxies, is very much lower in the Virgo cluster than elsewhere.

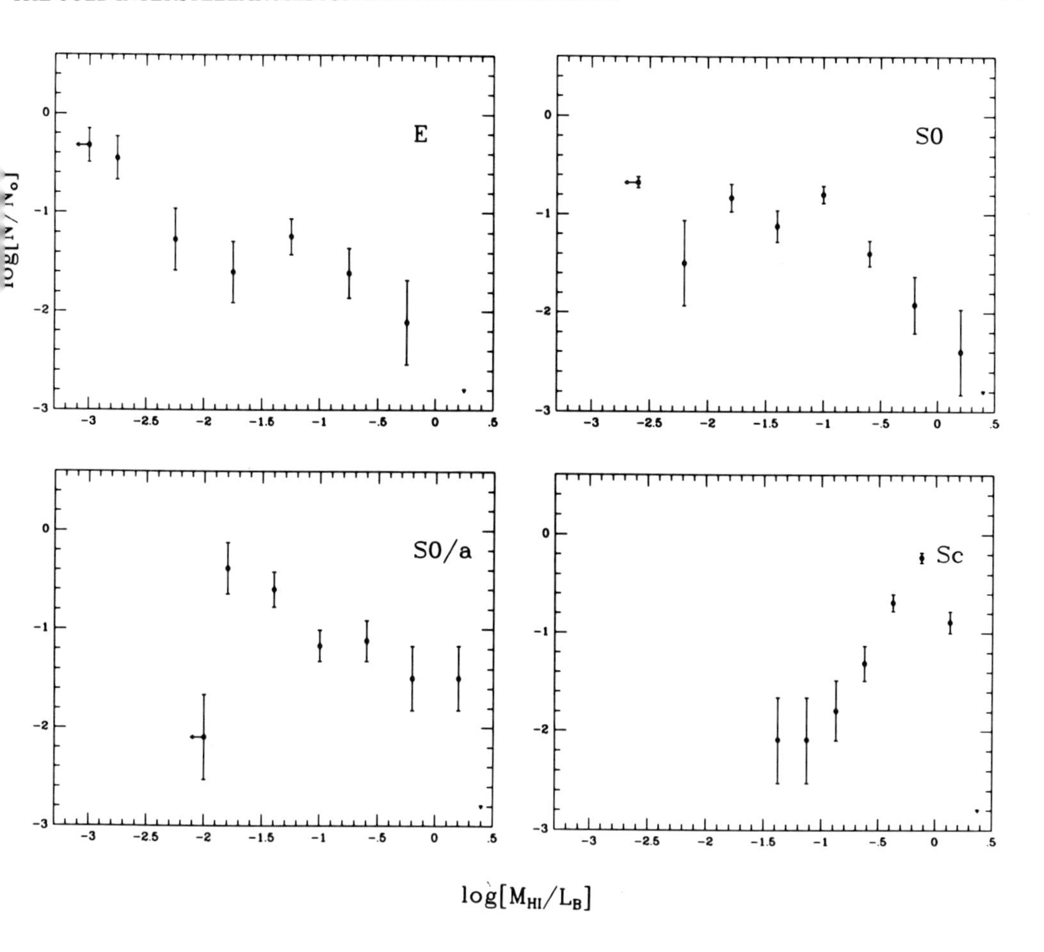

Figure 1. Distribution ϕ [$M(HI)/L_B$], where ϕ is the fraction of galaxies with a given value of $M(HI)/L_B$, for (a) E galaxies (upper left) (b) S0s (upper right) (c) S0/as (lower left) and (d) Scs (lower right). The HI observations for the Scs are by Shostak (1978).

3. HI DISTRIBUTION AND KINEMATICS

Several E and S0 galaxies contain enough HI that its distribution and kinematics can be mapped using synthesis arrays, i.e. the Westerbork array and the VLA. There is no space in this contribution to describe details of the individual galaxies [some of these are discussed by van Gorkom (this conference) and Dressel (this conference)]. The ensemble of results suggests:

(1) The HI emission is often distributed in large rings lying outside the main body of the galaxy. This is seen, for example, in the elliptical NGC4278 (Raimond *et al.* 1981 and in preparation) and in the S0s NGC2655 (Shane and Krumm, in preparation) NGC4203 (van Driel *et al.*, in preparation) and NGC5084 (Gottesman and Hawarden 1986).

There is no accompanying stellar disk in these galaxies or other galaxies of this type (c.f. Hawarden et al. 1981).

(2) The HI features are often highly inclined to the stellar galaxy, both structurally and kinematically. Inclinations of > 60° are seen, for example, for the ellipticals NGC4278 (Raimond et al. 1981) NGC1052 (van Gorkom et al. 1986) NGC5128 (van Gorkom, this conference) and NGC5666 (Lake et al. 1986) and for SOs NGC3998 (Knapp et al. 1985) and NGC2787 (Shostak 1986). Exceptions do exist - NGC5084 and NGC807 (Gottesman and Hawarden 1986; Dressel, this conference) have HI structures along the bulge major axis.

(3) The specific angular momentum for the gas is much greater than that of the stars; the gas is usually at larger radii than are the stars and appears to be rotationally supported.

(4) Some of the gas structures in E and SO galaxies show elongated and warped structures which are reminiscent of tidal tails (e.g. NGC1052, van Gorkom et al. 1986, and NGC1023, Sancisi et al. 1984, where the observations clearly show the interaction between a gassy dwarf and the massive early-type galaxy).

(5) Non-circular motions of the HI are apparent in the inner regions of some Es (e.g. NGC1052, van Gorkom et al. 1986a). In particular, some radio galaxies have HI absorption components which are redshifted with respect to the galaxy's systemic velocity.

These mapping results plus the results on the statistics of the global HI content described in the last section show that most of the HI seen in ellipticals is not produced by stellar mass loss, but was probably acquired from outside the galaxy. There are several ways in which this might happen. For example, the HI structures seen in the outskirts of some E and SO galaxies may be due to continued slow infall of primordial gas in a manner similar to the process thought to build the disks in spirals (Gunn 1982). This may be the case for isolated ellipticals like those discussed by Haynes and Giovanelli (1980). Another possibility is that the HI-rich ellipticals acquired their gas via mergers with gas-rich dwarfs or during a tidal interaction with a neighboring gas-rich spiral. Such processes are strongly suggested at least for some galaxies by the unsettled appearance of gas in the outer regions (e.g. NGC1052, NGC5128, NGC1023). The misalignments between the HI and optical structures also suggest that the gas was acquired fairly recently. Approximately equal numbers of E and SO galaxies have the gas - dust lanes parallel to the major (e.g. NGC5084) or minor (e.g. NGC5128) axes; a similar number have the lanes at large angles to either axis (e.g. NGC4278). These statistics suggest recent capture and settling into one of the two principal planes of the galaxy's potential.

The observation that the HI is often in rings well outside the optical galaxy suggests that in some cases HI may have been ablated from the volume of the galaxy occupied by stars by, for example, the action of stellar winds. In this context, the HI distribution in the large-bulge Sa galaxy NCG4594 is of interest (Bajaja et al. 1984). In this galaxy, the HI is found only in the outer parts of the stellar disk, suggesting that it has been removed from the inner regions of the disk by interaction with winds from the bulge stars. A similar fate may await cold gas accreted to the inner regions of E or SO galaxies.

4. DYNAMICS, MASSES AND MASS-TO-LIGHT RATIOS

The synthesis maps described above show in several cases that the projected circular velocities of the HI structures stay constant over several effective radii, i.e. that E and SO galaxies, like spirals, have "flat rotation curves". Such flat rotation curves are observed for the ellipticals NGC5128 and NGC1052 (van Gorkom, this conference), for NGC4278 (Raimond et al. 1981 and in preparation) and NGC5666 (Lake et al. 1986). Thus, in the outer parts of ellipticals, $\rho \sim r^{-2}$ and the mass-to-light ratio increases with radius. Because the HI is in rings rather than disks in many cases the rotation curve cannot be followed to the inner regions of the galaxy; however, some comparison of the dynamics of the inner and outer regions can be found from an examination of the stellar velocity dispersion and the HI circular velocity. For a spherical galaxy with "dark matter" density $\rho \sim r^{-2}$, the velocity dispersion is constant with radius, as is the circular velocity V_c, and $V_c = \sqrt{2}\,\sigma$, where σ is the one-dimensional velocity dispersion. If the stars in the galaxy have a density distribution $\rho_* \sim r^{-3}$ and the potential is dominated by the dark matter, then σ_*, the one-dimensional stellar velocity dispersion, is also constant with radius and $V_c = \sqrt{3}\,\sigma_*$.

Values of V_c at large radii (5-40 kpc) can be measured from HI maps if the inclination of the HI structure can be found. These estimates can be quite uncertain because of the irregularities in the HI distribution and the low signal-to-noise ratio of the observations. The stellar velocity dispersion in Es and spiral bulges is observed to be roughly constant with radius but often rises at the center of the galaxy (e.g. Kormendy and Illingworth 1982). In Figure 2 V_c(HI) is plotted versus the central velocity dispersion σ_* (from the compilation by Whitmore et al. 1985) for E and SO galaxies for which both measurements are available, plus NGC4594. All of the galaxies lie between the $\sqrt{2}$ and $\sqrt{3}$ lines except NGC3998, whose central velocity dispersion is very large for its luminosity and may indicate the presence of strong anisotropies or the presence of a massive central object. For the galaxy with the lowest value of σ_* and V_c in Figure 2 (NGC7743) no HI map is available; the value of V_c plotted is for the global HI profile width and is therefore a lower limit.

Figure 2 shows that elliptical galaxies have a roughly r^{-2} density dependence, i.e. a flat rotation curve, between approximately a few hundred pc and a few tens of pc. Thus the distribution of matter, both visible and invisible, in ellipticals is controlled by the dark matter distribution in a similar way to the distribution in spirals.

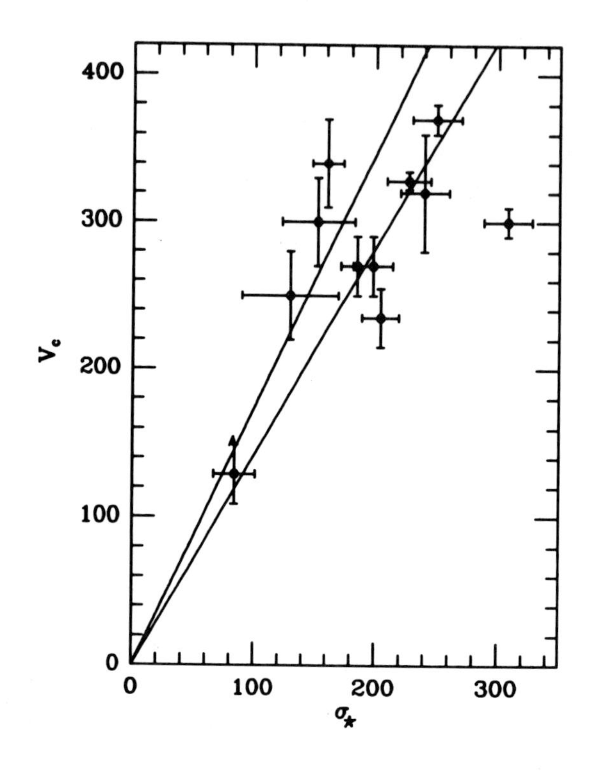

Figure 2. HI circular velocity V_c versus central stellar velocity dispersion σ_* for early-type galaxies. The lines correspond to $V_c = \sqrt{3}\ \sigma_*$ and $V_c = \sqrt{2}\ \sigma_*$.

The masses and mass-to-light ratios can now be estimated for some E and SO galaxies within the observed HI radius and are listed in Table 1. These values are calculated by the present author from values given in the cited literature. The assumed distance and the maximum radius R_m to which HI is detected are also given. The values of M/L_B are in

TABLE 1.

Galaxy	Type	D(Mpc)	R_m(kpc)	$M(M_\odot)$	$\frac{M}{L_B}\left(\frac{M_\odot}{L_\odot}\right)$	References
N4278	E2	8	7.5	$>1.8\times10^{11}$	>44	Raimond *et al.* 1981
N5084	SOp	21	45	$>1.1\times10^{12}$	>82	Gottsman and Hawarden 1986
N2787	SO	10	20	$>3.4\times10^{11}$	>85	Shostak 1986
N1052	E4	13	12	$>1.5\times10^{11}$	>17	van Gorkom *et al.* 1986
N4594	Sa	10	10	$>3.2\times10^{11}$	> 8	Bajaja *et al.* 1984
N5128	Ep	2	3.5	$>5.1\times10^{10}$	> 6	van Gorkom (this conference)

almost all cases very much higher than the value of ~5 expected for the stars in these systems. The above data show that, in a large-scale sense, E/SO and spiral galaxies are very similar to each other, except for the larger masses of the ellipticals.

It is of interest to compare these mass estimates with those derived by other methods. Currently, there is no overlap of early-type galaxies well observed in their shell distribution and in HI, and only two galaxies observed in both HI and X-rays, viz. NGC4594 and NGC5128. The HI results for these two galaxies are compared with the X-ray results of Forman, Jones and Tucker (1985) in Table 2. In making the comparison, the distances assumed by FJT were used. FJT calculate the

TABLE 2. Comparison of masses derived from X-ray and HI measurements for NGC4594 and NGC5128.

	NGC4594	NGC5128
Distance	17 Mpc	5 Mpc
R_m(X-ray)	32 kpc	20 kpc
M_T(X-ray)	1.7×10^{12} $M_\odot$	1.2×10^{12} $M_\odot$
"V_c"(X-ray)	480 km/sec	510 km/sec
V_c(HI)	370 km/sec	240 km/sec
R_m(HI)	18 kpc	9 kpc

mass within the maximum radial extent of the X-ray emission by assuming an isothermal gas sphere at the same temperature (1.2×10^7 K) for all galaxies. This is tantamount to assuming a circular velocity of ~500 km/sec at R_m(X-ray) for all galaxies, as shown in Table 2. This value is much higher than the value observed in HI. To be sure, R_m(HI) $< R_m$ (X-ray) for both galaxies, and there is no direct evidence against the amplitude of the HI rotation curve staying constant to R_m(HI) and then rising steeply somewhere between R_m(HI) and R_m(X-ray). Personally I regard this as unlikely and suspect that for these two galaxies at least the X-ray models overestimate the mass. A possible refinement might be made to the X-ray models by scaling the assumed temperature to the observed stellar velocity dispersion.

5. HI AND RADIO CONTINUUM

An intriguing idea suggested by the discovery of cold gas in some ellipticals is that the activity seen in the nuclei of some Es is fuelled by cold gas. In this scenario, most large Es contain a central engine, active in the early universe but now quiet because of lack of fuel. Such monsters can obtain a temporary new lease on life if the parent galaxy accretes more fuel at some later time (e.g. Gunn 1979). This hypothesis finds observational support in the perpendicularity of radio jets and gas/dust lanes (Kotanyi and Ekers 1979; Laing 1984) and in the presence of HI in the active Es NGC1052, NGC4278 and NGC5128 (in particular in the redshifted HI absorption components seen for some active nuclei - van Gorkom, this conference). The accumulation of large bodies of HI and continuum data for nearby E/SO galaxies allows this question to be adressed statistically (Knapp and Wardle 1986, in preparation). We find (1) the distribution of P_c/L_B (where P_c is the radio continuum power) is similar for E and SO galaxies, but different for spirals, which show the presence of disk continuum emission caused by star formation activity, (2) for high luminosity galaxies ($L_B > 3\times10^9$ $L_\odot$) the observed presence of HI significantly enhances the likelihood that the galaxy is a radio source, (3) low-luminosity E/SO galaxies ($L_B < 3\times10^9$ $L_\odot$) are not radio sources even when they contain cold gas; it is therefore likely that these galaxies do not contain central engines, (4) the radio power is not proportional to the HI content, i.e. the HI is in a reservoir and the accretion rate to the center does not depend on the total amount of HI present, and (5) we see no evidence in the data for disk emission from E and SO galaxies, i.e. of the presence of massive star formation (although the present limits are not very stringent).

6. THE GAS-TO-DUST RATIO

More and more observations are turning up dust lanes and patches in elliptical galaxies (e.g. Sadler and Gerhard 1985, Ebneter and Balick 1985). Gas-to-dust ratios are hard to estimate from such data because of the unknown relative distributions of the dust and stars. A different approach has been taken by Jura (1986) who found that some nearby Es are detected by the IRAS satelite at 100μ. The gas-to-dust ratios inferred from these observations and the HI data may be similar to that found in the solar neighborhood; a similar result is found for SOs (Jura and Knapp, in preparation). Much more work remains to be done on this subject; in particular the non-detections need to be included, but the current results provide another piece of evidence against the hypothesis that the cold gas is primordial, at least in the inner parts of the galaxies.

Finally, it is worth mentioning molecular gas. Little work has been done on this to date, but positive results include the detection of several species (e.g. H_2CO) in absorption against the nucleus of NGC5128 and the detection of CO emission from NGC185 reported by Wiklind and Rydbeck (1986).

ACKNOWLEDGEMENTS

I'm most grateful to Mark Wardle, Jacqueline van Gorkom, Mike Jura, Linda Dressel, Seth Shostak, John Bally and Tommy Wiklind for the use of their results before publication.

REFERENCES

Bajaja, E., van der Burg, G., Faber, S.M., Gallagher, J.S., Knapp, G.R., and Shane, W.W. 1984, A.&A. 141, 309.
Ebneter, K., and Balick, B. 1985, A.J. 90, 183.
Feigelson, E.D., and Nelson, P.I. 1985, Ap.J. 293, 192.
Forman, W., Jones, C., and Tucker, W. 1985, Ap.J. 193, 102.
van Gorkom, J.H., Knapp, G.R., Raimond, E., Faber, S.M., and Gallagher, J.S. 1986, A.J. 91, 71.
Gottesman, S.T., and Hawarden, T.G. 1986, M.N.R.A.S. 219, 759.
Gunn, J.E. 1979, in "Active Galactic Nuclei", ed. C. Hazard and S. Mitton, Cambridge University Press, p. 213.
Gunn, J.E. 1982, in "Astrophysical Cosmology", ed. H.A. Brück, G.V. Coyne and M. Longair, Pont. Acad. Script 48, 233.
Hawarden, T.G., Elson, R.A.W., Longmore, A.J., Tritton, S.B., and Corwin, H.G. 1981, M.N.R.A.S. 196, 747.
Haynes, M.P., and Giovanelli, R. 1980, Ap.J. (Letters) 240, L87.
Johnson, D.W., and Gottesman, S.T. 1983, Ap.J. 275, 549.
Jura, M. 1986, Ap.J. (in press).
Kent, S. 1985, Ap.J. Suppl. 59, 115.
Knapp, G.R., Turner, E.L., and Cuniffe, P.E. 1985, A.J. 90, 454.
Knapp, G.R., van Driel, W., and van Woerden, H. 1985, A.&A. 142, 1.
Kormendy, J. and Illingworth, G. 1982, Ap.J. 252, 460.
Kotanyi, C.G., and Ekers, R.D. 1979, A.&A. 78, L1.
Laing, R.A. 1984, in "Physics of Energy Transportation in Extragalactic Radio Sources", ed. A.H. Bridle and J.A. Eilek, NRAO Pub., p. 119.
Lake, G.B., Schommer, R.A., and van Gorkom, J. 1986, Ap.J. (submitted).
Raimond, E., Faber, S.M., Gallagher, J.S., and Knapp, G.R. 1981, Ap.J. 246, 708.
Sadler, E.M., and Gerhard, O.E. 1985, M.N.R.A.S. 214, 177.
Sancisi, R., van Woerden, H., Davies, R.D., and Hart, L. 1984, M.N.R.A.S. 214, 497.
Shostak, G.S. 1978, A.&A. 68, 321.
Shostak, G.S. 1986, A.&A. (in press).
Wardle, M., and Knapp, G.R. 1986, A.J. 91, 23.
Whitmore, B.C., McElroy, D.B., and Tonry, J.L. 1985, Ap.J. (Suppl.) 59, 1.
Wiklind, T., and Rydbeck, G. 1986, A.&A. 164, L22.

DISCUSSION

Norman: Could you have missed a substantial amount of gas with velocity dispersions of > 500 km/s?

Knapp: Yes, the observations made up to ~ 5 years ago used HI front– and back ends whose effective velocity coverage was ≤ 1600 km/s, so that very wide HI lines such as you have for the high–luminosity ellipticals might well have been missed. Recently, both Arecibo and NRAO have acquired equipment allowing velocity coverage of > 4000 km/s; the effectiveness of this in detecting HI from very massive gas has been demonstrated recently by Giovanelli and collaborators. So far, to my knowledge, no new HI detections have been made for high–luminosity ellipticals.
A related problem which makes the HI data incomplete is the great difficulty in observing ellipticals which are powerful radio emitters because of the poor spectral baselines caused by the enhanced noise temperature of the "on–source" observation.

Kochhar: We (cf. *Supernovae, their Progenitors and Remnants*, eds G. Srinivasan & V. Radhakrishnan, Bangalore, p. 47, 1985) have suggested that the presence of gas and radio activity in ellipticals is also correlated with the occurrence of supernovae in them. For example, in the Arecibo survey, whereas only 25% of observed galaxies are radiosources, 60% of the ellipticals with supernovae are radiosources.

Sarazin: The number of Type I SN in ellipticals is only about 15, so I don't think one could usefully constrain any SN versus HI correlation.

X-RAYS FROM ELLIPTICAL GALAXIES

A.C. Fabian & P.A. Thomas
Institute of Astronomy
Madingley Road
Cambridge CB3 OHA
UK

ABSTRACT. X-ray observations have shown that early-type galaxies contain a hot interstellar medium. This implies that the galaxies have a) a low supernova rate; b) high total gravitational binding masses and c) continuous star formation. Much of the gas in isolated galaxies is probably due to stellar mass-loss. The details of its behaviour are complex.

1. INTRODUCTION

Diffuse X-ray emission from individual elliptical galaxies in the Virgo cluster was discovered with the Einstein Observatory by Forman et al. (1979). Widespread X-ray emission due to intracluster gas was already well-known, but except for peaks on central galaxies such as M87 and NGC 1275, individual galaxies had not been previously detected. A 'plume' of emission from the fast-moving galaxy M86 (Forman et al. 1979), a diffuse component to Cen A (Feigelson et al. 1982) and studies of the X-ray spectrum (Forman, Jones & Tucker 1985; Trinchieri, Fabbiano & Canizares 1985; hereafter FJT and TFC, respectively) combine to make thermal emission (bremsstrahlung, line and recombination radiation) from hot gas at a temperature $\sim 10^7$K the most plausible origin for the X-rays. Low-mass X-ray binaries, such as found in the bulge of our Galaxy and in M31, are only expected to make a significant contribution in low luminosity ellipticals (FJT; Trinchieri & Fabbiano 1985).

Long & Van Speybroeck (1983) and Biermann & Kronberg (1983) showed that a wide range of nearby E and S0 galaxies are X-ray emitters and Nulsen, Stewart & Fabian (1984; hereafter NSF) found X-rays from a relatively isolated elliptical galaxy, NGC 1395. The X-ray emission is an intrinsic property of early-type galaxies and is not directly related to the environment. FJT carried out the first comprehensive X-ray study of a moderately large number (55) early-type galaxies and presented surface brightness profiles, X-ray spectra and mass estimates. These estimates were high enough to confirm that many E and S0 galaxies have dark haloes.

The work of FJT and the statistical analysis by Trinchieri &

T. de Zeeuw (ed.), Structure and Dynamics of Elliptical Galaxies, 155–165.

Fabbiano (1985) showed that the X-ray luminosity, L_X, correlates with the visual luminosity, L_V, so that

$$L_X \propto L_V^a \tag{1}$$

with $1.5 \lesssim a \lesssim 2$. This relation is explained by cooling of stellar mass loss and gravitational heating (NSF; TFC; Sarazin 1986; Canizares, Trinchieri & Fabbiano 1986, hereafter CTF). Further detailed studies have been carried out by TFC, CTF and Stanger & Warwick (1986). All of the data are from the *Einstein Observatory* in the energy band ∿ 0.2 to 4 keV and, apart from a few exceptions (e.g. Mason & Rosen 1985), no useful X-ray observations of individual early-type galaxies have been carried out since it became inoperational. Further X-ray observations must await the launch of ROSAT and of AXAF.

The mass of X-ray emitting gas in a typical E or S0 galaxy is between 10^9 and $10^{10} M_\odot$ and at a temperature of $\sim 10^7$K. This is consistent with accumulated stellar mass-loss heated by collisions and gravity. Given the simplicity of this picture, it is worth re-examining previous widely-held views that early-type galaxies are devoid of gas. Although some do contain significant quantities of HI and patches of star formation, this does not provide a general explanation for the removal of stellar mass loss. The main argument used by Faber & Gallagher (1976) against the retention of hot gas was that the gas is not observed optically (by Hβ emission, for example). We now know that the *central* gas density is much lower than they assumed. They concluded either that galactic winds powered by Type I supernovae sweep out the gas (Mathews & Baker 1971) or that star formation with a preference for low-mass stars operates. Although this last option was considered favourably by Jura (1977) and now appears to be correct, galactic winds were preferred and featured in many later discussions (e.g. MacDonald & Bailey 1981; White & Chevalier 1983; but see Norman & Silk 1979 for another view). In fact, the presence of detectable X-ray emission from E and S0 galaxies means that galactic winds are rare or even non-existent at the present time (see Mathews & Loewenstein 1986 for a discussion of the evolution of galactic winds). The mass-loss rate would otherwise far exceed the integrated stellar mass-loss rate (NSF). Consequently the supernova rate in early-type galaxies must be much lower than previously estimated (NSF; White & Chevalier 1984; Canizares 1986; Sarazin 1986; Thomas 1986). (Preliminary results of a large optical supernova search by Tammann and collaborators (private communication) indicate this is indeed the case and the rate is less than $(500\ \mathrm{yr})^{-1}$). Furthermore, the relatively short cooling time of the hot gas means that star formation is widespread in early-type galaxies.

In summary, the X-ray observations show that the mass of gas and star formation rate of E and S0 galaxies are not very different from those in spiral galaxies. The distribution of the hot gas can, in principle, provide accurate mass profiles and limits on total masses. In the next Section we argue that the total masses are large and that early-type galaxies possess massive dark haloes. Estimates of the rate at which gas is cooling throughout the galaxies and the consequent star formation and optical emission are then discussed. It appears that the gas contains a range of densities throughout and its flow is complicated. The much larger (10 - 1000 $M_{\odot}yr^{-1}$) cooling flows that are observed around many dominant central cluster galaxies are discussed only briefly; more detailed reviews are given by Fabian, Nulsen & Canizares (1984) and Sarazin (1986). Finally, some obvious implications for nucleus activity, jet propagation, cold discs etc. are outlined. It is clear to us that X-ray data are essential ingredients for a complete understanding of early-type galaxies.

2. GRAVITATIONAL MASSES

Large velocity flows approaching sonic speeds imply impossibly large mass rates in X-ray luminous galaxies (NSF) so it is safe to assume that hydrostatic equilibrium is a good approximation.

$$\frac{dP_{gas}}{dr} = - \rho_{gas} \frac{d\phi}{dr} \quad (2)$$

P_{gas}, ρ_{gas} and ϕ are the gas pressure and density and the gravitational potential, respectively. This can be rearranged, for an ideal gas, to give (Fabricant, Lecar & Gorenstein 1980);

$$\frac{d\phi}{dr} = - \left(\frac{kT}{\mu m}\right) \left(\frac{d \ln \rho_{gas}}{dr} + \frac{d \ln T_{gas}}{dr}\right) \quad (3)$$

The gravitational potential, and thus mass profiles can then be measured from observations of the gas density and temperature profiles. As the X-ray emission varies as ρ^2_{gas}, the density is usually well-determined, but the temperature is much less so. Where measured, T appears to increase outward (Forman, Jones & Tucker 1985). Estimates of the gravitational mass profiles of the best-studied galaxies are given by FJT, TFC, Thomas (1986), Canizares (1986) and Sarazin (these proceedings). There are uncertainties but it appears that many E and S0 galaxies have masses within $\sim$ 50 kpc that exceed $10^{12} M_{\odot}$. NGC 4472 (M49) appears to have a massive dark halo with a core radius of 10 - 60 kpc and a total mass profile similar to its neighbour NGC 4486 (M87; Thomas 1986).

A minimum lower limit to the total gravitational mass of a galaxy is obtained by assuming that the observed hot gas, of temperature T and pressure P_o at radius r_o, is confined by a convectively stable outer atmosphere through which the pressure decreases out to P_{∞} at r_{∞} (Fabian et al. 1986). The whole atmosphere sits at rest in the potential well of the galaxy. The total gravitational binding mass is minimised by the atmosphere with the steepest falling temperature gradient and, for convective stability, this follows an adiabat. The minimum total mass, M_T, is given by

$$M_T \geqslant \frac{5}{2} \frac{kT\, r_o}{G\mu m} \frac{[1 - (P_{\infty}/P_o)^{2/5}]}{(1 - r_o/r_{\infty})} \quad (4)$$

(Fabian et al. 1986). When applied to the galaxies well-studied by FJT and TFC, $\langle M \rangle > 5 \times 10^{12} M_{\odot}$ and $\langle M_T/L \rangle > 75$. Elliptical and S0 galaxies must have massive dark haloes in order to confine the observed X-ray gas.

We have assumed that $P_\infty \ll P_\sigma$, which is reasonable anywhere but near the centre of a cluster. The gas temperature is the most uncertain quantity. Measured values are an average and need not apply at r_o. However, the shape of the X-ray surface brightness profiles means that about half the flux originates from within $r_o/4$. A lowest limit is then obtained by dividing our straightforward limits by 4. The mean total mass-to-light ratio is now comparable with that for spiral galaxies, so we can definitely conclude that E and S0 galaxies have dark haloes at least as important as those of spirals. This lowest limit is very conservative and runs into some problems with interpretation due to the low temperature of the outer gas near r_o. The density and mass of gas necessary to produce the observed flux is then increased substantially, as is the cooling rate to many solar masses per year. The straightforward estimate (Eqn.4) is probably more realistic.

Our mass limit was obtained by assuming that all the mass lies within r_o. This leads to a contradiction between M_T and other estimates of $M(< r_o)$ in many cases. This is acceptable since the confining mass will be distributed over a region much larger than r_o. Whilst that has the effect of reducing $M(< r_o)$, it also increases M_T. An isothermal halo ($M(< r) \propto r$) typically requires that M_T is 3 times greater than our limit. Overall, the X-ray emission from E and S0 galaxies strongly suggests that their total masses exceed $10^{13} M_\odot$ and total mass-to-light ratios are some hundreds.

3. DENSITY PROFILES AND MASS DEPOSITION RATES

The radiative cooling time of most hot gas in ellipticals is shorter than a Hubble time. In the absence of a heat source, the gas cools and forms stars, settling inward and forming a cooling flow (NSF; Stanger & Warwick 1986; Thomas 1986; CFT). Even if there is some heating, such as from supernovae, it is not clear that the gas is prevented from being thermally unstable, as the densest gas will continue to cool.

The mass deposition rates of cooled gas in 18 E and S0 galaxies and the bulge of the Sombrero galaxy have been estimated by Thomas et al. (1986). The X-ray images are deprojected by dividing the counts into annular rings about the galaxy centre and then estimating count emissivities in concentric shells. These are related to density and temperature through the emission and detection processes. The optically measured velocity dispersion gives ϕ for the equation of hydrostatic support which further relates particle density, n, and T and enables n(r) and T(r) to be determined separately. Estimates of n(r) are particularly robust (Fig. 1), whereas T(r) depends sensitively on ϕ(r) and the value of pressure at one radius, typically the outer radius. We choose this pressure so that the overall spectrum from the gas is consistent with gas at 1.5×10^7 K.

As a first method (a) for estimating the mass deposition profile assume that no radial flows occur and all gas injected into a shell cools there. The mass deposition rate in that shell is given by

$$\Delta \dot{M}_a = \frac{\Delta L}{H}, \tag{5}$$

where ΔL is the X-ray luminosity of that shell and H is the enthalpy of the gas (5/2 kT/μm). $\dot{M}_a(<r)$ is then obtained as $\Sigma\Delta\dot{M}_a$.

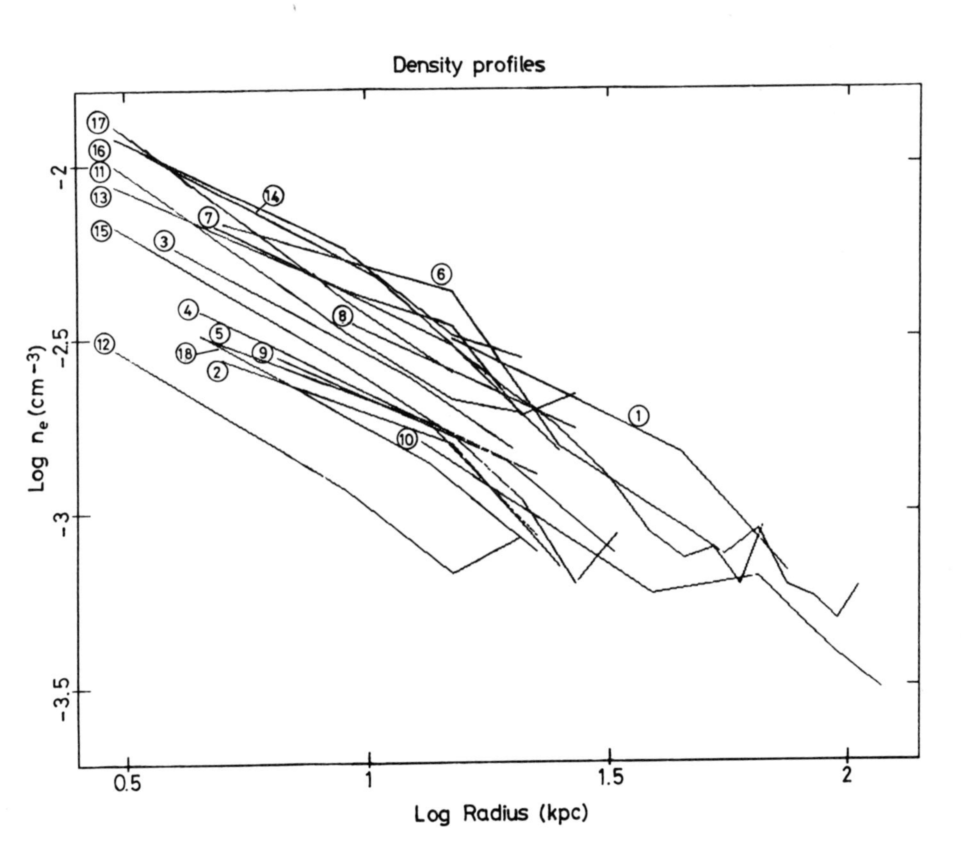

Figure 1. Electron density profiles of the hot gas in a sample of 18 E and S galaxies and the Sombrero galaxy. The individual galaxies are identified in Table 1 of Thomas _et al_. (1986).

A second method (b) allows radial flows with 2 gas phases. One phase flows across a shell whilst the other cools out there. There is no mass injection and

$$\dot{M}_b = \frac{\Delta L - \dot{M}\,(\Delta H + \Delta\phi)}{H} \tag{6}$$

The term $\dot{M}\,(\Delta H + \Delta\phi)$ represents energy losses or gains by the gas flowing across the shell.

Finally, (c), we allow radial flows with 2 phases and mass ($\Delta\dot{M}_+$) and energy (H_+) injection.

$$\Delta\dot{M}_c = \frac{\Delta L + \dot{M}\,(\Delta H + \Delta\phi) - \Delta\dot{M}_+\,(H_\mp - H)}{H} \qquad (7)$$

Both inflow and outflow are possible.

The first 2 methods run into significant problems. Estimating $\dot{M}_+$ at $1.5 \times 10^{-11} M_\odot yr^{-1} L_{\odot B}^{-1}$ (Faber & Gallagher 1976), we find $\dot{M}(< r_{cool}) << \dot{M}_+(r_{cool})$. In particular, we expected that $\dot{M}(r_e) = \dot{M}_+(r_e)$ but the discrepancy is even worse there (Fig. 2). In other words, the rate of mass deposition is much less than the expected rate of mass injection. There are ways of overcoming this problem. The stellar mass loss rate may not mix into the hot phase, so is not detected in X-rays (see White & Chevalier 1983 and Thomas 1986). A study of the evolution of red giant ejecta suggests that this is indeed likely and $\dot{M}$ estimated from X-rays is then a lower limit on the total $\dot{M}$. Alternatively, the rate of stellar mass loss may be overestimated. The mass of the resulting white dwarf is basic to global estimates of the mass loss from a red giant.

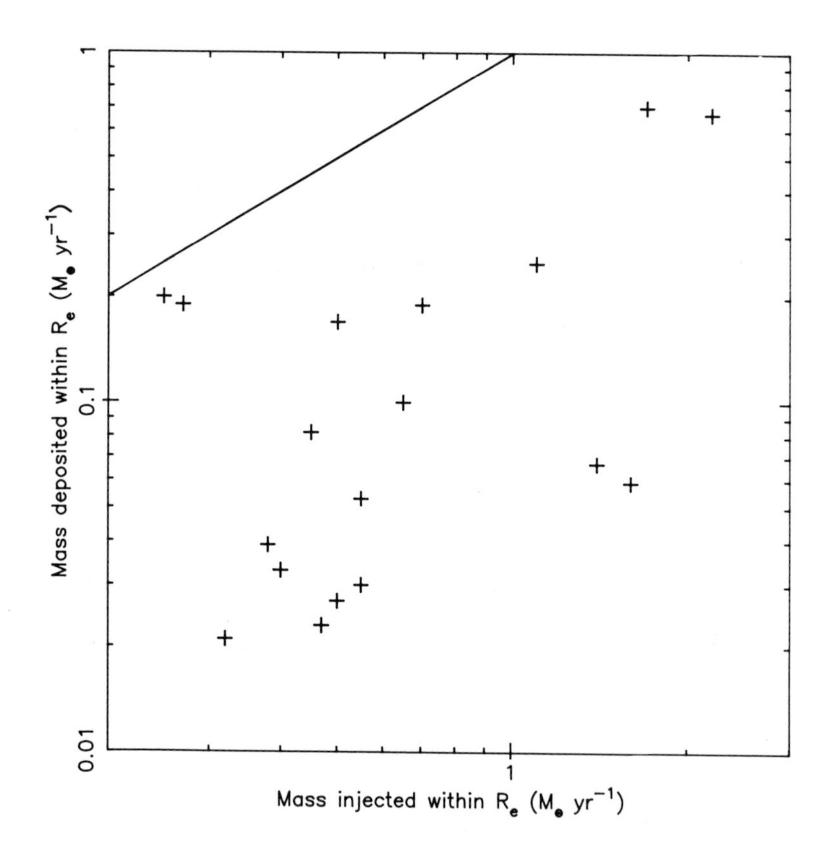

Figure 2. Mass deposited within the effective radius, $\dot{M}_a(r_e)$, in the 18 galaxies shown in Fig. 1, compared with the expected mass injection, $\dot{M}_+(r_e)$. Most of the points fall well below the solid line representing $\dot{M}_a(r_e) = \dot{M}_+(r_e)$.

Could white dwarfs in E and S0 galaxies be more massive ($\gtrsim 0.9\ M_\odot$) than those (0.6 - 0.7 $M_\odot$) than those in our Galaxy? Lastly, significant photoelectric absorption due to cooled gas (column densities $\gtrsim 10^{21} cm^{-2}$) could mean that we underestimate the X-ray luminosity and thus mass deposition rate. Coupled with a lower gas temperature we could perhaps force $\dot{M}_a(< r_e)$ to equal $\dot{M}_+(< r_e)$. Such column densities correspond to HI masses $\sim 10^8 M_\odot$ within r_e and are consistent with X-ray spectra (TFC).

Our last method can accommodate all these problems by allowing the inner gas to flow out a little way whilst the outer gas flows in. This is possible with multiphase gas provided that there is some heat source consistent with a low supernova rate. The system then resembles a complex galactic fountain, as discussed for our galaxy by Bregman (1980) and others. Gas at large galactic radii may have accumulated from earlier times when the galaxy was more active (Mathews & Loewenstein 1986).

In summary, we find that the hot gas in ellipticals is continually cooling at total rates of at least 0.02 to about 3 $M_\odot yr^{-1}$. The higher rates are associated with central galaxies in groups such as NGC 4472 in Virgo and NGC 1399 in Fornax. There is then gas at all temperatures from $\sim 1 - 2 \times 10^7 K$ down to 100 K and below. Much of the stellar mass loss is probably not shocked to the higher temperatures and is not heated much beyond $10^4 K$. Measurements and limits on HI in E and S0 galaxies suggest that most of the cooled gas rapidly and efficiently forms stars.

Much larger mass deposition rates of between 10 - 1000 $M_\odot yr^{-1}$ are inferred in 30 - 50 percent of the clusters studied with the _Einstein Observatory_ (Stewart _et al_. 1984). The cooled gas is deposited such that $\dot{M}(< r) \propto r$ over a range of radii in the central galaxy out to 100 - 200 kpc (see Fabian, Arnaud & Thomas 1986).

4. STAR FORMATION AND OPTICAL LINE EMISSION

The evidence collected so far indicates that gas is continually cooling throughout normal and S0 galaxies. Unless estimates of stellar mass loss are grossly incorrect, the total mass deposition rate from all gas is $\sim 1\ M_\odot yr^{-1}$. The star formation rate then approaches that for spiral galaxies (see e.g. Kennicutt 1983). Some fraction of the cooled gas probably forms stars with a 'normal' initial-mass-function (IMF) but the remainder could become 'dark' matter ('Jupiters' or whatever). As pointed out by Jura (1977), the pressure of any interstellar medium in ellipticals is high and this could depress the Jeans mass and skew the IMF to low masses. There will be little dust in the X-ray emitting phase due to sputtering (see Draine & Salpeter 1979). Cold blobs of gas from the mass loss of individual stars probably contain less than 1 $M_\odot$ and so cannot collapse to form OB stars unless there is considerable coalescence of blobs.

Some fraction of the cooled gas may accumulate as larger clouds in the galaxy core in some ordered form, particularly as a disk where relative velocities can be much less. Dust injected into cooled gas is not sputtered and so the hot phase spread throughout the galaxies is not inconsistent with a small dusty region in the core. Optical emission lines should occur from any region where massive stars form and (much weaker) from the cooling gas. There is, of course growing evidence that emission lines are common (Caldwell 1984; DeMoulin-Ulrich, Butcher & Boksenberg 1984; Phillips et al. 1986; Sadler these Proceedings).

Widespread cooled gas provides a ready explanation for observations of star formation in elliptical galaxies (Bertola et al. 1980; Gunn, Stryker & Tinsley 1982; Rose 1984; Pickles 1985 and these Proceedings; Veron & Veron 1986). The cycling of stellar mass-loss will cause chemical evolution of the stellar population and create colour gradients. The current rarity of supernovae in ellipticals discussed earlier means that the elements synthesised and ejected by giants will build up relative to iron.

Cooling flows are a large internal source of gas in elliptical galaxies. Mergers and other external causes need not always be invoked to explain observational phenomena. This is particularly relevant to the extensive optical line emission commonly observed in the central galaxy of cluster (and group) cooling flows (Kent & Sargent 1979; Ford & Butcher 1979; Heckman 1980; Fabian et al. 1982; Cowie et al. 1983; Hu, Cowie & Wang 1985). Johnstone, Fabian & Nulsen (1986) have found in such galaxies that the Hβ luminosity correlates inversely with the strength of the 4000 A break in the underlying stellar population. This shows that most of the line emission is due to photoionization by the radiation from massive stars and cooling gas. A small fraction (1 - 10 percent) of the mass deposited forms stars with a 'normal' IMF, the hotter members of which produce ionizing radiation and fill in the 4000 Å break. This fraction may be associated with the largest cooling blobs in the flow. Only 0.5 to 10 $M_\odot\,yr^{-1}$ of 'normal' star formation is then occurring in most dominant galaxies (see also Bertola et al. 1986 and O'Connell, these Proceedings), but the rate is 100 $M_\odot\,yr^{-1}$ in the most extreme case, PKS 0745-191. The remainder of the deposited matter presumably forms 'dark matter' (Fabian, Arnaud & Thomas 1986).

5. ACTIVITY, JETS AND SHELLS

A general interstellar medium has obvious consequences for central activity. Only a small fraction of the cooling flow need penetrate to a central mass to power a luminous active nucleus by accretion. The hot medium is pervasive and, if Bondi accretion takes place, the accretion radius adjusts such that $\dot{M} \approx (1 - 10)$ percent of $\dot{M}_{Edd}$ (Begelman 1986; NSF). Radio jets emerging from the nucleus are shaped by the surrounding interstellar medium (Sanders 1983).

A sudden global perturbation of the gas in an E or S0 galaxy, perhaps due to a nuclear outburst or a merger, could cause the star formation rate to increase briefly. Stars formed from matter that was in a highly subsonic flow are 'cold' in the sense of low velocity dispersion and will undergo phase-wrapping (Quinn 1984) and give rise to Malin-Carter (1980) shells. Cooled gas is a ready and widespread internal source of 'cold' stars.

6. THE EVOLUTION OF E AND SO GALAXIES

Substantial amounts of hot gas are an intrinsic part of all elliptical and SO galaxies. Only those galaxies which are kept fully-stripped in rich clusters are likely to be devoid of their own gas and even there the intracluster gas is present. The pressure of the gas is much higher than in spiral galaxies; nT is $\sim 10^5 - 10^6$ $cm^{-3}K$. The presence of hot gas in individual galaxies means that their total gravitational binding masses are large. Much of the gas is cooling, at rates of between 0.02 and 1000 $M_{\odot} yr^{-1}$, depending upon whether the galaxy is a small elliptical or a dominant cluster galaxy. Mass deposition from cooled gas is widely spread throughout the galaxies and 1 to 10 per cent of this gas form stars with a 'normal' IMF. Continuous star formation is thus important when modelling the evolution of early-type galaxies and in particular central cluster galaxies.

REFERENCES

Begelman, M.C., 1986. Nature, **322**, 614.

Bertola, F., Capaccioli, M., Holm, A.V. & Oke, J.B., 1980. Astrophys.J., **237**, L65.

Bertola, F., Gregg, M.D., Gunn, J.E. & Oemler, A., 1986. Astrophys.J., **303**, 624.

Biermann, P. & Kronberg, P.P., 1983. Astrophys.J., **268**, L69.

Bregman, J.N., 1980. Astrophys.J., **236**, 577.

Caldwell, N., 1984. Publs.astr.Soc.Pacif., **96**, 287.

Canizares, C.R., 1986. In 'Proc. IAU Symp. No.117'.

Canizares, C.R., Trinchieri, G. & Fabbiano, G., 1986. Astrophys.J., in press.

Cowie, L.L., Hu, E.M., Jenkins, E.B. & York, D.G., 1983. Astrophys.J., **272**, 29.

DeMoulin-Ulrich, M.-H., Butcher, H.R. & Boksenberg, A., 1984. Astrophys.J., **285**, 527.

Draine, B.T. & Salpeter, E.E., 1979. Astrophys.J., **231**, 77.

Faber, S.M. & Gallagher, J.S., 1976. Astrophys.J., **204**, 365.

Fabian, A.C., Arnaud, K.A. & Thomas, P.A., 1986. In 'Proc. IAU Symp. No.117'.

Fabian, A.C., Atherton, P.D., Taylor, K. & Nulsen, P.E.J., 1982. Mon.Not.R.astr.Soc., **201**, 17p.

Fabian, A.C., Nulsen, P.E.J. & Canizares, C.R., 1984. Nature, **310**, 733.

Fabian, A.C., Thomas, P.A., Fall, S.M. & White, R.E., 1986. Mon.Not.R.astr.Soc., **221**, 1049.

Fabricant, D., Lecar, M. & Gorenstein, P., 1980. Astrophys.J., **241**, 552.

Feigelson, E.D., Schreier, E.J., Delvaille, J.P., Giacconi, R., Grindlay, J.E. & Lightman, A.P., 1981. Astrophys.J., **251**, 31.

Ford, A.C. & Butcher, H., 1979. Astrophys.J.Suppl., **41**, 147.

Forman, W., Schwarz, J., Jones, C., Liller, W. & Fabian, A.C., 1979. Astrophys.J., **234**, L27.

Forman, W., Jones, C. & Tucker, 1985. Astrophys.J., **293**, 102.
Gunn, J.R., Stryker, L.L. & Tinsley, B.M., 1981. Astrophys.J., **249**, 48.
Heckman, T.M., 1980. Astrophys.J., **250**, L59.
Hu, E.M., Cowie, L.L. & Wang, Z., 1985. Astrophys.J.Suppl., **59**, 447.
Johnstone, R.M. & Fabian, A.C. & Nulsen, P.E.J., 1986. Mon.Not.R. astr.Soc., in press.
Jura, M., 1977. Astrophys.J., **212**, 634.
Kennicutt, R.C., 1983. Astrophys.J., **272**, 54.
Kent, S.M. & Sargent, S.M., 1979. Astrophys.J., **230**, 667.
Long, K.S. & Van Speybroeck, L., 1983. In 'Accretion Driven Stellar X-ray Sources', eds. Lewin & van den Heuvel, C.U.P.
MacDonald, J. & Bailey, M.E., 1981. Mon.Not.R.astr.Soc., **197**, 995.
Malin, D.F. & Carter, D., 1980. Nature, **285**, 643.
Mason, K.O. & Rosen, S.R., 1985. Space Sci.Rev., **40**, 675.
Mathews, W.G. & Baker, J.C., 1971. Astrophys.J., **170**, 241.
Mathews, W.G. & Loewenstein, M., 1986. Astrophys.J., **306**, L7.
Nulsen, P.E.J., Stewart, G.C. & Fabian, A.C., 1984. Mon.Not.R.astr.Soc., **208**, 185.
Norman, C. & Silk, J., 1979. Astrophys.J., **233**, L1.
Pickles, A.J., 1985. Astrophys.J., **296**, 340.
Phillips, M.M., Jenkins, C.R., Dopita, M.A., Sadler, E.M. & Binette, L., 1986. Astrophys.J., in press.
Quinn, P.J., 1984. Astrophys.J., **279**, 596.
Rose, J.A., 1985. Astrophys.J., **90**, 787.
Sanders, R.H., 1983. Astrophys.J., **266**, 73.
Sarazin, C.L., 1986. In 'Proc. IAU No.117'.
Sarazin, C.L., 1986. Rev.Mod.Phys., **58**, 1.
Stanger, V.J. Warwick, R.S., 1986. Mon.Not.R.astr.Soc., **220**, 363.
Stewart, G.C., Fabian, A.C., Jones, C. & Forman, W., 1984. Astrophys.J., **285**, 1.
Thomas, P.A., 1986. Mon.Not.R.astr.Soc., **220**, 949.
Thomas, P.A., Fabian, A.C., Arnaud, K.A., Forman, W. & Jones, C., 1986. Mon.Not.R.astr.Soc., in press.
Trinchieri, G., Fabbiano, G. & Canizares, C.R., 1986. Preprint.
Trinchieri, G. & Fabbiano, G., 1985. Astrophys.J., **296**, 447.
Veron, P. & Veron-Cetty, M.-P., 1986. Astrophys.J., **145**, 433.
White, R.E. & Chevalier, R.A., 1983. Astrophys.J., **275**, 69.
White, R.E. & Chevalier, R.A., 1984. Astrophys.J., **280**, 561.

DISCUSSION

Ebneter: You listed a number of active galaxies which contain cooling flows. All of these galaxies are dusty ellipticals (and an additional one is NGC 708, by the way). Do you think that the dust is being formed in the cooling flow?

Fabian: No. The dust is perhaps associated with stellar mass loss that was never heated to X-ray temperatures.

Ebneter: If the dust and gas *are* coming from the evolved stars in the galaxy, how does it end up in a nice, rotating disk of gas? All of the galaxies listed have very nice, regular dust lanes and rotating gas disks.

Fabian: I don't know, but the mass of the gas in disks is much smaller than the total stellar mass loss in those galaxies. Consequently, the small gas fraction that cools and has the angular momentum of stable orbits may remain there.

Sadler: Frogel and Whitford have recently used strong–lined M giants in the Galactic bulge to synthesize the IR stellar population in ellipticals. They derive a mass injection rate about a factor of ten *lower* than the Faber and Gallagher value.

Toomre: Did I hear you right in still entertaining the notion that the Malin/Quinn–like shells could possibly be features that somehow form spontaneously (or even otherwise!) in the *very hot* gaseous X–ray halos which certainly exist in, or surround, many of the ellipticals? Whatever gas–dynamical process do you feel could possibly give rise to such sharp–edged features in those hot gaseous spheres?

Fabian: Distributed star formation occurs in ellipticals from the hot gas. These stars are cold and could 'phase–wrap' as in Quinn's—and your—models.

King: It would be interesting to have comparable X–ray observations in the Coma cluster, where there would be a real confrontation. In Coma, dynamical–friction arguments set an *upper* limit to M/L that is considerably lower than the *lower* limit that you find.

Fabian: Presumably most of the mass that we infer to be present in elliptical galaxies is at large radii (hundreds of kpc). In a cluster such as Coma, these halves will overlap and I assume that they merge to form the general cluster potential.

Lauer: Looking at your list of lower limits to galaxy masses I noticed that isolated faint galaxies had the highest limits, while several bright ellipticals including those in Virgo had much more modest lower limits. Have you looked at M/L as $f(L)$ or as function of isolation?

Fabian: No, but I will do. I think that your comment strengthens our conclusion that the mass is associated with the ellipticals.

Binney: I think Cowie and his collaborators looked for a velocity gradient in the $H\alpha$ filament around NGC 1275 and found nothing significant. I recall reassuring myself that if the rotation parameter of the entire cluster has the canonical value $\lambda = 0.07$ (Efstathiou and Jones, 1979, *Mon. Not. R. astr. Soc.*, **186**, 133), and the gas in Perseus has lost from the cluster center a mass equal to the conventional $(M \leq 10^{12} M_\odot)$ mass of of NGC 1275, then the present rate of rotation of the cluster gas could not have been detected by Cowie *et al.*

Fabian fielding questions. Canizares looks on.

The audience listens attentively.

STAR FORMATION IN COOLING FLOWS

R. W. O'Connell
Astronomy Department, University of Virginia
P.O. Box 3818, Charlottesville, VA 22903
USA

ABSTRACT. Star formation, probably with an abnormal initial mass function, represents the most plausible sink for the large amounts of material being accreted by cD galaxies from cooling flows. There are three prominent cases (NGC 1275, PKS 0745-191, and Abell 1795) where cooling flows have apparently induced unusual stellar populations. Recent studies show that about 50% of other accreting cD's have significant ultraviolet excesses. It therefore appears that detectable accretion populations are frequently associated with cooling flows. The questions of the form of the IMF, the fraction of the flow forming stars, and the lifetime of the flow remain open.

1. INTRODUCTION: THE FATE OF GAS IN CLUSTER COOLING FLOWS

This is less a review than a prospectus for a process which, if it occurs, may play a fundamental role in the evolution of cD galaxies and perhaps many other ellipticals. The process is the conversion of material in cooling flows accreted by galaxies into stars--"accretion populations". The difficulty confronting this idea is that observational evidence has been scanty and contradictory. Recent studies, however, lend more credence to it and indicate that accretion populations may be a widespread phenomenon.

The X-ray evidence for cooling flows in clusters of galaxies has been thoroughly reviewed recently by Fabian, Nulsen, & Canizares (1984) and Sarazin (1986a), and I will not recover that ground. Estimated accretion rates, $\dot{M}$, range from $\sim 10\ M_{\odot}\ yr^{-1}$ for modest nearby cases such as M87, where sensitivity is good, to $500-1000\ M_{\odot}\ yr^{-1}$ for extreme cases like Abell 1126 and PKS 0745-191. The age of the flows cannot be accurately estimated, but since the cooling time in their centers is less than the Hubble time it is possible that they have continued for $t_{CF} \sim 10$ Gyr. At a typical accretion rate of $\dot{M} \sim 100\ M_{\odot}\ yr^{-1}$, $\sim 10^{12}\ M_{\odot}$ would then have been deposited, which is comparable to the stellar mass of a typical cD galaxy. (To avoid any dispute over nomenclature, I will use "cD" in this paper to designate the "centrally dominant" galaxy which is observed to be located in the center of cluster cooling

T. de Zeeuw (ed.), Structure and Dynamics of Elliptical Galaxies, 167–177.

flows.) Accretion from cooling flows could clearly have dramatic consequences for the evolution of cD galaxies.

There are not many viable repositories for this massive inflow (cf. Fabian et al. 1982, Sarazin & O'Connell 1983, White & Sarazin 1986). Accretion onto a compact massive object would generate extraordinary nonthermal activity at levels of 100 $M_{\odot}$ yr^{-1} and would also lead to excessive central velocity dispersions, which are not observed (e.g. Malumuth & Kirshner 1985). Limits on cool interstellar material in nearer accretors are orders of magnitude below the values implied by the accretion rates (e.g. Burns et al. 1981, Valentijn & Giovanelli 1982). This appears to leave star formation as the only plausible sink for accreted material.

Star formation with the initial mass function (IMF) prevailing in our Galaxy would be easily detectable at typical accretion rates, as shown by the following simple estimate. The fraction of the V light contributed by an accretion population is given by $f_V = (\dot{M}t_{CF})/[(M/L_V)_{AP} L_{tot}]$, where $(M/L_V)_{AP}$ is the visual mass-to-light ratio of the accretion population and L_{tot} is the V luminosity of the underlying galaxy (all values will be expressed in solar units). Typical accreting cD's have integrated $M_V = -23.5$ ($H_o = 50$), implying $L_{tot} = 2.1 \times 10^{11}$. Larson & Tinsley (1978) give convenient estimates for the mass-to-light ratios of populations with continuous, uniform star formation. (They employ an IMF, truncated at a lower limit of 0.1 $M_{\odot}$, which is slightly different from the best current estimate [Scalo 1986] for normal Galactic star formation but is certainly close enough for our purposes.) For a 10 Gyr-old population, they find $M/L_V = 2.0$. Such a population will have a broad-band (U-V) color -1.2 mag bluer than a normal gE galaxy and a (λ2500 - V) color -3.8 mag bluer than normal. For an accretion rate of 1 $M_{\odot}$ yr^{-1}, the accretion population would produce 2.4% of the integrated V light and UV excesses which would be only marginally detectable with current instrumentation. However, a rate of only 5 $M_{\odot}$ yr^{-1} would yield 12% of the V light and unambiguous excesses of -0.25 in (U-V) and -1.72 in (λ2500 - V). (These are integrated average values; localized measures--e.g. of the nucleus--could, of course, yield smaller values if little material were deposited there.) It turns out that because $(M/L_V)_{AP}$ declines quickly with t_{CF}, and colors become rapidly bluer, these excesses are only weakly dependent on the assumed t_{CF}, changing less than a factor of 2 for lifetimes between 10 Gyr and 0.1 Gyr.

Given the fact (see Sec. 2) that color anomalies in most accretors are small, the upper limit to the amount of mass being converted into stars with a normal IMF in typical accreting cD's is of order 5 $M_{\odot}$ yr^{-1}, independent of the lifetime of the cooling flow. This is only a small fraction of the typical accretion rate and a minuscule 1% of the more extreme rates. It is evident that if accretion populations are to be repositories for typical cooling flows, we require $(M/L_V)_{AP} \gg 2$. This has led to the proposal that star formation in cooling flows is preferentially weighted toward low mass stars and hence large mass-to-light ratios (Cowie & Binney 1977, Fabian et al. 1982, Sarazin & O'Connell 1983; see also Jura 1977). There are several theoretical reasons for expecting shifts of the IMF to lower masses, including the lower density of dust in the flows, higher shear, and, most importantly, the lower Jeans mass which is a consequence of the high gas pressure in the flows. None of these arguments are rigorous, however.

Sarazin & O'Connell (1983, hereafter SO) calculated colors for 10 Gyr-old accretion populations with power-law IMF's characterized by the index x (in the notation of Larson & Tinsley 1978), an upper mass limit M_U and a lower mass limit M_L. With $M_L = 0.1$, models with $M_U = 0.75$ or 1.0 yield $(M/L_V)_{AP}$ in the range 3-50 for $x \sim 0$-2, high enough to accommodate typical flows without large disturbances to optical and infrared colors. An interesting possibility to explore in the future is a bimodal IMF (Larson 1986) with a suppressed high mass mode.

Cooling flows, at much smaller rates, probably also exist in many normal elliptical galaxies not associated with cluster cores (Nulsen et al. 1984; Forman et al. 1985; Sarazin 1986b; Fabian, this conference). Arguments concerning the fate of material in these flows are similar to those above. Here, however, because many of the putative accretors are bright, nearby objects, more stringent limits on star formation with a normal IMF may be placed, perhaps as small as $SRF/L_V = 10^{-12}\ M_\odot\ yr^{-1}\ L_\odot^{-1}$ (O'Connell 1986). I will not discuss these cases further here.

Of course, a fundamental alternative to the accretion population interpretation is that the $\dot{M}$'s in cooling flows have been badly overestimated. Rates are estimated from quantities readily determined from the X-ray data ($\dot{M} \sim L_x/T$ in the absence of heating, where the values refer to the region inside the cooling radius). There is much independent X-ray evidence that cooling flows exist (cf. Sarazin 1986a), and further very strong support comes from the correlation between optical line emission and cooling times for cluster gas (Hu et al. 1985). However, there is ongoing debate over the importance of various heating mechanisms such as conduction (Bertschinger & Meiksin 1986), supernovae explosions (Silk et al. 1986), and galaxy motions (Miller 1986), among others. I leave this controversy to others, except to remark that plausible reductions in $\dot{M}$ from such processes may not be very large (e.g. supernova heating by a normal IMF reduces $\dot{M}$ by only a factor of 2, Silk et al. 1986).

2. OBSERVATIONAL EVIDENCE FOR ACCRETION POPULATIONS

Analyses of three reasonably solid detections of accretion populations have been published. The best studied, and perhaps most remarkable, case is that of NGC 1275 in the Perseus cluster (Abell 426) for which $\dot{M} \sim 300\ M_\odot\ yr^{-1}$. SO suggested that the unusual A-star absorption line spectrum and optical colors of NGC 1275 were consistent with an accretion population with $M_U \sim 3\ M_\odot$ and $x \sim 1.5$ if M_L were lowered to $0.001\ M_\odot$. Later (V-K) measures are consistent with this model (Romanishin 1986). Independent evidence for an absence of high mass stars comes from vacuum ultraviolet (Fabian, Nulsen & Arnaud 1984) and infrared (Gear et al. 1985) observations. Population synthesis of an optical spectrum by Wirth et al. (1983) finds no evidence for early B or O stars, and the contribution of the earliest type present (B5, 5 $M_\odot$) is probably overestimated because of contamination by the nuclear nonthermal source. While these studies all agree on the absence of massive stars in NGC 1275, this could be produced by a decrease in the star formation rate over the last few 10^7 yrs rather than a truncation of the IMF. Whatever M_U, however, the bulk of the 300 $M_\odot$ yr^{-1} being accreted must be going into objects of much lower mean mass than for a normal IMF.

Another spectacular accretor is PKS 0745-191 (Fabian et al. 1985), where the estimated accretion rate is ~1000 $M_\odot$ yr^{-1}. In this instance, optical data are not very good yet, but the spectrum shows strong emission lines and a very blue continuum, with evidence of a color gradient. Formation of stars well above 1 $M_\odot$ is evidently vigorous.

The third case is less dramatic than the two preceding. IUE and optical band spectrophotometry of NGC 6166, the accreting cD in Abell 2199, shows evidence of a small (~4% of the V light) young population (Bertola et al. 1986). In this case, the IUE spectrum clearly indicates the presence of O stars. In fact, a far UV upturn suggests the IMF is more abundant in O stars than is normal, though this could be caused instead by a recent surge in the star formation rate. As expected from the figures quoted in Sec. 1, Bertola et al. derive a total star formation rate for NGC 6166 of only a few $M_\odot$ yr^{-1} if the IMF is normal, which is to be compared to an estimated accretion rate of 110 $M_\odot$ yr^{-1} (Sarazin 1986a). In a case like NGC 6166, where only modest spectral anomalies are present, it would be important to distinguish the effects of a putative accretion population from those of the old, hot star population apparently responsible for the far-UV light in normal E's. It is not obvious that this can be done unambiguously yet (cf. Burstein et al. 1986), but Bertola et al. (1986) argue that anomalies at optical wavelengths support their young star interpretation.

As for the other known accreting cD's, published studies indicate they are outwardly normal in the optical region. This is actually a rather soft statement, since they are typically faint objects which have not received careful study. Various investigators have obtained broad band colors or spectra for nuclear velocity dispersion studies but not necessarily with the precision necessary to detect small departures in energy distribution or in the 3400-4500 A region where the largest effects might be expected. Romanishin (1986) made (V-K) observations to search for accretion population effects in 8 cases and, apart from NGC 1275, found no peculiarities. However, for M_U slightly below 1 $M_\odot$, the SO models have the interesting property that accretion populations hardly disturb the infrared energy distribution, so this is not necessarily a serious difficulty. Romanishin and Hu et al. (1985) found no evidence to support Valentijn's (1983) claim of unusual (B-V) color gradients in accretors, which apparently resulted from a slight miscalibration in his surface photometry.

Romanishin also compiled previously published U-V and g-r photometry on accretors, mostly taken with apertures over 20" in diameter. Known accretors again did not separate from non-accretors. However, there are puzzling inconsistencies here. Two of the objects with normal claimed g-r colors are NGC 6166 and Abell 1795. While the anomalies in the former (as described above) are confined to shorter wavelengths and might not be reflected in g-r colors, A1795 has manifestly unusual colors (see below). A number of other accretors have UV excesses which have not been previously reported (see below). It therefore appears that earlier studies may not constitute a suitable database for investigation of accretion population effects.

3. NEW SPECTROSCOPIC STUDIES

Two recent studies designed especially to search for such effects in the 3400-5000 A spectra of cD's with cooling flows indicate that accretion populations are frequently present. First, from IPCS measurements with the AAT of the 4000 A break and Hβ emission line strengths in 9 accretors, Johnstone et al. (1986) argue that a small fraction (0.1% to 10%) of the cooling flow forms into stars with a normal IMF which then help to maintain the ionization of the extended emission line filaments. They suggest that the bulk of the inflow is deposited into faint, low mass objects.

The other study (O'Connell et al. 1986, hereafter OSWM) is based on KPNO 4-m IIDS observations with a 5" diameter entrance aperture. Flux-calibrated spectrophotometry with 12 A resolution was obtained of 9 accretors and of a control sample of other cD's known not to have large cooling flows, which serve as templates of normal spectra. Template objects were chosen such that color differences with the accretor sample caused by luminosity or aperture effects should be minimized. The spectra were corrected for foreground reddening and registered by a cross-correlation technique. Ratio spectra were generated by dividing the accretor spectra by the template. Examples are shown in Figures 1 and 2, where the template is the sum of the two non-accreting cD galaxies MKW 1 and MKW 2. A color excess, δUV, between the galaxies and the template is based on the magnitude difference in 100 A bands centered at 3600 A and 4500 A. It is insensitive to emission lines, and excesses larger than about 0.05 mag are significant.

Five of the nine accretors in the sample exhibit significant δUV's: M87, A1795, A1991, A2052, and NGC 6166. All but A1795 have δUV ~ -0.2 to -0.3; the excess for NGC 6166 is apparently consistent with the data of Bertola et al. (1986). A1795, with δUV = -0.54 and an AF-type absorption spectrum, is a clearcut case of an accretion population (see Figure 1). A spectrum centered 10" (17.5 kpc) south of the nucleus is similar in appearance, with δUV = -0.33. The color anomaly is therefore unrelated to nuclear nonthermal activity. Johnstone et al. (1986) confirm this result. The unusually blue (B-V) of A1795 is also apparent in the spectra of Hu et al. (1985), though they do not comment on it.

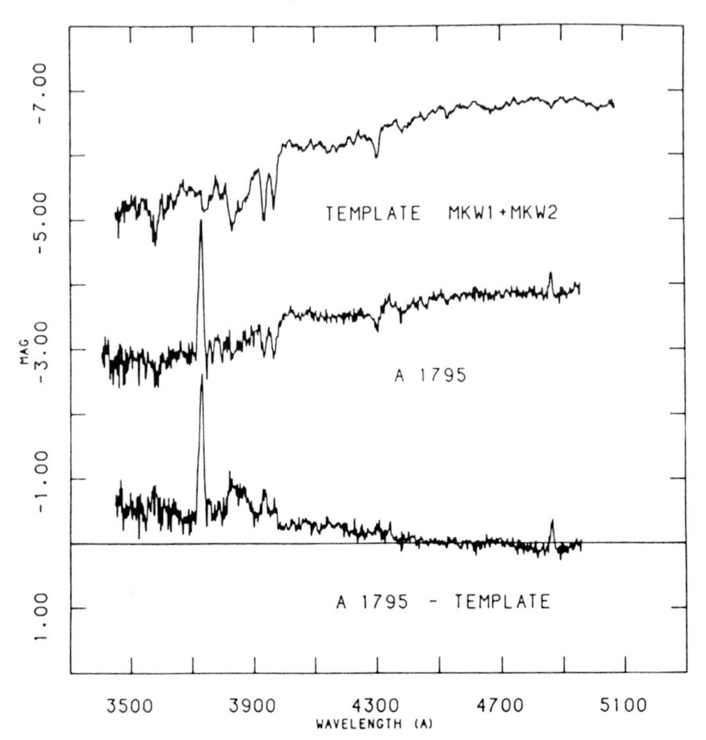

Figure 1: Middle panel: An unsmoothed spectrum (2 A per channel) of the accreting cD galaxy Abell 1795 in the restframe with arbitrary normalization, expressed in magnitudes. Upper panel: The mean spectrum of the non-accreting cD's MKW 1 and MKW 2, as for A1795. Lower panel: The ratio of the A1795 spectrum to the upper panel template, expressed in magnitudes and normalized at 4500 A. The large UV excess and weak stellar absorption features in the ratio spectrum are evidence of an accretion population. Strong emission lines of [O II] and hydrogen are also present.

The absence of significant color gradients in their data is further evidence of the unimportance of nuclear optical non-thermal radiation.

M87 is also worth comment. In this case, the template is the sum of the nuclear spectra (5" aperture) of the two Virgo E's NGC 4374 and 4552. As mentioned by de Vaucouleurs earlier in this conference, the nucleus of M87 exhibits a definite ultraviolet excess with respect to the template (δUV = -0.29) and strong, broadened emission lines. The stronger absorption lines show evidence of filling by a weak-lined continuum, which could be either a hot accretion population or non-thermal radiation. The UV excess is also present, though smaller (-0.10 to -0.20), at positions 10" and 20" north of the nucleus. This seems larger than would be produced by the normal decrease of metallicity with radius (cf. the U-R broad-band photometry of Davis et al. 1985), but without further analysis the issue cannot be decided.

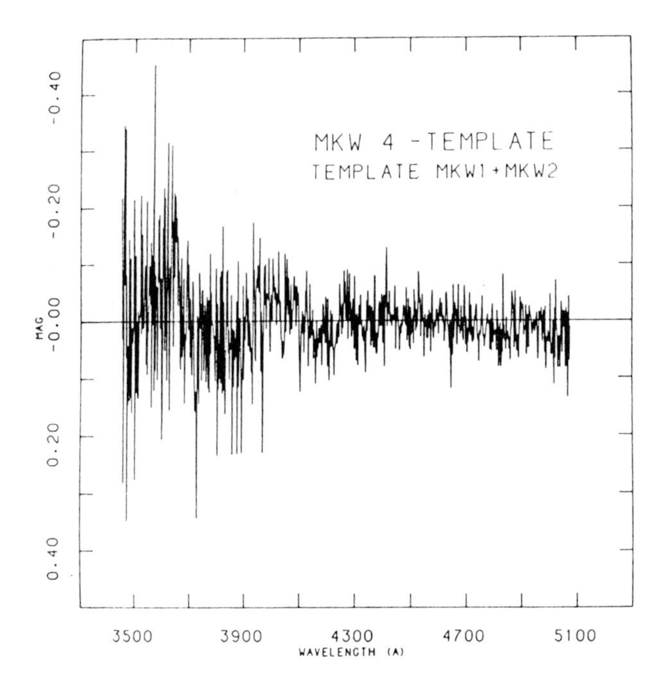

FIGURE 2: The ratio of the spectrum of the accreting cD MKW 4 to the MKW 1 + MKW 2 non-accretor template, as in Fig. 1 but with an enlarged vertical scale. Unlike A1795, there is no significant evidence for emission lines or unusual stellar populations in MKW 4.

Population synthesis of the objects in the OSWM sample is under way in an attempt to separate hot star effects from those of non-thermal radiation, internal dust, or metallicity. Our preliminary analysis suggests that accretion population models of the type described by SO, with $M_L = 0.1\ M_\odot$ and $M_U \sim 1\text{-}2\ M_\odot$ for $x \sim 1\text{-}2$, are compatible with the observed nuclear spectra for all the UV-excess objects except A1795. A fully consistent "conservative" accretion population interpretation such as this would also demand, however, that similar δUV values be observed at larger radii. (Note that absorption lines would be weakened by the fact that accreting gas appears to have sub-solar abundances [Sarazin 1986a] as well as by the contribution of stars hotter than normal for a gE.) Owing to the uncertain precision of the published photometry (see Sec. 2), it is unclear whether this is the case. Should further work not confirm an extended area of UV-excess, then the estimated M's would require that $M_L \ll 0.1\ M_\odot$ and/or $x > 2$. The data for A1795 already require such values (as in the case of NGC 1275, see Sec 2).

The other four accretors in the OSWM sample (AWM 4, MKW 3s, MKW 4, and A2029) do not exhibit significant UV excesses or other spectral anomalies (see Figure 2), apart from weak line emission in MKW 3s. The size of δUV in the OSWM sample is not correlated with the estimated global $\dot{M}$ nor with $\dot{M}/M_{stars}$ (derived assuming that the normal stellar population has constant M/L_V), which should be a good index of the expected effect of accretion on the integrated spectrum. In fact, two of the apparent non-detections (A2029 and MKW 3s) have larger values for these quantities than any of the detections except A1795.

The only positive correlation emerging from this relatively small sample is that all objects with significant δUV's have strong nuclear emission lines. Hu et al. (1985) have shown that such emission is found only in objects with cluster gas cooling times less than 7 $h^{-1/2}$ Gyr. However, there are many objects with short cooling times and little evidence of either optical emission lines or population anomalies. Abell 2029, an object which would "seem to be an ideal candidate for observing cooling flows" (Cowie et al. 1983) on the basis of X-ray evidence, is perhaps the best example.

4. SUMMARY AND CONCLUSION

There are three unambiguous cases (NGC 1275, PKS 0745-191, and Abell 1795) of unusual stellar populations apparently induced by cluster cooling flows. Approximately 50% of other accreting cD's in a relatively small sample exhibit nuclear ultraviolet excesses probably indicating accretion populations. The presence of UV excesses and other spectral anomalies is correlated with the presence of strong nuclear line emission but not with estimated global $\dot{M}$ or $\dot{M}/M_{stars}$.

Star formation with a normal (Galactic) initial mass function and the estimated $\dot{M}$'s would produce much larger color anomalies than observed in typical accretors. In most cases, less than 1-10% of the estimated $\dot{M}$ can be deposited in this form. This constraint is not substantially altered if one assumes that the cooling flow has a lifetime much shorter than a Hubble time.

Accretion population models with roughly normal values of M_L and x but lowered M_U's can possibly explain a minority of cases. But the optical properties of most accretors apparently demand that most of the estimated $\dot{M}$ is deposited in an "invisible", very low mass form (M_L << 0.1 $M_\odot$ and/or x > 2). In the case of IMF's with lowered M_U's, a reduced cooling flow lifetime does significantly relieve the observational constraints and permits less extreme parameters to be assigned.

The situation cannot be said to be satisfactory. Perhaps the least unpalatable of the various available alternatives is that the $\dot{M}$'s have been correctly estimated and that star formation with an abnormal IMF is the respository for the accreted material but that typical cooling flows have persisted for only a fraction of a Hubble time. The strong evolution of X-ray structure with redshift which that interpretation would require should be readily susceptible to test with AXAF.

REFERENCES

Bertola, F., Gregg, M.D., Gunn, J.E., & Oemler, A. 1986, Ap.J., **303**, 624.

Bertschinger, E., & Meiksin, A. 1986, Ap.J. (Letters), **306**, L1.

Burns, J.O., White, R.A., & Haynes, M.P. 1981, A.J., **86**, 1120.

Burstein, D., Bertola, F., Faber, S.M., & Lauer, T. 1986, in preparation.

Cowie, L.L., & Binney, J. 1977, Ap.J., **215**, 723.

Cowie, L.L., Hu, E.M., Jenkins, E.B., & York, D.G. 1983, Ap.J., **272**, 29.

Davis, L.E., Cawson, M., Davies, R.L., & Illingworth, G. 1985, A.J., **90**, 169.

Fabian, A.C., Arnaud, K.A., Nulsen, P.E.J., Watson, M.G., Stewart, G.C., McHardy, I., Smith, A., Cook, B., Elvis, M., & Mushotzky, R.F. 1985, M.N.R.A.S., **216**, 923.

Fabian, A.C., Nulsen, P.E.J., & Arnaud, K.A. 1984, M.N.R.A.S., **208**, 179.

Fabian, A.C., Nulsen, P.E.J., & Canizares, C.R. 1982, M.N.R.A.S., **201**, 933.

----- 1984. Nature, **310**, 733.

Forman, W., Jones, C., & Tucker, W. 1985. Ap.J., **293**, 102.

Gear, W.K., Gee, G., Robson, E.I., & Nolt, I.G. 1985. M.N.R.A.S., **217**, 281.

Hu, E.M., Cowie, L.L., & Wang, Z. 1985. Ap.J. Suppl., **59**, 447.

Johnstone, R.M., Fabian, A.C., & Nulsen, P.E.J. 1986, preprint.

Jura, M. 1977, Ap.J., **212**, 634.

Larson, R.B. 1986. M.N.R.A.S., **218**, 409.

Larson, R.B., & Tinsley, B.M. 1978, Ap.J., **219**, 46.

Malumuth, E.M., & Kirshner, R.P. 1985, Ap.J., **291**, 8.

Miller, L. 1986. M.N.R.A.S., **220**, 713.

Nulsen, P.E.J., Stewart, G.C., & Fabian, A.C. 1984. M.N.R.A.S., **208**, 185.

O'Connell, R.W. 1986. In Proc. of the STScI Conference on Stellar Populations, eds. C. Norman, M. Tosi, & A. Renzini (Cambridge: Cambridge Univ. Press), in press.

O'Connell, R.W., Sarazin, C.L., White, R.E., & McNamara, B. 1986. In preparation.

Romanishin, W. 1986. Ap.J., **301**, 675.

Sarazin, C.L. 1986a, Rev. Mod. Phys., **58**, 1.

----- 1986b. In Proc. Green Bank Workshop on Gaseous Haloes Around Galaxies, eds. J. Bregman & F. Lockman, in press.

Sarazin, C.L. & O'Connell, R.W. 1983, Ap.J., **268**, 552.

Scalo, J.M. 1986. In IAU Symposium No. 116, "Luminous Stars and Associations in Galaxies", eds C.W.H. de Loore, A.J. Willis & P. Laskarides, (Reidel, Dordrecht), p. 451.

Silk, J., Djorgovski, S., Wyse, R.F.G., & Bruzual, A.G. 1986, preprint.

Valentijn, E.A. 1983. Astr. Ap., **118**, 123.

Valentijn, E.A., & Giovanelli, R. 1982, Astr. Ap., **114**, 208.

White, R.E., & Sarazin, C.L. 1986, Ap.J. Supp., in press.

Wirth, A., Kenyon, S.J., & Hunter, D.A. 1983, Ap.J., **269**, 102.

DISCUSSION

Richstone: If cooling flows make heavy haloes, wouldn't cD's be completely at rest in cluster centers?

Sarazin: No. Gravitational interactions with other cluster galaxies will always give cD's velocities greater than 100 km/s.

White: I have two questions. First, could you or Andy Fabian comment on recent work by Bertschinger and Meiksin which claims that mass deposition rates drop by an order of magnitude when conductive effects are included? Second, what is known about the metallicity of the X-ray gas in elliptical cooling flows? If E's have "no" supernovae resulting from the star formation processes responsible for the bulk of the observed stars, where do the heavy elements come from?

Fabian: Large temperature gradients are observed in the hot gas so conduction is a possibility. However, the Bertschinger and Meiksin calculations, which reduce our inferred values for M by a factor of $\sim$ 10, do not appear to account for the low temperature ($10^7 K$) gas seen in X–ray lines by the Einstein Observatory SSS and FPCS. The emission measure of this gas is just what is expected from cooling.

O'Connell: My theoretical colleagues inform me they have two objections to the claims that conduction can realistically reduce estimated accretion rates. First, that conduction can be readily suppressed by tangled magnetic fields. And second, that only over a narrow range of conductive efficiency would you find both that conduction is important *and* the increased central densities apparently resulting from some degree of cooling in the gas. Efficient conduction would result in an isothermal halo. It seems unlikely that conduction can operate at just the right level in all the cluster X-ray sources which show some evidence of cooling, since these have a very wide range of physical characteristics. A final observational point is that you may need a factor of 100 reduction in the estimated flow rates to make the optical observations consistent with star formation with a normal IMF in some accretors.
Concerning metal abundances, the visible stars in most accretors appear to be normal E galaxy populations, presumably formed long ago with a normal IMF. The intracluster gas was evidently enriched in metals by such populations at an early stage of cluster evolution, ejected, and now is raining back onto the galaxies in the form of cooling flows. The abundances in the intracluster gas are about 1/2 of solar, so that stars forming from cooling flows will be a factor 2-10 times less metal rich than typical E galaxy stars. You are certainly right that if the IMF for accretion populations is truncated below the threshold mass for supernovae, no further enrichment will occur. I don't know what the situation is with regard to abundances of X-ray gas in normal E's.

Canizares: There is no information about abundances from the X-ray data for any ellipticals other than those at the centers of rich clusters.

Schechter: We've heard a lot about conservation of mass but little about conservation of angular momentum. Is there reason to think these systems won't form disks?

O'Connell: Not really. Cluster gas wil tend to have small but not, of course, zero angular momentum. Disklike structures might well form; but as conditions in the flow change, their orientation could change. Cowie, Hu and collaborators find the emission line filaments to be organized in thick disks. It would be very interesting to study the spatial distribution of color in systems with detected accretion populations in order to look for disklike structures.

N. Bahcall: Very large accretion rates of $10^{2-3} M_\odot/\mathrm{yr}$ that create low mass stars are likely to cause considerable deviations from the standard parameter correlations of elliptical galaxies, such as those observed between velocity dispersion, surface brightness, and effective radius (including the Faber-Jackson relation). Since the above relations for ellipticals are rather tight, those galaxies expected to have large accretion rates should deviate in a unique manner from the standard curves. Has this been checked, and, if so, what does it show?

O'Connell: Mass-to-light ratios have been looked at by Romanishin using data from Malumuth and Kirshner; he found no obvious effects. Non-accreting cD's have nuclear $M/L_v \sim 10$ but show a fairly large scatter. A typical accretion flow lasting for 10 Gyr would not push M/L_v beyond that scatter. Most M/L determinations are based on nuclear velocity dispersion measurements, which may not reflect the full effects of the accretion population if it is deposited preferentially at large radii. Dressler (1981, *Astrophys. J.*, **243**, 26) found an increasing velocity dispersion with radius in Abell 2029, which has a large cooling flow, and that could be regarded as evidence for an accretion population in its outer parts.

Kochhar: What does the gas do in the cases which do not show a UV excess?

O'Connell: Assuming that the mass accretion rates are correct, we presume that it forms into stars with a mean mass well below $1 M_\odot$, which will not have a significant effect on the optical energy distribution. For an upper mass cutoff around 0.7 $M_\odot$, the Sarazin and O'Connell models show that even infrared colors will be insensitive to the accretion population.

Craig Sarazin.

MASS DISTRIBUTIONS IN ELLIPTICAL GALAXIES AT LARGE RADII

Craig L. Sarazin*
Department of Astronomy, University of Virginia,
Charlottesville, VA 22903 and
Joint Institute for Laboratory Astrophysics, University of
Colorado and National Bureau of Standards, Boulder, CO 80309

ABSTRACT. Recently, x-ray observations have shown that elliptical galaxies generally contain large quantities of hot gas. Central dominant cluster ellipticals have even more gas, which they have accreted from the surrounding clusters. The mass distributions in these galaxies can be derived from the condition of hydrostatic equilibrium. M87, the best studied central dominant galaxy, has a massive, dark halo with a total mass of about 4×10^{12} $M_\odot$ within a radius of 300 kpc. The total mass-to-light ratio within this radius is at least 150 $M_\odot/L_\odot$. The x-ray observations of normal ellipticals also strongly suggest that they have heavy haloes, although the distribution of the mass is much less certain than in M87.

1. INTRODUCTION

Do elliptical galaxies have heavy haloes of optically dark matter? The rotation curves of spiral galaxies require that they possess such haloes (Rubin, Ford, and Thonnard 1980), but ellipticals generally lack extended disks of gas or stars which would allow a determination of their rotation curves at large radii. Recent x-ray observations show that these galaxies do generally have extensive coronae of very hot gas (Forman, Jones, and Tucker 1985; Trinchieri and Fabbiano 1985). If the profiles of gas density and temperature can be derived from the x-ray observations, then the mass distribution of the galaxy can be determined from the condition of hydrostatic equilibrium (Sec. 2). At least crudely, elliptical galaxies can be divided into two classes based on their x-ray properties. Many central dominant cluster galaxies are accreting large amounts of hot gas from their surrounding cluster, and are very luminous in x-rays. M87 in the Virgo

---

*1985-86 JILA Visiting Fellow; permanent address, University of Virginia.

T. de Zeeuw (ed.), Structure and Dynamics of Elliptical Galaxies, 179–188.

cluster is the best studied example of this class, and its mass distribution (and that of other central dominant galaxies) is discussed in Sec. 3. Normal elliptical galaxies have quantities of hot gas in reasonable agreement with the expected rates of stellar mass loss within the galaxies, and as a result have lower x-ray luminosities than the central dominant ellipticals. The less complete information we have on the mass distributions of normal ellipticals is described in Sec. 4.

All numerical values quoted in this paper assume a Hubble constant of H_o = 50 km/s/Mpc and a distance to the center of the Virgo cluster of 20 Mpc.

2. HYDROSTATIC EQUILIBRIUM

Masses for elliptical galaxies can be derived by assuming that the x-ray emitting gas is in hydrostatic equilibrium with the gravitational field of the galaxy. This is a reasonable assumption as long as the galaxy is stationary (the gravitational potential does not change on a sound-crossing time), other forces (magnetic fields, etc.) are negligible, and the gas motions are significantly subsonic. This latter condition requires, first of all, that rotation in the gas be dynamically unimportant. This is consistent with the fact that neither large elliptical galaxies nor clusters of galaxies rotate significantly (see, for example, the paper by Davies in this Proceedings), and that the x-ray isophotes of central dominant and normal ellipticals are fairly circular, except in their outermost regions. The gas in these galaxies cannot be in supersonic inflow or outflow; since the sound crossing time for a galaxy is much shorter than gas cooling time, supersonic motions would remove the gas faster than it could radiate, and a very large source of gas would be required to explain the observed x-ray emission. In particular, the cooling inflows that are commonly seen in central dominant and normal ellipticals (see Secs. 3 and 4 and the paper by Fabian in this Proceedings) are very subsonic, except possibly in their innermost regions.

Under these circumstances, the gas obeys the hydrostatic equation

$$\vec{\nabla} P = -\rho \vec{\nabla} \phi \tag{1}$$

where P and ρ are the gas pressure and density, and ϕ is the gravitational potential of the galaxy. The distribution of the total galaxy mass can then be derived from the variation of the gas pressure and density. This method has several advantages over stellar dynamic methods, using the radial velocity dispersion of stars or the individual radial velocities of globular clusters or companion galaxies, for determining the mass distributions of ellipticals at large radii. First, the gas is a collisional fluid, and the particle velocities are isotropically distributed. On the other hand, stars in galaxies (or globular clusters or companion galaxies) are collisionless, and uncertainties in the distribution of particle orbits can significantly influence the derived mass distribution (see, for example, Tonry 1983).

Second, the gas extends to large distances from the galaxy center, and it is easier to measure the density and temperature of the gas than it is to measure the velocity distribution of the stars in the outer regions of the galaxy. Third, the statistical errors in x-ray mass determinations are much smaller than those for globular clusters or companion galaxies. Finally, x-ray mass determinations are not very sensitive to the shape of the galaxy (Strimpel and Binney 1979; Fabricant, Rybicki, and Gorenstein 1984).

The first applications of x-ray distributions to derive mass distributions in galaxies were by Bahcall and Sarazin (1977) and Mathews (1978) in M87. The method has been developed extensively by Fabricant, Gorenstein, and collaborators (Fabricant, Lecar, and Gorenstein 1980; Fabricant and Gorenstein 1983; Fabricant, Rybicki, and Gorenstein 1984). Ideally, one would measure the spatially and spectrally resolved x-ray surface brightness of the galaxy $I_\nu(\vec{b})$, where $h\nu$ is the x-ray photon energy and $\vec{b}$ is the projected position relative to the center. This would be inverted to give the local x-ray emissivity $\varepsilon_\nu(\vec{r})$, where $\vec{r}$ is the position relative to the cluster center. This deconvolution is stable because the observed x-ray images of galaxies are quite smooth except in the outermost regions (Forman, Jones, and Tucker 1985; Trinchieri and Fabbiano 1985). To deconvolve the projected surface brightness one must assume that the actual gas distribution is spherical or spheroidal (Strimpel and Binney 1979) or, more generally, has an axis of symmetry in the plane of the sky (Fabricant, Rybicki, and Gorenstein 1984). However, the resulting mass distributions are not affected strongly by the shape. For a spherical cluster, the Abel integral inversion for $\varepsilon_\nu(\vec{r})$ is:

$$\varepsilon_\nu(\vec{r}) = -\frac{1}{\pi}\frac{d}{dr^2}\int_{r^2}^{\infty}\frac{I_\nu(b)\,db^2}{(b^2-r^2)^{1/2}} \quad . \tag{2}$$

The x-ray emissivity of a hot plasma depends on its electron density n_e, its temperature T, and its abundances (Sarazin and Bahcall 1977)

$$\varepsilon_\nu = n_e^2\,\Lambda_\nu\,(T,\ \text{abundances}) \quad . \tag{3}$$

Heavy elements mainly produce discrete line features in x-rays; the strength of these features determines the heavy element abundances. The x-ray continuum is exponential $\varepsilon_\nu \propto \exp(-h\nu/kT)$, and thus the spectral shape of the emissivity $\varepsilon_\nu(r)$ determines $T(r)$, while its normalization gives $n_e(r)$. Then the hydrostatic equation gives the total mass $M(r)$ interior to r as

$$M(r) = -\frac{k\,T(r)r}{\mu\,m_p\,G}\left\{\frac{d\log n_e}{d\log r} + \frac{d\log T}{d\log r}\right\} \quad , \tag{4}$$

where μ is the mean molecular weight. It is important to note that the mass depends only weakly on $n_e(r)$ (only on its logarithmic derivative), but depends strongly on the temperature profile $T(r)$.

3. M87 AND OTHER CENTRAL DOMINANT CLUSTER GALAXIES

Many central dominant cluster galaxies contain large amounts of hot gas, which they are accreting from the surrounding cluster. These cluster cooling flows have been reviewed by Fabian, Nulsen, and Canizares (1984), by Sarazin (1986a), and by Fabian in these Proceedings. The x-ray emission from this gas can be used to determine the distribution of mass around these central dominant galaxies.

M87 in the Virgo cluster is the nearest, brightest, and best studied example of a central dominant cluster galaxy with a cooling flow. (However, it has a much less x-ray luminous cooling flow than the typical, more distant, optically luminous central dominant cluster galaxy.) The most successful application of x-ray measurements for the mass determination of an elliptical galaxy has been to M87, because this galaxy has the best determined temperature distribution.

Originally, Bahcall and Sarazin (1977) and Mathews (1978) suggested that M87 must have a very high mass if it gravitationally binds its x-ray emitting gas. Bahcall and Sarazin found a mass of 1-6 × 10^{13} $M_\odot$ within several hundred kiloparsecs. However, the temperature distribution in the gas was not known at that time, and thus these estimates were very uncertain (Binney and Cowie 1981). Observations with the _Einstein_ x-ray observatory permitted a definitive determination of the mass profile in M87 (Fabricant, Lecar, and Gorenstein 1980; Fabricant and Gorenstein 1983; Stewart _et al_. 1984). The results are shown in Figure 1. The hatched area gives the allowed area for the x-ray derived mass from Fabricant and Gorenstein (1983), while the solid curve is the best-fit mass model of Stewart _et al_. (1984). The dots give the mass determined from the optical velocity dispersion measurements of Sargent _et al_. (1978), assuming isotropic stellar velocities. The Stewart _et al_. model was constrained to roughly fit the optical data, and involved a comparison to low ionization x-ray lines that may help constrain the mass distribution at intermediate radii $10 \lesssim r \lesssim 100$ kpc.

Figure 1 shows that M87 has a very massive halo, with a total mass

$$M \approx 3\text{-}6 \times 10^{13}\ M_\odot \left(\frac{r}{300\ \mathrm{kpc}}\right) \quad . \tag{5}$$

The material providing this mass is quite dark; beyond 100 kpc, the total mass-to-light ratio is $(M/L_B)_{tot} > 150\ M_\odot/L_\odot$, and the local value is $(M/L_B)_{loc} > 500\ M_\odot/L_\odot$.

The mass distribution in the outer regions of M87 is roughly consistent with $M \propto r$, implying an approximately constant velocity dispersion. However, the same proportionality between mass and radius that applies in the outer region overestimates the mass in the interior region ($r \lesssim 10$ kpc) by a factor of about 5. In other words, the implied radial velocity dispersion (assuming isotropic orbits) increases from ~300 km/s to ~600 km/s in the outer regions. This is a considerably different behavior than observed in spiral galaxies (Rubin, Ford, and Thonnard 1980), where the circular velocity remains roughly constant in the outer regions of the galaxy.

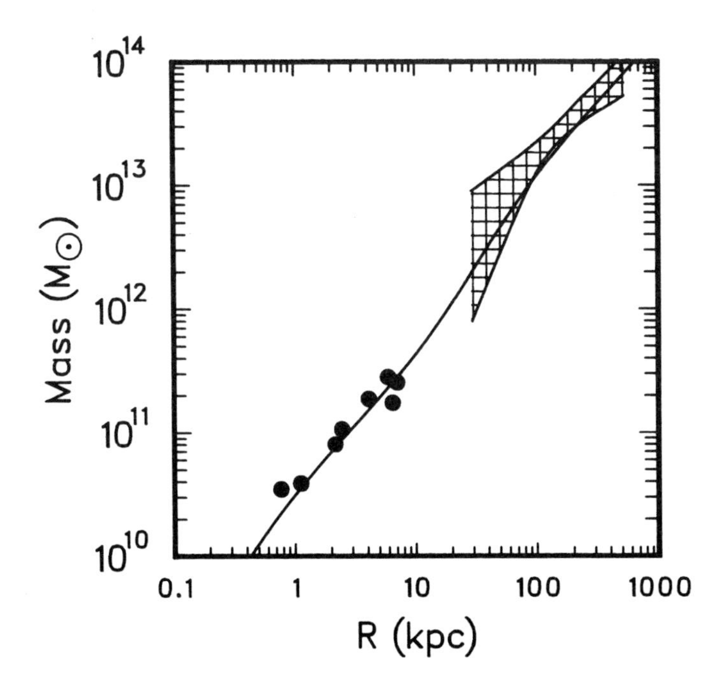

Fig. 1. The mass distribution in M87, assuming a distance of 20 Mpc. The hatched area gives the range of masses determined from the x-ray data by Fabricant and Gorenstein (1983). The solid line is the best fit model from Stewart et al. (1984). The filled circles are masses derived from the optical velocity dispersion measurements of Sargent et al. (1978), assuming isotropic stellar orbits.

The dark matter is strongly concentrated to the center of M87; Stewart et al. (1984) find that the core radius of the dark matter is less than 50 kpc. This is smaller than the typical core radius determined for clusters of galaxies of ~250 kpc. Moreover, there are no other bright galaxies within the region of the x-ray measurements and M87 provides the great majority of the optical luminosity from this region, so it seems reasonable to associate the dark matter with M87.

The x-ray observations of other central dominant cluster galaxies suggest that they also have dark, heavy haloes, although the temperature profiles are not very well determined in these other cases. Matilsky, Jones, and Forman (1985) find a total mass of 2×10^{13} $M_\odot$ with 200 kpc for NGC 4696 at the center of the Centaurus cluster. This central dominant galaxy and cluster are very similar in most x-ray and optical properties to M87/Virgo. Fabian et al. (1981) find a total mass of $\sim 7 \times 10^{13}$ $M_\odot$ within 300 kpc of NGC 1275 in the Perseus cluster. Central dominant galaxies in poor clusters also show evidence for massive, dark haloes (Kriss, Cioffi, and Canizares 1983; Malumuth and Kriss 1986), with typical total mass-to-light ratios of ~ 100 $M_\odot/L_\odot$ at radii of ~100 kpc.

Central dominant cluster galaxies have many special properties (see the paper by Tonry in this Proceedings). Although these galaxies do appear to have heavy haloes, the dark halo material might be associated with the location of these galaxies near the kinematic centers of clusters. Thus it is important to determine the mass profiles of more isolated, normal elliptical galaxies.

4. NORMAL ELLIPTICAL GALAXIES

Recent x-ray observations indicate that essentially all elliptical galaxies that are not in the cores of rich compact clusters have extended x-ray emission (Forman, Jones, and Tucker 1985; Trinchieri and Fabbiano 1985; Nulsen, Stewart, and Fabian 1984). These galaxies have x-ray luminosities of $L_x \sim 10^{39}$-10^{42} ergs/s, and sizes of typically $R_x \sim 50$ kpc. There is a strong correlation between the x-ray and optical luminosities of the galaxies $L_x \propto L_B^{1.6-2.0}$, where L_B is the blue luminosity. It seems most likely that the x-ray emission is thermal with typical gas temperatures $T \sim 10^7$ K. The gas densities in the inner parts of the coronae exceed 0.01 cm^{-3}, and vary roughly as $r^{-3/2}$. The total gas mass is typically $M_g \sim 10^9$-10^{10} $M_\odot$, and the ratio of gas mass to luminous stellar mass M_* is typically $M_g/M_* \sim 0.02$. (See the papers by Fabian and by Canizares, Fabbiano, and Trinchieri in this Proceedings.)

The x-ray observations show that elliptical galaxies are gas-rich systems. Apparently, they do not have the global galactic winds suggested by Mathews and Baker (1971). Instead, the observations are basically consistent with a simple cooling flow model in which the x-ray emitting gas is the result of mass loss by stars. This gas is heated by the motions of the mass losing stars and by infall in the galactic gravitational potential, cools radiatively, and slowly flows inward in the galaxy (Nulsen, Stewart, and Fabian 1984; White and Chevalier 1984; Canizares 1987; Fabian 1987; Sarazin 1986b, 1987; Sarazin and White 1986; Thomas <u>et al</u>. 1986; Trinchieri, Fabbiano, and Canizares 1986).

The presence of this x-ray emitting gas means that mass profiles for these galaxies can be derived from hydrostatic equilibrium. Unfortunately, there is essentially no direct information about the temperature profiles T(r) in these galaxies; even the average temperatures for the x-ray image are very poorly determined. The main reason the temperature profiles are so poorly known is that these normal elliptical galaxies are much less luminous in x-rays than M87 (10^{41} versus 10^{43} ergs/s), but are typically at about the same distance.

Forman, Jones, and Tucker (1985) estimated masses for several galaxies assuming that the gas temperature was isothermal at T(r) = 1.2×10^7 K. The hydrostatic equation was applied out to the largest radius at which x-ray emission was observed. Table 1 lists the results of their analysis, including the average temperatures derived from the x-ray spectra, the radius of the x-ray image, and the derived mass and mass-to-light ratios. Note that the mass-to-light ratios range from roughly 5-100 $M_\odot/L_\odot$, with an average value of ~40 $M_\odot/L_\odot$. Based on these results, Forman, Jones, and Tucker suggested that elliptical galaxies did indeed possess massive dark haloes.

Table 1. Masses of Normal Elliptical Galaxies from Forman, Jones, and Tucker (1985).

Galaxy	Temperature (keV)	Radius (kpc)	Mass (10^{12} $M_\odot$)	M/L_B ($M_\odot/L_\odot$)
N315		110	5.7	26
N1316		34	2.0	8
N1332		36	2.5	62
N1395	0.51-0.90	98	5.1	88
N2563		80	4.1	60
N4374	0.93-1.32	15	0.9	14
N4382	>0.80	48	2.5	37
N4406	0.94-1.03	88	4.6	63
N4472	1.07-1.30	80	4.6	33
N4594	>1.6	32	1.7	8
N4636	0.78-1.21	44	2.3	87
N4649	1.78-4.12	36	2.1	24
N5128		20	1.2	7

Trinchieri, Fabbiano, and Canizares (1986) have described the effect of uncertainties in the temperature distribution on the mass estimates. They derived the range of average gas temperatures from the x-ray spectra, and assumed a power-law temperature distribution $T \propto r^{\alpha}$ for $-0.5 \leq \alpha \leq 0.5$. Because the outermost portions of the x-ray images of many galaxies are irregular, they assumed smaller radii than Forman, Jones, and Tucker in most cases. For most galaxies, a range of total mass-to-light ratios of $(M/L_B)_{tot} \approx 10$ to $100\ M_\odot/L_\odot$ was found. Since this range extends down to the mass-to-light ratios found in the centers of elliptical galaxies from optical velocity dispersions, this analysis suggested that the present x-ray data do not necessarily require heavy haloes.

Fabian et al. (1986; see also the paper by Fabian in this Proceedings) have derived a lower limit on the total mass surrounding an elliptical galaxy. They assume that the temperature at the outermost x-ray detected radius is given by the average gas temperature derived from the integrated x-ray spectrum, and determine the pressure at this point from the density derived from the x-ray image. They assume that the gas pressure decreases beyond this radius, the gas being gravitationally bound. Finally, the temperature decrease is assumed to be sub-adiabatic, so that the gas beyond the x-ray radius is convectively stable. These assumptions lead to a lower limit on the total mass, which is typically $M \gtrsim 5 \times 10^{12}\ M_\odot$. This implies a mass-to-light ratio $(M/L_B) \gtrsim 75\ M_\odot/L_\odot$, and strongly supports the existence of massive haloes around elliptical galaxies. Unfortunately, this limit provides no information about the distribution of the dark matter. For example, in principle, the binding mass might be associated with the group or cluster in which the galaxy is located.

The mass determinations of Forman, Jones, and Tucker (1985) and Trinchieri, Fabbiano, and Canizares (1987) assumed very simple temperature profiles. More realistic profiles might come from detailed numerical models for cooling flows in elliptical galaxies. Sarazin and White (1986) have calculated a grid of spherical, steady-state cooling flow models for the x-ray emission in these galaxies. In these models, the source of the gas is stellar mass loss, and the gas is heated by the motions of the gas-losing stars and by infall in the galactic gravitational potential. These models are in good general agreement with the x-ray luminosities, x-ray versus optical luminosity correlation, and gas temperatures observed in normal elliptical galaxies. Sarazin and White find that these models fit the observed radial distribution of the x-ray surface brightness in ellipticals if ellipticals have heavy haloes with $M \propto r$. In models in which ellipticals lack heavy haloes, the predicted x-ray surface brightness profile is more centrally peaked than is observed.

There is a simple argument that indicates why this is so. I have shown that the energy equation for a steady-state cooling flow leads to approximate thermal equilibrium in the flow, and that this implies that (Sarazin 1986b, 1987; Sarazin and White 1986)

$$I_x(r) \propto I_B(r)\, \sigma_*^2(r)\, \frac{\Lambda_x}{\Lambda}(T(r)) \quad , \tag{6}$$

where I_x and I_B are the x-ray and (blue) optical surface brightnesses of the galaxy, σ_* is the stellar velocity dispersion, and (Λ_x/Λ) is the fraction of the radiative cooling that is emitted in the x-ray band. Now, if the velocity dispersion is roughly constant, then hydrostatic equilibrium implies that the temperature is nearly constant, and equation (6) implies that $I_x \propto I_B$. In fact, the observed x-ray surface brightness profiles fit this relationship quite well (Trinchieri, Fabbiano, and Canizares 1987). On the other hand, if elliptical galaxies lacked heavy haloes, $\sigma_*^2 \propto 1/r$ in the outer parts of the galaxy, and the x-ray surface brightness would drop with increasing radius at least as rapidly as I_B/r. This is much steeper than is observed.

Thus, the x-ray surface brightness profiles of normal elliptical galaxies strongly suggest that these galaxies have heavy haloes (Forman, Jones, and Tucker 1985; Fabian et al. 1986; Sarazin and White 1986), although this result is much less certain than in the central dominant galaxy M87 (Trinchieri, Fabbiano, and Canizares 1986). Moreover, we have at present very little information on the distribution of the dark matter in these heavy haloes.

ACKNOWLEDGMENTS

I thank Claude Canizares, Andy Fabian, and Pepi Fabbiano for communicating their results prior to publication, and for useful conversations. This work was supported in part by NSF Grant AST81-20260, by NASA Astrophysical Theory Center Grant NAGW-764, and by a Visiting Fellowship at JILA.

REFERENCES

Bahcall, J. N., and Sarazin, C. L.: 1977, Ap. J. (Letters) **199**, L89.
Binney, J., and Cowie, L.: 1981, Ap. J. **247**, 464.
Canizares, C. R.: 1987, in Proceedings of IAU Symposium No. 117: Dark Matter in the Universe, edited by J. Kormendy and G. Knapp, in press.
Fabian, A. C.: 1987, in Proceedings of IAU Symposium No. 117: Dark Matter in the Universe, edited by J. Kormendy and G. Knapp, in press.
Fabian, A. C., Hu, E. M., Cowie, L. L., and Grindlay, J.: 1981, Ap. J. **248**, 47.
Fabian, A. C., Nulsen, P. E., and Canizares, C. R.: 1984, Nature **310**, 733.
Fabian, A. C., Thomas, P. A., Fall, S. M., and White, R. E.: 1986, preprint.
Fabricant, D., and Gorenstein, P.: 1983, Ap. J. **267**, 535.
Fabricant, D., Lecar, M., and Gorenstein, P.: 1980, Ap. J. **241**, 552.
Fabricant, D., Rybicki, G., and Gorenstein, P.: 1984, Ap. J. **286**, 186.
Forman, W., Jones, C., and Tucker, W.: 1985, Ap. J. **293**, 102.
Kriss, G. A., Cioffi, D., and Canizares, C. R.: 1983, Ap. J. **272**, 439.
Malumuth, E. M., and Kriss, G. A.: 1986, preprint.
Mathews, W. G.: 1978, Ap. J. **219**, 413.
Mathews, W. G., and Baker, J. C.: 1971, Ap. J. **170**, 241.
Matilsky, T., Jones, C., and Forman, W.: 1985, Ap. J. **291**, 621.
Nulsen, P. E., Stewart, G. C., and Fabian, A. C.: 1984, M.N.R.A.S. **208**, 185.
Rubin, V. C., Ford, W. K., and Thonnard, N.: 1980, Ap. J. **238**, 471.
Sarazin, C. L.: 1986a, Rev. Mod. Phys. **58**, 1.
Sarazin, C. L.: 1987, in Proceedings of IAU Symposium No. 117: Dark Matter in the Universe, edited by J. Kormendy and G. Knapp, in press.
Sarazin, C. L.: 1986b, in Proceedings of the Greenbank Workshop on Gaseous Haloes and Galaxies. edited by J. Bregman and F. Lockman, in press.
Sarazin, C. L., and Bahcall, J. N.: 1977, Ap. J. Suppl. **34**, 451.
Sarazin, C. L., and White, R. E.: 1986, preprint.
Sargent, W. L., Young, P. J., Boksenberg, A., Shortridge, K., Lynds, C. R., and Hartwick, F. D.: 1978, Ap. J. **221**, 731.
Stewart, G. C., Canizares, C. R., Fabian, A. C., and Nulsen, P. E.: 1984, Ap. J. **278**, 536.
Strimpel, O., and Binney, J.: 1979, M.N.R.A.S. **188**, 883.
Thomas, P. A., Fabian, A. C., Arnaud, K. A., Forman, W., and Jones, C.: 1986, preprint.
Tonry, J. L.: 1983, Ap. J. **266**, 58.
Trinchieri, G., and Fabbiano, G.: 1985, Ap. J. **296**, 447.
Trinchieri, G., Fabbiano, G., and Canizares, C. R.: 1986, preprint.
White, R. E., and Chevalier, R. A.: 1984, Ap. J. **280**, 561.

DISCUSSION

Richstone: Since most galaxies are better fit by models without constant density cores, and since counts in clusters of galaxies are well fit by a similar model, I wonder if you would have still needed to assign the dark matter near M87 to the galaxy if you had assumed a sharper dark matter distribution for the cluster.

Sarazin: Of course, this is largely a semantic issue. The real point is just that the dark matter is concentrated towards the center of M87, and that there are no other galaxies with any significant luminosity in the same region.

Binney: One of my students, Ian Pollister, has modeled Perseus in detail with cooling flows in $r^{1/4}$ models, and gets excellent fits for reasonable cluster parameters and no galaxy contribution to the potential. In fact, the X–ray profiles of cooling clusters are remarkably featureless, showing no core, and no separation of central–galaxy and cluster components.

Mould: Information on the M87 potential between the radii available from the galaxy light and the X–ray gas can be obtained from the kinematics of the globular clusters as shown elsewhere in this volume (p. 451). We find $\sigma = 370 \pm 50$ km/s for the velocity dispersion, which is different from 800 km/s (your equivalent velocity dispersion for the X–ray gas), but not significantly different from 300 km/s. Although the error bars are large as yet, there is at least no sign that the velocity dispersion is turning down, and our inferred mass–radius relation seems to connect the corresponding relations from the galaxy light and the X–ray gas.

Sarazin: Globular clusters offer a very exciting opportunity to bridge the gap in mass determinations between optical velocity dispersions and X–ray data. The present data give a velocity dispersion which is slightly higher than the stellar velocity dispersion at a slightly larger radius. Because there are many globulars extending out into the X–ray region in M87, I hope we will hear more about these measurements in the future.

van Gorkom: Is it not a bit worrying that NGC 4472 is always mentioned as *the* prototype of a normal elliptical galaxy? It is the brightest elliptical in the Virgo cluster and probably the dominant central elliptical in a separate group. For how many real normal ellipticals are the radial distributions known and similar to NGC 4472?

Sarazin: Of course, it is the fact that NGC 4472 is bright that allows it to be studied in detail. Good X–ray surface brightness profiles are available for another half dozen to dozen ellipticals. However, X–ray emission is found in essentially all nearby ellipticals, with the same relationship between L_X and L_B. The few undetected Es have upper limits consistent with this relationship. So, the X–ray emission does seem to be a very general property of ellipticals.

LINE-STRENGTH GRADIENTS IN EARLY-TYPE GALAXIES

J. Gorgas
Departamento de Astrofisica
Universidad Complutense
28040 Madrid
Spain

G. Efstathiou
Institute of Astronomy
Madingley Road
Cambridge CB3 OHA
England

ABSTRACT. We have measured line-strength gradients in a sample of 15 early-type galaxies. The line-strength measures include the Mg_2 index and the equivalent widths of Hβ and two iron blends at 5270Å and 5335Å. In most of the galaxies we find gradients in the metallic line-strengths. However, the gradients vary markedly from object to object and do not correlate strongly with other parameters such as total luminosity, rotation, etc. A comparison of the line-strengths in the outer parts of these galaxies with galactic globular clusters suggests relatively modest abundance gradients in early-type galaxies.

1. INTRODUCTION

It is now clear that the stellar populations in globular clusters and elliptical galaxies do not simply represent a one parameter family in which metal abundance is the key variable (see Burstein, 1985 for a detailed review, Pickles 1986 in this volume). We mention here just a few of the developments which have led to this conclusion: (a) although in integrated colours galactic globular clusters appear as a one parameter family, their horizontal branch morphologies require at least two additional parameters (Freeman and Norris 1981); (b) M31 globular clusters which have similar metal line-strengths to those of galactic globular clusters generally have stronger Balmer lines (Burstein et. al. 1984, hereafter BFGK); (c) The Balmer lines in the nuclei of elliptical galaxies are stronger than expected in synthesis models of old (~15Gyr) metal rich stellar population (O'Connell 1976; Gunn, Stryker and Tinsley 1981) and do not match continuously with those of metal rich galactic globular clusters (BFGK). (d) M31 globular clusters have stronger CN features when compared to elliptical galaxy nuclei with similar values of magnesium line-strength (BFGK). The reasons for these differences are not well understood. Gunn, Stryker and Tinsley have suggested that the enhanced Balmer line-strengths in elliptical nuclei might be caused by a modest rate of current star formation; O'Connell (1980), Pickles (1985) and Rose (1985) argue that the stellar populations in elliptical galaxies contain a substantial intermediate

T. de Zeeuw (ed.), Structure and Dynamics of Elliptical Galaxies, 189–201.

age component (~5-10Gyr). However, other possibilities such as blue stragglers cannot be convincingly ruled out.

In this article, we investigate another aspect of old stellar populations, namely line-strength gradients in early-type galaxies. The existence of radial abundance variations in elliptical galaxies and the bulges of spirals has been suspected ever since early work on colour gradients (eg. de Vaucouleurs 1961, Tifft 1961). More recent work (see Gallagher 1984; Cohen 1986) has shown that the colours in the halos of giant elliptical galaxies are generally redder than those of metal poor galactic globular clusters. At face value, these colour gradients suggest relatively weak abundance gradients. Further progress in understanding these gradients requires spectroscopic data. As indicated above, spectroscopic studies have revealed a number of intriguing differences between spheroidal stellar populations. Line-strength variations **within** elliptical and other early-type galaxies provide a further test of the nature of these populations.

Previous line-strength studies have been reviewed by Faber (1977). They have provided unambiguous evidence of gradients in strong absorption features such as CN and Mg. However, only a few features, in a small sample of early-type galaxies, have been studied and the results have left a number of important questions unanswered (see below). This lack of data is not altogether surprising. High precision measurements at large distances from the centres of elliptical galaxies require long exposures and careful data reduction. A "two-dimensional" detector is essential if a large body of homogeneous data is to be obtained on a reasonable time scale. Our own results are based on spectra obtained with the Image Photon Counting System (IPCS) at the Anglo-Australian Telescope in a number of observing runs between 1979 and 1982. The observations were made primarily for dynamical studies of early-type galaxies and details may be found in various published papers (Efstathiou, Ellis and Carter 1980, 1982; Davies et. al., 1983; Carter et. al., 1985). Our sample spans a range of morphological types (ellipticals, S0's and cD's), absolute magnitudes and ellipticities and thus provides a starting point for a more systematic analysis of line-strength gradients.

We first outline a number of interesting problems which remain to be solved:

1. Are there any regularities in line-strength gradients? For example, do gradients correlate with parameters such as absolute magnitude, surface brightness, ellipticity, etc. Such correlations could provide important clues to the formation of elliptical galaxies (eg. Carlberg 1984). The available results are ambigous. Strom and Strom (1978) found that colour gradients vary strongly from galaxy to galaxy and presented evidence that gradients correlate with total luminosity. Burstein (1979) found indications of a correlation between central line-stength and gradients while Spinrad et. al. (1972) argued that CN gradients along the major axes of highly flattened galaxies are weaker than those in round galaxies. However, Faber (1977) found that gradients in CN, Mg and NaD in 8 early-type galaxies of widely different ellipticities were all roughly similar.

2. It is well known that line-strengths in the centres of ellipticals and S0's are strongly correlated with total luminosity, suggesting that the stars in the inner parts of giant galaxies are more metal rich than those in low mass galaxies (Faber 1973, Burstein 1979, Terlevich et al 1981, Tonry and Davis, 1981). If the stellar populations in ellipticals were governed by only one parameter, the mean metallicity of the system, then the gradients in all features within a galaxy should follow the same set of relations as those obeyed by the nuclei.

3. Do line-strength gradients correlate with morphological type? Wirth and Shaw (1983) find that colour gradients in spiral bulges are steeper than those in elliptical galaxies and argue that these systems must have formed in differents ways. It clearly important to check this result in more detail. There is now considerable speculation on the formation mechanism of cD galaxies. If galactic cannibalism (Hausman and Ostriker 1978) or cooling flows (Fabian et. al. 1984) are responsible for their high luminosities, it may be possible to detect anomalies in the spectral properties when compared to giant ellipticals.

4. Larson (1975) presented detailed models of the dissipative collapse of elliptical galaxies and predicted that the isoabundance contours ("isochromes") would be flatter than the isophotes. Results on this point remain ambiguous. Strom and Strom (1978) found supporting evidence from their study of colour gradients. Faber (1977) also measured such an effect but concluded that "the behaviour is quite variable from feature to feature and from object to object". Recently, Carlberg (1984) has presented a dissipative scheme for the formation of slowly rotating ellipticals which does not lead to a strong flattening of the isochromes.

In this article, we present results on line-strength gradients in 15 early-type galaxies. Our discussion will be focussed primarily on points (1) and (2) described above. An important aspect of our study is that we have adopted the line-strength system described by BFGK, thus allowing direct comparison with their extensive results for globular clusters and the nuclei of ellipticals. A more complete discussion of our work is given elsewhere (Gorgas and Efstathiou 1986).

2. Mg_2 GRADIENTS

We have measured the Mg_1, Mg_2, Hβ , Mgb, Fe5270 (henceforth Fe52) and Fe5335 (henceforth Fe53) indices defined by BFGK. We will not discuss the lengthy details of the data reduction here, but it is important for the reader to be aware that the the main sources of error in our line-strengths arise from a number of systematic effects associated with the instrument configuration. These include, S-distortion, non-uniformities in the response of the IPCS and saturation of the detector at high count rates. In addition, the Mgb, Fe52 and Fe53 indices are affected by velocity dispersion broadening which we have calibrated by convolving spectra of G and K giants with Gaussian distributions. Our line indices have been corrected to a fixed resolution corresponding to a Gaussian distribution of 220 km s^{-1} to match the resolution used by BFGK. The spectra were calibrated to an absolute scale using Oke (1974) flux standards observed during each run. This differs from the absolute calibration used by BFGK who calibrated with respect to a tungsten lamp. The absolute calibration is important in determining the zero point of the Mg_2 and Mg_1 indices, which have widely separated side bands, but is unimportant for the other indices. An important check that our results are on the same system as that of BFGK is provided by comparing our nuclear values with their relations for elliptical nuclei. This is shown in Figure 1 for all the ellipticals in our sample together with an additional 8 ellipticals for which short exposures were taken (Davies et. al., 1983). The nuclear values of BFGK were determined using a 1.4"x4" aperture, while we have binned our data to simulate an aperture of sx4" where s denotes the slit width in arcsec (between 2" and 3" depending on the observing run). The nuclear values shown in Figure 1 obey similar relations to those found by BFGK. In fact the scatter in the Fe lines is nearly twice as small as theirs. The scatter that we observe in Figure 1 is compatible with our estimated measurement errors with perhaps the exception of the Hβ-Mg_2 relationship. NGC 4742 provides an extreme example of a discrepancy since this galaxy has a central H equivalent width of 3.8Å which is nearly twice as large as the average value for an elliptical with similar nuclear metal line-strengths.

In Figure 2, we show how the major axis Mg_2 gradients vary for 10 ellipticals and 2 bulge dominated S0 galaxies. We have presented the Mg_2 gradients since this index is more accurately measured than the others. Gradients in the other indices will be discussed in Section 3. The triangles in Figure 2 show Burstein and Faber's measurements for NGC 4472 (Faber, 1977) which are in good agreement with those presented here.

Figure 2 shows that the Mg_2 gradients vary substantially from galaxy to galaxy. For example, NGC3904 appears to have a strong gradient while NGC 4478 does not possess a significant gradient. Our results do not support Faber's (1977) conclusion that early-type galaxies have similar gradients. We have fitted a straight line to each of the line-strength profiles shown in Figure 2 and tested for correlations between the slopes and other parameters by applying the Spearman rank-correlation test. The parameters chosen were: blue absolute magnitude, mean surface brightness within r_e, mean velocity dispersion within $0.5r_e$, ellipticity, ratio of maximum rotational velocity to central velocity dispersion, velocity dispersion gradient and central Mg_2 value. The strongest correlation (-0.50 for the 12 galaxies shown in Figure 2, -0.66 if the two S0's are excluded) is with velocity dispersion (in the sense that galaxies with large velocity dispersions have steeper gradients)- but the result is significant at less than the 95% level. We therefore conclude that Mg gradients do not correlate strongly with other parameters. The mean of the

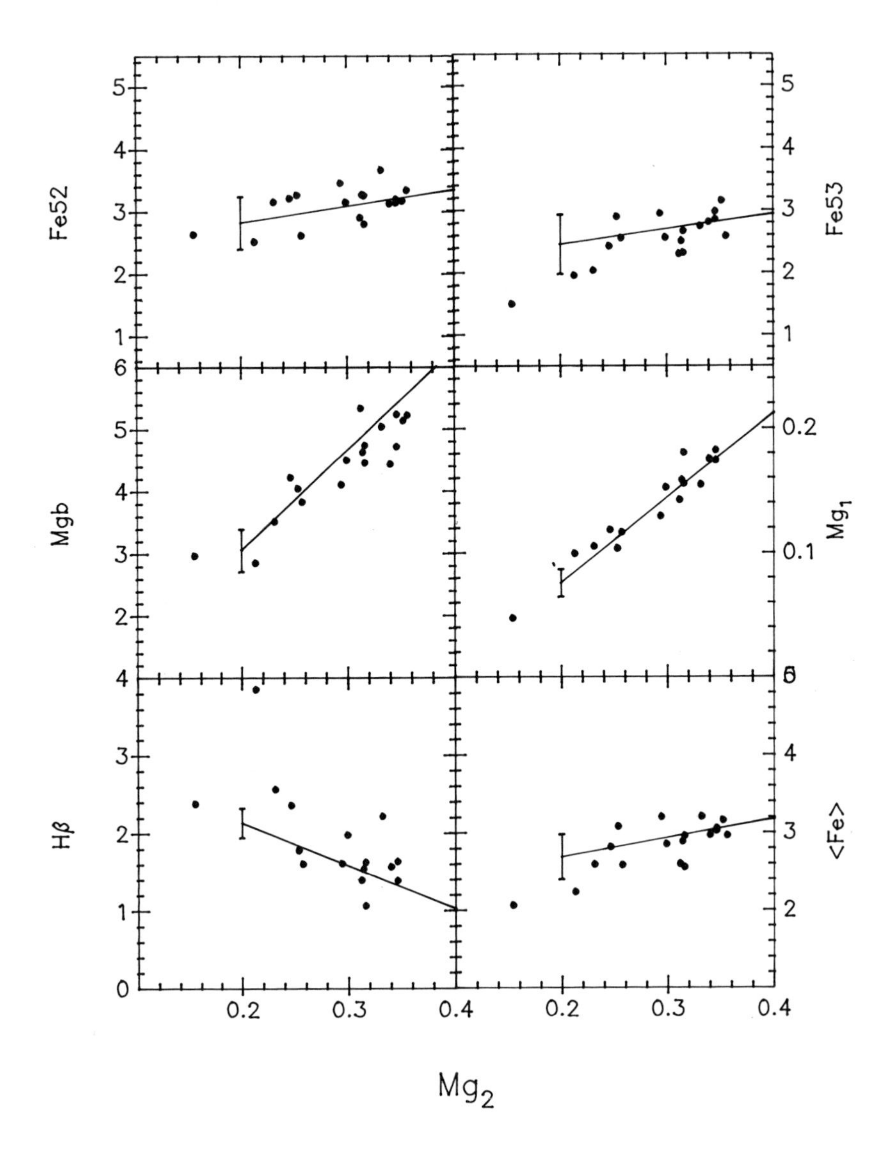

Figure 1. Comparison of nuclear line-strengths for elliptical galaxies in our sample with the mean relations (solid lines) determined by Burstein et. al. (1984). Their rms scatter is indicated by the error bars. <Fe> is the average of the two iron indices Fe52 and Fe53. The units of Fe52, Fe53, <Fe>, Mgb and Hβ are in Angstroms and those of Mg_1 and Mg_2 are in magnitudes.

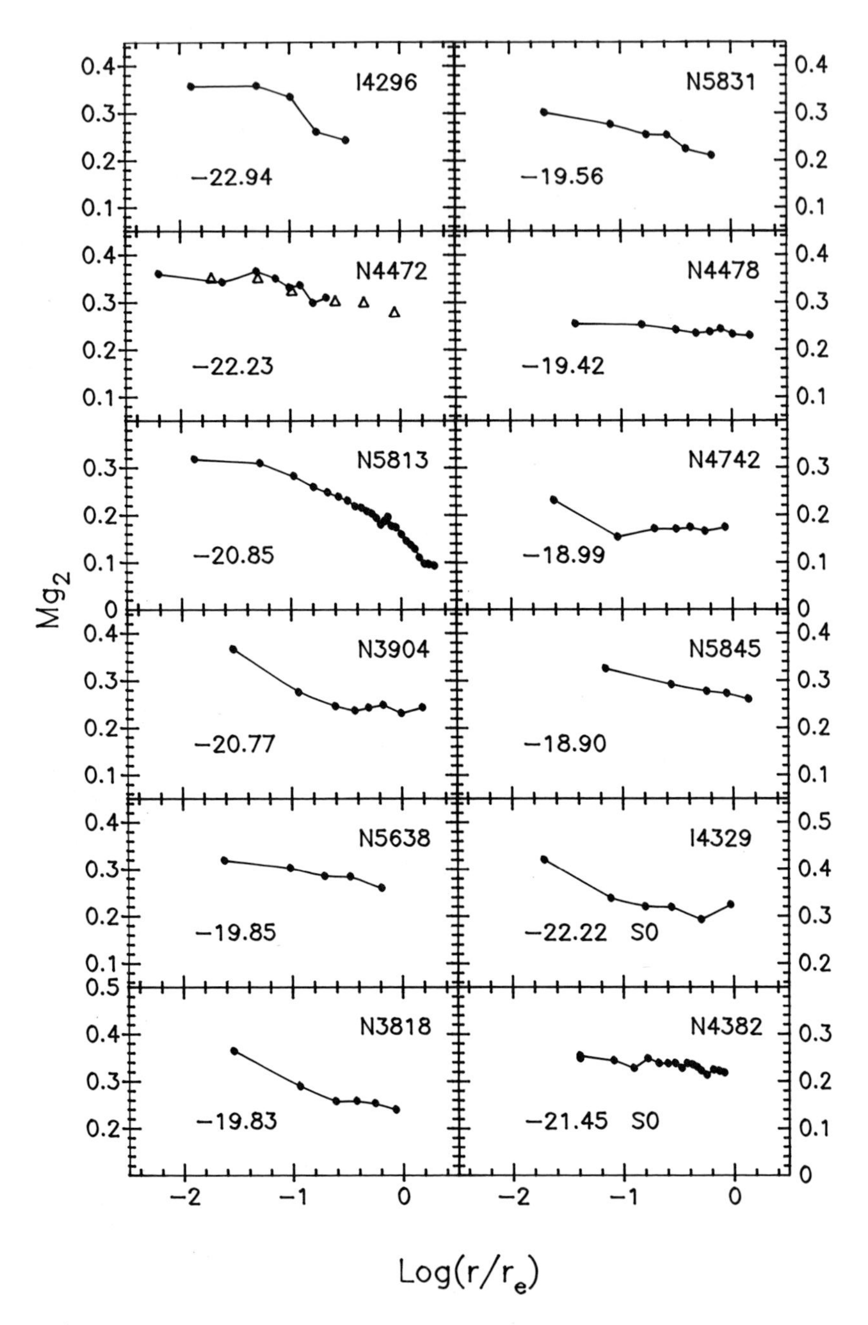

Figure 2. Major axis Mg_2 gradients for 12 early-type galaxies. We have plotted Mg_2 against $\log(r/r_e)$ where r_e is de Vaucouleurs' effective radius as listed in the Second Reference Catalogue of Bright Galaxies (de Vaucouleurs et. al. 1976). For each galaxy we give the absolute magnitude in the B_T system assuming a Hubble constant of H_o = 50 km s^{-1} Mpc^{-1}.

gradients for this sample is

$$Mg_2 = -(0.06 \pm 0.03) \log(r/r_e) + \text{constant} \qquad (1)$$

where the error represents one standard deviation. Converting such results into true metallicity gradients remains a major unsolved problem which we will not address in detail in this article (see the reviews by Burstein 1985, Pickles 1986 this volume, and references therein). In Figure 3, we show Mould's (1978) theoretical computation (as recalibrated by Terlevich et. al., 1981) for the relation between the Mg_2 index and [Fe/H] for an old (13Gyr) stellar population ([Fe/H] = $3.9Mg_2 - 0.9$). In addition, we have plotted Mg_2 values for galactic globular clusters measured by BFGK against estimates of [Fe/H] from Table 6 of Zinn and West (1984) (which gives [Fe/H] = $11.551\ Mg_2 - 2.274$). If Mould's calibration is correct then elliptical galaxies generally have weak metallicity gradients with [Fe/H] ~ $-0.22 \log(r/r_e)$.

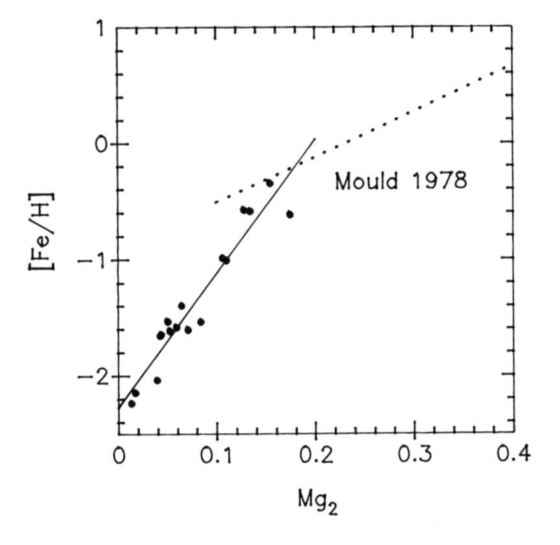

Figure 3. Mg values for galactic globular clusters (from BFGK) plotted against estimates of [Fe/H] from Zinn and West (1984). The solid line shows a least squares fit. The dotted line shows a calibration of the Mg_2 index at high metallicities for a 13 Gyr stellar population.

Gradients for three cD galaxies are presented in Figure 4. Our results show that these galaxies also possess modest Mg_2 gradients. The cD's all have shallower gradients than the mean given in equ. (1), but a larger sample is required to test whether this represents a systematic trend rather than a statistical fluctuation.

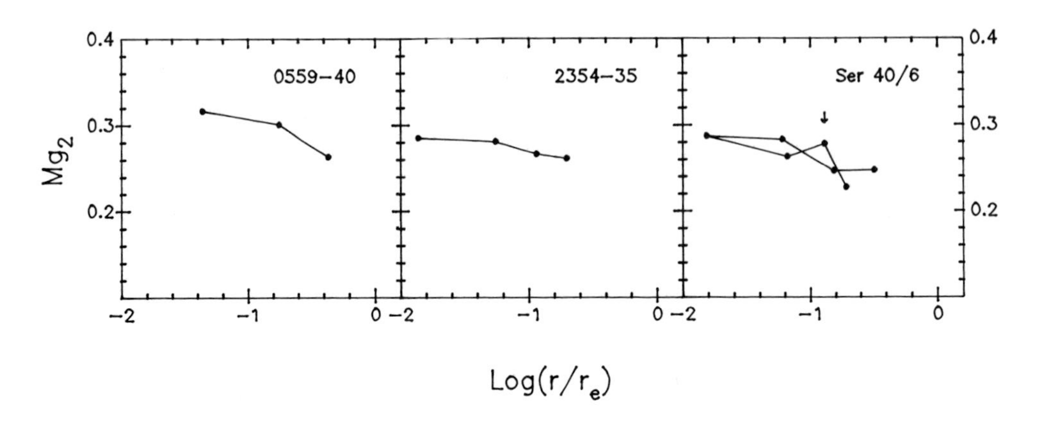

Figure 4. Major axis Mg_2 gradients in three cD galaxies. Sersic 40/6 is a dumbell galaxy and the position of the secondary nucleus is indicated by the arrow.

3. GRADIENTS IN OTHER FEATURES

For one galaxy, NGC 5813, we have particularly high signal-to-noise spectra extending out to large radii. We therefore present results for this galaxy separately in Figure 5 where we compare our line-strength estimates with those measured by BFGK for galactic globular clusters and for the nuclei of elliptical galaxies (see also Efstathiou and Gorgas 1985). This Figure shows that the line-strengths in the outer regions of NGC 5813 match continuously with those of metal rich galactic globular clusters. Notice that the Hβ index is nearly constant with radius outside the nuclear regions of the galaxy and thus the stellar population at large radii does not obey the same Mg_2-Hβ relationship as do the nuclei of elliptical galaxies. Previous studies of line-strength gradients have been based on changes in strong CN, Mg and Na features. Faber (1982) has stressed the importance of detecting changes in iron-peak features as a way of distinguishing between selective overenhancements in some elements and true abundance gradients. As Figure 5 shows, we do detect significant radial gradients in the iron features.

A similar comparison is shown in Figure 6 for elliptical galaxies from the sample of Davies et. al. (1983). The observational data for each of these galaxies are less extensive than for NGC 5813 so the line-strengths at large radii are less well determined. We have therefore averaged the results in the outer parts of each galaxy (typically between $0.5r_e$ to r_e). These points are plotted as the crosses in Figure 6. They have been connected to the nuclear line-strengths to give an impression of the gradients within each galaxy. These galaxies show similar trends to those found in NGC 5813.

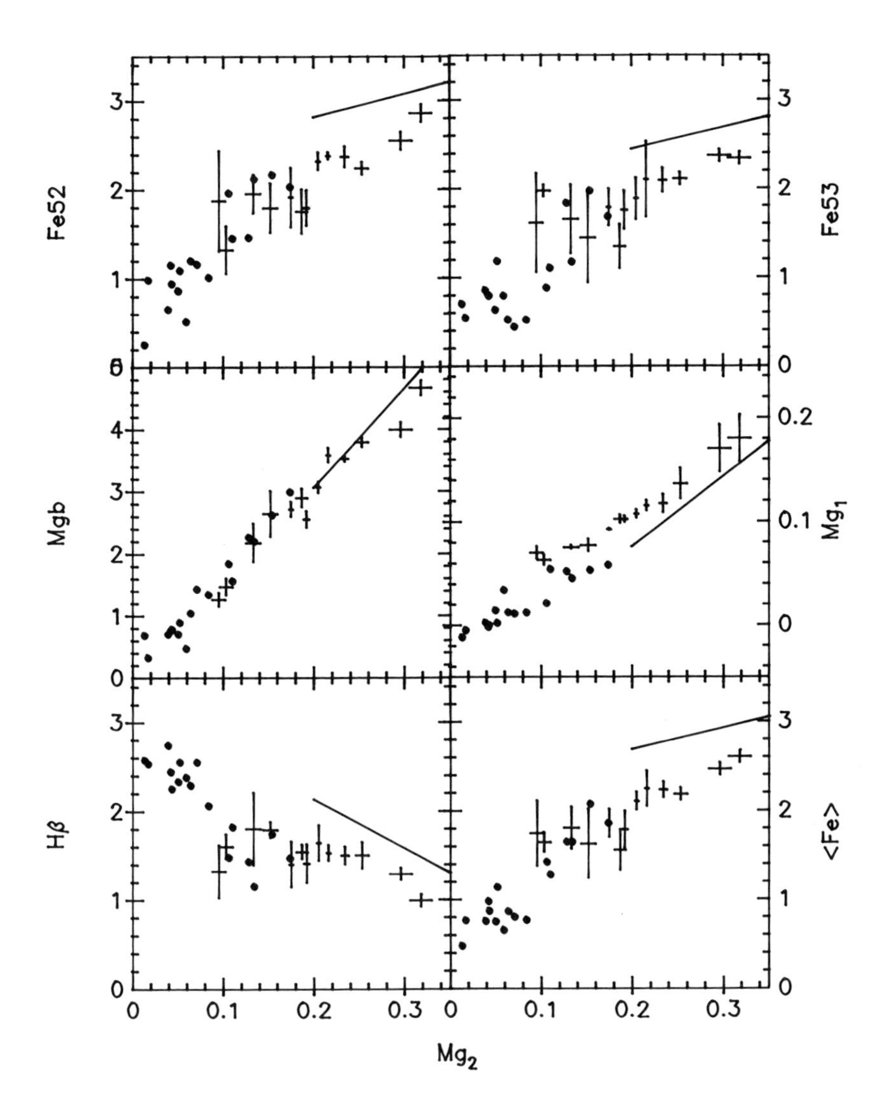

Figure 5. A comparison of the line-strengths in NGC 5813 (crosses) with those of galactic globular clusters (filled circles). The sizes of the crosses give estimates of 1σ random errors. We estimate that inaccuracies in sky subtraction could lead to systematic errors of up to 0.03-0.04 mag in Mg_2 for the outermost points. The solid lines show the mean relations deduced by BFGK for the nuclei of ellipticals.

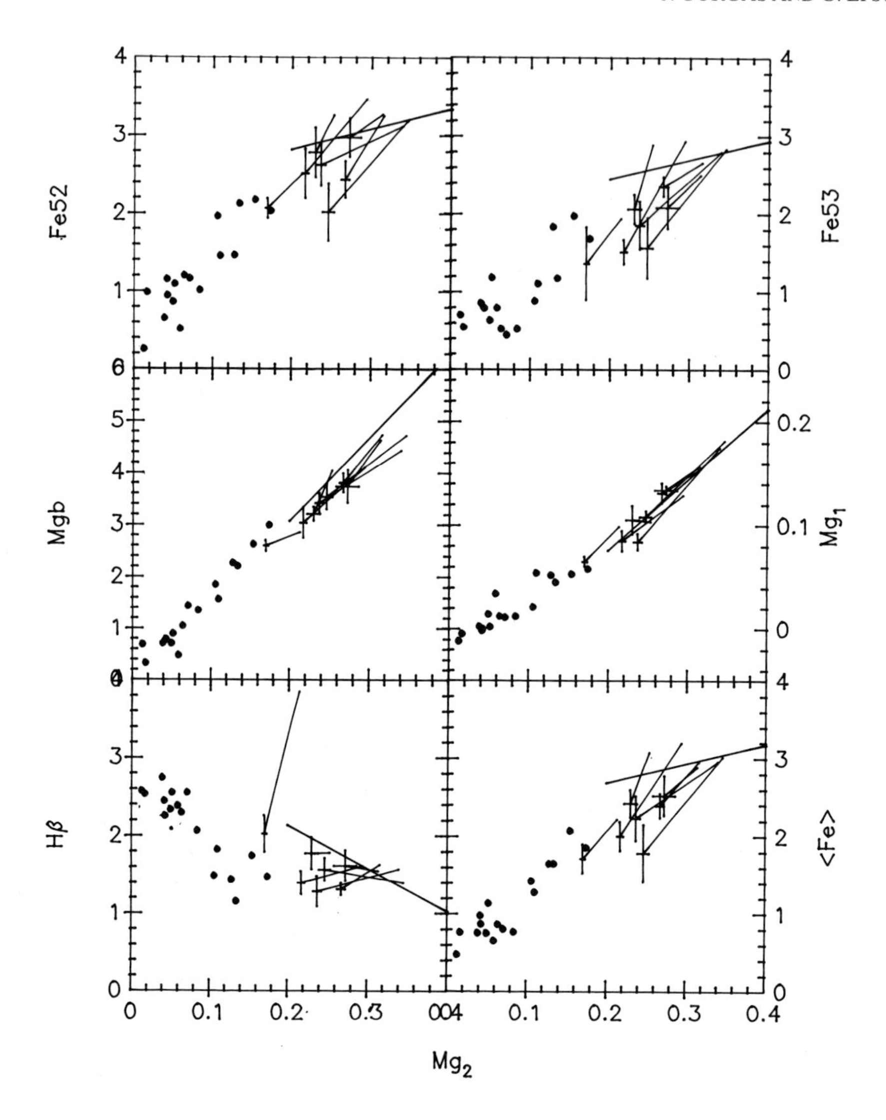

Figure 6. As for Figure 5 except we show our results for ellipticals from the sample of Davies et. al. (1983). The crosses show line-strengths in the outer parts of each galaxy (typically $0.5r_e$ - r_e). The crosses have been connected to the nuclear line-strengths by straight lines.

Hβ in the outer parts of these galaxies is lower than would have been inferred from the relation obeyed by elliptical galaxy nuclei. In addition the iron features, particularly Fe53, show quite steep gradients.

A detailed interpretation of these results in terms of stellar population models is beyond the scope of this article. The deviations shown in Figures 5 and 6 from a one parameter relation for the stellar populations in elliptical galaxies, especially in the Hβ-Mg_2 relation, may be indicative of variations in both age and metallicity. This possibility is explored further in the discussion following Pickles' contribution to these proceedings.

4. CONCLUSIONS

Our results show that most early-type galaxies, including cD's, possess line-strength gradients within one effective radius but the indices do not generally attain values as low as those of metal rich galactic globular clusters. This, in turn, suggests that typical abundance gradients are quite weak. The relative line-strengths in the outer parts of ellipticals appear to obey a different set of relations when compared to the nuclei of elliptical galaxies indicating that these stellar populations do not represent a one parameter family.

We have addressed only a few of the points raised in the introduction. Much more work needs to be done and we anticipate that this will be forthcoming as CCD/spectrograph combinations become more readily available on large telescopes. We have emphasised the inter-relationships between line-strength gradients and the variations seen amongst stellar populations in the nuclei of elliptical galaxies. More detailed work along these lines may help determine whether these relations can be understood in terms of a combination of age and metallicity variations.

REFERENCES

Burstein, D., 1979, Ap. J., **232**, 74.

Burstein, D., 1985, PASP, **97**, 89.

Burstein, D., Faber, S.M., Gaskell, C.M. and Krumm, N., 1984, Ap. J., **287**, 586.

Carlberg, R.G., 1984, Ap. J., **286**, 403.

Carter, D., Inglis, I., Ellis, R.S., Efstathiou, G. and Godwin, J.G., 1985, MNRAS, **212**, 471.

Cohen, J.G., 1986, Astron. J., in press.

Davies, R.L., Efstathiou, G., Fall, S.M, Illingworth, G. and Schechter, P.L., 1983, Ap.J., **266**, 41.

de Vaucouleurs, G., 1961, Ap. J. Suppl., **5**, 233.

de Vaucouleurs, G., de Vaucouleurs, A. and Corwin, H.G., 1976, "Second Reference Catalogue of Bright Galaxies", University of Texas, Austin.

Efstathiou, G. and Gorgas, 1985, MNRAS, **215**, 37p.
Efstathiou, G., Ellis, R.S. and Carter, D., 1980, MNRAS, **193**, 931.
Efstathiou, G., Ellis, R.S. and Carter, D., 1982, MNRAS, **201**, 975.
Faber, S.M., 1973, Ap. J., **179**, 731.
Faber, S.M., 1977, In "The Evolution of Galaxies and Stellar Populations" p157, eds. Tinsley, B.M., & Larson., R.B., Yale University Observatory, New Haven.
Faber, S.M., 1982, Highl. Astr.. **9**, 165.
Fabian A.C., Nulsen, P.E.J. and Canizares, C.R., 1984, Nature, **310**, 733.
Freeman, K.C., and Norris, J., 1981, Ann. Rev. Astr. Ap., **19**, 310.
Gallagher, J.S., 1984, in "Stellar Nucleosynthesis", p325, eds. Chiosi, C. and Renzini, A.. Reidel, Dordrecht.
Gorgas, J. and Efstathiou, G., 1986, in preparation.
Gunn, J.E., Stryker, L.L. and Tinsley, B.M., 1981, Ap. J., **249**, 48.
Hausman, M.A. and Ostriker, J.P., 1978, Ap.J., **224**, 320.
Larson, R.B., 1975, MNRAS, **173**, 671.
Mould, J.R., 1978, Ap. J., **220**, 434.
Oke, J.B., 1974, Ap. J. Suppl., **27**, 21.
O'Connell, R.W., 1976, Ap. J., **206**, 370.
O'Connell, R.W., 1980, Ap. J., **236**, 430.
Pickles, A.J., 1985, Ap. J., **296**, 340.
Rose, J.A., 1985, Astron. J., **90**, 1927.
Spinrad, H., Smith, H.E. and Taylor, B.J., 1972, Ap. J., **175**, 649.
Strom, K.M. and Strom, S.M., 1978, Astron. J., **83**, 73.
Terlevich, R., Davies, R.L., Faber, S.M. and Burstein, D., 1981, MNRAS, **196**, 381.
Tonry, J.L. and Davis, M., 1981, Ap. J., **246**, 666.
Tifft, W.A., 1961, Astron. J., **66**, 390.
Wirth, A. and Shaw, R., 1983, Astron. J., **88**, 2.
Zinn, R. and West, M.J., 1984, Ap. J. Suppl., **55**, 45.

DISCUSSION

White: Is it fair to say on the basis of your data that more than half of the stars in these relatively bright ellipticals (those outside r_e) have lower metallicities than the more metal-rich galactic globular clusters?

Efstathiou: We overlap with the metal-rich galactic globulars in one galaxy, NGC 5813, and here the inferred metallicity is about [Fe/H] $\sim$ -0.5 at $r \sim 1.5 - 2r_e$. The line-strengths for the rest of the galaxies in our sample are still larger than those of metal-rich galactic globulars at the outermost points that we have measured ($< r_e$). Data at larger radii are needed to fully answer your question. However, results on colour gradients, in particular recent work by Cohen, suggest that the halos of ellipticals are not extremely metal poor. I should add that we have checked the analysis of NGC 5813 and we do not think that errors in sky subtraction could eliminate the overlap with galactic globular clusters.

Davies: Sadler and I find weak emission filling in the $H\beta$ line in the centres of ellipticals, which causes the line–strength to fall in the centre. Thus the true central strength of $H\beta$ is stronger, and the need for a blue stellar component increased. We also find that the brighter galaxies have shallower Mg_2 gradients, however, given the work of Guinan and Smith (1984, *Publ. Astr. Soc. Pacific*, **96**, 354) and Faber, Friel, Burstein & Gaskell (1985, *Astrophys. J. Suppl.*, **57**, 711) showing the gravity sensitivity of this index, I see no reason to assume a linear relationship between Mg_2 and [Fe/H], especially as the gradients in Fe are so shallow and these are the only independent line–strengths measured.

Efstathiou: Let me take your points in turn: (i) your results on $H\beta$ are very interesting. We find that the $H\beta$ equivalent width falls in the very centre of NGC 5813 and the feature is essentially absent in the centre of the cD galaxy PKS2354–35, but in neither case can we tell whether this is caused by emission filling in the absorption feature (we find no evidence of [O III] emission lines. (ii) We do not find a statistically significant correlation between absolute magnitude and Mg_2 gradients. (iii) Our use of Mould's work is meant to be illustrative. A more extensive investigation between Mg_2 and metallicity would be worthwhile, though we have tried to stress the importance of looking at gradients in a variety of features. Our data do show substantial gradients in the Fe indices. In fact, the gradients in the two Fe blends are generally steeper than would have been inferred from the relations found for the nuclei of ellipticals.

Peletier: Regarding the dust seen in the centre of NGC 5813 by, e.g., Jedrzejewski, how do you think dust affects the line–strength gradients?

Efstathiou: The sidebands span such a small wavelength range that I don't think there could be a significant effect.

Burstein: The correlation of Mg_2 with metallicity is still somewhat misunderstood. Although the Mg index does increase in strength with metallicity, the giant branch also becomes much redder and Mg_2 increases strongly with decreasing temperature. Much of the change of Mg_2 with [Fe/H] in integrated spectra comes from this temperature dependence. Gravity effects in Mg_2, while important, are less than these temperature effects.

George Efstathiou, with an advertisement for the AAT.

Josh Barnes took all the photos, except this one.

POPULATION SYNTHESIS OF COMPOSITE SYSTEMS

Andrew Pickles
Kapteyn Laboratorium,
Postbus 800, 9700 AV Groningen,
The Netherlands

ABSTRACT: Metallicity and age dispersion among the stars comprising stellar composite systems are discussed, with emphasis on the present and future implications for the technique of population synthesis.

1. INTRODUCTION

In preparing this talk I have referred to a wealth of excellent material, too extensive to list here in full. In particular however, I have tried to avoid excessive repetition of topics already amply reviewed by Burstein (1985), O'Connell (1985) and Whitford (1985a,b).

The identification of two distinct populations by Baade (1944) laid the framework for the concept of discrete stellar populations within composite systems. In particular he identified M32 and the bulge of M31 as pure population II, analagous to the oldest and lowest abundance tracers in our Galaxy. This established the expectation that purely Stellar Composite Systems (SCS), with no gas or dust, would consist entirely of metal-weak and old stars formed in the earliest stages of galaxy formation.

The idea that stellar populations could be well represented by a few essentially discrete points within a four-fold continuum of age, metallicity, mass function and population dynamics implies a similar separability in star formation history. The initial expectation that spiral bulge and elliptical stars would prove to be metal-weak foundered early however with Morgan's (1956) observations of strong lines in the brighter systems, prompting an extensive reapraisal of stellar characteristics in these systems. Subsequent work established the morphological and photometric similarity between spiral bulges and ellipticals (de Vaucouleurs 1961, Johnson 1966a, Frogel *et. al.* 1978 - FPAM). Faber (1977) emphasised their similarity of line strength gradients, and Whitford (1978) identified the Galactic bulge as being spectroscopically similar to other spiral bulges and ellipticals.

Early narrow-band and scanner work to describe the integrated light of galaxies in terms of their stellar constituents was instrumental in

T. de Zeeuw (ed.), Structure and Dynamics of Elliptical Galaxies, 203–216.

establishing the high abundance of bright systems and the variation of mean metallicity with absolute magnitude (Wood 1966; McClure and van den Bergh 1968; Spinrad and Taylor 1971 - ST; Faber 1972, 1973). Faber (1977) summarised the evidence that mean metallicity decreasing with luminosity was the basis for the colour-magnitude relation observed in SCS (Visvanathan and Sandage 1977). Subsequent work confirmed this (FPAM; Aaronson *et. al.* 1978 - ACMM; Frogel, Persson and Cohen 1980).

At this point perhaps the safest interpretation of stellar populations in SCS was that they were of similar age (dynamics and mass function), but distinguished by their mean metallicity as a consequence of initially differing galaxy mass. By identifying the age as that of Galactic Globular Clusters (GGCs), this hypothesis has attraction as satisfying much of the observational data and being not too distant from Baade's original and powerful arguments.

There are problems with this picture however. It is a fact that no synthesis of an SCS has achieved a fit without at least one of: i) inclusion of a young component in addition to a basically old population (Gunn, Stryker and Tinsley 1981 - GST; Bertola *et. al.* 1986), ii) inclusion of a metal-weak component in addition to a basically metal-rich population (ACMM), iii) adoption of an intermediate age (roughly half that of GGCs) for the dominant population (Wood 1966; McClure and van den Bergh 1968; ST; Faber 1972, 1973; O'Connell 1976, 1980; Williams 1976; Pritchett 1977; Pickles 1985b; Rose 1985).

A comparison of synthesis fits shows them to be more remarkable for their similarities than differences. The problems of non-uniqueness in the presence of observational errors have been noted by several of the above and Peck (1980). But it is clear that this work has defined the general distribution of stellar numbers and light contributions within SCS, has converged to a relatively small range of acceptable mass function (power law) slopes, and mapped out the trend of mean metallicity with absolute magnitude.

I would suggest that a major source of residual disagreement comes not from observations or problems with synthesis techniques, but from attempts to derive global galaxy population parameters which are too restrictive. In essence I think the synthesis work tells us that SCS populations are truly composite, and should be interpreted in terms of distributions of (at least) metallicity and age.

More direct evidence for lack of metal-age uniformity in SCS comes from U and particularly uv observations of ellipticals (Nørgaard-Nielson and Kjærgaard 1981; Oke, Bertola and Capaccioli 1981; Wu *et. al.* 1980). The excess flux below 1800Å requires either a young or metal-weak component in an otherwise old and metal-rich system (Nesci and Perola 1985). Also the work of the Lick group indicates marked differences between Galactic globular clusters, M31 globulars and elliptical nuclei in terms of Hβ and CN line stengths, which can not be explained solely in terms of metallicity variations (Burstein *et. al.* 1984; Burstein 1985). Finally there has been a crescendo of direct work on stellar populations in local group systems, finding metallicity and age spread in globular clusters in the Magellanic clouds (Gascoigne 1980; Searle, Wilkinson and Bagnuolo 1980; Rabin 1982), in globulars and the halo of M31 (Burstein *et. al.* 1984, Mould and Kristian 1986), and evidence for metallicity spread (Whitford and Rich 1983, Terndrup, Rich and Whitford 1984) and young populations (Wood and Bessell 1983) in the Galactic bulge. This direct work more than any other perhaps is redefining the acceptable range of stellar populations in SCS.

I summarise below the evidence for both metallicity and age spread within SCS, and outline prospects for an expanded synthesis scheme based on available observations and theoretical data.

2. METALLICITY DISPERSION WITHIN SCS

2.1 Synthesis fits with metallicity dispersion

Several empirical population synthesis fits have included a range of stellar metallicities in their flux libraries. The strongest lined stellar components are usually required to match the strong lined composite spectra of bright systems such as the nuclei of M31 or bright ellipticals. Indeed a failure to completely match strong lines in galaxies has been taken as evidence for incompleteness in the stellar libraries at the high metallicity end. Fainter ellipticals have line strengths which can be well matched by existing libraries, but still with a ***range*** of metallicities from those available. Thus empirical models for M32 by McClure and van den Bergh (1968), ST and O'Connell (1980) utilise stellar components covering more than 1 dex in [Fe/H], from integrated globular cluster light to strong lined giant components. In fits to ellipticals covering a range of luminosities I also found most required a combination of different metallicity components (Pickles 1985b).

2.2 Direct evidence for metallicity dispersion

The most direct evidence for large metallicity spread within SCS comes from observations of stars in the Galactic centre and other local group systems. Whitford and Rich (1983 - WR) found a metallicity spread among Galactic bulge giants of more than 2 dex in [Fe/H]. Mould, Kristian and Da Costa (1983, 1984) found relatively small metallicity dispersions of 0.3 dex in NGC147 and 0.5 dex in NGC205 (dwarf elliptical companions to M31). Mould and Kristian (1986) find a dispersion of 1 dex or more in the halo of M31 however, and discuss the possibility of dispersion increasing with mean abundance.

2.3 Metallicity dispersion in evolutionary syntheses

Tinsley (1972) discussed the chemical enrichment process within ellipticals, and noted that steady enrichment during initial collapse and star formation could explain radial colour and and composition gradients. In general however, models by Tinsley and co-workers and by Bruzual (1983) have adopted a simpler scenario in which rapid star formation in the initial collapse produces a stellar population of roughly uniform composition, generally taken as solar for convenient application to existing isochrone and stellar data.

Arimoto and Yoshii (1986 - AY) have recently extended this work to explicitely allow for chemical enrichment during star formation within ellipticals. The effects included in their scheme are:

i) star formation rate (SFR) linearly proportional to the gas fraction until the thermal energy of the gas exceeds its binding energy, after which remaining (enriched) gas is expelled by a supernova driven galactic wind.

ii) metal enrichment based on star formation, nucleosynthesis and re-ejection via supernovae; numerical integration of the gas fraction and chemical enrichment equations occurs in time steps of 5.10^6 years.

iii) evaluation of star formation at each time (and gas metallicity) step, for a range of SFR and initial mass function (IMF) parameters.

iv) evolution of all stars along the appropriate (Sweigart and Gross 1978) tracks.

By explicitely including chemical enrichment during star formation, AY find that both the SFR ***and*** the IMF strongly affect the colours of galaxies. Metallicity enrichment is amplified and accelerated by having a giant dominated mass function, causing colour evolution (as well as luminosity evolution) to depend strongly on the IMF. These results can be compared with previous results (Tinsley 1972, Tinsley and Gunn 1976) which showed relatively small colour dependence on the IMF slope, but used single (solar) metallicity models.

The final (present day) stellar metallicity distribution derived by AY covers more than 2 orders of magnitude in [Fe/H]. This theoretically derived metallicity spread compares well with the distribution observed in the Galactic centre. I think the degree of correspondence with that observed by WR in a real SCS is both remarkable and gratifying. It is also very cautionary to all practitioners of the population synthesis technique.

AY also derive the variation of mean metallicity with total galactic mass. The variation with mass occurs because the supernovae driven thermal energy exceeds the gravitational binding energy earlier in lower mass galaxies, causing earlier (and proportionately larger) gas expulsion, and earlier termination of star formation.

3. DWARF TO GIANT RATIOS AND M/L IN SCS

A notable success of population synthesis has been the ability to determine mass to light ratios for the visible component to within quite high precision (factor of 2 roughly). Some early results gave mass to light ratios significantly higher than those determined dynamically for individual galaxies. These fits are now thought to be wrong for a combination of reasons; using too many late-type dwarfs to fit the strong NaI lines, and putting too much reliance on early infrared data. Later work showed much lower values were preferable if proper account was taken of the cool giant population (O'Connell 1976).

Since the lower main sequence dwarfs contribute so much mass and so little light, it is difficult to determine their contribution from medium resolution optical spectra. Detailed work on individual line strengths such as the FeH band at 990nm (Whitford 1977), and infrared features such as the CO band at 2.3μ and the H_2O band at 1.9μ (FPAM; ACMM; Aaronson, Frogel and Persson 1978) show the presence of late giants, cooler and brighter than found in GGCs, and indicating M/L values $\lesssim$ 10. This corresponds to a (power law) IMF slope less than or equal to the commonly used Salpeter (1955) value of x=1.35. Evolutionary syntheses based on fits to wide colour ranges have also converged to similar values (GST, AY). It is important to note that population synthesis is really fitting the dwarf:giant ratio, particularly in the infrared, and mass functions other than a simple power law can also fit.

One important aspect of this is that population synthesis, unlike dynamical determinations, gives M/L independent of distance. This has been a major impetus behind attempts to determine these values more precisely, by synthesising higher resolution near infrared spectra (Carter, Visvanathan and Pickles 1986) or CVF spectra of the CO band (Arnaud and Gilmore 1986, and this conference).

4. AGE DISPERSION WITHIN SCS

The most serious problem facing population synthesis is that of age determination of the stars forming the composite system. This is also the area in which there is most disagreement, real and percieved, between different approaches and between empirical as opposed to evolutionary synthesis. I would like to suggest that this is (in order of priority):

i) primarily due to ***external*** problems, particularly regarding isochrone placement in the HR diagram,

ii) partly due to interpretational problems, especially with regard to the tendency to summarise metallicity and age distributions in terms of single characterising parameters,

iii) partly due to real differences in synthesis techniques, specifically with respect to the relative weight placed on fitting continuum and spectral features, and

iv) only slightly due to problems with non-uniqueness, providing that the data quality is good (S/N $\gtrsim$ 25).

4.1 Isochrone placement

Figure 1 shows two interpolated Yale isochrones (solid lines) I have used for evolutionary synthesis. The parameters are (Z=0.017, Y=0.23, T=10 Gyr, α=1.6). The metallicity changes were accomplished by direct interpolation among the Yale isochrones (Ciardullo and Demarque 1977 - CD), and the ratio of mixing length to pressure scale height adjusted empirically (from α=1.0) to the more popular value by applying Twarog's (1980) algorithm - that is by increasing Log $T_{\bullet}$ by between 0.005 on the main sequence and 0.025 on the giant branch, without adjusting M_{bol}. The amplitudes of these shifts are indicated by arrows.

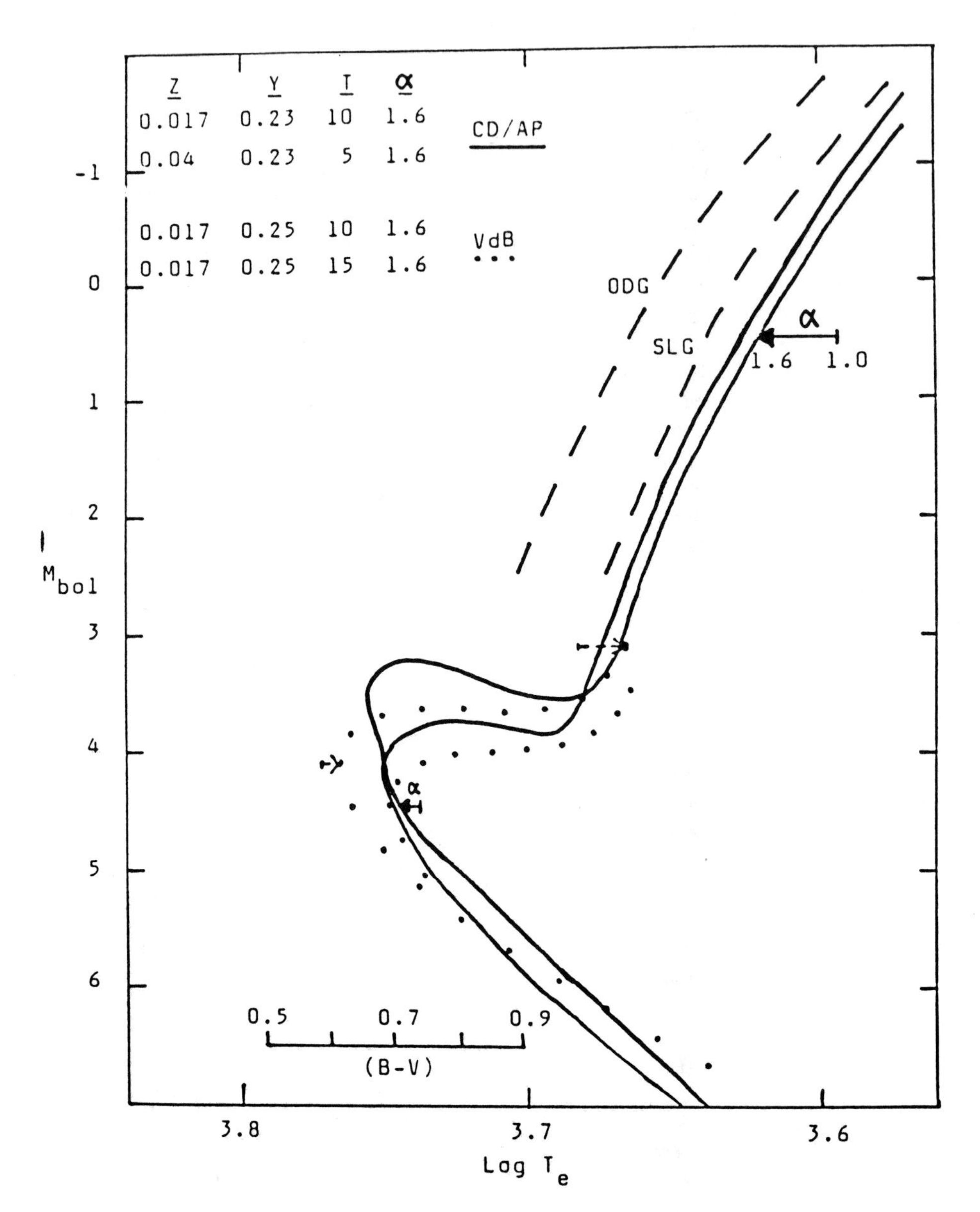

Figure 1: compares the position in the Mbol-LogTe plane of two interpolated and adjusted Yale isochrones (CD/AP, solid lines), and two shifted VandenBerg isochrones (VdB, dotted lines). The abundance, age and mixing length ratio parameters are given in the top LH corner. Arrows on the giant branch and main sequence indicate the empirical blueward shift applied to the Yale isochrones to achieve the higher mixing length ratio, and the redward colour shift applied to the VandenBerg isochrones. Also shown are the positions (dashed lines) of the Old Disk Giant and Strong Lined Giant branches used by Gunn, Stryker and Tinsley.

These isochrones are very close to the revised Yale isochrones (Green, Demarque and King 1985). At the bottom is a (**B**-**V**) colour scale from Johnson (1966b).

This diagram illustrates one problem concerning age dating via population synthesis. Empirical syntheses are not heavily constrained by ***a priori*** isochrone data, and typically give elliptical turnoff colours of (**B**-**V**)~0.7. This implies Yale ages of about 10 Gyr for a solar metallicity population, or younger for a metal-rich population. In evolutionary syntheses of Virgo and Coma ellipticals, GST preferred redder turnoff colours of (**B**-**V**)~0.8, which clearly imply older ages. O'Connell (1985) has pointed out one major difference between the two approaches which concerns giant branch placement. Dashed lines in the top part of figure 1 indicate the giant branches used by GST. Both the Old Disk Giant branch (ODG - used for GST's solar metallicity models) and the Strong Lined Giant branch (SLG - used for their metal-rich models) are substantially bluer than the Yale giant branches. This difference then results in a redder turnoff colour to compensate.

Clearly the question of isochrone placement is ***external*** to that of population synthesis. If we knew their correct placement this source of disagreement would disappear. Population synthesis is not a lottery; you can't get any answer you like, but you can get a range of answers depending on the framework in which you work - and this framework is ***not*** well defined at the higher metallicity end.

VandenBerg (1985) has recently calculated a new set of isochrones which explicitely allow for a higher mixing length ratio, aswell as using the latest opacity and nuclear reaction rate data; two of these are shown as dotted lines in figure 1. They have parameters (Z=0.0169, Y=0.25, T=10 and 15 Gyr, α=1.6), corresponding to solar metallicity. They have been shifted redward in colour according to the recipe given in VandenBerg (1985); this empirical shift is necessary to give reasonable fits to the sun with a 5 Gyr solar metallicity isochrone, and to GGCs with lower metallicity isochrones.

The shifted VandenBerg isochrones are bluer than the corresponding Yale isochrones (only partially due to the small helium abundance difference), implying older ages for a given turnoff colour. From the point of view of population synthesis there is a very rough correspondence in turnoff colour (at **B**-**V** ~ 0.7) between a) a Yale isochrone of 2½ times solar metallicity and age 5 Gyr, b) a Yale isochrone of solar metallicity and age 10 Gyr, and c) a VandenBerg isochrone of solar metallicity and age 15 Gyr. VandenBerg (and others) are continuing to work on metal-rich isochrones, and it seems clear that population synthesis should play as significant a role here in fitting theory to observations, as do direct observations of CM diagrams at lower metallicities. The point being that there are very few metal-rich populations accessible to direct CMD work.

4.2 Effects of metallicity dispersion

In view of the large metallicity dispersion present in galaxies, it is obviously pertinent to ask whether the blue turnoff colours indicated by empirical syntheses could be due to a large metal-weak component.

Rose (1985) has dealt with this problem quite extensively for the particular case of M32. He finds that a metal-poor contribution of 8% light at 4000Å provides a good fit to his (H_ϵ + CaII H)/CaII K index, whereas a much larger contribution of 20% would be needed to fit the colours if all stars were old. Notably a metal-weak fraction of ~8% is commensurate with that predicted by AY. In fact it is also that predicted by Pagel and Patchett (1975) for approximately the same choice of heavy yield parameter which reproduces the WR results in the Galactic centre (Pagel 1985; Rich 1986).

In comparing observations with theory it is certainly better at this stage to look at galaxies of solar mean abundance or less. Existing stellar libraries are then more complete, and comparison with both Yale and VandenBerg isochrones is possible. Figure 2 shows the situation pertaining to three Fornax ellipticals (NGC1336, JJ51, JJ79) of roughly solar mean abundance, which have been synthesised with a flux library containing a range of metallicities (Pickles 1985a,b). Note that the temperatures of the stellar groups are fixed by their observed colours, but that the absolute magnitudes can be adjusted slightly, particularly on the giant branch. Only solar metallicity dwarfs and subgiants were available for these fits, but the flux library did contain horizontal branch stars (spectral types B to F) and the fit adopts a 5% **V** light contribution from these stellar groups. The adjusted Yale isochrones shown (solid lines) are all of old age (15 Gyr, α=1.6), and are of low, solar and high metal abundance. Notable features of figure 2 are:

i) The giant branch can be fitted by a purely old (15 Gyr) population of solar mean abundance and some metallicity dispersion. The age resolution is poor on the giant branch however, and the same can be said for younger (10 Gyr or less) isochrones there.

ii) The main sequence requires a substantial contribution from blue stars. The numbers indicate that these would have to be ***all*** metal-weak to fit a Yale 15 Gyr isochrone. Younger isochrones provide a better fit to the main sequence and giant branch simultaneously.

iii) There is no compelling evidence from these (or other) synthesis fits to normal ellipticals for a substantial spread of ages. Most fits work well with a single (intermediate) age and a spread of metallicity.

What could be wrong here and is there any reasonable way to get a basically old population?

i) I could force in metal-weak giants and redden the turnoff colour slightly. This is not appropriate within the context of either the Yale or VandenBerg isochrones however.

ii) I could boost the horizontal branch contribution. But Rose (1985) finds an upper limit of 4% at 4000Å for these hot stars in both field and Virgo ellipticals, by fitting to the H_ϵ/CaII ratio. The horizontal branch fraction here contributes 5% at **V** or about 8% at 4000Å, so is already excessive.

iii) Rose's restrictions on hot stars apply to blue stragglers or uv bright stars aswell, so I cannot add these stars freely to provide the blue light.

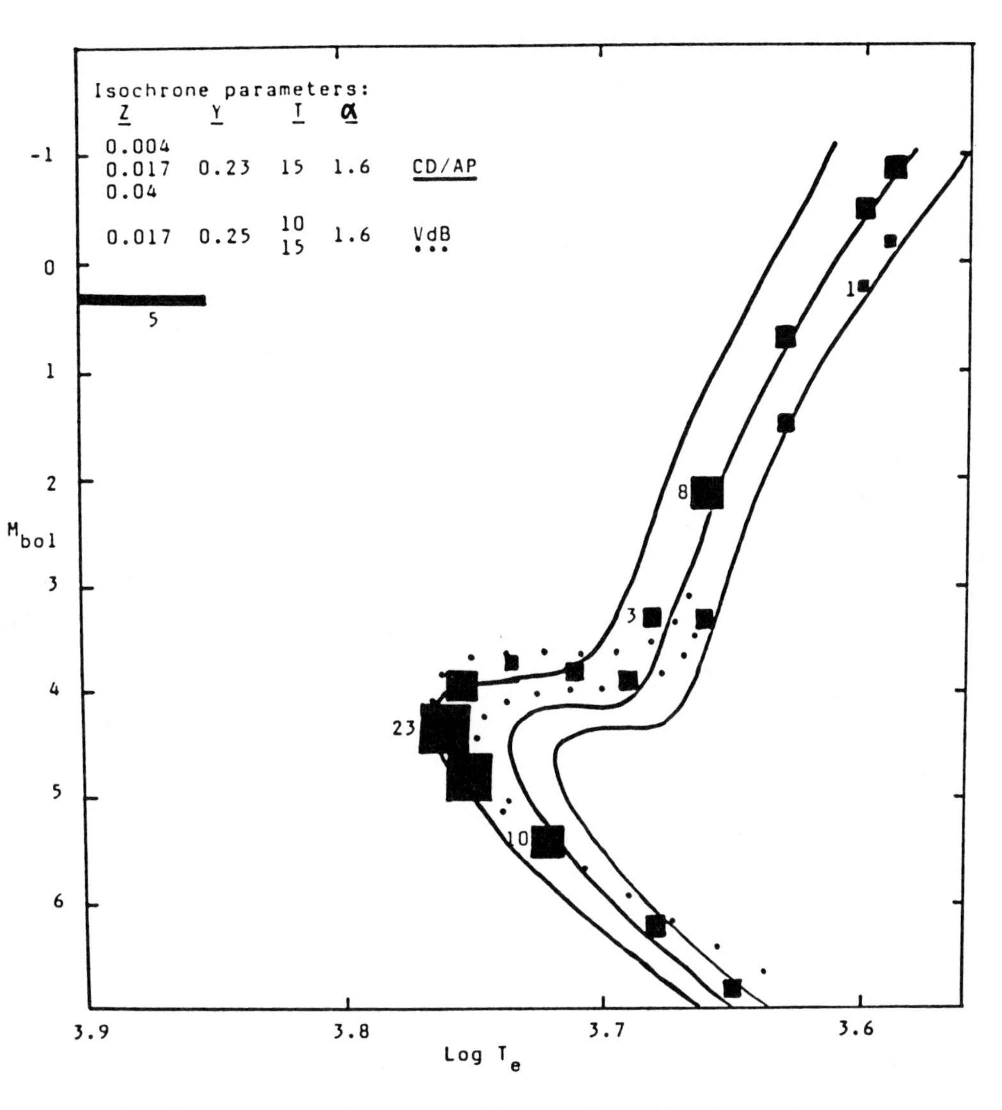

Figure 2: The mean synthesised light distribution of 3 Fornax ellipticals (solar mean abundance) is represented by discrete squares. Each square represents the location in the Mbol-LogTe plane of a stellar group used in the synthesis; their area is proportional to the percentage V light contributed by that group, with numbers given for representative groups. The horizontal branch contribution is represented by a rectangle extended in temperature.

The synthesised light distribution is compared with 3 interpolated and adjusted Yale isochrones (solid lines) of weak solar and rich metal abundance, all of age 15 Gyr. Also shown are Vandenberg's 10 and 15 Gyr, solar metallicity isochrones (dotted lines) from figure 1.

Use of the VandenBerg isochrones (dotted lines) clearly gives ages older than indicated by the Yale isochrones, but the light distribution is still ***not*** well fitted by purely old isochrones. The presence of a blue turnoff appears to be unavoidable within the context of either Yale or VandenBerg isochrones; this is strongly supported by Rose's higher resolution data which sets strong limits on the permissible contributions from hot stars which could otherwise compensate for a redder turnoff. A blue turnoff is even more pronounced for ellipticals like M32 (O'Connell 1980, 1985; Rose 1985). The prospects of satisfactorily fitting all or even some ellipticals with purely old isochrones just do not look good, even allowing for wide metallicity dispersion. Intermediate ages, or at least a significant age dispersion, appear to be a necessary component of these 'simple' systems.

4.3 Origin of age dispersion

Bruzual (1983) has calculated models of elliptical galaxy light distribution as a function of lookback time, which have been extensively used in the analysis of galaxy colours in clusters of high redshift. Bruzual's (solar metallicity) models are characterised by a parameter 'μ' which describes the exponential rate at which stars form out of the original gas. Without getting too deeply into this additional but related topic, it seems clear that observations of galaxies at large distances are mostly contained within the envelope specified by μ=1.0 (effectively instantaneous star formation) and μ=0.5 (half the gas converted to stars in 1 Gyr) - (Lilly and Longair 1984; Eisenhardt and Lebofsky 1986).

It is interesting to compare Bruzual's models, which are quite well known, with those of Arimoto and Yoshii, which are less so. The AY models use a parameter 'ν', which quantifies the SFR (per unit mass of residual gas) in units of the PDSFR in the solar neighborhood. Despite the fact that gas conversion to stars is not exactly an exponential in the AY models (and they explicitely reject the instantaneous recycling approximation), there is a rough analogy which is given here (Arimoto 1986 - private communication):

Arimoto and Yoshii ν	Bruzual μ
1	0.1
6	0.5
11	0.7
21	0.9
42	0.99

The star formation rate determined by AY to best fit present day giant elliptical colours, assuming star formation started 15 Gyr ago, is ν=42, corresponding to rapid star formation and a galactic wind occurring less than 1 Gyr after its onset. Their star formation rate parameter would decrease however if a higher mass function slope were used (they used x=0.95 in Tinsley's notation), and particularly if they adjust their models to a higher value of the mixing length ratio. They have used stellar colours derived from Sweigart and Gross (1978) evolutionary tracks for a mixing length ratio of α=1.0; increasing this makes the component stars bluer, necessitating a slower SFR to match present day galaxy colours. An estimate of the SFR parameter appropriate to a mixing length ratio of α=1.6 is ν=10 - 20 (Arimoto 1986, private communication), implying quite extended star formation terminating after 5-10 Gyr with the onset of a galactic wind. These calculations are being repeated for the higher ratio of mixing length to pressure scale height.

5. CONCLUDING REMARKS

It may be possible to reconcile all the observations and 'classical' concepts of elliptical galaxy formation with temporally extended star formation, culminating in a galactic wind which ejects hot, metal-enriched gas into the intergalactic medium. It seems to be possible to reproduce the colours this way; it should also be possible to construct synthetic spectra appropriate to the AY models for comparison with observations.

If an extended but declining star formation process cannot reproduce the observed spectra, then another source of gas for intermediate age star formation must be found. This could come from mergers, which qualitatively at least might explain the radial variation in clusters of the Elliptical to Spiral ratio. As a stochastic process they might also explain the colour range amongst dominant galaxies in clusters at large lookbacks. An alternative source of gas for late star formation might be infall of ejected gas, which might occur preferentially onto galaxies in the centre of the cluster potential well.

The key to unravelling these uncertainties possibly lies in a study of the radial variation of stellar populations in nearby ellipticals. This is necessary to enable quantitative discussion of population variations following dynamical interactions such as mergers or cooling flows. It is also necessary to rationally compare distant (whole) galaxies with nearby elliptical nuclei. Existing data indicate that radial variations are more complicated than a simple radial decline in mean abundance (Efstathiou; Peletier, Valentijn and Jameson; this conference).

Good spectra of distant galaxies at large lookbacks have been and can be obtained. These should provide a reliable check on the time evolution of stellar populations within galaxies. For synthesis however, it is necessary that the data be of sufficient quality to enable good deconvolutions.

Finally, since population synthesis is only a poor substitute for direct CMD work (but all that is possible outside the local group), every effort should be made to feed the expanding knowledge on local group populations into the synthesis process.

I thank the organisers and participants for a pleasant and memorable conference.

REFERENCES

Aaronson M, Cohen J.G, Mould J & Malkan M, (ACMM), 1978, *Ap. J.* **223**, 824
Aaronson M, Frogel J.A. & Persson S.E, 1978, *Ap. J.* **220**, 442
Arimoto N. & Yoshii Y, (AY), 1986, *Astr. Ap.* (in press)
Arnaud K.A. & Gilmore G, 1986, *MNRAS*, **220**, 759
Baade W, 1944, *Ap. J.* **100**, 137
Bertola F, Gregg M.D, Gunn J.E. & Oemler A, 1986, *Ap. J.* **303**, 624
Bruzual A.G, 1983, *Ap. J.* **273**, 105
Burstein D, 1985, *PASP*, **97**, 89
Burstein D, Faber S.M, Gaskell C.M. & Krumm N, 1984, *Ap. J.* **287**, 586
Carter D, Visvanathan N. & Pickles A.J, 1986, *preprint*
Ciardullo R.B. & Demarque P, (CD), 1977, *Trans.Astr.Obs*, Yale Univ, v33
Eisenhardt P.R.M. & Lebofsky M.J, 1986, *Ap. J.* (in press)
Faber S.M, 1972, *Astr. Ap.* **20**, 361
_________ 1973, *Ap. J.* **179**, 731
_________ 1977, in *The evolution of Galaxies and Stellar Populations*, B.M. Tinsley & R.B. Larson, eds. (New Haven, Yale Univ. Obs.), p177
Frogel J.A, Persson S.E, Aaronson M. & Matthews K, (FPAM), 1978, *Ap. J.* **220**, 75
Frogel J.A, Persson S.E. & Cohen J.G, 1980, *Ap. J.* **240**, 785
Gascoigne S.C.B, 1980, in IAU symposium 85, *Star Clusters*, ed. J.E. Hesser, Dordrecht, Reidel, p305.
Green E.M, Demarque P. & King C.R, 1985, *BAAS*, **16**, 51.06
Gunn J.E, Stryker L.L. & Tinsley B.M, (GST), 1981, *Ap. J.* **249**, 48
Johnson H.L, 1966a, *Ap. J.* **143**, 187
__________ 1966b, *Ann. Rev. Astr. Ap.* **4**, 193
Lilly S.J. & Longair M.S, 1984, *MNRAS*, **211**, 833
Morgan W.W, 1956, *PASP*, **68**, 509
Mould J. & Kristian J, 1986, *preprint*
Mould J.R, Kristian J. & Da Costa G, 1983, *Ap. J.* **270**, 471
________________________________ 1984, *Ap. J.* **278**, 575
McClure R.D. & van den Bergh S, 1968, *Astron. J.*, **73**, 313
Nesci R. & Perola G.C, 1985, *Astr. Ap.* **145**, 296
Nørgaard-Nielsen H.U. & Kjærgaard P, *Astr. Ap.* **93**, 290
O'Connell R.W, 1976, *Ap. J.* **206**, 370
____________ 1980, *Ap. J.* **236**, 430
____________ 1985, in *Spectral Evolution of Galaxies*, C. Chiosi & A. Renzini, eds. (Dordrecht, Reidel), p321
Oke J.B, Bertola F. & Capaccioli M, *Ap. J.* **243**, 453
Pagel B.E.J, 1985, in *Symposium on Cosmogonical Processes*, Boulder, Colorado
Pagel B.E.J & Patchett B.E, 1975, *MNRAS*, **172**, 13
Peck M.L, 1980, *Ap. J.* **238**, 79
Pickles A.J, 1985a, *Ap. J. Suppl*, **59**, 33
__________ 1985b, *Ap. J.* **296**, 340
Pritchett C, 1977, *Ap. J. Suppl*, **33**, 397
Rabin D, 1982, *Ap. J.* **261**, 85
Rich R.M, 1986, Conf. on *Stellar Populations*, Baltimore, USA
Rose J.A, 1985, *Astron. J.* **90**, 927

Salpeter E.E, 1955, *Ap. J.* **121**, 161
Spinrad H. & Taylor B.J. (ST), 1971, *Ap. J. Suppl*, **22**, 445
Sweigart A.V. & Gross P.G, *Ap. J. Suppl*, **36**, 405
Searle L, Wilkinson A. & Bagnuolo W, 1980, *Ap. J.* **239**, 803
Terndrup D.M, Rich R.M. & Whitford A.E, 1984, *PASP*, **96**, 94
Tinsley B.M, 1972, *Ap. J.* **178**, 319
Tinsley B.M. & Gunn J.E, 1976, *Ap. J.* **206**, 525
Twarog B.A, 1980, *Ap. J.* **242**, 242
VandenBerg D, 1985, *Ap. J. Suppl.* **58**, 711
de Vaucouleurs G, 1961, *Ap. J. Suppl*, **5**, 223
Visvanathan N. & Sandage A.R, 1977, *Ap. J.* **216**, 214
Williams T.B, 1976, *Ap. J.* **209**, 716
Whitford A.E, 1977, *Ap. J.* **211**, 527
__________ 1978, *Ap. J.* **226**, 777
__________ 1985a, *PASP*, **97**, 205
__________ 1985b, in *Spectral Evolution of Galaxies*, C. Chiosi & A. Renzini, eds. (Dordrecht, Reidel), p157
Whitford A.E & Rich R.M (WR), 1983, *Ap. J.* **274**, 723
Wood D.B, 1966, *Ap. J.* **145**, 36
Wood P.R. & Bessell M.S, 1983, *Ap. J.* **265**, 748
Wu C-C, Faber S.M, Gallagher J.S, Peck M. & Tinsley B.M, 1980, *Ap. J.* **237**, 290

DISCUSSION

FREEMAN: At the ST meeting last week, Demarque mentioned new work on evolution of stars more metal-rich than solar. This suggested that the incidence of evolved uv bright stars may be higher in very metal-rich systems. Would this help significantly in reducing the need for younger populations in ellipticals?

PICKLES: The observations do not reveal large numbers of hot evolved stars. Rose's observations in particular set strong limits on the numbers of A or B spectral type stars, so I would be surprised if they were shown to be expected in large numbers in metal-rich populations.

FREEMAN: i) Do you include red horizontal branch stars in your intermediate abundance components, and ii) do you think that a smoothly changing abundance could explain the constancy of $H\beta$ with Mg_2 shown earlier by George Efstathiou?

BURSTEIN: I think George Efstathiou's data may present a serious problem to 'closed system' models: George's data suggest that the Mg_2-Fe gradient ***within*** a galaxy is different from the Mg_2-Fe correlation ***between*** galaxies. If this difference is substantiated by future data, then the center of an elliptical is anything ***but*** a closed system.

PICKLES: i) The horizontal branch components are not extensive and include only metal-weak standards. ii) One possible explanation for the observed radial line strength variations is that stellar ages might increase radially outwards in ellipticals. The table shows (top rows) HI ($H\alpha+H\beta$), (middle rows) Mgb, and (lower rows) Fe+CN equivalent widths in angstroms, measured in a variety of synthetic spectra constructed to match selected Yale isochrones. There are problems with these synthetic spectra, but the trends in line strength should be correct.

<Fe/H> \ Age in Gyr	5	10	15	
0.4	4	4	3	($H\alpha+H\beta$)
	22	24	26	(Mgb)
	14	14	16	(Fe+CN)
0.0	5	4	4	
	17	21	24	
	9	11	12	
-0.6	7	4	4	
	8	13	16	
	2	5	7	

nuclear (vertical arrow down the 5 Gyr column); radial (diagonal arrow from 0.0/10 Gyr to -0.6/15 Gyr)

Observations show Mgb strength decreasing and $H\beta$ increasing with decreasing luminosity in elliptical nuclei. This implies that metallicity decreases with luminosity and favours young ages in the nuclei, so E nuclei may fall along the vertical line in the table. Mgb strength also decreases radially, whereas $H\beta$ stays constant. This would be the case if the mean stellar age increased outwards in addition to the mean metallicity declining: *ie.* the diagonal line in the table. The Fe+CN index closely parallels trends in Mg strength and it is not obvious what population variations would break this correspondence. Dwarf enrichment could give strong Mgb and NaD features, and mixing processes may give enhanced CN.

CANIZARES: It may be that winds have cleared out the gas in low mass galaxies, but we know that this cannot be true in high mass galaxies because we see their X-ray emission. Doesn't this make you suspicious about the validity of the particular scenario of your synthesis models?

PICKLES: We see hot, metal-enriched gas surrounding (particularly) the dominant galaxies in clusters. Arimoto and Yoshii's models offer a plausible early origin for this gas although they have not treated its subsequent progress through the cluster or cooling and infall. I don't see how purely low mass star formation would generate the observed metallicities in giant ellipticals. This needs supernovae which ***will*** expel residual gas from a ~spherical system when the thermal energy is high enough.

MOULD: (Comment) The need for supernova driven winds stems from the need to reproduce the mass-metallicity relation in elliptical galaxies. Without the wind to switch off the chemical evolution in low mass systems, we could not easily produce dwarfs that are metal poor.

THE INTRINSIC SHAPES OF ELLIPTICAL GALAXIES

P. L. Schechter
Mount Wilson and Las Campanas Observatories
813 Santa Barbara Street
Pasadena, CA 91101
USA

ABSTRACT. Distribution functions for the intrinsic shapes of elliptical galaxies are discussed, starting with the simplest and proceeding to the more complex. A variety of competing "proxy" observables, which can in principle be used to recover at least some of information lost in the projection of a galaxy onto the plane of the sky, are considered.

1. INTRODUCTION

A frequently repeated theme in these proceedings has been that elliptical galaxies span a wider range of observed properties, and could in theory span a wider range of physical properties, than has heretofore been appreciated. This increasing complexity is reflected in the growing number of observables, on one hand, and theoretical parameters, on the other, which can be used to describe ellipticals.

As the dimensionality of the space of elliptical galaxies grows, the difficulty in determining the distribution of intrinsic properties from the distribution of observed properties grows too. Not only are larger statistical samples and better data needed, but more sophisticated analyses are required. I shall argue that while there has been marked improvement in the observations, the machinery necessary for interpreting these have not kept pace.

In particular, there is now a considerable body of data on position angle twists and ellipticity variations in ellipticals. But while these data have been used to draw the qualitative conclusion that at least some ellipticals are triaxial, the data have not yet been used to produce a convincing joint distribution function for intrinsic ellipticity and triaxiality.

Just beneath the question of what ellipticals are lies the question of how they formed. We seek a statistical description of ellipticals because we imagine that this will allow us to discriminate among competing models for their formation. Aguilar's contribution to the

T. de Zeeuw (ed.), Structure and Dynamics of Elliptical Galaxies, 217–228.

present proceedings provides an example of the testable predictions of one particular model. Along the same lines, I suspect that a gradual collapse model for the formation of ellipticals, similar to the one often proposed for the formation of the spheroid of our own Milky Way, is unlikely to produce much triaxiality. As difficult as the determination of the intrinsic shapes of ellipticals may ultimately prove to be, the effort will not go entirely unrewarded.

But if the problem really is such a difficult one, which we might reasonably infer from the fact that it has not yet been treated satisfactorily, we must make every effort to cast it in as simple a form as possible. In the spirit of first approximation, we should be willing to disregard details which we suspect will only marginally influence our results.

There are two strategic questions which I would like to address, one more abstract, the other more practical. The abstract question concerns the choice of "interesting" parameters, and the choice of model distribution functions over those interesting parameters. I will argue that there are both observational and theoretical reasons to prefer some distribution functions over others. The practical question involves the choice of weapons. Since we lose one of the three dimensions of an elliptical galaxy to projection, we must find some other observable as a proxy. The candidate proxies include surface brightnesses, velocity dispersions, rotation velocities and position angle twists. I will consider in turn the relative strengths of each of these.

I will then briefly raise the question of the shape of the mass distribution in elliptical galaxies, which may be very different from the shape of the light distribution, and which requires a different observational approach.

2. DISTRIBUTION FUNCTIONS

2.1. $\phi = \phi(\varepsilon)$

If one makes the assumption that elliptical galaxies are axisymmetric, the problem of inverting the observed distribution of apparent ellipticities is then a straightforward one. Nonetheless, there are differences in the distributions derived from different data sets which bear on important physical questions.

Binney and de Vaucouleurs (1981) used Lucy's method to invert the distribution of apparent flattenings in the Second Reference Catalog, and found a peak in the distribution of intrinsic ellipticities at $\varepsilon = 0.38$, and a relatively flat distribution from $\varepsilon = 0.20$ to $\varepsilon = 0$. This plateau, if not an artifact of the inversion technique or the data, might be taken to indicate a process, perhaps a dynamical instability, which favors the formation of perfectly round galaxies. But Benacchio and Galletta (1980), who used data on cluster ellipticals from the work of Strom and Strom, found an extreme deficiency of very round systems.

A third data set is available, in a doctoral thesis by Djorgovski (1986). Taking the ellipticity at a representative isophote ($r = 20.5$ mag/arcsecond^2) yields a histogram of apparent ellipticities which is intermediate between that used by Binney and de Vaucouleurs and that used by Benacchio and Galletta. The paucity of very round systems in the Benacchio and Galletta data may be the result of their use of the maximum ellipticity for each galaxy rather than a mean ellipticity or the ellipticity at a fiducial isophote. I must apologize for not having gone through the exercise of inverting Djorgovski's distribution.

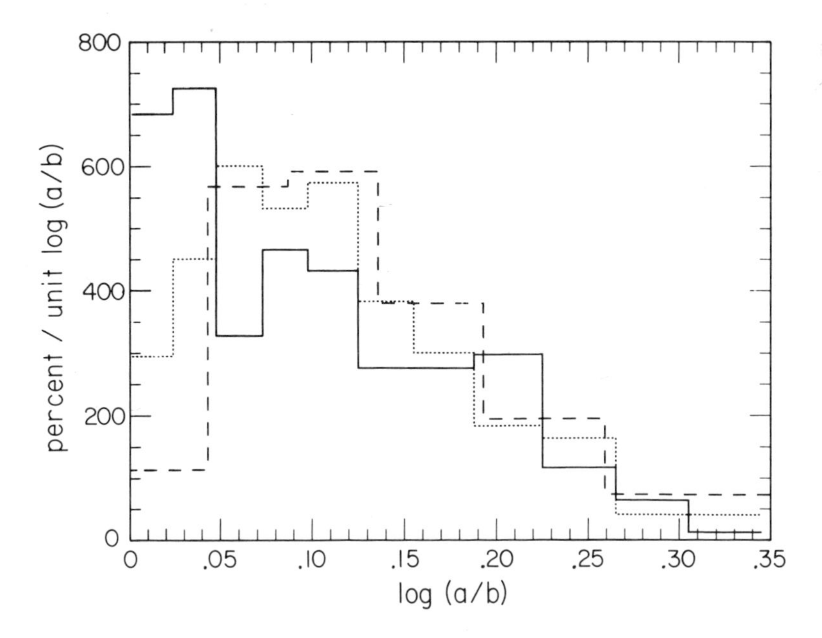

Figure 1. The distribution of observed axial ratios in the samples used by Binney and de Vaucouleurs (dashed line), Benacchio and Galletta (solid line), and Djorgovski (dotted line).

Djorgovski's data includes ellipticities at a wide range of radii. I was struck by the fact that among the systems which were very round at the chosen isophote, some were considerably flatter at other isophotes and others exhibited large position angle twists. Jedrzejewski (1986), who obtained photometry for a sample of southern ellipticals, finds that only 2 out of 49 galaxies have ellipticities everywhere less than 0.1. At the other extreme, de Vaucouleurs (1977) finds that there is an upper limit to the ellipticity of bona fide ellipticals, $\varepsilon < 0.55$.

2.2. $\phi = \phi(\varepsilon, \tau)$

The next level of complexity drops the assumption of axisymmetry and allows for a third axis intermediate in length between the longest and shortest axes. Such configurations, usually called triaxial, span the full range from oblate to prolate. If we adopt the convention $a > b > c$, and define a triaxiality parameter $\tau \equiv (a-b)/(a-c)$, then $\tau = 0$ for oblate galaxies and $\tau = 1$ for prolate galaxies. The wanted distribution is now a function of two variables, ε and τ.

This second variable greatly complicates matters. As a first approximation, some authors make the assumption that all ellipticals have the same value of τ, i.e. that the distribution in τ is a delta function. But while a single value of τ may be easier to deal with, it is easy to imagine physical processes which would give a range of values.

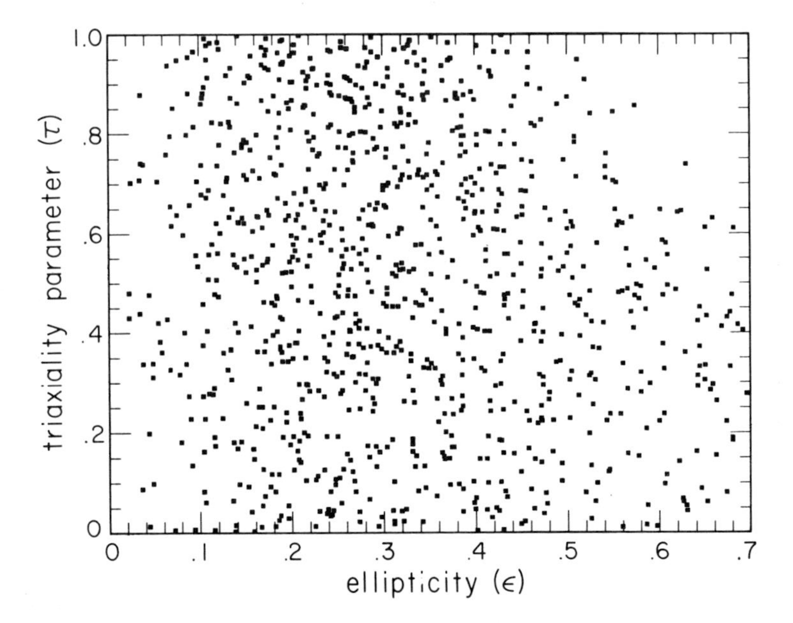

Figure 2. A 1000 point realization of a numerological model for the joint distribution of ε and τ.

For the sake of argument, consider a model for the distribution in ε and τ in which all three axes are mutually independent, with normal distributions about a common mean value. The distribution for such a numerological model is shown in Figure 2. The τ distribution spans the range from oblate to prolate almost uniformly. The ellipticity distribution peaks at an ellipticity of 0.3, and shows relatively few very round galaxies. The absence of round galaxies is explained by the low probability that all three axes will have the same length. While this model contains no physics, one might suppose that the velocity dispersion tensors in elliptical galaxies preserve some memory of the shape of the proto-objects from which they formed. Gaussian fluctuations about an average value might apply in such a case.

2.3. $\phi = \phi(\varepsilon, \tau, d\varepsilon/d\ln r)$

As bad as distribution functions over two variables may seem, we find ourselves driven to a third variable. One of the more convincing arguments for triaxiality comes from the observation of twisting isophotes. Twists occur if a) a galaxy is triaxial and b) its axial ratios vary as a function of radius. The effect results from the fact that the major and minor axes of the projection onto the plane of the sky are not the projections of the intrinsic major and minor axes.

A quantitative treatment of position angle twists therefore demands a model for the radial variations in ellipticity and triaxiality. Since the former are easier to determine, and since we wish to avoid introducing a fourth random variable, it seems reasonable to take τ to be constant throughout a galaxy.

One approach to ellipticity variations would be to assume a constant gradient in ε with respect to $\ln r$, assuming further that this gradient varied randomly about some mean value. While such an approach has the advantage of simplicity, it does considerable injustice to what we know about ellipticity variations. A more flexible but more complicated alternative would be to adopt an ellipticity autocorrelation function, from which one could compute the probability of observing a change in ellipticity $\Delta\varepsilon$ at distance $\Delta\ln r$ from a randomly chosen point. One might then compute predicted and observed position angle autocorrelation functions. The position angle autocorrelation would need to be studied as a function of observed ellipticity, since the biggest twists occur at the smallest apparent ellipticities.

3. OBSERVATIONS BEARING ON INTRINSIC SHAPES

Several different kinds of observations can be used as proxies for the unobservable third axis, and used to infer the the intrinsic shapes of ellipticals. The following table summarizes the different approaches:

PROXY	ASSUMPTIONS
surface brightness	$\mu = \mu(\varepsilon)$
velocity dispersion	$\sigma = \sigma(\varepsilon)$
minor axis rotation	shortest axis coincides with J vector
P.A. twists	none; ε gradient distribution needed
dust-lanes, radio jets	equilibrium configurations? self-gravitating disks?

Surface brightness and velocity dispersion measurements have thus far been inconclusive in helping to determine the intrinsic shapes of ellipticals. The underlying idea in both cases is simple. For an oblate galaxy, the surface brightness and velocity dispersion should be higher when viewed edge-on than when viewed pole-on. For a prolate system, the converse should apply.

Unfortunately, this simple situation is complicated by the possibility that surface brightness and velocity dispersion might be expected to vary with intrinsic axial ratio. Merritt (1982) found that given some latitude in the dependence of these quantities on intrinsic shape, and second, some latitude in the assumed scatter at a given intrinsic shape and luminosity, both oblate and prolate models could reproduce the available data. There has, however, been some improvement in our understanding of the dependence of velocity dispersion and surface brightness on intrinsic luminosity (Dressler *et al.*, 1987), and the time may be ripe for a new investigation of the subject.

Position angle twists, while the most reliable indicator of the existence of triaxiality, are rather difficult to analyze quantitatively. Benacchio and Galletta (1980) interpreted the isophote twists observed in the Strom's data in terms of a model with $\tau \approx 0.5$, the value for maximal triaxiality, and with ε varying with radius from zero to a maximum value which had a Gaussian distribution about $\varepsilon = 0.38$, with a dispersion of 0.13. They took the conservative view that only position angle twists greater than 10° would be considered significant.

With the advent of CCD's, position angle accuracies of 1-2° have become relatively easy to achieve. Leach (1981) obtained data for a sample of 32 galaxies, and interpreted them in terms of a triaxial model with Gaussian distributions in b/a, c/a, Δ(b/a), and Δ(c/a). While not unique, his adopted distribution function gives roughly equal numbers of nearly prolate ($2/3 < \tau < 1$) and triaxial ($1/3 < \tau < 2/3$) galaxies, with only half as many nearly oblate ($0 < \tau < 1/3$) ones.

Minor axis rotation offers a tool for studying triaxiality which does not require the presence of axial ratio variations. Binney (1985) has calculated the frequency distribution of the ratio of minor to major axis rotation for $\tau = 0.20$, 0.50, and 0.95. Working with data for only ten galaxies, he found that the extreme prolate hypothesis could be ruled out with considerable confidence. Most of the data used in his study were drawn from work by Gunn and myself, so I have no qualms about saying that much better data could be obtained today. Parallel efforts are underway by Franx and Illingworth and by Jedrzejewski and myself. Thus far, the evidence is that roughly equal numbers of galaxies do and don't show minor axis rotation at the 10 km/s level.

The picture is clouded by the discovery by Davies and Birkinshaw (1986) that NGC 4261, otherwise a relatively undistinguished elliptical, shows considerable minor axis rotation, and almost no major axis rotation. While their data are not inconsistent with the hypothesis of a nearly prolate object tumbling about its shortest axis, this demands an unlikely orientation to the line of sight. The alternative hypothesis, that the galaxy rotates about its long axis, violates the one explicit assumption in Binney's analysis.

Another observation which bears on intrinsic shapes is the relatively rapid rotation seen in intrinsically faint ellipticals (Davies _et al._, 1983). They are consistent with models of rotationally flattened oblate systems with isotropic dispersion tensors, and might therefore be thought to be oblate. If intrinsically faint systems are indeed oblate, then they ought not to exhibit position angle twists.

Dust-lane orientations have been frequently been taken as indicators of a galaxy's intrinsic shape (e.g. Hawarden _et al._, 1981; Bertola, these proceedings). Galaxies with dust-lanes along their minor axes have been called prolate, and those with dust-lanes along their major axes have been called oblate. But in a triaxial system, gas and dust can settle into equilibria about either the longest or the shortest axis.

The "skewed" dust-lane ellipticals are yet more puzzling. While small deviations from orthogonality could result from chance projections of triaxial ellipticals, larger angles are harder to explain. One possible explanation is self-gravity, which Sparke (1986) has invoked to explain the apparent stability of inclined "polar ring" S0s. Another possible explanation is that the skewed dust-lanes are transient phenomena. If they resulted from the disruption of a smaller galaxy, one might expect such systems to have a higher incidence of shells, tails, etc.

The alignments (or rather the misalignments) of the radio and optical axes of radio galaxies have also been used to test the intrinsic shapes of ellipticals. A recent effort on the subject, by Birkinshaw and Davies (1985), shows a frightening degree of misalignment. Heckman _et al._ (1985) have found that radio lobes show better alignment with the kinematic minor axes of the gas in ellipticals than they do with the optical isophotes. One might once again appeal to self-gravity of the gaseous disks, which might precess as a unit. One would then expect to see signatures of such precession in the radio emission from misaligned doubles.

Centaurus A, NGC 5128, deserves special mention, by virtue of the fact that it is both a dust-lane elliptical and a radio double, and by virtue of its proximity, which makes it the best studied of such systems. It shows modest rotation along its major axis, and little or none about its minor axis (Wilkinson _et al._, 1986).

4. SHAPES OF HALOS

There is considerable evidence that spiral galaxies are embedded dark halos which produce roughly logarithmic potentials. While less overwhelming, there is analogous evidence for ellipticals in their X-ray temperatures and profiles and in the velocities of their globular clusters and dwarf companions. Since the run of mass with distance from the center seems not to follow the run of light with distance from the center, one might think that the shape of the halo and the shape of luminous component could also be different.

It is inherently harder to determine the shapes of potentials than it is to determine the shapes of the luminous matter because the equipotentials of centrally condensed systems are rounder than their equidensity contours. A mass distribution with an ellipticity of 0.3 and a logarithmic potential will produce a potential with an ellipticity of only 0.1. We are therefore looking for subtler effects in potentials than in the mass itself.

Thermal bremsstrahlung from the X-ray coronae of ellipticals should trace the potential perfectly, since the pressure in the X-ray emitting gas cannot be anisotropic. The situation is rendered somewhat less clean by the possibility of ram-pressure distortions of X-ray coronae. I have examined the X-ray isophotes in the papers by Forman *et al.* (1985) and Trinchieri *et al.* (1986), and would only point to one case (NGC 720) where the X-ray isophotes look to be significantly out of round and yet sufficiently symmetric about the central elliptical to believe that one is seeing the shape of the potential.

If one permits discussion of the halos around non-elliptical galaxies, then there is more evidence. Whitmore, McElroy and Schweizer (1987) have observed polar rings around 3 S0 galaxies, measuring velocities both in the rings and in the central S0 components. The velocities in the two nearly orthogonal planes are close enough to permit interesting limits on the shapes of the potentials, with ellipticities of 0.2 or less. One might suspect, however, that such systems form preferentially in systems with rounder halos, in which case they might not give a fair representation of all halos. Moreover a potential with an ellipticity of 0.1 corresponds to a oblate mass distribution with an ellipticity of 0.3, which is just the typical intrinsic ellipticity seen in elliptical galaxies.

Another way to determine the shapes of spiral halos is to look for minor axis rotation in the disks of spirals. The effect is described in a paper by Binney (1978) called "Twisted and Warped Disks as Consequences of Heavy Halos." If halos are triaxial, then the apparent minor axis of a spiral galaxies would not necessarily coincide with the projection of its rotation axis. The effect is greatest in galaxies which are nearly face on, and demands careful measurements of position angles, which are made more difficult by the presence of spiral structure. This latter difficulty might be avoided by looking for the effect in the stellar disks of S0 galaxies.

5. SUMMARY

In preparing this review I was dismayed at how little can be said with any confidence about the intrinsic shapes of elliptical galaxies despite the substantial effort that has gone into studying them. I do not, however, believe that this effort has been wasted. There are promising opportunities to build on these earlier efforts which I suspect will yield the wanted answers.

A. Ellipticity and position angle data now exist for large samples of ellipticals. What is needed is a good statistical treatment of ellipticity variations and a good statistical treatment of position angle variations. One could then try to "predict" the observed position angle variations using the observed ellipticity variations and a variety of assumptions about the triaxiality distribution.

B. Minor axis rotation can be measured both in ordinary ellipticals and in those with dust-lanes. Either we will not see many repeats of the pathology of NGC 4261, and will be able to draw conclusions about triaxiality, or we will be forced to re-examine some cherished notions.

C. X-ray isophotes can give us the shapes of the potentials in which our ellipticals are embedded. Those of us who believe in triaxiality expect some twisting, and depending upon how ellipticals form, might expect some 90° misalignments.

REFERENCES

Benacchio, L. and Galletta, G. 1980, M.N.R.A.S., **193**, 885.
Binney, J. 1978, M.N.R.A.S., **183**, 779.
Binney, J. 1985, M.N.R.A.S., **212**, 767.
Binney, J. and de Vaucouleurs, G. 1981, M.N.R.A.S., **194**, 679.
Birkinshaw, M. and Davies, R.L. 1985, Ap.J., **291**, 32.
Davies, R.L. and Birkinshaw, M. 1986, Ap.J.(Letters), **303**, L45.
Davies, R.L., Efstathiou, G., Fall, S.M., Illingworth, G., and Schechter, P.L. 1983, Ap.J., **266**, 41.
de Vaucouleurs, G. 1977, in The Evolution of Galaxies and Stellar Populations, ed. B.M. Tinsley and R.B. Larson (New Haven: Yale University Observatory) p. 43.
Djorgovski, S. 1986, unpublished Ph.D. thesis, University of California, Berkeley.
Dressler, A., Lynden-Bell, D. Burstein, D., Davies, R.L., Faber, S.M., Wegner, G., and Terlevich, R. 1987, Ap.J., in press.
Forman, W., Jones, C., and Tucker, W. 1985, Ap.J., **293**, 102.
Hawarden, T.G., Elson, R.A.W., Longmore, A.J., Tritton, S.B., and Corwin, H.G. 1981, M.N.R.A.S., **196**, 747.

Heckman, T.M., Illingworth, G.D., Miley, G.K., and van Breugel, W.J.M. 1985, Ap.J., **299**, 41.
Jedrzejewski, R.I. 1987, M.N.R.A.S, in press.
Leach, R. 1981, Ap.J., **284**, 485.
Merritt, D. 1982, A.J., **87**, 1279.
Sparke, L.S. 1986, M.N.R.A.S., **219**, 657.
Trinchieri, G., Fabbiano, G., and Canizares, C. 1986, Ap.J., **310**, 637.
Whitmore, B.C., McElroy, D.B., and Schweizer, F. 1987, Ap.J., **314**, 439.
Wilkinson, A., Sharples, R.S., Fosbury, R.A.E., and Wallace, P.T. 1986, M.N.R.A.S., **218**, 297.

DISCUSSION

Aguilar: I want to mention the results of some dissipationless collapse calculations that David Merritt, Martin Duncan and I have made and that may explain the lack of round galaxies. We have run a series of collapses started from spherical, oblate, and triaxial initial conditions. It seems that only cold initial conditions ($2T/W < 0.1$) result in final models with realistic surface density profiles ($r^{1/4}$–laws) but whenever this happens the models develop an instability associated with the predominance of radial orbits. This instability produces prolate bars out of spherical initial conditions. Non-spherical initial conditions produce prolate and triaxial configurations but never oblate or spherical models. We should point out, however, that we have not yet included rotation in our simulations.

Binney: Sverre Aarseth and I played similar games, though starting from flattened initial conditions. We found that even the modest amount of net angular momentum pushed the final configuration towards oblate axisymmetry. We should not forget that practically all ellipticals do rotate at some level.

Capaccioli: Let me assume that the variations of ellipticity and position–angle do not occur in the same regions of the galaxy, i.e. that they are spatially not correlated. Would this influence the analysis?

Schechter: While I have not done such experiments myself, I would expect a correlation between the location of observed ellipticity variation and position angle variations. There are several galaxies in Djorgovski's sample which exhibit rapid 90 degree position angle twists. These occur near zero ellipticity.

Jarvis: Do you observe from your data, or expect on theoretical grounds, a correlation between the amount of rotation on the minor axis and the strength of the isophotal twisting?

Schechter: One of the strengths of the minor axis rotation test is that it is independent of ellipticity variations. Our sample was selected to avoid galaxies with large twists, which might conceivably introduce a selection effect.

King: At the cost of introducing still another complication, I would like to ask whether there are galaxies with good elliptical isophotes but where the twists are too large to account for with constant axial directions.

Williams: There are some potentially very worrisome systems observed, where there is a significant axis ratio change and position angle twist, while the isophotes are never very round. NGC 584 is an example. It is very difficult to do this by projection effects alone, without resorting to large three–dimensional axis ratio gradients, which will result in large deviations from elliptical isophotes (not observed). We will be forced to non–coaxial models, which I suspect can reproduce both twists and axis ratio changes at any flattening without so seriously distorting the individual isophotes.

Lauer: There are systems of strongly interacting elliptical galaxies that show strong twists; it seems that this is a likely case of galaxies with *non*–coaxial ellipsiods.

Schechter: I agree.

Porter: Preliminary indications are that brightest cluster ellipticals can show large twists simultaneously with large ellipticities. Whether they are too large to be projection effects, I'm not prepared to state yet.

Gerhard: Two comments. (i) Some years ago I came across an N–body model with *intrinsic* twists of the principal axes, which lasted for $\sim$ 15 dynamical times. While this may be a little too short to say much about the inner parts of ellipticals where dynamical time–scales are short, it may mean that intrinsic twists in the outer parts of these systems are dynamically possible (cf. 1983, *Mon. Not. R. astr. Soc.*, **203**, 198,). (ii) I would like to emphasize the value of gas disks in ellipticals for the deciphering of their intrinsic shapes. In a joint poster paper with Mario Vietri, it is shown that, if the circumstances are favorable, one may determine both the axial ratios from the geometry and the velocity field.

Schechter: Thank you for reminding us of your models with intrinsic twists. Under the intrinsic hypothesis, one might expect greater twisting in the outer parts than in the inner parts. Your method for determining the shapes of spheroidal components looks to be a powerful one. There is, of course, a rear guard which banishes from the class of all ellipticals all objects with any hint of a disk.

Williams: In three dimensions, for ellipsoidal figures, there are two potentially variable functions–the two axis ratios of the figure. The projection of these onto the sky results also in two variable functions: axis ratio changes and position angle twists. You have proposed a statistical analysis which admits only one variable three dimensional function to produce both projected functions. I suspect that you would be able to find projection angles which will allow this, but then you will introduce correlations between projection angles and the one variable function which do not exist in the actual objects. I am afraid this will very much complicate the analysis, but may well be unavoidable.

Schechter: You have considerably more experience in this matter than I do. Nonetheless, I would think one could place limits on such systematic effects through Monte Carlo simulations. The sense of the effect you describe would be due to the inferred distribution of triaxialities closer to maximal triaxiality than is really the case.

Davies: If NGC 4261 is to be oblate/triaxial as is shown to be possible in the posters of Statler and Levison, I think it is surprising that the $\Delta\theta = \theta_{rot} - \theta_{min}$- histogram that I showed yesterday has no entries with $30 < \Delta\theta < 80$. Do you agree? I believe this oblate-triaxial configuration to be possible but unlikely. If the orbits are populated in this particular way, it appears that we should expect to see the kinematic axes to be non-perpendicular. Observers wishing to test this need to use more than two position angles, at least four, I think.

Schechter: Perhaps galaxies populate either the short axis tubes or their long axis tubes, but not both.

Burstein: On a somewhat different topic, I note that the Strom & Strom and Djorgovski samples include elliptical galaxies with a wide intrinsic range of luminosity, while the RC2 sample, with its Malmquist-bias, will be dominated by high luminosity galaxies. Independent of errors or other kinds of selection effects, a plot of ellipticity vs. radial velocity for the RC2 sample might show if a real difference in these samples exists.

Schechter: I have used only Djorgovski's "sample I", which is magnitude limited and therefore has roughly the same luminosity distribution as the RC2 sample. The Stroms' sample is more nearly volume limited, and as such permits a test for a trend of ellipticity with absolute magnitude.

Valentijn: Responding to your request for more and uniform data on axial ratios and isophotal twists, I can announce that together with A. Lauberts a large two-dimensional photometric project is in progress at ESO. We have scanned all 16000 galaxies present in the ESO–Uppsala catalogue, on both red and blue original sky survey plates. In an automated mode, we are extracting magnitudes, colour gradients, axial ratios, isophotal twists, and various other parameters, as function of radius. Individual plates are calibrated using photo-electric data. As of May 1986, we have determined parameters of 6000 objects, and we envisage a full presentation of the results in the summer of 1988.

ORBITS

James Binney
Department of Theoretical Physics
Keble Road
Oxford OX1 3NP
England

ABSTRACT. Orbits that respect at least three isolating integrals of motion have very special structures in phase space. The main characteristics of this structure are reviewed, and the concrete examples that are provided by orbits in Stäckel potentials, are discussed. Many orbits in general potentials admit three approximate isolating integrals and closely resemble orbits in Stäckel potentials. If the potential is that of an elliptical galaxy with negligible figure rotation, the overall orbital stucture of the potential differs from that of a Stäckel potential only by the presence of a few unimportant families of resonant orbits. However, this elegant picture is shattered by the introduction of non-negligible figure rotation: though substantial regions of phase space may still be occupied by orbits that individually resemble orbits in Stäckel potentials, the overall orbital structure is radically changed by figure rotation, and in a rotating potential significant portions of phase space are given over to chaotic orbits, quite unlike orbits in Stäckel potentials.

1. INTRODUCTION

The emergence a decade ago of the view that elliptical galaxies are normally triaxial bodies, obliged us to discover what kinds of orbits are possible in a given triaxial potential. In this review I shall concentrate exclusively on this question, neglecting for example the extensive work of Richstone and his collaborators (see Richstone 1984), of Kent (1983) and others on the orbital structures of axisymmetric potentials.

The last decade has yielded a good basic understanding of orbits in potentials with figures fixed in inertial space, and this understanding has formed the basis of much recent work on galaxy models. Most of this review is taken up with a summary of the key results obtained in this area. Many of the concepts that arise from this work, especially the concept of orbital tori, are widely applicable in stellar dynamics. However, we should not lose sight of the likelihood that many orbits in elliptical galaxies are significantly influenced by rotation of the potential's figure, and therefore that we shall not be able to construct fully satisfactory galaxy models until we have cracked the much harder problem of motion in a potential with non-negligible figure rotation. I shall mention some pioneering work on this complex problem at the end of the review, but the space allocated to this topic in no way reflects its likely importance.

T. de Zeeuw (ed.), Structure and Dynamics of Elliptical Galaxies, 229–239.

Galaxies are three-dimensional, but two-dimensional orbits are much easier to study (not least because the power of Poincaré's surfaces of section). In the interests of brevity, results that are equally valid for two- and three-dimensional orbits, will be described in terms of n dimensions.

2. REGULAR ORBITS

Simple numerical experiments show that few, if any, orbits in galaxy-like potentials explore the whole "energy" hypersurface $H(\mathbf{x}, \mathbf{v}) = E$. The dimensionality of the phase-space subset to which a given orbit is confined can be elucidated by studying the range of velocity vectors $\mathbf{v}$ with which an orbiting particle passes by a particular place $\mathbf{x}$. On an *irregular* orbit, the range of velocities $\mathbf{v}$ at $\mathbf{x}$ is at least one-dimensional. The orbit is said to be *regular* if this range consists of a small number, typically 2–6, of isolated possibilities. In the case of two-dimensional orbits, it is immediately apparent from a simple tracing of the orbit, whether the orbit is regular or irregular; irregular orbits *look* messy [see, for example, Fig. 3 of Binney (1982)].

Since the value of $\mathbf{v}$ at a given point $\mathbf{x}$ on an n-dimensional regular orbit is determined up to a few-fold degeneracy, the $2n$ phase-space coordinates $(\mathbf{x}, \mathbf{v})$ of points on the orbit must be constrained by n relations of the form $H \equiv I_1(\mathbf{x}, \mathbf{v}) = i_1 \equiv E, \ldots, I_n(\mathbf{x}, \mathbf{v}) = i_n$, where the *isolating integrals* I_k are smooth single-valued functions of the phase-space coordinates. Conversely, along an irregular orbit, fewer functional relationships constrain the coordinates $(\mathbf{x}, \mathbf{v})$; One usually has $H = E$ and one or more inequalities $i'_k < I_k < i_k$.

The phase-space structure of regular orbits is strongly constrained by the nature of Hamilton's equations of motion. From the mere existence of the n isolating integrals $I_1, \ldots, I_n$, one may demonstrate the following (Arnold 1978):

(i) In $2n$-dimensional phase space the orbit lies on an n-dimensional surface which is topologically equivalent to an n-torus. In other words, a continuous one-to-one map exists of the orbital surface onto the unit cube of n-dimensional Euclidean space with opposite faces identified.

(ii) The *action integrals* $J_\gamma \equiv (2\pi)^{-1} \oint_\gamma \mathbf{v} \cdot d\mathbf{x}$ around a given orbital torus are equal for any two closed paths γ on the torus that can be continuously deformed into one another by motions confined to the torus.

(iii) It is possible to incorporate n of these action integrals into a system of *angle-action* coordinates $(\boldsymbol{\theta}, \mathbf{J})$ for that part of the phase space in which the I_k are integrals. In this portion of phase space, the action integrals label the orbital tori, while position on any torus is specified by the n angle variables $\theta_1, \ldots, \theta_n$. The coordinate system $(\boldsymbol{\theta}, \mathbf{J})$ is canonical. In particular, a small element of phase-space volume is $d^n\mathbf{x}d^n\mathbf{v} = d^n\boldsymbol{\theta}d^n\mathbf{J}$ and Hamilton's equations $\dot{\boldsymbol{\theta}} = [\boldsymbol{\theta}, H]$, $\dot{\mathbf{J}} = [\mathbf{J}, H]$ apply. The Hamiltonian H, being constant on orbital tori, is a function $H(\mathbf{J})$ of the actions only. Hence Hamilton's equations $\dot{\boldsymbol{\theta}} = \boldsymbol{\omega}(\mathbf{J}) \equiv (\partial H/\partial \mathbf{J})$ integrate immediately to $\boldsymbol{\theta}(t) = \boldsymbol{\theta}(0) + \boldsymbol{\omega}t$.

The Cartesian coordinates $(\mathbf{x}, \mathbf{v})$ are periodic functions of the θ_k with period 2π; $\mathbf{x}(\boldsymbol{\theta} + 2\pi\mathbf{m}, \mathbf{J}) = \mathbf{x}(\boldsymbol{\theta}, \mathbf{J})$ for any integer vector $\mathbf{m}$. Hence we may expand $\mathbf{x}$ as a Fourier series $\mathbf{x} = \sum_{\mathbf{m}} \mathbf{X}_{\mathbf{m}}(\mathbf{J}) \exp[i\mathbf{m} \cdot \boldsymbol{\theta}]$. Substituting for $\boldsymbol{\theta}(t)$, we obtain the position $\mathbf{x}$ of an orbiting particle as a *quasi-periodic* function of time: $\mathbf{x}(t) = \sum_{\mathbf{m}} \mathbf{X}'_{\mathbf{m}}(\mathbf{J}) \exp[i\mathbf{m} \cdot \boldsymbol{\omega}t]$, where $\mathbf{X}'_{\mathbf{m}} \equiv \mathbf{X}_{\mathbf{m}} \exp[i\mathbf{m} \cdot \boldsymbol{\theta}(0)]$. Thus the Fourier decomposition $\tilde{\mathbf{x}}(\omega)$ of the position vector $\mathbf{x}(t)$ along a regular orbit consists of a series of discrete lines. The frequencies $\mathbf{m} \cdot \boldsymbol{\omega}$ at which these lines occur are integer combinations of three fundamental frequencies ω_k, and by deducing the integer vector $\mathbf{m}$ associated with each line, one can reconstruct the angle representation $\mathbf{x}(\boldsymbol{\theta})$ from the time evolution $\mathbf{x}(t)$ (Binney & Spergel 1984). This reconstruction is useful, because $\mathbf{x}(\boldsymbol{\theta})$ contains much more information than $\mathbf{x}(t)$ (see also Ratcliff *et al.* 1984).

Regular orbits fall naturally into *families*. Each orbit family is parented by a sequence of stable closed orbits. In a realistic non-rotating triaxial potential there are three principal orbit families (Schwarzschild 1979): the box family, whose parents are the long-axis orbits; the short-axis tube family, which is parented by the closed short-axis loop orbits, and the long-axis tube family which has the closed long-axis loop orbits for its parents. (De Zeeuw (1985) additionally subdivides the long-axis tube family into inner- and outer-tubes.) At a given energy we may think of the orbital tori of each family as wrapped around the degenerate wire-like torus of the parent orbit of that energy, much as the insulator and sheath of a coaxial cable encircle the cable's central wire [see Fig. 1 of Lynden-Bell (1962)].

The orbits of each orbit family form a n-dimensional continuum. A useful graphical display of these continua is obtained by treating n independent action integrals over the orbits as Cartesian coordinates, and thus identifying the orbit that has actions $(J_1, \ldots, J_n)$ with the corresponding point in an n-dimensional *action space*. The orbits associated with neighbouring points in action space, occupy adjacent regions of phase space. Furthermore, the volume of phase space that is occupied by the orbits whose representative points lie with a volume element $d^n\mathbf{J}$ in action space, is $(2\pi)^n d^n\mathbf{J}$. Consequently, action space gives a fair representation of the *a priori* probability of a group of orbits.

If the frequencies ω_k are nearly everywhere incommensurable (as will usually be the case) a strengthened Jeans theorem applies: the distribution function of a steady-state galaxy in which almost all orbits are regular with incommensurable frequencies, may be presumed to be a function $f(\mathbf{J})$ of the actions only. Furthermore, the number of stars with actions in the range $d^n\mathbf{J}$ is $dN = (2\pi)^n f(\mathbf{J}) d^n\mathbf{J}$, so f is up to a constant, simply the density of stars in action space.

In general, the orbits of different orbit families have to be accomodated in different action spaces. Schwarzschild's principal families of orbits in a non-rotating potential form an exception to this rule, however; it is possible to define the actions of orbits of the principal families in such a way that the continua of all three families may be fitted together into a single action space, the principal action space. Any additional orbit family gives rise to a zone of missing actions in the principal action space. The volume of this zone is proportional to the phase-space volume occupied by the subfamily's orbits, but actions cannot be assigned to the family's orbits in such a way that they occupy the zone of missing actions in the principal action space (Binney & Spergel 1984).

2.1 Resonances and Extra Integrals

If the fundamental frequencies ω_k of a regular orbit are rationally related, that is, if we have $\mathbf{m} \cdot \boldsymbol{\omega} = 0$ for integer vector $\mathbf{m}$, then the orbit does not explore all of the torus $\{H = E, \ldots, I_n = i_n\}$ to which it is confined. The restriction of the orbit to a subset of its torus may be attributed to an extra isolating integral, $I_{n+1} \equiv \mathbf{m} \cdot \boldsymbol{\theta}$. If $n = 2$, the orbit is closed. If $n = 3$ the orbit closes only if a second rational relationship holds, $\mathbf{m}' \cdot \boldsymbol{\omega} = 0$ for $\mathbf{m}' \neq \mathbf{m}$. Familiar examples of these phenomena are furnished by motion in spherical potentials; the four isolating integrals (H, L_x, L_y and L_z) may be decomposed into three, say H, $|\mathbf{L}|$ and L_z that specify a torus, and a fourth, say L_x/L_y that arises because the orbit has only two independent frequencies, the radial frequency κ and the azimuthal frequency Ω. If κ and Ω happen to be rationally related, as in Kepler ($\kappa = \Omega$) or harmonic ($\kappa = 2\Omega$) motion, a fifth integral arises (the position angle of the apocentre), and the orbit closes.

2.2 Stäckel Potentials

Recently, de Zeeuw (1985) has shown that potentials studied a century ago by Jacobi and Stäckel provide analytic models of the most important features of the orbital structures of non-rotating elliptical-like potentials. In particular, (i) essentially all orbits in a Stäckel potential are confined to tori; (ii) motion on the tori is quasiperiodic; (iii) in realistic cases, all orbits belong to the same three orbit families as orbits earlier integrated numerically by Schwarzschild (1979). De Zeeuw's discovery of handy analytic models of orbits in non-rotating galaxy-like potentials, has opened up a rich vein of exploration. It is worth taking a little time to review the main features of Stäckel orbits.

Orbits in Stäckel potentials are intimately connected with systems of confocal ellipsoidal coordinates. In two dimensions these coordinates are most neatly expressed by writing ($x = \Delta_1 \sinh u \cos v$, $y = \Delta_1 \cosh u \sin v$), where (x, y) are the usual Cartesian coordinates, $\Delta_1 > 0$ is a constant, and u and v, which are constant on ellipses and hyperbolae respectively, are the new coordinates. In three dimensions, ellipsoidal coordinates ($\lambda \geq 0 \geq \mu \geq -\Delta_1{}^2 \geq \nu \geq -\Delta_2{}^2$) may be defined as the roots for τ of the cubic

$$\frac{x^2}{\tau} + \frac{y^2}{\tau + \Delta_1{}^2} + \frac{z^2}{\tau + \Delta_2{}^2} = 1 \tag{1}$$

where $0 \leq \Delta_1 \leq \Delta_2$ are constants. λ is constant on ellipsoids which at large $|\mathbf{x}|$ approximate spheres of radius $|\mathbf{x}| \simeq \sqrt{\lambda}$. In the (x, y) plane, $\lambda = \Delta_1{}^2 \sinh^2 u$, $\mu = -\Delta_1{}^2 \cos^2 v$. At large $|\mathbf{x}|$, μ and ν specify angular position, μ depending mainly on azimuth ϕ and ν depending most strongly on colatitude θ (de Zeeuw 1985, Appendix A).

Let p_τ be the momentum canonically conjugate to $\tau = \lambda$, μ or ν. Then the remarkable property of Stäckel potentials is that on an any orbit in one of these potentials (and these alone!), p_τ is a function of the corresponding coordinate alone. In fact

$$p_\tau^2(\tau) = 2\Big(E - \frac{i_2}{\tau} - \frac{i_3}{\tau + \Delta_2{}^2} + G(\tau)\Big)\Big/(\tau + \Delta_1{}^2), \tag{2}$$

where E, i_2 and i_3 are the values of the energy and two non-classical integrals on the orbit, and G defines the potential Φ through

$$\Phi(\mathbf{x}) = -\sum_{\lambda \to \mu \to \nu} \frac{\lambda(\lambda + \Delta_2{}^2)G(\lambda)}{(\lambda - \mu)(\lambda - \nu)}. \tag{3}$$

The real-space boundaries of the orbits are the curves on which one of the momenta vanishes; hence all orbits are bounded by coordinate surfaces. The number of

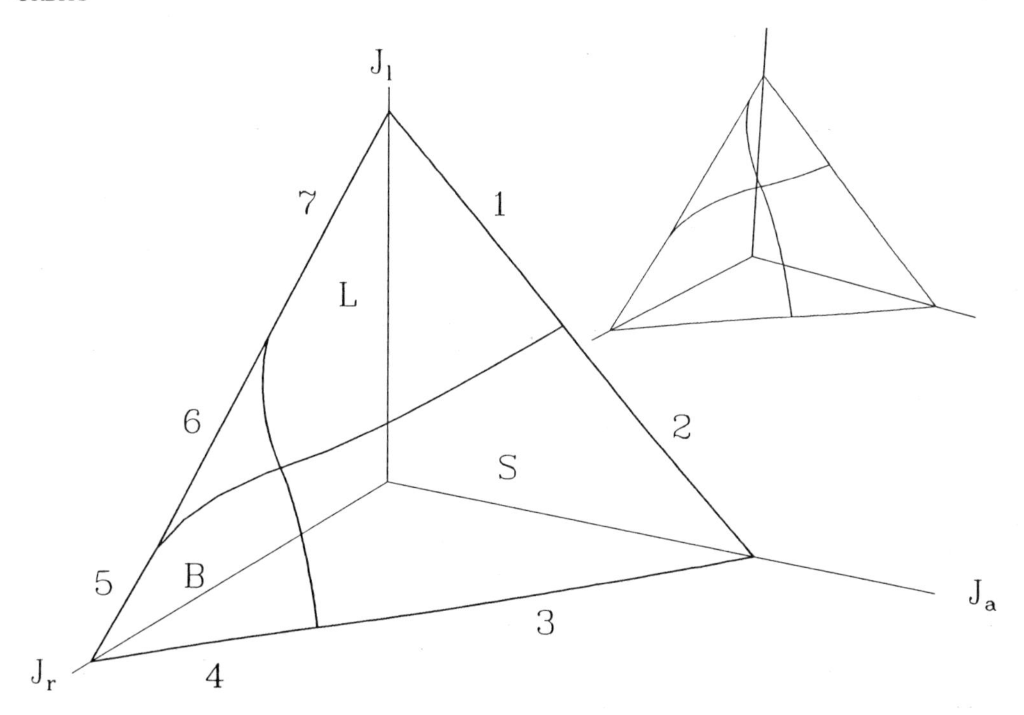

Figure 1. The partition of action space between orbit families in the potential of the perfect ellipsoid that has unit mass and axes in the ratios $b/a = 0.75$, $c/a = 0.5$. The small and large triangles are surfaces of constant energy; $E = -0.4$ and $E = -0.2$ respectively. They are plotted on the same scale and divided into the domains of the boxes (B), short-axis tubes (S) and the long-axis tubes (L). The quantity γ plotted in Fig. 2 is the ratio of the length of the portion of the edge of the triangle marked "3" to the sum of the lengths of "3" and "4".

possible momentum, and therefore velocity, vectors at any point $\mathbf{x}$ on the orbit ranges from 1–8 depending on how many of the p_τ change sign on the orbit. On box orbits, all momenta change sign, and eight velocity vectors are possible at any given point, while on tube orbits, only two momenta, p_λ and one other, change sign, so only four velocities occur at a point. The number of roots of the equations $p_\tau = 0$, and thus the family to which a given orbit belongs, depends on the values of the constants E, i_2 and i_3.

The action integrals which enable all orbits to be represented in a common action space, are

$$J_\tau(\mathbf{x},\mathbf{v}) = J_\tau(H, I_2, I_3) \equiv \frac{k}{\pi}\int_{\tau_{\min}}^{\tau_{\max}} |p_\tau(\tau)|\,\mathrm{d}\tau, \quad \text{where} \quad k = \begin{cases} 1 & \text{for } \tau = \lambda \\ 2 & \text{for } \tau = \mu \text{ or } \nu, \end{cases} \tag{4}$$

and $\tau_{\min}$ and $\tau_{\max}$ are the smallest and largest values of τ along the orbit. Unfortunately no comparably simple expressions are available for the angle coordinates θ_τ as functions of the phase-space coordinates.

While Stäckel potentials provide invaluable models of galaxy potentials, they are subject to significant limitations. The most important of these arise from the speed with which the isopotential surfaces become round at large $|\mathbf{x}|$. If the po-

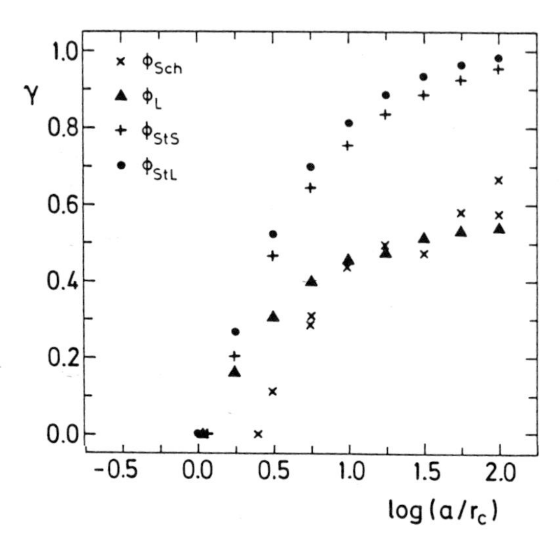

Figure 2. The quantity γ defined in the caption to Fig. 1 plotted for four potentials against $a(E)$, the distance along the potential's major axis at which the potential $\Phi = E$. r_c is the core radius of the body generating the potential. In their cores, the bodies generating the four potentials $\Phi_{\rm Sch}$, $\Phi_{\rm L}$, $\Phi_{\rm StS}$ and $\Phi_{\rm StL}$ all have axis ratios similar to those of the perfect ellipsoid for which Fig. 1 is plotted (see Gerhard & Binney 1985 for details). The potential $\Phi_{\rm Sch}$ is an approximation to Schwarzschild's (1979) potential ($\rho \propto r^{-3}$ for $r \gg r_c$), and $\Phi_{\rm StS}$ is the Stäckel potential that most closely approximates it. Similarly, $\Phi_{\rm StL}$ is a Stäckel approximation to the potential $\Phi_{\rm L}$ generated by a body with roughly constant axis ratios and density profile $\rho \propto r^{-2}$ at $r \gg r_c$.

tential were axisymmetric, all orbits would belong to one of the tube families. (Short-axis tubes if the potential were oblate, and long-axis tubes in the prolate case.) Hence the fraction of orbits of a given energy E that are boxes is a natural measure of the importance of the potential's triaxiality at radii characteristic of E. Figure 1 shows the action-space boundaries between the different orbit families for two values of E in a particular Stäckel potential. Notice (i) that the box-orbit fraction decreases quite rapidly with increasing energy, and (ii) that the shape of the box domain is almost independent of energy. Consequently, at any energy the box-orbit fraction is roughly proportional to $(1-\gamma)^2$, where $\gamma(E)$ is the fraction of orbits in the potential's equatorial plane that are loops. (These orbits fall in the (J_r, J_a) plane in action space.) In Figure 2 γ is plotted against a measure a of orbital energy for two Stäckel and two general potentials. One sees that with increasing energy, $(1-\gamma)^2 \to 0$ rather rapidly in the case of the Stäckel potentials, but slowly, if at all, for the other potentials plotted.

The concentration of the effects of triaxiality to the centres of Stäckel potentials is mirrored in conditions which the density profile $\rho(\mathbf{x})$ of a body must satisfy if it is to generate, via Poisson's equation, a Stäckel potential. In fact, if $\rho(\mathbf{x})$ does generate a Stäckel potential, then (i) then $\rho(\mathbf{x})$ must have a homogeneous core, *i.e.* $\rho(\mathbf{x}) \simeq$ constant for $|\mathbf{x}| < a$, a constant; and (ii) at $|\mathbf{x}| \gg a$, the non-spherical part of ρ, $\rho_2 \equiv \sqrt{\sum_m \left| \int \mathrm{Y}_2^m(\mathbf{\Omega})\rho(\mathbf{x})d^2\mathbf{\Omega}\right|^2}$, must fall off as $\rho_2 \propto |\mathbf{x}|^{-4}$. In particular, the ellipticity of a surface of constant ρ is independent of $|\mathbf{x}|$ only if $\rho \propto |\mathbf{x}|^{-4}$ for

$|\mathbf{x}| \gg a$, as in de Zeeuw's perfect ellipsoids. If we require that the density along the body's minor axis falls off as $\rho \propto |\mathbf{x}|^{-n}$, with $n \approx 3$ as is suggested by Hubble's law, or $n \approx 2$ as would be required to generate a flat rotation curve, then the body can generate a Stäckel potential only if it rapidly becomes round for $|\mathbf{x}| > a$. Thus bodies that generate Stäckel potentials are characterized by cores in which stars move nearly harmonically, and outer envelopes, in which the potential is either near Keplerian (if $n \geq 3$) or dominated by a locally spherical density distribution (if $n < 3$). Unfortunately there are many problems in galactic dynamics for which it is essential to consider models which have either singular central densities (see Gerhard, this meeting) or massive, strongly aspherical outer envelopes (*e.g.* Binney *et al.* 1986), and the application of Stäckel models to such problems can be frustrating.

3. ORBITS IN SLOWLY-ROTATING POTENTIALS

Rotation of the figure of a triaxial potential has far-reaching consequences for both regular and irregular orbits. Observationally, the most important effect of figure rotation is to imbue the vital triaxiality-supporting box orbits with a definite sense of circulation about the potential's rotation axis (usually assumed to coincide with the galaxy's shortest axis) (Schwarzschild 1982). The work of Merritt (1980) and Vietri (1985) has shown that strongly triaxial systems cannot achieve significant ratios v/σ of rotational and random velocities unless rotation of the potential has thus enabled the boxes to contribute to the overall stellar circulation.

Rotation affects the work of the galaxy modeler in two further ways: (i) in the presence of rotation the action spaces of Schwarzschild's principal orbit families no longer fit neatly together to form a single principal action space (Binney & Spergel 1984); (ii) some distance from the centre, figure rotation eliminates the majority of orbital tori in favour of a sea of stochastic orbits.

4. CHAOS

Since the frequencies ω_k are continuous functions of the phase-space coordinates, a general potential supports an infinite number of closed orbits. Consideration of the stability characteristics of these orbits gives some insight into the way in which the regular orbital structure exemplified by Stäckel potentials can dissolve into chaos.

If one linearizes the equations of motion around a closed orbit of period T, one obtains a set of coupled linear differential equations with coefficients that are periodic functions of time. By Floquet's theorem (*e.g* Margenau & Murphy 1956), any solution of these linear equations can be written as a sum of solutions of the form $Xe^{\mu t}P(t)$, where X and μ are constants, and P has period T: $P(t+T) = P(t)$. There are generally $(n-2)$ possible non-zero values of μ. The orbit is unstable if the real part of any of the μ's is positive. The orbit is stable if at least one μ is non-zero, and every such μ is pure imaginary. A stable closed orbit, is always a parent of an orbit family.

There are infinitely many closed orbits of any given energy E in a Stäckel potential, but *no* subsidiary orbit families, because all but 3–6 of these closed orbits are neutrally stable (every $\mu = 0$). This is a *very* special circumstance. In less special potentials, there are infinitely many of both stable and unstable closed orbits. In many cases of astrophysical interest, a few subsidiary families are able to push the principal tori back enough to gain a significant foothold in phase space, but the families parented by all other stable closed orbits are too small to be seen in a quick survey of possible orbits; these subsidiary families live like lice squeezed flat between the barely ruffled surfaces of the principal families. Fig. 7 of Binney (1982) shows an example of this phenomenon.

Now, just as stable closed orbits on the tori of a principal family give birth to subsidiary families, so stable closed orbits can arise on the tori of a subsidiary family, and call into being orbit families of the third generation. Closed orbits on tori of the third-generation spawn fourth-generation orbit familes, and so on *ad infinitum*— Gustavson (1966) shows a concrete example of this hierarchy. Little of the original regular phase-space structure survives the endless formation of orbit families unless the proportion of phase space that is claimed by families of any generation m is significantly less than that claimed by families of generation $(m-1)$.

At present there is no inexpensive way to estimate how much of phase space a particular resonance will seize. If a moderately triaxial potential is stationary in inertial space, resonances are unimportant, but resonances rapidly create havoc in phase space once we set the potential rotating. A partial explanation of this phenomenon is as follows. By setting the potential rotating at angular frequency Ω, we shift some of the orbital frequencies ω_k by $\pm\Omega$ while leaving others unchanged, and thus call into being a whole new set of resonances. Contopoulos & Mertzanides (1977), Athanassoula *et al.* (1983), Teuben & Sanders (1985) and others have studied these resonances in planar bars, while Binney (1981), Magnenat (1982), Mulder & Hooimeyer (1983) and Pfenniger (1984) have studied resonances that involve the motion perpendicular to the potential's equatorial plane. Evidently, these resonances of rotating potentials are much more effective at breaking up tori of the principal families than are the resonances of non-rotating potentials, but no simple mechanism has been identified to date.

The study of the generation of chaos by resonances, provides endless entertainment for analysts and computer enthusiast alike. In fact, the detailed structure of phase space at the onset of chaos is so fascinating, that before plunging deeply into its study, it is well to decide what we *need* to know about the chaotic regions of phase space before we can build serviceable galaxy models. My list would include:

(i) What are the characteristics of a potential Φ that determine whether its phase space is largely regular or largely chaotic? One possible answer to this question is "the distance of Φ from the nearest integrable potential Φ_I". Unfortunately, along any given orbit, *any* Hamiltonian lies arbitrarily close to an integrable Hamiltonian (Contopoulos 1963), so this is not a very satisfactory answer as it stands. Evidently the regular orbital structures of some integrable Hamiltonians are more easily disrupted than those of others. Also the effectiveness of a potential perturbation $\delta\Phi$ in introducing chaos depends on more than just the magnitude of $\delta\Phi$ (Gerhard 1985). Figure rotation seems to be an especially potent form of perturbation.

(ii) How should we incorporate stochastic orbits into our galaxy models? On closer examination, stochastic orbits often prove to be nearly quasi-periodic for several orbital periods at a time, and seem chaotic only because they switch abruptly from one quasi-periodic structure to another in an apparently random way. Are numerical experiments trustworthy here? What are the statistical characteristics of this switching? Can stochastic orbits be treated as linear combinations of regular orbits?

(iii) How many fundamentally different stochastic orbits are there at each energy? Goodman & Schwarzschild (1981) found that a single stochastic sea in Schwarzschild's (1979) model galaxy actually contains orbits that differ from one another on time-scales of a Hubble time. As Petrou (1984) and Pfenniger (1985) have pointed out, partial barriers, or "cantori" in phase-space can contain stochastic orbits for many orbital periods before suddenly releasing them into a wider volume of phase space. In the long run, this process gives rise to "Arnold diffusion" through phase space. How should we describe this sort of process mathematically, and what are its astronomical consequences?

REFERENCES

Arnold, V. I. 1978 *Mathematical Methods of Classical Mechanics*, (New York: Springer).

Athanassoula, E., Bienaymé, O., Martinet, L. & Pfenniger, D. 1983. *Astron. Astrophys.*, **127**, 349.

Binney, J. J. 1981. *Mon. Not. Roy. Astron. Soc.*, **196**, 455.

Binney, J. J. 1982. *Mon. Not. Roy. Astron. Soc.*, **201**, 1.

Binney, J. J. & Spergel, D. N. 1984. *Mon. Not. Roy. Astron. Soc.*, **206**, 159.

Binney, J. J., May, A. & Ostriker, J. P. 1986 *Preprint*, Princeton University.

Contopoulos, G. 1963. *Astrophys. J.*, **138**, 1297.

Contopoulos, G. & Mertzanides, C. 1977. *Astron. Astrophys.*, **61**, 477.

de Zeeuw, P. T. 1985. *Mon. Not. Roy. Astron. Soc.*, **216**, 273.

Gerhard, O. E. 1985. *Astron. Astrophys.*, **151**, 279.

Gerhard, O. E. & Binney, J. J. 1985. *Mon. Not. Roy. Astron. Soc.*, **216**, 467.

Goodman, J. & Schwarzschild, M. 1981. *Astrophys. J.*, **245**, 1087.

Gustavson, F. G. 1966. *Astron. J.*, **71**, 670.

Kent, S. 1983 *Preprint*, Center for Astrophysics.

Magnenat, P. 1982. *Celest. Mech.*, **28**, 319.

Margenau, H. & Murphy, G. M. 1956 *Mathematics of Physics and Chemistry*, (New York: Van Nostrand), §2.17.

Merritt, D. 1980. *Astrophys. J. Suppl.*, **43**, 435.

Mulder, W. A. & Hooimeyer, J. R. A. 1983. *Astron. Astrophys.*, **134**, 158.

Petrou, M. 1984. *Mon. Not. Roy. Astron. Soc.*, **211**, 283.

Pfenniger, D. 1984. *Astron. Astrophys.*, **134**, 373.

Pfenniger, D. 1985. *Astron. Astrophys.*, **150**, 97.

Ratcliff, S. J., Chang, K. M. & Schwarzschild, M. 1984. *Astrophys. J.*, **279**, 610.

Richstone, D. O. 1980. *Astrophys. J.*, **238**, 103.

Richstone, D. O. 1982. *Astrophys. J.*, **252**, 496.

Richstone, D. O. 1984. *Astrophys. J.*, **281**, 100.

Schwarzschild, M. 1979. *Astrophys. J.*, **232**, 236.

Schwarzschild, M. 1982. *Astrophys. J.*, **263**, 599.

Teuben, P. & Sanders, R. 1985. *Mon. Not. Roy. Astron. Soc.*, **212**, 257.

Vietri, M. 1985. *PhD thesis*, Princeton University.

DISCUSSION

Toomre: What is the best evidence you know that the shapes of triaxial galaxies (which I have finally gotten used to!) actually *revolve* appreciably in space?

Binney: In a non-rotating potential, the box orbits, which form the backbone of a triaxial galaxy, do not contribute to the circulation. Here one can get the stars to circulate only by weighting tube orbits, which are fundamentally opposed to the bar. The degree of their opposition grows with the strength of the bar. Merritt (1980, *Astrophys. J. Suppl.*, **63**, 435) found that for Schwarzschild's axis ratios $a : b : c = 1 : 0.625 : 0.5$, only modest rotation could be found in this way. Thus if rapidly-rotating galaxies are strongly triaxial near their core, their figures must rotate. I suspect that there are two main classes of rapidly rotating galaxies: (a) those which are strongly triaxial at small radii and have rapidly rotating figures which become axisymmetric well in advance of the Inner Lindblad Resonance; (b) those which are axisymmetric at small radii and become triaxial with low pattern speed far from the center.

Ostriker: Am I correct in thinking that the limitation (for non-rotating figures) to small regions of substantial triaxiality for Stäckel potentials is not fundamental. Real galaxies could have large (in units of core radius) triaxial parts but one simply could not compute the equilibrium analytically.

Binney: Yes. The logarithmic potential investigated by many people has a beautifully regular orbital structure to very large radii. But at large radii this orbit structure diverges from that characteristic of Stäckel potentials in that it supports a host of subsidiary families in the box domain. We still haven't figured out the details of galaxy building with potentials of this sort, but I see no fundamental problem.

Illingworth: My question relates to Alar Toomre's question. (a) What general constraints can be placed on the amount of figure rotation for a given (e.g., observed) level of rotation? (b) Could figures counter-rotate?

Binney: As far as I am aware, the literature contains no satisfactory study of the problem you raise in (a). Vietri's study of triaxial spheroids for our Galaxy is the best reference I know. As to (b), in principle figures can counter-rotate (Freeman's analytic models do) but never by very much and I rate the probability of such models very low—see Vietri for details.

White: We know that the bars in barred spirals are rapidly rotating triaxial systems. If nature can do the trick in this case, how confident can we be that it is not possible for more elliptical-like systems?

Binney: The density profiles of bars and ellipticals are quite different. Bars in disks have fairly constant surface density inside pretty sharp edges, while ellipticals have smooth, steep density profiles. Thus in a bar, the orbital frequencies can all be comparable with the pattern speed Ω, while in an elliptical the range in frequencies is very great. Furthermore, the sharp edges of bars can be identified with a resonance, perhaps corotation or the Inner Lindblad Resonance. Ellipticals seem to have no comparable characteristic radius.

Statler: Levison & Richstone find in the non-rotating logarithmic potential, strange orbits that look like boxes but do not line up with the symmetry planes. Are these resonances, or nearly-stochastic orbits, or something else?

Binney: Many resonances arise in the logarithmic potential when the semi-axes of the zero-velocity surfaces are much larger than the potential's core radius (Levison and Richstone *do* use a finite core contrary to what is said in their poster). Their weird orbits are almost certainly trapped, or partially trapped, around such resonances.

Sellwood: N-body simulations by Wilkinson & James, Gerhard etc. took a remarkably long time to settle to a steadily rotating potential. Is it likely that real elliptical galaxies have settled yet to uniform figure rotation?

Binney: No. In studying steady-state models we are surely doing no more than getting zeroth-order approximations to reality. As yet we know little of how a steady-state system will respond to perturbations. I expect many will be stable, and others will display only slowly-damped long-period oscillations.

Toomre asking a question, in his usual style.

James Binney explains that orbits are simple.

DYNAMICS IN THE CENTRES OF TRIAXIAL ELLIPTICAL GALAXIES

Ortwin E. Gerhard
Max-Planck-Institut für Astrophysik
Karl-Schwarzschild-Str. 1
8046 Garching bei München
West Germany

Abstract. Orbits in the inner *kpc* of a triaxial galaxy are discussed, taking into account the effect of a central density concentration like a massive black hole, a dense stellar nucleus, or a de Vaucouleurs-type cusp. Since the box orbits that form the backbone of a triaxial galaxy pass arbitrarily close to the centre after long enough time, they will eventually be subjected to large-angle deflections by a central point mass, and the triaxiality of the inner part of the system will thereby be destroyed. A $10^8 M_\odot$ black hole is estimated to affect box orbits out to $1kpc$ in a Hubble time, while a similar influence of the observed (extended) nucleus in $M31$ reaches out to $500pc$ in the bulge. Regular box orbits persist, however, in systems with singular central density profiles such as implied by carrying the $r^{1/4}$ law all the way to the centre. This result can be approximately understood in terms of the frequency ratio Ω_r/Ω_θ remaining close to the harmonic value of 2 for many orbits in the corresponding spherical potential. Finally, I discuss observable consequences of the box orbit scattering process and future work, and use the presence of isophote twists in the central parts of a number of elliptical galaxies to obtain approximate upper limits on the masses of the black holes these systems may contain.

INTRODUCTION

While most existing galaxy models are assumed to have homogenous cores, giant elliptical galaxies often have central brightness profiles that appear to be still rising at the smallest presently resolvable radii (Schweizer 1979, Lauer 1985, Kormendy 1985). Moreover, they may contain massive black holes at their centres (Lynden-Bell 1969, Rees 1984), or dense stellar nuclei such as are observed in the bulge of M31 (Light et al. 1974) and probably in the bulges of many other nearby spiral galaxies (Kormendy 1985). It is therefore important to understand the dynamical effects of such features on the system as a whole.

In a spherical or axisymmetric potential, at least one of the components of the angular momentum vector is conserved, and so most stars are kept far away from the centre. Since the total mass of the central object is small compared to the mass of the core region of the galaxy (which is typically $\sim 5 \times 10^9 M_\odot$), most stars in the *cores* of axisymmetric systems would also not be influenced much by the precise mass distribution at the very centre. However, there is increasing evidence

T. de Zeeuw (ed.), Structure and Dynamics of Elliptical Galaxies, 241–248.

for elliptical galaxies being in part triaxial systems (see Binney 1985 for a review, and Davies, this conference). In a triaxial galaxy the influence of a central compact object on the dynamics is much more profound, because a triaxial mass model depends critically on the existence and selective population of the box orbits, in particular the 'thin' box orbits (Schwarzschild 1979). As shown e.g. in Fig.1 of Gerhard & Binney (1985; hereafter GB) these orbits go arbitrarily close to the centre after sufficient time, and different box orbits of the same energy differ at the centre by the direction of their velocity vectors. Any feature in the centre of the potential that can scatter the box orbits will therefore redistribute them in angular action (or in the size of their waists) and thereby affect the very heart of the system.

In the following I shall discuss three kinds of deviations from a homogeneous core: (i) a central point mass with a typical mass of $10^8 M_\odot$, ie. $\sim 2\%$ of the galaxy's core mass, (ii) a dense but extended central star cluster such as that inferred for M31, and (iii) central cusps as eg. implied by carrying a de Vaucouleurs profile all the way in to the centre (then $\rho \propto r^{-3/4}$ asymptotically, Young 1976). All of these features give rise to a complicated phase-space containing resonant or stochastic orbits in place of some or most box orbits. However, the first two are localized, and the results of the numerical orbit integrations can to a fair degree be described by a simple 'box orbit scattering' model. This will be described next; then I shall briefly consider effects on the global dynamics of the galaxy and the orbit structure of cuspy potentials, and finally describe some observational consequences and future work.

BOX ORBIT SCATTERING

Numerical orbit integrations show (GB; Figs. 2-4) that when a softened point mass with about 2 percent of the galaxy's core mass is placed at the centre of a triaxial galaxy and is gradually hardened, the box orbit region in phase-space first breaks up into regions occupied by families of resonant orbits and then becomes more and more irregular. In the presence of a hard point mass, most box orbits at low and intermediate energies become highly stochastic. Inside the core radius some box orbits are transformed into loops; of the box orbits reaching out to several core radii only a small fraction can avoid disruption by being trapped in orbital resonances. Stochastic orbits at these energies can be described as sequences of segments of regular box orbits, separated by discrete scattering events.

These numerical results may be approximately understood by the following simple model. Denote the core mass of the underlying galaxy potential by M_c and the mass of the central compact object (CCO) by m; then for a typical $\mu \equiv m/M_c \simeq 0.02$ the radius of influence of the CCO - at which the forces due to the galaxy and the CCO are equal - is $r_h \simeq \mu^{1/3} r_c \simeq 0.3 r_c$, with r_c being the galaxy's core radius. Now imagine that stars from the galaxy enter the radius of influence r_h with a velocity v_h set by the underlying galaxy potential, and that inside r_h they may be considered to be on hyperbolic Keplerian orbits with asymptotic velocity v_h. Then large-angle scattering occurs for impact parameters

$$b < b_l = \frac{Gm}{v_h^2} = \mu r_c \left(\frac{v_c}{v_h}\right)^2, \tag{1}$$

where v_c is the circular velocity at the edge of the core. Empirically, one finds that box orbit scattering by a point mass becomes effective in destroying the regularity of an orbit when

$$b \leq 3b_l, \tag{2}$$

and by an extended nuclear star cluster when in addition the half-mass radius is

$$r_{1/2} \leq 2b_l. \tag{3}$$

The latter condition corresponds approximately to the condition $v_n^2/v_h^2 \leq 5$, where v_n is the characteristic velocity of the nucleus. Since box orbits come arbitrarily close to the centre after sufficient time, equation (2) implies that even a very small point mass in the centre will ultimately destroy the triaxiality of the *entire* system.

Thus the problem becomes one of time-scales. To estimate how long it takes a star on a given box orbit of specified energy to come sufficiently close to the centre, one needs to know the probability for it to go anywhere through the orbit's waist (i.e. the section $x = 0$ through the orbit). The simplest assumption is that the probability of going through any point in the orbit's waist is independent of where this point lies in this section; then $N(< b) \propto b^2$. The numerical integrations in GB verify the analogous assumption for planar orbits, $N_{pl}(< b) \propto b$, and in this case also the amplitude of $N_{pl}(< b)$ as determined by a suitable average over the angular actions for the box orbit segments agrees very well with the numerical results. Then one expects a star to be scattered after about (area of the waist)/$9\pi b_l^2$ passages through the centre, or (in the simplest case) after time

$$\tau_l = \frac{(2r_c)^2}{9\pi b_l^2}\tau_{dyn} \propto \mu^{-2}\left(\frac{v_h}{v_c}\right)^4 \tau_{dyn}. \tag{4}$$

τ_l is thus proportional to μ^{-2}, to the dynamical time, and to a dimensionless velocity factor that describes the energy of the orbit in a given mass model of the galaxy. Both this factor and τ_{dyn} increase with energy; thus equation (4) shows that *for any value of the mass ratio μ, there is a maximum radius out to which box orbit scattering is effective, and, conversely, for any radius there is a minimum mass required for scattering in a Hubble time.*

Assuming that the inner parts of the galaxy are approximately described by a modified Hubble law, one finds in this way that in a typical triaxial elliptical galaxy a central black hole of $\sim 10^8 M_\odot$ will scatter box orbits that reach out to $\sim 1kpc$ (GB). In the case of the extended nucleus of M31 (according to the Stratoscope II observations of Light et al. 1974 and the rotation measurements of Walker 1974, but see Kormendy, this conference) the maximum radius for box orbit scattering is *not* determined by the time-scale, but through the softness condition (3) (very approximately), and is numerically found to be $\sim 500pc$ (Gerhard 1986).

These results remain unchanged if the figure of the galaxy rotates slowly about its smallest principal axis. In a tumbling potential the closed long-axial orbits that sire the box orbit family become prograde orbits which no longer pass exactly through the centre of the system. The distance at which these orbits pass the centre as they cross the intermediate axis of the potential, increases both with orbital energy and with the angular speed at which the potential rotates. However, rotation can only prevent the dissolution of the triaxiality of the system by the box orbit scattering process if a substantial part of the box orbit family never comes close to the centre. I.e. the offset of the closed long-axial orbit along the intermediate axis has to be at least of order the maximum width of a box orbit in the absence of figure rotation or black hole. From this condition one estimates (GB) that only box orbits reaching out to at least a quarter of the corotation radius can escape being scattered, so that figure rotation cannot prevent the dissolution of box orbits in the inner few core radii of a slowly rotating elliptical galaxy, or in the inner 500*pc* of M31's bulge (which has an estimated corotation radius of at least 5*kpc*, cf. Gerhard 1986).

EFFECT OF BOX ORBIT SCATTERING ON THE GALAXY

The scattering of the box orbits by a compact object in the centre of a triaxial galaxy will tend to establish a uniform distribution of box orbits in their relevant angular action. Then the selective population of the thin box orbits cannot longer be maintained and hence the CCO must ultimately destroy the triaxiality of the system. What is less certain is whether the disruption of triaxiality occurs abruptly, or whether it proceeds gradually from the inside out. The latter possibility arises because the disruption time-scale (4) is a rapidly increasing function of energy, and so in a fixed potential the low-energy box orbits are scattered much before the high-energy ones and axisymmetry may spread gradually from the centre. However, if collective effects due to the dissolution of the box orbits in the core of the system are strong, they may alternatively lead to a rearrangement of a large part of the system relatively quickly.

To settle this question will require reliable N-body models of triaxial galaxies with a genuine central point mass. A first attack on this very difficult numerical problem was made by Norman, May & van Albada (1985); their simulations indicate, however, that individual orbits can currently not be integrated with sufficient accuracy for one to have confidence in results obtained from N-body simulations for this problem (see also van Albada, this conference).

The following simple argument suggests, however, that triaxiality will be lost gradually, from the inside out, rather than in the manner of an abrupt phase change. If one compares a prolate stellar system of constant ellipticity with one in which only the outer parts are prolate, but which has a spherical core, then the quadrupole moment in the outer parts is not very different between the two systems. Since it is this quadrupole moment (i.e. the tangential force) which every few dynamical times reverses the angular momentum of stars on box orbits, this suggests that at any time the galaxy will approach axisymmetry only at those radii at which the disruption time (4) is less than its age, and remain triaxial at larger radii.

CENTRAL DENSITY CUSPS

In recent years there has been much discussion of whether the stellar densities in the centres of elliptical galaxies continue to rise at the smallest radii that can be resolved from the ground, or whether the density finally flattens off in the very inner parts of the system (e.g. Schweizer 1979). In view of the ability of CCO's to destroy the triaxiality of at least the inner parts of the galaxy, it is therefore interesting to know how pronounced a density cusp has to be before it can have a similar effect. GB investigated the orbital structure in the equatorial planes of triaxial potentials, that correspond to power-law density distributions in the ellipsoidal radius a (actually, they studied orbits in the planes of prolate systems containing the respective long axis, but the difference to the triaxial case is small). A particular such case is that

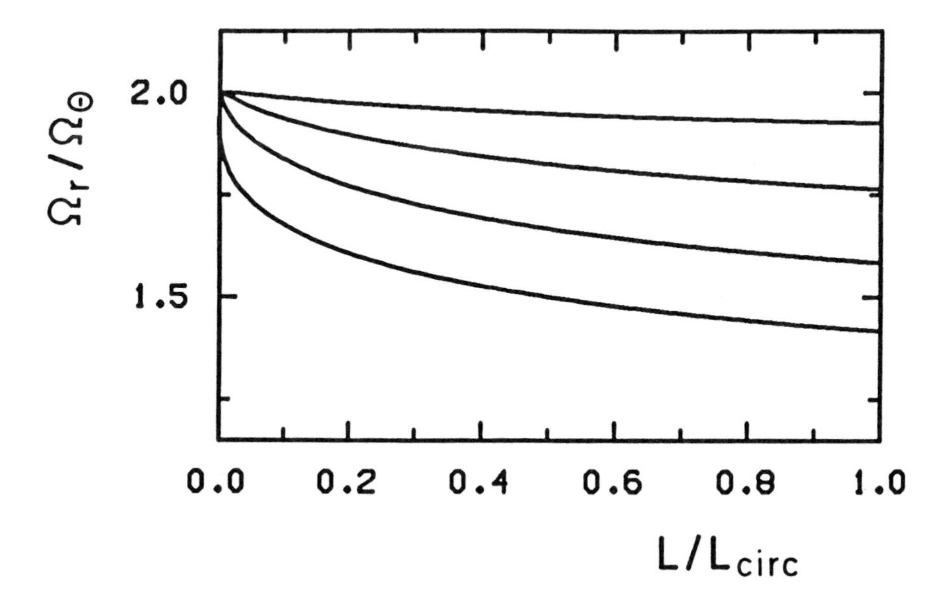

Fig.1: The frequency ratio Ω_r/Ω_θ in a number of spherical potentials corresponding to density profiles $\rho(r) \propto r^{-\gamma}$, as a function of the angular momentum of the orbit normalized to that of the circular orbit of the same energy. Curves are (from top) for $\gamma = 0.3$, $\gamma = 0.9$, $\gamma = 1.5$, $\gamma = 2.0$ (scale-free logarithmic potential).

obtained by carrying the de Vaucouleurs profile all the way into the centre; then for small a

$$\rho(a) = \rho_0 a^{-p}, \qquad a^2 = x^2 + \frac{y^2 + z^2}{q^2} \tag{5}$$

and $p = 0.75$ (Young 1976). In this case, both the force and the circular velocity tend to zero at small a, and the corresponding potential contains box orbits at all energies of interest, besides containing also a few resonant orbit families and some stochastic orbits. Indeed, the fraction of phase-space occupied by the box orbits is substantially larger in this potential than in Schwarzschild's (1979) potential at both low and high energies. Thus there is every reason to believe that self-consistent triaxial galaxy models can be constructed which have a de Vaucouleurs brightness profile.

The existence of box orbits in potentials with singular density profiles can be related to the range of values of the frequency ratio Ω_r/Ω_θ in the corresponding spherical potential. Fig. 1 shows this ratio for a number of spherical potentials, as a function of the orbit's angular momentum normalized by the angular momentum of the circular orbit. In a harmonic potential, $\Omega_r/\Omega_\theta = 2$ for all orbits; in the case corresponding to $\rho \propto r^{-\gamma}$ it is near the harmonic value for almost all orbits when $\gamma = 0.3$, and for many orbits still when $\gamma = 0.9$. Only when the potential approaches the scale-free logarithmic potential ($\gamma = 2.0$) is Ω_r/Ω_θ significantly different from 2 already for orbits with quite low angular momenta. Correspondingly, most of the phase-space in a genuinely scale-free logarithmic potential is occupied by loop orbits and by (in x-y) 1:2, 3:2, and 4:3 resonant orbits families, while box orbits play little or no role; cf. GB.

OBSERVABLE CONSEQUENCES, BLACK HOLE MASSES, FUTURE WORK

We have argued above that if box orbit scattering due to a central massive black hole or dense star cluster occurs in a triaxial galaxy, it will gradually axisymmetrize the system from the inside out. Then if the galaxy started out with constant ellipticities, it should now appear rounder in its inner region as compared to its outer parts. However, there is nothing that prevents a galaxy from being made with a rounder core in the beginning, so that interpreting such observations is inherently ambiguous. Nevertheless it is interesting to notice that galaxies which are fairly round inside a few core radii and become more flattened further out are common, eg. in Lauer's (1983) sample. Among his 42 galaxies, there are 21 which are consistent with having no isophote twist in the region outside of 4". Of these 10 have ellipticity profiles that rise from being nearly round at 4" to $\epsilon \sim 0.3$ further out (both before and after seeing corrections). This pattern is consistent with these systems being triaxial galaxies seen from near one of their principal planes, or with them being axisymmetric objects without constraint on the projection geometry. A Monte Carlo-type statistical analysis is needed to see whether one could rule out on probability grounds the hypothesis that the inner parts of these galaxies are triaxial. If so, this would be interesting from the point of view of the scattering process.

The argument may be turned around in the following way. If a galaxy shows isophote twists at some small radius a which are not caused by dust, and if one assumes that these are due to triaxiality and imply the existence of box orbits of scale a, one may set an approximate upper limit to the mass of the black hole this galaxy could contain, by estimating the box orbit disruption time and requiring that it should exceed a Hubble time at scale a. This has been done for the galaxies in Lauer's sample that show clear isophote twists outside 4" and for which the seeing correction left the position angle profiles nearly unchanged. Since the photometry of these galaxies is in the red, the effects of dust should not play a too important role, but some galaxies were discarded on Tod Lauer's advice. In Table 1 I list for the remaining galaxies with near-central twists (names in column 1) the radius at which the twist occurs (col. 2), the magnitude of the twist (col. 3), the ellipticity in the region of the twist (col. 4), the core radius (cols. 5,6), the line-of-sight velocity dispersion (col. 7), and the upper bound inferred for the mass of the central compact object (col. 8). This was obtained by equating to a Hubble time a more accurate expression for the box orbit scattering time τ_l (equation 25a of GB), assuming that the twists are due to box orbits at the scale of the twists. Typical upper limits range from a few 10^7 to a few 10^8 solar masses. If, less conservatively, one included a galaxy like NGC 1052 in the list, which may have a small twist at 3", one would derive $\sim 9 \times 10^6 M_\odot$ on the same assumptions. With the very best multicolour photometry one might push the limit down to $\sim 2 \times 10^6 M_\odot$ if isophote twists due to triaxiality were seen at one core radius of $\sim 1" \sim 100pc$.

On the theoretical side the most uncertain part is the collective response of the system as a whole to the box orbit scattering and subsequent evolution of its inner parts. The study of this important problem will require sophisticated N-body simulations, possibly making use of a hybrid approach that treats orbits in the vicinity of the central object specially. Alternatively, one might use that

NGC	at/"	$\Delta\phi/^{0}$	ϵ	rc/"	rc/pc	σ/kms^{-1}	$M/M_{\odot}$
596	6-9	20	.05	0.9	83	180	4×10^7
636	6-10	12	.06	1.39	128	186	5×10^7
741	4-5	10	.15	1.91	519	309	2×10^8
1407	4-6	15	.05	3.21	247	288	4×10^7
1453	5	5	.12	1.38	285	280	2×10^8
5846	4-16	20	.04	2.51	292	255	7×10^7
7626	4-18	9	.10	1.1	206	275	2×10^8
7768	7-10	5	.20	1.11	467	260	9×10^8
1052	3	10	.20	1.08	75	207	9×10^6
4474	2-4	5	.07	3.70	282	317	3×10^7
4649	3	10	.05	3.41	260	361	2×10^7
7785	3	8	.25	1.12	226	241	5×10^7

Table 1: Upper limits to the mass of a possible central black hole in a number of ellipticals, obtained from relating the scale of observed isophote twists to the corresponding box orbit scattering time-scale. See text for details. The first part of the table includes only galaxies with twists outside 4", while the second part lists a few galaxies with apparent twists at $\sim$ 3".

the evolution at each radius occurs on a time-scale much longer than the dynamical time-scale, so that the action variables are conserved except in the scattering events. Then, even though each star changes its angular action substantially when being scattered, the distribution function of the angular action may still be considered as slowly varying in a known way, so that the change in the mass distribution could be computed iteratively.

REFERENCES

Binney, J.J., 1985. *Mon. Not. R. astr. Soc.* **212**, 767.
Gerhard, O.E., 1986. *Mon. Not. R. astr. Soc.* **219**, 373.
Gerhard, O.E., Binney, J.J., 1985. *Mon. Not. R. astr. Soc.* **216**, 467.
Kormendy, J., 1985. *Astrophys. J. (Letters)* **292**, L9.
Lauer, T.R., 1983. *PhD thesis*, University of California, Santa Cruz.
Lauer, T.R., 1985. *Astrophys. J.* **292**, 104.
Light, E.S., Danielson, R.E., Schwarzschild, M., 1974. *Astrophys. J.* **194**, 257.
Lynden-Bell, D., 1969. *Nature* **223**, 690.
Norman, C.A., May, A., van Albada, T.S., 1985. *Astrophys. J.* **296**, 20.
Rees, M.J., 1984. *Ann. Rev. Astr. Astrophys.* **22**, 471.
Schwarzschild, M., 1979. *Astrophys. J.* **232**, 236.
Schweizer, F., 1979. *Astrophys. J.* **233**, 23.
Walker, M.F., 1974. *Publs. astr. Soc. Pacif.* **86**, 861.
Young, P.J., 1976. *Astron. J.* **81**, 807.

DISCUSSION

Norman: Two points. Firstly, the stochasticity produced by the black hole may help provide fuel for the hole. Unpublished calculations done with Andrew May indicate that stochasticity is significant in the nucleus with $0.001 < M_h/M_{core} < 0.1$ with the upper limit occurring because the hole dominates the nucleus and produces regular orbits again. This could be very important for luminosity evolution. Secondly, although as probably will be discussed by Tjeerd van Albada, the simulation we made was very difficult and needs to be greatly improved, the result was a significant roundening of isophotes by one or two ellipticity subclasses in a Hubble time. Radio galaxies are seen to be rounder–the latest on this is in a poster at this meeting. Although this occurs at larger radii than seen in our simulation, the effects may well be correlated.

Gerhard: To your first point: The cross section for tidal disruption of stars by the hole is always much smaller than the cross section for box orbit scattering, so that the triaxiality of the core, and the stochastic orbits introduced by the hole in it, disappear quickly. Consequently, the feeding rate is a rapidly decreasing function of time, and it seems unlikely to me that triaxiality can help feed powerful radio sources—let alone quasars—for longer than $\sim 10^6$ yr, even with very massive black holes (cf. Figure 9 in Gerhard & Binney, 1985, *Mon. Not. R. astr. Soc.*, **216**, 467). To the second point: I understand that there is some controversy about the observational effect you mention, but if radio galaxies are rounder at R_{25} than normal galaxies, *very* massive black holes and *very* small core radii are required if this is to be explained in terms of the box orbit scattering process. But I agree that it is important to understand better the collective effects that the changes in the inner kpc may induce.

Lake: Begelman, Blandford and Rees (1984, *Rev. Mod. Phys.*, **56**, 255) argue that radio galaxies emit at small fractions of the Eddington limit. NGC 1052 is a radio galaxy and is the galaxy whose black hole mass is most strongly constrained by your method. How does the mass from the Eddington limit compare?

Gerhard: NGC 1052 belongs to a class of compact-active E/SO galaxies for which $\sim 10^{40}$ erg/s is a characteristic luminosity (Jones, Wrobel & Shaffer, 1984, *Astrophys. J.*, **276**, 480). If this is 1% of the Eddington limit, the inferred mass would be $\sim 10^4 M_\odot$.

SHELLS AND THE POTENTIAL WELLS OF ELLIPTICAL GALAXIES

P.J.Quinn[1] and Lars Hernquist[2]
1. Space Telescope Science Institute, Baltimore, MD 21218.
2. Dept. of Astronomy, Univ. California, Berkeley, CA 94720.

ABSTRACT. A survey of the possible variety of sharp-edged, caustic features that may arise in the collision of galaxies with very different masses and sizes (Hernquist and Quinn 1986a) has shown that in general shells are morphologically very complex. It is therefore not easy to determine the history of the collision that produced the shells nor the properties of the galaxies involved. However, a small number of shell galaxies (notably NGC 3923) have a sufficiently simple and orderly shell distribution that we believe the shells were formed by a chance very symmetric and simple encounter. In such cases we are presented with a unique opportunity to investigate the potential well of an elliptical galaxy over a large range in radius ($\simeq 0.5r_e - 20r_e$). An analysis of the NGC 3923 shell system (Hernquist and Quinn 1986b) has shown that a large amount of dark matter is present ($M_{dark} \simeq 40M_{luminous}, r < 17r_e$).

1. SHELL HISTORY.

Over the past six years, our knowledge of shell systems has grown from a handfull of examples to a point where shells are now considered a common feature of field ellipticals. The word "shell" was first used by Arp (1966) to classify some of the features he saw in ARP 230. The Arp atlas has proved to be an invaluable source of information on the variety and abundance of interacting systems and has provided much of the fuel for the growing idea that galaxies may have been severly modified by mergers during their lifetimes (Toomre 1977). Further examples of shells were slow in coming after the Arp Atlas; Dave Malin's work on M89 being a notable addition (Malin 1979). The full implications of shells and their connection to interactions involving disk galaxies was first pointed out by Francois Schweizer in his work on Fornax A (Schweizer 1980). Schweizer discovered prominant sharp-edged features (which he called "ripples") in the outer optical envelope of Fornax A and concluded correctly that such sharp stellar features could not be formed from the "hot" stellar population that made up the main body of Fornax A. Rather, this material must have been introduced from another "colder" galaxy. At this point the connection between peculiar galaxies, mergers and shells had been made. However the real surprise was the keystone work of Malin and Carter (1980) who demonstrated the existence of shells in otherwise completely normal looking ellipticals. The discovery of a beautiful set of very symmetric shells around NGC 3923 was achieved by applying photographic enhancement techniques to U.K. Schmidt IIIaJ survey plates and A.A.T. prime focus plates. These techniques were largely

T. de Zeeuw (ed.), Structure and Dynamics of Elliptical Galaxies, 249–259.

developed for astronomical applications by Malin (Malin 1981,1982). The existence of low light level shells around otherwise normal ellipticals means that shells are possibly a very common and relatively long lived component of many ellipticals and implies mergers are a fundamental part of the history of elliptical galaxies.

Subsequent surveys for shells (Malin and Carter 1983, Schweizer and Ford 1985) have turned up over 200 examples of shell systems. Very few of these galaxies have been examined photometrically and spectroscopically but in the few cases that have, the general conclusion is that shells are indeed stellar; they have colors that are close to that of the parent and the parents themselves are kinematically and chemically like shell-less ellipticals (Carter, Allen and Malin 1982; Quinn 1982; Fort et. al. 1985; Pence 1986; Wilkinson, Sparks, Malin and Carter 1986). The proposition that some well studied peculiar ellipticals (like Cen. A) were the result of a merger gained support by finding telltale shells in their outer luminous envelopes (Malin, Quinn and Graham 1983).

On the theoretical side, Fabian, Nulsen and Stewart (1980) were the first to propose an origin for the shells reported by Malin and Carter (1980). Their theory called for shells being formed at shocks in an outflowing wind (a similar theory has also been proposed by Williams and Christiansen (1985)). At present, wind related theories seem to be less consistent with the observations than those involving mergers (see Athanassoula and Bosma 1986 for a review). Following Schweizer's suggestion of a merger related origin for shells, Quinn (1982, working with Agris Kalnajs) was able to produce sharp edged features in collisions involving disks and massive ellipticals that were qualitatively very similar to those found by Malin and Carter. Subsequent work by Toomre (1983), Quinn (1984), Dupraz and Combes 1986, and Hernquist and Quinn (1986 a,b) has shown that mergers and accretions can very naturally account for the diversity of the observed shell galaxies. Shell making is a very robust and inevitable consequence of the destruction and cannibalism of a small and/or cold companion galaxy by a massive parent. Further studies are reported elsewhere in this volume (James and Wilkinson 1986, Piran and Villumsen 1986).

2. LAYING TO REST SOME MISCONCEPTIONS ABOUT SHELLS.

Like all good prototypes in astronomy, NGC 3923 and the first sample of shell systems presented by Malin and Carter (1980) have proven to be atypical of the majority of the shell systems now known. Hence, by attempting to model the features of NGC 3923, the initial theoretical work of Quinn (1982, 1984) gave a rather narrow view of the way shell systems might be formed. In particular, Quinn found that a nearly radial orbit for the companion was prefered if the shells were to be aligned like those of NGC 3923. Nonradial companion orbits gave rise to complex and interpenetrating structures that were unlike the shells in NGC 3923. Also Quinn neglected the possibility that a small spheroidal companion could be as "cold" as the disk-like companions proposed by Schweizer to be responsible for the shells.

Hernquist and Quinn (1986a) have conducted a fairly complete survey of the possible parameter space taking into account progressive disruption of the companion, disk and spheroidal companions, and non-spherically symmetric parents. Their conclusion was that shell-like features can be formed during the accretion ($\Delta E_{orbit} = 0$) or merger ($\Delta E_{orbit} < 0$) events involving a fixed primary galaxy (either spiral or elliptical) and a small and/or cold companion. Companions have to have a small phase space volume relative to the primary in order to make many sharp shells. Hence the companion can be either spatially smaller than the primary, or have a smaller velocity width than the primary (colder) or both. Shell-making is therefore a very robust dynamical process and should be the rule rather than the exception.

The diversity of structures found in the Hernquist and Quinn models complements the observed diversity of morphologies in the Malin-Carter Catalogue (Malin and Carter 1983). The typical shell galaxy may have a small number of non-concentric, non-aligned, interpenetrating shells as well as radial filaments. In a case like NGC 3923 which has aligned, very symmetric, simple shells, the companion must have had a simple morphology and the orbit must have been very nearly radial (see Figure 1). Non-radial encounters involving either disks or elliptical companions and radial encounters involving disks, produce shells that are very different from those observed in NGC 3923 (see Figure 2). Hence NGC 3923-like systems should be rare amongst the shell galaxies (which they are) and they may afford us a particular simple case in which to test and use the phase-wrapping theory of shell formation (Quinn 1984).

3. MAKING SOME SENSE OUT OF THE DIVERSITY OF SHELL GALAXIES.

Shells were defined by Malin and Carter as "edge-brightened structures in the optical envelope or (more often) beyond the discernible limits of the elliptical galaxy". Most of the structures formed in the simulations of Hernquist and Quinn would certainly be classified as "shells" by the above definition, yet geometrically they may be totally unlike a shell. Clearly the label "shell" should not be taken in its literal sense but rather as a catch-all for a large class of sharp-edged stellar features which are not necessarily of constant curvature.

One clear systematic trend evident in the Hernquist and Quinn simulations is that the most complex looking structures are always associated with disk-like companions. The fact that the companion is sheet-like, permits folding of the companion in configuration space as well as phase space. The fold catastrophies that result when the distorted disk-like companion is projected onto the plane of the sky appear as extended, bright, sharp-edged features of the types classified by Arnold (1984). This would then imply that in those cases where we see a complex shell morphology (e.g. Cen. A, NGC 2685) we should also expect to see other signs of a disk-like intruder such as gas and dust. In all the examples known to the authors, this is observed to be the case.

The theory of catastrophic processes has direct implications for the photometry of shells. The fall-off of density behind a simple fold caustic is proportional to $x^{-\frac{1}{2}}$(where x is the distance from the caustic) regardless of the form of the potential well in which the system evolves. Hence for a "shell" produced by the spatial folding of a disk we would expect the luminosity to decline like $x^{-\frac{1}{2}}$ giving the shell a bright edge and rapid fall off towards the galaxy. If, however, the shell was produced by the phase folding of a spheroidal companion, then the $x^{-\frac{1}{2}}$ behaviour occurs when the complete phase space is projected on the spatial (radial) axis. When this distribution is projected on the sky, the resultant luminosity profile is nearly flat, giving the shell a plateau-like appearance. Both types of photometric behavior have been observed (Pence 1986, Fort et. al. 1985). We would expect the

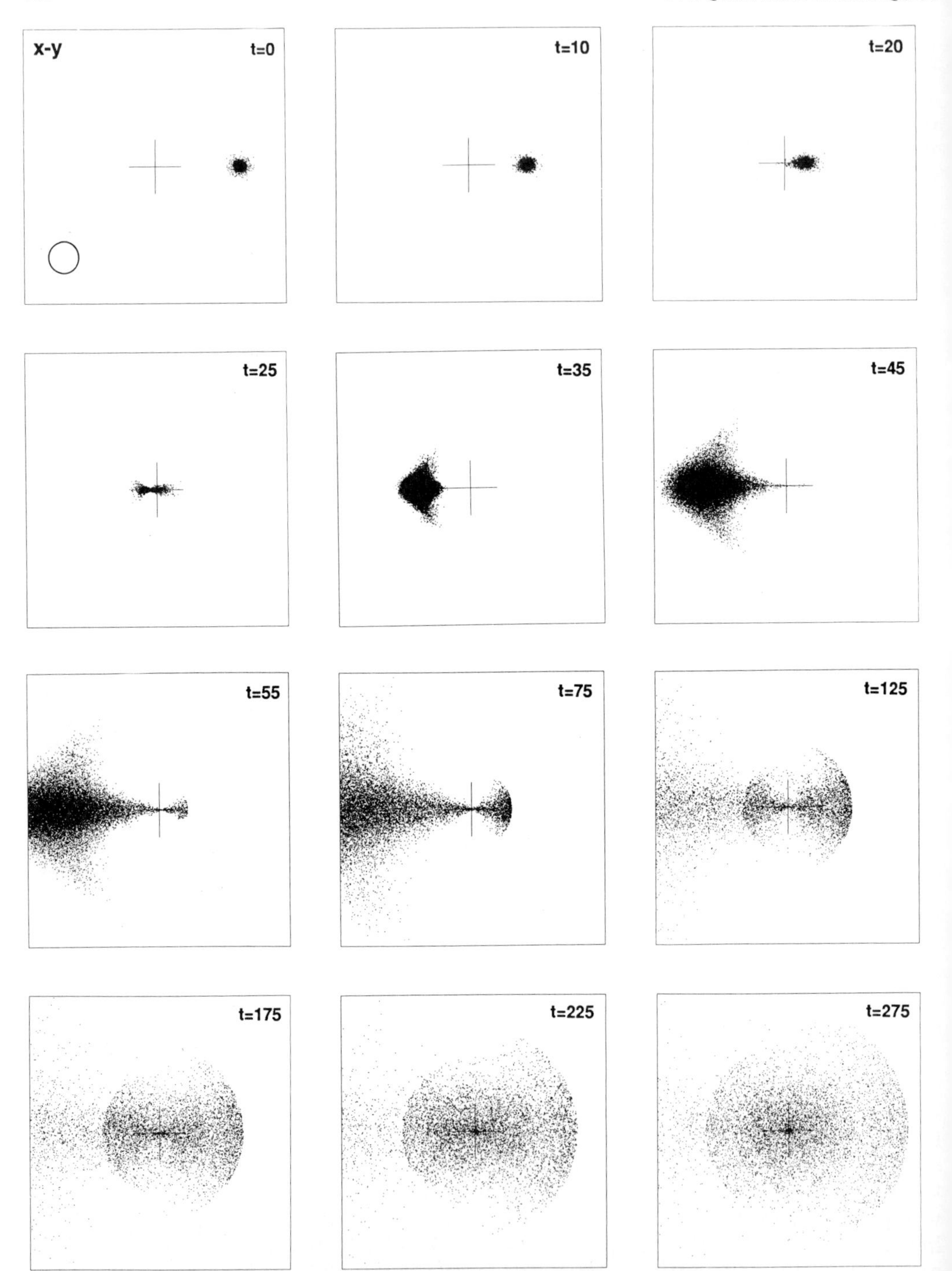

Figure 1. A radial encounter between a spherical Plummer primary and a spherical companion. The companion mass was 0.01 and its half-mass radius was 0.2 (both 1 for primary). The circle in the first frame indicates a spherical primary was used and the cross is at the center-of-mass.

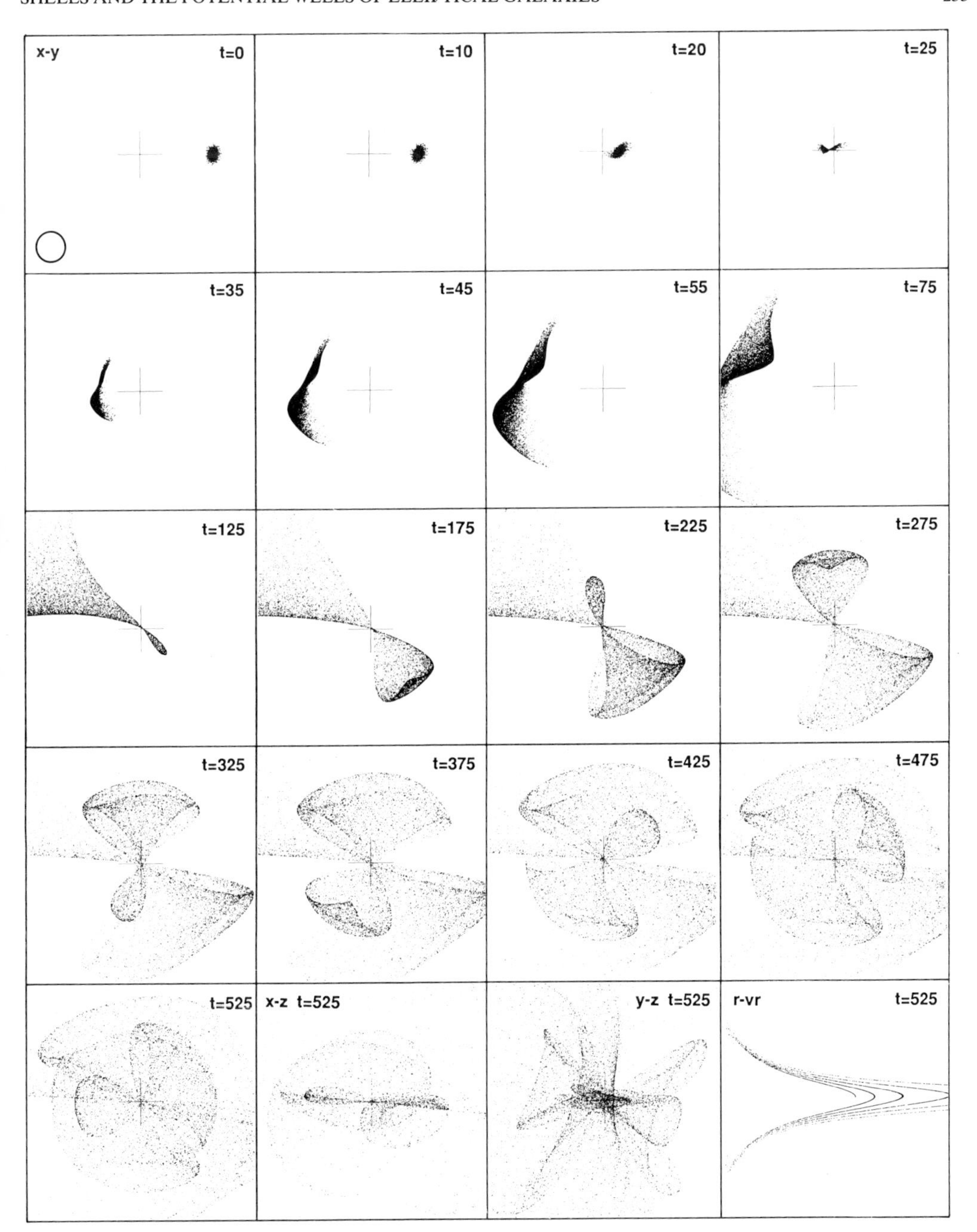

Figure 2. A radial encounter between a spherical Plummer primary and an exponential disk with an inclination of 45° to the x-y plane. The disk has a mass of 0.01. The final panel shows the phase space distribution of particles at $t = 525$. The box dimensions are 40×40 as in Figure 1. The time units are 4.3×10^6 years for a half-mass radius of 2 kpc. and a total mass of $10^{11} M_{\odot}$.

complex shell systems to show the strongly edge-brightened behaviour as it is most likely that these encounters involve disk-like companions.

Finally, if very little orbital energy is lost by the companion before it is destroyed (an accretion) then the spread in energies of the shell stars will reflect that initially present in the companion. If the orbital energies are close to zero then it is unlikely that shell stars would have turning-radii closer than a few half-light radii of the elliptical parent provided the companion is much less massive than the parent. Several shell galaxies have shells at radii as small as $0.5r_e$ which means that either the companion was quite massive (i.e. had a large spread in internal energies) or a significant fraction of the companions orbital energy was lost by dynamical friction on the parent before the companion was destroyed (a merger). The luminosity of the shell material is difficult to estimate but a reasonable guess seems to be less than 5% of the parent luminosity (Pence 1986). It would therefore seem likely that the companions involved were not very massive compared to the parent. This would then imply that dynamical friction had been at work in those systems with inner shells. Given a constraint on the luminous mass of the companion from the shell luminosity, the observed range in shell radii and some constraints on the orbit of the companion, limits can be placed on the work done by friction and hence the amount of dark material present. Clearly if one knew the number of shell systems with inner shells, the shell luminosities and morphologies, then useful constraints could be placed on dark matter in ellipticals. Most likely self-consistent simulations are called for to get the orbital evolution and disruption process right (James and Wilkinson, this volume).

4. NGC 3923 AS A SIMPLE SHELL GALAXY.

If the shells in NGC 3923 were produced by an encounter with a elliptical companion on a nearly radial orbit then the resultant shells are formed by the phase wrapping mechanism. In this process, the companion is drawn out and wrapped up in phase space by the difference in radial periods between the most bound and least bound stars. The number of wraps that occur by a time t after the companion is destroyed is given approximately by :

$$N \sim t(\frac{1}{P_{min}} - \frac{1}{P_{max}})$$

If we measure the time passed in units of the maximum period $\tau = \frac{t}{P_{max}}$ then

$$N \sim \tau(\frac{P_{max}}{P_{min}} - 1)$$

where P is the radial period. If we model NGC 3923 as a $r^{\frac{1}{4}}$ law mass profile then given an outer shell radius $d_{max} \sim 16r_e$ and an inner shell radius $d_{min} \sim 0.6r_e$, the above equation would predict $N_{r^{\frac{1}{4}}} \sim 106$ if τ has its minimum value. However the observed number of shells between these two radii is ~ 20. We could solve this discrepancy by either decreasing P_{max} or increasing P_{min}. Since NGC 3923 does follow an $r^{\frac{1}{4}}$ law luminosity profile we are confident about P_{min}. So we are forced to decrease P_{max} by a factor of order 6 or equivalently we must increase the mass at large radius by a factor $\sim 6^2$ since $P(r) \sim M(r)^{-\frac{1}{2}}$. Making the system older (τ larger) only increases the discrepancy. Hence the relatively small number of shells in the outer envelope of NGC 3923 given the spread in radius, implies a large amount of dark matter ($M_{dark} \sim 40 M_{lum.}$) out to a radius $\sim 100h^{-1}kpc$.

A more detailed analysis of the shell distribution has been conducted by Hernquist and Quinn(1986b). They conclude that if the halo has a density distribution like that of a non-singular, isothermal sphere then the halo mass out to $100h^{-1}kpc$. is $40M_{lum.}$ and the core radius $\gamma \sim 3r_e$. There are uncertainities and difficulties with the analysis of NGC 3923. Firstly to correctly establish the relative amount of dark matter present, the age of the shell system (τ) must be known. Allowing τ to be greater than its minimum value increases the amount of dark matter required. Fortunately τ for NGC 3923 is constrained from above by requiring the luminous material to be self-gravitating (see Hernquist and Quinn 1986 for details). Secondly, there are missing shells in NGC 3923. Gaps are expected since the shell brightness distribution depends on the spatial and velocity distributions in the companion and the shape of the parent potential. Irregularities in position or velocity in the companion or in the potential well of the parent will be reflected as missing or faint shells. Thirdly, what influence has dynamical friction had on the shell distribution? To answer this completely will require detailed, self-consistent simulations. It is hoped that by avoiding the very innermost shells, we are dealing with a set of shells that were all produced from the same disruption event rather than from multiple passages of a single companion as its orbit decays or multiple companions. Bearing these difficulties in mind, simple shell galaxies like NGC 3923 present us with a unique opportunity to probe the potentials of ellipticals.

5. THE HALOS OF ELLIPTICALS AND SPIRALS

Given that the dark halos of spirals and ellipticals behave like isothermal spheres at large radii, we can characterize the halos by either their asymptopic rotational velocity or the slope of their mass profile :

$$\frac{M}{r} = \frac{V_A}{G} = constant = \Xi$$

The constant Ξ is directly proportional to the depth of the halo potential and the mean halo density within some given radius. For NGC 3923 we know that approximately 40 times as much dark matter as bright is contained within the outer shell radius ($r \sim 100h^{-1}kpc.$) giving $\Xi \sim 6 \times 10^7 M_\odot pc^{-1}$. A similar number can be obtained from the X-ray sample of early type galaxies by Forman, Jones and Tucker (1985). Assuming that the maximum radius at which X-rays are detected is well outside the core radius of the dark matter then $\langle\Xi\rangle \sim 5 \times 10^7 M_\odot pc^{-1}$ in good agreement with the shell data. For spirals, we can estimate Ξ from the Bahcall and Casertano sample (Bahcall and Casertano 1985) or the Tully-Fisher relationship again assuming that the velocities truly represent the asymptopically flat values. We obtain : $\langle\Xi\rangle \sim 0.7 \times 10^7 M_\odot pc^{-1}$ (see Hernquist and Quinn 1986c, Figure 3, this volume). This would then imply that ellipticals may have typical halo densities which are perhaps an order of magnitude larger than that of spirals. Some caution is necessary however, since the overlap in luminosity between the X-ray sample and the spirals is very small. Hence it is difficult to compare spirals and ellipticals at the same luminous mass until X-ray data can be taken on less luminous ellipticals.

However, since the luminosity density of ellipticals is approximately ten times that of spirals, an indication of a similar ratio for their dark matter would favour theories in which the halos of galaxies strongly influence the type of galaxy that forms within them.

REFERENCES

Arnold, V.I., 1984, *Catastrophe Theory*, Springer Verlag.
Arp, H., 1966, *Atlas of Peculiar Galaxies*, *Astrophys. J. Suppl.*, **123**, 1.
Athanassoula, E., & Bosma, A., 1985, *Ann. Rev. Astr. Astrophys.*, **23**, 147.
Bahcall, J.N., & Casertano, S., 1985, *Astrophys. J. Letters*, **293**, L7.
Carter, D., Allen, D.A., & Malin, D.F., 1982, *Nature*, **295**, 126.
Dupraz, C., & Combes, F., 1986, *Astr. Astrophys.*, **166**, 53.
Fabian, A.C., Nulsen, P.J., & Stewart, G.C., 1980, *Nature*, **287**, 613.
Forman, W., Jones, C., & Tucker, W., 1985, *Astrophys. J.*, **293**, 102.
Fort, B.P., Prieur, J.L., Carter, D., Meatheringham, S.J., & Vigroux, L., 1986, *Astrophys. J.*, **306**, 110.
Hernquist, L., & Quinn, P.J., 1986a, *Astrophys. J.*, submitted.
Hernquist, L., & Quinn, P.J., 1986b, *Astrophys. J.*, in press.
Hernquist, L., & Quinn, P.J., 1986c, this volume, p. 467.
James, R.A., & Wilkinson, A., 1986, this volume, p. 471.
Malin, D.F., 1979, *Nature*, **277**, 279.
Malin, D.F., 1981, *J. Photo. Sci.*, **29**, 199.
Malin, D.F., 1982, *J. Photo. Sci.*, **30**, 87.
Malin, D.F., & Carter, D., 1980, *Nature*, **285**, 643.
Malin, D.F., & Carter, D., 1983, *Astrophys. J.*, **274**, 534.
Malin, D.F., Quinn, P.J., & Graham, J.A., 1983, *Astrophys. J. Letters*, **272**, L5.
Pence, W., 1986, STScI preprint.
Piran, T., & Villumsen, J.V., 1986, this volume, p. 473.
Quinn, P.J., 1982, *Ph.D. Thesis*, Australian National University.
Quinn, P.J., 1984, *Astrophys. J.*, **279**, 596.
Schweizer, F., 1980, *Astrophys. J.*, **252**, 303.
Schweizer, F., & Ford, W.K.Jr., 1985, in *New Aspects of Galaxy Photometry*, Lecture Notes in Physics No. 232, ed. J.-L. Nieto (Springer Verlag), p. 145.
Toomre, A., 1977, In *The Evolution of Galaxies and Stellar Populations*, eds B. Tinsley & R. Larson, (Yale Univ. Obs.), p. 401.
Toomre, A., 1983, in IAU Symposium No. 100, *The Internal Kinematics and Dynamics of Galaxies*, ed. E.O. Athanassoula, (Reidel, Dordrecht), p. 319.
Wilkinson, A., Sparks, W.B., Malin, D.F., & Carter, D., 1986, this volume, p. 465.
Williams, R.E., & Christiansen, W.A., 1985, *Astrophys. J.*, **291**, 80.

DISCUSSION

Ostriker: i) Do I recall correctly that at last year's dark matter symposium you said that in order to fit the observations you required that there be *no* massive halo, that the mass given by X–ray observations be wrong, too high? Now I understand you to say that the X–ray observations confirm the model. ii) If the shells add up to 10% of the light, then from the Faber–Jackson relation, a velocity dispersion of the captured object would be 50% of the cannibal. Is this cold enough to give you the observed sharp features?

Quinn: As regards the paper I gave with Wojcek Zurek and John Salman at IAU 117 on Dark Matter, we concluded that it was *not* possible to fit the shell distribution in NGC 3923 without dark matter. What was in doubt at that time was the distribution of the dark matter at large radii. Lars Hernquist and I have now clarified this in a recent preprint. Whether or not a given companion can produce a given number of shells before smearing sets in, depends not only on the velocity dispersion in the companion relative to the primary, but also on its physical size. In other words, the phase–space volume of the companion has to be small compared to the primary. Hence a small hot companion can produce shells. One should also note that shells sharpen as time goes by due to the system being stretched in phase. I believe that a 50% velocity dispersion ratio is not a problem, provided the sizes have a ratio of greater than about ten to one, which does not seem to be unreasonable. It is my impression from the work of Bill Pence on NGC 3923, that the shells represent more like 2% of the total luminosity of this galaxy.

Fabian: I think that perhaps Alar Toomre and you misunderstand my comment yesterday on producing shells in cooling flows. The stars produced there are very cold ($v \sim 10$ km/s) and much colder than any galaxy.

Quinn: Yes, such cold stars could produce shells, but if they formed with the body of the main galaxy they could not produce shells at large radii.

Binney: Did I understand that in your shell–counting argument you assumed that all shells started at the same instant? I should have thought that the outer shells got a head start by one free–fall time, since they would form as the victim started to fall in, while the inner shells would form only after the victim had been dragged, kicking and screaming, to the center.

Quinn: The shell counting argument is not very sensitive to the distribution of initial orbital phases. If the companion disrupts on the first pass then differences in phase between lightly and loosely bound stars are quickly forgotten as the most bound stars complete a large number of periods in one crossing time of the least bound stars. Even if I forbid the most tightly bound stars from making shells at all for a time comparable to the period of the least bound stars, the corrections to the shell numbers are $\sim 2/3$ which still leaves us with a factor of 5 discrepancy between the observed shell distribution and that predicted by the Kepler formula.

Bland: Would you not expect the shells to be interleaved in radius with respect to the deep potential and is this observed for the most part in the Malin & Carter catalogue?

Quinn: NGC 3923 does show interleaving as do a few others I know of (NGC 3051, Arp 230 etc.). As far as the whole catalogue goes we will have to wait for Althea Wilkinson to reduce her beautiful new CCD data.

Nieto: Some of your simulations show very complex inner structures that are very reminiscent of the structure of dusty or box-shaped ellipticals, once the low luminosity gradient has been taken off. Do you think that this could mean a relationship between dust and ripples? In other words, do you think that the dust and/or the box-shapes in ellipticals may be a remnant of a merger implying a spiral galaxy?

Quinn: Certainly I would expect complex (crossing, nonconcentric) shells to be associated with the distruction of a disk-like companion. Hence in such cases dust and gas should be present. In the case where the primary is axisymmetric, the orbital plane of the companion precesses about the symmetry axis and can produce X-like structures as have been seen in spiral bulges. "Boxy-ness" in ellipticals may be due to contributions from debris in an X-shaped configuration.

Freeman: You showed a run with a triaxial potential that gave nice well aligned shells like those of NGC 3923. If there were a high incidence of triaxial potentials, might one expect more shell systems like NGC 3923's?

Quinn: The problem is that even when the parent potential is strongly triaxial, the form of the shells depends also on the type of the companion and its orbital parameters. For example, an inclined disk-like companion or even a radial orbit does not produce a clean set of aligned shells when thrown into a triaxial parent. Similarly a spheroidal companion falling down the short axis produces shells which are completely misaligned. It is therefore dangerous to draw conclusions about parent shapes from the shell alignment alone.

Whitmore: If your need for a heavy halo in NGC 3923 is solely reliant on being able to identify all $\approx$ 120 shells you predict in your model with no halo, the observational difficulty of actually identifying that many very faint features will probably be the biggest problem. However, if the qualitative look of the shells and their distribution in radius is also better fit by the heavy halo model, this would be more convincing.

Quinn: If there were even twice as many shells present as observed, I'm sure they could be easily seen on the plates. I agree that the shell distribution is the telling factor for the dark matter.

Bertin: Is the self-gravity of the infalling disks included in your simulations?

Quinn: Not in the simulations with Hernquist. In my Ph.D. thesis I did some self-gravitating simulations and showed that it was not important for shell making. Shell making (phase wrapping) is a test particle process, self-gravity just regulates the rate of release of free particles into the primary potential.

Combes: Simulations show that around a prolate galaxy, shells are usually seen aligned along the apparent major axis. The prototype case is NGC 3923. But when a strong figure rotation is introduced (tumbling period short in front of companion stars oscillation periods), the alignment disappears. The prolate potential is time–averaged in an oblate potential for the shells and one can see randomly spread shells in the plane perpendicular to the rotation axis— this means that NGC 3923 is tumbling very slowly, if at all.

Quinn: I agree. The period of stars in the outermost shells is very long ($\geq 5 \times 10^8$ yrs) for NGC 3923. Hence even a very slow tumble rate could be ruled out ($t_{tumble} \geq 5 \times 10^9$ yrs). Observations along the major axis of NGC 3923 I made in my thesis showed no apparent rotation at the ± 30 km/s level.

Peter Quinn.

Aguilar, with Meylan on his right, and on his left Schechter and Toomre.

Lauer, with next to him Vandervoort.

SPHERICAL GALAXIES: METHODS AND MODELS

Douglas O. Richstone
Department of Astronomy
University of Michigan

ABSTRACT. Over the last 5 years, considerable progress has been made in our ability to construct self-gravitating stellar equilibria. One of these new methods is essentially a variant of Eddington's (1916) method. Two other key approaches are logical extensions of Schwarzschild's Linear Programming method, and can be applied to nonspherical models as well. These methods are reviewed below.

The application of these methods to galaxies has yielded a few very interesting results within the last year or two. The methods described below unambiguously establish M/L's for M87 and M32 within about 30 arc seconds. They strongly support Tonry's contention that the nucleus of M32 contains a large invisible mass, possibly a $10^7 M_\odot$ black hole. They also suggest that observational recovery of the projected velociy distribution might permit the observer to distinguish between a massive halo and an increasingly tangential velocity distribution function.

1. METHODS

We begin with Jeans's theorem, which states that the phase space distribution function $f(\mathbf{r}, \mathbf{v})$ is a function only of the isolating integrals. For truly spherical symmetry these are the energy (E) and the square angular momentum (J^2). Any dependence on the individual components of **J** other than through J^2 would create a prefered dynamical direction. In a sense, Jeans's theorem solves the problem of constructing spherical systems: we simply choose a form of the phase space distributioin function $f(E, J)$, integrate over all velocities to find the density ρ, and then solve the Poisson Equation for the potential Φ, which is contained implicitly in f and ρ.

This approach in fact describes the first class of methods — which we term King type methods (see Michie and Bodenheimer 1963 and King 1965). It has been recently applied to the dynamics of clusters of galaxies by Kent and Gunn (1982). The method requires a reasonably simple choice of the functional form of f, so that the integral for the density can be performed in terms of the potential. Its major limitation is that there is no guarantee that the assumed distribution function approximates the system under study.

Moreover, under normal circumstances we have a density profile $\rho(r)$ and wish to recover the distribution function which gives rise to that spatial density. This problem is underdetermined, since we are deriving a function of two variables (E and J) from a function of one variable. Eddington (1916) provided one possible solution

T. de Zeeuw (ed.), Structure and Dynamics of Elliptical Galaxies, 261–269.

by showing that if $f = f(E)$ it can be recovered from $\rho(r)$. A distribution function which is only a function of energy is isotropic and contains no free parameters. Merritt (1985) has shown that the distribution function can also be recovered from the density if it is a function of $Q_{\pm} = E \pm J^2/r_a^2$. This choice of functions has two advantages over Eddington's: it permits the modelling of velocity anisotropy and it contains an adjustable parameter. For many choices of $\rho(r)$, it will yield physically reasonable (nonnegative) distribution functions. Even so, its virtue is also its fault: the explicit restriction to a particular combination of the two integrals sharply restricts the solutions which can be found. It also will happily produce distribution functions which are negative in some regions of phase space, although it is immediately obvious when this occurs.

A third useful approach to this problem was invented by Binney and Mamon (1982, see also Tonry 1983), in their study of M87. Binney and Mamon used the density profile, observed velocity dispersion profile, and the equation of stellar hydrodynamics to recover the internal radial and tangential velocity dispersions. Unfortunately, this method does not work well when the observed dispersion is available over a small range in radius. It also has the important limitation that the dispersions constructed by this method need not correspond to a nonnegative distribution function, and it may in fact not be easy to discover that. For example, it is possible to use this technique to construct a de Vaucouleurs law model with purely radial orbits by setting the tangential dispersions everywhere equal to zero; such a system does not have a nonnegative distribution function everywhere (see Richstone and Tremaine 1984).

A fourth kind of approach to this problem was pioneered by Schwarzschild (1979). He noted that each time averaged orbit in any specific potential is a solution of the collisionless Boltzmann equation, and that these orbits can be summed to find a solution of the Poisson equation for that potential. He chose to use linear programming to perform the sums, since it guarantees a set of orbit occupation numbers with nonnegative weights. The formal validity of this approach has been demonstrated by Vandervoort (1984). In the spherical case, this approach can be described as writing the distribution function as a sum of products of delta functions of the isolating integrals: $f(E, J^2) = \sum_i [\delta(E - E_i)\delta(J^2 - J_i^2)]$. It is immediately clear that one can in fact express f as any sum of functions of isolating integrals. In this regard Merritt's functions may be particularly useful, since any weighted sum of his functions (say, with different r_a's) will remain a valid solution for that particular $\rho(r)$.

Although linear programming has been used extensively for modelling spherical galaxies (Richstone and Tremaine 1984, 1985), there are at three other methods which can be used to combine orbits (or families of phase space distribution functions) to produce models. Each of these methods must solve (or approximate) a set of equations of the form

$$M_i = \sum_i m_{ij} w_j, \tag{1}$$

$$w_j \geq 0,$$

which states that the sum of the products of the mass distributions of the orbits (m_{ij}) with their occupation numbers (w_j) is equal to the mass distribution (M_i) of the galaxy under consideration. One such method is advocated by Pfenniger (1984) and described in detail in Lawson and Hanson (1974). It uses a pivoting method (like linear programming) to minimize the sum of squared residuals from eqn (1) above.

A second method is the application of Lucy's (1974) method due to Newton and Binney (1984). Lucy's method solves equation 1 by starting with some guess for w_j with no zero or negative components, and iterating as follows:

$$Q_{ij} = \frac{m_{ij} w_j^{(g)}}{\sum_i m_{ij} w_j^{(g)}}, \tag{2}$$

$$w_j^{(g+1)} = \sum_i M_i Q_{ij}. \tag{3}$$

Lucy's method is motivated by Bayes's theorem on conditional probabilities. This method turns out to work extremely well for good first guesses. Its great virtue relative to linear programming is that it tends to produce smooth distribution functions. It is also fairly easy to code.

Another method which has been used to construct galaxy models is a maximum entropy method using Lagrange multipliers to stay on the constraints (Richstone and Tremaine 1986). Any entropy of the form $\sum_j S_j(w_j)$ can be maximized subject to those constraints by solving

$$S_j' - \sum_i \lambda_i m_{ij} = 0. \tag{4}$$

subject to equation 1, where $S_j' = \partial S_j / \partial w_j$. This can be accomplished by guessing w_j , and expanding eq (4) as a Taylor series to get

$$\Delta w_j = \frac{\sum_i \lambda_i m_{ij} - S_j'}{S_j''}, \tag{5}$$

for Δw_j and multiplying by m_{kj} and summing to get

$$\sum_i \lambda_i \Big(\sum_j \frac{m_{ij} m_{kj}}{S_j''} \Big) = \sum_j \frac{S_j'}{S_j''} m_{kj} + \Delta M_k, \tag{6}$$

for λ_i. Eqn 6 is solved first for the λ_i and eqn 5 is then solved for the Δw_j . This method converges very rapidly, but requires the solution of a set of linear equations (for the λ_j) at each step.

Statler (unpublished) has shown that the Binney-Newton-Lucy method leads to a variety of different solutions in the solution space. We have compared the operation of this method and the maximum entropy method for a toy problem and found that the maximum entropy method usually finds a solution in about 1/5 the number of iterations of the BNL method, but then spends more time finding the particular solution with maximum entropy. The virtue of the maximum entropy method is that it produces a solution to the problem which is well defined in terms of some principle.

It does not seem appropriate to regard the classical entropy $f \ln(f)$ as having any particular significance for galaxies. They are, after all, only violently relaxed (see Tremaine, Henon, and Lynden-Bell 1986). Their further diffusion in phase space occurs on a timescale much longer than the age of the Universe. In this view, the entropy serves as a device to avoid negative occupation numbers and produce a smooth distribution function. For this purpose, other choices of an 'entropy' function would suffice, and we have explored other choices. All models displayed below were constructed using the classical definition.

2. MODELS

Various authors have employed these methods to construct spherical galaxy models during the last few years. Here I first want to display two maximum entropy models (using the classical entropy) with mass distributions of a Plummer Model and of an $r^{1/4}$ law. Both models were rescaled in mass to have the same observed velocity dispersion. Note that in each case the dispersion profile slightly favors tangential anisotropy at large radii. In these cases, the maximum entropy method described above converges to a solution in less than 4 iterations, and finds a maximum to reasonable precision in 10 iterations.

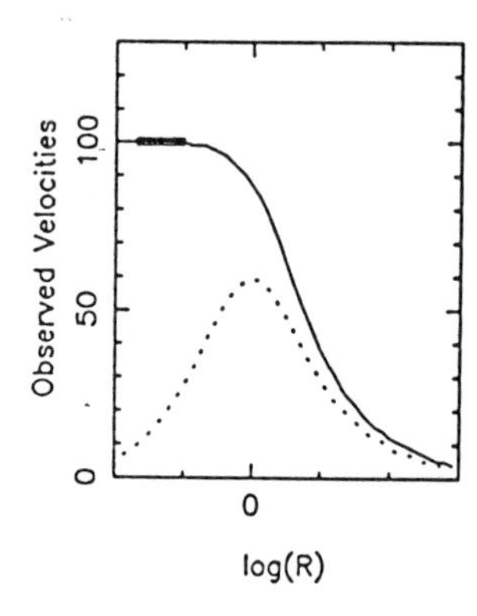

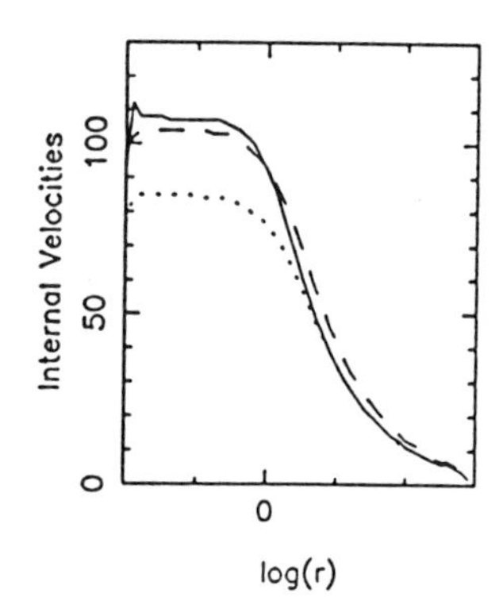

Figure 1. – Dynamical properties of a maximum entropy Plummer model. Left box shows the observed dispersion for a nonrotating model (solid line), and the maximum rotation rate as described in the text (dotted line) as a function of projected radius. The right box show the σ_r, σ_t for a nonrotating model and the maximum rotation rate as a function of radius.

One important use of these methods is to improve our understanding of galactic nuclei. Since the most popular theory of quasars uses supermassive black holes for energy production, the demonstration that they are present in galactic nuclei would provide strong support for that view. Starting with Young et. al. (1978) and Sargent et. al. (1978) various investigators have addressed these issues. Tremaine and I (1985) have recently reviewed the history of work on M87. For that particular galaxy, only one additional footnote seems appropriate here. Not only (as Duncan and Wheeler(1980) and Binney and Mamon pointed out) is it possible to construct constant M/L models for the galaxy. It is even (especially if Dressler's(1980) results for the dispersion near the center are correct) easy to make a variety of constant M/L models consistent with the observations, including models with only mildly anisotropic velocity dispersions. Such a model is displayed below.

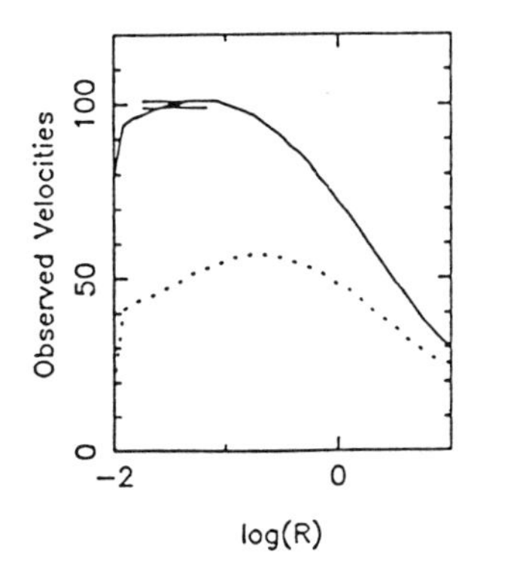

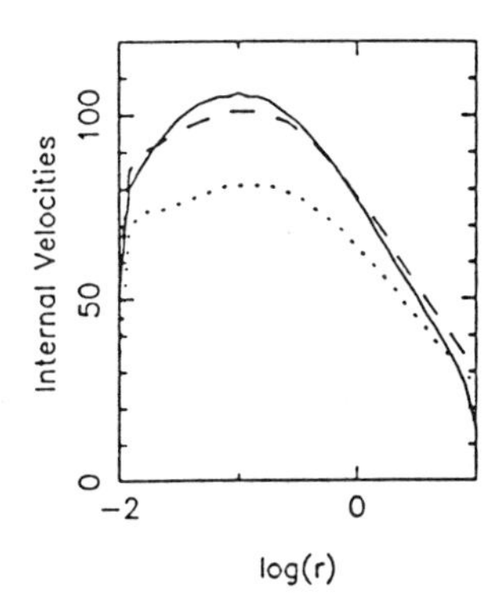

Figure 2. – Same as Figure 1 for a deVaucouleurs law mass distribution.

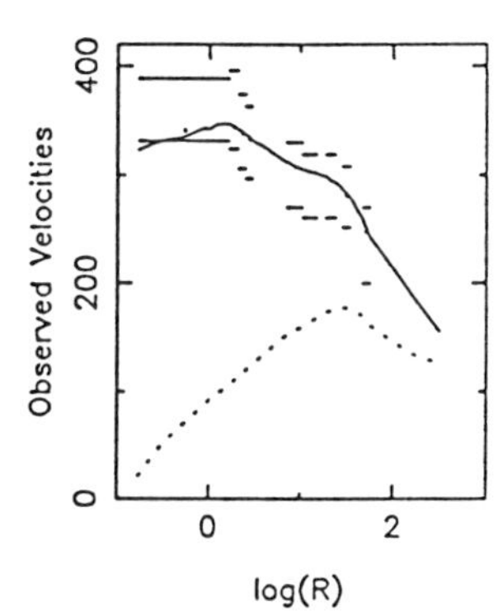

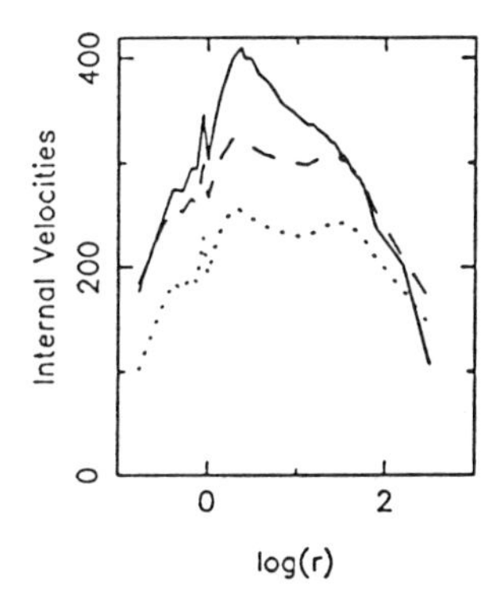

Figure 3. – A dynamical model for M87. The double horizontal lines in the left box are the 1σ upper and lower error bars from observations. Otherwise the quantities plotted are the same as in Figures 1 and 2. The units are km/sec and arcseconds.

The situation is much more interesting in the case of M32, which has been studied by Tonry (1984) and Dressler (1984) and now by Kormendy (this meeting). M32 displays a sharp jump in velocity dispersion, which may be unresolved rapid rotation. Tonry remarked that the rotation curve of M32, which reaches about 40 km/sec at about 3 arcseconds (12 pc) from the center, must be lowered by seeing. Seeing has a particularly significant effect on rotation curves, since it carries light from the wrong side of the minor axis into the slit. Tonry modeled the rotation curve with various profiles, concluding that the true rotation at about 3 arcseconds was about 70 km/sec and that it must rise still further at smaller radii. He stated that this established a higher M/L inside 2 arcseconds than outside 2 arcseconds.

I have attempted to verify this statement by using the maximum entropy method to construct M32 models consistent with the observed velocity dispersions with the maximum amount of rotation at the center consistent with a spherical model. The maximum rotation rate for a given orbit at a given point can be shown to be $2/\pi(J_i^2/r^2)^{1/2}$. The M32 model with maximum rotation near 1 arcsecond is shown below. Note that the projected rotation velocity near 1-3 arcseconds is only 60 km/sec . I felt that this discrepancy of 60 vs 70 km/sec was too small to justify Tonry's claims for this object, but now that Kormendy has demonstrated that the rotation velocity of the galaxy must continue to rise inside 2 arcseconds it is clear that the M/L must rise above the 2.4 ± 0.1 (red) that characterizes larger radii in this galaxy. This makes M32 a prime candidate for a black hole in the $10^7 M_\odot$ range.

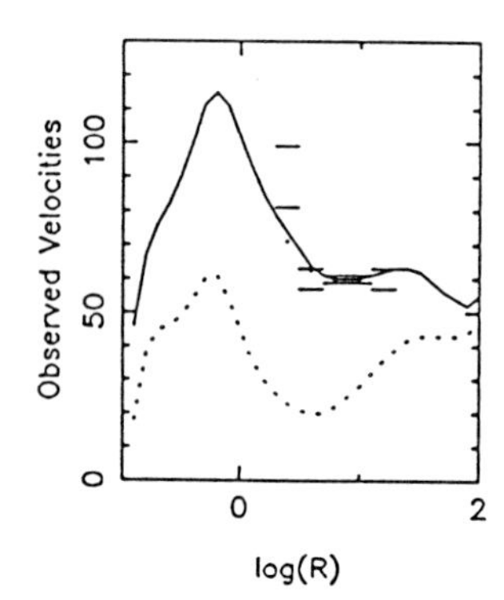

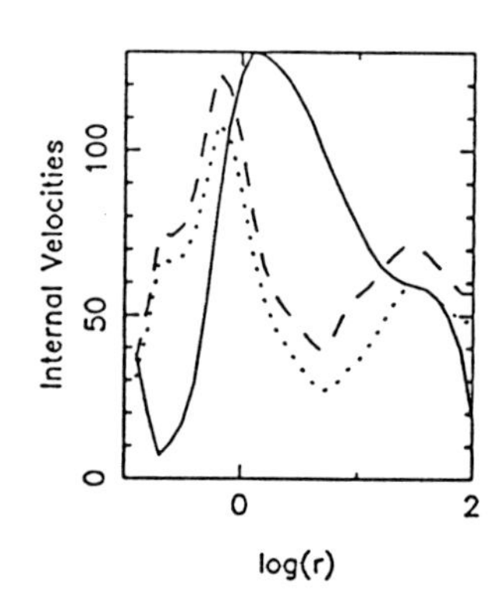

Figure 4. – Same as Figure 3 for M32.

In retrospect, M31 and M32 should have been studied all along, since the maximum radius at which such an object can make its influence felt is defined by

$$r = \alpha G M_h / v^2, \tag{7}$$

where M_h is the mass of the black hole, v characterizes the velocities of the stars near the center of the galaxy, and α is of order unity. So, in order to 'see' the black hole in a galaxy, it must have a minimum mass of order

$$M_h = \frac{\sigma^2 D\theta}{\alpha G}, \tag{8}$$

where D is the distance to the galaxy, θ is the observational spatial resolution and σ is the system's velocity dispersion. M32, with its dispersion of 60 km/sec and its distance of 700 kpc, presents a minimum detectable black hole mass near $10^6 M_\odot$, while M87, at 15Mpc and $\sigma = 300$ km/sec has a minimum mass near $1.5 \times 10^9 M_\odot$, (for $\theta = 1$ arcsec) so a $10^8 M_\odot$ object in M87 will never be dynamically detectable from the ground.

An interesting sidelight of the careful work on M87 and M32 is that the systemic M/L ratios for these systems are now quite well constrained, from about 2 arcseconds to about 30 arcseconds, assuming that the mass follows the light over that range in radius. In that case, M/L for M87 is 10 ± 2 (in B), while for M32 it is 2.4 ± 0.1 in Gunn red. This real difference in M/L is not an artifact of using different bandpasses. It cannot be wriggled out of via anisotropy. It must reflect a real difference in the stellar populations of these galaxies or of the ratio of stellar to dark matter in them. It seems to me to support Kormendy's (this meeting) statement about a trend in M/L with luminosity, although it is clearly desirable to carry out a rather more detailed analysis of more than two systems.

One other interesting recent development in spherical galaxies has been DeJonghe's calculation of the distribution of projected velocities in a Plummer model. DeJonghe showed that the profile is bimodal if the dispersion tensor is very tangentially elongated. Below we illustrate this effect for power law light distributions in a logarithmic potential with the circular velocity everywhere unity. An isotropic distribution function would have a Gaussian $f(v_p)$ at all radii, with dispersion given by $\sigma^2 = v_c^2/K$, where $K = -d\ln\epsilon/d\ln r$. The observed profiles are decidedly not Gaussian. Further investigation is required to see if this effect can be observed after convolution with a stellar template. If so, it offers a possible approach to breaking the degeneracy between velocity dispersion and anisotropy in these models.

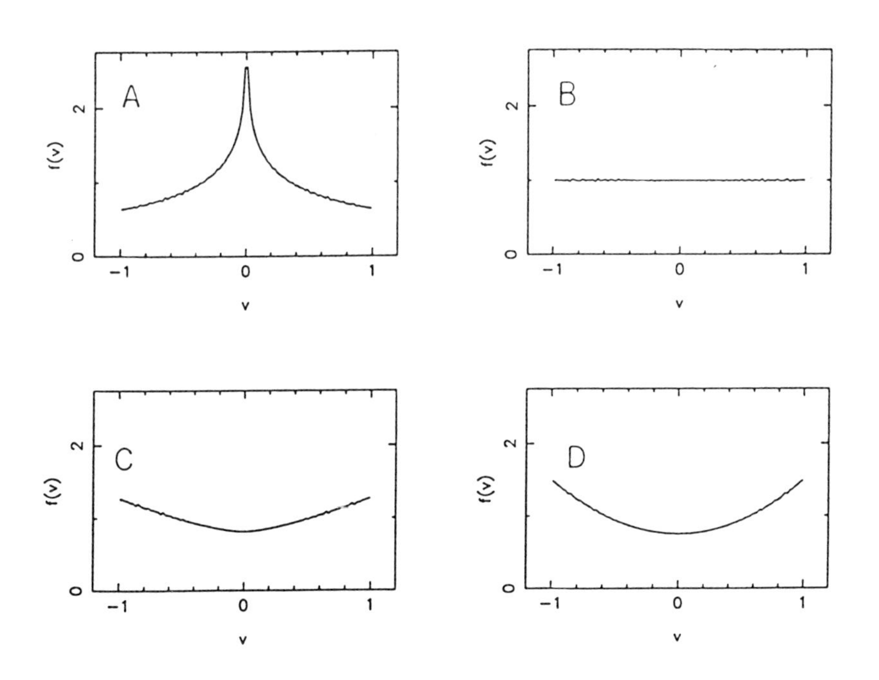

Figure 5. – Observed velocity distributions for power law models described in text. All models have $v_c = 1$. Values of K were as follows: A: 2, B: 3, C: 4, D: 5.

REFERENCES

Binney, J. and Mamon, G. A. 1982, *Mon. Not. Roy. Astr. Soc.* **200**, 361.
DeJonghe, H. 1987, *Mon. Not. Roy. Astr. Soc.* **224**, 13.
Dressler, A. 1980, *Astroph. J. Letters*, **240**, 111.
Dressler, A. 1984, *Astroph. J.*, **286**, 97.
Duncan, M. J. and Wheeler, J. C. 1980, *Astroph. J.* **237**, L27.
Eddington, A. 1916, *Mon. Not. Roy. Astr. Soc.* **76**, 572.
Kent, S. M. and Gunn, J. E. 1982, *Astron. J.* **87**, 945.
Lawson, C.L. and Hanson, R.J. 1974, *Solving Least Squares Problems*, (Prentice Hall, Englewood Cliffs, New Jersey).
Merritt, D. 1985, *Astron. J.* **90**, 1027.
Michie, R. W. and Bodenheimer, P. H. 1963, *Mon. Not. Roy. Astr. Soc.* **126**, 269.
Newton, A. J. and Binney, J. 1984, *Mon. Not. Roy. Astr. Soc.* **210**, 711.
Pfenniger, D. 1984, *Astron. Astroph.* **141**, 171.
Richstone, D. O. and Tremaine, S. 1984, *Astroph. J.* **286**, 27.
Richstone, D. O. and Tremaine, S. 1985, *Astroph. J.* **296**, 370.
Richstone, D. O. and Tremaine, S. 1986, in preparation.
Sargent, W. L. W., Young, P. J., Boksenberg, A., Shortridge, K., Lynds, C. R., Hartwick, F. D. A., 1978, *Astroph. J.* **221**, 731.
Schwarzschild, M. 1979 *Astroph. J.* **232**, 236.
Tonry, J. 1983, *Astroph. J.* **266**, 58.
Tonry, J. 1984, *Astroph. J. Letters* **283**, L27.
Tremaine, S., Henon, M., and Lynden-Bell, D. 1986, *Mon. Not. Roy. Astr. Soc.* **219**, 285.
Vandervoort, P. O. 1984, *Astrophys. J.* **287**, 475.
Young, P. J., Sargent, W. L. W., Boksenberg, A., Lynds, C. R. and Hartwick, F. D. A. 1978, *Astroph. J.* **222**, 450.

DISCUSSION

Jaffe: Is looking for non–gaussian velocity profiles a useful way to resolve the ambiguities in solving for the distribution function?

Richstone: Dressler and I tried to do that for the center of M32 with a very extreme distribution function. After convolution with a standard star the observed line profile is nearly indistinguishable from a Gaussian convolved with a star. We were discouraged and did not proceed further. Dejonghe's work *suggests* that it is worthwhile to pursue this question at large radii.

Lupton: In view of the short relaxation time in the core of M32 ($\sim \frac{1}{40} H_0^{-1}$), does a one–component, constant M/L model make any sense?

Richstone: At $\sim 1''$ the relaxation time is about $\frac{1}{2} H_0^{-1}$. The traditional logic in searching for black holes or dark matter is to see first whether they are mandated by a failure to produce a constant M/L model. In this case, it seems to be impossible to fit the observed velocities with a constant M/L model.

Kormendy: The core radius of M32 is $<< 1''$. At $r_c = 1.3$ pc, the best limit I can get from my observations, the relaxation time is $\sim \frac{1}{30}$ Hubble time. The relaxation time becomes equal to a Hubble time at about 2–3″ radius.

Burstein: Over what range of radii do your M/L–estimates for M32 and M87 apply?

Richstone: About 3″ to about 20″ in M32. 0″ to about 60″ in M87. In both cases M/L was assumed to be constant over those ranges.

King: The M/L that is derived is not global. It will apply to the region where the velocities are observed.

Binney: A short sermon on entropy. E. T. Jaynes (*e.g. Papers on probability, statistics and statistical physics*, ed. R. D. Rosenkrantz, Reidel 1983; also Dejonghe, H., 1987, *Astroph. J.*, in press) argues eloquently that the Gibbs–Boltzmann entropy $S = -\sum_i p_i \ln p_i$ enjoys a special place amongst the many convex functionals that can be used for rating probability distributions p_i. Jaynes contends that S is a purely subjective quantity that enables us to decide which of two probability distributions, that are both compatible with the available data, is more plausible. The structure of S is enforced by the laws of probability theory, and *has nothing whatever to do with physics.* In particular, the additivity of information on which Shannon's uniqueness proof rests, is not connected with the kind of physical additivity that arises in extensive thermodynamic systems. Shannon's additivity is to do with the variation with n of amount of information communicated by telling someone how many stars are in each of a series of phase-space cells of size $n\tau$. We obtain the most plausible probability distribution $p_i \equiv f(w_i)d^6w$ (and the only consistent interpretation of the distribution function is as a *probability* density—attempts to interpret f as some kind of stellar *density* generate only confusion) by maximizing S subject to *all* available constraints. As Jaynes emphasizes, attacks on the maximum-entropy procedure generally consist in obtaining manifestly absurd results by failing to include an important piece of prior information in the constraints, and we should beware of falling into this trap.

We still do not know what are the essential constraints for stellar systems. Usually people impose values of E and M, and sometimes a constraint derived from a theory of galaxy formation, such as the maximum phase-space density. One can think of many other constraints, e.g., the value of a tidal radius or some kinematic data (see also Dejonghe, *ibid.*). The consistent application of the maximum entropy procedure would involve repeatedly maximizing S subject to an ever-lengthening list of constraints, until further constraints do not significantly improve the agreement with observation.

Finally, let me remark that the following mathematical fact is an important source of confusion. If f_n maximizes S subject to constraints $C_1, \ldots, C_n$, then the distribution f_{n+1} that maximizes S subject to $C_1, \ldots, C_{n+1}$ is the same as the distribution f that maximizes $S' \equiv -\int d^6w f(w) \ln[f(w)/f_n(w)]$ subject to the single constraint C_{n+1}. In other words, one can consolidate a long list of constraints into a "prior" f_n. Later, we tend to forget about the constraints $C_1, \ldots, C_n$ and imagine that we are simply maximizing a new entropy S' subject to a single constraint. There is no objection to this way of thinking when S' *has* been properly founded at an earlier stage, but we should not listen to people who pull "entropies", or H–functions (Tremaine *et al. M.N.R.A.S.* **219**, 285), out of thin air, and claim that they enjoy the same status as S.

Part of the Dutch connection: Reynier Peletier (with his back to the camera), Andrew Pickles, Peter Teuben, Marijn Franx & Myriam Hunink.

DYNAMICAL MODELS FOR AXISYMMETRIC AND TRIAXIAL GALAXIES

Tim de Zeeuw
The Institute for Advanced Study
Princeton, NJ 08540, USA

ABSTRACT. Non-spherical dynamical models for galaxies, and the methods for their construction, are reviewed. The theory for two-integral axisymmetric models is reasonably well developed. Stäckel models give considerable insight in the structure of both three-integral axisymmetric models and non-rotating triaxial systems. Triaxial galaxies with appreciable figure rotation require much further study. Applications to elliptical galaxies and the bulges of disk galaxies are discussed.

1. THE FUNDAMENTAL PROBLEM OF STELLAR DYNAMICS

The structure and dynamics of a collisionless stellar system are determined completely by specification of the distribution function $f(\mathbf{r}, \mathbf{v}, t)$, which gives the distribution of the stars in the system over position $\mathbf{r}$ and velocity $\mathbf{v}$ as function of the time t. The distribution function satisfies the collisionless Boltzmann equation

$$\frac{\partial f}{\partial t} + \mathbf{v} \cdot \nabla f - \nabla V \cdot \frac{\partial f}{\partial \mathbf{v}} = 0, \tag{1}$$

where V is the potential in which the stars move, and we have used Newton's equations of motion $\dot{\mathbf{v}} = -\nabla V$. If the stellar system is in a steady state then $\partial f / \partial t = 0$. The density $\rho(\mathbf{r})$ of the system is the integral of f over the velocities

$$\rho(\mathbf{r}) = \int \int \int f(\mathbf{r}, \mathbf{v}) \, d^3\mathbf{v}. \tag{2}$$

In a self-gravitating system, V is the gravitational potential of the density itself, and is connected to ρ via Poisson's equation

$$\nabla^2 V = 4\pi G \rho. \tag{3}$$

In order to obtain a *dynamical model* for a self-gravitating stellar system in equilibrium, equations (1), (2) and (3) have to be solved simultaneously. A solution corresponds to a dynamical model only if $f \geq 0$. The problem of finding f for a stellar system in equilibrium is the fundamental problem of stellar dynamics (Chandrasekhar 1942). It is often referred to as the *self-consistent* problem.

T. de Zeeuw (ed.), Structure and Dynamics of Elliptical Galaxies, 271–290.

In many cases, Jeans' theorem can be used in order to reduce the number of variables in the problem. It states that f depends on the phase–space coordinates $(\mathbf{r}, \mathbf{v})$ only through the isolating integrals of motion admitted by the potential (Jeans 1915; Lynden–Bell 1962*b*). Any such function automatically satisfies equation (1). Hence one is left with equations (2) and (3).

Two main approaches towards a solution of the self–consistent problem may be identified. In the first method one specifies f as a function of the integrals of motion. Then equations (2) and (3) combined form an integro–differential equation for V. The alternative is to specify the density and/or the potential, and then to solve the integral equation (2) for f. Non–consistent models are obtained in a similar way. Since in this case Poisson's equation does not have to be satisfied, one usually chooses a potential and then either gives f and calculates ρ from equation (2), or one specifies ρ and solves equation (2) for f.

As an example, consider spherical galaxies. A spherical potential admits four isolating integrals. In addition to the energy E, the three components L_x, L_y and L_z of the angular momentum vector $\mathbf{L}$ are conserved as well. As a result, the general distribution function of a spherical galaxy must be of the form $f(E, \mathbf{L})$. If f is spherical in all its properties, it can depend only on the magnitude of $\mathbf{L}$, but not on its direction, so that $f = f(E, L^2)$. Such models have anisotropic velocity distributions. For $f = f(E)$ the velocity distribution is isotropic. Eddington showed in 1915 that, for a given $\rho(r)$, it is always possible to invert equation (2) explicitly in order to obtain a unique $f(E)$. If $\rho(r)$ falls off with radius sufficiently rapidly (cf. Hunter 1974) this $f(E)$ is nowhere negative, and represents the unique isotropic solution. Many anisotropic solutions $f(E, L^2)$ exist. They can sometimes be found by analytic inversion techniques (e.g., Dejonghe 1987*a*), but they are usually constructed by assumption of a special functional form for f, or by numerical techniques. For recent reviews, see Binney (1982*a*), Binney & Tremaine (1987), and Richstone (1987).

In this paper, we review dynamical models that are not spherically symmetric. In §2 we discuss axisymmetric models with distribution functions that depend on two integrals. It has recently become clear that such models apply at most to the spheroidal bulges of disk galaxies only. Elliptical galaxies and box–shaped bulges require three–integral models. These may be axisymmetric, but most likely they are triaxial and have slow figure rotation. Due to lack of knowledge regarding extra integrals of motion, few such models exist to date. They are described in §§3 and 4. §5 is devoted to the so–called Stäckel models, for which three exact integrals are known in closed form. They can be used to construct realistic three–integral models of axisymmetric galaxies, and also of non–rotating triaxial systems.

2. AXISYMMETRIC MODELS WITH $f = f(E, L_z)$

Let (ϖ, ϕ, z) be cylindrical coordinates. An axisymmetric potential $V(\varpi, z)$ always admits two exact isolating integrals of motion, the energy E and the component of the angular momentum vector that is parallel to the symmetry axis, L_z. It is therefore natural to consider dynamical models with $f = f(E, L_z)$. Although such models are not as widely applicable to galaxies as was once believed, it is nevertheless useful to review the various methods for their construction. We shall mainly discuss three–dimensional models; for a review of circular disks, see Kalnajs (1976).

2.1 Inversion

Fricke (1952) showed that in a potential $V(\varpi, z)$, a distribution function f of the form

$$f(E, L_z) = \sum a_{ij} E^{i-j-3/2} L_z^{2j}, \tag{4}$$

corresponds to a density ρ that is given by

$$\rho(\varpi, z) = \sum b_{ij} V^i \varpi^{2j}, \tag{5}$$

where the summation is over i and j, and the a_{ij} and b_{ij} are constants that are related one to one. Thus, if a given density $\rho(\varpi, z)$ in a potential $V(\varpi, z)$ can be expressed as $\rho(\varpi, V)$, and can be expanded in the form (5), then the unique distribution function $f(E, L_z)$ that is consistent with it follows from equation (4). The corresponding velocity moments can be given as series expansions in a form similar to equation (5). However, the velocity dispersions can often be derived more rapidly by direct integration of the stellar hydrodynamical equations (see §2.3). A major drawback of Fricke's method is that the convergence of the series (4) is not guaranteed for all values of E and L_z.

An early application of Fricke's method was given by Kuzmin & Kutuzov (1962), who calculated $f(E, L_z)$ for a family of oblate mass models found previously by Kuzmin (1956). The models are nearly spheroidal, and form a continuous sequence between Hénon's (1960) spherical isochrone and Kuzmin's (1953) disk.

Miyamoto & Nagai (1975) presented a remarkable (ρ, V)–pair that describes a sequence of flattened models that connects the spherical Plummer (1911) model with Kuzmin's disk. At large radii, the density falls off as ϖ^{-3} in the equatorial plane, but it decreases as z^{-5} everywhere else. As a result, the models appear to have both a bulge and a disk. Nagai & Miyamoto (1976) derived $f(E, L_z)$ for their models by Fricke's method, and delineated the kinematic properties.

Lynden–Bell (1962*a*) generalized Eddington's (1915*b*) isotropic spherical inversion formula to axisymmetric models. His method for the actual calculation of f is not easy to apply in practice. It requires taking a Laplace transform, and then two inverse Laplace transforms, and hence is restricted to very special densities. Alternative formulations in terms of Stieltjes and Mellin transforms (Hunter 1975*b*; Kalnajs 1976) avoid this problem. However, just as in Fricke's method, these different versions of the inversion method all require that ρ is given explicitly as $\rho(\varpi, V)$. Only very few (ρ, V)–pairs with this property are known. Consequently, only a small number of exact models have been constructed.

The inversion method has been applied to various modifications of the Plummer model, both oblate and prolate (Lynden–Bell 1962*a*; Hunter 1975*b*; Lake 1981*a*). Lynden–Bell's generalized Plummer models have a finite total mass; their distribution function is the sum of two terms of the form (4), so that it could have been calculated easily with Fricke's method as well.

Hunter (1975*a*) calculated $f(E, L_z)$ for a homogeneous spheroid of finite radius by Lynden–Bell's method, and showed that it is not everywhere positive. He also derived an approximate form for the distribution function of a mildly inhomogeneous spheroid, in the limit of small flattening, and found that $f \geq 0$.

Recently, Dejonghe (1986) rediscussed the inversion method in detail, with emphasis on the use of Mellin transforms. He gives a very useful list of explicit distribution functions $f(E, L_z)$ that correspond to a variety of fairly general functions

$\rho(\varpi, V)$. These can be used as convenient building blocks for dynamical models. Dejonghe's list includes all previously known cases, and many new ones. In particular, he derived $f(E, L_z)$ for a density of the form $\rho = V^a(1 - \varpi^2 V^2)^b$, with a and b constants. This allowed him to give the distribution function for the Miyamoto & Nagai (1975) model in a more convenient form, and to show that it is positive everywhere. By using the same building block, Dejonghe & de Zeeuw (1987) found the exact distribution function for the Kuzmin & Kutuzov (1962) model.

Dejonghe discusses the connection between the inversion problem for axisymmetric systems and the corresponding problem for anisotropic spherical models, and shows explicitly that the direct inversion method is numerically unstable. As a result, in practice the method is limited to *analytic* (ρ, V)–pairs for which, as we have seen, it must be possible to write $\rho = \rho(\varpi, V)$ in closed form. In addition to the six cases already mentioned, not counting the homogeneous spheroid, only one further (ρ, V)–pair with this property is known. These are Satoh's (1980) $n = \infty$ models, which form a continuous sequence connecting the Kuzmin disk with the point mass. The models resemble those of Miyamoto & Nagai (1975), but have a weaker disk component. Satoh calculated the kinematic properties of these models (see also §2.3), but he did not derive $f(E, L_z)$. This can be done by Fricke's method, and most likely also by direct inversion.

Although the above results seem to indicate that the calculation of $f(E, L_z)$ for a given density $\rho(\varpi, z)$ is limited to a few special cases, which have been found in a haphazard way, the situation is not quite desperate. Dejonghe (1986) has developed a method to generate many (ρ, V)–pairs with the desired property in a systematic fashion. In an earlier paper the same method was used to generate new exact spherical models (Dejonghe 1984). Construction of axisymmetric models via this technique promises to yield many new and useful models.

Finally, we remark that ρ constrains only the part of the distribution function that is even in the velocities, i.e., it can be used to determine $f(E, L_z^2)$. The odd part of f can be found by exactly the same techniques as the even part, by inversion of $\varpi\rho\langle v_\phi\rangle$ (cf. §2.3), instead of ρ (Hunter 1975*a*; Dejonghe 1986).

2.2 Assumed Form for f.

The alternative to direct inversion is to assume a functional form for $f(E, L_z)$, and then to find the potential–density pair (ρ, V) that is consistent with it. This approach has produced realistic spherical models (e.g., Michie 1963; King 1966; Binney 1982*b*). Prendergast & Tomer (1970) developed an efficient numerical technique to do this for axisymmetric models. After adopting a simple form for $f(E, L_z)$, they calculate ρ by equation (2), and substitute the result in equation (3). This produces a non–linear equation for the potential V, which is solved by iteration. The authors then employed this *self–consistent field method* for the construction of models with a variety of flattenings and radial density profiles.

Wilson (1975) extended the work of Prendergast & Tomer, with a smoother form for f, and constructed a small number of oblate models designed to fit the observations of the elliptical galay NGC 3379. He was able to reproduce the radial density profile, the isophote shape and even the inner part of the rotation curve. Further exploration showed that the models have a variation of eccentricity with radius that is always peaked, and that models flatter than about E4 are unrealistic. Lake (1981*b*) used another form for f, and was able to find prolate models flatter than E4. His models have eccentricity profiles which are decreasing with radius, and can have streaming, i.e., rotation, around the long axis only.

The most recent, and succesful, application of the self-consistent field method is by Jarvis & Freeman (1985*a*, *b*). They constructed models for the bulges of disk galaxies. A new aspect of their models is that the potential is taken as the sum of a bulge potential and a disk potential. During the iteration, the latter is kept fixed, and is set equal to the potential of a Miyamoto & Nagai model. Thus, the resulting bulge models are not self-consistent, which—for this application—is exactly as it should be. Jarvis & Freeman find that, with only a small contribution from the disk, their models accurately reproduce all available photometric and kinematic properties of spheroidal bulges, but not of the box-shaped ones.

A difficulty with the self-consistent field method is that the shape of a dynamical model is not related to the form of f in a very transparent way. This is evident from the differences between the above mentioned models, and also from the work of Ruiz & Schwarzschild (1976), who constrained the form of $f(E, L_z)$ and the shape of the density distribution simultaneously, and found that their conditions overspecified the model. This problem was studied in detail by Hunter (1977), who derived the conditions that f must satisfy in order to produce density distributions with roughly constant eccentricity profiles.

A special case where a particular choice of $f(E, L_z)$ leads to an analytic solution of the self-consistent problem was given by Toomre (1982). He considered scale-free axisymmetric density distributions with, in spherical coordinates, $\rho(r, \theta) = S(\theta)/r^2$, so that $V \propto \ln[r + P(\theta)]$. Here $S(\theta)$ and $P(\theta)$ describe the shape of the model, and are still to be determined. Toomre assumed that f is given by

$$f_n = c_n L_z^{2n} e^{-E}, \tag{6}$$

with c_n a constant. Combination of equations (2) and (3) then produced a nonlinear equation for $P(\theta)$ which, remarkably, can be solved in closed form for all $n \geq 0$. $S(\theta)$ then follows by differentiation. Thus, the form (6) describes a one-parameter family of exact scale-free models. Two special cases are the standard isothermal halo ($n = 0$) and Mestel's (1963) disk of infinite extent ($n = \infty$).

The scale-free models are not realistic, since they have infinite total mass, central density, and central potential. Also, for $n > 0$ the surfaces of constant density are tori with vanishingly small central holes. However, this last defect can be remedied by considering two-component models. Toomre showed that the structure of a model consisting of an infinite Mestel disk and a halo with distribution function f_n can also be found analytically. The special case $n = 0$ had already been discussed in detail by Monet, Richstone & Schechter (1981). Furthermore, two-component models with $f = f_n + f_m$ and $0 \leq n \leq m < \infty$ can have a bulge-disk structure, and can be given in closed form for $m = 2n + 1$. The particular case $n = 0$, $m = 1$ is the model used by Richstone in his extensive study of scale-free models with three integrals of motion (Miller 1982; §3.3 below).

2.3 Stellar Hydrodynamics

The kinematic properties of a dynamical model can be derived from the velocity moments of f. Let v_ϖ, v_ϕ and v_z be the velocity components in the cylindrical coordinates (ϖ, ϕ, z), and denote an average over all velocities by $\langle \enspace \rangle$. By symmetry, it follows that $\langle v_\varpi \rangle = \langle v_z \rangle \equiv 0$, so that there can be mean streaming around the symmetry axis only, with velocity $\langle v_\phi \rangle$ (which is a function of ϖ and z).

The second order moments of f are related to ρ and V by the Jeans equations, also referred to as the equations of stellar hydrodynamics (Jeans 1922; Chandrasekhar 1942). By symmetry, $\langle v_\varpi v_\phi \rangle = \langle v_\phi v_z \rangle = 0$. For $f = f(E, L_z)$, we also have $\langle v_\varpi v_z \rangle = 0$, $\langle v_\varpi^2 \rangle = \langle v_z^2 \rangle$, and the Jeans equations reduce to

$$
\begin{aligned}
\frac{1}{\rho}\frac{\partial \rho \langle v_\varpi^2 \rangle}{\partial \varpi} + \frac{\langle v_\varpi^2 \rangle - \langle v_\phi^2 \rangle}{\varpi} &= -\frac{\partial V}{\partial \varpi}, \\
\frac{1}{\rho}\frac{\partial \rho \langle v_z^2 \rangle}{\partial z} &= -\frac{\partial V}{\partial z}.
\end{aligned}
\tag{7}
$$

Given a density ρ in a potential V, the moments $\langle v_\varpi^2 \rangle = \langle v_z^2 \rangle$ and $\langle v_\phi^2 \rangle$ can be determined by direct solution of equations (7), without knowing the explicit form of the corresponding f. In fact, even if f is known, it is often easier to calculate the moments in this way, rather than by averaging of v_ϖ^2 and v_ϕ^2 over velocity space. However, a solution of the Jeans equations may give velocity dispersions that are unphysical, since the implied f is not guaranteed to be non–negative.

As already mentioned, specification of ρ does not fix the part of f that is odd in the velocities. Hence $\langle v_\phi \rangle$ is undetermined. In many cases $\langle v_\phi \rangle$ is chosen such that the local velocity dispersions are equal in all three directions, i.e., $\langle v_\varpi^2 \rangle = \langle v_z^2 \rangle = \langle v_\phi^2 \rangle - \langle v_\phi \rangle^2$. These solutions are referred to as *isotropic*. It is also possible to determine $\langle v_\phi \rangle$ via an entropy argument (Dejonghe 1986).

The Jeans equations have played a central role in Galactic Dynamics (e.g., Oort 1965). They have also been put to good use for spherical galaxies (Binney & Mamon 1982), but few applications to general axisymmetric models exist. Bagin (1972) considered rather special density distributions. A very interesting set of solutions was given by Nagai & Miyamoto (1976). They derived an infinite set of axisymmetric (ρ, V)–pairs, each of which connects the Plummer model with one of Toomre's (1963) disks. The Miyamoto & Nagai (1975) models mentioned earlier are given by the first of these pairs, since Toomre's $n = 1$ disk is identical to Kuzmin's disk. The authors were able to integrate equations (7) explicitly for all these generalized Toomre models, so that the velocity dispersions could be given in closed form. In the same fashion, Satoh (1980) presented another infinite set of three–dimensional models, connecting the Kuzmin disk with various generalizations of the Plummer model. He solved the Jeans equations for his $n = \infty$ model only, and compared the results with observations of NGC 4697.

Hunter (1977) used the Jeans equations in his investigation of the relations between the functional form of $f(E, L_z)$, the anisotropy of the velocity dispersions, and the shape of the density distribution in self–consistent models (cf. §2.2). He also showed that equations (7) can be solved by simple quadrature in case $\rho = \rho(\varpi, V)$ explicitly. Since both the Nagai & Miyamoto $n = 1$ models and the Satoh $n = \infty$ models have densities with this property, it is not surprising that for them the Jeans equations can be solved exactly.

2.4 Applications

The standard example of an axisymmetric system is our own Galaxy. However, it has long been known that $\langle v_\varpi^2 \rangle \neq \langle v_z^2 \rangle$ in the solar neighbourhood (e.g., Oort 1928). It was concluded that the distribution function of the Galaxy must depend on a third argument, i.e., there must be a third isolating integral of motion, I_3. Although special potentials which admit a third integral were considered by various authors (Oort 1932; Lindblad 1933; Kuzmin 1953, 1956; Hori 1962) it was established only through the analytic work of Contopoulos (1960) and the numerical work of Ollongren (1962) that in realistic galactic potentials indeed most stellar orbits have an effective third integral (see also Martinet & Mayer 1975). No simple expression is known for I_3, and to date no satisfactory dynamical model for the Galaxy exists.

Although a two-integral model was known to be inadequate for the Galaxy, it was thought for a long time that elliptical galaxies have a simpler dynamical structure, and are oblate axisymmetric systems with $f = f(E, L_z)$ and an isotropic velocity distribution, so that their flattening is caused by rotation (Freeman 1975). This premise generated most of the work described in the previous sections. However, at about the time that Wilson (1975) produced his two-integral models, spectroscopic observations indicated that large elliptical galaxies rotate much slower than expected (Bertola & Capaccioli 1975; Illingworth 1977).

All $L_z \neq 0$ orbits in an oblate galaxy have a definite sense of rotation around the symmetry axis, but both clockwise and counter-clockwise motion may occur. By reversing the direction of motion for an arbitrary fraction of stars in each orbit, and hence making the velocity distribution anisotropic (§2.3), we may obtain as small a mean streaming velocity $\langle v_\phi \rangle$ (i.e., observed rotation) as desired, even with $f = f(E, L_z)$. However, by using the tensor virial theorem to connect the ratio of the observed maximum rotational velocity and the central velocity dispersion with the apparent flattening, Binney (1978*a*) concluded that the then available kinematic observations of elliptical galaxies were best represented by models with $\langle v_\varpi^2 \rangle \neq \langle v_z^2 \rangle$. This is supported by more recent studies (§3.4). As a result, it is generally accepted that the most natural models for elliptical galaxies have all three velocity dispersion components unequal, and hence must have distribution functions that depend on three integrals.

In smaller elliptical galaxies the velocity anisotropy decreases, and rotation becomes more important (Davies *et al.* 1983). The still smaller bulges of disk galaxies, which resemble elliptical galaxies in many respects, all rotate nearly as fast as the oblate isotropic models (Kormendy & Illingworth 1982). Jarvis & Freeman (1985*b*) have shown that, by inclusion of the effect of the disk, the *spheroidal* bulges are completely consistent with $f(E, L_z)$-models. However, the *box*-shaped bulges have velocity fields that need three-integral models.

3. AXISYMMETRIC MODELS WITH $f = f(E, L_z, I_3)$

Most orbits in realistic axisymmetric potentials are tubes around the symmetry axis, and have an effective third integral. The remainder is generally made up of a host of minor orbit families and irregular orbits. The latter do not have a third integral of motion. Binney (1982*c*) has argued that Jeans' theorem is not valid for potentials that support irregular orbits (see also Pfenniger 1986). Hence, no true equilibrium solutions may exist for such systems. In many cases of interest, however, the fraction of irregular orbits is small. On time scales of the order of a Hubble time these orbits are nearly indistinguishable from regular ones (e.g., Goodman & Schwarzschild 1981). For practical purposes, one may probably still use Jeans' theorem for these systems, and construct approximate equilibrium models. Since a given $\rho(\varpi, z)$ determines a unique $f(E, L_z)$, generally many different distribution functions $f(E, L_z, I_3)$ are consistent with it.

3.1 Exact Models

An exact third isolating integral of motion is known for the special axisymmetric potentials for which the Hamilton–Jacobi equation separates in spheroidal coordinates (Stäckel 1890; Lynden–Bell 1962*c*). In a classic paper, Kuzmin (1956) showed that many such Stäckel potentials correspond to realistic axisymmetric mass models. For such models Jeans' theorem is strictly valid. We shall consider exact solutions based on these potentials in §5.

3.2 Approximate Distribution Functions

For nearly spherical systems, the third integral is related closely to the square of the total angular momentum. This is suggested by the form of I_3 in the Stäckel potentials, most clearly in the limiting case where $V(r,\theta) = F(r) + G(\theta)/r^2$, with $F(r)$ and $G(\theta)$ arbitrary functions. This is the Eddington (1915*a*) potential, which admits $I_3 = L^2 - 2G(\theta)$ as exact integral. Numerical orbit calculations in more realistic potentials show that L^2 indeed does not vary much along an orbit (e.g., Saaf 1968; Innanen & Papp 1977). This fact can be used to include the dependence of f on I_3 in an approximate way.

Lupton (1985) simply used L^2 as third integral, and constructed realistic models of globular clusters with an assumed form for $f(E, L_z, I_3)$. A more elaborate treatment was given by Petrou (1983*a*). She used an approximation of the form $I_3 = L^2 - 2G(r,\theta)$, with $G(r,\theta)$ a simple function of the potential. This I_3 turned out to be constant along individual orbits to better than a few percent. She then modified the lowered Maxwellian form of the distribution function $f(E, L_z)$ used by Prendergast & Tomer (1970), by inclusion of a factor that depends on I_3, chosen such that the orbits that do not come close to the center are all depopulated, independent of their inclination. The potential and density that correspond to this f were then determined by the self–consistent field method (§2.2). This produced quite realistic models for nearly round elliptical galaxies. Due mainly to the chosen L_z-dependence of f, the models all have peaked rotation curves, and cannot become much flatter than E2.3. In order to remedy this defect, Petrou (1983*b*) modified f in such a way that the non–radial orbits are depopulated predominantly outside the equatorial plane, which results in flatter models. She presented detailed results for an E3.5 model. It has a nearly flat rotation curve, a velocity ellipsoid that becomes radially aligned at large distances from the center, a realistic density profile, and elliptical isophotes.

A somewhat different approach was taken by Binney & Petrou (1985). They proposed a particular form for the three–integral distribution function for box–shaped bulges, and showed that the corresponding density in the potential of the spherical isochrone has the correct photometric and kinematic properties. It is likely that more realistic models can be obtained by taking their distribution function, and applying the self–consistent field method after inclusion of a disk potential, just as was done for spheroidal bulges by Jarvis & Freeman (1985*a, b*).

3.3 Scale–Free Models

As we have seen in §2.2, scale–free models suffer from some defects. They can, however, provide considerable insight in the structure of more realistic models. Richstone (1980, 1982, 1984) made a detailed study of the possible three–integral models that are consistent with a particular axisymmetric scale–free model. He took the case where the potential is logarithmic, and constant on similar oblate spheroids. The associated density has dimples near the symmetry axis. Richstone studied a particular example with a flattening similar to an E6 galaxy. The great majority of orbits in this potential belongs to one family, that of tubes circling the short axis. He did not encounter any stochastic orbits in his survey. Thus, all orbits have an effective third integral.

Richstone used Schwarzschild's method (§4.2) to compute numerical distribution functions by reconstructing the given density with individual orbit densities. Because the model is scale–free, this reconstruction needs to be done at one radius only, so that the problem is effectively one–dimensional. By optimizing the values of the total angular momentum and the total second velocity moments, Richstone found a large variety of different dynamical models.

Subsequently, Levison & Richstone (1985*a*, *b*) produced non–consistent models in the same potential, by using the orbits to reproduce a density $\rho(r) \sim r^{-3}$, instead of $\rho(r) \sim r^{-2}$, so that the mass–to–light ratio $M/L \propto r$. They obtained the somewhat unexpected result that the kinematics of these different models are very similar. If this is true for more realistic models as well, then the chances of finding evidence for dark matter in elliptical galaxies when only observations of stellar kinematics are available, are very slim.

3.4 Stellar Hydrodynamics

Solutions of the Jeans equations for anisotropic axisymmetric models relevant for elliptical galaxies have been given only recently. Bacon, Simien & Monnet (1983), and Bacon (1985) considered the case where $\langle v_\varpi^2 \rangle \neq \langle v_z^2 \rangle$, but with the restriction that the velocity ellipsoid is radially aligned everywhere, so that $\langle v_r v_\theta \rangle = 0$. By analogy with the spherical case (Binney & Mamon 1982), they derived the formal solution of the Jeans equations for an assumed functional form of the anisotropy parameter $\beta = 1 - \langle v_\theta^2 \rangle / \langle v_r^2 \rangle$. They considered the case $\beta = \beta(r) = kr/(1 + ar)$, with k and a constants, in detail. This corresponds to a velocity distribution that is isotropic in the center, and, for $0 < k \leq 1$, predominantly radial at large radii, as suggested by N–body simulations (van Albada 1982). The authors evaluated their solutions numerically for density distributions that are stratified on similar oblate spheroids and have, in projection, a de Vaucouleurs profile. Comparison of the results with observations of a small number of well–observed elliptical galaxies revealed that the larger ones are indeed best fit with anisotropic models.

A similar investigation was done by Fillmore (1986), who solved the Jeans equations by an iterative numerical technique for three different assumptions for the shape of the velocity ellipsoid. He employed a mass model identical to the one used by Bacon, and showed that observations of both major and minor axis dispersion profiles can put strong constraints on the form of the velocity distribution.

4. TRIAXIAL MODELS

Since models with $f = f(E, L_z, I_3)$, and hence with anisotropic velocity distributions, seem to be required for the majority of elliptical galaxies, the assumption that these galaxies are oblate, i.e., with two of the three axes exactly equal, becomes rather artificial. It is much more natural to assume that these galaxies are triaxial

(Binney 1978*a, b*). This hypothesis is supported by a number of observational lines of evidence (e.g., Schechter 1987).

A general triaxial potential admits only one isolating integral, the orbital energy E. Although there are three planes of reflection symmetry, there is no symmetry axis, and no component of the angular momentum is conserved. Schwarzschild (1979, 1982) has shown by numerical means that in realistic triaxial potentials most stellar orbits possess two effective integrals I_2 and I_3 in addition to E. Just as for the third integral in axisymmetric systems, in the general case no simple expressions are known for these extra integrals.

4.1 Exact Solutions

There exists a large class of triaxial mass models with *non–rotating* figures that have a potential of Stäckel form, and admit three exact integrals of motion. Self–consistent models of this kind will be discussed in §5. For triaxial systems with *rotating* figures, only two analytic solutions exist, one exact, the other approximate.

Freeman (1966) constructed dynamical models for homogeneous triaxial ellipsoids of arbitrary axis ratios, rotating at a critical frequency such that the centrifugal force exactly balances the gravitational attraction on the long axis. The distribution function f of his models depends on the energy integral in the corotating coordinate system only. However, shape and rotation are related in a unique way, which is in contradiction with the observations of elliptical galaxies. Hunter (1974) proved that for general axis ratios, and a general figure rotation, no solutions exist. In the special case of a homogeneous spheroid, solutions exist that depend on more integrals (Bisnovatyi–Kogan & Zeldovich 1970), although the model with $f(E, L_z)$ is unphysical (§2.1).

Vandervoort (1980*a, b*) constructed approximate solutions for rotating polytropic models in which the density is a power of the potential, and is only mildly concentrated towards the center. The distribution functions found by Vandervoort depend on the energy only, and the velocity dispersions are isotropic. This illustrates that even triaxial models can have an isotropic velocity distribution. However, these models again require a unique rotation speed. Vandervoort & Welty (1981, 1982) developed an analytic—and approximate—version of the self–consistent field method (§2.2), in which the iterative solution is terminated after the first one–half iteration. They used this method to construct a more general set of polytropic models, with a variety of rotation speeds. They considered distribution functions that depend on an approximate second integral, which is only valid in nearly homogeneous systems. The resulting anisotropic models can be considered as the stellar dynamical counterparts of the S–type Riemann fluid ellipsoids (Chandrasekhar 1969). Bohn (1983) used the same method to construct prolate models.

4.2 Schwarzschild's Method

A completely new approach to the self–consistent problem, which sidesteps our ignorance of two of the three arguments of f, was introduced by Schwarzschild (1979). He specified a mass model and derived the gravitational potential by integration of Poisson's equation. Then he calculated a large number of stellar orbits by numerical means, and computed their individual density distributions on a grid of cells, by determining the average time spent in each cell by each orbit. He then used linear programming to find a combination of orbital densities that reproduces the original density distribution, with all occupation numbers non-negative.

If this procedure is succesful, one has obtained a numerical representation of a distribution function that is consistent with the assumed density. In each cell the number of stars going through it are known. Since the orbits have been computed, the velocities of the stars are known at each position. Thus, the orbits and their occupation numbers give the distribution of the stars over position and velocities. This is, by definition, the distribution function (§1). Schwarzschild (1979) applied this method to a nearly ellipsoidal triaxial mass model with axis ratios 1 : 5/8 : 1/2, and a realistic density profile, and was able to find a solution for f. This established the existence of realistic triaxial equilibrium models.

Most orbits in Schwarzschild's models have three effective integrals of motion, and belong to four major families: box orbits, which have no net average angular momentum, tube orbits around the short axis and two kinds of tubes circling the long axis. Each star in a tube orbit has a definite sense of rotation around the appropriate axis, but both clockwise and counter-clockwise motion may occur on the same orbit. The orbital shapes and their individual density distributions are determined by the values of the integrals. As a result, the derived occupation numbers depend on the integrals. Thus, without knowledge of the explicit forms of two of the integrals, still a distribution function f is found that depends on all three of them. We remark that instead of linear programming, one can also use Lucy's (1974) method (Newton & Binney 1984), non-negative least squares (Pfenniger 1984), or a maximum-entropy method (Tremaine & Richstone 1987).

Schwarzschild constructed his model with boxes and short axis tubes only. Its internal structure, and the observable properties, were studied by Merritt (1980). Later work showed that long axis tubes can be included in the model as well, indicating that different equilibrium models can exist for the same density distribution. Levison & Richstone (1987) used Schwarzschild's method to survey the solutions for two different triaxial scale-free models, one nearly oblate, and the other nearly prolate. These models contain the same major orbit families as found in Schwarzschild's model (Levison, priv. comm.). The equilibrium models can have mean streaming about the long axis and about the short axis (but see below), and show a large variety in observable properties.

4.3 Figure Rotation

Schwarzschild (1982) constructed two distinct equilibrium solutions for his original model, but now with a slow rotation of the figure around its short axis. This showed that realistic triaxial dynamical models exist whose shape is not determined by the (figure) rotation.

In the rotating models, the Coriolis force distinguishes direct and retrograde motion, and each of the tube orbit families mentioned above splits up in two branches, with different shapes. The long axis tubes tip out of the plane that contains the intermediate and short axes, with the direct and retrograde branches tipping in opposite directions. In order to obtain a model with triaxial symmetry, each branch of this family has to be populated equally, so that no net streaming about the long axis can occur. The streaming about the short axis can be quite complicated, since in addition to the short axis tubes, also the boxes and the tipped long axis tubes have a net average angular momentum with respect to this axis.

Vietri (1986) used Schwarzschild's method to construct rotating triaxial models for the bulge of the Galaxy: He assumed that the bulge counter–rotates with respect to the disk, and heavily populated the tube orbits that are retrograde in a frame that corotates with the bulge. If this counter–streaming is large enough, then the bulge will seem to rotate in the normal sense to an external observer. This unorthodox approach was inspired by the possibility that in this way the Liszt & Burton (1980) tilted HI disk can be explained as a stable dynamical phenomenon. Unfortunately, Vietri found that the counter–streaming is so small that the net rotation of the bulge (figure rotation minus counter–streaming) would be contrary to Galactic rotation, which is not observed.

Vietri's result does not rule out the existence of slowly rotating triaxial elliptical galaxies with counter–streaming in their central parts, even although van Albada (1987) could not simulate them in N–body experiments.

5. STÄCKEL MODELS

It is evident from the two preceding sections that the collection of realistic three–integral models for galaxies, both axisymmetric and triaxial, is still limited. Many questions regarding the structure of these systems remain. What are the extra integrals? How many f's are consistent with a given ρ? What is the full variety of equilibrium figures? What observations do we need in order to determine the intrinsic dynamical structure of an elliptical galaxy?

These questions, and many others, cannot be answered easily by numerical methods alone. Each model has to be constructed, or simulated, separately. This involves considerable labour and expense. We now show that many general aspects of the dynamics of triaxial and axisymmetric galaxies can be understood by analytic means, through a study of the Stäckel models.

5.1 Stäckel Potentials

Many triaxial mass models exist with a gravitational potential of Stäckel form, for which the Hamilton–Jacobi equation separates in ellipsoidal coordinates (Kuzmin 1973; de Zeeuw 1985*b*). Every orbit in such a model enjoys three exact isolating integrals of motion, E, I_2 and I_3, which are known explicitly, and which are quadratic in the velocities. I_2 and I_3 are related to the angular momentum integrals in the axisymmetric and spherical limits.

An individual orbit can be considered as the sum of three motions, one in each ellipsoidal coordinate. The stars are thus constrained—by the integrals of motion—to move between coordinate surfaces. Thus, all possible orbital shapes can be found by inspection of the ellipsoidal coordinate system in which the motion separates. It turns out that all centrally concentrated triaxial mass models of this kind have four families of orbits: boxes, short axis tubes, and two families of long axis tubes. These are exactly the four major orbit families that occur in Schwarzschild's (1979) nonrotating model, and also in Levison & Richstone's (1987) scale–free models. The orbital structure of the Stäckel models is *generic* for all moderately flattened triaxial systems without figure rotation (de Zeeuw 1985*a*; Gerhard 1985).

The prototypical triaxial Stäckel model is the perfect ellipsoid. It has a density distribution given by

$$\rho = \rho(m^2) = \frac{\rho_0}{(1+m^2)^2}, \qquad m^2 = \frac{x^2}{a^2} + \frac{y^2}{b^2} + \frac{z^2}{c^2}, \tag{8}$$

with $a \geq b \geq c$. The oblate case, $a = b$, was discovered by Kuzmin (1956), in his classic study of separable models of the Galaxy. He also obtained the general case (Kuzmin 1973), and showed that it has four major orbit families. The perfect ellipsoid was rediscovered by de Zeeuw & Lynden–Bell (1985), who showed that it is the only inhomogeneous triaxial mass model with a Stäckel potential in which the density is stratified exactly on similar concentric ellipsoids. The orbital structure was delineated in detail by de Zeeuw (1985*b*).

The general form of a Stäckel potential contains a free function of one variable. The associated ellipsoidal coordinate system is determined by specification of two parameters. As a result, there is a large variety of mass models with a Stäckel potential. They have remarkable properties.

Kuzmin (1956, 1973) showed that in a Stäckel model the density $\rho(x, y, z)$ at a general point is related to that on the short axis, $\rho(0, 0, z)$, by a very simple formula, and that $\rho(x, y, z) \geq 0$ if and only if $\rho(0, 0, z) \geq 0$. This makes it possible to choose a short–axis density profile, and the values of the central axis ratios of the model, and find the complete mass model that has a Stäckel potential and this density profile, by one integration in one variable. A second integration gives this potential explicitly (de Zeeuw 1985*c*). This means that Poisson's equation can be integrated in closed form for the whole class of Stäckel models.

De Zeeuw, Peletier & Franx (1986) constructed many different mass models, and delineated their general properties. Models with a singular density in the centre only do not exist. The density cannot fall off more rapidly than r^{-4} as $r \to \infty$, except on the short axis. Models in which ρ falls off less rapidly than r^{-4} become spherical as $r \to \infty$. The only models that have surfaces of constant density that approach a finite flattening at large radii are those with $\rho \sim r^{-4}$.

With the exception of the perfect ellipsoid, on projection the triaxial Stäckel models have isophotes that are not exact ellipses. Their ellipticity changes with radius, but they do not show twisting isophotes (Franx 1987).

5.2 Equilibrium Models

The individual orbit densities in a Stäckel model are known in analytic form, and hence are evaluated easily. This makes it straightforward to construct self–consistent models by means of Schwarzschild's method, while avoiding laborious numerical orbit integrations. This is true not only for the triaxial models, but also for the various limiting cases with more symmetry, and a simpler orbital structure. These cases are depicted schematically in Figure 1, which shows the axis ratio plane for the perfect ellipsoid, and hence gives the layout of "Ellipsoid Land". Elliptical galaxies are no flatter than at most E6, and hence occupy the upper part of the triangle. This coincides with the area where the Stäckel models provide an adequate description of the orbital structure.

Bishop (1986) considered the perfect oblate spheroids. All orbits in them are short axis tubes. He first constructed one–dimensional continua of orbits, effectively specifying the dependence of f on one integral, and solved for the remaining unknown part of f by an algebraic technique (Vandervoort 1984). Prolate models have not yet been constructed. Since they can have streaming about the long axis only, they are probably less relevant for galaxies.

For $c = 0$ the models reduce to elliptic disks. These contain flat box orbits, and flat short axis tubes. Teuben (1987) used Schwarzschild's method to construct equilibrium models for nine different axis ratios, both with minimum and with

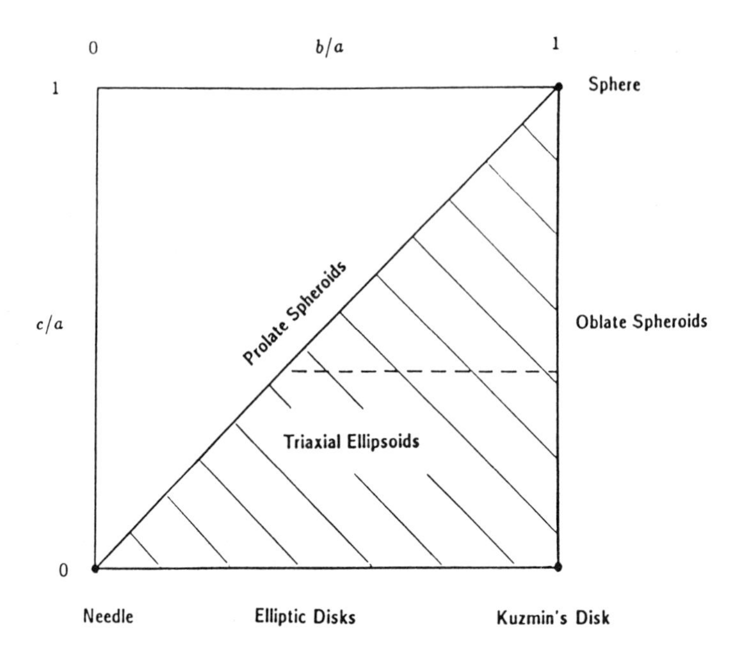

Figure 1. Plane of axis ratios for the perfect ellipsoid. The area above the dashed line corresponds to axis ratios relevant for elliptical galaxies.

maximum possible streaming, and discussed the kinematic properties. De Zeeuw, Hunter & Schwarzschild (1987) used analytic methods to prove rigorously that exact equilibrium models exist, and showed that there is a two–dimensional continuum of different f's that all are consistent with the same ρ. They also constructed the model with maximum possible streaming in nearly explicit form, by using all the box orbits, but only the thinnest tubes, i.e., the elliptic closed orbits. The fundamental reason for the non–uniqueness of the dynamical solutions is the existence of more than one major orbit family, so that orbits can be exchanged while keeping the density fixed and $f \geq 0$. Schwarzschild (1986) showed that when the perfect elliptic disk is truncated inside the foci of the elliptic coordinate system in which the motion separates, and hence only box orbits can be used, the dynamical solution is unique. The circular limit of the perfect elliptic disk is Kuzmin's disk, for which many authors have derived self–consistent dynamical models (cf. Kalnajs 1976). The opposite limit is a one–dimensional needle. Its curious properties are discussed by Tremaine & de Zeeuw (1987).

The triaxial case was investigated by Statler (1987). He considered 21 different perfect ellipsoidal mass models covering a regular grid in Figure 1. For each of these he constructed one solution by Lucy's method, thus establishing existence, and 15–20 distinct solutions by means of linear programming. In these, he optimized either the streaming around the short axis, or the streaming around the long axis, or a combination of the two. The properties of the models all vary smoothly with the axis ratios. Statler delineated the kinematic properties of the resulting models in detail, and discussed how observations might distinguish them.

The fundamental result of Statler's study is that many different dynamical models can be constructed with the same density distribution, as already suggested by the earlier work discussed in §4.2. By analogy with the above mentioned results for the perfect elliptic disk, this freedom is expected to be that of an arbitrary function of three variables (see also Dejonghe 1987*b*).

5.3 Exact Solutions

Many properties of the Stäckel models can be given in analytic form. It is therefore natural to ask whether it is possible to give exact distribution functions $f(E, I_2, I_3)$. This is generally not to be expected since, after all, every spherical potential is of Stäckel form, but only few analytic spherical models exist. However, in the last year some special solutions have been obtained, and more are likely to follow soon. A formal solution of the inversion problem has been obtained by Dejonghe (1987).

For any oblate Stäckel potential the special equilibrium model that contains the thinnest short axis tubes only (i.e., tubes without radial epicyclic motion) can be found by simple inversion of a one-dimensional Abel equation (de Zeeuw 1987). As a result, the corresponding distribution function can be given explicitly, and the kinematic properties can be calculated very easily. These models have the maximum possible streaming around the symmetry axis. Numerical solutions of this kind have been given by Bishop (1987). The similar prolate models require more work, due to the presence of the two types of long axis tubes. Exact distribution functions have recently been found by Park, de Zeeuw & Schwarzschild (1987).

It is likely that triaxial models with maximum streaming, containing thin tube orbits of the three families as well as boxes, can also be constructed in nearly explicit form, by generalizing the analysis of the perfect elliptic disk by de Zeeuw, Hunter & Schwarzschild (1987). This would give insight in the relative importance of the different orbit families, and in the detailed behaviour of the solutions near the focal curves of the ellipsoidal coordinates. This would resolve questions that Statler's models cannot answer, due to lack of numerical accuracy that results from the finite grid of cells.

Dejonghe & de Zeeuw (1987) have succeeded in generalizing Fricke's method (cf. §2.1) to axisymmetric three-integral models, by writing $f(E, L_z, I_3)$ as a triple series in powers of E, L_z and I_3. For each term the corresponding density in a given Stäckel potential can be calculated. Expansion of a given density in a series of these terms then gives f by comparison of coefficients. In practice, this last step may have to be done by numerical means, analogous to Schwarzschild's method, but now using these density components instead of individual orbit densities. The same method can be applied to triaxial systems also, and should give smooth solutions.

In order to obtain fully analytic solutions, Dejonghe & de Zeeuw took a slightly different approach. The Kuzmin & Kutuzov (1962) model (cf. §2.1), has a potential that is of Stäckel form. The authors reconstruct part of the density distribution of this model with three-integral components, and represent the remaining density by an $f(E, L_z)$ found via the two-integral inversion method.

All the above efforts have been directed at finding f for a given ρ. We have seen that for axisymmetric models often the approach that assumes a form for f was taken. This was motivated by the fact that in these cases many solutions exist for a given density, so that a plausible choice for f is likely to lead to an equilibrium model. When it became evident that elliptical galaxies are probably triaxial, virtually nothing was known concerning the existence of non-axisymmetric equilibrium models, let alone regarding their non-uniqueness. Hence, guessing a plausible distribution function seemed rather difficult. Therefore, the second approach was adopted, notably by Schwarzschild.

Now we know that many equilibrium solutions are likely to exist for triaxial models as well, it should not be too difficult to guess a plausible form for f. It seems very useful to attempt to construct dynamical models with this first method as well, in particular for the Stäckel models. A first step in this direction has been taken recently by Stiavelli and Bertin (1985). It should be noted that in order to obtain self-consistent models, f should not be chosen to obey the Ellipsoidal Hypothesis, i.e., $f = f(Q)$, where Q is any linear combination of the three integrals E, I_2 and I_3 (e.g., Chandrasekhar 1942). Although such an assumed form for f has produced interesting spherical models (Eddington 1914; Osipkov 1979; Merritt 1985), no self-gravitating axisymmetric of triaxial models of this kind exist (Eddington 1915*a*; Camm 1941; Fricke 1952).

5.4 Stellar Hydrodynamics

Eddington (1915*a*) already knew that in a Stäckel potential, the principal axes of the velocity ellipsoid are locally always aligned with the coordinate system in which the equations of motion separate. For triaxial models this means that the velocity ellipsoid is always aligned with the principal axes of the model in the central regions, and always aligned nearly radially at large distances. This is exactly what is also seen in the non-separable models (Merritt 1980). Furthermore, the Jeans equations have a simple form in ellipsoidal coordinates (Lynden-Bell 1960). Solution of them gives the three non trivial second velocity moments for any density in a given Stäckel potential. Since many different distribution functions are consistent with the same density, the danger of finding non-physical solutions of the equations is not very severe. For axisymmetric models it turns out that the equations are formally equivalent to those already solved by Bacon (1985) in a slightly different context (§2.3). Thus, by using his solutions, kinematic properties of axisymmetric Stäckel models can be derived easily. The triaxial case is under study by Wyn-Evans (1987, priv. comm.).

6. CONCLUDING REMARKS

It is fair to say that in the last decade considerable progress has been made in the construction of equilibrium models. However, Hunter's (1977) remark that *determining appropriate distribution functions for elliptical galaxies [is] a problem that deserves more study from fundamental stellar dynamic considerations than it has yet received*, has not lost its value.

For moderately flattened axisymmetric and non-rotating triaxial galaxies, the Stäckel models provide an adequate description of the internal dynamical structure. Construction of realistic three-integral models of this kind should evidently be pursued. Methods for doing this are already available. These models are well suited to establish which observations would have to be done in order to determine the intrinsic shape and structure of elliptical galaxies.

Much work remains to be done for triaxial systems with rotating figures, not only for moderately flattened elliptical galaxies, but also for the rapidly rotating nearly flat bars. It is well possible, although not proven, that no realistic rotating potentials with exact integrals, in addition to the energy in the corotating frame, exist. As a result, construction of such systems may have to be done largely by numerical means.

It is evident that the family of triaxial equilibrium structures is very rich. The most important next step in the study of these systems is the delineation of the models that are dynamically stable. If we then can establish which of the models are favored by the elliptical galaxies, we will have a much better understanding of the formation of these systems.

It is a pleasure to thank Herwig Dejonghe and Martin Schwarzschild for many stimulating discussions, and a careful reading of the manuscript. This research was supported in part by an RCA Fellowship, and by NSF Grant PHY 82–17352.

REFERENCES

Bacon, R., 1985*a*. *Astr. Ap.*, **143**, 84.
Bacon, R., Simien, F., & Monnet, G., 1983. *Astr. Ap.*, **128**, 405.
Bagin, V.M., 1972. *Astr. Zh.*, **49**, 1249 (translated as 1973. *Sov. Astr.*, **16**, 1003).
Bertola, F., & Capaccioli, M., 1975. *Ap. J.*, **200**, 439.
Binney, J.J., 1978*a*. *M.N.R.A.S.*, **183**, 501.
Binney, J.J., 1978*b*. *Comments Ap.*, **8**, 27.
Binney, J.J., 1982*a*. *Ann. Rev. Astr. Astrophys.*, **20**, 399.
Binney, J.J., 1982*b*. *M.N.R.A.S.*, **200**, 951.
Binney, J.J., 1982*c*. *M.N.R.A.S.*, **201**, 15.
Binney, J.J., & Mamon, G., 1982. *M.N.R.A.S.*, **200**, 361.
Binney, J.J., & Petrou, M., 1985. *M.N.R.A.S.*, **214**, 449.
Binney, J.J., & Tremaine, S.D., 1987. *Galactic Dynamics*, Princeton University Press, in press.
Bishop, J., 1986. *Ap. J.*, **305**, 14.
Bishop, J., 1987. Preprint.
Bisnovatyi-Kogan, G.S., & Zeldovich, Ya.B., 1970. *Astrofisika*, **6**, 387 (translated as 1973. *Astrophysics*, **6**, 207).
Bohn, C., 1983. *Ap. J.*, **268**, 646.
Camm, G.L., 1941. *M.N.R.A.S.*, **101**, 195.
Chandrasekhar, S., 1942. *Principles of Stellar Dynamics*, University of Chicago Press, Chicago.
Chandrasekhar, S., 1969. *Ellipsoidal Figures of Equilibrium*, Yale University Press, New Haven.
Contopoulos, G., 1960. *Zs. Ap.*, **49**, 273.
Davies, R.L., Efstathiou, G., Fall, S.M., Illingworth, G., & Schechter, P.L., 1983. *Ap. J.*, **266**, 41.
Dejonghe, H., 1984. *Astr. Ap.*, **133**, 225.
Dejonghe, H., 1986. *Physics Reports*, **133**, 218.
Dejonghe, H., 1987*a*. *M.N.R.A.S.*, **224**, 13.
Dejonghe, H., 1987*b*. Preprint & this volume, p. 495.
Dejonghe, H., & de Zeeuw, P.T., 1987. Preprint & this volume, p. 497.
de Zeeuw, P.T., 1985*a*. *M.N.R.A.S.*, **215**, 731.
de Zeeuw, P.T., 1985*b*. *M.N.R.A.S.*, **216**, 273.
de Zeeuw, P.T., 1985*c*. *M.N.R.A.S.*, **216**, 599.
de Zeeuw, P.T., 1987. Preprint.
de Zeeuw, P.T., Hunter, C., & Schwarzschild, M., 1987. *Ap. J.*, in press.
de Zeeuw, P.T., & Lynden-Bell, D., 1985. *M.N.R.A.S.*, **215**, 713.

de Zeeuw, P.T., Peletier, R.F., & Franx, M., 1986, *M.N.R.A.S.*, **221**, 1001.
Eddington, A.S., 1914. *M.N.R.A.S.*, **75**, 366
Eddington, A.S., 1915*a*. *M.N.R.A.S.*, **76**, 37.
Eddington, A.S., 1915*b*. *M.N.R.A.S.*, **76**, 572.
Fillmore, J.A., 1986. *A. J.*, **91**, 1096.
Franx, M., 1987. *M.N.R.A.S.*, in press.
Freeman, K.C., 1966. *M.N.R.A.S.*, **134**, 1.
Freeman, K.C., 1975. In *Galaxies and the Universe*, Stars and Stellar Systems Vol. 9, p. 409, eds A. Sandage & J. Kristian, University of Chicago Press.
Fricke, W., 1952. *Astr. Nachr.*, **280**, 193.
Gerhard, O.E. 1985, *Astr. Ap.*, **151**, 279.
Goodman, J., and Schwarzschild, M. 1981, *Ap. J.*, **245**, 1087.
Hénon, M., 1960. *Ann. d' Astrophys.*, **23**, 474.
Hori, G., 1962. *Publ. Astr. Soc. Japan*, **14**, 353.
Hunter, C., 1974. *M.N.R.A.S.*, **166**, 633.
Hunter, C., 1975*a*. in IAU Symposium No. 69, *Dynamics of Stellar Systems*, ed. A. Hayli, (Dordrecht, Reidel), p. 195.
Hunter, C., 1975*b*. *A. J.*, **80**, 783.
Hunter, C., 1977. *A. J.*, **82**, 271.
Illingworth, G., 1977. *Ap. J.*, **218**, L43.
Innanen, K.A., & Papp, K.A., 1977. *A. J.*, **82**, 322.
Jarvis, B.J., & Freeman, K.C., 1985. *Ap. J.*, **295**, 314.
Jarvis, B.J., & Freeman, K.C., 1985. *Ap. J.*, **295**, 324.
Jeans, J.H., 1915. *M.N.R.A.S.*, **76**, 71.
Jeans, J.H., 1922. *M.N.R.A.S.*, **82**, 122.
Kalnajs, A., 1976. *Ap. J.*, **205**, 751.
King, I.R., 1966. *A. J.*, **71**, 64.
Kormendy, J., & Illingworth, G., 1982. *Ap. J.*, **256**, 460.
Kuzmin, G.G., 1953. *Tartu Astr. Obs. Teated*, **1**.
Kuzmin, G.G., 1956. *Astr. Zh.*, **33**, 27.
Kuzmin, G.G., 1973. in *Dynamics of Galaxies and Clusters*, Materials of the All–Union Conference in Alma Ata, ed. T.B. Omarov, (Alma Ata, Akademiya Nauk Kazakhskoj SSR), p. 71 (translated in this volume, p. 553).
Kuzmin, G.G., and Kutuzov, S.A., 1962. *Bull. Abastumani Astroph. Obs.*, **27**, 82.
Lake, G., 1981*a*. *Ap. J.*, **243**, 111.
Lake, G., 1981*b*. *Ap. J.*, **243**, 121.
Levison, H., & Richstone, D.O., 1985*a*. *Ap. J.*, **295**, 340.
Levison, H., & Richstone, D.O., 1985*b*. *Ap. J.*, **295**, 349.
Levison, H., & Richstone, D.O., 1987. *Ap. J.*, **314**, 476.
Lindblad, B., 1933. *Handbuch der Astrophysik* (Springer, Berlin), Vol. **5**, p. 937.
Liszt, H.S., & Burton, W.B., 1980. *Ap. J.*, **236**, 779.
Lucy, L.B., 1974. *A. J.*, **79**, 745.
Lupton, R., 1985. In IAU Symposium No. 113, *Dynamics of Star Clusters*, eds J. Goodman & P. Hut, (Dordrecht, Reidel), p. 327.
Lynden–Bell, D. 1960, *Ph.D. Thesis*, Cambridge University.
Lynden-Bell, D., 1962*a*. *M.N.R.A.S.*, **123**, 447.
Lynden-Bell, D., 1962*b*. *M.N.R.A.S.*, **124**, 1.
Lynden-Bell, D., 1962*c*. *M.N.R.A.S.*, **124**, 95.
Martinet, L., & Mayer, F., 1975. *Astr. Ap.*, **44**, 45.
Merritt, D.R., 1980. *Ap. J. Suppl.*, **43**, 435.

Merritt, D.R., 1985. *A. J.*, **90**, 1027.
Mestel, L., 1963. *M.N.R.A.S.*, **126**, 553.
Michie, R.W., 1963. *M.N.R.A.S.*, **126**, 466.
Miller, R.H., 1982. *Ap. J.*, **254**, 75.
Miyamoto, M., & Nagai, R., 1975. *Publ. Astr. Soc. Japan*, **27**, 533.
Monet, D.G., Richstone, D.O., & Schechter, P.L., 1981. *Ap. J.*, **245**, 454.
Nagai, R., & Miyamoto, M., 1976. *Publ. Astr. Soc. Japan*, **28**, 1.
Newton, A.J., & Binney, J.J., 1984. *M.N.R.A.S.*, **210**, 711.
Ollongren, A., 1962. *Bull. Astr. Inst. Netherlands.*, **16**, 241.
Oort, J.H., 1928. *Bull. Astr. Inst. Netherlands.*, **4**, 269.
Oort, J.H., 1932. *Bull. Astr. Inst. Netherlands.*, **6**, 249.
Oort, J.H., 1965. In *Galactic Structure*, Stars and Stellar Systems Vol. 5, p. 455, eds A. Blaauw & M. Schmidt, University of Chicago Press.
Osipkov, L.P., 1979. *Pis'ma Astr. Zh.*, **5**, 77 (translated as *Sov. Astr. Lett.*, **5**, 42).
Park, Ch., de Zeeuw, P.T., & Schwarzschild, M., 1987. In preparation.
Petrou, M., 1983*a*. *M.N.R.A.S.*, **202**, 1195.
Petrou, M., 1983*b*. *M.N.R.A.S.*, **202**, 1209.
Pfenniger, D., 1984. *Astr. Ap.*, **141**, 171.
Pfenniger, D., 1986. *Astr. Ap.*, **165**, 74.
Plummer, H.C., 1911. *M.N.R.A.S.*, **71**, 460.
Prendergast, K.H., & Tomer, E., 1970. *A. J.*, **75**, 674.
Richstone, D.O., 1980. *Ap. J.*, **238**, 103.
Richstone, D.O., 1982. *Ap. J.*, **252**, 496.
Richstone, D.O., 1984. *Ap. J.*, **281**, 100.
Richstone, D.O., 1987. This volume, p. 261.
Ruiz, M.T., & Schwarzschild, M., 1976. *Ap. J.*, **207**, 376.
Saaf, A., 1968. *Ap. J.*, **154**, 483.
Satoh, C., 1980. *Publ. Astr. Soc. Japan*, **32**, 41.
Schechter, P.L., 1987. This volume, p. 217.
Schwarzschild, M., 1979. *Ap. J.*, **232**, 236.
Schwarzschild, M., 1982. *Ap. J.*, **263**, 599.
Schwarzschild, M., 1986. *Ap. J.*, **311**, 511.
Stäckel, P., 1890. *Math. Ann.*, **35**, 91.
Statler, T.S., 1987. *Ap. J.*, in press.
Stiavelli, M., and Bertin, G., 1985. *M.N.R.A.S.*, **217**, 735.
Teuben, P. 1987. *M.N.R.A.S.*, in press.
Toomre, A., 1963. *Ap. J.*, **138**, 385.
Toomre, A., 1982. *Ap. J.*, **259**, 535.
Tremaine, S.D., & de Zeeuw, P.T., 1987. This volume, p. 493.
Tremaine, S.D., & Richstone, D.O., 1987. Preprint.
van Albada, T.S., 1982. *M.N.R.A.S.*, **201**, 939.
van Albada, T.S., 1987. This volume, p. 291.
Vandervoort, P.O., 1980*a*. *Ap. J.*, **240**, 478.
Vandervoort, P.O., 1980*b*. *Ap. J.*, **241**, 316.
Vandervoort, P.O., 1984. *Ap. J.*, **287**, 475.
Vandervoort, P.O., & Welty, D.E., 1981. *Ap. J.*, **248**, 504.
Vandervoort, P.O., & Welty, D.E., 1982. *Ap. J.*, **263**, 654.
Vietri, M., 1986. *Ap. J.*, **306**, 48.
Wilson, C.P., 1975. *A. J.*, **80**, 175.

DISCUSSION

Whitmore: I would like to add a note of caution about making the firm conclusion that bulges are completely explained by the oblate rotator models. In a paper by Whitmore, Rubin, and Ford two years ago (1984, *Ap. J.*, **287**, 66), we reanalysed Kormendy and Illingworth's observations along with some of our own. After taking into account several effects, the most important of which is contamination in the spectrum by light from the disk, we find that the spiral bulges fall about 30–40% below the oblate spheroid line. It has been stated during this conference that taking into account the flattening of the bulge by the disk potential makes the agreement nearly perfect. This is only about a 10% effect at best, so I would urge people to keep an open mind about this question.

Illingworth: In response to Brad Whitmore's comment that the bulges of disk galaxies are not fully consistent with being oblate rotators, I would like to note that the detailed comparisons made by Jarvis and Freeman do not support that contention. These authors compared the Kormendy/Illingworth bulge kinematical data with their isotropic dispersion oblate rotationally–flattened models (which include a superimposed disk potential) and found excellent agreement.

Jarvis: In relation to Brad's comment I would like to point out that at least in the case of one of the galaxies that we modeled, NGC7814, there is a negligibly small *luminous* disk. This means that contamination of bulge light by disk light would have been insignificant, leading us to believe that the kinematic observations are reliable and truly reflect the kinematics of the bulge *alone.*

Vietri: I would like to add my voice to the cautionary note rung by Whitmore on the nature of bulges (and small ellipticals). In fact, the only bulge for which we have a good de–projection, M31, has been known for $\sim$ 30 years to be inconsistent with being oblate (Lindblad 1956, *Stockholm Obs. Ann.*, **19**, No. 2). It can only be triaxial, or at most prolate. Furthermore, Zaritsky & Lo (1986, *Ap. J.*, **303**, 66) found variations of ellipticities and major axis position angles in all 12 bulges they observed.

Binney: A comment on the structure of Ellipsoid–Land. I believe there are two non–trivial state holders in Ellipsoid–Land: Statler & Bishop. Though nobody has yet proved this, I believe that if you choose a distribution function at random, you have a finite chance of landing on Bishop's frontier province, and a finite chance of landing in Statler's interior. By contrast, there is no need to obtain a visa for any of Ellipsoid–Land's lower frontier provinces.

de Zeeuw: In the "low countries" the orbital structure that is found in the Stäckel potentials is not generic; a small perturbation will produce a markedly different phase–space structure. Hence it is very likely that this part of the diagram is populated by unstable equilibrium models that are of minor interest only. I would think that the chance of landing on the oblate axisymmetric models is considerably smaller than ending up in the triaxial province.

N-BODY SIMULATIONS OF ELLIPTICAL GALAXIES

T.S. van Albada
Kapteyn Astronomical Institute
P.O. Box 800
9700 AV Groningen
The Netherlands

ABSTRACT. N-body simulations are a useful tool for constructing equilibrium models of elliptical galaxies and for the exploration of their kinematical properties, in particular the tumbling rate of the figure about some axis and the internal streaming. As yet little is known about these, except that there is a large variety of possible equilibrium models. It is easy to make triaxial systems that tumble about the short axis, with internal streaming aligned with the rotation axis of the figure. Attempts to construct systems with figure rotation and internal streaming in opposite directions have not been successful. Use of current simulation codes for detailed studies of particle orbits is limited to several (about 10) dynamical times due to non-physical fluctuations in the force field.

1. INTRODUCTION

Papers presented at this symposium make it clear that our view of the dynamics of elliptical galaxies has changed greatly during the last decade. Much of this is due to the realization that elliptical galaxies are probably triaxial in shape rather than spheroidal. The insignificant rotation of many flattened ellipticals is the main basis for this insight, but a distinct contribution has also been made by N-body simulations. In an early paper on the subject, Binney (1976; see also Aarseth and Binney 1978) showed that violent relaxation from irregular initial conditions does not wipe out all memory of the initial state, and leads to triaxial equilibrium systems. The non-spherical structure is supported by an anisotropic velocity distribution. Later work on dissipationless formation of elliptical galaxies has concentrated on the shape of the radial density profile. Results indicate that violent relaxation starting from cold, irregular initial conditions inescapably leads to a density law that is close to the observed $r^{1/4}$ law. For a review see White (1987 – this volume). It is not surprising that N-body simulations are useful for these problems because they probably are the best available tool for exploring the *terra incognita* of galaxy dynamics. The absence of collisions in simulations of elliptical galaxies is achieved by calculating the force field from a smoothed-out density distribution. Existing codes use either Fourier series or spherical harmonics expansions; see Miller (1978), Wilkinson and James (1982), Villumsen (1982), van Albada (1982), White (1983),

T. de Zeeuw (ed.), Structure and Dynamics of Elliptical Galaxies, 291–299.

McGlynn (1984), and Bontekoe (1987).

N-body simulations can also be used to construct equilibrium models. Usually the latter are also made by violent relaxation from non-equilibrium conditions, but less significance is then given to the particular choice of initial conditions. In this way, Miller (1978), Miller and Smith (1979), and Hohl and Zang (1979) constructed tumbling bars from rotating homogeneous spheres. Several aspects of the dynamics of elliptical galaxies, such as orbital structure (Wilkinson and James 1982) and stability (Merritt 1987 – this volume, Stiavelli 1986), can conveniently be studied with N-body simulations. Here, however, a weakness of N-body simulations appears: dynamical principles are not made explicit, e.g. detailed numerical information on orbits is not sufficient to tell *why* orbits should have a particular shape. Fortunately, more abstract mathematical tools are now available to make up for this deficiency, for example: spectral stellar dynamics (Binney and Spergel 1982), the study of orbits in separable potentials (de Zeeuw 1985), and the orbital density method for constructing equilibrium systems (Schwarzschild 1979).

Somewhat surprisingly, not much work has been done on the exploration of the full range of possible equilibria with N-body simulations. In this paper I will concentrate on work in that area, but first I will discuss briefly some of the limitations of N-body simulations for studies of the dynamics of collisionless systems. (For some comments on N-body methods, see van Albada 1986.)

2. LIMITATIONS OF COLLISIONLESS N-BODY SIMULATIONS

All properties of equilibrium models, and of time dependent systems as well, depend on the microscopic structure of the system, that is on the orbits. It is therefore important to look into the quality of N-body orbits, even if one is only interested in global properties. Since orbits in collisionless systems are smooth, numerical procedures can be much simpler than in collisional systems, but the question remains whether present techniques are perhaps too approximate. The fact that the quality of orbits in collisionless N-body systems is hardly ever discussed may indicate that things are not as good as they should be.

As an example, consider the orbits in the equilibrium model discussed by Norman, May and van Albada (1985). This is a nearly prolate equilibrium configuration ($N = 20000$), with $b/a = 0.65$ and $c/a = 0.64$. Orbits in this model have been classified by monitoring the components of the angular momenta of the individual particles. Nearly all orbits can be classified in this way as belonging to one of the major orbit families (see, however, problem of type switching below). In the above model we find 67% X-tubes, 6% Z-tubes and 27% boxes (including stochastic orbits and misclassified tubes). The large fraction of X-tube orbits reflects the nearly prolate structure of the system. The model is a nicely stationary configuration as far as the global properties are concerned: energy is conserved to within 0.4% over 25 crossing times and the radius of the 90% mass shell changes less than 1% in this period. However, this near constancy of global properties masks considerable changes in the individual particle orbits. Secular changes in the binding energy of individual particles of about 10 - 20% occur over a time interval of 50 crossing times. (Similar variations have been observed by Wilkinson, private communication.) These are caused by 'noise' in the force field, due to the limited number of particles, and by truncation errors that are hard to avoid. The presence of noise implies that orbits close to a boundary between families may pass the boundary

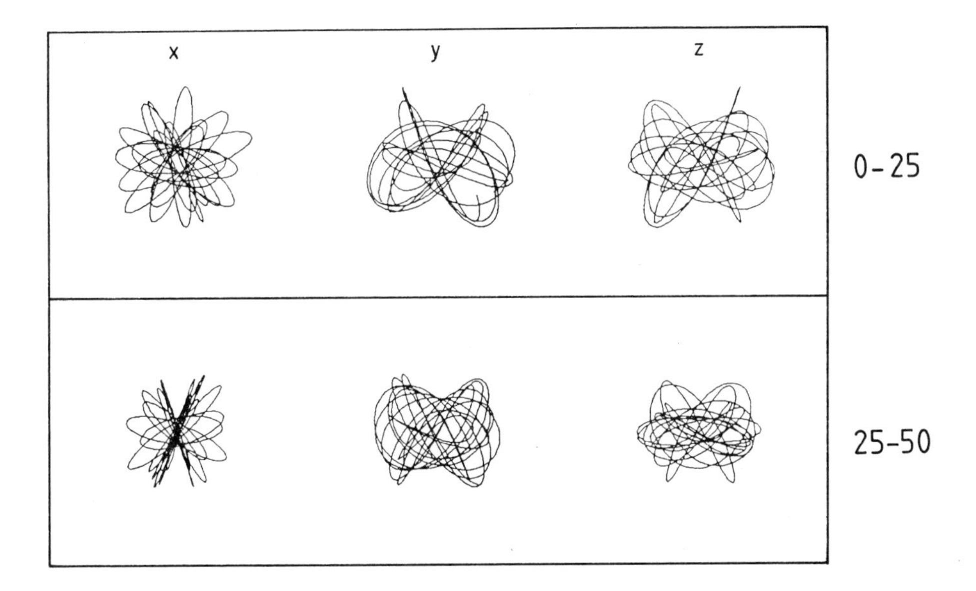

Figure 1. Three projections of an orbit in a triaxial galaxy; from left to right, views along X, Y, and Z axes respectively. Upper panel: first 25 time units; the orbit is a long-axis tube. Lower panel: second 25 time units; somewhere in this time interval the orbit switches from a long-axis tube to a box orbit.

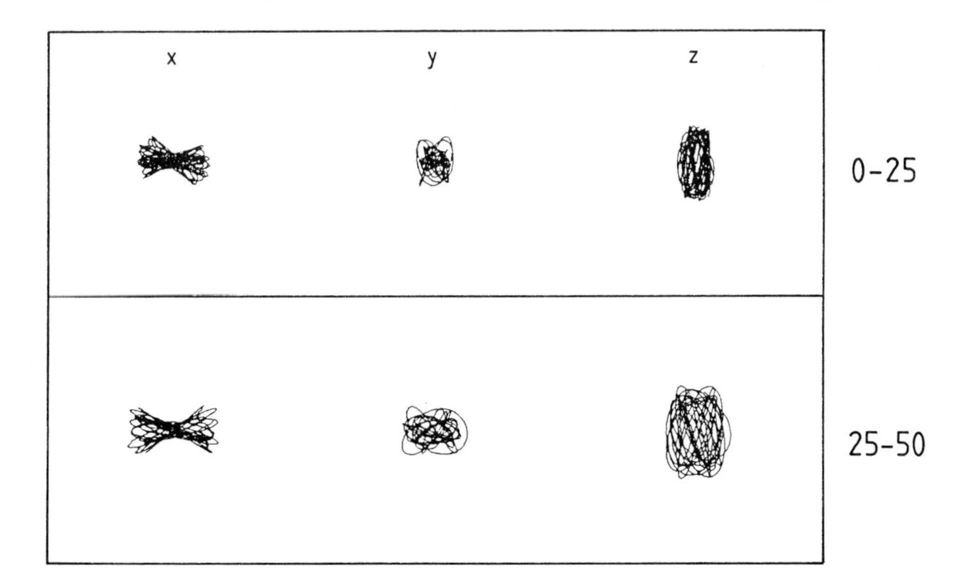

Figure 2. As Figure 1, for box orbit in central region. Note difference in size of orbit for the two time intervals.

and change family. Thus, the stationary state of a model is an equilibrium in a statistical sense only.

These problems are illustrated in Figures 1 and 2, which show projections of orbits for two time intervals (0-25 and 25-50 crossing times after equilibrium has been reached). The orbit of the particle in Figure 1 is an X-tube during the first 25 crossing times, but somewhere during the second time interval it changes to a box orbit. Note that the 'hole' around the X-axis is rather small so it is relatively easy to pass the center on the 'wrong' side. Most tube orbits have larger angular momenta than the case shown in Figure 1 and for those the problem of type switching does not occur. Figure 2 shows an orbit in the central region. This orbit grows in size as time goes on; thus, the orbital energy is not strictly conserved.

One may conclude that the N-body orbits in collisionless systems are basically correct, even though their shapes and sizes vary somewhat with time. The orbital structure of an equilibrium model should be correct for a period of several (about 10) crossing times, and the study of processes like violent relaxation should be no problem. On the other hand, due to the non-physical variations of the properties of individual particle orbits, it is not yet possible to study evolution caused by slowly changing external force fields, or secular internal evolution.

3. INTERNAL STREAMING AND TUMBLING

As demonstrated by Miller (1978), Milller and Smith (1979,1980), Hohl and Zang (1979), and Wilkinson and James (1982), triaxial equilibrium models can be made that tumble about the short axis with constant angular velocity. A large range in tumbling speeds is possible. In addition to tumbling, also called figure rotation, there is internal streaming with respect to the figure in these models. In principle the kinematics of triaxial systems need not be limited to motions about the short axis. By a suitable alignment of the directions of motion of orbits circulating about one of the principal axes (i.e. tube orbits) one can produce internal streaming about an axis that need not coincide with the axis of figure rotation. Both of these axes can, perhaps, have an arbitrary orientation with respect to the principal axes of the system. Many parameters are therefore needed for a complete description of a given configuration. It seems unlikely that all of these can be derived from observation for a given system, even though kinematical models (Binney 1985, Davies and Birkinshaw 1986) show that one can derive interesting constraints.

In this situation N-body simulations can help to restrict the range of possibilities by finding out which configurations can exist as systems in dynamical equilibrium. One way to conduct such experiments is to take a stationary triaxial system in dynamical equilibrium and to induce tumbling by applying an external torque for a limited period of time, allowing the system to relax thereafter. Note that the initial effect of a torque is internal streaming; only when the torque is applied over a sufficiently long time does it induce tumbling. By varying the orientation and strength of the torque one can generate different equilibrium systems. In this way one can determine, for example, whether a triaxial sytem can tumble about its intermediate principal axis or only about the longest or shortest principal axes. (Rotation about the intermediate principal axis is unstable for a solid figure, cf. Goldstein (1980).) Internal streaming about a preferred direction can also be introduced by monitoring the sense of circulation of the particles about the principal axes and by reversing the direction of motion of some of the particles which

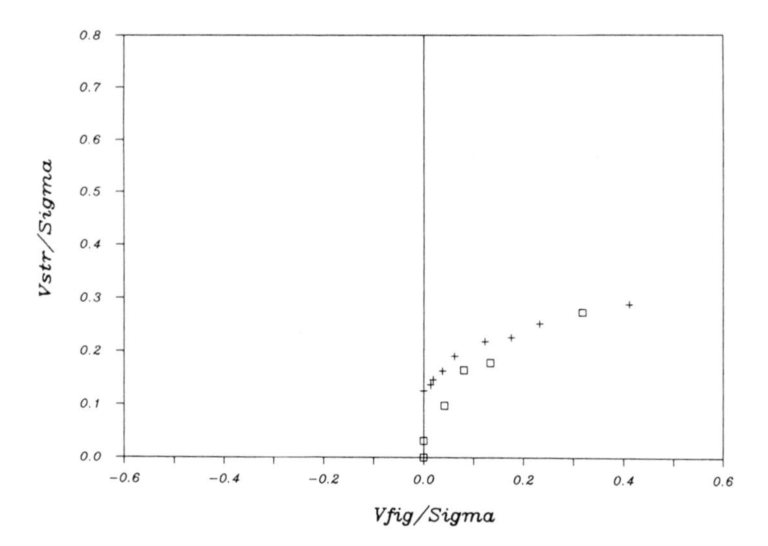

Figure 3. Relation between tumbling and streaming velocities for two sequences of triaxial models discussed in text.

circulate in the 'wrong' sense.

Experiments of this type were carried out by the author for a nearly prolate equilibrium model, similar to the one described in section 2. Only the simple case that both figure rotation and internal streaming are about the short axis was considered. Thus, the final models are tumbling bars. In this sense our models are similar to those of Miller and Smith (1980), who have already emphasized the large variety of possible equilibrium models. Two sequences of tumbling models were made. For one sequence the starting model had no internal streaming (open squares in Figures 3 and 4). For the other sequence the maximum amount of internal streaming was introduced by aligning the senses of circulation of all Z-tube orbits in the initial configuration (crosses in Figures 3 and 4). However, because the fraction of Z-tube orbits in the model is small the net streaming velocity in the starting model is limited. For both sequences application of the external torque not only introduces tumbling about the Z-axis but also causes additional internal streaming.

In Figure 3 the relation between tumbling and streaming in the final models is represented in terms of v_{fig}/σ_0 and v_{str}/σ_0, where v_{fig} and v_{str} are respectively the figure rotation and streaming velocity at the half-mass radius (as seen by an observer in the XY-plane) and σ_0 is the central velocity dispersion in one dimension. Each symbol in this Figure represents one possible equilibrium model. The main conclusions can be summarized as follows:

(*i*) Stationary, i.e. non-tumbling, systems with internal streaming are possible equilibrium configurations (models indicated by cross and square on vertical axis). This may seem a trivial result, but it is nice to see that such models are stable.

(*ii*) All tumbling systems in Figure 3 posses internal streaming. It is not clear whether this result is a general property of triaxial systems. It could perhaps be a result of the special way in which tumbling was introduced in these models, namely

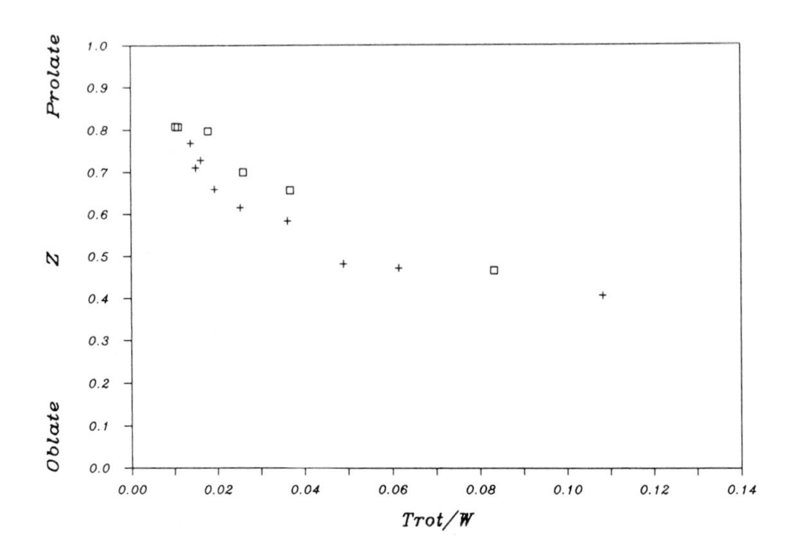

Figure 4. Shape versus amount of rotational kinetic energy, T_{rot}/W, for triaxial models discussed in text. Shape is measured by the triaxiality parameter $Z = (1 - b/a)/(1 - c/a)$. Note increasing oblateness with increasing T_{rot}/W.

through the application of an external torque.
(*iii*) Tumbling in a direction opposite to the internal streaming does not appear possible. This is evident from the absence of models in the left part of Figure 3. Attempts were made to make such models by applying a torque in a direction opposite to the internal streaming, but without success. What happens in such experiments is that the torque reduces the internal streaming; it does not induce tumbling as long as internal streaming is opposed to the torque. This behaviour is quite different from that when torque and internal streaming are parallel. Then the system responds promptly by increasing its tumbling rate. A similar conclusion on the alignment of figure rotation and internal streaming has been reached by Vietri (1986), who tried, without success, to make a specific triaxial model with figure rotation opposed to internal streaming using Schwarzschild's orbital-density method.

Some other features of these models are worth noting. First, a large range in tumbling speeds about the short axis is possible. For the models in Figure 3 the tumbling periods range from 4.2×10^8 to 75×10^8 yrs (adopting a total mass of $2 \times 10^{11} M_\odot$ and a half-mass radius of 4 kpc). Second, the models in Figure 3 vary in shape (see Figure 4); systems become increasingly oblate if the strength of the torque is increased. Finally, there is a tight correlation between the amount of figure rotation and internal streaming for the models in Figure 3. In a general sense this agrees with earlier results from N-body experiments: rapidly rotating systems develop a rapidly tumbling bar (Hohl and Zang 1979, Miller and Smith 1979), while slowly tumbling triaxial systems have a modest amount of internal streaming (Wilkinson and James 1982). On the other hand, the tightness of the correlation in Figure 3 is no doubt a consequence of the use of a single (nearly prolate) starting configuration, with at most a small amount of internal streaming.

More oblate models have a larger fraction of short-axis tube orbits (Statler 1986), and by aligning the directions of motion of these orbits one can probably make a non-tumbling system with a fairly large amount of internal streaming.

The above discussion shows that N-body simulations form a versatile tool to study the range in stable equilibria of elliptical galaxies, but present work still has the character of an initial exploration. One should keep in mind that real elliptical galaxies form a subset of the stable N-body systems due to additional constraints acting during the formation phase; the physics of galaxy formation determines the details of the collapse and subsequent violent relaxation phase, and thereby specifies the resulting equilibrium system.

I am indebted to Andrew May for carrying out most of the analysis presented in section 2 and to Linda Sparke for comments.

REFERENCES

Aarseth, S.J. and Binney, J.J., 1978. *Mon. Not. Roy. Astr. Soc.*, **185**, 227.
Binney, J., 1976. *Mon. Not. Roy. Astr. Soc.*, **177**, 19.
Binney, J., 1985. *Mon. Not. Roy. Astr. Soc.*, **212**, 767.
Binney, J. and Spergel, D., 1982. *Astrophys. J.*, **252**, 308.
Bontekoe, T.R., 1987. Ph.D. Thesis Groningen University, in preparation.
Davies, R.L. and Birkinshaw, M., 1986. *Astrophys. J.*, **303**, L45.
de Zeeuw, T., 1985. *Mon. Not. Roy. Astr. Soc.*, **216**, 273.
Goldstein, H., 1980. *Classical Mechanics*, 2nd*ed.*, p. 210, Addison-Wesley.
Hohl, F., and Zang, T.A., 1979. *Astron. J.*, **84**, 585.
McGlynn, T.A., 1984. *Astrophys. J.*, **281**, 13.
Merritt, D., 1987. This volume, p. 315.
Miller, R.H., 1978. *Astrophys. J.*, **223**, 122.
Miller, R.H. and Smith, B.F., 1979. *Astrophys. J.*, **227**, 407.
Miller, R.H. and Smith, B.F., 1980. *Astrophys. J.*, **235**, 793.
Norman C.A., May, A. and van Albada, T.S., 1985. *Astrophys. J.*, **296**, 20.
Schwarzschild, M., 1979. *Astrophys. J.*, **232**, 236.
Statler, T.S., 1986. Ph.D. Thesis Princeton University.
Stiavelli, M., 1986. Ph.D. Thesis Scuola Normale Superiore, Pisa.
van Albada, T.S., 1982. *Mon. Not. Roy. Astr. Soc.*, **201**, 939.
van Albada, T.S., 1986. In: *The Use of Supercomputers in Stellar Dynamics*, eds. P. Hut and S. McMillan, Springer Verlag. p. 23.
Vietri, M., 1986. *Astrophys. J.*, **306**, 48.
Villumsen, J.V., 1982. *Mon. Not. Roy. Astr. Soc.*, **199**, 171.
White, S.D.M., 1983. *Astrophys. J.*, **274**, 53.
White, S.D.M., 1987. This volume, p. 339.
Wilkinson, A. and James, R.A., 1982. *Mon. Not. Roy. Astr. Soc.*, **199**, 171.

DISCUSSION

Quinn: If streaming around a figure were present and comparable to the orbital velocities of stars in the galaxy, then when the galaxy appears to be nearly round on the sky we would expect to see a large V/σ for a small ellipticity ϵ. There appear to be very few large V/σ small ϵ galaxies among observed ellipticals. This may then constrain the contribution of streaming velocities to V.

van Albada: If I understand correctly, the situation you describe is that of a nearly prolate galaxy, with internal streaming about the short axis, which is seen end–on. Such an orientation has a rather low probability. One should probably take a better look at the statistics to see whether the observations give useful constraints.

Vietri: (i) I do not understand why you cannot set up counter–streaming motions for sufficiently low values of Ω_p. Suppose in fact $\Omega_p = 0$. Then you can get stars streaming in either direction. Now give a gentle tug to the figure. I would expect the model to show some degree of counterstreaming. Can you say why you do not find such solutions? (ii) Gravitational torques work more subtly than one would think naively. Tremaine and Weinberg (1984, *Mon. Not. R. astr. Soc.*, **209**, 729) showed that the average star is not spun up at any given time. Only orbits momentarily in resonance are spun up. This is different from what you do, since you spin up all orbits, at the same time. Can you say whether your results would be effected by including this effect?

van Albada: Perhaps the N–body simulations are too rough. Tumbling can only be measured when it exceeds about one revolution per 10^{10} years. As regards your second point, the temporal behavior of the force seen by a star in the torque experiments is indeed quite different from that caused by a satellite in circular orbit. I don' t know what would happen in the latter case.

Bishop: For the N–body simulation, how do you specify the initial conditions and how well does the distribution function retain its initial form?

van Albada: The equilibrium model I showed was generated from non–equilibrium initial conditions, as described in Norman, May and van Albada (1985). Once equilibrium has been reached, the shape of the distribution of individual particle energies is well conserved.

Lake: Freeman's models (1966, *Mon. Not. R. astr. Soc.*, **133**, 47; **134**, 1, 15) counterstream. They certainly aren't stable, but one could try them as an initial state.

van Albada: Yes, that could be tried.

White: I found your pictures of the long term evolution of orbits to be very instructive but I think your conclusions may be too pessimistic. It is important to bear in mind that present N–body techniques were not set up primarily to attack problems of this kind and we may be able to do significantly better by some careful attention to the algorithms.

van Albada: I agree that that should be done.

Schwarzschild: The last speaker is, I feel, too terribly modest regarding the accuracy and completeness he has reached. From his results as they stand one can, I think, draw some quite secure general conclusions. Specifically I would conclude from his results—bolstered by some recent work of Mario Vietri—that if one observes an elliptical to rotate with a substantial velocity, say $V/\sigma > 0.2$, then the figure rotation, if any, must be in the same sense as the observed net streaming, i.e. no strong counterstreaming.

Spergel: While N–body techniques may be robust when there are not many resonant orbits, resonant orbits might introduce spurious stochasticity. This is not a problem for loop orbits. However, when there are many resonant tori, the fluctuations in the N–body code might scatter the orbits from one torus to a nearby torus. The Liapunov coefficients for particles with finite separation might be quite large. Energy conservation is not a good diagnostic of this problem, since the neighboring tori may have similar energies but visit very different regions of phase space.

van Albada: The point I wanted to make is that even the necessary condition that individual particle energies are conserved, is not fulfilled in present N–body simulations. I agree that this condition would not be a sufficient one for strict equilibrium.

Statler: I hope that Dave Spergel's comment does not discourage people from looking at the stellar orbits in N–body models. We can use Schwarzschild's method and linear programming to find what kind of self-consistent solutions can *exist*, but we need the N–body models to tell us which of them can actually be *made*. So I would like to encourage the N–body experts, especially those who simulate mergers, to do what Althea Wilkinson has done, and calculate distribution functions for their models, or at least find the mass fractions in the major orbit families.

Petrou: From the computation of orbits in two–dimensional potentials I have done, I know that in order to study high order resonances and their effect you need at least 7–8 significant figure accuracy in the energy conservation. With a 0.4% accuracy in the energy you can only see the effect of the very simple resonances.

Gerhard: (i) Can I interpret your results to the effect that the observed rotation velocity in an edge–on bulge at given radius R and highest slit position z, gives an upper limit to the figure rotation velocity at R, and hence a lower limit to the corotation radius, if this bulge were triaxial? In this case, corotation in NGC 4565 would be at least at $\sim$ 15 kpc (using data from Kormendy and Illingworth 1982, *Astrophys. J.*, **265**, 637). (ii) Concerning the role of resonant orbits: Is the orbit sample you showed representative of all orbits in this system, and do you have a surface of section for the potential of this model to see whether resonant orbits are important in its phase-space and/or preferentially populated?

van Albada: (i) Rejecting the possibility of streaming and tumbling in opposite directions, I would indeed interpret any observed rotation as an upper limit to the figure rotation. (ii) I mainly showed orbits with "problems," leaving out well-behaved tube orbits. No further analysis of the orbits has been made.

Tjeerd van Albada.

STABILITY OF ELLIPTICAL GALAXIES. THEORETICAL ASPECTS

V.L. Polyachenko
Astrosovet
Ul. Pyatnitskaya 48
109017 Moscow Sh–17
U.S.S.R.

1. INTRODUCTION

Stars, which contain more than 90% of the visible matter of the Universe, are organized into systems comprising a hierarchy of different masses and scales: from galactic stellar clusters to clusters of galaxies themselves. Most of the systems in this hierarchic sequence are collisionless: the characteristic time between collisions is much more than typical dynamical times (e.g., the period of oscillations). Various conditions of formation of collisionless gravitating systems leave traces on their main characteristic parameters—such as the angular momentum, the degree of anisotropy, the kind of dependence of the stellar density on coordinates, and so on. The lifetime of a system with a certain set of parameters depends essentially on whether these parameters belong to a region in which the system is stable, or not. If they do, the system may then exist practically without changes for an arbitrarily long time (in terms of the dynamical time scale); if they do not, it must quickly adjust to another, stable state.

Thus, the main question for each special system of whether it is stable or in a state of dynamical relaxation is answered by asking whether the parameters of the system belong to stable or unstable regions. Determination of the boundaries between these two regions is the main problem for stability theory. The importance of the problem is sufficiently evident when studying various collisionless systems, in particular, elliptical galaxies, which have determined in many respects the development of contemporary astrophysics. It is important to note that the requirement of stability imposes actually very essential limits on permissible stationary models of collisionless gravitating systems (it follows from the results below; for details see Fridman & Polyachenko 1984, hereafter FP).

2. REVIEW OF EARLIER WORK

Many of the main results on the stability of collisionless ellipsoidal systems are already published in English and so are easily accessible to Western readers (first of all, see FP). I shall briefly remark on these results and the history of the problem at the beginning of the talk, and devote the main part of the talk to some new results which are not yet available in English translation.

T. de Zeeuw (ed.), Structure and Dynamics of Elliptical Galaxies, 301–314.

Studies of the stability of stellar systems already have a sufficiently interesting history of their own. But when we discuss the origins of such work, we should keep in mind the theory of figures of equilibrium and the stability of different geometrical shapes of an incompressible fluid (sphere, ellipsoid, ring, etc.) which was created by many outstanding scientists. The results of their research have been summarized in a large number of reviews and monographs. It is clear, however, that the incompressible fluid approximation (and the corresponding mathematical formalism), generally speaking, is not suitable for describing collisionless systems.

It is quite natural that some concepts and methods used in the equilibrium and stability theory of gaseous gravitating systems proved to be relevant for describing stellar clusters. We should mention here, first of all, the concepts connected with the name of Jeans, who was the first to investigate the stability of a homogeneous distribution of compressible matter: the Jeans instability, the Jeans critical wavelength, etc. The similarity between stellar and gaseous systems was confirmed in the pioneering investigations of collective effects in a collisionless gravitating medium carried out by Antonov (1960), Lynden-Bell (1962), and others. For example, the search for stationary states of stellar systems is equivalent in some cases to solving the corresponding problem in the equilibrium theory of gaseous spheres. Moreover, it turns out that the problem of the stability of such collisionless systems can also be reduced to the stability of certain analogous hydrodynamical systems. However, it should be noted that this "hydrodynamical analogy" (strictly proved by Antonov 1962, and Lynden-Bell & Sanitt 1969) occurs only for a very special class of *isotropic* collisionless systems with distribution functions depending on the total energy of a star only. In such a case it is really possible to draw an analogy between the characteristics of the two systems under consideration (for instance, the stellar velocity dispersion then is analogous to the thermal particle velocity of the gaseous system, etc.).

But the uniqueness of collisionless systems consists just in the possibility of *anisotropy*: the velocity dispersions along different directions may differ strongly from each other. Anisotropy often arises in a natural way—for instance, at the origin of a system; it is typical for various real systems (stellar clusters, galaxies, clusters of galaxies). At the same time, a tradition of applying the hydrodynamical analogy had led to the opinion that stellar clusters are as stable or even more stable than gaseous systems. However, this is correct only for isotropic distributions where the "collisionlessness" reveals itself in the behaviour of perturbations on very small scales only. In real (anisotropic) collisionless systems, not only "old" instabilities (characteristic also of the gaseous medium—for instance, the Jeans instability) are strongly modified, but also new types of instabilities appear which are due to the anisotropy of the velocity distribution of particles, the occurrence of beams, etc.

In the late sixties, after the discovery of quasars, there arose the idea to construct models of spherical stellar clusters with a large redshift. In this connection Mikhailovsky, Fridman and others have carried out a series of investigations on the study of the stability of stellar systems with purely circular orbits (Michailovskii, Fridman & Epel'baum, 1970; Fridman & Shukhman 1972; Shukhman 1973; Synakh, Fridman & Shukhman 1972). Those systems were apparently the first examples of essentially anisotropic systems whose stability was investigated in detail and for which in some cases exact spectra of oscillations were obtained. It follows from these papers that stellar clusters with circular orbits and an outwardly decreasing density are probably stable.

The situation with respect to stability theory of stellar systems in the early seventies may be summarized as follows (restricting ourselves to the example of spherical systems). By that time the stability of systems with isotropic distribution functions had been proved and, besides, it had been shown that systems with circular orbits were apparently also stable (excluding a few unphysical ones). Thus there emerged the opinion that all spherical systems should be stable. Such an opinion looked natural, especially if we recall the traditional view that the spherical shape is the most stable one.

However, in 1972 Schukhman and I noticed that systems with nearly radial orbits should be unstable and stated the following problem: to derive the critical anisotropy separating stable and unstable clusters (Polyachenko & Shukhman, 1972; Polyachenko 1972). That problem was resolved in our subsequent papers, the main results of which were presented in FP. The instability is due to the lack of velocity dispersion in transversal directions and, consequently, has the Jeans nature. If we consider any narrow cone with its center in the center of a sphere and imagine that at the initial time we have slightly compressed it into a new cone by bringing its generatrices together, the inevitability of further Jeans contraction is evident if one takes into account that the particles cannot escape from the perturbation region. A study of the stability of systems with such an anisotropy is important since, on the one hand, it arises naturally at the center of stellar systems, and, on the other hand, stellar orbits in spherical and elliptical galaxies are actually strongly radially elongated.

Due to the importance of these problems I shall return to them below in order to consider the radial instability in a new light and to present more exact stability computations for certain realistic models of spherical stellar clusters, and to outline possible further work in this field.

All the problems mentioned above are concerned with the stability of stationary spherical collisionless systems. But it is no less interesting to ask what role possible instabilities may play in the evolution of systems initially far from equilibrium. For example, one of the most important problems in the theory of galaxy or cluster formation is that of the collapse of a spherical cloud of collisionless particles. In the case of a cloud that is sufficiently cold at the initial moment, practically all the energy released during the contraction converts into radial motion. Consequently, a strong radial orbit instability may develop. As a matter of fact, the conclusion that an ellipsoidal deformation should arise during the collapse of such a cloud was trivial after the "bar-like" instability of corresponding stationary systems were proved; it was made earlier in the author's paper (Polyachenko 1981), dealing, mainly, with instabilities of stationary systems (some more details are presented in Polyachenko 1985). Dr. D. Merritt has recently determined the minimum dispersion of initial transverse velocities in a collisionless cloud that is necessary to ensure its stability (see his paper in this volume).

Then I shall consider the stability of spherical systems with the opposite kind of anisotropy (*i.e.*, with nearly circular orbits). Investigations of such systems were at the beginning mainly of methodological interest. However, the situation has now changed. For instance, it is important to study the processes occurring in the stellar component of dense galactic nuclei, collapsed objects (e.g., black holes), whose vicinities should, evidently, be rich with stars having nearly-circular orbits: stars with elongated orbits having been "absorbed" by a central body. For these systems it was possible to obtain the dispersion relation describing short wavelength perturbations (of the type of the well-known Lin-Shu dispersion relation for disk-like galaxies). The physical mechanism of instability which is evident for systems with nearly radial orbits (of course, of the Jeans nature) is not yet so evident for the nearly–circular orbits systems. One may give some arguments that this instability has a "beam" nature.

The above listed results were obtained by the same method, using the reduction procedure which will now be proved. This procedure reduces the stability problem for arbitrary spherical collisionless systems to a more simple problem on the stability of the corresponding cylindrical systems relative to flute–like perturbations (when the component of the wave vector along the axis of the cylinder equals zero).

3. REDUCTION PROCEDURE

The kinetic equation in the variables r, θ, φ, v_r, $v_\perp$ and α (where r, θ, φ are the standard spatial spherical coordinates, v_r is the radial velocity component, $v_\perp^2 = v_\theta^2 + v_\varphi^2$ with v_θ, v_φ are the velocity components in the θ and φ directions, and $\alpha = \arctan(v_\varphi/v_\theta)$) has the form:

$$\frac{\partial f}{\partial t} + \frac{v_\perp}{r}\left[\cos\alpha\frac{\partial f}{\partial\theta} + \frac{\sin\alpha}{\sin\theta}\frac{\partial f}{\partial\varphi} - \sin\alpha\cot\theta\frac{\partial f}{\partial\alpha}\right]$$
$$+ v_r\left(\frac{\partial f}{\partial r} - \frac{v_\perp}{r}\frac{\partial f}{\partial v_\perp}\right) + \left(\frac{{v_\perp}^2}{r} - \frac{\partial\Phi}{\partial r}\right)\frac{\partial f}{\partial v_r}$$
$$- \frac{1}{r}\left(\cos\alpha\frac{\partial\Phi}{\partial\theta} + \frac{\sin\alpha}{\sin\theta}\frac{\partial\Phi}{\partial\varphi}\right)\frac{\partial f}{\partial v_\perp} - \frac{1}{rv_\perp}\left(\sin\alpha\frac{\partial\Phi}{\partial\theta} - \frac{\cos\alpha}{\sin\theta}\frac{\partial\Phi}{\partial\varphi}\right)\frac{\partial f}{\partial\alpha} = 0.$$

Consider small perturbations of the sphere with the equilibrium distribution function $f_0 = f_0(E, L^2)$, where E is the stellar energy, $E = \frac{1}{2}({v_r}^2 + {v_\varphi}^2) + \Phi_0(r)$, with Φ_0 the gravitational potential, $L^2 = r^2 v_\perp^2$ is the square of the stellar angular momentum. Such a system is evidently non–rotating. Then the linearized kinetic equation can be presented as

$$\frac{\partial f_1}{\partial t} + \frac{v_\perp}{r}\hat{L}f_1 + \hat{D}f_1 = \frac{\partial\Phi_1}{\partial r}\frac{\partial f_0}{\partial E}v_r + \frac{1}{r}\hat{L}\Phi_1\left(\frac{\partial f_0}{\partial E}v_\perp + \frac{\partial f_0}{\partial L^2}2v_\perp r^2\right), \tag{1}$$

where we introduced the operators:

$$\hat{L} = \cos\alpha\frac{\partial}{\partial\theta} + \frac{\sin\alpha}{\sin\theta}\frac{\partial}{\partial\varphi} - \sin\alpha\cos\theta\frac{\partial}{\partial\alpha}, \tag{2}$$

$$\hat{D} = v_r\frac{\partial}{\partial r} - \frac{v_r v_\perp}{r}\frac{\partial}{\partial v_\perp} + \left(\frac{{v_\perp}^2}{r} - \frac{\partial\Phi_0}{\partial r}\right)\frac{\partial}{\partial v_r}. \tag{3}$$

The operator $\hat{L}$ has the standard form of the infinitesimal rotation about the y–axis (being expressed through the Euler angles).[1] In the case of full spherical

[1] Possibly, it was first noted, in the aspect interesting to us, in the unpublished work: M.Ya. Pal'chik, A.Z. Patashinskii, V.K. Pinus and Yu.G. Epel'baum, *Institut Yudarnoi iziki Sibir Otd. Akad. Nauk SSSR*, 99–100, Novosibirsk, 1970 (In Russian).

symmetry, the angular part of the potential may be separated in a form proportional to the spherical harmonics: $\Phi_1 \sim Y_\ell^m(\theta,\varphi)$. The following calculations may also be considered as the formal justification of this natural statement.

It is natural to solve the kinetic equation in a system of coordinates in which the operator $\hat{L}$ is diagonal, *i.e.*, corresponds to a rotation around the z'-axis of the turned coordinate system. In this system we present the perturbation of the distribution function in the form of the expansion:

$$f_1 = \sum_s f_s(r, v_r, v_\perp) T^\ell_{ms}(\varphi', \theta', \alpha'), \tag{4}$$

where the functions

$$T^\ell_{ms}(\varphi_1, \theta, \varphi_2) = e^{-im\varphi_1 - is\varphi_2} P^\ell_{ms}(\cos\theta), \tag{5}$$

are introduced, and the $P^\ell_{ms}(\cos\theta)$ are the three–index functions (Vilenkin, 1968; see also FP), in particular the $P^\ell_{m0}(\cos\theta)$ functions, identical (except for constant coefficients) to the associated Legendre functions. Therefore it is convenient to write the potential in the form

$$\Phi_1 = \chi(r,t) T^\ell_{m0}(\varphi, \theta, \alpha), \tag{6}$$

or, in the "primed" system,

$$\Phi_1 = \chi(r,t) \sum a^\ell_s T^\ell_{ms}(\varphi', \theta', \alpha'), \tag{7}$$

where the a^ℓ_s are the coefficients of the rotation transforming the y-axis into the z-axis. Thus, in the "primed" system we have an independent equation for each function of the expansion (4). Taking into account the relation $\hat{L} f_s = is f_s$, these equations can be written as

$$\frac{\partial f_s}{\partial t} + \frac{v_\perp}{r} is f_s + \hat{D} f_s = \frac{\partial \Phi_s}{\partial r}\frac{\partial f_0}{\partial v_r} + \frac{is\Phi_s}{r}\frac{\partial f_0}{\partial v_\varphi}. \tag{8}$$

Equation (8) can be presented in a form identical to the standard equation for the response of a cylindrical (or disk) system f_s to the flute–like perturbation of the potential, $\Phi_s(r,\varphi) = \Phi_s(r) e^{is\varphi}$ (where r and φ are the corresponding cylindrical coordinates); if we denote for uniformity $v_\varphi \equiv v_\perp$, then

$$\begin{aligned} \frac{\partial f_s}{\partial t} + v_r \frac{\partial f_s}{\partial r} + \frac{v_\varphi}{r}\frac{\partial f_s}{\partial \varphi} + \left(\frac{{v_\varphi}^2}{r} - \frac{\partial \Phi_0}{\partial r}\right)\frac{\partial f_s}{\partial v_r} - \frac{v_r v_\varphi}{r}\frac{\partial f_s}{\partial v_\varphi} = \\ \frac{\partial \Phi_s}{\partial r}\frac{\partial f_0}{\partial v_r} + \frac{1}{r}\frac{\partial \Phi_s}{\partial \varphi}\frac{\partial f_0}{\partial v_\varphi}. \end{aligned} \tag{9}$$

Thus, for solving the initial "spherical" problem, it is sufficient to find solutions of the corresponding "cylindrical" problem (9) for all integers s with $-\ell \le s \le \ell$ that are even or odd (see below).

All the above forms the first part of the reduction procedure considered in this section. The second part of this procedure consists in the recipe for calculating the full density perturbation ρ_1.

Let us suppose that equations (9) for f_s are solved. Then for the calculation of ρ_1, it is convenient to turn in the expansion (4) again to the initial ("non–primed") system, which may be done by means of the formula

$$T^{\ell}_{ms}(\varphi', \theta', \alpha') = \sum_{s'} T^{\ell}_{ms'}(\varphi, \theta, \alpha)\bar{a}^{\ell}_{s'}. \tag{10}$$

Since the expression for the density perturbation $\rho_1 = \int f_1 v_\perp dv_\perp dv_r d\alpha$ includes the integration over the angular variable α, only one term (corresponding to $s' = 0$) remains from each sum (10), and we obtain the desired formula in the form

$$\rho_1^{(\ell)} = T^{\ell}_{m0}(\varphi, \theta, \alpha) \sum_s \alpha_s^{(\ell)} \int f_s(r, v_r, v_\perp) v_\perp dv_\perp dv_r, \tag{11}$$

where we have defined

$$\alpha_s^{(\ell)} \equiv |P^s_\ell(0)|^2. \tag{12}$$

In order to derive, e.g., the dispersion relation (in the self–consistent problem) we need only to solve the Poisson equation. This gives the potential that corresponds to the density ρ_1. A comparison with the initial expression will then give the dispersion relation.

I shall illustrate this reduction procedure with a few examples.

4. SPHERES WITH RADIAL ORBITS

Let us derive, with the help of the reduction procedure, the equations describing the Jeans instability of a spherical system with radial orbits only. In this case, it is possible to derive a simpler set of equations. Since the equilibrium distribution function has the form

$$f_0 = \delta(L^2)\phi_0(E) = \frac{1}{r^2}\delta(v_\perp^2)\phi_0(E), \tag{13}$$

i.e., contains a delta–function (E is the energy of radial motion, $E = v_r^2/2 + \Phi_0(r)$), the perturbation of the distribution function may be written[2]

$$f_s = a_s \delta(v_\perp{}^2) + b_s v_\perp \delta'(v_\perp^2), \tag{14}$$

where δ' is the derivative of the δ-function, and the a_s and b_s functions depend only on r and v_r (and, of course, on the time t): $a_s = a_s(r, v_r, t)$, $b_s = b_s(r, v_r, t)$. For these functions we obtain the set of equations:

$$\frac{\partial a}{\partial t} + v_r \frac{\partial a}{\partial r} + \frac{2v_r}{r} a - \frac{\partial \Phi_0}{\partial r}\frac{\partial a}{\partial v_r} - \frac{1}{r}\frac{\partial b}{\partial \varphi} = \frac{1}{r^2}\frac{\partial \Phi_s}{\partial r}\frac{\partial \phi_0}{\partial v_r}, \tag{15}$$

[2] An analogous procedure of δ–expansions was first applied by Michailovskii, Fridman and Epel'baum (1970).

$$\frac{\partial b}{\partial t} + v_r \frac{\partial b}{\partial r} + \frac{3v_r}{r} b - \frac{\partial \Phi_0}{\partial r} \frac{\partial b}{\partial v_r} = \frac{2is\Phi_s}{r^3} \phi_0. \tag{16}$$

The density perturbation is

$$\rho_1 = 4\pi \sum_s \alpha_s^{(\ell)} \int dv_r a_s. \tag{17}$$

If we substitute $b_s = 2isF_1$ and take into account that

$$\sum_s \alpha_s^{(\ell)} = 1, \qquad \sum_s \alpha_s^{(\ell)} s^2 = \frac{1}{2}\ell(\ell+1),$$

then we obtain the following set of equations for the functions F_1, $a_1 \equiv \sum_s \alpha_s^{\ell} a_s$ and $\chi(r)$ [with $\Phi_s = \chi(r) e^{is\varphi}$ and $\hat{D} \equiv v_r \partial/\partial r - (\partial\Phi_0/\partial r)\partial/\partial v_r$]:

$$\frac{\partial a_1}{\partial t} + \hat{D} a_1 + \frac{2v_r}{r} a_1 + \frac{\ell(\ell+1)}{r} F_1 = \frac{\partial \chi}{\partial r} \frac{1}{r^2} \frac{\partial \phi_0}{\partial v_r}, \tag{18}$$

$$\frac{\partial F_1}{\partial t} + \hat{D} F_1 + \frac{3v_r}{r} F_1 = \frac{\chi}{r^2} \phi_0, \tag{19}$$

$$\frac{1}{r^2} \frac{d}{dr}\left(r^2 \frac{d\chi}{dr}\right) - \frac{\ell(\ell+1)}{r^2} \chi = 4\pi G \int a_1 dv_r, \tag{20}$$

coinciding with the set in FP (derived in another way).

Further standard transformation of the set (18)–(20) consists of the natural substitution $a_1 r^2 = a$ and $F_1 r^3 = F$ in the expansions

$$\begin{aligned} a &= a_+ \delta(v_r - v_0) + a_- \delta(v_r + v_0) + b_+ \delta'(v_r - v_0) + b_- \delta'(v_r + v_0), \\ F &= F_+ \delta(v_r - v_0) + F_- \delta(v_r + v_0), \end{aligned} \tag{21}$$

where $v_0 \equiv \sqrt{2E_0 - 2\Phi_0(r)}$ and $\phi_0(E) = \int \phi_0(E_0)\delta(E - E_0) dE_0$. This gives

$$\frac{\partial a_\pm}{\partial t} \pm v_0 \frac{\partial a_\pm}{\partial r} \pm v_0' a_\pm - \frac{\partial b_\pm}{\partial r} + \frac{\ell(\ell+1)}{r^2} F_\pm = 0, \tag{22}$$

$$\frac{\partial b_\pm}{\partial t} \pm v_0 \frac{\partial b_\pm}{\partial r} \pm 2v_0' b_\pm = \frac{1}{2v_0} \frac{\partial \chi}{\partial r}, \tag{23}$$

$$\frac{\partial F_\pm}{\partial t} \pm v_0 \frac{\partial F_\pm}{\partial r} \pm v_0' F_\pm = \frac{1}{2v_0} \chi. \tag{24}$$

In particular, for the WKBJ–perturbations with $\ell \gg r(\partial/\partial r)$ the following symmetric set of equations can be obtained from (22)–(24) (Antonov, 1973):

$$\hat{D}_\pm \xi_\pm = \frac{\varsigma_\pm \pm v_0 \xi_\pm}{r}, \tag{25}$$

$$\hat{D}_+\varsigma_+ = \hat{D}_-\varsigma_- = 2\pi G r \int dE_0 \rho_{E_0}(\xi_+ + \xi_-), \tag{26}$$

where $\hat{D}_\pm = (\partial/\partial t) \pm v_0(\partial/\partial r) = (d/dt)_\pm$ are the operators of the derivative over time along the unperturbed radial orbits of the particle "flow" with energy E_0, $\rho_{E_o} = \phi_0(E_0)/v_0 r^2$, $\varsigma_\pm = -v_0 F_\pm$ and $\xi_\pm = \frac{1}{\ell^2} r v_0 a_\pm$. For perturbations of the form $\phi \sim Y_\ell^m(\theta,\varphi)\chi(r)$ the quantities $\xi_\pm$ and $\varsigma_\pm$ may be thought of as the corresponding linear displacements in the equatorial plane in the azimuth φ and the angular momentum (rv_φ) of the particle. Proceeding from the system (25)–(26), instability is proved by constructing the Lyapunov functional (Antonov, 1973; in English see FP).

The principal point in the derivation of the set of equations (25)–(26) was the use of the WKBJ–approximation in the angular variables [the condition $\ell \gg r(\partial/\partial r)$]. Such an approximation would be justified if the eigenfunctions of such a type exist. Strictly speaking this point requires an additional justification. One may recall, for example, that eigenfunctions of such a kind are absent in the case of the exactly solved problem of perturbations of uniformly rotating disks with arbitrary elliptical orbits. On the qualitative level, the basis for the existence of such solutions in the case of systems with radial orbits requires a repetition of the arguments similar to those given above under the qualitative description of the instability mechanism.[3]

Proofs of the instability based on the application of equations (25)–(26) describing the short–wavelength perturbations say nothing about how systems with nearly radial orbits evolve in reality. For instance, the answer to the question of whether the initial spherical shape of the system changes depends on the stability or instability of just the largest scale modes. The answer to this question was obtained in the paper of the author mentioned above (Polyachenko 1981). It was shown that an initially spherical system with radial orbits converts in its non–linear evolution into an ellipsoidal one. At the same time in that paper (and also in those of our works where the boundary between stable and unstable systems was found) the radial–orbit instability itself was finally proved (without any approximations).

The problem of finding the eigenmodes ($\sim e^{-i\omega t}$) in analytic form is in the general case rather complicated, even for the simplified set (25)–(26). However, in the particular case (important from the point of view of astronomical applications) when the system of particles is in a given external field (created, for example, by a more massive central condensation or by the "halo" which itself is not subjected to perturbations of such a type),[4] we can advance rather far in the way of an analytical solution. The point is that in this case the first approximation is evidently the perturbation of the radial–orbit system in the given external potential, which can easily be determined. The gravitational interactions of particles with each other (their self–gravitation which causes the instability) may be accounted for in the next approximation. Thus, there appears the possibility of a good perturbation theory

[3] Possibly all the systems may be separated into two classes: 1) those with nearly circular orbits (and "joining" to them) and 2) systems "joining" to those with radial orbits. For systems of the first class, the WKBJ–eigenfunctions with $\ell \ll r\partial/\partial r$ occur, and for systems of the second class, the eigenfunctions with $\ell \gg r\partial/\partial r$.

[4] For example, the role of a "halo" may be played by a massive spherical system with an isotropic velocity distribution function of stars.

in the small ratio $M/M_h \ll 1$ where M_h is the "halo" mass, and M is the mass of the radial–orbit system. Let us consider perturbations for which particles from one radial orbit (more exactly, a certain "flow" of particles with a fixed energy E) pass onto a nearby identical radial orbit. It is obvious that without taking into account the self–gravitation such perturbations correspond to a certain new equilibrium, *i.e.*, in this approximation their frequency $\omega = 0$. But the consistency condition for the solutions of the first and second (with self–gravity) approximations of the set (25)–(26) leads to the integral equation

$$\omega^2 F(E) = -4\pi G \int_0^{r_E} \frac{dr}{v_0(r,E)} r^2 \int dE_0 \rho_{E_0}(r) \frac{F(E_0)}{F_1(E_0)}, \tag{27}$$

where

$$F_1 \equiv \int_0^{r_E} r^2 \frac{dr}{v_0(r,E)}, \qquad F(E) = \omega^2 F_1(E)\xi_E.$$

In this case, the displacement $\xi(r,E)$ is proportional to the radius: $\xi = r\xi_E$, with r_E the maximal radius for a particle with the energy E: $\Phi_0(r_E) = E$. Thus, $F(E)$ (or ξ_E) must be a solution of the integral equation (27), and the square of the eigenfrequency, ω^2, is the eigenvalue of the corresponding integral operator.

Equation (27) can be easily transformed to the standard form of an integral equation with a symmetric kernel

$$-\omega^2 \bar{F}(E) = \int_{\Phi_0(0)}^{E_{max}} \bar{K}(E,E_0)\bar{F}(E_0), \tag{28}$$

where we introduced the new function

$$\bar{F}(E) = F(E)\sqrt{f(E)}, \qquad \left(f(E) \equiv \frac{2\pi G \Phi_0(E)}{F_1(E)}\right), \tag{29}$$

and the symmetric kernel is

$$\bar{K}(E,E_0) = \sqrt{f(E)f(E_0)} \begin{cases} \displaystyle\int_0^{r_{E_0}} \frac{dr}{\sqrt{E-\Phi_0(r)}\sqrt{E_0-\Phi_0(r)}}, & E_0 < E, \\ \displaystyle\int_0^{r_E} \frac{dr}{\sqrt{E-\Phi_0(r)}\sqrt{E_0-\Phi_0(r)}}, & E_0 > E. \end{cases} \tag{30}$$

In particular, it follows from (28)–(30) that for sign–definite eigenfunctions (for instance, $\bar{F}(E) \geq 0$ everywhere), $\omega^2 < 0$ and $\omega^2 \simeq -GM/R^3$, *i.e.*, instability with a growth rate of such an order takes place. For the simplest case of a homogeneous halo, $\bar{F} = \sqrt{4G}\sqrt{\phi_0(E)/E}\,F(E)$ and

$$\bar{K}(E,E_0) = 4G\sqrt{\frac{\phi_0(E)}{E}}\sqrt{\frac{\phi_0(E_0)}{E_0}} \begin{cases} \displaystyle\int_0^{\frac{\pi}{2}} \frac{d\varphi}{\sqrt{E-E_0\sin^2\varphi}}, & E_0 < E\,, \\ \displaystyle\int_0^{\frac{\pi}{2}} \frac{d\varphi}{\sqrt{E_0-E\sin^2\varphi}}, & E_0 > E, \end{cases} \tag{31}$$

where the potential of a halo was taken as $\Phi_h = r^2/2$. The integral equation (30) with kernel (31) was solved numerically for systems with various distribution functions $\phi_o(E)$. For instance, for $\phi_o(E) \propto E(E_{max} - E)^2$ the growth rate of the most unstable mode (which corresponds to sign–constant ξ_E with the maximum at $E = 0$) proved to be $\gamma \simeq 2.05\sqrt{GM/R^3}$.

The integral equation (28) can be solved easily numerically for an arbitrary potential of a "halo" $\Phi_k(r)$; when the external field is that of a point mass, the integral equation (28) reduces to the following form (assuming $GM = 1$):

$$-\omega^2 \bar{\chi}(x) = \int\limits_{y_0}^{\infty} \bar{K}(x,y)\bar{\chi}(y)dy, \tag{32}$$

where

$$\bar{K}(x,y) = \frac{8\pi G}{Cx^2y^2}\sqrt{\phi_0(x)x^{9/2}}\sqrt{\phi_0(y)y^{9/2}} \begin{cases} \dfrac{x^2+y^2}{2xy}\ln\dfrac{x+y}{y-x} - 1, & y > x, \\ \dfrac{x^2+y^2}{2xy}\ln\dfrac{x+y}{x-y} - 1, & x > y, \end{cases}$$

and

$$\bar{\chi}(x) = \sqrt{\frac{8\pi G}{C}}\sqrt{\phi_0(x)x^{9/2}}\chi(x),$$

with

$$C = \int\limits_{-\infty}^{\infty} \frac{t^6 dt}{(1+t^2)^4} = \frac{15\pi}{16}.$$

5. SPHERICAL SYSTEMS WITH A FINITE DISPERSION OF TRANSVERSE VELOCITIES

Stability criteria for spherical stellar systems with general distribution functions were considered first in the papers by the author and Shukhman (1981; the main results are described in detail in FP). In these works the supposition was made that the stability (or instability) of systems with radially elongated orbits is determined by the value of the parameter of "global anisotropy" $\xi = 2T_r/T_\perp$ where T_r and $T_\perp$ are the total kinetic energies corresponding to the radial and transversal degrees of freedom. For several series of distribution functions which differed greatly from each other, critical values of $\xi = \xi_c$ proved to lie in a rather narrow interval: between 1.4 and 2.0. The most interesting result is the fact that the anisotropy necessary for instability need not be too large: already for $\xi \gtrsim 2$ the system is usually unstable. (By the way, one sometimes considers models with a mean anisotropy $\xi \sim 5$–8, for instance, in some models of the stellar surroundings of massive black holes).

However, it is obvious that for a system immersed in a "halo" or containing a massive central cluster, *i.e.*, in essence, compound systems, the criterion needs some improvement: this requires introducing the energy of interaction between an inner cluster (or a "halo") and surrounding stars in which the instability may develop.

It is analogous to the necessity of accounting for the energy of interaction between disk stars and a "halo" in the Ostriker–Peebles criterion. Taking into account the Jeans nature of the instability, one may postulate a generalized instability criterion of the form:

$$T_r > \xi(T_\perp/2) + \eta U, \tag{33}$$

where U is the modulus of the interaction energy between the stars in a "corona" and a central condensation or a halo (a more accurate definition of which is given below), ξ is the critical anisotropy determined earlier, and η is a number of order unity. Inequality (33) is, in essence, the condition for the transverse Jeans instability $\omega_0^2 \simeq 4\pi G\bar{\rho} > \kappa_\perp^2 \overline{v_\perp^2}$, slightly transformed by using the virial theorem, for the largest scale mode ($\kappa \sim 1/R$), which may be written in energetic terms as

$$T_\perp < \alpha|U_1|, \tag{34}$$

where U_1 is the gravitational energy of the system under consideration ("corona"), and α is some constant. The gravitational interaction between the "corona" and the external gravitational field having been accounted for, the virial theorem gives

$$2(T_r + T_\perp) = |U_1| + \int \rho(r\frac{d\Phi_0}{dr})\, dV, \tag{35}$$

where ρ is the mass density of the system, Φ_0 is the external potential, and $dV = 4\pi r^2 dr$. U in the inequality (33) is, strictly speaking, simply the magnitude of the second term on the right hand side of the last formula (the "external" part of the virial): $U = \int \rho r(d\Phi_0/dr)dV$. Particularly, for a homogeneous "halo", writing $\Phi_0 = \Omega_h^2 r^2/2$ (with $\Omega_h^2 = const.$) we have $U = 2\int \rho\Phi_0 dV$, and for a point mass in the center, when $\Phi_0 = -GM/r$, $U = \int \rho|\Phi_0|dV$. Expressing $|U_1|$ in equation (35) in terms of T_r, $T_\perp$ and U, we obtain

$$2T_r > (\frac{1}{\alpha} - 2)T_\perp + U, \tag{36}$$

i.e., the inequality (33), with $\xi = (1/\alpha) - 2$, and $\eta = 1$.

It should be kept in mind that the simplest criteria for compound systems, such as inequality (33), operate with the kinetic energies T_r and $T_\perp$ of the "corona" stars only, *i.e.*, by our definition, of the component of a whole system that is responsible for the instability: only this component responds effectively to the most important perturbations (the fundamental mode, first of all—see footnote 3). If a finite—not necessarily massive—subsystem with nearly radial orbits may be distinguished in the whole system, it may be unstable, while the mean anisitropy of the whole compound system may be negligible (e.g., the ratio $2T_r^{tot}/T_\perp^{tot}$ for the total system happens to be arbitrarily close to unity). This follows clearly from physical arguments, and, formally, from the inequality (34). In this case, the growth rate must be small, being of order of the Jeans frequency ω_0 determined by the "corona" matter density.

We always have adopted just this interpretation of the instability conditions. This may be seen, for instance, in our critical comments on the paper by Duncan & Wheeler (1980), who examined models of spherical systems with large anisotropy (see FP, Vol. 2, p. 151–153). Given a sufficiently small mass for the central anisotropic part of the system, the anisotropy of the whole system may be very small, $2T_r^{tot}/T_\perp^{tot} \sim 1$. Nevertheless, perturbations localized near the centre (with a radial scale of order the size Δr of the anisotropic region) must be unstable.[5]

[5] This is valid with some reservations—see FP.

This conclusion has recently been confirmed by Merritt and Aguilar (1985).

Thanks to Dr. Merritt, the paper by Palmer and Papaloizou presented at this Symposium became available to me. In essence, their result, which seems plausible, is that small angular momentum stars can form a distinct unstable subsystem—in the sense discussed above—if the distribution function $f_0(E, L^2)$ of the whole system becomes unbounded as $L \to 0$, for all E. However, the numerical results for the generalized polytropes given in this paper become invalid for the most interesting—fundamental—mode, as they lie far beyond the framework of the approximations used. The stability of these models was studied earlier by us (Polyachenko 1983; in more detail, see FP): we showed that a pronounced instability, which develops in a few crossing times and leads to an easily perceptible ellipsoidal deformation of the whole system (and not only a small portion of it), occurs for anisotropies $2T_r/T_\perp \gtrsim 1.4$. This result was obtained both by solving the corresponding matrix eigenvalue problem, and by N-body simulations. The latter have recently been repeated by Barnes, Goodman and Hut (1986), using a different technique, and showing the same result.

It should be remarked that the last paper—as well as the paper by Palmer and Papaloizou—contains some critical comments resulting from the same misunderstanding, discussed above. These criticisms arise from a somewhat too blunt application of the instability criterion for the parameter $\xi = 2T_r^{tot}/T_\perp^{tot}$, which is assumed to be absolutely rigorous. Of course, this criterion is not valid in some cases, which we have already seen from general, reasonable considerations. One should realize that even physically more relevant criteria, like (33), may have only a preliminary, and very approximate character. They are useful, in our opinion, as a first step. Rigorous criteria are still absent. It is clear, however, that, when present, such criteria will be very complex, so that their applicability to the general case will still be questionable. Our criteria are simple, and cannot be "strongly" incorrect, since they are based on a simple and fairly obvious physical picture of the instability.

A way for isolating an "unstable subsystem" may be evident in some cases from the structure of the system considered. In other cases, this can be done only with some degree of uncertainty. However, for systems with a sufficiently smooth phase-space structure, one may apply the instability criterion ($\xi \gtrsim 2$) directly for the value $\xi = 2T_r^{tot}/T_\perp^{tot}$ without any substantial error.

Concerning real star clusters and galaxies, they often have a strongly inhomogeneous structure, with a high concentration of matter in the center. The velocity distribution of stars in the central region is usually assumed to be isotropic. Such an isotropic region plays—roughly speaking—the role of a central mass which (cf. equation [33]) exerts a stabilizing influence on perturbations in an anisotropic "corona".[6]

---

[6] One may consider an isotropic (or any other) subsystem as an immovable halo when the stellar density response of this subsystem is negligible in comparison with that of an "unstable subsystem" of stars with elongated orbits. It is easy to write down the corresponding condition suitable for rough estimates. Conceptually, this problem is quite similar to that of the derivation of the condition under which it is possible to neglect perturbations in the spherical and other non-flat components of spiral galaxies when studying waves in the disk component (see, e.g., Marochnik & Suchkov 1974). By analogy, here we may write down a similar condition in the form: $(\rho^{(h)}/\rho^{(c)}) \cdot (c_\perp^{2(c)}/c_\perp^{2(h)}) \ll 1$, where $\rho^{(h)}$ and $\rho^{(c)}$ are the densities of a halo

It is natural in such a situation to expect different critical values of anisotropy than for most of the models studied by us earlier. The simplest and rather realistic density distribution gives the well-known Plummer–Schuster law $\rho(r) = \rho(0)/(1 + r^2/r_0^2)^{5/2}$. The corresponding phase models were suggested by Kuzmin and Veltmann (1968). Their stability was investigated by us earlier, but the results presented, for example, in FP, had a rather large dispersion. The value $\xi_c \sim 1.6$ for three series of these models was obtained by means of a too-great extrapolation beyond the "large" growth rate values which were determined with confidence. At the same time, a smaller extrapolation, but taking into account rather small growth rates determined with less confidence, had led to values of ξ_c closer to $\xi_c \sim 2$ (however, with a large dispersion). The restriction on ξ_c from above in the form $\xi_c \lesssim 1.95$ arose just as a consequence of the latter determinations. Here we present the results defined more precisely (to be published in more detail elsewhere). It turned out that all the ξ_c values lie within the narrow interval between ~ 2.05 and 2.10 for all physically acceptable models by Kuzmin–Veltmann (with phase densities positive everywhere).

6. SPHERICAL SYSTEMS WITH NEARLY CIRCULAR ORBITS

In the above we considered the systems with radially elongated orbits. The opposite type of anisotropy is that of clusters with nearly circular orbits. For such systems, one can obtain the dispersion relation describing short wavelength perturbations (like the well-known dispersion relation by Lin–Shu for disk galaxies):

$$ux = 1 - \sum_k \alpha_k^\ell \frac{\nu_k \pi}{\sin \nu_k \pi} F_x(\nu_k), \tag{37}$$

where

$$\alpha_k^\ell = |P_k^\ell(0)|^2 = (\ell + k)!(\ell - k)!/\left[(\frac{\ell + k}{2})!(\frac{\ell - k}{2})!\,2^\ell\right]^2,$$

and P_k^ℓ are the associated Legendre functions, $\nu_k = \nu - \mu k/2$, $\nu = \omega/\kappa$, ω is the eigenfrequency, κ the epicyclic frequency, $\mu = 2\Omega/\kappa$, Ω is the angular velocity of the circular orbit, ρ_0 is the density, $u = \kappa^2/4\pi G\rho_0 = 4/(4 - \mu^2)$ for the self-consistent equilibrium,

$$F_x(\nu) = \frac{1}{2\pi} \int_{-\pi}^{\pi} e^{-x(1+\cos s)} \cos \nu s ds,$$

$x = k^2 c^2/\kappa^2$, c is the radial velocity dispersion, and k is the radial wavelength number. In the derivation of equation (37) the distribution function over v_r was assumed to be Maxwellian. Branches of oscillations described by equation (37) for various ℓ resemble very much the cyclotron branches of plasma oscillations. There is a separate set of branches for each pair of (ℓ, μ).

---

and a "corona", and $c_\perp^{(h)}$, $c_\perp^{(c)}$ are the transverse velocity dispersions. One might expect that this condition would be weakened if the "halo" occupied only a small central region.

Considering systems with small values of x and $\mu \simeq 4/3$ for $\ell = 3$ one can find the small root: $\nu^2 \simeq cx(\mu - 4/3)$, with the constant c equal to $\frac{25}{192}$, which implies instability if $\mu < 4/3$. Analogous resonant conditions are fulfilled also for other values of ℓ. Moreover, generally speaking, layers with arbitrary μ are unstable for one or another ℓ. Thus, there exist mechanisms for isotropisation not only in systems with an increasing function $\Omega(r)$ when $\mu < 1$ (this was known earlier) but also in normal systems with decreasing $\Omega(r)$.

REFERENCES

Antonov, V.A., 1960. *Astr. Zh.*, **37**, 918 [1961. *Soviet Astronomy*, **4**, 859].

Antonov, V.A., 1962. *Vestn. Leningrad Univ.*, **19**, 96.

Antonov, V.A., 1973. in *Dynamics of Galaxies and Star Clusters*, p. 139, ed. Omarov, T.B., Nauka (Alma-Ata) (in Russian).

Barnes, J., Goodman, J., & Hut, P., 1986. *Astrophys. J.*, **300**, 112.

Duncan, M.J., & Wheeler, J.C., 1980. *Astrophys. J.*, **237**, L27.

Fridman, A.M., & Polyachenko, V.L. 1984. *Physics of Gravitating Systems*, Vols. 1 and 2, Springer Verlag (FP).

Fridman, A.M., & Shukhman, I.G., 1972. *Dokl. Akad. Nauk SSSR*, **202**, 67 [*Sov. Phys. Dokl.*, **17**, 44].

Kuzmin, G.G., & Veltmann, Yu I.K. 1968. *Publ. Tartu Astroph. Obs.*, **36**, 470.

Lynden-Bell, D., 1962. *Mon. Not. R. astr. Soc.*, **124**, 279.

Lynden-Bell, D., & Sanitt, N., 1969. *Mon. Not. R. astr. Soc.*, **143**, 167.

Marochnik, L.S., & Suchkov, A.A., 1974. *Usp. Fiz. Nauk*, **112**, 275 [*Sov. Phys. Usp.*, **17**, 85].

Merritt, D.R., & Aguilar, L.A., 1985. *Mon. Not. R. astr. Soc.*, **217**, 787.

Michailovskii, A.B., Fridman, A.M., & Epel'baum, Ya.G. 1970. *Zh. Eksp. Teor. Fiz.*, **59**, 1608 [1971. *Sov. Phys. JETP* **32**, 878].

Polyachenko, V.L., 1972. Kandidatskaja dissertatsija, Leningrad.

Polyachenko, V.L., 1981. *Pis'ma Astr. Zh.*, **7**, 142 [*Soviet Astr. Lett.*, **7**, 79].

Polyachenko, V.L., 1985. *Astron. Tsirkular*, **1405**, 1.

Polyachenko, V.L., & Shukhman, I.G., 1972. *Inst. Zemn. Magn. Ionosf. Rasprostr. Radiovoln Sibir. Otd. Akad. Nauk SSSR*, 2–72, Irkutsk.

Polyachenko, V.L., & Shukhman, I.G., 1981. *Astr. Zh.*, **58**, 933 [*Soviet Astronomy*, **25**, 533].

Shukhman, I.G., 1973. *Astr. Zh.*, **50**, 651 [*Soviet Astronomy*, **17**, 415].

Synakh, V.S., Fridman, A.M., & Shukhman, I.G., 1971. *Dokl. Akad. Nauk SSSR*, **201**(4), 827 [1972. *Sov. Phys. Dokl.*, **16**(12), 1062].

Vilenkin, N.Ya. 1968. *Special Functions and Group Theory*, Nauka, Moscow (in Russian).

STABILITY OF ELLIPTICAL GALAXIES. NUMERICAL EXPERIMENTS

David Merritt
Canadian Institute for Theoretical Astrophysics
University of Toronto
Toronto, Ontario M5S 1A1
Canada

1. INTRODUCTION

The idea that dynamical instabilities might play an important role in determining the equilibrium structure of elliptical galaxies is a startling one, especially to those of us who are accustomed to associating instabilities with rapidly-rotating systems like disk galaxies. The shock is even greater when we learn that these instabilities have been taken seriously by workers in the Soviet Union for a long time. As Dr. Polyachenko points out in his talk, instabilities affecting spherical, non-rotating galaxies were being discussed by Soviet astronomers as long ago as 1972. Much of this work has recently become more accessible through the publication of an English-language version of Fridman and Polyachenko's monograph, *Physics of Gravitating Systems* (Fridman and Polyachenko 1984). In the West, it appears that only two people were prescient enough to systematically test the stability of spherical models before learning of the Soviet work (Hénon 1973; Barnes 1985). In particular, Barnes discovered independently that a spherical system composed of predominantly radial orbits evolves rapidly into a bar. Subsequent work has demonstrated that even some mildly anisotropic models can be unstable in this way.

Why did this important class of instabilities remain undetected for so long? There are probably two reasons. First, it is easy to construct stable equilibrium models for elliptical (or at least spherical) galaxies. Indeed, a number of mathematical proofs, beginning with Antonov's remarkable papers of 1960 and 1962, demonstrated (though not in a physically very intuitive way) that isotropic spherical systems are guaranteed to be stable as long as their distribution functions satisfy certain reasonable constraints. By contrast, stable disk models are notoriously difficult to construct. The second reason is more subtle, and, in retrospect, rather ironic. When an N-body experimenter wants to make a strongly anisotropic galaxy model, he generally does so by relaxing an out-of-equilibrium set of initial coordinates and velocities chosen in such a way as to give him roughly the final state he is looking for. This procedure is obviously much easier than constructing an exact equilibrium solution to the collisionless Boltzmann equation. But a model galaxy formed in this way will never be unstable, because if it were, the instability would have acted during the relaxation to produce a different (and stable) final state.

T. de Zeeuw (ed.), Structure and Dynamics of Elliptical Galaxies, 315–329.

What is ironic is that it is clear, from a careful re-reading of past work on galaxy formation, that instabilities very similar to the ones now known to afflict equilibrium systems were active in many of the published collapse simulations. It takes only a little imagination to wonder whether instabilities of the sort talked about by Polyachenko and Barnes might have played an active role during the formation of real galaxies.

This talk is divided into four parts. The first part summarizes what has been learned so far about the stability and instability of spherical equilibrium models. Unfortunately, nothing definite is known yet about the stability of triaxial models. The second part discusses how dynamical instabilities might be used to constrain the dynamics of particular well-observed galaxies. The third part describes some preliminary work on the question of whether instabilities could have played an active role during galaxy formation. The fourth part presents an efficient new algorithm for testing the stability of spherical and triaxial models.

The total number of published papers on this topic is still quite small, and this talk should be seen less as a review then as an introduction to a rapidly developing field.

2. STABILITY OF SPHERICAL MODELS

Orbits in spherical potentials are characterized by four integrals of motion, the energy E and the three components of the angular momentum $\mathbf{J}$. Since $\mathbf{J}$ is conserved, every orbit lies in a plane. According to Jeans's theorem, equilibrium models can be constructed from distribution functions f that depend on the phase-space coordinates $\mathbf{r}$ and $\mathbf{v}$ through E and $\mathbf{J}$ alone. If the model is to exhibit the same symmetries as the potential, then f must be a function of r, which means that $f = f(E, J^2)$. In fact this is not quite correct: one can imagine adding to f a term that is odd in $\mathbf{J}$ and contributes nothing to the total density. Such a term effectively specifies what fraction of the stars revolve clockwise or counterclockwise on each orbit. In what follows, however, only non-rotating models will be considered.

Consider first the case of velocity isotropy, $f = f(E)$. Most of what we know about the stability of isotropic systems was first discovered by Antonov (1960, 1962). Antonov considered systems for which $df/dE < 0$, and found a necessary and sufficient condition for stability in the form of a complicated variational principle. He went on to derive a number of simpler, *sufficient* conditions for stability. The most important of these are:

I. A spherical system with $f = f(E)$ and $df/dE < 0$ is stable to all non-radial (i.e. non-spherically-symmetric) perturbations;

II. A spherical system with $f = f(E)$, $df/dE < 0$ and $d^3\rho/d\Phi^3 \leq 0$ is stable to all perturbations. Here ρ and Φ are the density and potential, respectively.

Antonov was able to show that the family of "stellar dynamical polytropes" defined by

$$f(E) \propto (E_0 - E)^{n-3/2}, \quad E \leq E_0, \tag{1}$$

is stable for $n \geq 3/2$, i.e., for all values of n such that $df/dE \leq 0$. Antonov's theorems may be used to verify that many of the isotropic models that resemble real galaxies, such as Hénon's (1959) isochrone or the isotropic Michie-King (1966) models, are also stable.

Antonov's proofs leave open the question of the stability of systems whose distribution functions do not satisfy conditions I or II. The first systematic search for unstable, isotropic models was carried out by Hénon (1973), who used a spherical N-body code to check the stability of polytropes with $n < 3/2$ (i.e., $df/dE > 0$). He found that all were stable (at least to the spherical modes permitted by his computer code), with the possible exception of the $n = 1/2$ model, which appeared to oscillate at a level slightly above the noise. (The case $n = 1/2$ turns out to be a peculiar one, since all the stars in this model have exactly the same energy; smaller values of n are not allowed.) Barnes, Goodman and Hut (1986) later showed that these polytropic models are stable to non-spherical modes as well.

As of now, Hénon's is the *only* candidate for an unstable isotropic system. It is certainly possible that other unstable isotropic models exist, especially when $df/dE > 0$. On the other hand, it seems very unlikely that any isotropic model resembling a real galaxy will ever be found to be unstable. This is because an observed density profile $\rho(r)$ implies a unique isotropic distribution function $f(E)$, and the isotropic models corresponding to many of the standard galaxy surface density profiles are known to be stable.

The situation is very different, and much more interesting, for *anisotropic* systems. A number of attempts have been made to generalize Antonov's sufficient stability criteria to systems with $f = f(E, J^2)$. These proofs (e.g. Doremus and Feix 1973; Gillon, Doremus and Baumann 1976; Sygnet et al. 1984) are mostly still controversial, and at least one (Gillon, Doremus and Baumann 1976) appears to be contradicted by the numerical experiments described below. A more fruitful approach has been to search for particular, unstable models. Hénon (1973) tested the radial stability of the "anisotropic polytropes" defined by

$$f(E, J^2) \propto J^{2m}(E_0 - E)^{n-3/2}. \tag{2}$$

Models generated from equation (2) have velocity ellipsoids with fixed axis ratios $\sigma_r^2/\sigma_t^2 = (1+m)^{-1}$, where σ_r and σ_t are the radial and tangential components of the velocity dispersion tensor. Hénon found that the oscillatory instability that seemed to be present in the isotropic model with $n = 1/2$ became stronger as the velocity ellipsoid was made more prolate.

At about the same time that Hénon published his paper on instabilities in polytropic models, a number of Soviet workers had begun to apply techniques of perturbation theory to spherical systems. So far, two new classes of instabilities have been identified in this way. One class, associated with systems dominated by circular orbits, is probably not very relevant to elliptical galaxies. The other class, affecting systems dominated by radial orbits, almost certainly is.

2.1. Spherical Systems Dominated by Nearly-Circular Orbits

A disk galaxy in which all the stars move along exactly circular orbits is violently unstable to small-scale axisymmetric modes, i.e. clumping in rings (Toomre 1964). It is natural to ask whether a spherical galaxy composed of circular orbits is similarly unstable. The simple answer is "no": unlike a disk, the gravitational force in a sphere is due entirely to the interior mass, and this fact is sufficient to insure that a sphere will not clump into shells (Bisnovatyi-Kogan, Zel'dovich and Fridman 1968). Non-spherical perturbations can be unstable, however. For a spherical system composed purely of circular orbits, a sufficient condition for instability to non-spherically-symmetric modes is

$$\frac{d\rho}{dr} > 0 \tag{3}$$

(Fridman and Polyachenko 1984, Vol. 1, p. 179). Barnes, Goodman and Hut (1986) explored this type of instability numerically using N-body models generated from the anisotropic polytrope distribution function (2). For $m > 0$, equation (2) implies a preponderance of circular orbits, and a density profile that peaks at nonzero r. In the limit of large m, all the matter lies on a thin shell. Oscillatory instabilities are easy to understand for such an extreme system, since all the orbits have nearly the same period, and perturbations tend to recur naturally after a fraction of an orbital period. Barnes, Goodman and Hut (1986) found that all models with $m \gtrsim 1/2$ exhibited quadropole oscillations with rapidly increasing amplitude, and achieved stability only after a substantial rearrangement of matter.

Tangentially anisotropic models generated from equation (2) could never be mistaken for real galaxies because of their peculiar density profiles. Recently Polyachenko (1985) has suggested that systems with $d\rho/dr < 0$ might exhibit similar instabilities. This hypothesis is an important one to check. At present, however, there is no numerical evidence to suggest that this class of instability is relevant to systems that look like real elliptical galaxies.

2.2. Spherical Systems Dominated by Eccentric Orbits

If a circular-orbit model seems an unlikely one for an early-type galaxy, the opposite extreme, a galaxy consisting largely of radial orbits, seems much more natural: after all, collapse from cold initial conditions tends to build in very elongated orbits, at least at large radii. Antonov suggested as early as 1973 that a purely radial-orbit model would be unstable to clumping of particles around any radius vector. There is presently some uncertainty about the validity of Antonov's proof. Nevertheless the instability exists. It was first verified numerically by Polyachenko (1981), who used a direct-summation N-body code to follow the evolution of a 200-particle radial-orbit model with density profile $\rho \propto r^{-2}$. The initially equilibrium model evolved rapidly into a bar. The instability was rediscovered independently by Barnes (1985), who used a more sophisticated mean-field N-body code to test the stability of the anisotropic polytropes of equation (2).

Figure 1 illustrates the development of the radial-orbit instability in a model with an initial surface-density profile that is essentially identical to a de Vaucouleurs $r^{1/4}$ law; the initial velocities were chosen from a distribution function that gives increasingly radial orbits at large radii, similar to the models produced in collapse simulations (Merritt and Aguilar 1985).

What is the physical mechanism behind the radial-orbit instability? One simple interpretation (Fridman and Polyachenko 1984, Vol. 2, p. 148; Barnes, Goodman and Hut 1986) is based on the well-known Jeans (1929) instability of a uniform, self-gravitating medium. Jeans showed that any such medium is unstable to gravitational clumping on length scales

$$\lambda > \lambda_J = \sqrt{\frac{\pi}{G\rho}}\sigma \tag{4}$$

where ρ and σ are the density and the (isotropic) velocity dispersion. The *radial* velocity dispersion in a radially-anisotropic system of size R and mean density ρ must be of order $\sigma_r^2 \approx G\rho R^2$ to satisfy the virial theorem. Since the *tangential* velocity dispersion in such a system is much lower, equation (4) suggests that clumping might be expected in a tangential direction, i.e. in a cone around any radius vector. Since the growth time for the Jeans instability is not strongly dependent on λ, one might expect an unstable model to evolve on all azimuthal length scales greater than λ_J simultaneously. This means that small-scale clumping should not be sufficient to stabilize a model before a large-scale, or bar, mode has permanently destroyed its spherical symmetry. These predictions are consistent with the numerical experiments.

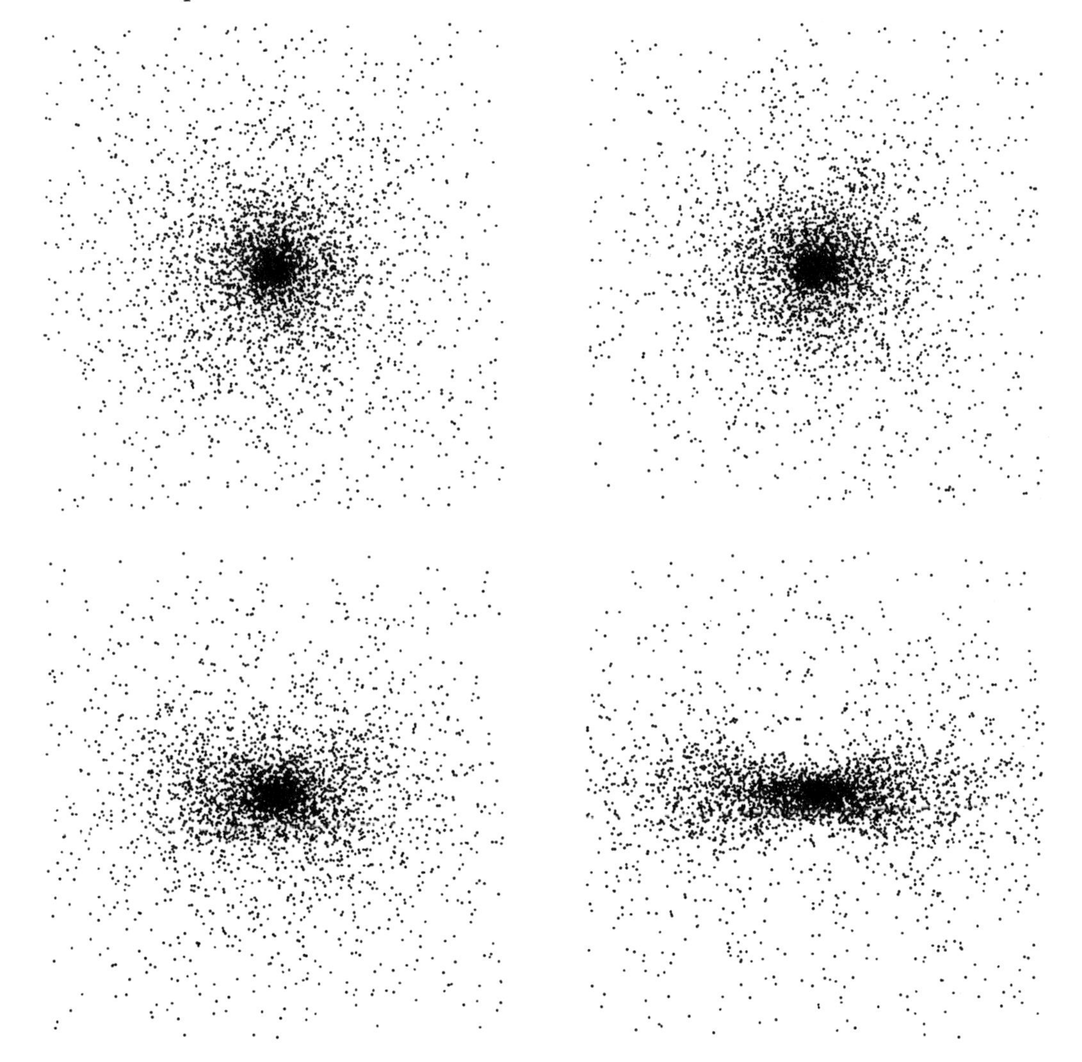

Figure 1. Radial-orbit instability in an anisotropic spherical model. The initial density and velocity dispersion satisfy equations (5) and (6), with anisotropy radius $r_a = 0.1 r_0$. This is an axisymmetric, mean-field N-body calculation with 5000 particles.

On the other hand, it would be a mistake to associate this class of instability too closely with the simple instability discussed by Jeans. Any system dominated by radial orbits must be very inhomogeneous. Since the time scale for growth of the Jeans instability and the orbital time scale are both roughly equal to $(G\rho)^{-1/2}$, an unstable mode would scarcely begin to grow before the particles contributing to it had moved far away from their initial positions, to regions of very different density and velocity dispersion. Furthermore, since the unperturbed trajectories are not closed, there is no guarantee that a perturbation that is initially confined to a small angular region will not quickly be spread over a much larger one.

A better description of the physical mechanism underlying the radial-orbit instability is given, in a slightly different context, by Lynden-Bell (1979). In a spherical potential, every orbit is a rosette, with an angle between apocenters that lies between π and 2π. For very eccentric orbits, this angle is close to π, and orbital precession is very slow (cf. Figure 2a). Now suppose that the potential is modified by the addition of a weak bar-like perturbation. If the minimum of the bar potential lies ahead of the orbit, the net torque will be in the same direction as the orbital motion. What Lynden-Bell showed was that for certain orbits—and, in particular, for very eccentric ones—a positive torque leads to a greater precession rate, causing the orbits to align with the bar and oscillate about it (Figure 2b). Note that in Lynden-Bell's picture, the growth rate of the instability is determined primarily by the orbital precession rate, and not by the dynamical time.

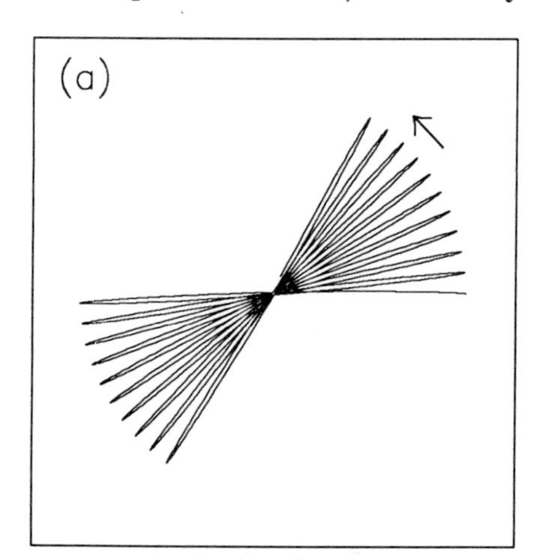

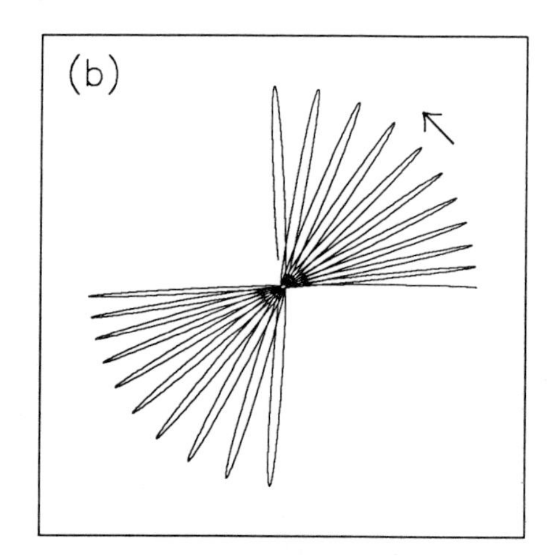

Figure 2. Attraction of eccentric orbits by a bar. (a) Spherical potential; orbit precesses at a constant rate. (b) Spherical potential plus a weak bar (oriented vertically). The orbit accelerates in the direction of the bar.

The most surprising thing about the radial-orbit instability is how *little* anisotropy is required for a model to be unstable. Barnes, Goodman and Hut (1986) found that the final ellipticity of spherical models generated from equation (2) increased smoothly as m was reduced below zero, suggesting that the instability is present even for slight departures from isotropy. Merritt and Aguilar (1985) found a similar behavior in two families of models with a density profile closer to that of real galaxies. One of these families was derived from a distribution function of the form $f(E, J^2) \propto J^{2m} g(E)$, similar to equation (2); the second was defined as the superposition of an isotropic and a purely radial model, and thus contained a finite number of particles with zero angular momentum. Note that all three of these distribution functions diverge as $J \to 0$. Although the ability of the N-body codes to distinguish between stable and unstable models is limited by particle noise, these numerical simulations suggest that models with velocity anisotropies as small as $\sigma_r/\sigma_t \approx 1.2$ can be bar-unstable.

This surprising conclusion has recently been verified analytically by Palmer and Papaloizou (1987; this volume). Their analysis shows that any spherical system characterized by a distribution function that diverges as fast as $J^{-\alpha}$ as J goes to zero, for all E, is guaranteed to be unstable to clumping on *all* angular scales. Since a model with $\alpha \approx 0$ has $\sigma_r/\sigma_t \approx 1$, Palmer and Papaloizou's work verifies that large anisotropies are not required for instability.

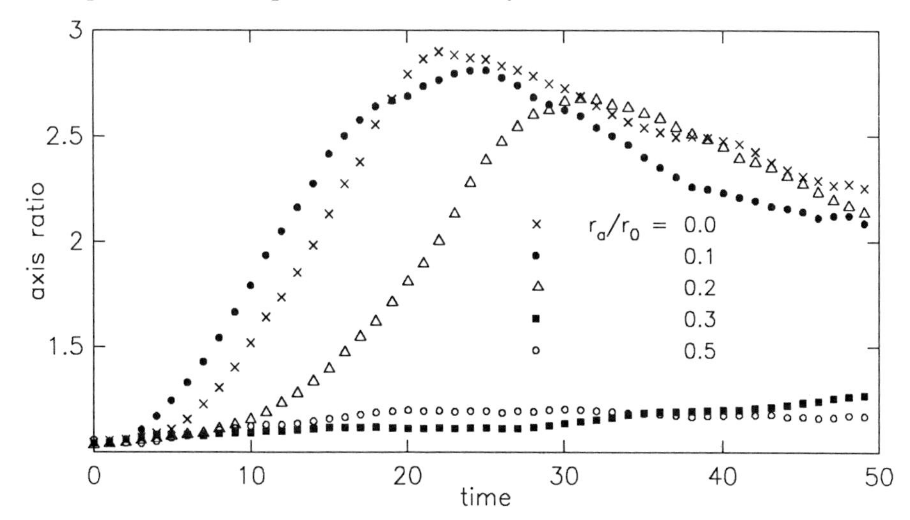

Figure 3. Evolution of the ellipticity of a family of anisotropic spherical models. r_a/r_0 is the ratio of anisotropy to half-mass radii (cf. eqs. [5], [6]). From a set of mean-field, 5000-particle N-body simulations by Merritt and Aguilar (1985).

Less is known about the stability of anisotropic models whose distribution functions do not diverge as $J \to 0$. Only one such family has been analyzed numerically. (Stability boundaries for two other non-singular families have been calculated by a normal-mode technique [Fridman and Polyachenko 1984, Vol. 1., p. 207], but there is some reason to doubt the accuracy of these calculations, which have not been checked numerically; see Polyachenko [1985] and Sec. 5 below.) Merritt and Aguilar (1985) tested the stability of a family of models with density profile

$$\rho(r) = \rho_0(r/r_0)^{-2}(1 + r/r_0)^{-2}, \tag{5}$$

suggested by Jaffe (1983) as an approximation to real galaxies, and a distribution function of the form $f = f(E + J^2/2r_a^2)$; here r_0 is the half-mass radius, and r_a is a free parameter that determines the degree of velocity anisotropy through the relation

$$\frac{\sigma_r^2}{\sigma_t^2} = 1 + \frac{r^2}{r_a^2}. \tag{6}$$

These models obtain their anisotropy essentially by excluding all high-angular-momentum orbits from the region $r > r_a$. Merritt and Aguilar (1985) found that this family undergoes a sudden transition to instability when r_a is reduced below $\sim 0.3r_0$ (Figure 3). Models near the stability boundary have an average anisotropy $\langle\sigma_r^2\rangle^{1/2}/\langle\sigma_t^2\rangle^{1/2} \approx 1.6$. Thus, at least for this family, instability does seem to require a sizable anisotropy.

The obvious next question is whether models with reasonable (i.e. non-divergent) distribution functions, and small overall anisotropies, can be bar-unstable. If the answer is "yes", then the stability of virtually *every* radially anisotropic model will have to be verified before it can be used to describe an equilibrium galaxy. A good starting point for such a study might be the families of models derived by Dejonghe (1986).

3. CONSTRAINING THE DYNAMICS OF OBSERVED GALAXIES

No galaxy can remain in an unstable state for more than a few crossing times. In the case of disk galaxies, this fact is often used to infer the presence of heavy halos, massive bulges, or some other source of "rigid" gravity capable of inhibiting the tendency to bar formation. The situation is obviously very different for elliptical galaxies. The bar instability discussed above will destroy the spherical symmetry of an initially spherical model; but since elliptical galaxies are often very elongated, there is no reason to suppose that this instability did not act at some time in the past. In fact, it will be argued below that the radial-orbit instability may have been partly responsible for producing the flattenings of observed galaxies.

Nevertheless the existence of instabilities means that a theorist has less freedom than he might otherwise have had in constructing equilibrium models for particular galaxies. Consider the simplest case, a spherical galaxy with constant mass-to-light ratio. Complete knowledge of the surface-brightness profile $\mu(r)$ is sufficient to yield a unique space density $\rho(r)$. The number of different distribution functions $f(E, J^2)$ consistent with a given $\rho(r)$ is very large; the only constraint derives from the fact that a very radial distribution of orbits implies a density profile that diverges as fast as

$$\rho \propto r^{-2}|\ln r|^{-1/3} \tag{7}$$

at small radii (Agekyan 1961). Since real galaxies appear to have cores, they cannot be constructed out of purely radial orbits. However one can come very close; the "maximally anisotropic" models corresponding to observed galaxy luminosity profiles are almost completely radial (Richstone and Tremaine 1984; Merritt 1985). This is where instabilities can play a useful role: a sequence of models with fixed density profile will often become bar-unstable before the limit of maximum anisotropy is reached.

In practice, more information is often available than just $\mu(r)$. But even an exact determination of the projected velocity dispersion profile $\sigma_p(r)$ still leaves a formally infinite set of possible distribution functions (Dejonghe 1987). In the case of the most thoroughly modelled elliptical galaxy, M87, theorists have so far been content to stop after finding just one or two distribution functions consistent with the observed profiles (Newton and Binney 1984; Richstone and Tremaine 1985). It would be interesting to construct a *family* of models for M87, each member of which has the *same* surface brightness and velocity dispersion profiles, and see whether some members can be ruled out on the basis of instability.

A first step in this direction would be to test the stability of the handful of published models of M87. These models appear to be excellent candidates for instability since their central regions are dominated by radial orbits. Unfortunately the fraction of the total mass contained within the strongly anisotropic region is only $\sim 10^{-3}$, making these models very difficult to treat with a standard N-body code. Preliminary calculations (Merritt 1986), based on a mean-field code with 10^4 particles, suggest that the model of Newton and Binney (1984) is unstable.

4. THE RADIAL-ORBIT INSTABILITY DURING GALAXY FORMATION

Polyachenko (1981) was the first to make the point that, if an elliptical galaxy is to avoid forming in an unstable state, the instability must somehow manifest itself during the formation process. He later verified (Polyachenko 1985) that collapse starting from very cold and spherical initial conditions can be bar-unstable. In retrospect, it appears that quite a few people narrowly missed discovering the radial-orbit instability in this way. For instance, Aarseth and Binney (1978), in their study of collapse from flattened initial conditions, found that "[the] flattest final configuration...started from the least flattened initial configuration!" Polyachenko's work demonstrates that there need be no simple relation between the initial and final ellipticities of a galaxy that forms via collapse, as long as the collapse is sufficiently "strong" that it produces a significant fraction of nearly-radial orbits which can then clump into a bar.

It is easy to estimate roughly how hot a spherical, proto-galactic cloud must be to avoid making a bar. Simulations of radial collapse (e.g. van Albada 1982) show that galaxies formed in this way tend to have isotropic "cores" and radially-anisotropic "envelopes." The total squared angular momentum of such a galaxy–defined as the sum of the squared angular momenta of all the stars–is roughly $M^2 r_c^2 \sigma^2$, where r_c is the radius of the isotropic "core" and σ is the central velocity dispersion. In a spherical collapse, angular momentum is conserved; thus

$$M^2 r_c^2 \sigma^2 \approx M^2 R_i^2 \sigma_i^2 \approx 2MR_i^2 T_i \approx 8MR^2 T_i \tag{8}$$

where the subscript i refers to the unrelaxed state, T is the kinetic energy, and R is the radius. Energy conservation further requires

$$M\sigma^2 \approx 2|W_i| \tag{9}$$

where W is the potential energy. Combining relations (8) and (9),

$$\frac{r_c^2}{R^2} \approx \frac{4T_i}{|W_i|}. \tag{10}$$

The family of equilibrium models studied by Merritt and Aguilar (1985) is bar-unstable when the "anisotropy radius" r_a (cf. equation [6]) is less than ~ 0.3 times the half-mass radius. Equating r_a with r_c and $r_{1/2}$ with R gives

$$\frac{2T_i}{|W_i|} \gtrsim 0.05 \tag{11}$$

to insure that the galaxy which is formed will not be bar-unstable.

This admittedly crude estimate turns out to be roughly correct: Merritt and Aguilar (1985) find, for a particular set of smooth and spherical initial conditions, that collapses with $2T_i/|W_i| \lesssim 0.1$ are bar-unstable. This is a potentially important result because, as van Albada (1982) and McGlynn (1984) have shown, collisionless collapse from smooth initial conditions is *only* capable of producing objects resembling real galaxies if the initial state is roughly this cold.

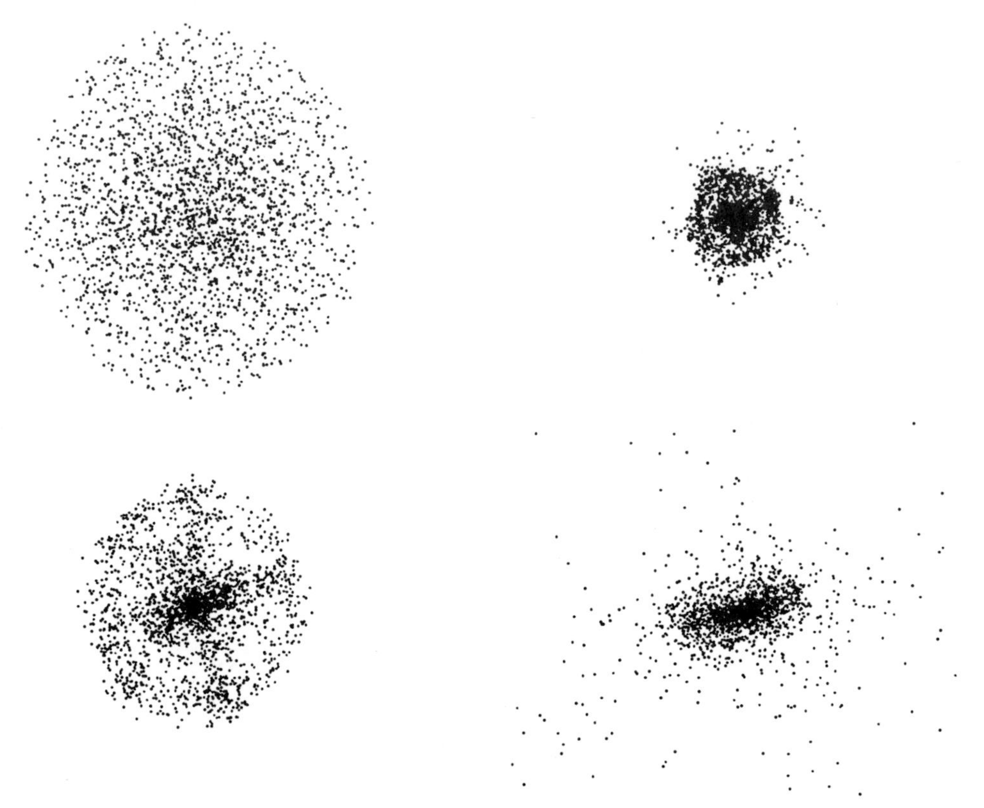

Figure 4. Collapse of a cold, initially oblate cloud, as seen from the direction of the initial symmetry axis. This is a 2500-particle, direct-summation N-body simulation (from M. Duncan).

How the radial-orbit instability manifests itself in more general collapses, which are unlikely to be smooth, spherical, or even very "cold", is an open question, and one that will be very difficult to answer numerically. All that is known so far (Aguilar, Merritt and Duncan, this volume) is that initial conditions that are close to oblate are susceptible to bar-formation in much the same way that spherical initial conditions are (cf. Figure 4). It would be hasty to conclude from this that oblate galaxies are impossible to make, however, since other processes, such as mergers or tidal-torquing, naturally produce oblate systems.

5. NUMERICAL TECHNIQUES

Broadly speaking, algorithms for evaluating the stability of equilibrium models can be divided into two classes: *linear* techniques that are based on the linearized (i.e. small-perturbation) form of the collisionless Boltzmann equation; and *nonlinear* techniques that are based on the actual equations of motion. Most of the results discussed above were obtained using N-body codes, which are fully nonlinear and relatively easy to program. Linear techniques, on the other hand, are far more accurate and efficient when searching for the exact boundary between stability and instability. The standard method for solving the linearized equations (cf. Kalnajs 1977) consists of expanding an initial perturbation in terms of a complete set of basis functions, computing the linear response density, and requiring the potential generated by the response to be equal to that imposed. In this way one obtains the "normal modes" of the system as well as the frequencies at which they grow.

In practice, normal mode calculations can be very difficult, and only a handful of stellar dynamical models (mostly disks) have been analyzed in this way. A feeling for the difficulties involved may be gotten by comparing the normal mode calculations of Fridman and Polyachenko (1984, Vol. 1, p. 219) for the anisotropic polytropes of equation (2), to those of Palmer and Papaloizou (1987) for the same family of models. The two sets of authors arrive at rather different conclusions: Fridman and Polyachenko find a clear-cut stability boundary at $m \approx -0.3$ (corresponding to $\sigma_r/\sigma_t \approx 1.2$), whereas Palmer and Papaloizou find instability for all $m < 0$ ($\sigma_r/\sigma_t > 1$) . The N-body experiments based on this family (Barnes, Goodman and Hut 1986) are not sufficiently accurate to decide which result is more correct, since the instability growth rate becomes very small near the stability boundary, and any evolution is swamped by noise due to the finite number of particles.

It would clearly be useful to develop a new technique that has the computational simplicity of an N-body code, and the (potential) accuracy of a normal mode analysis. Recently S. Tremaine and I have begun to investigate such a technique. The linearized, collisionless Boltzmann equation may be written

$$\frac{D_0 f_1}{Dt} = \frac{\partial \Phi_1}{\partial \mathbf{x}} \cdot \frac{\partial f_0}{\partial \mathbf{v}} \tag{12}$$

where f_0 and f_1 are the equilibrium and perturbed distribution functions, Φ_1 is the perturbed potential, and $D_0 f_1/Dt$ denotes the rate of change of f_1 along an unperturbed trajectory defined by Φ_0:

$$\frac{D_0 f_1}{Dt} = \frac{\partial f_1}{\partial t} + \mathbf{v} \cdot \frac{\partial f_1}{\partial \mathbf{x}} - \frac{\partial \Phi_0}{\partial \mathbf{x}} \cdot \frac{\partial f_1}{\partial \mathbf{v}}.$$

Equation (12) is the starting point for any calculation of the linear response (cf. Polyachenko, this volume). One way of understanding this equation is to note that it gives the change with time, evaluated along an *unperturbed* orbit, of the perturbed phase-space density. Thus

$$f_1(\mathbf{x}, \mathbf{v}, t) = f_1(\mathbf{x}_0, \mathbf{v}_0, 0) + \int_0^t \frac{\partial \Phi_1}{\partial \mathbf{x}} \cdot \frac{\partial f_0}{\partial \mathbf{v}} dt, \tag{13}$$

where $(\mathbf{x}, \mathbf{v})$ are the coordinates at time t of a particle initially at $(\mathbf{x}_0, \mathbf{v}_0)$, and the integral is understood to extend along the trajectory containing the initial and final points. If we imagine dividing up phase space at time zero into a set of N regions with volumes ΔV_i, Liouville's theorem states that motion along the unperturbed trajectories will leave these volumes unchanged. Equation (13) then predicts the change with time of the perturbed mass m_i associated with each volume element ΔV_i:

$$m_i(t) = m_i(0) + \Delta V_i \int_0^t \frac{\partial \Phi_1}{\partial \mathbf{x}} \cdot \frac{\partial f_0}{\partial \mathbf{v}} dt. \tag{14}$$

Equation (14) is a prescription for a Monte-Carlo integration of the linearized equations. Its implementation requires only (a) the unperturbed equations of motion, i.e. $\nabla \Phi_0$; (b) the first velocity derivatives of the unperturbed distribution function f_0; and (c) an algorithm for evaluating the potential Φ_1 generated by a set of mass points m_i. Since this technique assigns *all* of the particles to the perturbation, none are "wasted" in reproducing the underlying equilibrium as in a standard N-body code. Also, since the technique is derived from the linearized equations, the solution is guaranteed to remain in the linear regime at all times. This fact makes the search for normal modes much easier than in a standard N-body code (cf. Sellwood 1983).

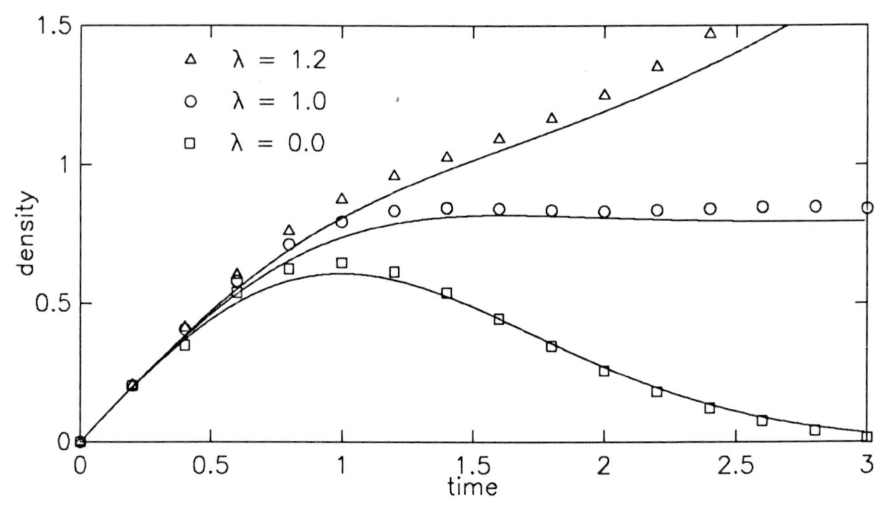

Figure 5. Evolution of the perturbed density of an initially homogeneous, Maxwellian medium. The parameter $\lambda = k_J/k$ is the ratio of the wavelength of the perturbation to the Jeans wavelength. Solid lines: exact solutions; symbols: Monte Carlo integrations.

Figure 5 shows a linearized Monte-Carlo calculation of the response of an infinite, homogeneous, Maxwellian distribution of particles to a perturbation of the form

$$\Phi_{ext}(\mathbf{x}, t) = \delta(t) \cos(\mathbf{k} \cdot \mathbf{x}),$$

i.e. an impulsive plane wave. The exact solution to the linearized equations is easy to obtain in this case (A. Toomre, private communication), and is also shown in Figure 5. The Monte-Carlo technique reproduces the exact solution quite well with only 1000 particles.

This Monte-Carlo technique can easily be applied to systems with any symmetry, as long as the equilibrium distribution function can be specified sufficiently smoothly. It should be a useful tool for evaluating the stability of both spherical and triaxial models.

REFERENCES

Aarseth, S. J., and Binney, J., 1978. *Mon. Not. R. Astron. Soc.*, **185**, 227.
Agekyan, T. A., 1961. *Leningrad University Bulletin*, **1**, 152.
Antonov, V. A., 1960. *Astr. Zh.*, **37**, 918 (translated in *Soviet Astron.*, **4**, 859).
Antonov, V. A., 1962. *Vestnik Leningrad Univ.*, **19**, 96 (translated in *Princeton Plasma Physics Lab. Rept. PPl-Trans-1*).
Antonov, V. A., 1973. In: *The Dynamics of Galaxies and Star Clusters*, p. 139, ed. G. B. Omarov, Nauka, Alma Ata (in Russian).
Barnes, J., 1985. In: *Dynamics of Star Clusters, IAU Symposium No. 114*, p. 297, ed. J. Goodman and P. Hut, Reidel, Dordrecht, The Netherlands.
Barnes, J., Goodman, J., and Hut, P., 1986. *Astrophys. J.*, **300**, 112.
Bisnovatyi-Kogan, G. S., Zel'dovich, Ya. B., and Fridman, A. M., 1968. *Doklady Akademii Nauk SSSR*, **182**, 794 (translated in *Soviet Physics-Doklady*, **13**, 969).
Dejonghe, H., 1986. *Physics Reports*, **133**, 217.
Dejonghe, H., 1987. *Mon. Not. R. Astron. Soc.*, **224**, 13.
Doremus, J. P., and Feix, M. R., 1973. *Astron. Astrophys.*, **29**, 401.
Fridman, A. M., and Polyachenko, V. L., 1984. *Physics of Gravitating Systems*, volumes 1 and 2, Springer-Verlag, New York.
Gillon, D., Doremus, J. P., and Baumann, G., 1976. *Astron. Astrophys.*, **48**, 467.
Hénon, M., 1959. *Annales d'Astrophysique*, **22**, 126.
Hénon, M., 1973. *Astron. Astrophys.*, **24**, 229.
Jaffe, W., 1983. *Mon. Not. R. Astron. Soc.*, **202**, 995.
Jeans, J. H., 1929. *Astronomy and Cosmogony, 2nd ed.*, Cambridge Univ. Press, Cambridge, England.
Kalnajs, A. J., 1977. *Astrophys. J.*, **212**, 637.
King, I. R., 1966. *Astron. J.*, **71**, 64.
Lynden-Bell, D., 1979. *Mon. Not. R. Astron. Soc.*, **187**, 101.
McGlynn, T. A., 1984. *Astrophys. J.*, **281**, 13.
Merritt, D., 1985. *Astron. J.*, **90**, 1027.
Merritt, D., 1986. In preparation.
Merritt, D., and Aguilar, L. A., 1985. *Mon. Not. R. Astron. Soc.*, **217**, 787.
Newton, A. J., and Binney, J., 1984. *Mon. Not. R. Astron. Soc.*, **210**, 711.
Palmer, P. L., and Papaloizou, J., 1987. This volume, p. 515.
Polyachenko, V. L., 1981. *Pis'ma Astron. Zh.*, **7**, 142 (translated in *Soviet Astron. Lett.*, **7**, 79).
Polyachenko, V. L., 1985. *Astronomicheskii Tsirkuliar*, No. 1405 (in Russian).
Richstone, D. O., and Tremaine, S., 1984. *Astrophys. J.*, **286**, 27.
Richstone, D. O., and Tremaine, S., 1985. *Astrophys. J.*, **296**, 370.
Sellwood, J. A., 1983. *J. Comp. Phys.*, **50**, 337.
Sygnet, J. F., Des Forets, G., Lachieze-Rey, M., and Pellat, R., 1984. *Astrophys. J.*, **276**, 737.
Toomre, A., 1964. *Astrophys. J.*, **139**, 1217.
van Albada, T. S., 1982. *Mon. Not. R. Astron. Soc.*, **201**, 939.

DISCUSSION

Lauer: According to Gerhard, even a small black hole will destroy box orbits. Is it possible to have a black hole large enough to prevent formation of a bar, but small enough to still permit anisotropies large enough to explain cusps and rises in velocity dispersion? Can a black hole prevent formation of a bar?

Merritt: Box orbits are not a prerequisite for velocity anisotropy, so the destruction of box orbits does not necessarily imply the elimination of cusps or high central velocities. There are three ways that a central black hole might suppress instabilities. 1) A black hole fixes part of the gravitational field. 2) Reflection of radial orbits around a black hole would probably reduce the growth rate of a bar mode. 3) Adding a black hole to a galaxy with a known central velocity dispersion entails a modification of the orbital distribution near the center so that the dispersion remains constant. This modification will usually go in the direction of more circular orbits, which increases stability.

de Zeeuw: What are the smallest axis ratios one can obtain via the radial–orbit instability?

Duncan: The initially oblate collapse that Dave showed was run using a direct summation N–body code and the results agree well with the results of the quadrupolar code. Secondly, I have run oblate and prolate configurations with initial flattenings as large as 6:1. The final states are always triaxial with largest axial ratios of 2.5 to 1. This is consistent with suggestions that flatter non-rotating systems are unstable.

Villumsen: Some years ago I made some collapse calculations from non–spherical rotating initial conditions. The most extreme results were triaxial E8 systems with low v/σ. They also had beautiful $r^{1/4}$ profiles.

King: You said that an isotropic core would stabilize an anisotropic envelope. In the case $(\sigma_r/\sigma_t)^2 = 1 + (r/r_a)^2$, how big a core do you need for stability? One reason why I ask is that Meylan's poster paper at this meeting (p. 449) shows that in Omega Centauri r_a is 2 or 3 times the core radius. Is Omega Centauri unstable?

Casertano: Josh Barnes and I have run N–body simulations for a King model with a concentration parameter ~ 1.5. We can see instability (on a scale of ~ 5 half–mass dynamical times) if $r_a \leq 2r_c$; no instability can be seen of $r_a \geq 4r_c$.

Merritt: Aguilar and I found that models with r_a greater than about 0.3 times the half–mass radius were stable. Meylan's inferred value for the anisotropy radius of Omega Centauri is roughly equal to its half–mass radius, so his models are probably stable.

Djorgovski: Can you tell whether the instabilities will occur if the stars move in a pre–existing, perhaps spherical, probably isothermal dark halo?

Merritt: I believe that Dr. Polyachenko addresses this question in his presentation.

Richstone: The next question one might ask is what the dynamical appearance of a galaxy is after the instability has run its course. In particular, suppose you take a model of M87 with $\sigma_r >> \sigma_T$ near $r = 0$ and it makes a bar. If you observe the bar, does it still have a large σ_{obs} as $r \rightarrow 0$? If so, maybe you don't need to put the black hole back in.

Merritt: The answer will depend on the direction from which the bar is observed. As seen from the long axis, the galaxy might well appear circular, with a high central velocity dispersion.

Palmer: In some of the models for which we calculated growth rates for the unstable modes, we found very little difference between the growth rates of the $n = 2, 4, 6$ & 8 terms in the spherical harmonic expansion of the potential. How well do you believe that your code can describe the evolution of the instability in these models?

Merritt: Mean–field codes will, of course, have difficulty following the growth of very small–scale modes. However, even in the models you analyzed, the largest scale ($n = 2$) modes were always the fastest growing ones. Furthermore, J. Barnes has shown that both mean–field and "exact" N–body codes give similar results for a number of unstable models.

Duncan: The diagram that Dave showed contains several spoke–like structures just as the bar is forming. They are real, but the bar–mode instability is dominant and one bar wins in a rather short time.

Palmer: Recently I have been simulating systems with very radial orbits using a direct–summation code. I find higher instabilities as well as the original $l = 2$ bar instability; however, in the long run these higher–l features seem to become weaker, leaving only the bar–deformation.

Binney: Andrew May and I have a note coming out (1986, *Mon. Not. R. astr. Soc.*, **221**, 13P) in which we point out that there is a natural method of testing for the stability of any model whose distribution function can be written in terms of the action integrals, $f = f(J_i)$: One distorts the model's potential and asks whether the density over– or under–responds. We used this method to determine the stability of anisotropic isochrone spheres with less than $1/500$ of the computer time required by a typical N–body simulation.

Merritt: The method is a very clever one and should be explored. Your comparison of computing times neglects the fact that the same N–body code can be used for a variety of models. Also, the adiabatic deformation technique is inherently approximate, whereas the accuracy of N–body tests can always be improved, *e.g.*, by increasing the number of particles.

David Merritt.

FORMATION AND SECULAR EVOLUTION OF ELLIPTICAL GALAXIES

George Lake
University of Washington
Dept. of Astronomy
Seattle, Washington 98195
U.S.A.

ABSTRACT. Three ways have been proposed to make elliptical galaxies: cooling by gas dynamical processes at late epochs, Compton cooling at early times and merging. These theories must address a variety of observational constraints, the most severe being the problems of slow rotation and high central phase densities. I look at some aspects of all three theories with particular attention to key numerical simulations and observations that can distinguish between the scenarios.

1. INTRODUCTION

To my mind, the most unique feature of this conference has been the dazzling array of talks on the secular evolution of elliptical galaxies. There's been nothing like this before. There were at least three talks about galaxy formation (Ivan King sat through all three). Star formation in cooling flows is slowly becoming well-established, though puzzling. Elliptical galaxies are influenced by cannibalism, merging and accretion; the resultant shells are seen for a long time. The centers of galaxies may have their integrals slowly broken by the central massive object leading to stochasticity and feeding.

The bottom line is that I find my charge to review secular evolution puzzling. I thought that I might talk a bit about galaxy formation (the most certain of all secular events) covering a couple of topics that have had little attention at this conference.

Galaxy formation theory has not advanced as briskly as our understanding of their internal structure. Rees (1977) has lamented the hazards of the roads to be traveled and likened the process to searches for probes of the Universe between a redshift of 3.5 (quasars) and 1000 (the microwave background). The one success of the theory has been the calculation of the characteristic masses of galaxies as the scale where dissipation breaks the hierarchy. While this idea has an irresistible appeal, there are problems with the way it has been calculated.

Here, I will examine three scenarios for the formation of elliptical galaxies. These are: 1) late dissipative when a large fraction of the binding energy is radiated by gas cooling 2) early dissipative when Compton cooling off the microwave background is important and 3) merging. I will show that there are key observations that will help discriminate between the three. In keeping with my charge to

T. de Zeeuw (ed.), Structure and Dynamics of Elliptical Galaxies, 331–338.

describe secular evolution, I'll also mention some long lasting dynamical effects that may still be visible.

At this conference, Carlberg has stressed the constraints that the core properties of ellipticals place on their formation. An important part of the discussion will be the ways the three scenarios address this problem and the longer standing problem of slow rotation (cf Illingworth 1977).

2. DISSIPATIONAL GALAXY FORMATION

Normally, it's said that the criterion for gas dynamics to have had significant effect is that the cooling time is less than the dynamical time. Contrary to the usual interpretation, this would seem like a good criterion for gas dynamics to **stop** restructuring galaxies. Once the cooling time is shorter than the dynamical time, the gas should be thermally unstable. The denser regions cool faster, the hot diffuse phase pushes on them and clouds form with large density contrasts and low temperatures. We must then ask if star formation occurs within a collapse time. In short, does the collapse rapidly become dissipationless? Avoiding this may be a strong argument against galaxies cooling too quickly. On the other hand, if they don't cool there won't be any contrast between galaxies and their environment. Galaxies have clearly pinched off the clustering hierarchy and gas dynamics is certainly the cause. Either the gas must follow the ridge line $\tau_{cool} = \tau_{ff}$ as closely as possible or we must find ways to loose energy after the formation of clouds.

Dark matter changes most of the details of galaxy formation in a manner that hasn't been well chronicled in the literature (though see White and Rees 1978). A classic argument by Ostriker and Rees (1977) is that cooling due to bremsstrahlung sets a characteristic scale of $\sim$ 80 kpc. The gas is assumed to be at a virial temp,

$$T_{gas} = 5.2 \times 10^5 K\ \mathbf{M}_{11}\ \mathbf{R}_{10}^{-1}$$

where $\mathbf{R}_{10}$ is the radius in units of 10 kpc and $\mathbf{M}_{11}$ is the total mass in units of $10^{11} M_{\odot}$, with a gas density

$$\rho = 1.6 \times 10^{-24} g\ cm^{-3}\ \mathbf{F}\ \mathbf{M}_{11}\ \mathbf{R}_{10}^{-3},$$

where $\mathbf{F}$ is the fraction of the matter that can dissipate. The bremsstrahlung cooling time for this gas is

$$\tau_{brem} = 2.2 \times 10^{14} s\ \mathbf{M}_{11}^{-1/2}\ \mathbf{F}^{-1}\ \mathbf{R}_{10}^{3/2}.$$

By comparison, the free fall time is given by

$$\tau_{ff} = 7.3 \times 10^{14}\ \mathbf{M}_{11}^{-1/2}\ \mathbf{R}_{10}^{3/2}$$

equating τ_{brem} to τ_{ff}, we find a characteristic radius: $R_{cool} = 73$ Fkpc. With $\mathbf{F} = 1$, the radius 73 kpc rings true as the radius that galaxies collapse from, but when we take the currently popular figure (Faber 1983, Gunn 1983) $\mathbf{F} = 0.07$ for the ratio of light to dark matter, the radius 5 kpc rings equally true for the final radius of collapse where dissipation **stops** being important.

Of course, at the virial temperatures of galaxies, cooling by radiative recombination dominates over bremsstrahlung. While in principle, this should lead to horrid complications, over the range of T = 10^4-10^6, the value of the cooling constant Λ for a zero-metal gas is nearly constant at a value of 2×10^{-23}, so that

$$\tau_{rad} = 5.5 \times 10^{13} s\ \mathbf{R}_{10}^2\ \mathbf{F}^{-1}.$$

For a value of **F** equal to .07, we find that thermal instabilities should occur at a radius of 10 kpc.

The properties of the clouds that form are easily estimated if the clouds are in pressure balance. The hot medium will be nearly fully ionized and have a temperature of 1-2 $\times$ 10^6 K, while the clouds will rapidly cool to approximately 10^4 K. The ratio of the densities $\mathbf{f} = \rho_{cool}/\rho_{hot}$ is just $2 \times T_{hot}/T_{cool}$. These clouds have several attractive properties. First, since we assumed that the protogalaxy is collapsing nearly quasistatically until it becomes thermal unstable, this means that we have one Jeans mass in one Jeans length. The Jeans mass in the clouds will be roughly $(T_{cool}/T_{hot})^2\ \mathbf{F}\ \mathbf{M}_{11}$ or $10^{-5}\ \mathbf{F}^{-1}\ \mathbf{M}_{11}$. Fall and Rees (1985) associate this mass with that of the globular clusters. While it is high for a globular cluster, they point out that there will be mass loss. I regret that I had not seen their paper at the time of the Princeton meeting. They argue that various other physical processes favor clouds of this size. I have a simpler argument that shows that clouds this size form. Since a large fraction of the gas will be tied up in the clouds, their filling factor will be roughly **f**, previously defined as the ratio of the densities. For illustrative purposes, let's consider N clouds of fixed size. The mean free path of the clouds is $N^{1/3}\ \mathbf{f}^{-2/3}$ times the radius of the protogalaxy. Clouds collide in a dynamical time if there are more than $\sim 10^5$ of them. If there are less than this number, they are gravitationally unstable and form something, presumably star clusters. If there are more, then when the clouds collide, the shock will be isothermal and lead to a large density increase in the postshock gas. For a shock velocity of 300 km s^{-1}, the densities rise by over two orders of magnitude and the Jean's mass decreases by an order of magnitude. So if there are $\sim 10^6$ of them, they collide in a dynamical time and immediately become Jean's unstable. If there are still more, they coagulate on a time scale that's short compared to a dynamical time and eventually reach the Jean's threshold in a few dynamical times.

Will these clouds with densities contrasts of 200-400 at a radius of 10 kpc evolve to form an elliptical galaxy with a radius of 3 kpc and a dense core? Well, maybe. Each time that the clouds collide, they can only reduce their relative kinetic energy. If they have to radiate enough to collapse by a factor of three, it takes about 15 collisions (Carlberg, Lake and Norman 1986). If the clouds start with masses of $\sim 10^3 M_{\odot}$, then it looks like this could work. But, there's some three card Monty going on here. A **F** of 0.07 is small for a scale of 10 kpc. Let's suppose that the dark matter in ellipticals has the same distribution that we see in spiral galaxies. Then at these radii, a more appropriate value would be 0.5. If ellipticals are surrounded by halos with the run of densities seen in spiral galaxies, they become thermally unstable at a radius of roughly 40-50 kpc. Without any halo, the collapsing cloud becomes thermally unstable at a radius approaching a Mpc. At either of these radii, a diffuse iceberg forms rather than an elliptical galaxy unless the clouds suffer many collisions.

It's easy enough to scheme up a way for the clouds to keep dissipating. They start small, bust up, reform, etc. until they've radiated away a lot of energy. It may happen that way. It stops looking very natural. We may just have to accept that all these collisions do occur. The good news is that there is a lot of action with clouds busting up and reforming during galaxy formation. It's a good epoch for hungry monsters. Star formation is occuring quickly, which bodes well for searches for primeval galaxies. The best news is that the physics of dissipational collapse is easy to simulate. The energy is being radiated by cloud-cloud collisions with mean free paths that are $\mathbf{O}(10^{-1})$ the size of the system. This is exactly the circumstance that Carlberg and I are now simulating using the San Diego Supercomputers (cf. Carlberg, Lake and Norman 1986). I now think that there is little reason to be apologetic about our dissipation scheme of bouncing particles with a coefficient of restitution. This captures the essential physics and means that we will learn a lot from these simulations. So far, we've learned a lot about the angular momentum problem. We find that when we collapse warm perturbations (velocity dispersion 30% of the virial value); the gas settles gently into a disk. If we collapse cold perturbations ($\sigma < 0.2$), there is a lot of angular momentum transfer and a dense, slowly rotating elliptical galaxy forms. The collapse of clouds is a good way to make all kinds of galaxies.

Before leaving this section, let's take a last look at the protoglobular clusters. If you accept the universal ratio of light to dark mass, you find that the Jean's mass of the clouds at the time of thermal instability is dependent on a a fairly high power of the mass of the protogalaxy. This deserves attention to see if it's really a viable theory for the origin of globulars.

3. BUT ELLIPTICALS ARE COMPTON COOLED

The previous section considered only the role of radiative cooling. In the early Universe, there is another mechanism– Compton cooling (cf. Ostriker and Rees 1977). This cooling owes to the scattering of the microwave (then infrared) background photons off electrons, robbing the latter of energy. It is independent of density, $\mathbf{F}$ and temperature (as long as $T_{electron} >> T_{radiation}$). The Compton cooling timescale for a completely ionized gas is given by

$$\tau_{comp} = 3.8 \times 10^{19}(1+z)^{-4}s,$$

where z is the redshift. As perturbations evolve in the Universe, they experience turnaround when their mean density is 5.5 times the cosmological density. The two times τ_{comp} and τ_{ff} are equal for perturbations that are turning around at a z of 10. At this redshift, both times are 5.5×10^{15} s. By comparison, T_{rad} is $8.8 \times 10^{14}\ \mathbf{M}_{11}^{2/3}\ \mathbf{F}^{-1}$ s.

If Comptonization is the only cooling mechanism, there isn't any thermal instability. Dense regions don't cool faster. Jeans and shape stabilities may still be important. It's important to know when Compton cooling dominates over radiative cooling. For an object just turning around, this occurs when $(1+z) = 25(\mathbf{F}\ \mathbf{M}_{11})^{-1/2}$ If $\mathbf{F}$ is 1, this yields the surprising result that Compton cooling isn't dominant until a redshift of 25 rather than 10.

What is $\mathbf{F}$ for perturbations turning around at redshifts of 10-20? We can gain some insight from the central densities of halos of spiral galaxies. From the data and analysis for NGC3198 (van Albada *et al.* 1985), we find a core halo density of $3.5 \times 10^{-25}\ g\ cm^{-3}$. This density sets an upper limit to the turnaround redshift of roughly 10. We find similar values for the sample of Carrignan and Freeman

(Carrignan 1983, Carrignan and Freeman 1985). It would seem that dark matter can't participate in galaxy formation before a redshift of $\sim$ 10. So **F** is probably **O**(1) at redshifts of order 20.

Even if thermal instabilities don't exist, the efficiency of Compton cooling leads to shape instability. The shocking pancake will produce stars *but very little of the binding energy of the galaxy is radiated.* If the protogalaxy pancakes at a time when its binding energy is appropriate for an elliptical galaxy. The high phase densities in the resultant sheet of stars is ideal for making the cores. The most likely time that such a sheet formed is at a redshift of 20. At this time, Compton cooling keeps the collapsing cloud at low temperatures and there are barely thermal instabilities.

The best argument that I can think of for galaxy formation in the Compton cooled era is that it affords the best chance of hiding baryons. There are several important "estimates" of Ω, the ratio of the mean mass density of the universe to the critical density. Ω_{stars} is about 5×10^{-3}. $\Omega_{baryons}$ determined from calculations of nucleosynthesis is roughly 0.1 (Olive *et. al.* 1981). Cosmic virial theorems yield $\Omega_{cvt} = 0.1$. Finally, the New Religion (or New Inflation) says that $\Omega_{true} = 1$. In the preceding list, science is accurate to a factor of two, while religion is exact. Even with these uncertainties, there are two discrepancies, one between stars and baryons and the other between religion and the universe. Its important to simulate the compton cooled era to see if elliptical galaxy formation can work in detail and if there is some way to hide baryons. An example of the interesting physical questions that await analysis is the importance of compton drag in slowing rotation and preventing fragmentation.

4. MERGERS

I'm amazed at how little discussion there has been of the merger hypothesis at this meeting. The two most difficult hurdles for the merger hypothesis are the L-σ relation (cf. Ostriker 1980, Tremaine 1981) and the extremely high central phase densities of ellipticals (Carlberg, this conference).

Recently, Dressler and I (1986) measured the velocity dispersion of objects from the Arp (1966) and the Arp-Madore (1986) Atlases that have characteristics of recently merged galaxies. Accurate velocity dispersions of objects dominated by Balmer lines in the blue were measured using the uncontaminated Ca triplet feature in the extreme red (8400-8700 Å).

What did we expect? Most mergers will occur when a pair of galaxies is just bound (White 1979). This is easy to see as the phase space of bound orbits is largest at marginal binding. Galaxies that are just unbound may merge if the orbital parameters are just right; they become bound by ejecting a fraction of their mass with positive energy. We consider the merger of spirals, since we selected galaxies with streamers or tails — clear indications of an initially cold component. Roughly half of our selected sample shows signs of star formation; the initial systems must have had a considerable gas content.

For mergers at zero orbital energy, the binding energy of the remnant is the sum of the binding energies of the initial spirals. The luminosity in the H-band of the remnant will be the sum of the luminosities in the H-band of the original spirals. Our last assumption is that the merging spirals have the same H-luminosity. This biases our result toward high luminosities at a fixed dispersion velocity, but the bias is small. We can now take the H-band Fisher-Tully relation for spirals and construct a model Faber-Jackson relation for ellipticals.

Aaronson, Huchra and Mould (1980) give a Fisher-Tully relation using aperture magnitudes of:

$$H^{abs}_{-0.5} = -21.23 - 10[\log \Delta V_{20} - 2.5].$$

Using their growth curves, we find a mean correction of 0.5 magnitudes for total magnitudes yielding

$$H^{abs}_{spirals} = -21.7 - 10[\log \Delta V_{20} - 2.5].$$

The quantity ΔV_{20} is twice the circular velocity in the disk which is in turn $\sqrt{2}\sigma$, where σ is the velocity dispersion. Using these transformations and our luminosity and binding energy sum rules, we expect

$$H^{abs}_{hypoth.merger} = -24.9 - 10[\log(\sigma) - 2.3].$$

We compare this to the Faber-Jackson (1976) relation for ellipticals (de Vaucouleurs and Olson 1982, with a mean B-H = 4):

$$H^{abs}_{E} = -23.4 - 10(\log(\sigma) - 2.3).$$

Our expectation is that there will be a 1.5 magnitude difference between merger remnants and normal ellipticals. Given the steepness of luminosity-velocity dispersion relation, this means that the velocity dispersions of merger remnants should be 30% lower than that observed in elliptical galaxies. Similar conclusions were reached by Veeraraghavan and White (1984) employing blue magnitudes and fading the disk and bulge separately.

What did we find? Our sample of merger remnants had a $\sigma(M_B = -21)$ of 184 km s^{-1} as compared to 184 for Sargent *et al.* (1977), 237 for Faber and Jackson (1976), 201 for Schechter and Gunn (1979), 237 for Whitmore *et al.* (1979) and 220 for Terlevich *et al.* (1981). The merger remnants are on the low side, but not outside the range seen in samples of elliptical galaxies.

We used blue luminosities. Many of the galaxies had bluer colors than normal ellipticals. If we fade the blue galaxies using starburst models (an extreme case, Larson and Tinsley 1978), the resultant $\sigma(M_B = -21)$ is 193 km s^{-1}.

Two of the three galaxies with blue colors (Arp 224 and Arp 226) were on Toomre's (1977) list of "prospects for ongoing mergers". A better observation technique for these and some other candidates (notably NGC 3256 and Arp 243) would be to use H-magnitudes which are much less sensitive to the recent bursts of star formation. An ideal follow-up to our study is one using measurements of the velocity dispersion from the Calcium triplet together with H-band luminosities.

The second problem of large core densities is unsettled. The bulges of spiral galaxies have even large phase densities than ellipticals. Carlberg pointed out a galaxy that had too small a core radius (large phase density) and associated it with a merger event. In using central phase densities to constrain galaxy formation scenarios, we assume that the cores are formed with the rest of the galaxy. If instead, they are secondary, formed by subsequent gas infall then their formation is decompiled from initial conditions. Space Telescope can make some key observations here. First, it can image merger remnants in the I-band in the same way that Kormendy (1985) has resolved the cores of nearby ellipticals. This way, we'll know *empirically* if phase space densities are a problem for the merger hypothesis. Also, if the cores of ellipticals (and/or merger remnants) are made by secondary infall, ST data will show that they rotate rapidly. I would have bet against this before I saw Kormendy's spectacular data (this conference) on the core of M31. That core certainly evolved after the galaxy formed.

Finally we may get a better understanding of all these issues using numerical simulation. There's no reason not to simulate the merger of two galaxies each with 100,000 particles. We can follow the dissipational collapse of galaxies. Simulations of collapsing Compton cooled clouds are interesting for both galaxy formation and for pointers to the mystery of Pop III and the missing baryons.

REFERENCES

Aaronson, M., Huchra, J. and Mould, J. 1980, *Ap. J.*, **237**, 655.

Arp, H. 1966, *Atlas of Peculiar Galaxies* (Pasadena: California Institute of Technology).

Arp, H. C. and Madore, B. F. 1985, *A Catalogue of Southern Peculiar Galaxies and Associations*, (Toronto:Clark Irwin), in press.

Carlberg, R., Lake, G. and Norman, C. 1986, *Ap. J. (Letters)*, **300**, L1.

Carrignan, C. 1983, Ph.D. thesis, Australian National University.

Carrignan, C. and Freeman, K. 1985, *Ap. J.*, **294**, 494.

de Vaucouleurs, G. and Olson, D. 1982, *Ap. J.*, **256**, 346.

Faber 1982, in *Astrophysical Cosmology*, eds. H. A. Bruck, G.V. Coyne and M.S. Longair, (Vatican: Pontificia Academia Scientarum).

Faber, S. M. and Jackson, R. 1976, *Ap. J.*, **204**, 668.

Fall, S. M. and Rees, M. J. 1985, *Ap. J.*, **298**, 18.

Gunn, J.E. 1982, in *Astrophysical Cosmology*, eds. H. A. Bruck, G.V. Coyne and M.S. Longair, (Vatican: Pontificia Academia Scientarum).

Illingworth, G. 1977, *Ap. J. (Letters)*, **218**, L43.

Lake, G. and Dressler, A. 1986, *Ap. J.*, in press.

Larson, R. B. and Tinsley, B. M. 1978, *Ap. J.*, **219**, 46.

Olive, K. A., Schramm, D. N., Steigman, G., Turner, M. S. and Yang, J. 1981, *Ap. J.*, **246**, 557.

Ostriker, J. P. 1980, *Comm. Astr. Ap.*, **38**, 177.

Ostriker, J. P. and Rees, M.J. 1977, *M.N.R.A.S.*, **119**, 81.

Rees, M. J. 1977 in *The Evolution of Galaxies and Stellar Populations*, eds. B. M. Tinsley and R. B. Larson (New Haven: Yale University Observatory).

Sargent, W. L. W., Schechter, P. L., Boksenberg, A. and Shortridge, K. 1977, *Ap. J.*, **221**, 731.

Schechter, P. and Gunn, J. E. 1979, *Ap. J.*, **229**, 472.

Terlevich, R., Davies, R. L., Faber, S. M. and Burstein, D. 1981, *M.N.R.A.S.*, **196**, 381.

Toomre, A. 1977 in *The Evolution of Galaxies and Stellar Populations* , eds. B. M. Tinsley and R. B. Larson (New Haven: Yale University Observatory).

Tremaine, S. D. M. 1981, in *The Structure and Evolution of Normal Galaxies*, eds. Fall, S. M. and Lynden-Bell, D. (Cambridge: Cambridge University Press).

van Albada, T. S., Bahcall, J. N., Begeman, K. and Sancisi, R. 1985, *Ap. J.*, **295**, 305.
Veeraraghavan, S. and White, S. D. M. 1985, preprint.
Whitmore, B. C., Kirshner, R. P. and Schechter, P. L., *Ap. J.*, **3234**, 68.
White, S. D. M. 1979, *M.N.R.A.S.*, **189**, 831.
White, S.D.M. and Rees, M.J. 1978, *M.N.R.A.S.*, **183**, 341.

DISCUSSION

Fall: Martin Rees and I studied many of the processes you discussed. Our main point was that a collapsing proto–galaxy would be thermally unstable and would develop a two–phase structure: cold clouds at $T \sim 10^4 K$ compressed by the surrounding hot gas at $T \sim 10^6 K$. The clouds with masses exceeding some critical value of order $10^6 M_\odot$ would then be gravitationally unstable; we identified these objects as the progenitors of globular clusters. In contrast to your claim, the clouds must cool slowly at temperatures just below $10^4 K$ to imprint a characteristic mass of order $10^6 M_\odot$. The condition for this to happen is that the sound crossing times of the clouds be *shorter* than their cooling times. A complete account of our work was published last year (1985, *Astroph. J.*, **298**, 18).

DISSIPATIONLESS FORMATION OF ELLIPTICAL GALAXIES

Simon D.M. White
Steward Observatory
University of Arizona
Tucson, AZ 85721
U.S.A.

ABSTRACT. Dissipationless formation mechanisms envisage elliptical galaxies as arising from the collective relaxation of an aggregate of stars. Their key ingredients are thus a set of initial conditions derived from consideration of prior evolution, and a treatment of the relaxation process. I review numerical studies of violent relaxation carried out over the last decade and purely theoretical treatments going back twice as far. Relaxation is always incomplete, and as a result the final structure of a "galaxy" depends sensitively on the initial conditions assumed. The viability of dissipationless formation thus rests on the identification of plausible stellar initial conditions which relax to the present structure. I discuss the extent to which such initial conditions are compatible with current ideas on the origin of structure in the universe.

I. INTRODUCTION

Present elliptical galaxies appear to approximate smooth equilibrium solutions of the gravitational N-body problem. However, we believe that at high redshift their material was in the form of hot uniformly distributed gas. The transition between these two states clearly involved gravitational, hydrodynamical, radiative and perhaps hydromagnetic processes. As a result, a purely stellar dynamical model for elliptical formation is useful only if star formation occurred before the final phase of relaxation. Any dissipationless model should therefore start from initial conditions in which stars could already have formed. The main idea motivating dissipationless models is the hope that gravitational evolution will cause convergence to an equilibrium distribution which depends weakly on the initial conditions. It turns out, however, that this hope is only partially fulfilled.

Numerical experiments show that the gross equilibrium properties of an isolated system of stars are established in the first two or three free fall times of its evolution. During this phase, evolution can be thought of as resulting from an interaction between the time-dependent gravitational potential and the orbits of the stars responsible for that potential. Qualitatively, one can say that the greater the inhomogeneity of the initial conditions, the broader the spectrum of potential fluctuations excited during collapse, and the greater the randomisation of stellar orbits. In his classic paper of 1967 Lynden-Bell analysed this process by assuming

T. de Zeeuw (ed.), Structure and Dynamics of Elliptical Galaxies, 339–351.

the maximum randomisation possible; in addition he coined the oxymoron "violent relaxation" to describe it. This work and extensions of it remain the main theoretical treatments of collisionless relaxation.

In the next section, I review Lynden-Bell's work highlighting the logic which underpins it, some aspects of its predictions, the contradictions to which it leads, and its connection to notions of entropy in stellar dynamics. Much of this discussion draws on the recent paper of Tremaine, Henon and Lynden-Bell (1986). I follow it by a review of the numerical evidence that, in practice, violent relaxation does not lead to full randomisation, and that elliptical galaxies retain considerably more information about their initial conditions than is required by conservation constraints alone. Some structural properties of dissipationless formation models can, however, be derived on quite general grounds. In Section IV I give a simple argument (also presented by Jaffe at this meeting) which shows that in the outermost regions of a galaxy the stellar density is expected to decline as the inverse fourth power of radius. I also give a heuristic explanation for the power-law structure found in numerical experiments involving very strong relaxation. In a final section I return to the question of initial conditions and I consider how dissipationless formation might plausibly occur in a cosmological context. The conclusion of this discussion is that current models for the evolution of structure predict initial conditions which are sufficiently inhomogeneous that the formation process is best characterised as the collision and merging of pre-existing virialised stellar systems.

II. VIOLENT RELAXATION

As formulated by Lynden-Bell (1967), violent relaxation describes the dynamical equilibration of a system of stars which is so large that the fundamental graininess of the stellar distribution can be ignored. It can thus be thought of as the relaxation of a system which is completely described by the coupled Collisionless Boltzmann and Poisson equations:

$$\frac{\partial f}{\partial t} + \mathbf{v} \cdot \frac{\partial f}{\partial \mathbf{x}} - \nabla \phi \cdot \frac{\partial f}{\partial \mathbf{v}} = 0; \tag{1}$$

$$\nabla^2 \phi = 4\pi G \int m f d^3 v dm; \tag{2}$$

where $f(\mathbf{x}, \mathbf{v}, m, t)$ is the single particle phase space density. The first equation is simply a statement of mass conservation for each type of star, while the second asserts that the stars are themselves the source of the (Newtonian) gravitational potential in which they move. Because $-\nabla\phi(\mathbf{x})$ is the acceleration of all stars at $\mathbf{x}$, the operator in equation (1) is just the standard convective time derivative along the stellar trajectories. The Collisionless Boltzmann equation can thus be recognised as the equation of motion of an incompressible fluid in a 6-dimensional space. Following the motion of each star, f as a function of m is conserved. We can therefore drop the mass dependence of f and consider it as the total stellar mass density at each point in phase space. In addition, since surfaces of constant f are convected with the fluid, the total volume and the total mass they contain are independent of time. Thus if we define $m(f)df$ as the mass of stars in regions with phase space density in the range $(f, f + df)$, $m(f)$ is conserved as the system relaxes. Following Tremaine, Hénon and Lynden-Bell (1986; hereafter THLB) we can then define the mass and volume at phase space densities exceeding f by,

$$M(f) = \int_f^\infty m(f')df'; \quad V(f) = \int_f^\infty m(f')\frac{df'}{f'}. \tag{3}$$

Since both these functions depend on $m(f)$ alone, they remain constant during violent relaxation. A third similar quantity is

$$S = -\int_0^\infty df' m(f') \ln(f'), \tag{4}$$

which may be recognised as the analogue of the Boltzmann entropy for a collisional gas. Hence if f is used to define an entropy, violent relaxation is adiabatic.

The second important ingredient in the theory of violent relaxation is the notion that although f is conserved, the structure of its phase space distribution becomes more and more convoluted as evolution proceeds. At late times f is no longer an observable quantity; rather the "interesting" information about the structure of a system is contained in a coarse-grained distribution, f_c, obtained by smoothing f over some finite volume. As a system evolves, initially adjacent elements of phase density become separated by arbitrarily large distances. This orbital divergence is caused by differential scattering off fluctuations in the potential, and by phase-mixing as a result of the slightly different periods of neighboring orbits. (Notice that the latter process continues even after violent potential fluctuations have died down.) Thus at late times phase space elements from very different parts of the initial conditions may contribute to the same value of the coarse-grained phase space density. Only because of the implicit assumption that such smoothing has occurred can it be said that violent relaxation causes an elliptical galaxy to "forget" the details of the initial conditions from which it formed. This information loss differs from the corresponding effect in a gas. In the latter case our failure to keep track of the positional correlations between molecules results in a loss of the details of individual collisions. For a gas, an entropy defined using the fine-grained distribution function is increased by collisions.

Lynden-Bell's analysis of violent relaxation is based on a maximal mixing hypothesis. He assumes that the evolution randomises the position of elements of f completely, subject to conservation of total energy and total angular momentum. If every averaging volume contains a large number of such elements, he shows that the most probable coarse-grained distribution function satisfying the constraints is approximately

$$f_c(\mathbf{x}, \mathbf{v}) = \int_0^\infty df' A(f') \exp(-f'\beta\epsilon), \tag{5}$$

with

$$\epsilon = v^2/2 + \phi(\mathbf{x}),$$

and where the energy and mass constraints require

$$E = \int d^3x d^3v \; f_c \epsilon; \quad m(f) = A(f) \int d^3x d^3v \exp(-f\beta\epsilon). \tag{6}$$

Equation (5) is simpler than the form given by Lynden-Bell, because I have assumed that $f'\beta\epsilon >> 1$ for all significant contributions to the integral, and that the total angular momentum of the system is negligible. The first condition is equivalent to assuming that mixing is sufficiently strong for f_c to be everywhere small compared to typical values of f in the initial conditions. Numerical experiments suggest that this condition is violated for a small fraction of the mass at the centre of a violently relaxed system; however, it appears to hold for the bulk of the stars. Equation (5) then gives f_c as a superposition of Maxwellians, one for each value of f present in the initial conditions. The weighting function $A(f)$ is related to the mass of stars with initial phase density f through equation (6); their final velocity dispersion is inversely proportional to the value of f.

It is sometimes stated that as a result of violent relaxation the central regions of an elliptical galaxy should resemble an isothermal sphere. However, the distribution function of equation (5) will resemble a single temperature Maxwellian only if the initial distribution of f satisfies rather restrictive requirements, for example if all stars are initially in regions of the same phase space density so that $m(f)$ and $A(f)$ are delta functions. In general a rather wide range of initial phase space densities is to be expected; the shape of f_c at low energies, and thus the theoretical prediction for the core structure of the resulting galaxy, then depends sensitively on initial conditions. In addition, as noted above, numerical experiments suggest that in many situations the central value of f_c in the final galaxy is not much smaller than the largest values of f_c in the initial conditions. Thus mixing of phase space elements appears to be far from complete. Failure of the maximal mixing hypothesis is also apparent in the angular momentum distributions found experimentally in systems formed by violent relaxation. Lynden-Bell shows that his hypothesis predicts the most probable distribution to be in solid-body rotation. However, the rather sparse experimental data available suggest that a typical final state has a rotational velocity at each radius which is roughly proportional to the local velocity dispersion (White 1979; Farouki and Shapiro 1982; Villumsen 1982; Frenk *et al.* 1985).

Lynden-Bell's derivation of a most probable distribution involves finding a stationary point of an entropy defined using the coarse-grained phase space density,

$$S = -\int d^3x d^3v f_c \ln(f_c). \tag{7}$$

The solution (5) has the unpleasant property that although it is derived using conservation of mass and energy as constraints, it implies an infinite mass and energy for an isolated, unconfined system. In fact, it is easy to see that there is no distribution which maximises the entropy (7) for a given finite mass and energy. A small fraction of the mass of any system can always be put into an arbitrarily weakly bound outer halo which extends to arbitrarily large distances and makes an arbitrarily large contribution to the integral in (7) (see THLB, this argument is originally due to J. Binney). Thus not only is maximal mixing never realised in practice, it does not even lead to a definite prediction for finite systems (see, however Section IV below).

If we abandon the idea of defining the endpoint of violent relaxation as a maximum entropy state, we may ask whether we can, at least, say that violent relaxation always leads to an entropy increase. A proof of this statement, originally due to Tolman (1938), runs as follows:

$$\begin{aligned} S(t_f) &\equiv \left[-\int f_c \ln(f_c) d^3x d^3v\right]_{t_f} \\ &\geq \left[-\int f \ln(f) d^3x d^3v\right]_{t_f} \\ &= \left[-\int f \ln(f) d^3x d^3v\right]_{t_i} \\ &= \left[-\int f_c \ln(f_c) d^3x d^3v\right]_{t_i} \equiv S(t_i), \end{aligned} \tag{8}$$

where t_i and t_f denote the initial and final times respectively. The inequality in the second line is a result of the averaging implicit in the definition of f_c, together with the fact that for

$$C(f) = -f \ln(f), \tag{9}$$

$\overline{C(f)} > C(\overline{f})$ for any set of unequal values of f. The equality in the third line of (8) holds because, as we have already seen, an entropy defined using equation (4) is conserved during violent relaxation. The equality in the final line of (8) reflects the assumption that in the initial conditions the fine-grained phase space density is a slowly varying function of phase space coordinates so that $f = f_c$. The result that the final entropy exceeds the initial entropy is a direct consequence of this assumption. This has some peculiar corollaries. Since equations (1) and (2) are time reversible we can imagine integrating backwards in time to find the entropy at some $t_o < t_i$. The above argument then implies $S(t_o) > S(t_i)$. In addition, if t_1 and t_2 are both later than t_i, we cannot say anything about the relative sizes of $S(t_1)$ and $S(t_2)$. Thus we have not proved that violent relaxation causes S to increase with time; merely that there is, in general, less information in the coarse-grained distribution than in the fine-grained distribution. If we assume the two to be identical at some epoch, then this time is picked out as special.

As THLB note in their discussion of Tolman's proof, the analysis only requires $C(f)$ to satisfy the inequality following equation (9). Thus the proof does not prefer the particular functional form given in that equation; any convex function $C(f)$ could replace $-f\ln(f)$ in the definition (7). Once again the situation differs from that for gaseous systems where the standard proof of the H-theorem shows that equation (4) is the only possible definition for which entropy always increases. In collisionless stellar dynamics there are infinitely many H-functions corresponding to all possible choices of C. THLB show that there is actually a one-to-one correspondence between the choice of C and the distribution function $f_c(\epsilon)$ (satisfying $df_c/d\epsilon < 0$) given by Lynden-Bell's maximisation procedure. Thus, although Lynden-Bell's distribution (5) is the unique most probable state implied by his counting rules for assigning statistical weights to macrostates, Tolman's analysis of evolution does not assign it any special status. The only clear prediction would seem to be that violent relaxation should lead to isotropic distribution functions in which the phase space density is a declining function of energy. Unfortunately even this very general prediction is not well borne out by experiment. Numerical models show the outer regions of "galaxies" formed by violent relaxation to be highly anisotropic; they are composed mainly of stars on near radial orbits which have been ejected from the central regions.

It is clearly desirable to find some way to define the effect of evolution which does not make reference to any specific choice of H-function. THLB give a simple formalism which does this by utilising the functions $M(f)$ and $V(f)$ defined in equation (3). Eliminating the fine-grained phase space density leads to a new function, $M(V)$. This specifies the maximum mass which collisionless evolution could ever conceivably pack into any phase space volume of size V. If we now imagine using the coarse-grained phase space density of the system to define similar functions $M_c(f_c)$, $V_c(f_c)$ and $M_c(V)$, then, because of the averaging implicit in the definition of f_c, $M_c(V)$ must be less than $M(V)$ for all V; the two can be equal only if $f_c = f$ throughout the system. Thus if we are willing to assume that $f_c = f$ at some particular time (say the initial time), $M_c(V)$ at all other times, earlier or later, is bounded above by $M_c(V)$ at that time. This result can be expressed, following THLB, by saying that the system must always appear more mixed than at the chosen time. Thus if we are willing to assume that the initial conditions for violent relaxation are fully specified by f_c, we can restrict the class of equilibrium distributions for the final state to those which are more mixed than the initial conditions. Notice that there is still no sense in which we have picked out an arrow of time. Our assumptions promote one particular time to special status, but we have no way to order the properties of the system at any other two times. A well-known consequence of this mixing constraint (obtained in the limit $V \to 0$) is the fact that the maximum value of f_c in the final state must be less than its maximum value in the initial conditions. A version of this argument was used by Tremaine and Gunn (1979) to put upper limits on the central density of the galaxy halos which might form in a neutrino dominated universe.

Finally, I think that it is important to note that despite the formal difficulties discussed at length above, the conceptual framework of violent relaxation theory is of considerable use. In most situations the collapse and relaxation of an inhomogeneous and relatively cold system does indeed lead to an equilibrium with typical phase space densities much lower than in the initial conditions. This final state will almost always have a significantly higher coarse-grained entropy.

III. NUMERICAL STUDIES OF DISSIPATIONLESS FORMATION

A full treatment of evolution under equations (1) and (2) can, in general, be carried out only by numerical simulation, and, indeed, any large-scale N-body simulation of gravitational evolution from nonequilibrium initial conditions can be said to be a simulation of violent relaxation. Typically if the total energy of a system is negative, it settles down after a few dynamical times to an approximately steady state. A small fraction of the mass may escape to infinity. The structure of the final state depends quite strongly on the initial conditions. Its maximum phase space density cannot exceed that of the initial conditions, but is often found to be almost equal to it (Melott 1982; Farouki, Shapiro and Duncan 1983; May and van Albada 1984). The overall shape of the final state usually reflects the geometry of the initial conditions from which it relaxed (White 1976, 1979; Aarseth and Binney 1978). In particular, Aarseth and Binney showed that relaxation from sheet-like initial conditions can lead to equilibria which are considerably more flattened than observed elliptical galaxies. Finally, systematic series of experiments by White

(1979) and van Albada (1982) showed that while violent relaxation from a wide range of initial conditions leads to density profiles which are generally similar to each other and to the luminosity profiles of elliptical galaxies, it is easy to trace systematic trends in the concentration of the final state to details of the structure of the initial conditions.

The strong concentration of elliptical galaxy luminosity profiles is one of their most striking features. For any system in virial equilibrium one can define a characteristic phase space density using its mass, its gravitational radius and its velocity dispersion as,

$$f_o = 3M/(8\pi r_g^3 < v^2 >^{3/2}) = 3(-2E)^{3/2}/(\pi G^3 M^{7/2}), \tag{10}$$

where E is the total energy of the system. For the model which King (1966) fitted to the galaxy NGC 3379 the central phase space density is about 100 f_o. Thus for violent relaxation to stand a chance of reproducing the structure of this galaxy it must occur from initial conditions where a significant fraction of the mass has phase space density a hundred times that defined by the total mass and energy. If density variations in the initial conditions are of order unity, then random velocities must be much smaller than the characteristic virial values. On the other hand, if the initial conditions are highly inhomogeneous, stars in dense regions can have moderately large random velocities and still have high phase space density. Thus cold collapse and the merging of previrialised subunits provide two possible routes for dissipationless galaxy formation. It turns out that both routes can lead to systems which resemble observed ellipticals. White (1979) and van Albada (1982) give examples which have density profiles very similar to de Vaucouleurs (1948) empirical density law, and which formed by merging and from a cold collapse, respectively. However, there is nothing magic about the $r^{1/4}$-law. The early spherical models of Gott (1973) show that if too much symmetry is imposed on a collapse, randomisation processes are too weak to produce a concentrated galaxy. More recent work shows that both merging (Villumsen 1982; Farouki, Shapiro and Duncan 1983) and cold collapse (McGlynn 1984) can also lead to systems which are more concentrated than observed galaxies. In such systems the density profiles approximate power laws of slope -3 over many decades in density, a property for which I give a tentative explanation in the next section. The observed uniformity of luminosity profiles cannot, therefore, be explained by violent relaxation alone; it must also reflect some uniformity in the initial conditions.

IV. POSSIBLE EXPLANATIONS FOR OBSERVED DENSITY PROFILES

In this section I present two simple arguments which give some insight into the results of empirical studies of violent relaxation. It seems intuitively clear, at least with hindsight, that the final equilibria should be ellipsoidal, centrally concentrated systems with an extensive outer envelope. (In fact, intermediate states with less symmetry can be surprisingly long lived, see Gerhard (1983)). The asymptotic structure of the outer envelope can be derived as follows. In a system which undergoes strong violent relaxation, stellar energies in the central active region are greatly modified over a period of a couple of dynamical times. Subsequently, stars which have been scattered into weakly bound orbits move out to populate the outer envelope, and a significant number of them escape (typically between 2 and 20 percent of the total mass). Thus if we denote by $N(E)dE$ the number of stars in the final system with energies in the range $(E, E + dE)$, we expect $N(E)$ to be continuous and nonzero in the neighborhood of $E = 0$. In the outer envelope the galaxy potential is approximately Keplerian, so that the size of an orbit is inversely proportional to E. The density profile of the outer envelope follows immediately, since

$$\rho(r) \propto 1/r^2 N(GM/r)\, d(GM/r)/dr \propto r^{-4}. \tag{11}$$

Hoffman and Shaham (1985) found an analogous result for the outer structure of the clumps which form in an open universe. Aguilar and White (1986) applied this argument to the outer parts of tidally stripped galaxies, and Jaffe (this volume) has given it in the present context. Since the stars which populate the outer envelope were all ejected from the central regions, they must be on primarily radial orbits. The asymptotic behaviour of the radial and tangential velocity dispersions is then easily derived from the hydrostatic equilibrium equation and from conservation of angular momentum:

$$\sigma_r^2 = GM/3r >> \sigma_t^2 \approx GMr_i/r^2, \tag{12}$$

where r_i is the scale of the strong interaction region from which stars are ejected. The effective value of r_i is smaller for systems which form by cold collapse than for merger products. As a result the region of strong radial anisotropy encompasses a larger fraction of the mass in the former case (compare the results of White (1979) and van Albada (1982)).

The asymptotic profile just derived applies, if at all, only to the extreme outer parts of real elliptical galaxies. The slope of the $r^{1/4}$-law luminosity profile steepens to -3 at 6 effective radii; this is near the outer limit of photometry in most systems. It would be desirable to find some explanation for the power-law structure of the main body of a violently relaxed system. The following heuristic argument suggests that this structure may perhaps be understood as the maximum entropy member of a restricted class of distributions.

As a toy model for the result of violent relaxation, consider a system with a density profile of the following form:

$$\rho(r) = \begin{cases} \rho_o(r/r_c)^{-\gamma}; & r_c < r < cr_c, \\ 0; & \text{otherwise.} \end{cases} \tag{13}$$

For given values of the shape parameters γ and c, the scale factors ρ_o and r_c are determined by the total mass and energy of the system. At each radius let us estimate the coarse-grained phase space density as

$$f_c(r) = \rho(r)/(G\rho(r)r^2)^{3/2}. \tag{14}$$

If we substitute this expression into the standard definition of entropy (equation 7), the result, for fixed total mass and energy, is a function of γ and c alone. To within an additive constant it can be written as,

$$S(\gamma, c) = \ln\left[(3-\gamma)^2\left(\frac{c^{5-2\gamma}-1}{5-2\gamma} - \frac{c^{2-\gamma}-1}{2-\gamma}\right)^{3/2}(c^{3-\gamma}-1)^{-7/2}\right] - \frac{6-\gamma}{6-2\gamma} + \frac{6-\gamma}{2}\frac{c^{3-\gamma}}{c^{3-\gamma}-1}\ln c. \tag{15}$$

This expression turns out to be a remarkably weak function of its arguments. For $0 < \gamma < 5$ and $3 < c < 10^6$, S varies by less than 35% around the value -1.5, except for a sharp and well defined region of high values for $\gamma = 2.9 \pm 0.3, c \geq 10^4$. This peak arises because for $2.5 < \gamma < 3$ the mass of the distribution is concentrated at large radii while the binding energy is concentrated at small radii. Thus the half-mass radius can exceed the gravitational radius by a large factor and a significant fraction of the mass of the system can be at phase space densities much lower than the characteristic value of equation (10).

From this argument we see that the profile attained by systems which undergo strong violent relaxation is close to the maximum entropy solution for a system of fixed mass and energy and with a truncated power-law density profile. Note, however, that equation (15) has no true maximum, but diverges logarithmically as $c \to \infty$ for $2.5 < \gamma < 3$. This reflects the the lack of any global entropy maximum for a finite isolated system. It is amusing that equation (15) is minimised for moderately large c and for $\gamma \approx 2$. Thus, contrary to intuition, a standard bounded isothermal halo would appear unlikely to arise from violent relaxation. Finally note that the profiles of elliptical galaxies deviate systematically and consistently from single power-laws. As already remarked above, these deviations must reflect a consistent incompleteness of the relaxation process, and thus must retain information about the initial conditions from which the galaxies formed, if indeed they formed in a dissipationless manner.

V. INITIAL CONDITIONS

Under what circumstances might elliptical galaxies form the bulk of their stars before they relax to their present state? An obvious possibility is elliptical formation by the merging of pre-existing galaxies. Nobody doubts that such mergers occur—a number of clear examples of currently merging systems are known (Toomre 1977; Schweizer 1986). These systems will end up as ellipsoidal star piles, and computer simulations suggest that their density structure and kinematic properties will be quite similar to those of real ellipticals (see White 1983 for a review). Criticism of the hypothesis that this process formed all ellipticals (Toomre and Toomre 1972) has focused on whether the systematic properties of the observed elliptical population can be a result of mergers between observed spirals (e.g. Tremaine 1981). This is not a mandatory requirement, since in most cosmological models mergers occur predominantly at early times when spirals presumably differed considerably from their present day counterparts. However, recent observational work has tended to lessen the force even of arguments against formation from present-day spirals (see Schweizer 1982,1986; Dressler and Lake 1986). The data presented by Djorgovski at this conference suggest that the observed parameters of ellipticals spread over much of the range allowed by the constraints of a fixed density profile and a fixed stellar population. These latter regularities are the only ones expected on the merger hypothesis.

Let us consider whether dissipationless formation is otherwise plausible within current cosmogonies. The latter may be broadly divided into those in which galaxies form by collapse from a coherent shock front of galactic or supergalactic scale, and those in which they form by the hierarchical aggregation of material into larger and larger systems. The old "adiabatic" and "isothermal" formation pictures of the 1970's are the basis of this classification (Sunyaev and Zel'dovich 1972; Peebles

1974; White and Rees 1978). Among more recent theories, neutrino dominated universes (Bond, Efstathiou and Silk 1980: Doroshkevich *et al.* 1980) and explosive formation theories (Ostriker and Cowie 1981; Ikeuchi 1981) belong to the first category; cold dark matter universes (Blumenthal *et al.* 1984; Davis *et al.* 1985) and string-induced galaxy formation (Vilenkin 1985) belong to the second. It is important to note that in all of these theories the characteristic scale of galaxies is set by the cooling properties of the baryonic gas. Strictly dissipationless formation (except by mergers at late times) is thus implausible in any of them.

The results of Aarseth and Binney (1978) show that if stars were formed by fragmentation of a large-scale shock front, subsequent collapse could produce flattened systems with the concentration of elliptical galaxies. In fact, it could form objects which are much flatter than any known galaxy. Thus pancake theories might in principle have difficulty explaining why elliptical galaxies are so round. However, it turns out that in most situations the shocks fragment, if at all, into smaller objects than galaxies; these must then merge into the systems we see. In hierarchical clustering models each object forms from smaller objects which collapsed earlier. Galaxy formation is thus intrinsically a merger process. However the merging systems are mixtures of stars, of any pre-existing dark matter, and of gas that may later make stars. It is not clear that a dissipationless model is appropriate for this situation. However, to the extent that it is, highly inhomogeneous initial conditions are clearly implied. Among hierarchical models, the standard cold dark matter theory predicts the minimum possible level of substructure in collapsing protogalaxies. Nevertheless, the pictures of Frenk *et al.* (1985) show that even in this case large objects form primarily by merging. This bears on the question of whether the radial orbit instability discussed by Merritt at this conference could play a role in the formation of real ellipticals. In the initial conditions predicted by currently popular cosmological theories, inhomogeneities are probably too large to allow a highly radial collapse. The strongly unstable models which have so far been the main focus of investigation thus seem unlikely to be relevant. For more plausible initial conditions it will be difficult to tell if flattening results directly from asymmetries in the initial conditions or from a gentler radial orbit instability.

I would like to thank Herwig Dejonghe for fruitful discussions of entropy in stellar dynamics. This research was supported by NSF Presidential Young Investigator Award AST-8352062.

REFERENCES

Aarseth, S.J. and Binney, J., 1978. *Mon. Not. R. astr. Soc.*, **185**, 227.
Aguilar, L.A. and White, S.D.M., 1986. *Astrophys. J.*, **307**, 97.
Blumenthal, G.R., Faber, S.M., Primack, J.R. and Rees, M.J., 1984. *Nature*, **311**, 517.
Bond, J.R., Efstathiou, G. and Silk, J.I., 1980. *Phys. Rev. Lett.*, **45**, 1980.
Davis, M., Efstathiou, G., Frenk, C.S. and White, S.D.M., 1985. *Astrophys. J.*, **292**, 371.
de Vaucouleurs, G., 1948. *Ann. d'Astrophys.*, **11**, 247.
Doroshkevich, A.G., Khlopov, M.Yu., Sunyaev, R.A., Szalay, A.S. and Zel'dovich, R.A., 1980. *Ann. N.Y. Acad. Sci.*, **375**, 32.
Dressler, A. and Lake, G., 1986, preprint.
Farouki, R.T. and Shapiro, S.L., 1982. *Astrophys. J.*, **259**, 103.

Farouki, R.T., Shapiro, S.L. and Duncan, M.J., 1983. *Astrophys. J.*, **265**, 597.
Frenk, C.S., White, S.D.M., Efstathiou, G. and Davis, M., 1985. *Nature*, **317**, 595.
Gerhard, O.E., 1983. *Mon. Not. R. astr. Soc.*, **203**, 19P.
Gott, J.R., 1973. *Astrophys. J.*, **186**, 481.
Hoffman, Y. and Shaham, J., 1985. *Astrophys. J.*, **297**, 16.
Ikeuchi, S., 1981. *Pul. Astr. Soc. Jap.*, **33**, 211.
King, I.R., 1966. *Astron. J.*, **71**, 64.
Lynden-Bell, D., 1967. *Mon. Not. R. astr. Soc.*, **136**, 101.
May, A. and van Albada, T.S., 1984. *Mon. Not. R. astr. Soc.*, **209**, 15.
McGlynn, T.A., 1984. *Astrophys. J.*, **281**, 13.
Melott, A.L., 1982. *Phys. Rev. Lett.*, **48**, 894.
Ostriker, J.P. and Cowie, L.L., 1981. *Astrophys. J. Lett.*, **243**, L127.
Peebles, P.J.E., 1974. *Astrophys. J. Lett.*, **189**, L51.
Schweizer, F., 1982. *Astrophys. J.*, **252**, 455.
Schweizer, F., 1986. *Science*, **231**, 227.
Sunyaev, R.A. and Zel'dovich, Ya.B., 1972. *Astron. Astrophys.*, **20**, 189.
Tolman, R.C., 1938. *The Principles of Statistical Mechanics*, Clarendon, Oxford.
Toomre, A., 1977. In: *Evolution of Galaxies and Stellar Populations*, eds. Larson, R.B. and Tinsley, B.M., Yale University Observatory.
Toomre, A. and Toomre, J., 1972. *Astrophys. J.*, **178**, 623.
Tremaine, S.D., 1981. In: *Structure and Evolution of Normal Galaxies*, p. 67, ed. Fall, S.M. and Lynden-Bell, D., Cambridge Univ. Press.
Tremaine, S.D. and Gunn, J.E., 1979. *Phys. Rev. Lett.*, **42**, 407.
Tremaine, S.D., Hénon, M. and Lynden-Bell, D., 1986. *Mon. Not. R. astr. Soc.*, **219**, 285.
van Albada, T.S., 1982. *Mon. Not. R. astr. Soc.*, **201**, 939.
Vilenkin, A., 1985. *Phys. Rep.*, **121**, 263.
Villumsen, J.V., 1982. *Mon. Not. R. astr. Soc.*, **199**, 493.
White, S.D.M., 1976. *Mon. Not. R. astr. Soc.*, **177**, 717.
White, S.D.M., 1979. *Mon. Not. R. astr. Soc.*, **189**, 831.
White, S.D.M., 1983. In: *Internal Kinetics and Dynamics of Galaxies*, p. 337, ed. Athanassoula, E., Reidel, Dordrecht.
White, S.D.M. and Rees, M.J., 1978. *Mon. Not. R. astr. Soc.*, **183**, 341.

DISCUSSION

Luwel: i) Severne and I derived an evolution equation for the coarse grained distribution function which has the Lynden–Bell distribution as a stationary state. We also have an H–theorem for that evolution equation. ii) What is your opinion on Shu's (1978, *Astroph. J.*, **225**, 83) result on violent relaxation?

White: i) It is interesting that you were able to obtain a result equivalent to the classical result that a Maxwellian is a stationary state of the Boltzmann equation for a collisional gas. However, I do not see how this can be applied to real systems where relaxation appears always to be incomplete and where global constraints preclude the Lynden–Bell distribution.
ii) Shu's arguments rely on examining the limit imposed on the validity of the collisionless Boltzmann equation by the discreteness of the stars. However, the

definition of violent relaxation which I have been using and which was, I believe, intended by Lynden-Bell, is that it is the dynamical relaxation undergone by a system which obeys the collisionless Boltzmann equation exactly. This would seem to be a valid approximation for most situations of interest. For example, if a large galaxy forms by the merger of two, ten or even a hundred subgalactic stellar systems it seems obvious that the discrete nature of the stars will play no role in determining how the final state is related to the properties of the initial set of systems.

Gott: You mentioned that the entropy would increase to both the future and past of the initial conditions. This occurs in any system. If you find an ice cube on a stove and ask what is the most probable configuration at a later time it is a partially melted ice cube. If you assume that it is an *isolated* system then the most probable configuration at an earlier time is also a partially melted ice cube. In a truly isolated system the ice cube is just a statistical fluctuation and to find a still larger ice cube at earlier times would be still more improbable.

White: Fair comment!

Ostriker: You argue that if there are three times $t_2 > t_1 > t_i$, then an appropriately defined entropy S would have the property such that $(S_2 > S_i, S_1 > S_i)$ but that nothing could be said concerning $(S_2 >< S_1)$. But could we not redefine t_1 to be a new "initial" time with a newly defined "fine grained" distribution, so that $S_2 > S_1$ and an arrow of time is defined?

White: I agree that the choice of the initial time is, to some extent, arbitrary. The assumption that the coarse and fine–grained distributions are identical at the initial time is essentially an expression of our ignorance about the fine details of the initial distribution. However, if we consider the evolution of any specific system we can consistently apply this assumption at one time only. The structure at any other time is related to that at the chosen time by the equations of motion which may well require specific kinds of small scale structure in the distribution function. Thus only the chosen time is picked out as having lowest entropy.

Ostriker: One difference between collapse from a sheet and a three–dimensional collapse is that the effect of overall expansion is different. The result is that infall continues in a three–dimensional collapse but is cut off at finite mass for two–dimensional collapse. Will this difference produce a different final structure in any measurable way?

White: Yes, I think collapse from a sheet would lead to an asymptotic outer profile with $\rho \propto r^{-4}$ as I described. Cosmological infall in a *flat* universe leads to profiles which, in general fall off less steeply with radius.

Djorgovski: You can do a good deal better than simply comparing the radial density profiles. After all, there are correlations, which need to be preserved: take the fundamental plane of physical properties of ellipticals, discovered by Faber *et al.*, myself, and indicated by others. It is not clear to me whether dissipationless mergers will keep the galaxies on this plane or not. This is not a question for the formation epoch only—you may ask whether, if you merge two present day ellipticals, the product will stay on the plane?

White: This is an important point which is not addressed by dissipationless formation models. In such models the distribution of characteristic properties over the population of ellipticals is a result of the distribution of the initial conditions for collapse (or merging). This can only be specified in the context of a more comprehensive theory of galaxy formation. It is possible for the remnant of a merger between two present day ellipticals to lie on the "fundamental plane" you refer to. Whether it does so depends on the orbit from which the galaxies merge and on whether they are surrounded by dark halos.

Burstein: I also wish to refer back to the real world. Recent observations indicate that almost all ellipticals have line–strength gradients that extend to, or beyond, their effective radii, (i.e., they are not just due to the central peak in line–strength). What are your current thoughts on this constraint on dissipationless formation processes?

White: The line–strength gradients do provide a constraint on dissipationless models, but they are relatively easy to understand in the context of models which have at least what I would consider to be the minimum plausible amount of initial inhomogeneity. This is so because the violent relaxation process roughly preserves the ordering of stars in binding energy. (See White 1978, *Mon. Not. R. astr. Soc.*, **184**, 185; van Albada 1982, *Mon. Not. R. astr. Soc.*, **201**, 939). Thus the most bound stars in the final galaxy tend to be those that were most bound in the smaller clumps in which they originally formed. One would naturally expect these to also be the most metal rich stars.

Palmer: If I understand you correctly, the only reason why these "entropies" increase is because you coarse–grain the distribution. The actual physics of the evolution *conserves* these functions. This is in marked contrast to the $f \ln f$ entropy for collision dominated systems, where physical collisions mix up the phase space distribution.

White: It is true that the $f \ln f$ entropy defined using the fine–grained single particle distribution function always increases in collisional systems, provided that the two–particle distribution function which appears in the Boltzmann collision term can be assumed to be a simple product of single particle distributions.

Barnes: You've shown how the splashing of stars out to infinity gives a characteristic profile to the outer part of an E galaxy. How is this argument modified by the presence of heavy halos which merge before the galaxies do?

White: My argument will probably not apply in this (quite plausible) situation. In particular, if the highest stellar energy generated during violent relaxation is less than the escape energy from the halo, the final stellar distribution will be bounded.

White and Dejonghe discuss entropy.

DISSIPATION AND THE FORMATION OF GALAXIES

R.G. Carlberg
York University and CITA, Toronto
and
Johns Hopkins University, Baltimore.

ABSTRACT. The evidence for dissipation in elliptical galaxies indicates neither the epoch of formation nor the rate of radiation. The hypotheses for the formation of ellipticals include mergers of pre-existing, mostly stellar, disk galaxies; accumulation of gassy fragments that subsequently turn into stars; and the dynamical collapse of a distinct protogalactic gas cloud with simultaneous star formation. Mergers of purely stellar disks seem unlikely, because the phase space density of disks is everywhere far below that of the cores of normal ellipticals. Allowing a few percent of the mass of the galaxy to dissipate into the core and turn into stars could remove this difficulty. In the Hubble sequence of galaxies, ellipticals are characterized by their low angular momentum content. As a start to understanding the general problem for galaxy formation and angular momentum acquisition in the presence of dissipation, a cosmological N-body experiment containing both a dominant collisionless component and an isothermal gas is described. The collisionless component clusters in the usual hierarchical manner appropriate to the spectrum of fluctuations. In contrast, the gas fragments only when the Jeans mass drops below the turnaround mass. The fragments subsequently shrink, becoming distinct entities with relatively low chances of being quickly incorporated in a larger unit. Gravitational torques transfer angular momentum outward in the dissipating gas, placing most of the gas angular momentum at large radii in the protogalaxy. The distant, high angular momentum gas has a relatively long infall time onto the galaxy. The gas may continue to rain down for some time if the galaxy remains undisturbed, or, the growth of clustering may strip the gas off, leaving a low angular momentum system.

1. THE EVIDENCE FOR DISSIPATION

The case for dissipation being a dominant process in the formation of elliptical galaxies largely rests on circumstantial evidence. Galaxies are an apparent break in the clustering hierarchy, since their internal luminosity densities are 1000 times higher than the extrapolation of the correlation function to the sizes of galaxies would predict (Efstathiou, Fall and Hogan 1979). Either galaxies suffered a radial contraction of a factor of 10, or galaxy formation and galaxy clustering arise independently. Whether dissipation was an intrinsic part of the formation of ellipticals is less clear. If dissipation and star formation take place in a common process they leave behind radial chemical abundance gradients, such as those measured (in this volume see papers by Efstathiou, and Davies and Sadler). Metallicity gradients

T. de Zeeuw (ed.), Structure and Dynamics of Elliptical Galaxies, 353–366.

do exist in elllipticals, but in contrast to the manifestly dissipated disk galaxies, ellipticals do not appear to have distinct chemical-kinematic stellar populations. In any case, the observation that large ellipticals are not rotationally flattened (Illingworth 1977, Davies, *et al.* 1983) means that they cannot be formed in dynamical equilibrium from a gaseous protogalaxy.

A case for radiative cooling is made using the average mass densities and virial temperatures of elliptical galaxies, that together imply gas cooling times less than a free fall time (Binney 1977, Rees and Ostriker 1977, Silk 1977). A possible caveat is that the cooling is calculated under the optically thin assumption, whereas the high average surface density of an elliptical implies an optical depth greater than unity to electron scattering alone (Fish 1964).

2. DISK GALAXY MERGERS AND DISSIPATION

Collisions between disk galaxies quite plausibly produce "star piles" having characteristics similar to ellipticals (Toomre 1977). An extreme version of this hypothesis is open to observational test. The phase space density cannot increase unless there is dissipation or two body relaxation. Comparison of the phase space densities in disk galaxies to those in ellipticals should find that the ellipticals have lower values.

The phase space density is estimated as the product of the volume density and a gaussian velocity distribution. The peak phase space density in the core of an elliptical is then,

$$\begin{aligned} f_c &= \frac{M}{L}\frac{\rho_L}{(2\pi\sigma_c^2)^{3/2}} \\ &= \frac{9\sigma_c^2}{2\pi G I_c r_c}\frac{I_c}{2r_c}\frac{1}{(2\pi\sigma_c^2)^{3/2}} \\ &= \frac{9}{2(2\pi)^{5/2}G}\frac{1}{\sigma_c r_c^2} \end{aligned} \tag{1}$$

where ρ_L is the central luminosity density, I_c is the central surface brightness, r_c is the radius at which the brightness drops to half of I_c, and σ_c is the central velocity dispersion. The constant is 10.5 $M_\odot$ $pc^{-3}(km\,s^{-1})^{-3}$ when σ_c is measured in $km\,s^{-1}$ and r_c in parsecs.

The central phase space densities of disks are calculated in a similar manner. Using the locally isothermal, constant scale height, model advocated by van der Kruit and Searle (1982) gives,

$$f_c = \frac{5.04}{(\frac{M}{L}\rho_L)^{1/2} z_0^3} M_\odot pc^{-3}(km\,s^{-1})^{-3}. \tag{2}$$

where ρ_L is the midplane luminosity density at the centre, and z_0 is the scale height. An M/L of 7 is adopted, as derived in van der Kruit and Freeman (1984), and suggested in van der Kruit and Searle (1982). The comparison of the central values is reasonable on the basis that the most dense, most bound material likely sinks to the centre of the merging system (although this has not been established), and it is desirable that any outward decline of metallicity within a disk be preserved in the elliptical.

Figure 1 shows the ellipticals as circles and hexagons (data taken from Lauer 1985 and Kormendy 1985, 1986) and the disks as squares, with the Galaxy as an open square. The disk luminosities are calculated from the disk masses of van der Kruit and Searle 1982 using their M/L. On the basis of Figure 1, purely stellar mergers of bulgeless disk galaxies cannot make the cores of normal ellipticals. Mergers are allowed for the most luminous ellipticals, $M_B \lesssim -22$. Note that the density deficiency is two orders of magnitude for $M_B = -20$ ellipticals.

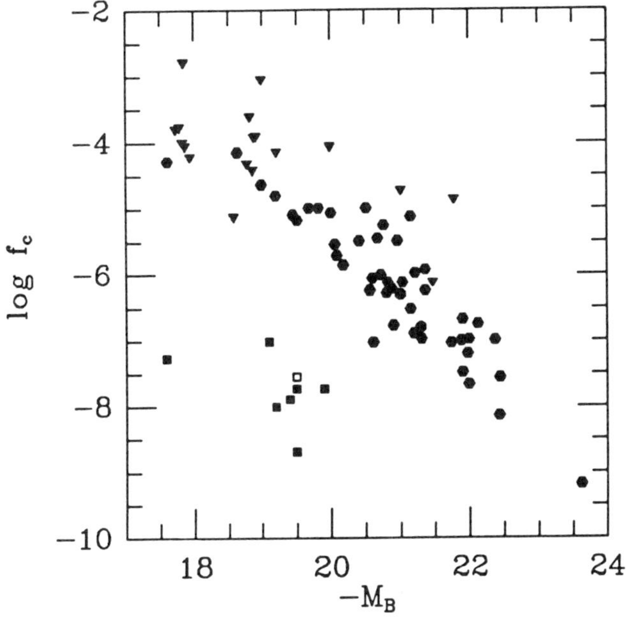

Figure 1. The central phase space densities of disks are shown as squares. Bulges and ellipticals are hexagons (Kormendy 1986) and circles (Lauer 1985). The inverted triangles are lower limits. $H_0 = 75$ for these measurements. Dissipationless merging can never move points upward.

A bulge in a disk galaxy completely overwhelms the phase space density of the disk. Bulges, being hot stellar systems, likely merge conserving their phase space density (Duncan, *et al.* 1983, Carlberg 1986). Since bulges appear to be part of the low luminosity extension of ellipticals in Figure 1, the merger of disks with bulges would move points horizontally forward placing them above the elliptical line in Figure 1. Candidates for old merged disk galaxies are the few normal ellipticals where the phase space densities are exceptionally high for their luminosity, possibly, those with only lower limits in Figure 1.

There are two conclusions that could be drawn from this comparison of disk galaxies and ellipticals. If mergers of disks make ellipticals, then the cores of ellipticals must be almost completely composed of stars subsequently added to the core, possibly through gas sinking to the centre and turning into stars. In this case, the cores could have different kinematics, for instance, more rotation; and possibly different metal abundances than the rest of the galaxy. The core buildup must be done in such a way that the fairly tight correlations between core radius, core surface brightness, and total luminosity arise (Kormendy 1985, Lauer 1985). A more extreme conclusion, and one that is premature in the absence of a better understanding of gas and star formation, is that mergers do not make most normal ellipticals.

3. GALAXY FORMATION IN A DISSIPATIVE COSMOLOGY

Dissipative formation models for individual ellipticals have been examined by Larson (1975), Miller and Smith (1981) and Carlberg (1984a,b). These experiments fairly successfully reproduce the typical elliptical's density profile, the consequence of the very rapid star formation allowing a nearly dissipationless collapse of the stellar component. The experiments of Carlberg (1984a, 1985) show that similar parameters for star formation and gas cooling can give rise to ellipticals, bulges, and disks depending on the ratio of core to envelope mass in the protogalaxy as it turns around from the Hubble flow.

The uncertainties in numerical experiments of galaxy formation are manifold. Simple gas physics can be adequately simulated, leaving the main uncertainties in the initial conditions. All the models reported above chose nearly uniform density and velocity fields as starting points and evolved the objects in isolation. Cosmological experiments show that the initial conditions are likely to be very nonuniform. Furthermore, since present day galaxies interact substantially with one another and their environment, at earlier times the protogalaxies probably did so even more.

To understand the relation of galaxies to large scale structure it is essential to have a basic understanding of the physics of a universe containing collisionless dark matter and gas. To this end, exploratory numerical experiments of the growth of galaxies within a two component universe are reported below. The gas physics and cosmological model are deliberately kept as simple as possible in order to highlight the general principles that are expected to hold for various initial conditions and more elaborate gas physics. The present analysis focuses on the growth of angular momentum in the two components.

3.1 The Experiment

Galaxy formation within a group or cluster of galaxies is an interesting situation, particularly since elliptical galaxies tend to be found in such high density neighbourhoods (Dressler 1980). The presence of considerable dark matter in the clusters is well established, and, in many, X-ray data suggest that the dark matter is smoothly distributed (Jones *et al.* 1979). If the gas and dark matter were initially similarly distributed, then the current presence of galaxies as discrete units argues that they must shrink enough through dissipation to lower their collision cross-sections (White and Rees 1978).

The experiment starts with an isolated expanding sphere of 2×10^5 particles, randomly distributed. Since the particles are so numerous the amplitude of density fluctuations on large scales is small. The particle positions are further perturbed with an imposed white noise spectrum, added as several hundred randomly oriented waves. The velocities are left as pure Hubble flow, with an expansion velocity such that the initial $\Omega = 1.28$. Half of the particles are collisionless and only interact via gravity, while the other half are subject to both gravity and collisions. One collisionless particle has a mass 10 times one gas particle. Initially the two particle distributions are identical. Gravity is evaluated on a grid 64 on a side, using an FFT method. The particles are advanced with a second order accurate method.

An isothermal gas is the simplest that loses energy under compression. Adopting this equation of state has the advantage of bypassing the complexities of cooling. The thermodynamic and transport properties of isothermal particles are fairly well defined, as opposed to, say, "sticky particles". Of course detailed realism has been sacrificed, but the aim of this exploratory study is to highlight the essential differences between a cooling gas and the collisionless background. The particles identified as gas are turned over to an algorithm that searches for nearest neighbours and tests them for approaching velocities. If there is a collision, then the particles are brought to their centre of mass and momentum, and assigned new relative velocities drawn from a Maxwellian distribution. This method is quick, has no problem with particle interpenetration, and is a fairly accurate reflection of the kinetics of gas atoms. Useful measurements require many particles to blur the particle microphysics simulation into an acceptable realization of the desired physical experiment. For these experiments there are usually 1000 or more particles per fragment, which is adequate for this investigation.

The problem of star formation is sidestepped here. These experiments are mainly investigating the properties of the protogalactic gas clouds that fragment out of the expansion. Adding star formation at this point would mostly confuse the issues. The fragments here are subject to considerable stirring from halo merging that would likely keep any stars that did form fairly well mixed with the gas. If the experiment was less overdense, $\Omega \leq 1$, some gas fragments would escape immediate inclusion in a cluster, and would ultimately dissipate down to flat, angular momentum supported, gas disks. To have any spheroids in such a situation it would be necessary to introduce star formation into the experiment.

The important physical length in this problem is the Jeans length,

$$\lambda_J = \frac{2\pi}{\sqrt{3}} R^{3/2} c_s,$$

in units where $G = M = R_0 = 1$, and where R is the current size of the sphere, and c_s is the sound speed of the isothermal gas. Experiments were done with c_s of 0.01 and 0.03 giving 10 to 30 Jeans lengths across the experimental volume.

The particle viscosity is of considerable importance when angular momentum is an issue. In these experiments the effective Reynolds number of a fragment is,

$$\Re = \frac{V_0 \lambda_J^2}{\frac{1}{3} \ell c_s}.$$

where ℓ is the mean interparticle seperation, and V_0 is the initial expansion velocity, 1.25 in the units here. The Reynolds number across a fragment is about 100, which should be safe from viscous effects for the duration of these experiments, only a few dynamical times. As a check, collisions were turned off for particle separations less than 0.01 units, thereby stopping any gas viscosity in regions with overdensities at maximum expansion greater than about 1000. No significant change was found in the results. It seems safe to conclude that angular momentum transfer in these experiments is dominated by gravitational torques.

3.2 The Growth of Clustering in the Two Components

Figure 2 shows an x-y projection of the distribution of collisionless and gas particles at a time near maximum expansion. The collisionless particle distribution is essentially identical in both the cold and warm gas experiments, hardly surprising since the collisionless particles contain most of the mass. The gas particles have a strikingly different distribution, dependent on the temperature, they are much more compact and the number of gas fragments is strongly dependent on temperature. The small sizes of the gas fragments are of course due to dissipation, although their core sizes are largely determined by the softening of gravity (there are about 60 softening lengths across the pictures).

The collisionless particles cluster in the hierarchical manner appropriate to the prescribed fluctuation spectrum. Small mass fragments first appear and then merge together to form fewer larger groups. Eventually nearly all of the particles here will be incorporated into one large group. The gas particles cluster rather differently. Gas fragments do not appear until the size scale that is turning around exceeds the Jeans length. The fragment masses are rather broadly distributed around the Jeans mass. Following turnaround they collapse and dissipate on a dynamical time scale. The number of gas fragments decreases only a little with time, going from 24 to 18 over the time interval from 3.5 to 7.0 in a warm experiment. In contrast the number of collisionless particle fragments decreases from 71 to 33 over the same time interval. The shrinkage of the gas fragments is enough to "break the hierarchy", and reduce their merger rate below the undissipated rate (Aarseth and Fall 1980). Unfortunately the shrinkage in these two experiments is inadequate to allow the entire cluster of galaxies to survive the epoch of entire cluster collapse without considerable merging.

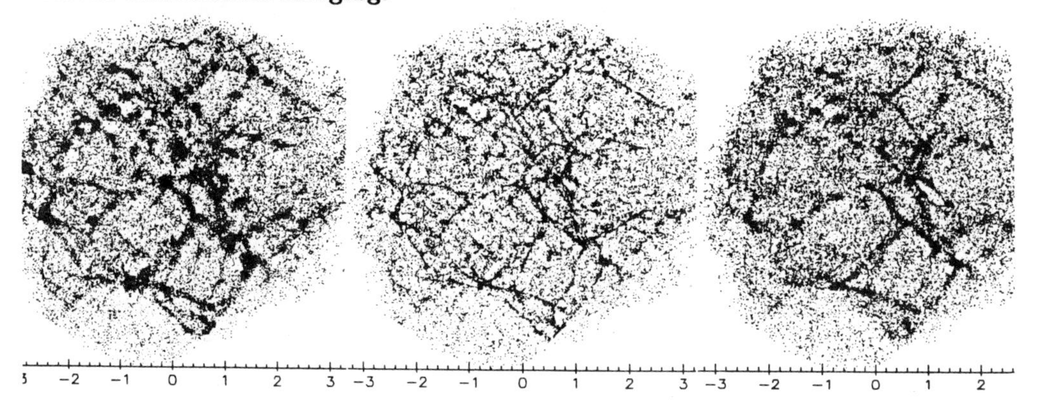

Figure 2. Projection on the x-y plane of the particles at t=4. From left to right, collisionless particles, $c_s = 0.01$ gas, and $c_s = 0.03$ gas.

3.3 Angular Momentum in the Fragments

Angular momentum is acquired when the quadropole moment of the protogalaxy couples to the tidal field produced by neighbouring galaxies (Hoyle 1949). Elliptical galaxies have roughly a factor of six times less angular momentum than comparable mass disks (Fall 1983) so their spins are not compatible with being the tail of the general spin distribution. Some process is required to either pick out fragments that are likely to have low angular momentum (Blumenthal *et al.* 1984) or remove or dilute the angular momentum that is generated by tidal torques (Fall 1983, Frenk *et al.* 1985). The experiments reported here find an additional effect; the protogalactic gas has a disproportionately large amount of angular momentum at large radii where it is susceptible to stripping during the build up of clustering.

To identify groups for analysis, a "friends of friends" algorithm is used that joins all particles that are within a specified minimum distance of each other. The standard separation is 0.02 units, or about 1/5 of the mean interparticle separation. The group radii and angular momenta are somewhat dependent on the choice of group density since the outer parts are constantly being disturbed, and the inner regions communicate with the outer parts via gravitational coupling. The chosen interparticle separation for inclusion in a group gives fragment properties that are well defined, and are repeatable at other seperations after suitable scaling.

To compare the properties of the collisionless matter with the gas a moment is chosen when the two types of fragments have nearly equal numbers of particles in them. Thus, in the warm ($c_s = 0.03$) experiment, the time 3.5, near maximum expansion, is chosen as the standard for analysis. Fragments are analyzed only if they contain at least 200 particles. About 80% of the collisionless particle fragments (halos) contain gas fragments (galaxies). In fact there is a close correspondence between the two types of particles, it just happens that some of them have insufficient numbers in their groups to be analyzed together.

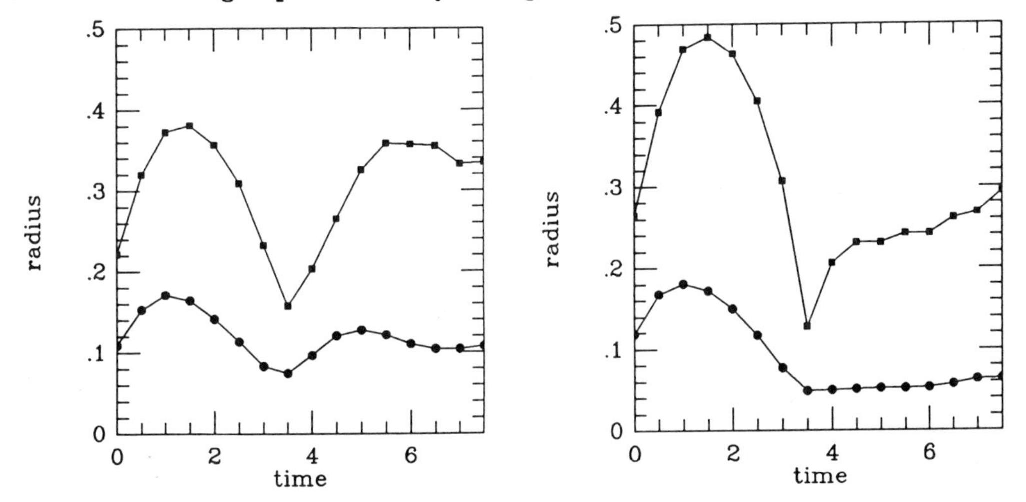

Figure 3. Average (circles) and maximum (squares) radii of groups in the collisionless (left) and gas (right) particles.

The positions of the particles that eventually come together to make the fragments identified at time 3.5 can be traced forward and backward in time. Figure 3 shows both the average radii and maximum radii of the fragments with time. The gas groups contain slightly more particles on the average so they are larger during the expansion phase. The separation of the gas fragments from the general clustering can be seen in Figure 3, since their average radius collapses to a well defined size, and stays there nearly to the end of the measurement time. The collisionless particles collapse, only to immediately re-expand as the original fragments are diluted by mergers with other groups. It is impressive that a little dissipative contraction (about a factor of 2 here) on top of a dissipationless collapse is sufficient to shut off a lot of merging in the gas component. The data given in §3.2 indicates the merging rate of the collisionless fragments is twice that of the gas fragments.

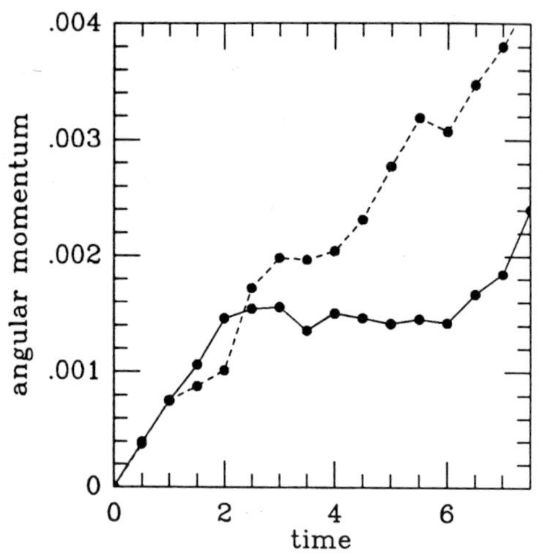

Figure 4. Average specific angular momentum vs time in the gas fragments (solid line) and collisionless fragments (dashed line) for particles in fragments identified at t=3.5.

The average specific angular momentum of the gas and collisionless particle fragments are displayed in Figure 4. The average specific angular momentum in the gas fragments is 70% of that in the collisionless fragments. The decoupling of the gas fragments from re-expansion and further spin up in larger units is again evident in Figure 4.

The spin of the gas is mostly acquired during the expansion phase, up to time 2.0. It is interesting to note that during the expansion phase both the gas and collisionless fragments gain angular momentum at the same rate, even though the gas groups are somewhat larger in size (see Figure 3) during this time. Since both components are experiencing the same tidal field the difference must be a result of different quadropole moments. Figure 5 shows the shape, measured as the geometric mean of the ratios of the two smaller principal moments of inertia to the largest one. The gas fragments stay significantly rounder during their expansion, leading to a smaller acquired angular momentum. The shapes are essentially the same after collapse. The difference in shape is not particularly large here, since most of the mass is in the collisionless component. If the gas is completely self-gravitating the effect is stronger. The relatively more spherical shape is a manifestation of the stabilization of the Lin-Mestel-Shu (1965) shape instability. A dynamical collapse of a constant density ellipsoid is stabilized against further flattening by a pressure of 3/5 of the gravitational field on the short axis (Lynden-Bell 1964, 1979). Some help may also come from density gradient (see Figure 8) present in the fragments before collapse (Goodman and Binney 1983).

The most remarkable feature of the gas angular momentum is the relatively large fraction at large radii in the gas fragment. Figure 6 gives shows $M(h)$ curves, that is, the particles in a fragment are sorted by radius, then the cumulative mass is plotted against the cumulative angular momentum, both normalized to their maximum values. The $M(h)$ are shown at two times, the time of maximum expansion of the groups and at the time of group identification. Most of the gas angular momentum is at large radii. In fact, about 50% of the angular momentum lies in the outer 20% of the mass, and 80% lies in the outer 50% of the mass. At no time does the angular momentum distribution closely resemble that of a uniformly torqued constant density sphere, for reasons discussed with respect to Figure 8.

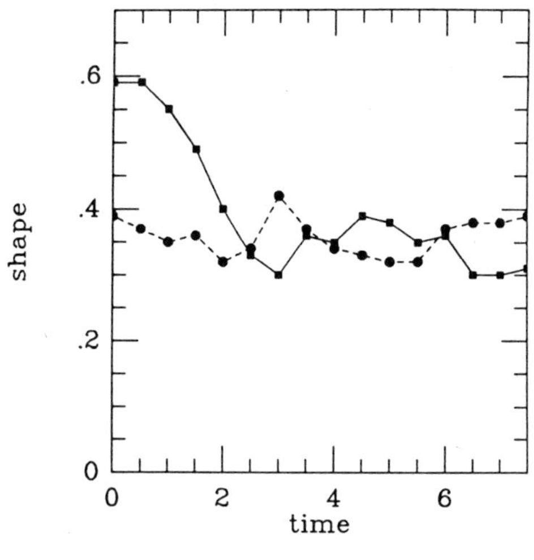

Figure 5. Shape vs time in the gas (solid line) and collisionless (dashed line) particle fragments. Perfectly spherical fragments would have a shape of unity. Most of the gas spin is acquired before time 2.0.

The angular momentum transfer is a consequence of the gravitational coupling between the dissipationless material and the cooler, dissipating gas. The transfer of angular momentum is only very weakly dependent on the gas temperature, if the sound speed in the gas is lower than about 50% of the virial velocity within the fragment. If the gas sound speed is near the sound speed, or equivalently, if the gas is adiabatic then the gas and the collisionless particles have nearly identical angular momentum distributions.

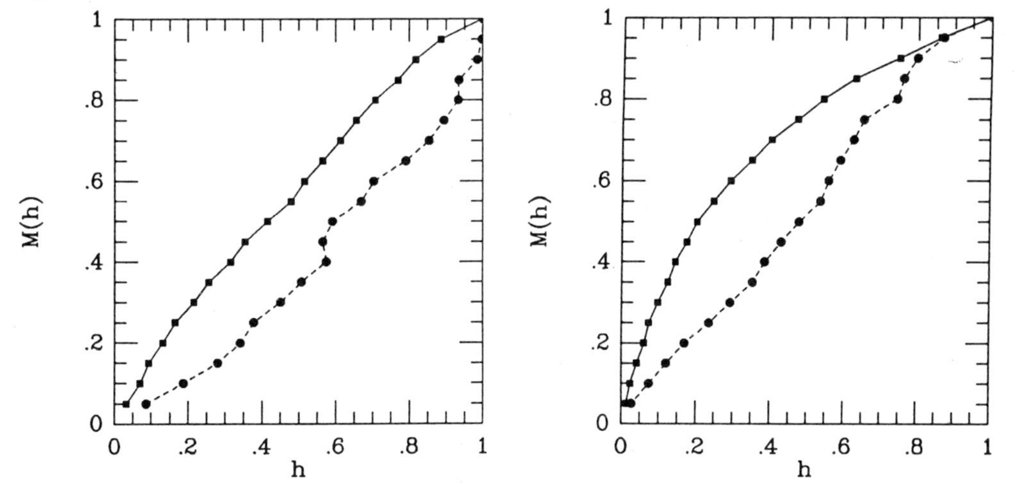

Figure 6. $M(h)$ distributions in the gas (solid line) and collisionless (dashed line) particles at time 1.5, maximum expansion, and time 3.5, when the fragments were identified.

The average characteristic velocities of the gas particles are shown in Figure 7. The circular, radial infall, and circular streaming velocities are measured as spherical averages over shells. At the outermost radius of the gas its streaming velocity is 0.31 of the circular velocity, whereas the collisionless particles have a streaming velocity of only 0.18 at the same radius. This outermost gas can only collapse by a factor of 3 in radius before it is completely supported by angular momentum. A general collapse is also evident if Figure 7, but with a relatively low velocity At this rate it will take five dynamical times, as measured at the outside edge, for the material at the edge to fall in. This outermost gas must be allowed to settle in a quiet environment if it is to make it into a disk.

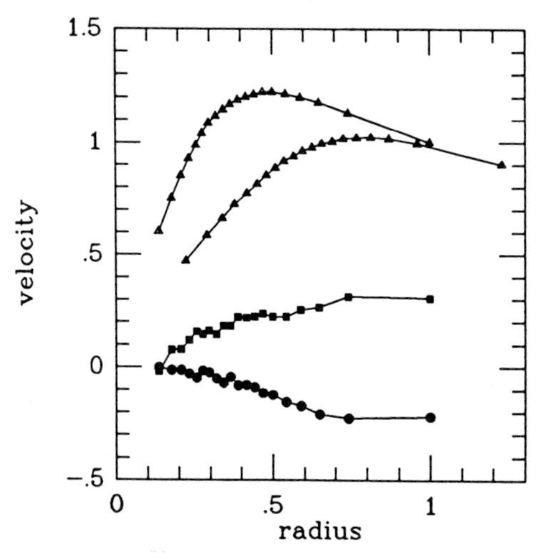

Figure 7. Characteristic velocities measured at time 3.5 normalized to the circular velocities at the edge of the fragments. Circles give the average radial velocity and squares the average azimuthal streaming velocity in the gas fragments. The upper line with triangles shows the circular velocity, $\sqrt{(GM/R)}$ assuming the gas is completely self-gravitating. The lower line with triangles shows the true normalized circular velocity from the collisionless particle mass distribution.

The objects that are collapsing are not cold, constant density spheres. Rather, the density declines outwards from the centre, so that the collapse time outside the core increases with radius. The radial density profile of the fragments at an early time and at the time of "turnaround" is shown in Figure 8. The points are at 5% intervals of the total mass.

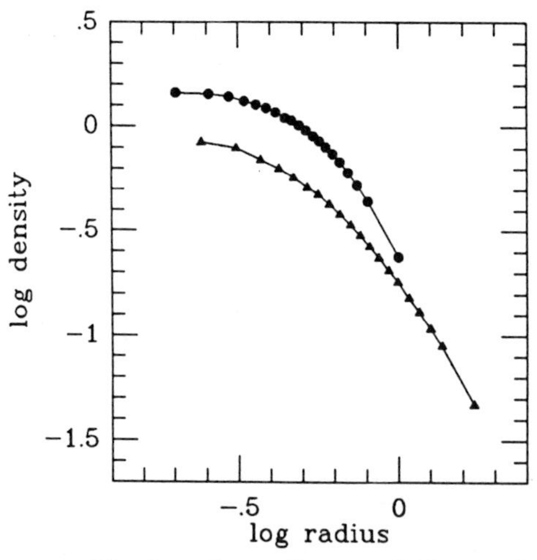

Figure 8. Average density profiles of the particles in the gas fragments identified at $t = 3.5$, measured at the earlier times $t = 0$, the beginning, and $t = 1.5$, turnaround.

During the collapse of the entire system, the gas fragments more or less all merge together to form a "hypergalaxy". Softening limits the collapse available to the fragments, that would core collapse leaving an envelope with $\rho \propto r^{-2}$ behind (Larson 1969). It is interesting to ask what might have happened had the cores been able to shrink more. Larson, Tinsley, and Caldwell (1980) provide a likely suggestion. The extended dark halos must come off the gas fragments to avoid too much mass segregation (White 1976), and metal enriched gas must come out of the galaxies to fill the cluster. Removal of half the gas mass would take 80% of the angular momentum. Furthermore the objects left behind have been subject to a great deal of violent relaxation from ongoing halo merging, making it difficult to form a very thin disk. It is tempting to identify these putative stripped gas fragments with bulges and ellipticals. A consequence would be that the Hubble type of an object would not be determined until the time of clustering of groups of galaxies.

4. CONCLUSIONS

The formation of ellipticals is likely to be strongly dependent on environmental influences, such as the clustering environment and interactions with other galaxies. Ellipticals are dynamically distinct as having relatively low angular momentum content, and high random velocities. The spin of a galaxy is generated by tidal torques, reduced by stripping, and diluted by merging. These three mechanisms have been investigated for their relation to dissipation. Mergers of present day disk galaxies would not make normal elliptical galaxies in the absence of some dissipative star formation to build up the core density. However the most luminous ellipticals have sufficiently large cores that there is no objection to mergers. For normal ellipticals it is not known whether gas resident in a disk could be brought to the centre and turned into stars to build up the core density to the observed values. The angular momentum imparted to a protogalaxy by tidal torques is slightly reduced in the presence of gas pressure, by about a one third. The reduction is due to the decreased quadropole moment as a result of partial pressure support of the protogalaxy and suppression of the Lin-Mestel-Shu instability. The main effect of dissipation is to allow gravitational torques associated with violent relaxation to transport the bulk of the angular momentum in the protogalaxy to large radii, where it could be stripped off during the growth of clustering in the collisionless component.

REFERENCES

Aarseth, S. J. and Fall, S. M. 1980, *Ap. J.*, **236**, 43.

Binney, J. J. 1977, *Ap. J.*, **215**, 483.

Blumenthal, G. R., Faber, S. M., Primack, J., and Rees, M. J. 1984, *Nature*, **311**, 517.

Carlberg, R. G. 1984a, *Ap. J.*, **286**, 403.

Carlberg, R. G. 1984b, *Ap. J.*, **286**, 416.

Carlberg, R. G. 1985, in *The Milky Way Galaxy*, I.A.U. Symp. No. 106, ed. H. van Woerden, p. 615.

Carlberg, R. G. 1986, *Ap. J.*, **309**, in press.

Davies, R. L., Efstathiou, G., Fall, S. M., Illingworth, G., and Schechter, P. L. 1983, *Ap. J.*, **266**, 41.

Dressler, A. 1980, *Ap. J.*, **236**, 351.

Duncan, M. J., Farouki, R. T., and Shapiro, S. L. 1983, *Ap. J.*, **271**, 22.

Efstathiou, G., Fall, S. M., and Hogan, C. 1979, *M. N. R. A. S.*, **189**, 203.

Fall, S. M. 1983, *in Internal Kinematics and Dynamics of Galaxies*, I.A.U. Symp. 100, ed. E. Athanassoula, (Reidel: Dordrecht).

Fish, R. A. 1964, *Ap. J.*, **139**, 284.
Frenk, C., White, S. D. M., Efstathiou, G., and Davis, M. 1985, *Nature*, **317**, 670.
Goodman, J. and Binney, J. 1983, *M. N. R. A. S.*, **203**, 265.
Hoyle, F. 1949, in *Problems in Cosmical Aerodynamics* (Central Air Documents Office: Dayton, Ohio), p. 195.
Illingworth, G. 1977, *Ap. J. (Letters)*, **218**, L43.
Jones, C., Mandel, E., Schwarz, J., Forman, W., Murray, S. S., and Harnden, F. R., Jr. 1979, *Ap. J. (Letters)*, **234**, L21.
Kormendy, J. 1985, *Ap. J.*, **295**, 73.
Kormendy, J. 1986, *private communication.*
Larson, R. B. 1969, *M. N. R. A. S.*, **145**, 271.
Larson, R. B. 1975, *M. N. R. A. S.*, **173**, 671.
Larson, R. B., Tinsley, B., and Caldwell, N. 1980, *Ap. J.*, **237**, 693.
Lauer, T. R. 1985, *Ap. J.*, **292**, 104.
Lin, C. C., Mestel, and Shu, F. H. 1965, *Ap. J.*, **142**, 1431.
Lynden-Bell, D. 1964, *Ap. J.*, **139**, 1195.
Lynden-Bell, D. 1979, *Observatory*, **99**, 89.
Miller, R. H. and Smith. B. F. 1981, *Ap. J.*, **244**, 467.
Toomre, A. 1977, in *Evolution of Galaxies and Stellar Populations*, ed. B. M. Tinsley and R. B. Larson, (Yale Observatory: New Haven).
Rees, M. J. and Ostriker, J. P. 1977, *M. N. R. A. S.*, **179**, 451.
Silk, J. 1977, *Ap. J.*, **211**, 638.
van der Kruit, P. C. and Freeman, K. C. 1984, *Ap. J.*, **278**, 81.
van der Kruit, P. C. and Searle, L. 1982, *Astr. Ap.*, **110**, 61.
White, S. D. M. 1976, *M. N. R. A. S.*, **174**, 19.
White, S. D. M. and Rees, M. J. 1978, *M. N. R. A. S.*, **183**, 341.

DISCUSSION

Lake: In using phase–space densities to constrain merging, you compare central phase densities. Shouldn't you use peak phase densities?

Carlberg: The peak phase–space density in an elliptical is in the core. In a disk the phase–space density increases away from the centre, peaking someplace in the outer disk. At the present time it is completely unclear what disk material will actually end up at the centre of the merged system. The higher phase–space density outer parts of a disk are very cool and likely very susceptible to phase–space dilution. If the disks were test particles imbedded in halos, we know that the central halo particles of the merged system come largely from the centres of the precursor objects. In any case, the peak phase–space density is only a factor of 10 larger than the central phase–space density, so the argument is not greatly affected.

White: Could you clarify your conclusion about the viability of forming ellipticals from "real" spirals which almost always have bulges whose phase–space density is inferred to exceed that required in the elliptical remnant?

Carlberg: Bulges are viewed as the low luminosity extension of the spheroidal sequence. As shown in Figure 8 the protogalaxy has a core of nearly constant density that likely collapses with considerable violent relaxation. The size of the core in a particular protogalaxy would then determine the size of the spheroid. This is clearly extremely speculative, and needs further investigation with higher resolution experiments.

Gerhard: If I wanted to merge disk galaxies to make ellipticals, I would try to do this at early times. Two ways then come to mind by which your phase–space constraint could be circumvented. (i) At early times, the disks contain more gas, which would in a merging event go to the center and increase the central phase–space density of the forming elliptical. (ii) The diffusion process increasing the velocities in disk stars, as discussed by Wielen (1977, *Astr. Astrophys.*, **60**, 263), would not have had time to operate. To show how sensitive the phase–space density of disk material is to this process, let me make the extreme assumption that all stars had only one third of their present peculiar velocity dispersion, *i.e.*, 10 km/s instead of 30 km/s in the solar neighborhood. Then the disk z–scalelength is reduced by a factor of 10, and the phase–space density increased by a factor of 100, according to your formula.

Carlberg: Certainly disks would have had higher gas contents and phase–space densities at earlier times. Mergers of very young gas disks are of course completely dominated by dissipative processes. The phase–space densities of the stellar component for a young disk can be estimated by imposing the constraint that the disk have a Q–value between 1 and 2. At constant Q, the phase–space density increases as the inverse cube of the surface density. Two or three orders of magnitude increase in phase–space density would require a surface density down by about a factor of ten. That means that the disks must be very young, implying a similarly reduced stellar mass and the presence of a great deal of gas in the vicinity.

Kochhar: Why do ellipticals have to come from a merger of spirals?

Carlberg: Making ellipticals from spirals has a number of attractive features (Toomre 1977, *Ann. Rev. Astr. Astroph.*, **15**, 437). In particular, merging avoids having two very different sites for star formation: very rapid star formation in a collapsing spheroid, and slow formation in disks. Mergers are of course seen to occur, and it is hypothesized that the merged systems resemble ellipticals.

Djorgovski: Is M32 a merger of two spirals?

Carlberg: M32 has such an enormous central phase–space density (around 0.1 units) that it doesn't even fit on Figure 1. There are no obvious precursors that have the appropriate properties to make M32.

White: In your talk you didn't mention dissipative models for elliptical formation which are analogous to observed cooling flows. In such systems stars may have formed during subsonic settling of gas in a temporarily static potential well. There has been done recent work in this area stimulated by observations of cooling flows. Would you or someone in the audience like to comment on it?

Carlberg: To some degree, all dissipative models for galaxy formation resemble cooling flows. It is a matter of knowing the Jeans mass of the cooling region. The extreme possibilities are that the entire galaxy is a hot gas of one Jeans mass that is in nearly hydrostatic equilibrium and cooling on a free fall time scale only in its central regions, as in a cooling flow. The other possibility is that the galaxy is made up of many Jeans masses, each of which is a little cooling flow. The time scales in the many small subgalactic clumps may be sufficiently short that most of the star formation would occur in them, and subsequently merge together to form the entire galaxy. Of course, the cooling time argument suggests that galaxies sit right near the dividing line of the two regimes.

Richstone: I was impressed that violent relaxation always gives you $\sigma_r > \sigma_z$ at large r. Can you make a similar statement about models where dissipation is important?

Carlberg: The velocity dispersion in the gas is isotropic, so that it is unlikely that the velocity ellipsoid can ever have a larger tangential component than radial component. In fact, in the models of Carlberg (1984a) it was found that the kinematics of dissipative models were generally similar to those of violently relaxed dissipationless models.

Schechter: The two key processes in your prescription for galaxy formation are merging and stripping. The merging occurs naturally in your simulations but the stripping is done in an ad hoc manner. Is there any prospect of a more realistic treatment of stripping?

Carlberg: The prospects for a more realistic treatment of stripping are excellent, provided that computer time continues to be available in abundance. It is mostly a matter of improving the gas physics. Based on the experience gained here improved methods are currently under development.

Ray Carlberg explains galaxy formation.

SUMMARY

Scott Tremaine
Canadian Institute for Theoretical Astrophysics
University of Toronto
Toronto, Ontario
M5S 1A1, Canada

This is the first IAU symposium devoted specifically to elliptical galaxies. I would like to congratulate the organizers for initiating a symposium on such a well-defined topic with so many interesting puzzles. I have been impressed at this meeting by the great strides made in this subject over the last ten years, both in the quality and quantity of data and in the sophistication of the theoretical models. In addition, I'm sure that I express the sentiments of the great majority of the participants when I say what a pleasure it was to attend a meeting on galaxies in which there was not a single paper on spiral structure!

Rather than attempting to give a general summary, I have chosen to concentrate on four specific topics which reflect some of my personal impressions of the meeting.

NOTATION

Before discussing science I wanted to comment briefly on the vexing issue of the names of the fundamental equations which describe the evolution of stellar systems. If $f(\vec{x}, \vec{v}, t)$ is the density of stars in 6-dimensional phase space, we know that in a collisionless system

$$\frac{\partial f}{\partial t} + \sum_{i=1}^{3} \left[v_i \frac{\partial f}{\partial x_i} - \frac{\partial \Phi}{\partial x_i} \frac{\partial f}{\partial v_i} \right] = 0.$$

This is usually referred to as the "Vlasov equation", but Hénon (1982) has stressed that it is more properly called the "collisionless Boltzmann equation", since it is merely a simplified version of an equation which Boltzmann derived in 1872, long before Vlasov's work. A second equation involves the probability density f_N of an ensemble of N-body systems in $6N$-dimensional phase space. This satisfies

$$\frac{\partial f_N}{\partial t} + \sum_{i=1}^{3N} \left[v_i \frac{\partial f_N}{\partial x_i} - \frac{\partial \Phi}{\partial x_i} \frac{\partial f_N}{\partial v_i} \right] = 0,$$

which is usually called the "Liouville equation". However, Liouville neither derived this equation nor even worked in statistical mechanics, and as far as I know it was first written down in a short abstract by Gibbs (1884), who states that "the object

T. de Zeeuw (ed.), Structure and Dynamics of Elliptical Galaxies, 367–373.

of this paper is to establish this proposition (which is not claimed as new, but which has hardly received the recognition which it deserves) and to show its applications to astronomy and thermodynamics". Although one wishes Gibbs had published more detail, it seems clear that he deserves priority (particularly in astronomical applications), and it might therefore be appropriate for us to replace "Liouville equation" by "Gibbs equation". Finally, the equations of stellar hydrodynamics or "Jeans equations"

$$\frac{\partial \overline{v}_i}{\partial t} + \sum_{j=1}^{3} \overline{v}_j \frac{\partial \overline{v}_i}{\partial x_j} = -\frac{\partial \Phi}{\partial x_i} - \frac{1}{\nu} \sum_{j=1}^{3} \frac{\partial \sigma_{ij}}{\partial x_j},$$

where n, $\overline{v}_i$, and σ_{ij} are the number density, mean velocity and dispersion tensor respectively, were first derived by Maxwell (1866). Unfortunately, he already has a set of eponymous equations, and since Jeans was the first to use the equations in stellar dynamics, the name "Jeans equations" is probably a reasonable compromise. My personal vote, then, is for "collisionless Boltzmann equation", "Gibbs equation" and "Jeans equations".

TRIAXIALITY

The modern era in the study of elliptical galaxies began with the paper by Bertola and Cappacioli (1975) showing that NGC 4697 had a rotation speed far smaller than would be expected if it were a rotating gaseous spheroid. This remarkable result was confirmed and considerably generalized by Illingworth's (1977) observations of thirteen ellipticals, most of which had rotation curves which fell far below the curves expected for rotating gaseous masses. At the same time, Binney (1976, 1978) stressed that slow rotation was a natural consequence of pancake theories of galaxy formation, which produce flattened or triaxial galaxies with little or no figure rotation. The hypothesis that elliptical galaxies may be triaxial has proved to be an extremely fertile one, and much of the theoretical and observational work over the last decade has been devoted to testing and elaborating on this idea.

On the observational side, the progress has been slow. It is discouraging that we still have no conclusive evidence and no clear consensus on whether most ellipticals are triaxial. It is true that observers can cite specific examples which are difficult to explain without triaxiality (see paper by Davies). However, I suspect that this issue, like many others in astronomy, will only be decided by statistical analysis of large samples of high-quality data, and one of the most encouraging features of this meeting was that such samples—CCD surface brightness distributions of dozens to hundreds of galaxies—are beginning to be reported (papers by Djorgovski, Jedrzejewski, and Kormendy). I echo Schechter's comment that these surveys deserve to be analyzed carefully to see whether some clues to the distribution of intrinsic shapes can be disentangled from projection effects.

On the theoretical side, the last decade has produced giant leaps in our understanding of the structure of triaxial galaxies—so much so, that it would be a real shame if the observers found that they were not triaxial after all! Here the turning point was Schwarzschild's (1979) construction of a realistic triaxial galaxy using linear programming. My belief is that the most lasting and important accomplishment of this paper was not the introduction of linear programming, which has by now been superseded in many cases by other approaches, such as Lucy's algorithm and maximum entropy methods (see paper by Richstone). Rather, it was

Schwarzschild's recognition that the traditional construction of models by what might be called "integral-based methods"—using Jeans' theorem and the integrals of motion in a given potential—could be replaced by "orbit-based methods", in which the model builder simply integrates a selection of orbits, computes the fractional time each orbit spends in each spatial grid cell, and combines the orbits so as to make a self-consistent model. For the first time, orbit-based methods enable us to construct galaxy models with realistic triaxial potentials, for direct comparison with both N-body experiments and real galaxies.

MERGERS

One striking aspect of this meeting has been the impressive display of evidence that mergers are a common and ongoing process in ellipticals. Although the importance of this process was clearly recognized by Toomre and Toomre (1972), it is only in the last few years that the "smoking gun"—evidence for recently completed mergers in single galaxies—has been found. Some of the most convincing candidates for recent mergers are the radio galaxies Fornax A (Schweizer 1980) and Centaurus A (Malin *et al.* 1983), and the "double core" elliptical NGC 5813 (Kormendy 1984). The features suggestive of recent mergers in many other candidates include (i) the presence of warped or misaligned HI disks in ellipticals (see papers by Knapp and Schweizer), in particular the presence of over $10^{10}\mathrm{M}_\odot$ of HI in the giant elliptical NGC 807 (Dressel); (ii) the presence of dust and ionized gas (Bertola and others); (iii) ripples and shells (Quinn and others); (iv) polar rings; and (v) the IRAS starburst galaxies. A nice recent review of this subject is given by Schweizer (1986).

VIOLENT RELAXATION

N-body experimenters now have codes which can reliably follow the collapse and violent relaxation of galaxies from cold initial conditions (see papers by van Albada and White). It appears that whenever the relaxation is violent enough, the final state is largely independent of the initial conditions and has an $R^{1/4}$ surface density profile. It would be nice to have a simple analytic distribution function which both reflects the physics of violent relaxation and fits the results of these numerical simulations. I have been inspired by two of the papers presented here to offer a heuristic derivation of such a function.

Violent relaxation redistributes the energies and angular momenta of stars through the strong potential fluctuations in the central core of the collapsing galaxy. Any relaxation process of this form tends to produce a Maxwell-Boltzmann distribution with temperature inversely proportional to energy, $f(\vec{x},\vec{v}) \propto \exp(-\beta E)$, where E is the energy per unit mass. However, violent relaxation is not complete, since the potential fluctuations only last for a limited time $t_{\rm eff}$, which is of order one or two crossing times. Thus highly eccentric orbits, whose radial orbit period t_r exceeds $t_{\rm eff}$, will be underpopulated, by a factor of order $t_{\rm eff}/t_r$. We may crudely account for this effect by replacing the Maxwell-Boltzmann distribution by $f \propto \min(1, t_{\rm eff}/t_r)\exp(-\beta E)$. Since the minimum orbit period is of order $t_{\rm eff}$, we can simplify this result to $f \propto t_r^{-1}\exp(-\beta E)$. Indeed, since the correction is most important for highly elongated orbits, and since these orbits are nearly Keplerian with $t_r \propto |E|^{-3/2}$, we can simplify again, to obtain the distribution function $f \propto |E|^{3/2}\exp(-\beta E)$. Furthermore, the potential fluctuations are effective only in a limited region, and hence cannot place stars on orbits whose angular momenta are so large that their pericenters lie outside this region. We must therefore expect that the distribution function exhibits a cutoff at large angular momenta, which we may incorporate by a multiplicative factor $\exp(-\alpha J^2)$, to arrive at our final guess at the distribution function arising from violent relaxation:

$$f(\vec{x},\vec{v}) = K|E|^{3/2}\exp(-\beta E - \alpha J^2).$$

This turns out to be precisely the distribution function which Bertin and Stiavelli (1984, and this conference) have advocated for violently relaxed systems, although on the basis of quite different arguments. They have already shown that it can provide a good match to the $R^{1/4}$ profile.

Although this distribution function ought to provide a good fitting formula, its exact functional form is only a guess—the arguments above only show that the distribution function should have the general form

$$f \propto \begin{pmatrix} \text{function which is} \\ \text{proportional to} \\ t_r^{-1} \text{ when } t_r \text{ is large} \end{pmatrix} \times \begin{pmatrix} \text{cutoff in peri-} \\ \text{center or } J^2 \end{pmatrix} \times \begin{pmatrix} \text{decreasing function} \\ \text{of } |E| \text{ which is} \\ \text{non-zero at } E = 0 \end{pmatrix}.$$

The condition that the third function is non-zero at $E = 0$ arises because the potential fluctuations have no way to recognize the special role of the escape energy, and hence must populate energy space smoothly around $E = 0$. Jaffe (this meeting) has used similar arguments to show that the asymptotic density distribution in a violently relaxed galaxy must be $\rho \propto r^{-4}$, a condition which the Bertin-Stiavelli distribution function satisfies.

The Bertin-Stiavelli distribution function has one free parameter once the total mass and energy are fixed. This parameter reflects the size of the region in which the potential fluctuations occur. It would be most interesting to compute a family of these models and to compare them in detail with a sequence of violent relaxation simulations.

This completes my detailed comments. Of course, the selection of topics I have made leaves out many of the most interesting aspects of the meeting. A more balanced treatment would spend much of its effort on the topic of X-rays from ellipticals and cooling flows, which was the subject of some of the most lively and controversial discussions of the meeting. Unfortunately I do not feel that I can add anything useful to the reviews already given here by Fabian, O'Connell and Sarazin.

Let me close with one final remark. The Hubble classification for spirals is useful because many properties of spirals (gas content, color, spiral arm morphology, bulge prominence, etc.) all correlate with Hubble time. By contrast, almost nothing correlates with the elliptical Hubble sequence E1 to E7. In view of the rapid increase in quality and detail of our data on ellipticals, is it not time for someone to come up with a new classification scheme for ellipticals to replace Hubble's?

REFERENCES

Bertin, G., and Stiavelli, M. 1984. *Astron. Astrophys.*, **137**, 26.
Bertola, F., and Cappacioli, M. 1975. *Astrophys. J.*, **200**, 439.

Binney, J. J. 1976. *Mon. Not. Roy. Astron. Soc.*, **177**, 19.
Binney, J. J. 1978. *Mon. Not. Roy. Astron. Soc.*, **183**, 501.
Gibbs, J. W., 1884. *Proc. Am. Assoc. Adv. Sci.*, **33**, 57.
Hénon, M., 1982. *Astron. Astrophys.*, **114**, 212.
Illingworth, G. 1977. *Astrophys. J. Lett.*, **218**, L43.
Kormendy, J. 1984. *Astrophys. J.*, **287**, 577.
Malin, D. F., Quinn, P. J., and Graham, J. A. 1983. *Astrophys. J. Lett.*, **272**, L5.
Maxwell, J. C., 1866. *Phil. Trans. Roy. Soc. London*, **157**.
Schwarzschild, M. 1979. *Astrophys. J.*, **232**, 236.
Schweizer, F. 1980. *Astrophys. J.*, **237**, 303.
Schweizer, F. 1986. *Science*, **231**, 227.
A. Toomre and J. Toomre, 1972. *Astrophys. J.*, **178**, 623.

DISCUSSION

Ostriker: You were looking for a classification parameter for ellipticals to replace Hubble's shape parameter. Perhaps the simplest such parameter would be the luminosity which several authors have shown correlates with metallicity, color, rotation, core radius etc. This parameter has the added virtue that, when applied to bulge luminosity in spirals, it provides a classification parallel to the Hubble sequence.

Whitmore: Since we seem to be talking about objective classification systems, I'd like to comment that although you state that there are many correlations for spirals and only a few correlations for ellipticals, a principal component analysis shows that there are two basic parameters for spirals 1) luminosity and/or radius 2) bulge/disk ratio and/or color (Whitmore 1984, *Astrophys. J.*, **278**, 61). I am therefore not so sure that we would have to treat ellipticals and spirals completely differently.

White: I think your suggested prescription for obtaining the distribution function of a violently relaxed system is incomplete because the parameters are not specified. They must depend on the initial conditions of the relaxation phase. It seems to me that a function with only one shape parameter will not be able to accommodate the results both of cold collapses and of mergers between fully formed stellar systems. Although the density profiles can look similar in the two cases, their anisotropy radii are very different.

Tremaine: The simple models I proposed have three parameters—energy, mass and an anisotropy radius. Of course, at some level they are incomplete, because they are spherical, while some initial conditions lead to non-spherical final states. However, they do seem to be much more appropriate than King–Michie models, isotropic de Vaucouleurs models, or the other models which have been traditionally compared to numerical simulations of violent relaxation.

Jaffe: Another way to formulate Simon White's comment is to ask for a quantitative explanation for where or how the transition from core to halo occurs.

Tremaine: The core–halo transition presumably occurs at the boundary of the region where the violent relaxation takes place.

Mamon: I would like to give a word of caution to the builders of self–consistent models of spherical galaxies. Many of you have been building your models on the model that James Binney and I had developed for M87. That model is clearly wrong, for at least three reasons. First, our model was constructed with no central mass in M87, whereas I don't see how the large jet could be present without a very large black hole on the center. Second, our models were constrained by the velocity dispersion in the very center of the galaxy, as measured by Dressler. However, his measurement is very difficult to interpret because of finite slit size and atmospheric seeing. The 'true' line–of–sight velocity dispersion in the core of M87 may rise much faster than what we assumed. Finally, Kormendy's surface brightness profile of M87 obtained with the excellent seeing in Hawaii shows a much smaller core, and a naive application of the Binney–Mamon algorithm would yield a radial anisotropy peak occuring 3 times deeper inside the galaxy. So I think people should base their self–consistent studies on a different galaxy model. Unfortunately, I don't know of any galaxy for which a decent model exists.

Goodman: The observational results that most surprised me were the X–ray results, particularly those for cooling flows in cluster ellipticals. If, as Dr. Fabian tells us, 100 to 1000 solar masses per year of hot gas is forming very low mass stars in M87, then even if these stars can't burn hydrogen, they should release gravitational binding energy comparable to the X–ray luminosity of the hot gas. (At least, this is what Binney and I estimated during the coffee break). Can't we detect these low–mass stars at infrared or millimeter wavelengths?

Gilmore: It is possible to measure the mass function in low mass stars directly by spectroscopy of the gravity sensitive 2-3mm CO absorption band (Arnaud & Gilmore, 1986 *Mon. Not. R. astr. Soc.*, **220**, 759; this volume, p. 445). We have done this for a sample of ellipticals with cooling flows and a large number of ellipticals with no cooling flows. These observations exclude the possibilty that the mass in the flow is forming stars with mass greater than $\sim$ 0.2 solar mass. If the mass is locked up in very low mass objects, they will radiate their binding energy during formation at a temperature of $\sim 1500 - 2000K$, and they will be detectable only as a low surface brightness excess in the near infrared. Such data await suitable array detectors. The available spectroscopic data show no differences between X–ray and non X–ray galaxies. They are also consistent with the stellar mass function in elliptical galaxies being very similar to that of the solar neighbourhood.

Schweizer: In their relation to other morphological types in the Hubble sequence, ellipticals seem to be the closest to what one might call dead galaxies. They barely form stars anymore, whereas the later Hubble types do. I visualize the E's as being in the central graveyard of a little city, whose surrounding inhabitants are the more lively spiral and irregular galaxies. In the past, we have concentrated our research on the central part of that graveyard, figuring out some properties of the oldest skeletons. During recent years and especially at this meeting, we have begun exploring the outer regions of the graveyard, where the fresh graves are. There has been a lot of excitement about ellipticals with shells, ripples, and inclined gas disks, presumably the aftermaths of some recent collisions involving disk galaxies. It seems to me that we stand to gain even more by turning our attention to where the open graves are, watching for the arrival of fresh victims of head–on collisions. We need to study the most luminous IRAS galaxies, if we wish to learn about elliptical formation. Having simulated this formation in a rather abstract manner from mergers in various cosmological models with given sets of rules, we should now try to model in much more detail the collisions occurring in IRAS galaxies like Arp 220 and NGC 6240, collisions that produce abundant molecular gas and vigorous star formation and that seem likely to lead to elliptical remnants. A first promising attempt in that direction has been made by Negroponte and White (1983, *Mon. Not. R. astr. Soc.*, **205**, 1009).

Toomre: Could we ask Martin Schwarzschild for the last comment?

Schwarzschild: A good portion of what we have discussed in this symposium was triggered by the decisive early radial velocity observations in ellipticals by Bertola and by Illingworth. The challenge of their surprising data was picked up by Binney and other theoreticians—I think with much success. But now—as this symposium has surely shown—we will again be in trouble if new key observational data can not be obtained. Brilliant progress is being made in superprecise photometry as well as in velocity dispersion measurements. The greatest difficulties are encountered, it seems, in the observations of mean streaming patterns in the inner portions of big ellipticals. But such mean stellar velocities are a most powerful diagnostic—if they can be obtained with an accuracy of about ± 5 km/s, which is a severe task in view of the typical velocity dispersions of 300 km/s. Some recent observational mean velocity data still appear beset by unidentified systematic errors of about ± 15 km/s which spoil their full diagnostic power. Here then lies a decisive challenge for new observational inventiveness.

While Toomre, Pfenniger, Porter & Nieto look on, de Zeeuw receives his lost fountain pen from Tremaine, who had it cast in lucite in order to prevent it from disappearing again.

Scott Tremaine discusses a famous equation.

Toomre, the chairman of the session, approves.

POSTERS

Discussions during a coffee break. In the foreground: Gilmore, Wyse, Hut and Bacon. Behind them: Mamon, Statler, Lacey, Gerhard and Vietri.

MORPHOLOGICAL PROPERTIES OF ELLIPTICAL GALAXIES

S. Djorgovski
Harvard–Smithsonian Center for Astrophysics
60 Garden St.
Cambridge, MA 02138, USA.

ABSTRACT. In the poster as presented at the meeting, I described global morphological properties of elliptical galaxies, based on the data from a CCD surface photometry survey of $\sim$ 200 ellipticals and $\sim$ 50 S0's (Djorgovski 1985). In this brief summary, I emphasize two points: (1) there is a very weak and very noisy trend of radial shape with luminosity, in the sense that more luminous galaxies are less concentrated, and (2) there is no preference for low-luminosity ellipticals to show boxy isophotes, and they differ in that respect from the bulges.

There is a wide variety of shapes of surface brightness profiles, which implies any formula or model which do not contain *at least one* shape parameter (e.g., the $r^{1/4}$ law, Binney, or Jaffe models) cannot describe satisfactorly all (or any?) elliptical galaxies. There are no obvious systematics in this variety, except for the very slight trend with luminosity, illustrated in Figure 1. The trend is in the sense that more luminous galaxies are less concentrated, or have shallower surface brightness profiles; for example, it is well known that cD's behave in that way, with respect to lower luminosity ellipticals. Schombert (1986) argued that this trend is well defined and that one can construct a "standard" surface brightness profile at any given luminosity. I agree that there is a trend, but it is by far too noisy, and at any given luminosity there is too much diversity in profile shapes.

The galaxies show also a wide variety of ellipticity profiles. There is a tendency toward positive ellipticity gradients, but there are exceptions. The gradients usually flatten into a constant ellipticity at some radius, and sometimes reverse after that. The isophotal twists are common, but not too large, typically a few degrees per decade in radius. The ellipticities, ellipticity gradients and isophotal twist rates are not mutually correlated, and none of them is correlated with luminosity, radial shape of light distribution, or any of the kinematical quantities.

Davies *et al.* (1983) demonstrated that the low luminosity ellipticals and bulges are similar in their dynamical properties. They suggested that, in analogy with boxy or peanut-shaped bulges, the low luminosity ellipticals may show a higher incidence of boxy isophotes. Figure 2 demonstrates that this is not the case: there is no trend of "boxiness" (or, for that matter, "diskiness") with luminosity. The small elliptical and bulges are not equivalent in all of their properties.

T. de Zeeuw (ed.), Structure and Dynamics of Elliptical Galaxies, 377–378.

REFERENCES

Davies, R., *et al.* 1983, *Astrophys. J.* **266**, 41. (DEFIS)
Djorgovski, S. 1985, Ph.D. Thesis, University of California, Berkeley.
Schombert, J. 1986, *Astrophys. J. Suppl. Ser.* **60**, 603.

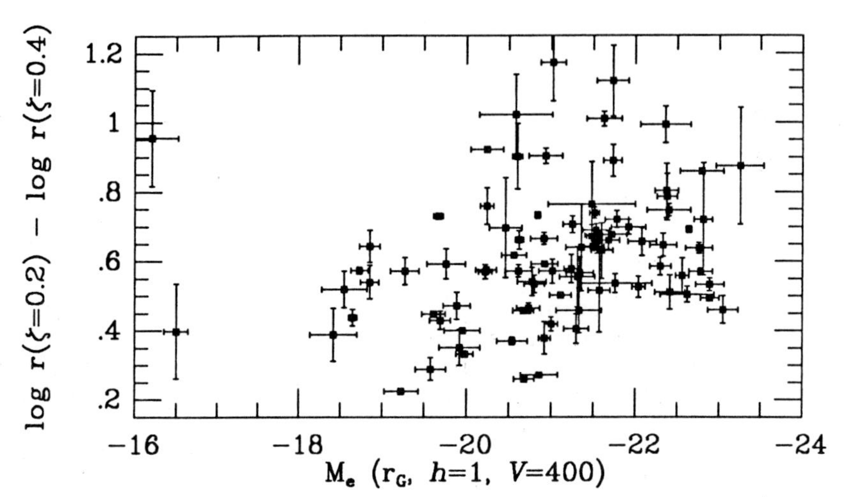

Figure 1: Ratio of two fiducial radii (in a logarithmic form), plotted *vs.* the luminosity. The more luminous galaxies tend to have shallower light profiles, and thus larger radial ratios. The very discrepant point at M_e =-16.2 is M32, which may be a post-core-collapse galaxy. A similar trend is also seen when a magnitude difference in two fiducial apertures is plotted against the luminosity.

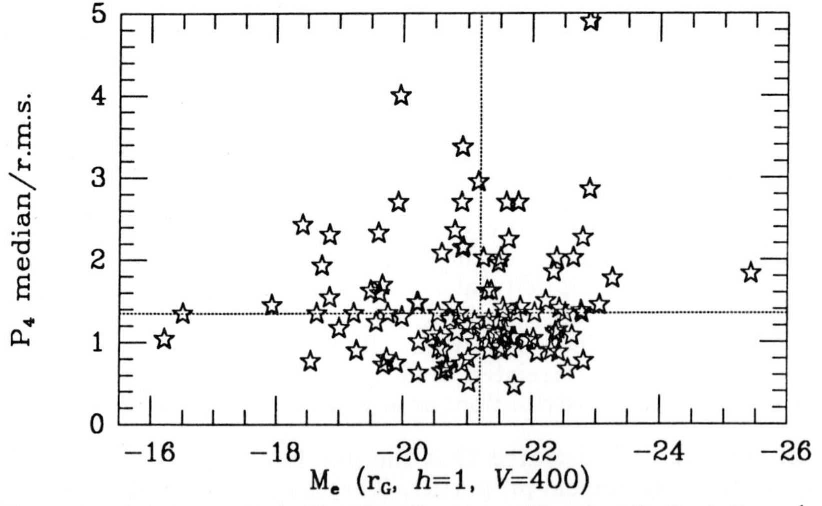

Figure 2: A parameter indicating the mean "boxines" of a galaxy (residual Fourier 4-wave amplitude, renormalized by the error-bar) plotted *vs.* the luminosity. The dotted lines indicate sample medians in both coordinates. There is no trend with luminosity. A Virgocentric infall model with V_{inf} = 400 km s^{-1}, and h = 1 were used in computing the absolute luminosities.

CORRELATIONS BETWEEN $r^{1/4}$-LAW PARAMETERS FOR BULGES AND ELLIPTICAL GALAXIES

Masaru Hamabe[1,2] and John Kormendy[1]
[1]Dominion Astrophysical Observatory
[2]Tokyo Astronomical Observatory

The correlation between the effective radius r_e and surface brightness μ_e for elliptical galaxies is a fundamental scaling law that theories of galaxy formation must explain. When r_e and μ_e are derived by fitting two-parameter fitting functions such as the de Vaucouleurs $r^{1/4}$ law to brightness profiles, the errors in the parameters are strongly coupled. The purpose of this paper is to rederive the $\mu_e(\log r_e)$ relations for ellipticals and bulges, taking account of the coupling in the errors and using only high-accuracy CCD data. Our preliminary conclusions are: (1) The coupled errors are too small to affect significantly the correlation derived for elliptical galaxies. (2) The correlation for bulges is not very different from that for ellipticals, but the galaxy sample is small and the errors in the parameters are large due to the inherent uncertainty in bulge-disk decomposition.

Brightness profiles of elliptical and early-type disk galaxies are taken from all published CCD photometry and from new CCD photometry obtained with the Canada-France-Hawaii Telescope (Kormendy 1986; these proceedings). Distances were derived using group velocities and Schechter's (1980) $\gamma = 2$ Virgo Cluster infall model, assuming a Local Group infall velocity of 300 km s^{-1} and a Virgo distance of 15.7 Mpc. All magnitudes were converted to the V bandpass.

For elliptical galaxies, the $r^{1/4}$-law parameters were obtained by least-squares fits to the mean profiles. We used only the parts of the profiles that are well described by $r^{1/4}$ laws. The parameters of the bulges of disk galaxies were obtained by simultaneous decomposition of the profiles into bulge and disk components; we assumed that disks are exponential. Galaxies with non-exponential disks or small bulge-to-disk ratios cannot be decomposed in this way and were discarded.

Because of strong coupling between the two $r^{1/4}$-law parameters, the error bars in the $\mu_e - \log r_e$ diagram are not perpendicular to the coordinate axes. For each galaxy we have calculated ellipses of constant $\chi^2 = \nu^{-1}\Sigma_i(\mu_i - \mu_{fit})^2/\sigma^2$, ν = number of degrees of freedom, as follows. The point plotted for each galaxy is that of lowest χ^2. Moving out from this point along 92 radial lines in the plane, the points were found at which χ^2/ν had increased by amounts corresponding to 99 % confidence levels. An ellipse was fitted to the resulting points. The main uncertainty is in ν; this was the reason why we adopted a strict confidence level.

The $\mu_{eV} - \log r_e$ diagram for 89 ellipticals is shown in Figure 1a; μ_{eV} and $\log r_e$ are tightly correlated. The best-fitting straight line is similar to relations derived previously (e.g., Kormendy 1982). The χ^2 ellipses are small enough that this relation is securely derived. Fig. 1b shows the same diagram with the parameters for 30 bulges added. The bulges do not obviously deviate from the parameter correlation for ellipticals. However, the fitting errors are larger than for ellipticals; a larger sample of measurements is needed.

T. de Zeeuw (ed.), Structure and Dynamics of Elliptical Galaxies, 379–380.

We thank P. Schechter for emphasizing the importance of the coupling in the errors in μ_e and r_e. The method used for the derivation of the χ^2 ellipses was originally developed by T. Boroson. The CFHT is operated by NRC (Canada), CNRS (France) and the University of Hawaii.

REFERENCES

Kormendy, J. 1982, in *Morphology and Dynamics of Galaxies*, ed. L. Martinet and M. Mayor (Sauverny: Geneva Observatory), p. 113.
Schechter, P. 1980, *A. J.*, **85**, 801.

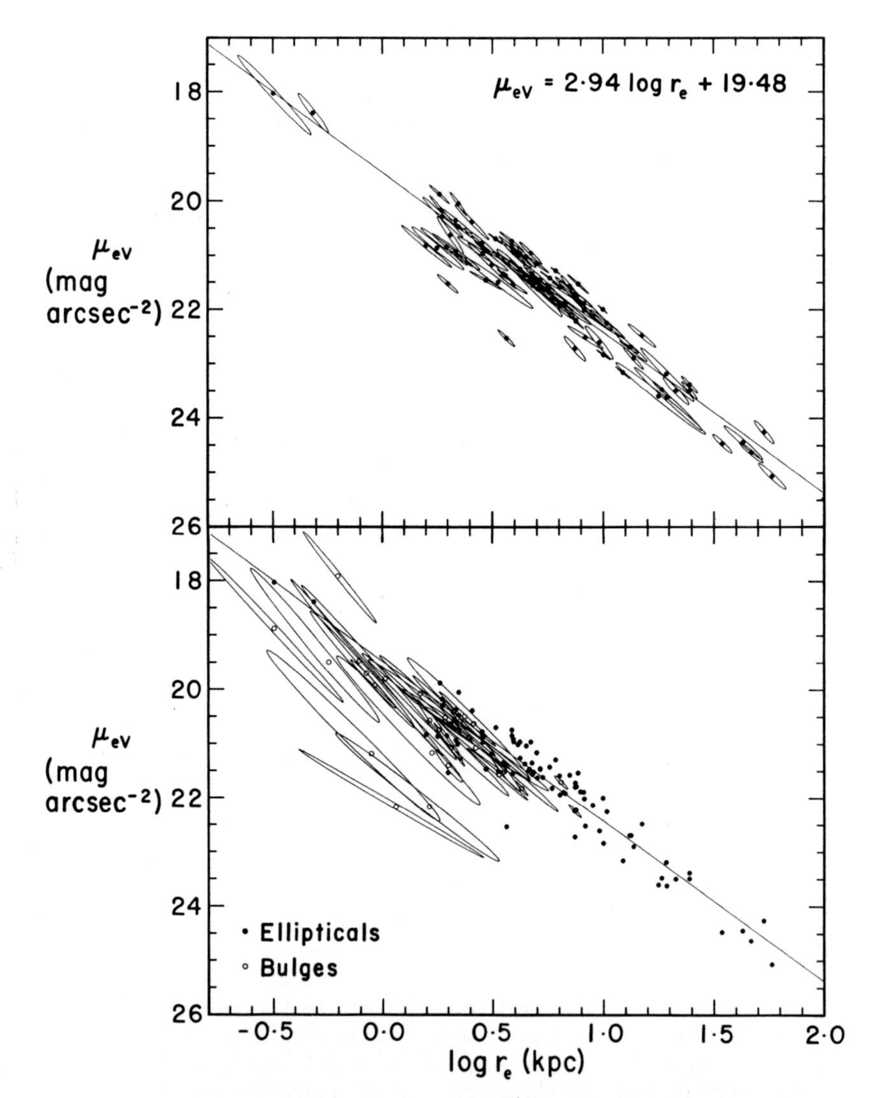

Fig. 1 – μ_{eV} *vs* log r_e with χ^2 ellipses for $r^{1/4}$-law fits to (a, upper) elliptical galaxies and (b, lower) bulges of early-type disk galaxies. The *straight line* is a least-squares fit to the points for elliptical galaxies.

THE CORE PROPERTIES OF ELLIPTICAL GALAXIES

Tod R. Lauer,
Princeton University Observatory
Princeton, NJ 08540, USA

ABSTRACT. The core structure of elliptical galaxies is determined by two parameters: total luminosity L and central luminosity density ρ_C. Scatter in ρ_C at any L may imply dissipational formation of core structure; once this scatter is accounted for, however, the pure L dependence of the core parameters can be isolated and perhaps used as a metric distance indicator.

1. DISCUSSION

Lauer (1985) and Kormendy (1985) find similar core parameter relationships as a function of luminosity. Brighter galaxies tend to have bigger and dimmer cores – specifically I get $r_C \propto L^{1.2}$ and $I_0 \propto L^{-1.0}$, where r_C and I_0, are the core radius and central surface brightness. There is large cosmic scatter in both relationships, however, but with tightly correlated residuals; at any luminosity excessively large cores are dim in the center and excessively compact cores are brighter; this can be attributed to even larger scatter in core luminosity densities. Over an order of magnitude variation is observed in ρ_C at any luminosity; this implies that L alone is not the sole parameter determining a galaxy's core structure.

I believe the scatter in ρ_C suggests dissipational formation of the cores. Cores of giant ellipticals contain only $10^8 - 10^9 M_\odot$, a small fraction of their total masses. Further, since the cores sit at the bottom of their galaxies' potential wells they are sensitive to the central accretion of even small amounts of matter which would otherwise have no effect on the global structure of their galaxies - it is interesting to note that there appears to be no relationship between the ratio R_E/r_C and galaxy luminosity. This argues against the formation of elliptical galaxies by any sort of dissipationless hierarchy of merging, since simulations of such processes predict R_E/r_C to increase as a strong function of L. Merging *with* dissipation might be expected to produce even more compact cores for a given final effective radius; however, variations in the amount of "stuff" allowed to sink to the center of the final merger product might obscure any basic dependence of R_E/r_C on L.

Once the scatter in r_C and I_0 at any luminosity is understood as being caused by variations in ρ_C, it is possible to isolate the pure luminosity dependence of the core parameters. A principal components analysis gives $L \propto r_C^{2.4} I_0^{1.7}$. As argued by Dressler *et al.* (1986) central velocity dispersion σ correlates better than L with global structural properties of the galaxies; this also appears to be true with the core

T. de Zeeuw (ed.), Structure and Dynamics of Elliptical Galaxies, 381–382.

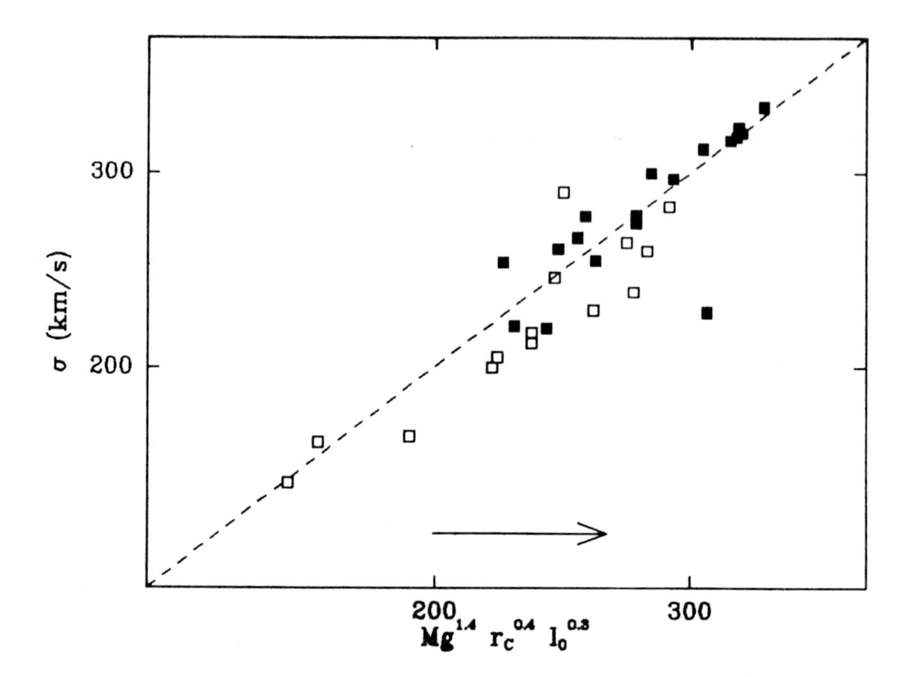

Figure 1: Central velocity dispersion plotted against that calculated from a fit to r_C, I_0 and Mg_2. Solid symbols are for galaxies with resolved cores. The arrow shows the change in calculated σ for a factor–of–two change in r_C.

parameters and works best with the central Mg_2 index tossed in. Figure 1 shows the results of a fit involving central velocity dispersion, σ, r_C, I_0, and Mg_2 on the assumption that the later three variables determine σ. The resulting fit using just the well resolved galaxies gives:

$$\sigma \propto r_C^{0.4} I_0^{0.3} 10^{1.4Mg_2}. \tag{1}$$

Dispersions predicted from the core parameters of the resolved galaxies fall very close to their true values. Note that r_C is the only distance dependent parameter in (1); the goodness of the fit suggests that cores might be used as a metric distance indicator if (1) can be confirmed with a larger sample of galaxies with resolved cores.

2. REFERENCES

Dressler, A., Lynden-Bell, D., Burstein, D., Davies, R. L., Faber, S. M., Terlevich, R. J., Wegner, G. 1986, *preprint.*

Kormendy, J. 1985, *Ap. J.*, **295**, 73.

Lauer, T. R. 1985, *Ap. J.*, **292**, 104.

SURFACE PHOTOMETRY OF BRIGHT ELLIPTICALS

James M. Schombert
Palomar Observatory
California Institute of Technology

ABSTRACT. Surface brightness profiles of bright ellipticals ($M_V < -17$, $H_o = 100$) in rich clusters and the field are reduced to structural parameters. Most fitting functions are found to be inadequate in describing the overall shape of profiles with only the $r^{1/4}$ law providing a good match over the range of surface brightness from 19 to 24 mag arcsec^{-2}. The systematic deviations in structure of first-ranked ellipticals from normal ellipticals (i.e. enlarged characteristic radii and shallow profile slopes) are well explained by comparison to the N-body simulations of merging systems from Duncan, Farouki, and Shapiro (1982).

I. PROFILE MORPHOLOGY

Photometry for this study (Schombert 1986) consists of 277 galaxies obtained in 71 rich and poor clusters plus 27 field ellipticals. The photometry is in the Johnson V band bounded by an inner radius of 2 kpc and light levels corresponding to 28-29 mag arcsec^{-2}.

The structures of the morphology types gE, D, and cD are clearly defined from their profile shapes. D and cD galaxies are diffuse and enlarged in characteristic radii compared to normal ellipticals (Thuan and Romanishin 1981, Malumuth 1984) and cD galaxies also have a characteristic extended envelope (Oemler 1976) (see also article by Tonry in this collection). Classification by profile occasionally differs from visual estimates, especially for cD systems whose envelopes are often below the 24 mag arcsec^{-2} level. NGC 6034 in Hercules is an example of a cD galaxy which was classified as a normal elliptical because it is not a brightest cluster member (BCM), nor at the center of a cluster system.

II. TEMPLATE PROFILES AND THE $r^{1/4}$ LAW

The structure of normal ellipticals was best described by a set of template profiles which were drawn from the entire sample, excluding BCMs, and displayed a smooth change in structure with luminosity (Schombert 1986). From inspection of the templates, the following comments can be made about the $r^{1/4}$ law and bright ellipticals: (1) the $r^{1/4}$ law is only valid for ellipticals with $M_V < -21$, (i.e. there are systematic deviations from $r^{1/4}$ at low luminosities; see also Binggeli, Sandage,

T. de Zeeuw (ed.), Structure and Dynamics of Elliptical Galaxies, 383–384.

and Tammann 1984), (2) and is valid only for intervals of $19 < \mu < 24$ mag arcsec^{-2}.

In Figure 2 the effective surface brightness (μ_e) and effective radius from $r^{1/4}$ fits under the above restrictions are plotted along with the relation found by Kormendy (1980). Notice that BCMs lie to the right of the relation indicating that they are more diffuse than normal ellipticals (i.e. larger characteristic radii).

III. MERGER EFFECTS

A large number of recent theoretical studies (May and van Albada 1984) have made similar predictions about the structure of merger remnants (i.e., larger characteristic radii, $r^{1/4}$ profile shapes, and anisotropic velocity distributions). These are the same properties seen for brightest cluster members in this sample. Even in the more extreme cD galaxies with "nested" satellite companions, there exists a fairly uniform $r^{1/4}$ region in their inner profile.

Shallow profiles for brightest cluster members, as found by Malumuth (1984), are confirmed by this study. This result, together with enlarged characteristic radii and the shape of D and cD profiles, matches in form and size the profiles predicted by the N-body merger simulations of Duncan, Farouki, and Shapiro (1982). As much as 16 L_* growth is estimated for some of the more extreme cD systems such as A1413 and A2670.

REFERENCES

Binggeli, B., Sandage, A., and Tarenghi, M. 1984, *A. J.*, **89**, 64.
Duncan, M.J., Farouki, R.T., and Shapiro, S.L. 1982, *Ap. J.*, **271**, 22.
Kormendy, J. 1980, *Proc. ESO Workshop on 2-D Photometry*, p. 191.
Malumuth, E.M. 1984, *Ph.D. thesis, Univ. of Michigan.*
May, A. and van Albada, T.S. 1984, *M.N.R.A.S.*, **209**, 15.
Oemler, A. 1974, *Ap. J.*, **194**, 1.
Schombert, J.M. 1986, *Ap. J. Suppl.*, **60**, 603.
Thuan, T.X. and Romanishin, W. 1981, *Ap. J.*, **248**, 439.

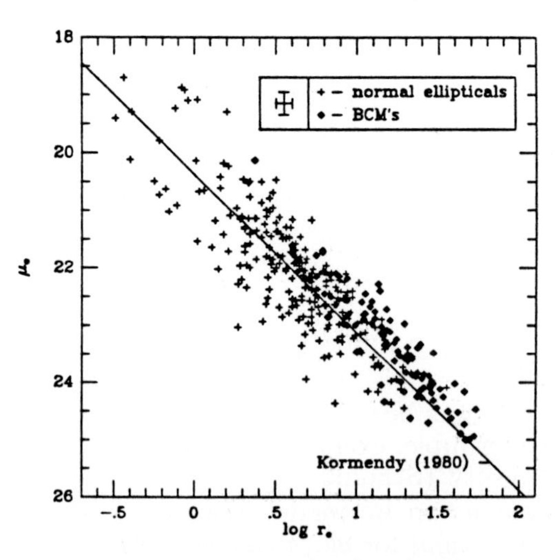

Figure 1. Effective radius versus effective surface brightness for all bright ellipticals (Schombert 1986) with linear fit from Kormendy 1980.

ISOPHOTOMETRY OF BRIGHTEST CLUSTER ELLIPTICALS

A.C.Porter, Caltech 105-24, Pasadena, CA 91125
D.P.Schneider, IAS, Princeton, NJ 08540
J.G.Hoessel, Univ. of Wisconsin, Madison, WI 53706

ABSTRACT

What are the two-dimensional distributions of projected luminosity in the brightest ellipticals in Abell clusters (hereafter called "E1"s)? If we treat isophotes as ellipses, how do their properties vary as functions of surface brightness? Are the isophote parameters correlated with a galaxy's global properties? Are they correlated with any properties of the surrounding cluster? Are E1s morphologically similar to other cluster or field ellipticals? This paper describes work in progress to the answers to these questions, and illustrates an interesting result.

1. OBSERVATIONS AND REDUCTIONS

We have obtained Gunn r CCD images of nearly 200 E1s from the samples of Hoessel et al (1980) and Schneider et al (1983). Most of the data were taken with a 500 x 500 Texas Instruments chip and the PFUEI optics (Gunn and Westphal 1981) on the Palomar 1.5 meter reflector. These pictures have a scale of 0.548 arcsec/pixel, giving a 4.5 arcmin field of view. Most were 500 second exposures in good (1") to moderate (2") seeing.

We are presently engaged in the laborious task of removing foreground stars and neighboring galaxies from these images and fitting ellipses to the isophotes of the E1s by standard least squares techniques. Details and limitations of this process will be presented later. Suffice it to say here that for each E1, we derive centroids (x, y), major axes (a), eccentricities (e), and position angles (θ) for isophotes at half magnitude intervals to an average depth of μ_r = 25 magnitudes/arcsec^2.

2. PRELIMINARY RESULTS

47 images have been processed so far. The following results are apparent:
1) Some E1s have complex and interesting eccentricity and position angle curves ($e(\mu)$, $\theta(\mu)$) despite having one-dimensional surface brightness profiles which appear normal.

T. de Zeeuw (ed.), Structure and Dynamics of Elliptical Galaxies, 385–386.

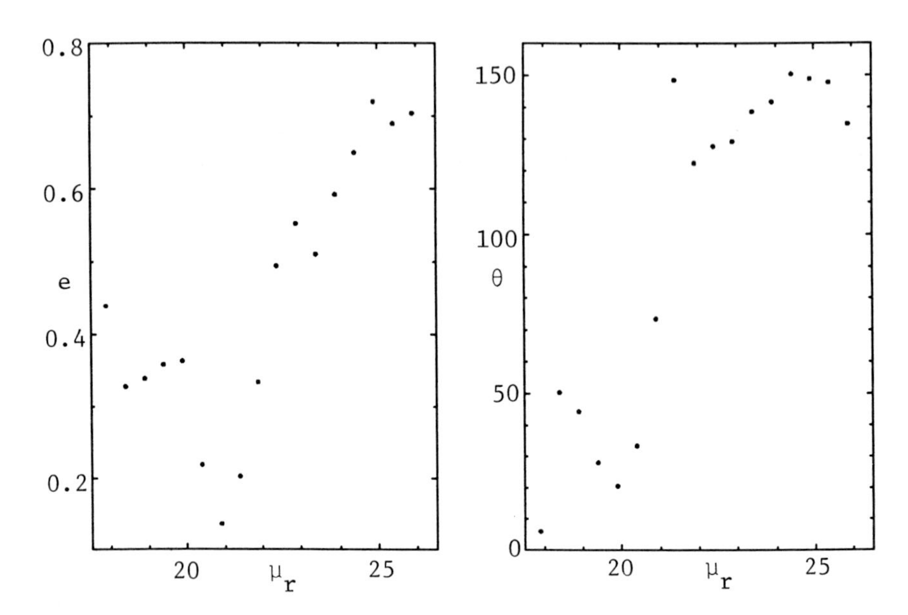

Figure 1. Eccentricity and position angle profiles of A1228.

More specifically, at least 2 galaxies (A1228, A1809) show pronounced rounding of isophotes between regions where the position angle of the galaxy's major axis has changed by significant amounts. The figure above illustrates this behavior in A1228. This suggests that we are observing two superposed, misaligned components of the central "galaxy".

2) Els differ from field ellipticals in that their isophotes become flatter with increasing size. This has been noted before (e.g. diTullio 1979), but we are struck by the generality of the rule: only 4 of the galaxies reduced so far do not have monotonically increasing e(a).

Indeed, the faintest, outermost isophotes of these galaxies are quite elongated. In only 5 galaxies do they remain rounder than e = 0.6, and in 19 galaxies, they have e $\geq$ 0.8. This strongly suggests to us that the outer regions of Els are not spheroidal: for a spheroid to appear flat, it must *be* flat *and* have its axis of symmetry near the plane of the sky. If the outer parts of Els were spheroidal, we would expect their axes to be randomly oriented, causing more of them to look rounder. They must be either extremely prolate, triaxial, or perhaps even irregular.

When the remaining Els have been profiled, we will run quantitative simulations to test these hypotheses. The questions in the abstract will also be reexamined in light of the complete results. The answers will tell us much about the nature and nurture of Els.

REFERENCES

diTullio, G.A. (1979) Ast.Ast.Supp.Ser. 37 591
Gunn, J.E., and Westphal, J.A. (1981) Proc.SPIE 290 16
Hoessel, J.G., Gunn, J.E., and Thuan, T.X. (1980) ApJ 241 493
Schneider, D.P., Gunn, J.E., and Hoessel, J.G. (1983) ApJ 268 476

INVESTIGATING THE SCATTER IN THE $V_{26} - \log \sigma$ RELATION

Michael Gregg and Alan Dressler
Mt. Wilson and Las Campanas Observatories
813 Santa Barbara St.
Pasadena, CA 91101, USA

ABSTRACT. The nature and existence of a second parameter needed to characterize the family of normal elliptical galaxies has been much discussed. The need for a second parameter has been demonstrated by the correlation of the residuals from the well-known magnitude-velocity dispersion relation for ellipticals with other observables such as ellipticity, mass-to-light ratio and Mg line strength or metallicity. Here, evidence for a correlation between residuals in the $V_{26} - \log \sigma$ relation and the strength of $H\beta$ is presented, suggesting that variations in the stellar populations in Virgo elliptical cores may be an important secondary parameter.

Dressler (1984) measured velocity dispersions for Virgo and Coma galaxy cluster members from reticon spectra taken through 16″ and 4″ apertures, respectively. The Virgo objects were also measured through a 4″ aperture, and velocity dispersions determined, though never published. As part of a detailed analysis of the absorption line strengths of Dressler's (1984) data, the strength of $H\beta$ has been measured for all objects in the sample, including the unpublished 4″ Virgo spectra.

Residuals from linear least-squares fits to the $V_{26} - \log \sigma$ relations for the three data sets are plotted against $\log(H\beta)$ in Figure 1. Galaxies lying far from the mean $\log \sigma$ relations were not included in the fits. A weak anti-correlation exists in the large aperture Virgo data. A much stronger trend is present in the small aperture data. Coma exhibits no such relation. In Figure 2, residuals from the tighter $\log D_\Sigma - \log \sigma$ relation are plotted against $H\beta$. $\log D_\Sigma$ is a quantity defined by Dressler et al. (1986) to be the log of the radius in arcseconds at which the surface brightness drops to 20.75 mag/arcsec^2. The tight correlation of $\log D_\Sigma$ with $\log \sigma$ is attributed to $\log D_\Sigma$ being dependent upon both core surface brightness and size, quantities more directly linked to the central potential than V_{26} (Dressler et al. 1986). The same trends are seen in Figure 1.

These relations cannot be interpreted in any precise way, though they suggest variations of the stellar population in the cores of the Virgo galaxies. The addition of a stellar component with a blue continuum and enhanced $H\beta$ would cause the velocity dispersion to be measured too low for its V_{26} or $\log D_\Sigma$ as well as enhancing the central surface brightness of the parent galaxy. Both effects would contribute to the anti-correlations seen here. This is demonstrated qualitatively by the location of NGC 4742 in the large aperture Virgo data and D107 in the Coma sample. Both galaxies have very strong Balmer lines and low $\log \sigma$. Such an effect could in

T. de Zeeuw (ed.), Structure and Dynamics of Elliptical Galaxies, 387–388.

principle be caused by young, intermediate age, or metal poor populations as well as variations in M/L (or equivalently, the slope of the IMF). That the correlation of $\log(H\beta)$ with $\delta \log \sigma$ is stronger in the Virgo 4×4 data than in the 16×16 data indicates that the population is concentrated at the centers of the galaxies. This would seem to favor the young population, as recent star formation in normal ellipticals is expected to occur where mass loss from stars would concentrate.

Whatever it is that drives the relation in Virgo is absent in the Coma cluster. If it is recent star formation that produces the Virgo correlation, then perhaps the Coma galaxies are kept clean by some gas evacuation mechanism such as ram presure from intracluster gas. It is also possible that the relation is suppressed by the somewhat lower signal–to–noise data for Coma as well as the larger effective size of the 4×4 aperture at the distance of Coma compared the the 16×16 aperture at Virgo. 0.7 arcsecond aperture measurements of Coma galaxies would be needed to obtain data equivalent to the Virgo 4×4 data.

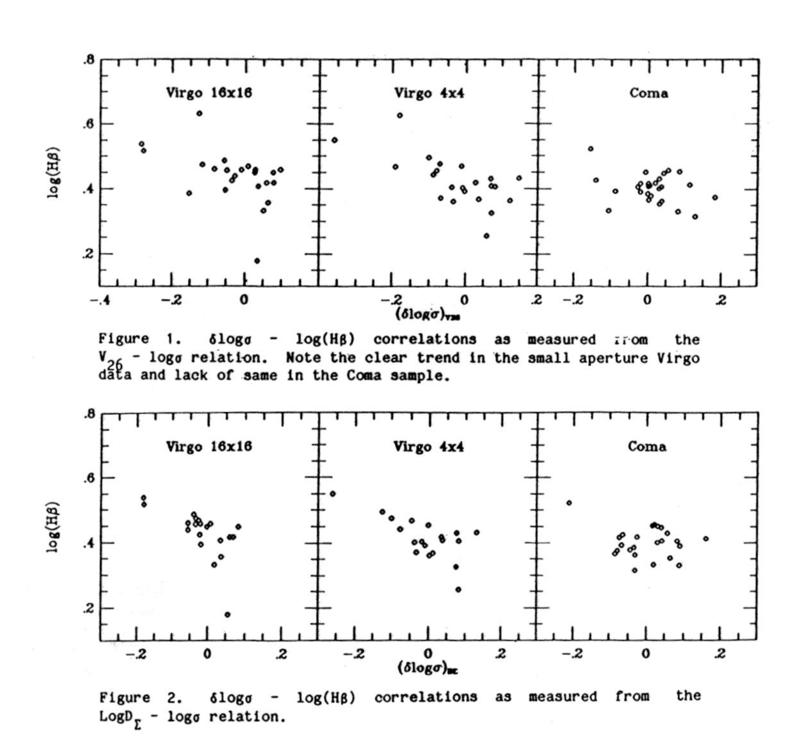

Figure 1. $\delta\log\sigma$ - $\log(H\beta)$ correlations as measured from the V_{26} - $\log\sigma$ relation. Note the clear trend in the small aperture Virgo data and lack of same in the Coma sample.

Figure 2. $\delta\log\sigma$ - $\log(H\beta)$ correlations as measured from the $\mathrm{Log}D_{\Sigma}$ - $\log\sigma$ relation.

REFERENCES

Dressler, A., 1984. *Ap. J.*, **281**, 512.

Dressler, A., Lynden–Bell, D., Burstein, D., Davies, R.L., Faber, S.M., Wegner, G., & Terlevich, R., 1986. "Spectroscopy and Photometry of Elliptical Galaxies. I. New Distance Estimator", preprint.

ON THE RELATION BETWEEN RADIUS, LUMINOSITY AND SURFACE BRIGHTNESS IN ELLIPTICAL GALAXIES

E. Recillas-Cruz and A. Serrano P.G.
Instituto de Astronomía, Universidad Nacional Autónoma de México, Apdo. Postal 70-264, México 04510, D.F., México.

ABSTRACT. We have analyzed luminosity profiles of E galaxies in six clusters of galaxies. We have found a relationship between radius, luminosity and surface brightness for galaxies in each of the clusters. Moreover, it seems that there is a dependence of the zero point of the relation with environment. This relationship implies that there is not a universal luminosity profile for elliptical galaxies.

We have used values for effective radius r_e and M_V for each galaxy in six clusters of galaxies (Strom and Strom 1978a,b,c,) to construct a linear regression of log r_e versus M_V. The residuals of these fits are correlated with the surface brightness in the R band, μ_e, we therefore made a regression of the residuals versus μ_e which has a dispersion four times smaller. However, galaxies that deviated most from the mean relationship were found to be the most and least luminous in each sample meaning that the dependence on M_V was not properly removed. Therefore a simultaneous regression was made of the form:

$$\log r_e = A + B\,M_V + C\,\mu_e. \qquad (1)$$

This relation has an even smaller dispersion than above mentioned relations. Moreover, differences in the values for the coefficients B and C (which are equal within a few sigma) between different clusters, are much smaller than corresponding differences for correlation of residuals. Assuming then that there is a universal value for B and C, we looked for those coefficients B and C that minimize the dispersion between the different clusters. Resulting values are:

$$\tilde{B} = -0.178\ (\pm 0.002),\ \text{and}\ \tilde{C} = 0.165\ (\pm 0.002).$$

Further, a value for $\tilde{A}$ as in Table 1 was found, from a least-squares fit of log r_e versus $\tilde{A} + \tilde{B}\,M_V + \tilde{C}\,\mu_e$, using the above values for $\tilde{B}$ and $\tilde{C}$. In Figure 1 we plot the estimated effective radius $\tilde{A} - 0.178M_V + 0.165\mu_e$ versus log r_e with values for coefficients taken from Table 1.

Equation (1) can also be expressed as $r_e \approx L^{\alpha}\, I_e{}^{\beta}$ where $\alpha = 0.445$ ($\pm$0.005) and $\beta = -0.413$ ($\pm$0.005). A universal luminosity profile will give instead $\alpha = 0.5$ and $\beta = -0.5$. Values found for α, β imply that more luminous galaxies have shallower profiles. A further discussion on the effects of environment on relation (1) can be found in Recillas-Cruz and Serrano (1986).

T. de Zeeuw (ed.), Structure and Dynamics of Elliptical Galaxies, 389–390.

TABLE 1. Fits of log r_e versus $\tilde{A} + \tilde{B}\, M_V + \tilde{C}\, \mu_e$ for $\tilde{B} = -0.178$ and $\tilde{C} = 0.165$.

Cluster	$\tilde{A}$	(std. error)	log r_e^{*a} (kpc)	log r_e^{**b} (kpc)
Coma Center	-6.71	(±0.04)	0.422	0.34
Coma West	-6.66	(±0.04)	0.470	0.41
Perseus Center	-6.68	(±0.02)	0.449	0.41
Perseus Outer Region	-6.66	(±0.03)	0.469	0.46
Perseus LSB	-6.61	(±0.01)	0.522	....
Abell 2199 Center	-6.62	(±0.02)	0.510	0.45
Abell 2199 Outer Region	-6.59	(±0.03)	0.542	0.53
Abell 1367	-6.60	(±0.04)	0.525	0.56
Abell 1228	-6.57	(±0.05)	0.554	0.52
Hercules	-6.54	(±0.04)	0.587	0.55

a. Log r_e^* is evaluated from equation (1) at $M_V = -21.5$ and $\mu_e = 20$.
b. Log r_e^{**} is the value for the effective radius taken from log r_{26} by Strom and Strom (1978d), and estimated at $\mu_e = 20$.

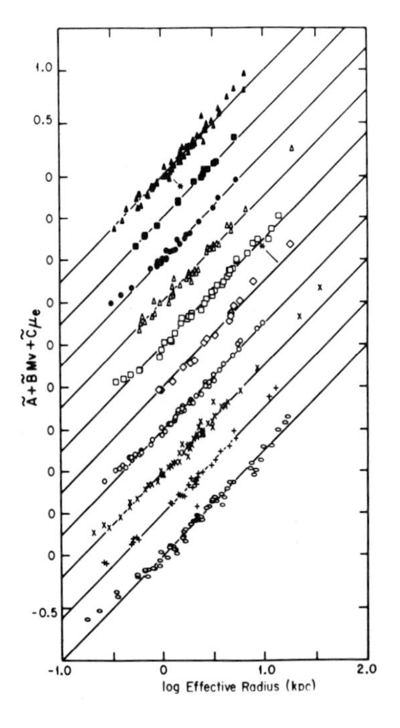

Fig. 1. Plot of the estimated efefective radius $A - 0.178M_V + 0.165\mu_e$ versus the true radius r_e. Lines correspond to the identity. Each cluster has its origin shifted as indicated to avoid overcrowding of data points. Filled symbols correspond to central regions of Coma (Δ), Perseus ( ) and Abell 2199 (O). Open symbols (Δ, , O) correspond to outer regions of same clusters. Other symbols are: ( ) for Perseus LSB, ( ) for Abell 1367 (+) for Abell 1228 and (O) for Hercules.

REFERENCES

Recillas-Cruz, E., and Serrano, P.G., A. 1986, preprint.
Strom, K.M., and Strom, S.E. 1978a, *Astron. J.*,83, 73.
Strom, S.E., and Strom, K.M. 1978b, *Astron. J.*,83, 732.
Strom, K.M., and Strom, S.E. 1978c, *Astron. J.*,83, 1293.
Strom, S.E., and Strom, K.M. 1978d, *Astrophys. J.*, 225, L93.

THE ISOPHOTAL STRUCTURE OF ELLIPTICAL GALAXIES

T.B. Williams and Bidushi Bhattacharya
Department of Physics and Astronomy
Rutgers University
P.O. Box 849
Piscataway, NJ 08854, USA

We have obtained photographic surface photometry for a large sample of elliptical galaxies. The plates were obtained using the KPNO 0.9 m telescopes with Carnegie image intensifiers and IIIaJ plates. The f/13.5 secondaries were used to obtain a large plate scale: 0.33 arcsec per 20 micron pixel. In addition to the galaxy images, flat fields on cloudy skies and geometrical distortion calibration images of astrometric clusters were obtained. All plates were calibrated with the standard KPNO spot sensitometers. The sample of galaxies contains 75 ellipticals of total m_B greater than 13 and north of -30^0 declination. The plates were digitized on the KPNO PDS microdensitometer by Dr. E. Malumuth. At the present time about 50% of the sample is reduced.

The reduction procedure consists of converting photographic densities to intensities, flagging plate flaws and foreground stars to be rejected, and fitting the remaining points in elliptical annuli. A least squares fitting procedure is employed which gives the flux, centers, position angle, axis ratio, and flux gradient for each isophotal annulus. We can determine these parameters accurately down to flux levels of about 10% of the sky. At fainter levels, systematic effects begin to enter the results.

The results for 37 galaxies can be summarized as follows:

Normal–No major axis twist, constant axis ratio		
2 of 37 certain,		
3 of 37 uncertain	total	14%
Twists– Change of the major axis position with radius		
8 have no twist,		
9 of 37 uncertain or noisy	total	46%
13 have monotonic twist,		
7 have oscillating twist	total	54%
Axis ratio variations with radius		
4 constant, 1 uncertain	total	14%
10 increase with radius, 4 decrease,		
12 have minima or maxima,		
5 have oscillations,		
1 shows major or minor axis reversal	total	86%

T. de Zeeuw (ed.), Structure and Dynamics of Elliptical Galaxies, 391–392.

Thus we see that most galaxies have dissimilar isophotes. The axis ratio variations could be explained by spheroidal three–dimensional models with changing axis ratios, but those with twists require ellipsoidal structures (or even more complex models). Any models which explain the observed dissimilarities also predict that the isophotes cannot be exactly elliptical. In the simplest models, those of dissimilar three–dimensional ellipsoids, these deviations from exact ellipses are certainly large enough to be observable, and may seriously constrain the range of acceptable structures for elliptical galaxies.

PARENT STRUCTURES IN E AND SO GALAXIES ?

R.Michard (1) and F.Simien (2)
(1) Observatoire de Nice,
B.P. 252, F-06007 Nice Cedex, France
(2) Observatoire de Lyon,
F-69230 Saint Genis Laval, France

ABSTRACT. Isophote shape and flattening are being used for studying a similarity between elliptical and lenticular galaxies.

One of us (Michard 1984) found a similarity between the ellipticity profiles of E galaxies flatter than E3 and those of SO's of equivalent apparent flattening. It was suggested that Ellipticals might have a two -component structure akin to the classical SO picture.

With the aim of investigating this, we have begun to parametrize the shape of isophotes for a sample of E and SO galaxies (B-band data from Michard 1985 together with new unpublished photographic material). Following Carter (1978), we describe an isophote by its best fitting ellipse and parameters which quantify the deviation from this ellipse. An isophote is represented by:

$$x = a\,(1 + e4 \cos 4t + e6 \cos 6t + \ldots) \cos t$$
$$y = c\,(1 + e4 \cos 4t + e6 \cos 6t + \ldots) \sin t,$$

where x,y are cartesian coordinates, a,c are the axes of the ellipse, and t is a polar angle; e4,e6 ... are the shape parameters, e4 being by far the most significant. A box-shaped isophote gives $e4<0$, a pure ellipse $e4=0$, and a "pointed" isophote $e4>0$. A position angle of major-axis is also obtained. We use algorithms and codes developed by Dr. A.Bijaoui starting from the work of Kuhl and Giardina (1982).

For many SO's, the departures from ellipses are obvious, and Figure 1 shows a typical example: e4 is positive at intermediate radii, and further out c/a eventually rises, where the bulge is thought to become dominant again.

E-classified galaxies exhibit a variety of behaviours; because of space limitation, only one is displayed here (Figure 2). There is often a strong similarity with SO's, for both c/a and e4. This is evidence that there exists, in these Ellipticals, a structure locally much flatter than indicated by the minimum value of c/a alone.

Only a dozen galaxies have been reduced so far, and we are processing more data in order to obtain significant statistics. This

T. de Zeeuw (ed.), Structure and Dynamics of Elliptical Galaxies, 393–394.

should eventually provide insights on the transition between E and SO galaxies, at least we hope so.

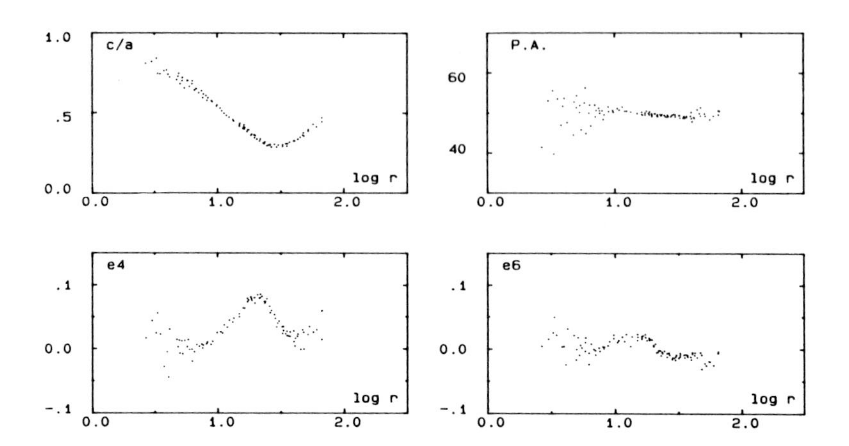

Figure 1. NGC 4417 (type SO): plots of axial ratio c/a, position angle P.A. of major-axis, and coefficients e4 and e6, versus log r ($r=\sqrt{ac}$, in arcsec). In this particular case, e6 also is significantly positive, near log r=1.2 .

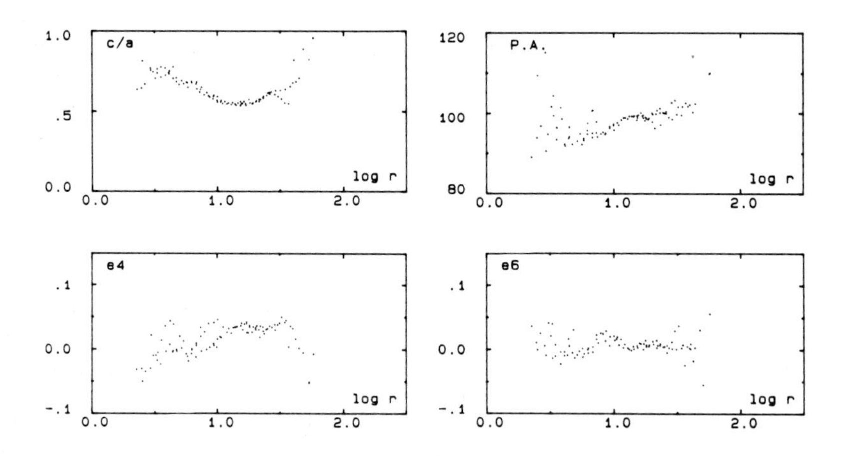

Figure 2. NGC 4660 (type E5): same plots as for Figure 1.

REFERENCES

Carter, D. 1978, Monthly Notices Roy. Astron. Soc. **182**, 797
Kuhl F.P., Giardina C.R. 1982, Comput. Graph. and Image Proc. **18**, 236
Michard, R. 1984, Astron. Astrophys. **140**, L39
Michard, R. 1985, Astron. Astrophys. Suppl. **59**, 205

INTRINSIC SHAPES OF ELLIPTICAL GALAXIES FROM A STATISTICAL COMPARISON OF TWO DIFFERENT ISOPHOTES

G. Fasano
Astronomical Observatory
Vicolo dell'Osservatorio 5
35122 Padova
Italy

Several statistical attempts were already made to discriminate between oblate forms vs. prolate forms (OF/PF) in elliptical galaxies (Marchant & Olson, 1979; Lake, 1979; Olson & de Vaucouleurs, 1981; Capaccioli *et al.*, 1984). They fell in with the fundamental problem that the observed quantities (*surface brightness, velocity dispersion, flattening etc.*) whose projection should depend on the assumed shapes (OF/PF), could be intrinsically correlated each other. This can mask the simple *projection–effect* (Richstone, 1979; Merritt, 1982).

The approach we have tried consists of two different tests. the first one (T1) is also exposed to the above mentioned problem. It works with both semimajor and semiminor axes of the *effective isophote*, whose behaviour as a function of the line of sight obeys the simple geometrical projection laws, depending on the assumed shapes. We have removed the obvious dependence of the semi–axes on the luminosity and analyzed the distribution of the residuals Δ_e vs. apparent flattenings (Fig.1a,b). Photometric and geometric data are taken from an homogeneous sample of elliptical galaxies in the central region of the Virgo Cluster (Liller, 1960, 1966; Capaccioli & Rampazzo, 1984). The statistical comparison between the distributions of Fig.1 and the model predictions are performed with a Montecarlo technique by using three different two–dimensional statistics: *Correlation Coefficient*, *Maximum Likelihood* and *Kolmogorov–Smirnov*. The T1 test is able to separate the acceptability ranges of the two hypotheses (OF/PF) on the basis af a parameter γ which defines the possible intrinsic correlation between linear size and true flattening (see the curves marked with T1 in Fig.2).

The second test (T2) works in the same way with the semi–axes of the isophote corresponding to the surface brightness $\mu' = 25$, whose behaviour as a function of the line of sight turns out to be almost independent on the choice of the shapes. This test allows us to estimate *a priori* the value of γ. In particular, the distribution of the residuals suggests that the possible correlation between linear size and intrinsic flattening must be small (see the curves marked with T2 in Fig.2).

By comparing the tests T1 and T2, we conclude that the oblate hypothesis is favoured in comparison with the prolate one. In fact, for $.05 < \gamma < .75$, the oblate hypothesis is consistent, at 90% of significance, with both T1 and T2 test, being the most likely value of $\gamma \approx .3$. On the contrary, at the same significance level, there is no range of γ where the prolate hypothesis is simultaneusly consistent with the two tests. More precisely, the prolate hypothesis can be rejected at 96% of significance level.

T. de Zeeuw (ed.), Structure and Dynamics of Elliptical Galaxies, 395–396.

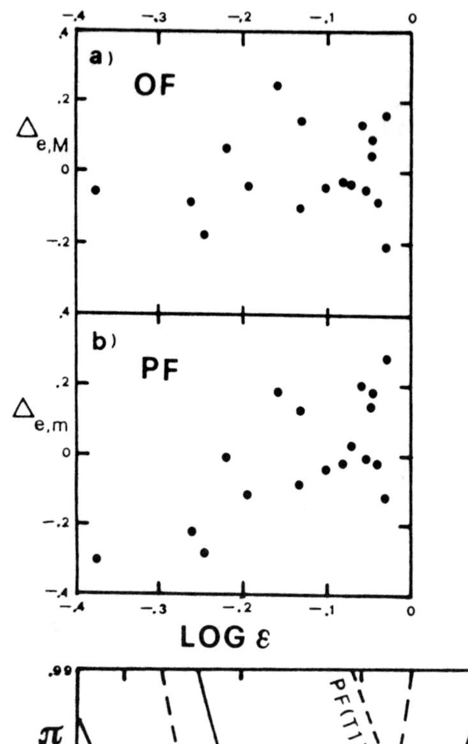

Fig.1a(b) - *Test T1. Residuals $\Delta_{e,M}$ ($\Delta_{e,m}$) of the major (minor) axes vs. apparent flattenings, in the oblate (prolate) hypothesis.*

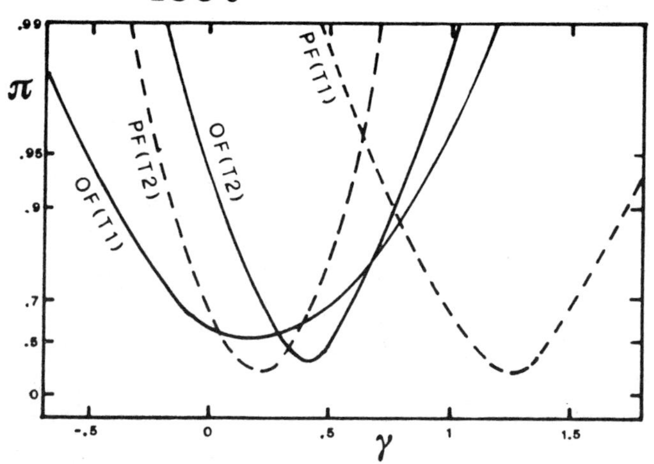

Fig.2 - *Behaviour of the rejection probability π of the two hypotheses (OF continuous lines; PF dashed lines) as a function of the parameter γ, for both T1 and T2 tests.*

REFERENCES

Capaccioli,M., Fasano,G. & Lake,G. 1984. *Mon.Not.R.astr.Soc.*, **209**, 317

Capaccioli,M. & Rampazzo,R. 1984. In: *New Aspects of Galaxy Photometry*, p.275, ed. Nieto,J.L., *Lecture Notes in Physics* n. 232

Lake,G. 1979. In: *Photometry, Kinematics and Dynamics of Galaxies*, p. 381, ed. Evans,D.S., Univ. Texas, Austin

Liller,M.H. 1960. *Astrophys. J.*, **132**, 306

Liller,M.H. 1966. *Astrophys. J.*, **146**, 28

Marchant,A.B. & Olson,D.W. 1979. *Astrophys. J.*, **230** , L157

Merritt,D. 1982. *Astron. J.*, **87**, 1279

Olson,D.W. & de Vaucouleurs,G. 1981. *Astrophys. J.*, **249**, 68

Richstone,D.O. 1979. *Astrophys. J.*, **234**, 825

DEPROJECTION OF GALAXIES: HOW MUCH CAN BE LEARNED?

George B. Rybicki
Harvard-Smithsonian Center for Astrophysics
60 Garden Street
Cambridge, MA 02138

ABSTRACT. A general discussion, based on the "Fourier Slice Theorem," is given for the problem of deprojecting the observed light distribution of galaxies to obtain their intrinsic three dimensional light distribution or "shape." Several results are obtained: 1) A model-independent deprojection of an axially symmetric galaxy is shown to be possible only if the symmetry axis lies in the plane of the sky. 2) A simple criterion is given to test whether two different galaxies can have the same intrinsic shape, based solely on their observed projections. 3) It is shown that a homogeneous class of galaxies can be deprojected using a sufficiently large number of projections of random perspective.

THE FOURIER SLICE THEOREM

The intrinsic shape of a galaxy, defined by its light distribution in space $\epsilon(x, y, z)$, determines the brightness projection, $I(x, y) = \int \epsilon(x, y, z)\, dz$, onto the x-y plane of the sky. An important question is to what extent this projection can be used to determine the intrinsic shape of the galaxy, what is usually called the *deprojection* problem. In principle other kinds of information (such as velocity data) might help to determine the intrinsic shape, but the discussion here is confined to the pure deprojection problem, using only projected brightness data.

It is convenient to discuss properties of projections in terms of Fourier transforms. Denoting the three-dimensional Fourier transform of $\epsilon(x, y, z)$ by $\tilde{\epsilon}(u, v, w)$ and the two-dimensional Fourier transform of $I(x, y)$ by $\tilde{I}(u, v)$, it is easily shown that $\tilde{I}(u, v) = \tilde{\epsilon}(u, v, 0)$. The generalization of this result to projections in arbitrary directions is that the values of the two-dimensional Fourier transform of the projection are equal to the values of the three-dimensional Fourier transform of the light distribution along a planar "slice" through the origin and normal to the direction of the projection (Bracewell and Riddle 1967; Crowther *et al.* 1970). There are several immediate consequences of this "Fourier slice theorem":

1) *Axially symmetric Galaxies.* Since the intrinsic shape of an axially symmetric galaxy and its projection are both determined by functions of two variables, it is tempting to believe that an inversion formula might be found for this case. To investigate this, one notes that the one observed projection yields information in the Fourier slice normal to the line of sight. Now the axial symmetry implies that exactly the same view would be seen at another perspective related to the original

T. de Zeeuw (ed.), Structure and Dynamics of Elliptical Galaxies, 397–398.

one by a simple rotation about the symmetry axis. Thus we can rotate the original Fourier slice about the symmetry axis, yielding information about other parts of the Fourier domain. If the symmetry axis lies in the plane of the sky, the entire Fourier space is covered in this rotation. However, if the symmetry axis is at an angle $\theta > 0$ to the plane of the sky, then this rotation of the Fourier slice leaves a "cone of ignorance," of half-angle θ, such that Fourier components within this cone remain completely unknown. Therefore if $\theta > 0$ the inversion problem for axially symmetric galaxies has no solution.

2) *A Test of Identity.* Consider two projections of a single galaxy. Using the Fourier slice theorem, one obtains the values along the two slices of the Fourier space normal to the two directions of projection. Since these two planes intersect in a line, the complex values along this line must be identical for the two projections (Crowther *et al.* 1970). This fact can be used as a test to determine if two different galaxies seen in projection can have the same intrinsic shape. After Fourier transforming the two projections, one looks for a line through the origin of each of these Fourier planes (in general at different angles) such that the Fourier values are identical along them. If such lines are found it is at least *possible* for the intrinsic shapes to be indentical. Otherwise, the possibility of having the same intrinsic shape is ruled out.

In practice the relative scaling of two projections might be difficult to determine, so it is appropriate to include the possibility of a simple rescaling the data by a uniform stretching or shinking in order to match the two lines of Fourier data. The identity of two functions is such a strong condition that it is not substantially weakened by allowing such rescalings.

3) *Deprojection of a Homogeneous Class of Galaxies.* Suppose there existed a homogeneous class of galaxies, all with the same intrinsic shape. From the preceding discussion, the Fourier transforms of any two projections would have a line of values in common (to within a simple rescaling). The corresponding Fourier slices could be fitted together along that line, but the angle between the planes would be undetermined. Now taking a third projection, one could locate the lines of common values with each of the first two projections, and this third plane could be fitted in, yielding a rigid structure for the three planes. Further planes could similarly be inserted, and, if the observed projections were sufficiently numerous and varied, eventually the Fourier space would be filled in enough for an inversion to be performed.

REFERENCES

Bracewell, R. N., and Riddle, A. C. 1967, *Astrophys. J.*, **150**, 427.

Crowther, R. A., DeRosier, D. J., and Klug, A. 1970, *Proc. Roy. Soc. Lond.* A. **317**, 319.

DISK ROTATION CURVES IN TRIAXIAL POTENTIALS

Ortwin E. GERHARD[1] and Mario VIETRI[2]

[1]Max-Planck-Institut für Astrophysik, Garching, West Germany
[2]Osservatorio Astrofisico, Firenze, Italy

ABSTRACT. The apparent rotation curve of cold gas in a triaxial potential often differs from that in an axisymmetric system in characteristic ways. The data argue for triaxial bulges in our Galaxy and several others.

Gas rotation on elongated closed orbits. A significant fraction of elliptical galaxies contain neutral or ionized gas (Knapp and Sadler, this volume). In the absence of shocks, this gas can be treated as a collisionless fluid orbiting on stable, closed, non-self-intersecting orbits (Sanders & Huntley 1976). Here we consider motion in the equatorial plane of a non-rotating triaxial galaxy. Then the relevant closed orbits are the short axis loops, which are elongated along the intermediate axis of the potential. When viewed down its major axis (the potential's intermediate axis), the tangent point velocity of such an orbit is higher than that of the circular orbit at the same radius in a corresponding axisymmetric potential (Lake & Norman 1982), while the reverse effect is seen when the orbit is viewed down its short axis. Thus in a triaxial potential, the apparent rotation curve is determined by the run with radius of M/r **and** orbital ellipticity (as well as geometry). E.g. in a de Vaucouleurs-type profile or in the density profiles of galactic bulges the closed orbits become rapidly round outside a turnover radius in the density distribution, introducing characteristic features in the rotation curve. We now describe how this applies to our own Galaxy (Gerhard & Vietri 1986; hereafter GV).

Model of the Galaxy. We use a triaxial bulge with $\rho(a) \propto a^{-1.8}$ inside $800pc$ as determined by infrared observations, and $\rho(a) \propto a^{-3.5}$ for $a > 800pc$ as is consistent with these data and the local density of spheroid stars. We also include an exponential disk with scale-length $3.5kpc$ and local surface density $64M_{\odot}/pc^2$ (see GV). Fig. 1 shows the apparent rotation curves from loop orbits in three bulge models, spherical, axisymmetric, and non-axisymmetric, with the same peak velocity, as seen by an observer on the intermediate axis of the potential. Also shown is the Galactic rotation curve from HI and CO terminal velocities. It is clear that **the observed galactic rotation curve is consistent with the $2.4\mu m$ bulge density profile and the local density of spheroid stars only if the bulge is non-axisymmetric.** In this model, the orbital ellipticity is $\sim 20\%$ inside $\sim 600pc$ and then decreases rapidly with r. Thus the peak in the rotation curve is not at the knee in the density profile as in the axisymmetric models, but closer to the centre ($r = 500$ pc). The Sun need not be aligned with the bulge intermediate

T. de Zeeuw (ed.), Structure and Dynamics of Elliptical Galaxies, 399–400.

axis exactly. Fig. 2 shows the apparent tangent velocity curves for observers at various angles ϕ *wrt* this axis. A range $-20^0 \lesssim \phi \lesssim 20^0$ seems acceptable for the solar position (with scaled M/L). Fig. 2 also shows that from a viewing direction along the potential's major axis the rotation curve rises unusually slowly.

Bulges and elliptical galaxies. The characteristic morphology of a peak and subsequent near-Keplerian fall-off in the rotation curve is also shown by NGC 3200 and UGC 2885. The two galaxies NGC 2708 and NGC 3054 have such slowly rising rotation curves, that the fits by an axisymmetric model based on CCD photometry appear to require negative M/L ratios for their bulges (Kent 1986). With triaxial bulges this is expected for lines-of-sight in the plane containing the major and minor axes of the potential ($\phi = 90^0$ in Fig. 2). Thus triaxiality of the bulge is probably important in modelling the rotation curves of at least a fraction of spiral galaxies, and may significantly affect the derived mass-to-light ratios.

In elliptical galaxies with a de Vaucouleurs profile the transition from elongated to nearly circular orbits occurs in the range between a fraction of an r_e and two or three r_e. The closed orbits in the inner part of the system are significantly more elongated than in the Galactic model of the same ellipticity; thus if gas is found on these orbits, the expected effects are significantly larger. Again the characteristic morphology indicating triaxiality is either a peak-fall-off structure or a very slow rise of the rotation curve, both inconsistent with constant M/L inside $\sim 2r_e$.

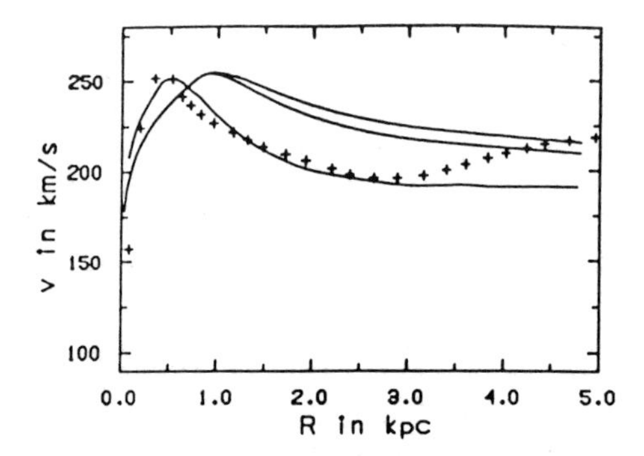

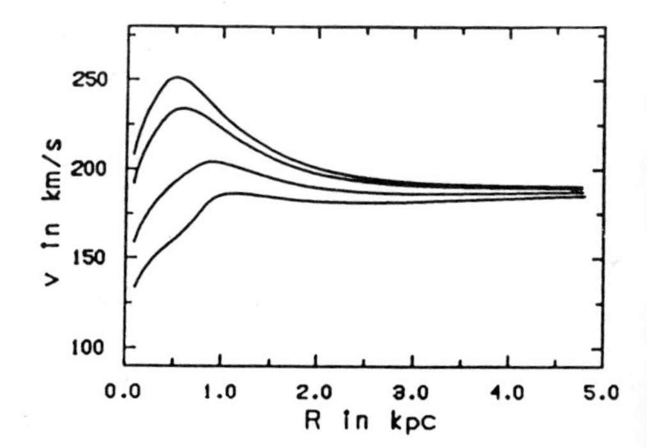

Fig. 1 (left): Apparent rotation curves along the intermediate axis for three galactic bulge models plus galactic disk, and compared with observations. Crosses are taken from a smooth curve drawn through the data points (Clemens 1985). Curves from top to bottom at large radii: Spherical, oblate (1:1:0.4), and prolate-barred (1:0.6:0.6) bulge.
Fig.2 (right): Tangent rotation curves in a prolate bulge (1:0.6:0.6) plus disk model, for observers at ∞ and with different position angles ϕ relative to the bulge's intermediate axis. Curves are (from top) for $\phi = 0^o, 20^o, 45^o, 90^o$.

REFERENCES

Clemens, D.P., 1985. *Astrophys. J.* **295**, 422.
Gerhard, O.E., Vietri, M., 1986. *Mon. Not. R. astr. Soc.*, **223**, 377.
Kent, S.M., 1986. *Astron. J.* **91**, 1301.
Lake, G., Norman, C., 1982. in *The Galactic Center*, A.I.P. Conference no. 83, ed. G.R. Riegler, R.D. Blandford, p.189.
Sanders, R.H., Huntley, J.M., 1976. *Astrophys. J.* **209**, 53.

A METHOD TO DETERMINE THE INTRINSIC AXIAL RATIOS OF INDIVIDUAL TRIAXIAL GALAXIES

Ortwin E. GERHARD[1] and Mario VIETRI[2]

[1]Max-Planck-Institut für Astrophysik, Garching, West Germany
[2]Osservatorio Astrofisico, Firenze, Italy

ABSTRACT. We present a method, partly geometrical and partly dynamical, to determine the intrinsic axial ratios of a triaxial galaxy. The method can be applied to bulges of spiral galaxies and to ellipticals containing sufficiently extended and kinematically regular gas disks. Required observational inputs are multi-colour surface photometry and the gas velocity field.

Motivation: It is now believed that giant elliptical galaxies are often triaxial systems (e.g. Binney 1985). There is increasing evidence that, despite their fast rotation, the bulges of some spiral galaxies are also triaxial (Zaritsky & Lo 1986; M31: Lindblad 1956; the Galaxy: Gerhard & Vietri 1986; barred galaxies: Kormendy 1982). Because of projection effects, the axial ratios of a triaxial galaxy are not readily determined and most attempts have therefore been statistical. Here we describe how in favourable circumstances the axial ratios of an individual triaxial system can be obtained directly from observations.

Input and assumptions: If applied to a triaxial bulge, our method requires photometry of the bulge, in several colours to remove the effects of dust, and the velocity field of the gas in the disk plane, at distances such that the potential is dominated by the bulge. Then, if we assume that the minor axis of the bulge and the spin axis of the disk are the same, that the disk is not flattened in its major plane, and that the bulge is non-rotating, and if the gas is in regular motion, the axis ratios of the bulge follow. In the case of an elliptical galaxy, the method requires the presence of a regular gas disk in one of the principal planes, with some gas inside r_e and some gas well outside r_e, and a regular velocity field. Examples may be NGC 1097 (Davies, this conference) and Cen A (Bland, this conference).

The method: The axial ratios of a bulge in a disk galaxy can be obtained as a function of one (unknown) parameter, e.g. the true angle ϕ of the bulge intermediate axis *wrt* the disk apparent major axis (Stark 1977). The method uses as constraints θ, the inclination angle of the disk, and ψ, the angle between the major axis of the isophotes of the bulge and disk ($= 10^\circ$ for M31). In an elliptical with extended gas disk, the outer parts replace the optical disk in the spiral to fix θ and ψ. - Here, we have computed gas velocity fields in various potentials, among which an illustrative de Vaucouleurs (dV) model with $r_e = 800$ pc and axial ratios $q = 0.6$, assuming that the gas moves on closed orbits. Generally, the non-circular gas streamlines in the plane of the galaxy cause a tilted velocity field when

T. de Zeeuw (ed.), Structure and Dynamics of Elliptical Galaxies, 401–402.

projected onto the sky, in which the velocity contours cluster around the projected intermediate axis of the potential and significant minor axis rotation is present. Fig.1 shows, for the above dV-model, one of the diagnostics that allow us to fix the remaining free parameter of the projection through the gas velocity field, and thus to obtain the intrinsic axial ratios of the triaxial body. In this case the velocities to be measured are of order $200 kms^{-1}$ on the major axis, and large (up to $100 kms^{-1}$) minor axis velocities occur. In an axisymmetric bulge, the rotation curve on the major axis determines the projected velocity anywhere. We can use this fact to test for a triaxial component in the potential, by plotting the velocity difference between the axisymmetric and the triaxial model, again as a funcion of position angle along the same ellipse as for Fig.1. Then one finds a typical $\cos 2PA$ component, and this will tell us whether the application of Fig.1 to a given bulge is justified.

Conclusions: Regular gas velocity fields can be used to infer the existence of a triaxial component in the potential and to measure its intrinsic axial ratios. If the profile is de-Vaucouleurs-like, the effects are easily measurable for $q \leq 0.9$; more centrally concentrated models have more circular orbits; then flattenings of order $q \leq 0.7$ are measurable.

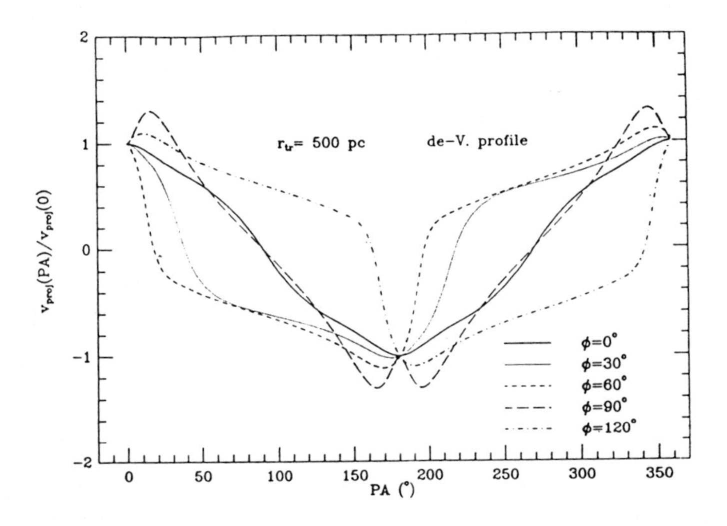

Fig.1: Gas velocity in the de Vaucouleurs model with q=0.6, projected along the line of sight, as a function of position angle along an ellipse on the plane of the sky that corresponds to a circle with r=500pc in the plane of the galaxy, and for various choices of the unknown angle ϕ. These velocities are normalized to the projected velocity measured on the disk major axis, so that these curves are independent of the choice of M/L.

REFERENCES

Binney, J.J., 1985. *Mon. Not. R. astr. Soc.* **212**, 767.
Gerhard, O.E., Vietri, M., 1986. *Mon. Not. R. astr. Soc.*, **223**, 377.
Kormendy, J., 1982. *Astrophys. J.* **257**, 75.
Lindblad, B., 1956. *Stockholm Obs. Ann.* **19**, no.2.
Stark, A.A., 1977. *Astrophys. J.* **213**, 368.
Zaritsky, D., Lo, K.Y., 1986. *Astrophys. J.* **303**, 66.

SETTLING OF GAS DISKS IN ELLIPTICAL GALAXIES

T.Y. Steiman-Cameron[1] and R.H. Durisen[2]
[1]J.I.L.A., U. of Colorado, Boulder, CO 80309
[2]Astronomy Dept., Indiana U., Bloomington, IN 47405

Prominent gas disks and dust lanes are found in a number of elliptical galaxies. It is widely believed that these disks are the result of the tidal capture of material from a nearby galaxy or the accretion of a gas rich companion into a larger galaxy. While a captured gas disk will ultimately settle into a steady-state orientation ("preferred plane"), the disk will initially be time varying due to precessional and viscous forces. We have developed methods for modeling the evolution of such inclined dissipative galactic gas disks (Steiman-Cameron 1984, Ph.D. dissertation, Indiana U.; Steiman-Cameron and Durisen 1986, Ap.J., in press, hereafter SCD), and we here present some of the results of this work.

While the full set of differential equations describing disk evolution must, in general, be solved numerically, certain limiting assumptions allow for analytic solutions, in particular, when: 1) the viscous time scale for inflow is much longer than the time scale for settling; 2) the precession rate is dominated by gravitational forces with viscous forces being a minor perturbation; 3) the initial inclination of the disk relative to a preferred plane is small, and 4) settling takes place in a nearly axisymmetric galactic potential. These conditions are generally met for parameters typical of real galaxies (see SCD). If i_0 is the initial disk inclination (i=0 being the steady-state orientation), Ω_p the precession rate due to the non-sphericity of the gravitational potential, and ν the coefficient of kinematic viscosity due to cloud-cloud collisions, then, according to the analytic solution in SCD, the inclination of the disk as a function of time and radius goes as

$$i \cong i_0 \exp\left[-(t/\tau_e)^3\right], \tag{1}$$

where the time constant τ_e is given by

$$\tau_e = \left[\nu(d\dot{\Omega}_p/dr)^2/6\right]^{-1/3}. \tag{2}$$

Note that the settling time scale is a weak function of ν, going only as $\nu^{-1/3}$. In view of the inadequacies of our understanding of cloud-

T. de Zeeuw (ed.), Structure and Dynamics of Elliptical Galaxies, 403–404.

cloud collisions, this result gives us confidence that we can nonetheless derive realistic settling times.

For galaxies with extended halos, as appears to be the case for luminous ellipticals, a useful gravitational potential is the self-similar logarithmic potential (Gunn 1979, in Active Galactic Nuclei, eds. C. Hazard and S. Mitton, p. 213.). Model galaxies with this potential have similar concentric isodensity surfaces, a zero-order density profile of $\rho(r) \propto r^{-2}$, and a flat rotation curve. Then

$$\tau_e = r^{4/3} (\nu/6)^{-1/3} (\tfrac{1}{2}\varepsilon v_c \cos i_o)^{-2/3} \tag{3}$$

where ε is the ellipticity of the model galaxy and v_c is the constant circular velocity. Since the maximum radial distance over which a cloud-cloud collision can transport momentum is the epicyclic amplitude of its orbit, the maximum value of ν occurs when the cloud mean free path is equal to its epicyclic amplitude. The value of ν_{max} is then a function only of the cloud velocity dispersion and galactic potential. Settling times determined using ν_{max} represent minimum settling times. Substitution of the expression for ν_{max} that we derive in SCD yields a minimum settling time of

$$(\tau_e/\tau_p)_{min} = 0.404 \, (v_c/v_{rms})^{2/3} (\cos i_o)^{-1/3}, \tag{4}$$

where v_{rms} is the cloud rms velocity. The precession period $\tau_p = 4\pi r/(\varepsilon v_c \cos i_o) = 2\tau_o/(\varepsilon \cos i_o)$, where τ_o is the orbit period. We therefore find that the normalized minimum settling time for the self-similar scale free potential is also self-similar and scale free. Calculations for a modified Hubble potential, which has often been used in modeling elliptical galaxies, show very similar results. Substitution of reasonable numbers into (4) shows that settling is slow, with unsettled disks at radii $\gtrsim$ 10 kpc being extremely old, in contrast to earlier work by Simonson (1982, Ph.D. dissertation, Yale U.) which we feel must be in error.

We have also developed numerical solution methods for parameters less restrictive than the analytic solution presented above. These allow us to test the limits of the analytic results and to further explore the characteristics of disk evolution. We find the following: 1) there is good agreement between the analytic solution and numerical simulations, even for moderate ($i_o = 40°$) initial inclinations; 2) settling is essentially a local process, depending primarily on local properties of the potential and gas disk; 3) minimum settling times for reasonable galaxy parameters are on the order of two or three orbit precession periods or, equivalently, ten to a few tens of orbit periods in galaxies with intrinsic flattenings representative of E3 to E7 ellipticals; 4) the actively settling region is somewhat narrow, with a radial width of about 30% of its outer radius; 5) radial inflow is enhanced during the active settling period.

DUST IN EARLY TYPE GALAXIES OBSERVED AT THE CFHT

J. Kormendy[1] and J. Stauffer
Dominion Astrophysical Observatory

A program of CCD imaging of early-type galactic nuclei carried out at the Canada-France-Hawaii Telescope (Kormendy, these proceedings) reveals dust in an unusually large fraction of the galaxies. Often the dust is in a ring or disk whose radius is comparable to the core radius r_c; since r_c is small, our detection rate benefits from the excellent CFHT seeing (median FWHM = $0\rlap{.}''7$; scale = $0\rlap{.}''22$ pixel^{-1} for an RCA CCD used at the Cassegrain focus). We plan to prepare a photographic atlas of unsharp-masked images, and to study correlations between dust properties and nuclear emission lines, HI and X-ray gas content, radio continuum jets and the overall light distributions of the program galaxies.

Two techniques of "unsharp masking" are used to illustrate dust distributions. In each case the galaxy image is divided by a dust-free model image. The simplest model image is a smoothed version of the original: each pixel is replaced with the mean in a square box surrounding it. Alternatively, we calculate the galaxy profile using the VISTA image processing system and construct a model galaxy having this profile and exactly elliptical isophotes. The first technique (Fig. 1d) is preferable for dust rings, the second (Fig. 1a-c) for radial or large-scale dust features.

Figure 1 shows some of the variety of dust structures seen. The panels are arranged in order of increasing regularity, from chaotic dust structure in an HI elliptical that may be a recent merger remnant (Raimond *et al.* 1978) to more "settled" distributions in NGC 1199 and the two radio-jet galaxies. The good CFHT seeing has allowed us to detect significant amounts of dust in many galaxies where it was not previously known (e. g. NGC 4621, NGC 4636, NGC 5846). For example, nearly half of King's (1978) sample of normal ellipticals show dust.

Dust lanes are especially common in radio galaxies. The relative alignment of the dust lane and the optical major axis is a diagnostic of the 3-D shape of the galaxy; comparison of the dust lane orientations with the radio jet axes has implications for the formation of the jet and its interaction with the local ISM. Figure 2a shows the comparison between dust and optical major-axis position angle; the tendency for $\triangle$PA(dust-opt) to have values near 0° and 90° suggests that most ellipticals are oblate while a few may be prolate. Like Kotanyi and Ekers (1979), we find that $\triangle$PA(dust-jet) tends to have values near 90°. This may be expected for jets produced by accretion disks, although jet/disk models do not require the ISM at large radii (where we observe it) to be aligned with the dust at small radii.

[1]Visiting Astronomer, CFHT, operated by NRC(Canada), CNRS(France), and the University of Hawaii.

REFERENCES

King, I. 1978, *Ap. J.*, **222**, 1.
Kotanyi, C. and Ekers, R. 1979, *A. A.*, **73**, L1.
Raimond, E., Faber, S., Gallagher, J., and Knapp, G. 1978, *Ap. J.*, **246**, 708.

T. de Zeeuw (ed.), Structure and Dynamics of Elliptical Galaxies, 405–406.

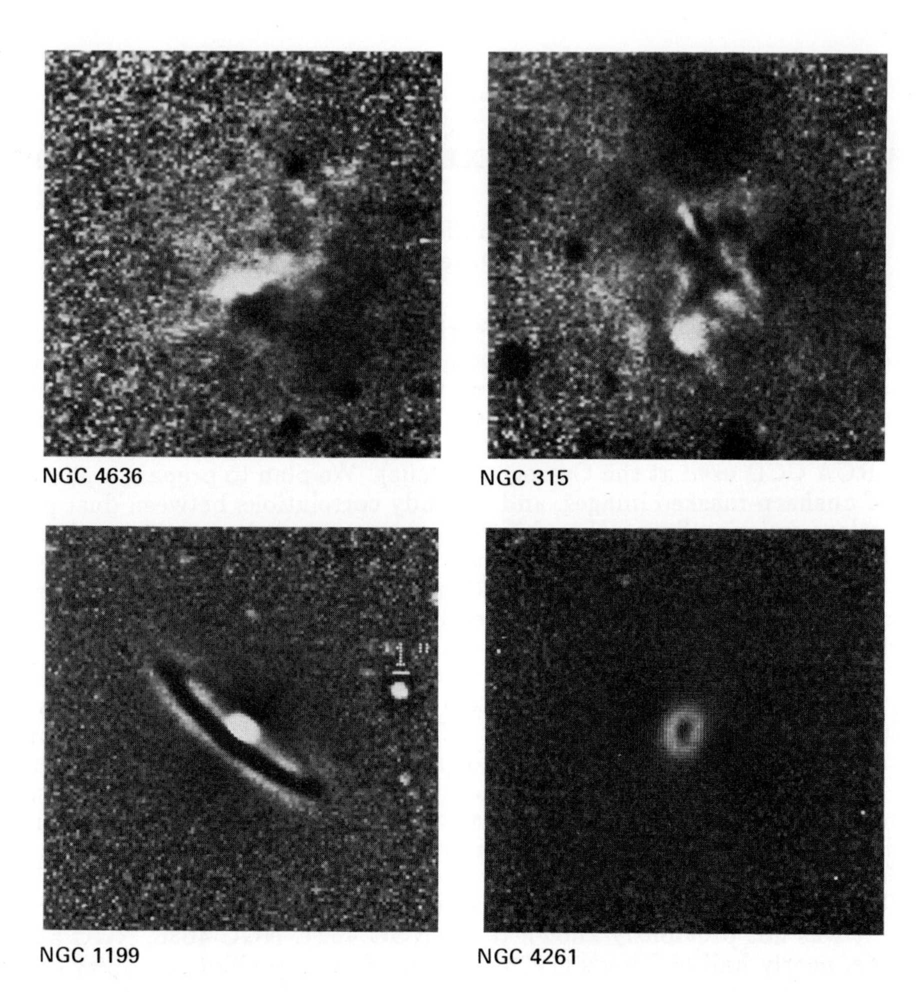

Figure 1 (above): Unsharp-masked images of four galaxies observed at the CFHT.
Figure 2 (below): Histograms of relative position angles of dust lanes, radio jets and optical major axes.

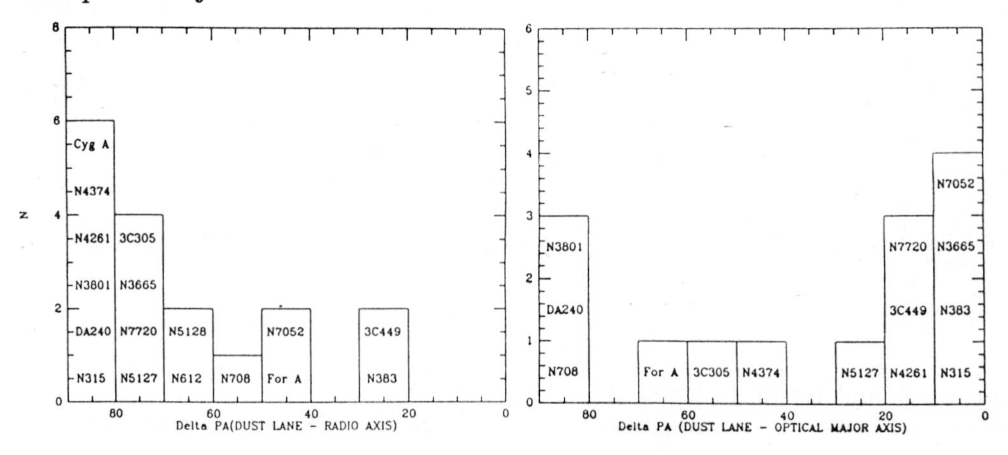

DETAILED SURFACE PHOTOMETRY OF THE DUST-LANE ELLIPTICAL NGC 6702

E. Davoust[1], M. Capaccioli[2], G. Lelièvre[3], J.-L. Nieto[4]

1: Observatoire de Besançon, 25000 Besançon, France
2: Istituto di Astronomia, Università di Padova, 35100 Padova, Italy
3: Canada-France-Hawaii Corporation, p.o. Box 1597, Kamuela HI 96743, USA
4: Observatoire du Pic du Midi et de Toulouse, 14 Avenue Edouard Belin, 31400 Toulouse, France

We present preliminary results of a detailed photometric study of NGC 6702, from high resolution photographs taken at the Cassegrain focus of the Canada-France-Hawaii (CFH) telescope. The luminosity distribution of the galaxy follows an $r^{1/4}$ law ($re^* = 11.6''$, $\mu_e^* = 22.17$). The axis ratio is 0.75 (corresponding to the morphological type E3) and the position angle of the major axis is 60° and fairly constant. Fig. 1 is an image in intensity of the galaxy, from a 30mn exposure in B. The galaxy is crossed by a dust lane, reported first by Capaccioli et al. (1984).

To reveal the structure of the dust, we have subtracted from the previous image the model of an elliptical galaxy having the photometric and geometric parameters quoted above. The result of this numerical unsharp masking is shown in Fig. 2. The dust lane is in position angle 320°, that is 10° West of the minor axis. The isophotes brighter than 20 magarcsec^{-2} are pear-shaped, and indicate that the dust obscures the North-Eastern half of the inner regions of the galaxy. The dust lane is 20" long (4.4 kpc at the distance of 45 Mpc given by de Vaucouleurs and Olson, 1984) and 3" wide (650 pc). Its dimensions and relative orientation are comparable to those of the dust lanes in NGC 5128 (5.5 x 1.1 kpc at 3 Mpc) and NGC 5266 (5.6 x 1.2 kpc at 24 Mpc). The origin of the dust lane must be sought elsewhere.

NGC 6702 is not a strong radiosource. It has not been detected at 21 cm (Knapp, Turner and Cunniffe, 1985), and its flux at 6 cm is less than 1 mJy (Impey, Wynn-Williams and Becklin, 1986).

The companion elliptical galaxy NGC 6703, located 10' to the South-East is a foreground galaxy at 17 Mpc. Its isophotes are fairly round (c/a = 0.95) and show no signs of tidal interaction.

T. de Zeeuw (ed.), Structure and Dynamics of Elliptical Galaxies, 407–408.

Figure 1: NGC 6702 from a 30mn blue exposure at the Cassegrain Focus of the CFH telescope.

Figure 2: The dust lane of NGC 6702 after removal of the luminosity gradient of the galaxy in the above image

REFERENCES

Capaccioli, M., Davoust, E., Lelièvre, G., Nieto, J.-L. : 1984. IAU colloquium n° **78**, p. 393.

de Vaucouleurs, G., Olson, D.W. : 1984. Astrophys. J. Suppl. Ser. **56**, 91.

Impey, C.D., Wynn-Williams, C.G., Becklin, E.E. : 1986, preprint.

Knapp, G.R., Turner, E.L., Cunniffe, P.E. : 1985, Astron. J. **90**, 454.

BOXY ISOPHOTES AND DUST LANES IN BRIGHT VIRGO ELLIPTICALS

C. Möllenhoff*, R. Bender*
Landessternwarte
Königstuhl
D-6900 Heidelberg
Fed. Rep. Germany

The CCD camera of the Landessternwarte Heidelberg was used at the 1.2 m Calar Alto Telescope for a V, R, I - survey of ~70 dusty and non-dusty elliptical galaxies. We report here about morphological studies of ten bright elliptical galacxies in the Virgo cluster. After the usual CCD data reduction the following procedures were carried through:

1. The CCD-frames were filtered and centered to an accuracy of ~0.2 pixels. By division we obtained V-R, V-I, R-I color index images which allow to separate dust absorption and gaseous emission from the stellar distribution. Five of the 10 bright Virgo ellipticals evidently show dust/gas, two other objects have weak features. The dust lanes in NGC 4261 (cf. Fig. 1), NGC 4365 and NGC 4552 were detected for the first time. A dust lane must be visible in all three color index frames in order to avoid fake identifications.

2. The isophotes in all three colors were fitted by least square ellipses. From these fits we obtained in dependence from the radius: color profiles, color index profiles, ellipticity profiles, isophote twists (cf. Fig. 2). The deviations of the isophotes from ideal ellipses were studied by a Fourier series analysis. The fourth coefficient is the main indicator for such deviations. Six of ten galaxies show box-shaped isophotes of an amount between 0.3 % and 1.5 % of the semi major axis (cf. Fig. 3), two galaxies show disc components (0.8 %, 1.5 %), only one galaxy is a perfect elliptical up to an error of ±0.2 %. The deviations are equal in all three colors and therefore are not produced by dust.

Conclusion: Dust/gas and box-shaped isophotes seem to be quite common among the bright Virgo ellipticals. Both phenomena may be interpreted as a consequence of merging or accretion processes (Binney, Petrou 1985).

----

* Visiting Astronomers, German Spanish Observ., Calar Alto

T. de Zeeuw (ed.), Structure and Dynamics of Elliptical Galaxies, 409–410.

Results of Isophote Analysis and Color Index Images:

NGC	Dust/Gas	4·cos/a	Δ(P.A.)	Δ(b/a)
4261	dust lane, P.A. ~0°	-1.2 %	5°	0.1
4278	patches, complex	±0.5 %	18°	0.05
4365	major axis dust lane	-1.0 %	5°	0.05
4374=M84	dust lane, P.A. ~90°	-0.4 %	3°	0.1
4382=M85		+0.8 %	10°	0.2
4406=M86	H_α pointsource	-0.5 %	7°	0.11
4472=M49		-0.3 %	4°	0.14
4552=M89	patches, minor axis	(-0.5 %)	30°	0.05
4621=M59		+1.5 %	0°	0.25
4636	gas in center	0.0 %	16°	0.26
3379	(not Virgo)	0.0 %	5°	0.1
errors		±0.2 %	$\pm 1^\circ$	±0.01

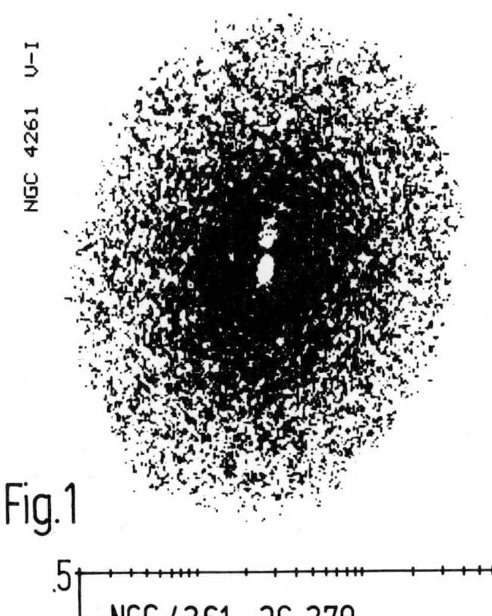

Figure 1. V-I color index image of NGC 4261. The dust lane is orthogonal to the radio jet (Birkinshaw, Davies 1985)

Figure 2. Due to the dust lane the ellipticity in V decreases faster towards the center.

Figure 3. The negative 4th Fourier coefficient indicates a color-independent 1.5 % boxiness of NGC 4261.

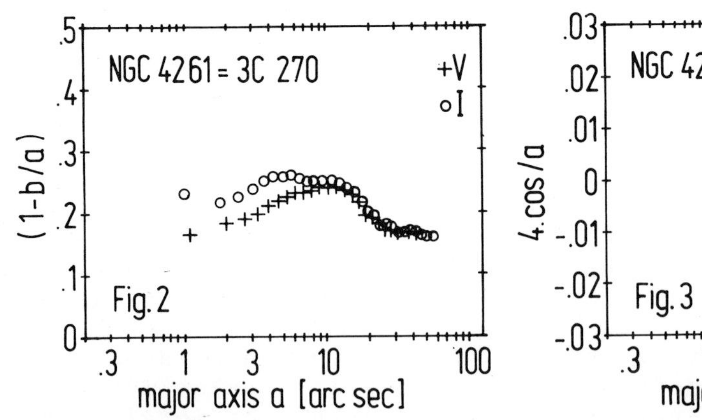

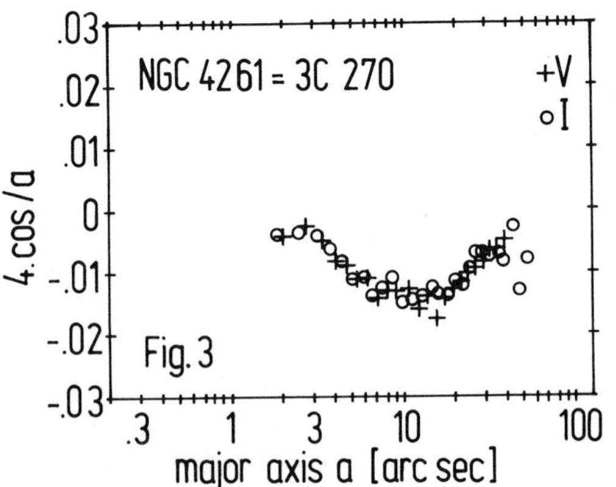

REFERENCES

Binney, J.J., & Petrou, M., 1985. *Mon. Not. R. astr. Soc.*, **214**, 449.
Birkinshaw, M., & Davies, R.L., 1985. *Astrophys. J.*, **291**, 32.

THE UNUSUAL BOX-SHAPED ELLIPTICAL(?) GALAXY IC 3370

Brian Jarvis
Geneva Observatory
CH-1290 Sauverny
Switzerland

ABSTRACT. Photometic and kinematic observations are presented for the box-shaped "elliptical" galaxy IC 3370, that show that this galaxy has characteristics more akin to an S0 galaxy than an elliptical and should be classified S0pec.

1. INTRODUCTION

The existence of a significant proportion ($\simeq$ 1 percent) of disk galaxies which possess box- or peanut-shaped bulges is now clear (Jarvis, 1986). However, until recently, similar morphologies in elliptical galaxies have gone unreported. IC 3370 is a galaxy classified as an elliptical in the RC2 and RSA catalogues but shows a strong box shape. Recent attempts to model disk galaxy bulges with box- or peanut-shapes have required large amounts of rotation ($V_m/\sigma_0 > 1.0$, May, van Albada and Norman 1985; Binney and Petrou 1985). However, most bright elliptical galaxies (say $M_B < -21.0$) rotate slowly (Davies *et al*, 1983). Hence, any *bright* "elliptical" galaxy (i.e. no disk) found to have a box or peanut shape with cylindrical rotation, would be unusual indeed for the reasons mentioned above.For these reasons, the bright box-shaped southern "elliptical" galaxy IC 3370, promised to be an interesting study.

2. OBSERVATIONS and RESULTS

Fig. 1 shows the isophotes of IC 3370 in the Johnson B band taken from a CCD image. The V and R isophotal maps are similar. The box shape is very noticeable over more than 3.5 magnitudes of surface brightness to the lowest surface brightness measured ($\mu_B = 25.0$ mag.arcsec^{-2}) and is strongly suggestive that it extends much further. The minor axis luminosity profile (not shown because of space restrictions) closely follows the de Vaucouleurs $r^{1/4}$ law. There is convincing evidence that IC 3370 is an S0 galaxy and not an elliptical since the major axis luminosity profiles, color profiles and fitted ellipses to the isophotes in all colors show a faint stellar disk, visible to a distance of about $15''$ on both sides of the nucleus. Fig. 3 shows the isophotes from the central $20''$ of IC 3370 in R (least affected by dust) with fitted ellipses. Clear evidence is seen for an excess of light on the major axes of these

T. de Zeeuw (ed.), Structure and Dynamics of Elliptical Galaxies, 411–412.

ellipses indicative of a stellar disk embedded within a much more luminous bulge. Further support is provided by the kinematical observations shown in Fig. 2. The maximum rotational velocity on the major axis is $V_m = 98 \pm 15$km.s^{-1}, more akin to S0 galaxies than ellipticals. Furthermore, IC 3370 shows strong cylindrical rotation, unlike any elliptical galaxy known at present. The central velocity dispersion was measured to be σ_0=220 $\pm$20 km.s^{-1}, giving V_m/σ_0=0.45, a value in good agreement with theoretical models of oblate models supported by rotation alone and greater on average than ellipticals of similar luminosity ($M_B = -22.2$ assuming a distance of 54.3 Mpc ($H_0 = 50$ km.s^{-1}.Mpc^{-1})) and observed flattening.

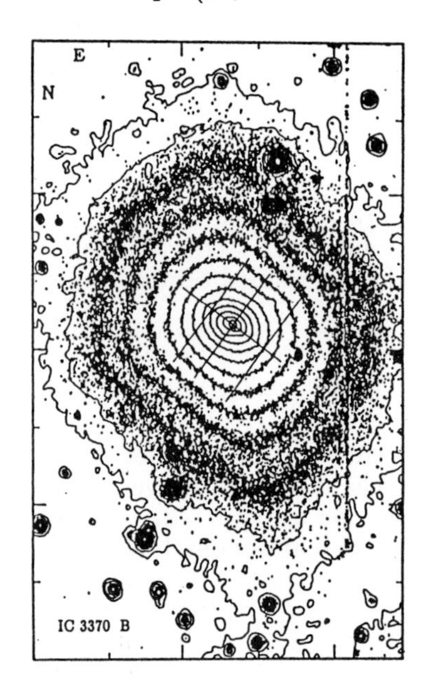

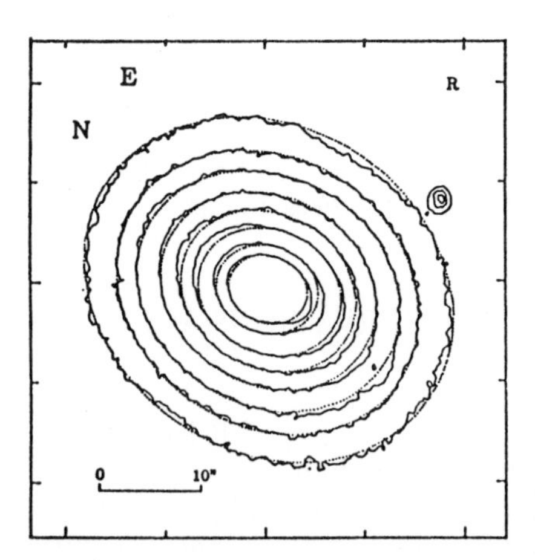

Fig.1 (Top left) Isophotal contour map of IC 3370 in B. The contour interval is 0.5 mag.arcsec^{-2} with the faintest contour at μ_B=25.0 mag.arcsec^{-2}.
Fig.2 (Top right) Rotation curves of IC 3370 along the major axis (top) and perpendicular to the major axis (bottom) but offset 10" from the minor axis.
Fig.3 (Bottom right) Fitted ellipses to inner 20" of IC 3370 in R. Note the disk shown by an excess of light on the major axes of the ellipses.

REFERENCES

Binney, J., and Petrou, M., 1985, *M.N.R.A.S.*,**214**, 449.
Davies, R.L., Efstathiou, G., Fall, S.M., Illingworth, G., and Schechter, P.L., 1983, *Ap.J.*,**266**,41.
Jarvis, B.J., 1986, *A.J.*,**91**, 65.
May, A., van Albada, T. S., and Norman, C. A., 1985. *M.N.R.A.S.*, **214**, 131.

AM2020–5050: AN ELLIPTICAL GALAXY WITH AN OUTER RING

Bradley Whitmore,
Doug McElroy
Space Telescope Science Inst.
3700 San Martin Drive
Baltimore, MD 21218

François Schweizer
Dept. of Terrestrial Magnetism
Carnegie Inst. of Washington
5241 Broad Branch Rd. NW
Washington, DC 20015

ABSTRACT. Photometric and spectroscopic observations show that the inner component of AM2020–5050 is an elliptical galaxy, unlike other polar-ring galaxies which have an S0 disk at the center. A comparison of the central velocity dispersion with the rotational velocity in the ring suggests the presence of a nearly spherical gravitational potential. The inner component has a rapidly rotating core with rotational velocities at 3″ substantially higher than at 8″. Although the optical ring is quite narrow, H_α emission is observed all the way through the center of the galaxy, indicating the presence of an extended gaseous disk.

OBSERVATIONS AND RESULTS

Observations have shown that the central components of polar-ring galaxies tend to be highly inclined S0 disks, with a large rotational velocity and small velocity dispersion. Photographs of the polar-ring galaxy AM2020–5050 led us to believe that the central component in this case might be an elliptical galaxy. We obtained photometric and spectroscopic observations using the 4 meter telescope at CTIO to check this hypothesis.

Photometry in the B and V bands with a CCD detector at prime focus shows that the luminosity profile follows an $R^{\frac{1}{4}}$ law characteristic of an elliptical galaxy (see Figure 1), rather than the exponential law seen in most disk galaxies. The ring is bluer than the central component with $B - V = 0.6$ compared to $B - V = 0.9$.

Long-slit spectra obtained in the blue (4100 - 5600 Å) show that the central component rotates quite slowly at 8″, unlike S0 disks which rotate rapidly. The ratio of the rotational velocity to the central stellar velocity dispersion is $V/\sigma_o = 0.4$, characteristic of an elliptical galaxy. However, the inner 3″ are rotating substantially faster, showing the presence of a rapidly rotating core.

Spectra obtained at H_α measure the rotation of the gas in the polar ring. The rotation curve has a constant gradient in the inner region, followed by a well defined turnover at about 10″. Beyond this radius the rotation curve is flat or slightly rising. Perhaps the most surprising fact is that while the ring appears to be a narrow feature at about 12″ radius, H_α emission is actually observed all the way through the nucleus. This suggests that the ring may represent the locus of most intense star formation within an underlying gaseous disk.

T. de Zeeuw (ed.), Structure and Dynamics of Elliptical Galaxies, 413–414.

DISCUSSION

In most polar-ring galaxies, a comparison between the rotational velocities in the ring and in the S0 disk provides a measurement of the flattening of the gravitational potential. In AM2020−5050, however, the low value of V/σ_o shows that the inner component is an elliptical galaxy rather than an S0 disk. Since most of the dynamical support comes from the velocity dispersion rather than rotation, a comparison between σ_o and V_{ring} is more relevant.

For bulge-dominated spiral galaxies, Whitmore and Kirshner (1981) find that $V_{disk}/\sigma_o = 1.54 \pm 0.06$. The value for the polar ring of AM2020−5050 is approximately the same, $V_{ring}/\sigma_o = 1.5 \pm 0.1$. This suggests that the rotational velocity of the ring is about the same as it would be in a hypothetical gaseous disk of the central elliptical component. Therefore, the gravitational potential of AM2020−5050 seems to be nearly spherical, even though the shape of the luminous matter is that of an E4 galaxy.

REFERENCES

Whitmore, B.C., McElroy, D., and Schweizer, F., 1987. *Ap. J.*, submitted.
Whitmore, B.C., and Kirshner, R.P., 1981. *Ap. J.*, **250**, 43.

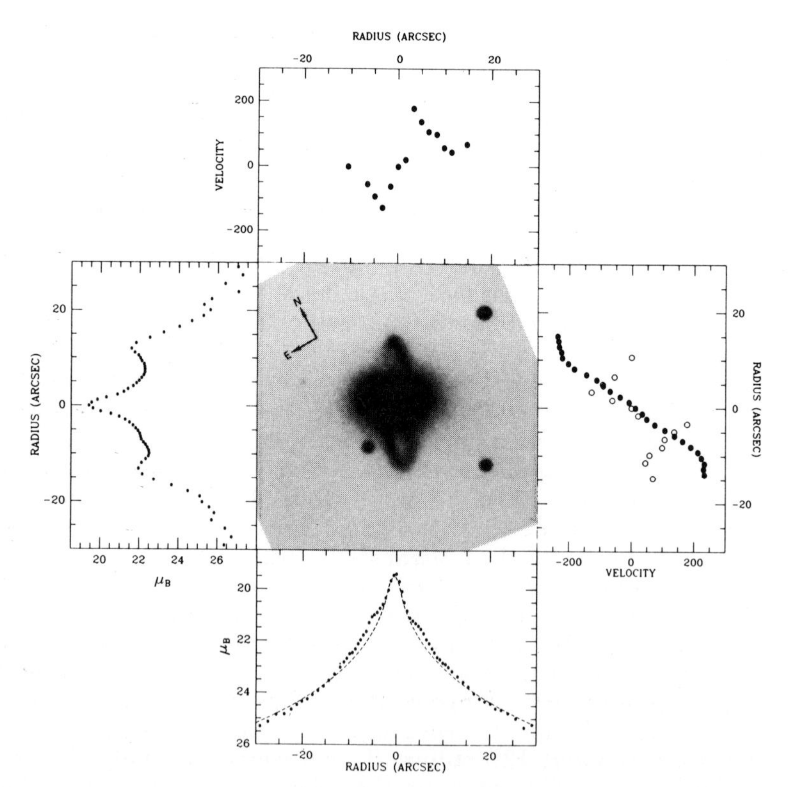

Figure 1. Photometric and Kinematic Observations of AM2020-5050.

TAURUS OBSERVATIONS OF S0 POLAR RING GALAXIES

R.A. Nicholson[1], K. Taylor[1], W.B. Sparks[1] & J. Bland[1,2]

[1] Royal Greenwich Observatory, Hailsham, Sussex, England
[2] Inst. for Astron., Univ. of Hawaii, Honolulu, USA

By use ot the TAURUS imaging Fabry–Perot interferometer (Taylor & Atherton 1980) we have obtained seeing limited two–dimensional velocity, line width and line flux maps of the ionised gas in two polar ring systems.

The Hα TAURUS maps of NGC 4650A, obtained at the AAT, show clearly an extended (110″) differentially rotating gas disc orthogonal to the projected minor axis of the galaxy. The gas warps in a manner consistent with it being in a polar orbit about an S0 galaxy, and stabilised by its self gravity (Sparke 1986).

NGC 2685 was observed with the 2.5–m Isaac Newton Telescope at the Observatory del Roque de Los Muchachos, La Palma. This has revealed the distribution of HII regions surrounding the stellar component. Provisional results show no sign of the complicated central warp structure seen in NGC 4650A. Evidence is found for possible severe warping at radii greater than 4 kpc (H_0 = 50 km/s/Mpc).

OBSERVATIONS/RESULTS

Figure 1 shows the velocity contour map obtained from the TAURUS data cube on NGC 4650A. The velocity field is found to be far more complex in structure than that expected from a laminar disc undergoing differential rotation. Figures 2a, b show the line of nodes and rotation curve obtained by sampling this velocity field with a series of concentric inclined circular annuli. The rotation and line of node curves were then used in modelling the velocity field expected from a thin warped disc. The resultant model reproduces well the double kinked 'M' type perturbations seen in the central regions of the contoured data map, Figure 3. Lausten and West (1980) suggested that the observed intensity distribution might be the result of the disc warping across our line of sight: this is supported by the thin disc model.

Preliminary analysis of NGC 2685 reveals a gas disc inclined at 15 degrees to the minor axis of the stellar system. Evidence for warping in the outer regions of this disc, and HII regions associated with the outer equatorial HI ring (Shane 1980) is found. The existence of two nearly orthogonal gas geometries may be explanable in terms of a severe (~ 90 deg) warp. Such a warp would give a natural explanation to the peculiar asymmetric distribution of dust bands seen across the face of the stellar component.

T. de Zeeuw (ed.), Structure and Dynamics of Elliptical Galaxies, 415–416.

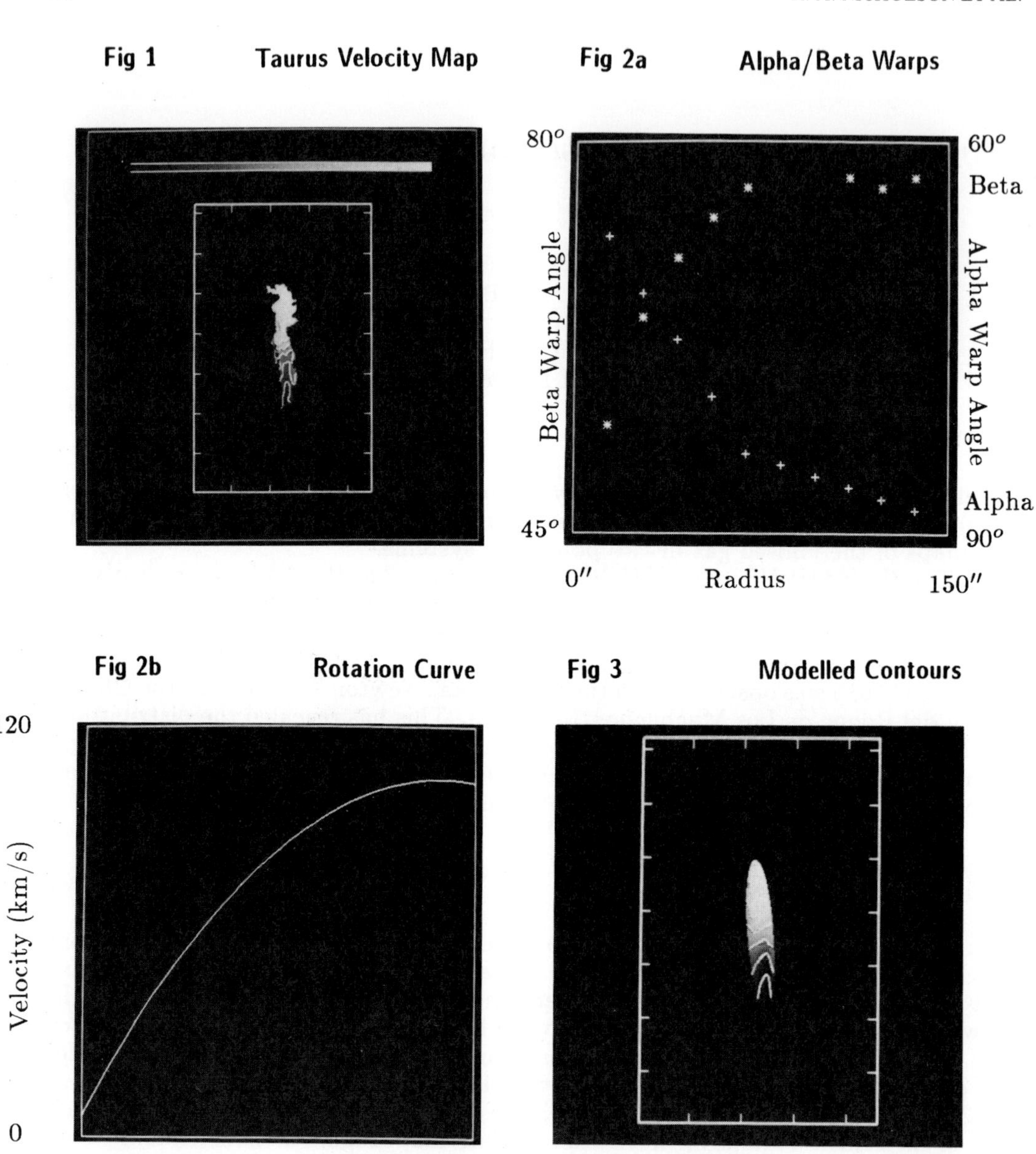

REFERENCES

Lausten, S., & West, R.M., 1980. *Journ. Astrophys. Astron.*, **1**, 177.
Shane, W.W., 1980. *Astr. Astrophys.*, **82**, 314.
Sparke, L.S., 1986. *Mon. Not. R. astr. Soc.*, **219**, 657.
Taylor, K., & Atherton, P.D., 1980. *Mon. Not. R. astr. Soc.*, **191**, 675.

THE STRUCTURE AND KINEMATICS OF THE IONIZED GAS IN CENTAURUS A

J. Bland[1,2] K. Taylor[1] P. D. Atherton[3]

1) Royal Greenwich Observatory, Hailsham, Sussex, England
2) Institute for Astronomy, Honolulu, Hawaii
3) Queensgate Instruments, London, England

ABSTRACT. The TAURUS Imaging Fabry-Perot System (Taylor & Atherton 1980) has been used with the IPCS at the AAT to observe the ionized gas within NGC 5128 (Cen A) at [NII]λ6548 and Hα. Seven independent (x,y,λ) data cubes were obtained along the dust lane at high spectral resolution (30 km/s FWHM) and at a spatial resolution limited by the seeing (~1"). From these data, maps of the kinematics and intensities of the ionized gas were derived over a 420" by 300" region. The maps are the most complete to date for this object comprising 17500 and 5300 fitted spectra in Hα and [NII]λ6548 respectively. The dust lane system is found to be well understood in terms of a differentially rotating disc of gas and dust which is warped both along and perpendicular to the line-of-sight.

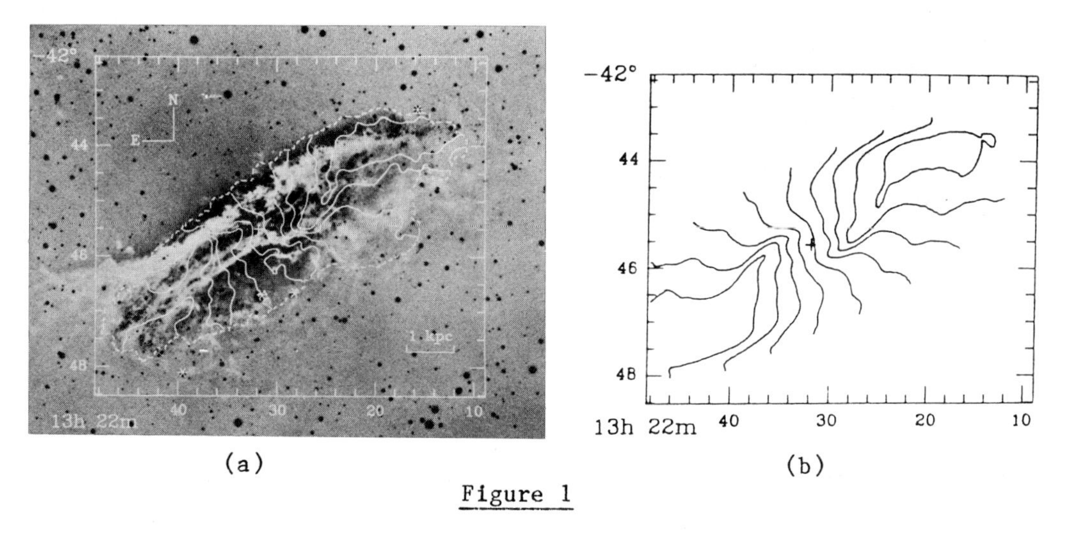

Figure 1

(a) The TAURUS Hα velocity field is overlayed on an unsharp masked image of Cen A (courtesy of D.F.Malin); the contours are drawn in steps of 50 km/s beginning at 300 km/s in the SE. (b) The model velocity field where contours are drawn in the same way.

T. de Zeeuw (ed.), Structure and Dynamics of Elliptical Galaxies, 417–418.

The TAURUS observations are discussed in detail elsewhere (Bland et al. 1986). In Figure 1(a), the derived isovelocity contours have been superimposed on an image of the dust lane system. The well-ordered velocity field is particularly striking in that the kinematic line-of-nodes is complex and not confined to a single position angle. A variety of kinematic models has been explored in order to interpret these new results (Bland and Taylor 1986). It is found that, to first order, the ionized gas is optically thin at Hα so that the observed velocity field is the result of seeing throughout the entire gas system. In contrast, the observed dust distribution is due primarily to dust in front of the plane of the sky (assumed to pass through the nucleus of the galaxy), highlighted against the bulk of starlight from the elliptical body. The ionized gas and dust appear to be physically mixed out to a radius of 250"; beyond here, only dust is seen. The proposed warped disc model explains successfully

i) the velocity field of the gas (see Figure 1(b)),
ii) the distribution of HII regions within the dust lane,
iii) the striking appearance of the dust lane system: the familiar dust bands are tangential points of the warp, and
iv) the distribution of broadened, skewed and split line profiles observed within the TAURUS data.

The distribution of line splitting indicates that the disc is thin ($\ll$ 1 kpc) normal to the surface of the warp. The rotation curve implies that the ionized gas moves primarily under the influence of a deep potential well; there is evidence for neither radial nor non-planar components of motion greater than 30 km/s within the system. There are further clues that the gas configuration is at least partially relaxed:

a) the velocity field is well-ordered and bisymmetric,
b) the entire system rotates about a common centre which is coincident with the nucleus of the old stellar population, and
c) its systemic velocity of 536 km/s is identical to that of the stellar component.

On the other hand, it has not been possible to understand the warped disc in the light of recent work on the stability of orbits in both stationary and tumbling, spheroidal and triaxial potentials (van Albada et al. 1982; Steiman-Cameron and Durisen 1984). Thereby, attempts to infer the 3D form of the old stellar component within Cen A have been inconclusive.

REFERENCES.

Bland J., and Taylor K., 1986, submitted to M.N.R.A.S.
Bland J., Taylor K., and Atherton P.D., 1986, M.N.R.A.S., in press
Steiman-Cameron T.Y., and Durisen R.H., 1984, Ap.J. **276**, 101
Taylor K., and Atherton P.D., 1980, M.N.R.A.S. **191**, 675
van Albada T.S., Kotanyi C.G., and Schwarzschild M., 1982, M.N.R.A.S. **198**, 303

FABRY-PEROT OBSERVATIONS OF CEN A

J.J.E. Hayes, R.A. Schommer* and T.B. Williams
Department of Physics and Astronomy
Rutgers University, Piscataway, NJ 08854, USA

*Visiting astronomer, NOAO-CTIO, operated by AURA
under contract from the NSF

We present Fabry–Perot spectrophotometry of the well–known peculiar galaxy Cen A (NGC 5128). The observations were carried out using the Rutgers Fabry–Perot system and a CCD as a detector. We scanned the $H\alpha$ and [NII] (λ6583) emission lines. From these data we were able to construct maps of the continuum, line emission, velocity and velocity dispersion. The velocity maps in both $H\alpha$ and [NII] have smooth gradients and twists in the line of nodes. The deprojected emission maps strongly resemble emission maps of face-on spirals. We speculate that Cen A is a merger between an elliptical and a spiral.

1. THE DATA AND REDUCTION

Our data were obtained at CTIO using the 1.5 m telescope, a GEC CCD and the Rutgers Fabry–Perot system on the nights of 8 March 1985 ($H\alpha$) and 11 March 1985 ([NII]). The image scale was 1.4″/pixel. We used an etalon with a passband of 2.5 Å. Twenty-four images were taken of Cen A as we scanned through the $H\alpha$ emission line, from 6555–6585 Å. The same procedure was used in obtaining the [NII] data, where we obtained 24 images as well. Each image had an integration time of 600s. The data were reduced on the Rutgers Physics VAX 11/780. The data were calibrated in the same fashion as is outlined in Williams et al. (1984). To produce our maps, we have fit 65,000 $H\alpha$ line profiles and 50,000 [NII] line profiles. The internal accuracy from pixel to pixel is about 5 km/s.

2. DISCUSSION

The deprojected $H\alpha$ emission map is shown in Figure 1. We have used an inclination angle of 72° to deproject the emission map, a value that agrees with Graham's work (1979). Figure 1 strongly looks like the spiral structure of an Sc galaxy (see M101 in the "Hubble Atlas" (Sandage 1961).

In the velocity maps the apparent line of the nodes twists from almost parallel to the dust lane in the outer parts of the galaxy ($r > 20'', 0.5kpc$), to being almost perpendicular to the dust lane in the inner portions. Bland et al.'s data (1987, these proceedings) also shows this peculiar behavior.

T. de Zeeuw (ed.), Structure and Dynamics of Elliptical Galaxies, 419–420.

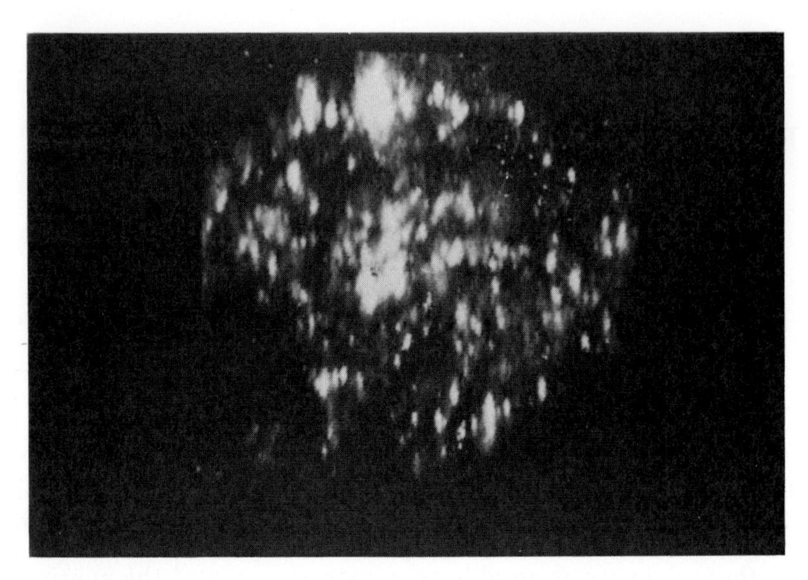

Figure 1. Deprojected Hα emission of Cen A. North is to the left, and West is to the top. The scale is 1.4″/pixel.

3. CONCLUSIONS

We conclude that Cen A is merger between an elliptical, making up the bulge, and a spiral making up the gas and dust lane, agreeing with Graham's (1979) work.

Wilkinson et al. (1986) have shown that there are no stable oblate, prolate or triaxial models which are consistent with the present appearance of Cen A. They state that there are, however, two cases (one oblate, one prolate) where the morphology of Cen A can be reproduced, both arguing for a recent encounter. Hayes (1985) has found that when a rotating stellar disk encounters a spherical potential, the disk loses its structure in only a few times 10^7 years. The twists in the line of nodes also argues for a merger, as it can be interpreted as a differentially precessing disk; a non-stable configuration. However, the uniformities seen in the inner portions of the velocity field argue that we are seeing stable inner orbits, and this would imply a triaxial potential (van Albada et al., 1982). We are left to conclude that Cen A is the product of a recent merger, with the outer disk still relaxing, a few times 10^8 to 10^9 years ago.

This research was carried out under the auspices of NSF grant 83-19344, and the observations were carried out in collaboration with M. Phillips and N. Caldwell. E. Malumuth assisted with some of the data reduction.

REFERENCES

van Albada, T.S., Kotanyi, C.G. & Schwarzschild, M., 1982. *Mon. Not. R. astr. Soc.* **198**, 303.
Bland, J., Taylor, K. & Atherton, P.D., 1987. This volume, p. 417.
Graham, J.A., 1979. *Ap. J.* **232**, 60.
Hayes, J.J.E., 1985. Ph.D. Thesis.
Sandage, A.R., 1961 *Hubble Atlas of Galaxies*, Carnegie Institution, Washington.
Wilkinson, A., Sharples, R.M., Fosbury, R.E.A. & Wallace, P.T., 1986. *Mon. Not. R. astr. Soc.* **218**, 297.
Williams, T.B., Caldwell, N. & Schommer, R.A., 1984. *Ap. J.* **281**, 579.

HI IMAGING OF RADIO ACTIVE ELLIPTICALS

J.H. van Gorkom
Princeton University Observatory
Peyton Hall
Princeton, NJ 08544
USA

ABSTRACT. Statistics on HI absorption in radio galaxies suggest that accretion of gas into the center might be a general phenomenon. Images of the HI emission from these galaxies show that the gas must have an external origin. The rotation curves derived from the HI data are flat.

High resolution observations of HI absorption against the unresolved radio cores exist for 4 radio galaxies. All show absorption at velocities redshifted with respect to the systemic velocities, by 100 to 400 km/s. Three show absorption at the systemic velocity as well, none shows blueshifted absorption. Examples are shown in Fig. 1 for NGC 1052 (van Gorkom et al, 1986) and NGC 5363 (van Gorkom and Laing, in prep.). The high velocity gas is only seen in absorption against the cores, not against extended structures, thus it must be close to the center. Either the gas is falling in or it is on non-circular orbits, if that is true no preference should be found for redshifted gas in a larger sample.

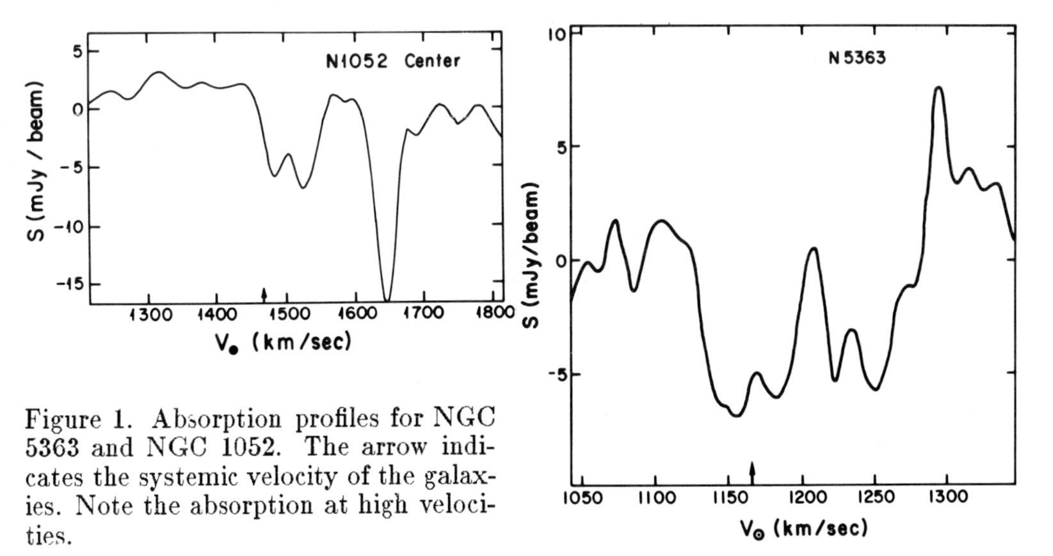

Figure 1. Absorption profiles for NGC 5363 and NGC 1052. The arrow indicates the systemic velocity of the galaxies. Note the absorption at high velocities.

T. de Zeeuw (ed.), Structure and Dynamics of Elliptical Galaxies, 421–422.

An overlay of the total HI emission on an optical photograph is shown in Fig. 2 for NGC 1052 and 5128 (van Gorkom et al., 1986, 1987). The galaxies are very similar: the HI has settled into a regularly rotating disk, but the rotation axes of gas and stars are almost perpendicular and the specific angular momentum of the gas is much larger than of the stars. The gas in the outer parts is not on stable orbits as can be inferred from the direction of rotation of gas and stars and the orientation of the warps. In the inner parts gas on non-circular orbits is seen, in NGC 1052 it is a regular flow pattern, showing up in emission and absorption, while in NGC 5128 it looks like individual clouds falling in. The gas to dust ratios are close to galactic for both galaxies.

From the above we conclude that the gas must have an external origin, as opposed to stellar mass loss. Most likely it comes from another galaxy, since the gas to dust ratios are normal.

The rotation curves (Fig. 3) stay flat out to the last measured point, giving an M/L of 15 at 16kpc and of 5 at 4 kpc for NGC 1052 and 5128 respectively. A comparison with the mass of NGC 5128 as derived from X-ray observations leads to a discrepancy of a factor 4 (Knapp, 1987).

REFERENCES

Knapp, G.R., 1987. This volume, p. 145.
van Gorkom, Knapp, Raimond, Faber, Gallagher, 1986. *A. J.*, **91**, 791
van Gorkom, van der Hulst, Haschick, Tubbs, 1987. in preparation

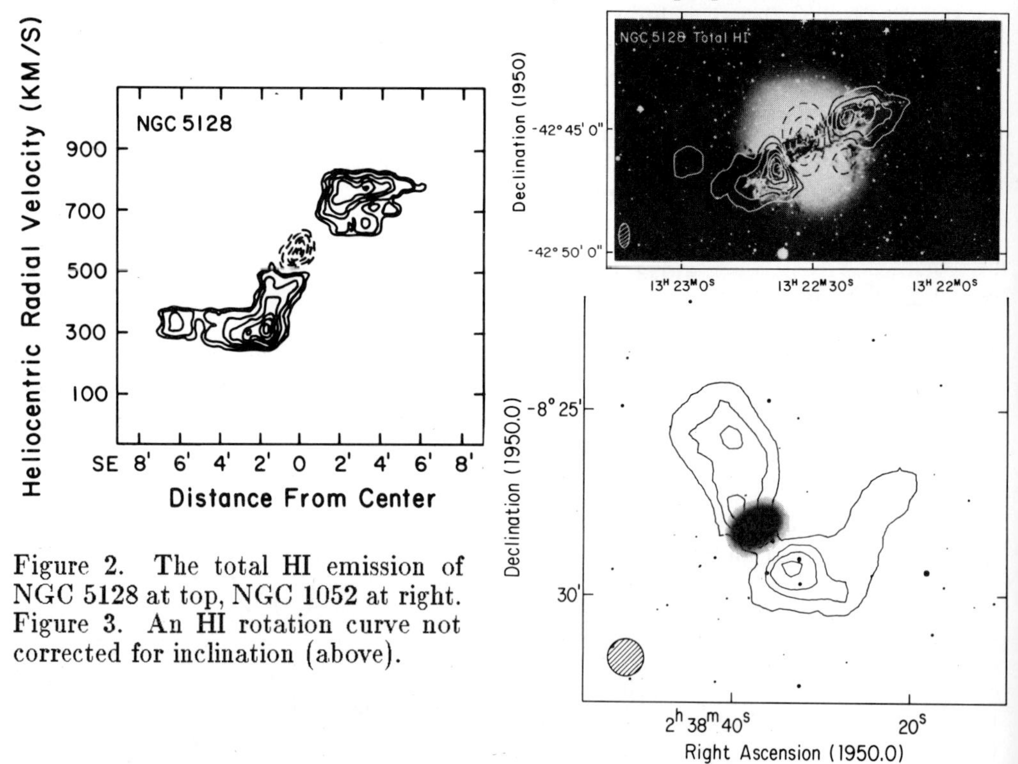

Figure 2. The total HI emission of NGC 5128 at top, NGC 1052 at right.
Figure 3. An HI rotation curve not corrected for inclination (above).

THE DISTRIBUTION AND KINEMATICS OF NEUTRAL HYDROGEN IN NGC 807

L. L. Dressel
Department of Space Physics and Astronomy
Rice University
P. O. Box 1892
Houston, TX 77251
U.S.A.

ABSTRACT. I have detected 21 cm line emission from neutral hydrogen in the giant elliptical galaxy NGC 807 at Arecibo Observatory, and I have mapped this emission with the VLA. Unlike the active and dwarf ellipticals that have been mapped thus far, NGC 807 has a fairly regular disk of gas rotating about the apparent optical minor axis. Combined with observations of active ellipticals, this observation suggests that two classes of HI-rich ellipticals may exist: ellipticals which have accreted gas and become active recently, and quiescent ellipticals which have either produced gas internally or accreted it so long ago that it has reached dynamical equilibrium.

1. INTRODUCTION

Because of the extremely low (usually undetected) 21 cm line flux of elliptical galaxies, few ellipticals have been detected or mapped at 21 cm. The maps of NGC 1052 (van Gorkom et al. 1986), NGC 4278 (Raimond et al. 1981), and NGC 5903 (Appleton et al. 1985) reveal irregular spatial distributions and kinematics, suggestive of tidal disturbance of internal gas or accretion of external gas. The first two galaxies are well-studied prototypes of the class of ellipticals with active nuclei, and the third galaxy is known to be a weak radio source. More maps must be made of both active and normal giant ellipticals, so that we can decipher and compare the evolution of the gaseous component in the two classes (or phases).

2. DATA

I have made 21 cm line observations of 66 elliptical galaxies with m(pg) $\leq$ 14.5 mag at Arecibo Observatory. One of the three detected galaxies is NGC 807, a normal giant elliptical (M(pg) = -21.8 for H_0 = 50 km s^{-1} Mpc^{-1}) with an inactive nucleus (S(2380 MHz) = 3 $\pm$ 4 mJy, Dressel and Condon 1978; W_λ(3727A) < 1A, O'Connell and Dressel 1978). I have observed NGC 807 for 24 hours in the D array of the VLA with 42

T. de Zeeuw (ed.), Structure and Dynamics of Elliptical Galaxies, 423–424.

km s^{-1} resolution. Using two different weighting functions for the baselines ("uniform" and "natural"), I have made surface brightness maps and isovelocity contour maps with resolutions of 0.6 and 1.0 arcmin.

3. RESULTS

NGC 807 has a double-horned 21 cm line profile indicative of a rotating disk. The width of the profile at the 20% intensity level is 510 km s^{-1}. A grid of Arecibo observations around the galaxy shows that the source is resolved by the 3.2 arcmin beam, and indicates a total HI mass of 1.4×10^{10} $M_{\odot}$. The VLA maps show that the HI is centrally concentrated, possibly with a central depression, and extends to at least three times the optical diameter. The apparent HI and optical axial ratios are similar, and the HI and optical position angles (Djorgovski 1986) agree to within a few degrees. The velocity field is consistent with pure rotation about the minor axis.

4. DISCUSSION

NGC 807 has the most massive and "normal" rotating gaseous disk found in an elliptical galaxy to date. Unlike the active giant ellipticals that have been mapped at 21 cm, it does not appear to have undergone tidal disturbance or accretion recently. It has no companions that could serve as likely donors of ~10^{10} $M_{\odot}$ of gas. The gas was apparently produced internally or accreted so long ago that it has reached dynamical equilibrium. The rotation axis of the gas implies that the galaxy is an oblate spheroid or quasi-oblate triaxial galaxy. Maps of other "inactive" giant ellipticals are needed to show whether NGC 807 is typical of this class.

5. ACKNOWLEDGEMENTS

This research was supported by a Cottrell Research Program grant from the Research Corporation and by a grant from Rice University.

6. REFERENCES

Appleton, P. N., Sparks, W. B., Pedlar, A., and Wilkinson, A. 1985, M.N.R.A.S., **217**, 779.
Djorgovski, S. 1986, private communication.
Dressel, L. L., and Condon, J. J. 1978, Ap. J. Suppl., **36**, 53.
O'Connell, R. W. and Dressel, L. L. 1978, Nature, **276**, 374.
Raimond, E., Faber, S. M., Gallagher, J. S., and Knapp, G. R. 1981, Ap. J., **246**, 708.
van Gorkom, J. H., Knapp, G. R., Raimond, E., Faber, S. M., and Gallagher, J. S. 1986, A. J., **91**, 791.

GAS, DUST AND RADIO EMISSION IN ELLIPTICAL GALAXIES

W. B. Sparks
Royal Greenwich Observatory
Herstmonceux Castle, Hailsham,
East Sussex, BN27 1RP,
England

ABSTRACT. CCD data are used to compare the isophotes, colour and dust-content of radio ($10^{21}<P_{5GHz}<10^{24}WHz^{-1}$, $H_o = 75kms^{-1}Mpc^{-1}$) and radio-quiet ellipticals. Radio ellipticals are round but not spheroidal, reddened and occasionally have disturbed dust lanes. X-ray emission correlates with both radio emission and shape. Detailed investigations of dust in NGC 1316 reveal a possible nuclear gas disc orthogonal to the radio jet.

RADIO ELLIPTICALS - ROUND BUT TRIAXIAL: Disney et al (1984) showed that radio ellipticals are significantly rounder than radio-quiet ellipticals. Highly accurate ellipticity data from CCD observations (Sparks et al. 1985, S85) confirm this. Fig. 1 shows the distribution of median axial ratios of radio ellipticals and comparison distributions. A lack of exactly E0 galaxies and the existence of isophotal twists both indicate that the true shape is not spheroidal - radio ellipticals are tri-axial with varying axis ratios or are somewhat distorted. 69% show twists > 2° (typically 10°), compared to 33% of the radio-quiet ellipticals, a difference significant at the 5% level.

DUST: S85 described a powerful new method for searching for dust in elliptical galaxies, the (B-model I) method. Some 25% of ellipticals have dust lanes, which tend to have a disturbed appearance when found in radio-ellipticals. An overall reddening of the radio ellipticals was found, which could not be attributed to discrete dust features.

Fornax A: Fig. 2 shows a (B-model I) map of NGC 1316. Notice the inner dust feature essentially perpendicular to the radio jet. Hitherto, it was thought that NGC 1316 violated the correlation that radio structure lies perpendicular to any dust lane present (Kotanyi and Ekers 1979). Physical properties of individual dust clouds are similar to the averaged properties of the cool interstellar medium of our galaxy, and are consistent with the dust arising from the disrupted remains of a merged spiral galaxy.

COLOUR: Sparks (1983, S83) found radio-ellipticals are redder than radio-quiet ellipticals. By ranking the galaxies of S83 in optical

T. de Zeeuw (ed.), Structure and Dynamics of Elliptical Galaxies, 425–426.

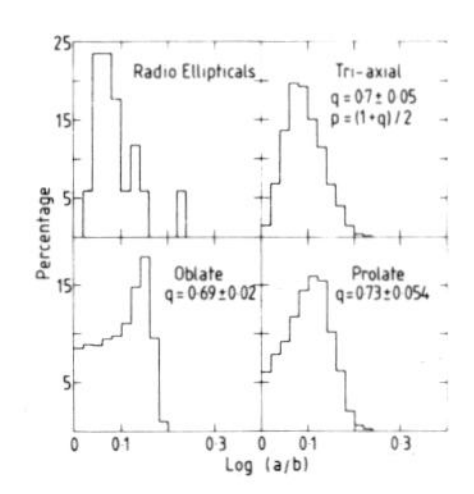

Fig. 1: Axial ratio distributions for 17 radio ellipticals and ensembles of triaxial ellipsoids, intrinsic axis ratios 1:p:q ($1 \leq p \leq q$).

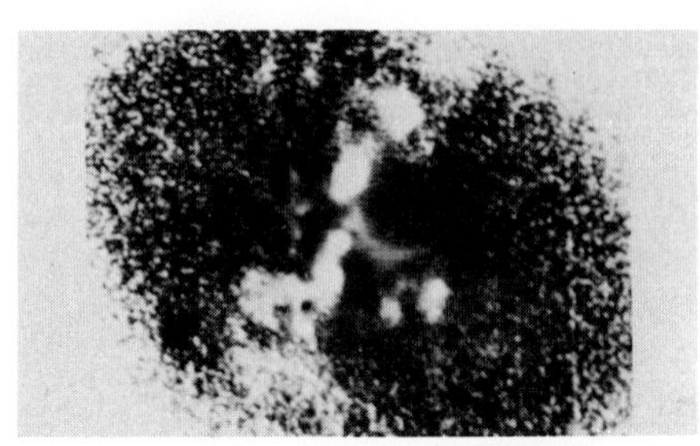

Fig. 2: (B-model I) image of dust in NGC 1316. Areas of extinction and reddening appear bright.

luminosity and applying the Sign test, it may be seen that absolute magnitude effects do not cause this. The sequence SO_1 to SO_3 is a sequence of early-type galaxies with increasing dust content. Classifications of the **Revised Shapley-Ames Catalog of Bright Galaxies** were used to find that mean values of A (see S83) for the SO_1, SO_2 and SO_3 galaxies are 0.04±0.01, 0.07±0.03 and 0.16±0.03 respectively. The difference between the dusty SO_3 and the others is much the same as between radio and radio-quiet elliptical galaxies, supporting the hypothesis that dust causes the reddening.

X-RAYS: X-ray satellites have provided direct observations of hot gaseous atmospheres within elliptical galaxies. Various correlations have been examined between X-ray and other properties for the galaxies studied by Forman et al (1985). Rounder galaxies are more X-ray luminous and there are correlations between X-ray and optical luminosity, radio power and $(B-V)^0_T$. Partial correlation (Macklin 1982) indicates that the correlation between X-ray and radio luminosity may not be due entirely to a mutual correlation with optical luminosity. Controlling for this and distance, the significance level of the X-ray/radio correlation is 5% (24 galaxies, all ellipticals). Similarly the correlation between X-ray luminosity and shape for ellipticals has a 1% significance level.

CONCLUSION: The data suggest that activity in elliptical galaxies is caused by accretion of gas onto the nucleus. The massive, more spherical radio ellipticals contain denser, more massive gaseous atmospheres which may be thermally unstable. Cooling gas could provide fuel for the active nucleus and, if dusty, redden the galaxy. A less common fuelling mechanism is gas accreted from outside the galaxy, recognized by the disturbed dust lanes of some radio ellipticals.

Disney, M.J., Sparks, W.B., & Wall, J.V., 1984. M.N.R.A.S., **206**, 899.

Forman, W., Jones, C., & Tucker, W., 1985. Ap. J., **293**, 102.

Kotanyi, C.G., & Ekers, R.D., 1979. Astron. Astroph., **73**, L1.

Macklin, T., 1982. M.N.R.A.S., **199**, 1119.

Sparks, W.B., 1983. M.N.R.A.S., **204**, 1049 (S83).

Sparks, W.B., Wall, J.V., Thorne, D.J., Jorden, P.R., van Breda, I.G., Rudd, P.J., & Jorgenson, H.E., 1985. M.N.R.A.S., **217**, 87 (S85).

THE DISCOVERY OF BLAZAR-TYPE NUCLEI IN TWO NEARBY RADIO ELLIPTICALS

W B Sparks
Royal Greenwich Observatory
Herstmonceux Castle, Hailsham,
East Sussex, BN27 1RP,
England

J C Hough, C Brindle
Physics Department
Hatfield Polytechnic
Hatfield, Herts AL10 9AB
England

J Bailey
Anglo-Australian Observatory
P O Box 296, Epping
NSW 2121, Australia

D J Axon
NRAO, University of Manchester
Jodrell Bank, Macclesfield
Cheshire, England

Near infrared polarimetry of Centaurus A and IC5063 has revealed the existence of a steep spectrum highly polarized source in the nuclei of both galaxies. The position angle of polarization is perpendicular to the radio position angle. We interpret this polarized emission as synchrotron radiation. This, together with a luminosity of 5 x 10^{41}erg s^{-1}, suggests the galaxies are low luminosity blazars and that such nuclei may be common in elliptical galaxies.

Low luminosity BL Lac objects (e.g. Mkn 421, 501) are clearly associated with the nuclei of elliptical galaxies. Many nearby E/S0 galaxies have similar flat radio spectra suggesting they too may contain blazar type nuclei. Axon, Baily and Hough (1982) reported the discovery of a very red near infrared source in the nucleus of IC5063, however it was not then possible to distinguish between thermal and non-thermal radiation mechanisms. Polarimetric observations offer the chance to do this, since high polarization is the main characteristic of blazar type objects. By going into the infrared, the behaviour may be followed as the contribution from starlight falls. All observations were made with the Hatfield Optical-IR Polarimeter (Bailey and Hough 1982) on the 3.9-m Anglo-Australian Telescope. Figures 1 and 2 present polarization levels and position angles for various aperture sizes and as a function of wavelength for the two galaxies.

1. Centaurus A: The results are described in detail in Bailey et al. (1986). The data are readily understood if the nucleus has its own intrinsic polarization with a different p.a. to that caused by the dust lane. A good fit is obtained with a power-law nucleus (dominant at longer wavelengths) with wavelength independent polarization of 9% in p.a. 147 degrees, spectral index -5.3 and flux 3mJy at J, together with

T. de Zeeuw (ed.), Structure and Dynamics of Elliptical Galaxies, 427–428.

a stellar component and polarization due to the dust lane applied to both (a standard interstellar law).

2. IC5063: The polarization decreases from B to J and then rises sharply towards longer wavelengths. In the smaller apertures, the polarization increases at H and K, and the p.a. of polarization is wavelength independent and perpendicular to the east-west extension of the radio source centred on the galaxy (Danziger et al. 1981). The spectral index as calculated using the polarized flux between H and K is -4.6 ± 0.4, the same as from photometry alone (Axon et al. 1982). We therefore conclude that the infrared excess arises from a non-thermal source whose intrinsic polarization is $(17.4 \pm 1.3)\%$, typical for a BL Lac object. The origin of the optical polarization is unknown although a reasonable fit may be obtained by polarization due to the passage of radiation through aligned grains (see Fig. 2).

Axon D.J., Bailey, J. & Hough, J.H., 1982. Nature, 299, 234
Bailey, J. & Hough, J.H., 1982. Publ. Astr. Soc. Pacific, 94, 618
Bailey, J., Sparks, W.B., Hough, J.H., & Axon, D.J., 1986. Nature, 322, 150
Danziger, I.J., Goss, W.M. & Wellington, K.J., 1981. Mon. Not. R. astr. Soc., 196, 845

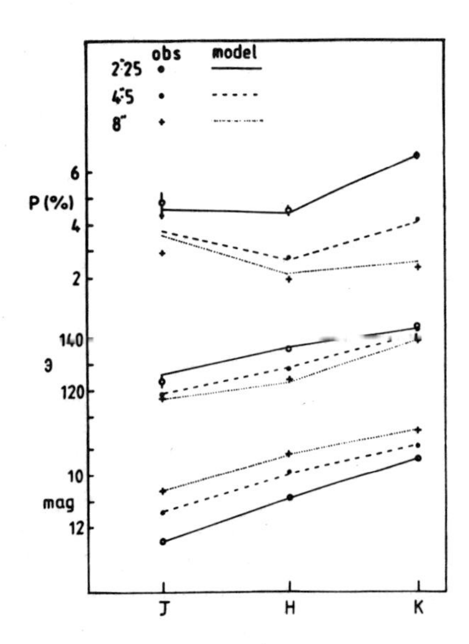

Fig. 1. Polarization, position angle and magnitude of NGC5128 (Cen A) compared to the model described in the text.

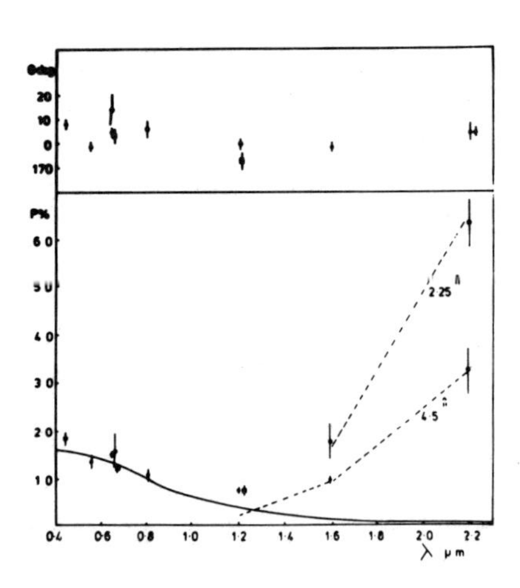

Fig. 2. Polarization and position angle for IC5063. Solid line is an interstellar curve with maximum polarization 1.5% at 0.40μm. Dashed line connects polarization levels calculated for a non-thermal source with spectral index -4.5, polarization 17.4% and flux 4.5mJy at K.

STELLAR DYNAMICS OF RADIO ELLIPTICAL GALAXIES

A. Sansom
University of Sussex, Falmer, Brighton, U.K.
J.V. Wall, W.B. Sparks
Royal Greenwich Observatory, E. Sussex, U.K.

ABSTRACT. Stellar kinematical and dynamical results are presented for 34 radio ellipticals. The radio galaxies we observed were brighter than m(B)=16. The results show that these radio ellipticals are not generally more rapidly rotating than their non-radio counterparts. Evidence for some rotation about the major axis is seen in two cases. These radio ellipticals do not appear to obey the luminosity, velocity dispersion trend seen for normal ellipticals.

Long slit spectra of 34 radio ellipticals were obtained on the AAT. Stellar rotation curves and velocity dispersions were derived from these spectra using cross-correlation techniques. Two of the resulting rotation curves and velocity dispersion profiles, showing major and minor axis rotation are in Fig.1.

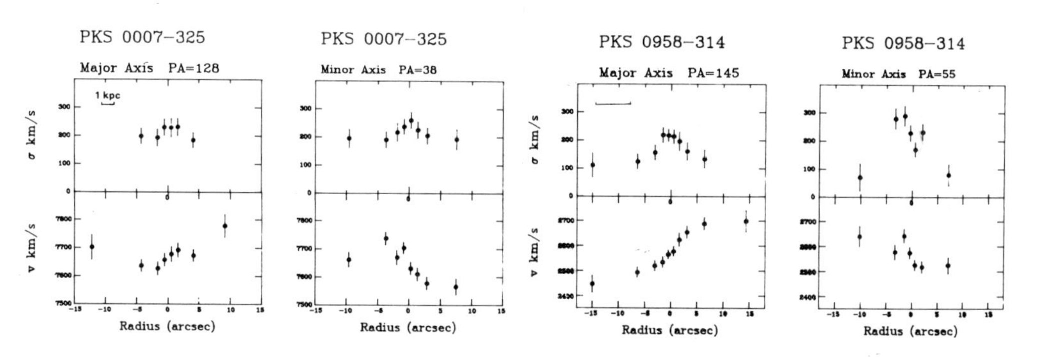

Fig.1. Stellar kinematic of two radio galaxies.

The ratio of measured rotation about the apparent minor axis to the central stellar velocity dispersion was plotted against ellipticity. Comparison with normal ellipticals from Davies et.al.(1983) shows that the radio galaxies, like normal ellipticals, are predominantly slowly rotating, in agreement with Heckman et.al. 1985.

T. de Zeeuw (ed.), Structure and Dynamics of Elliptical Galaxies, 429–430.

ABSOLUTE LUMINOSITY, VELOCITY DISPERSION RELATION

Observations, and accurate velocity dispersion measurement methods have been made for both radio and non-radio ellipticals, the most recent results of these are shown in Fig.2, together with the current data for the 34 radio galaxies (circles). Dressler's observations of Coma and Virgo ellipticals show a fairly tight correlation, for which $L \sim \sigma^{3.5}$ (for $\log\sigma > 2$).

The radio galaxies do not follow the $L \sim \sigma^{3.5}$ law. They lie systematically below the line and are different from this power law at the 99.9 percent confidence level. The spread of $\log\sigma$ at a given luminosity is larger for the radio galaxies (Shaver et.al. and Heckman et.al. samples) than it is for the galaxies in the Virgo and Coma clusters. Tests using PKS 0958-314 showed that the effect of the galaxies being at different distances and having decreasing velocity dispersion with radius is insufficient to produce the observed deviations in σ of the radio galaxies from the line in Fig.2.

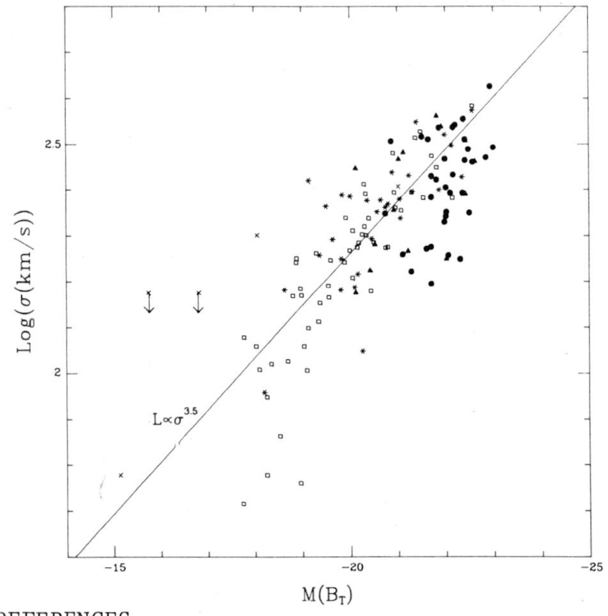

Fig.2.
Absolute B magnitude against velocity dispersion.

x=Faber and Jackson (1976)
□=Dressler(1984)
●=Shaver et.al.(1982)
▲=Heckman et.al.(1985)
*=Davies et.al.(1983)

REFERENCES

Davies, R.L., Efstathiou, G., Fall, S.M., Illingworth, G.D., & Schechter, P.L., 1983. Ap. J., **266**, 41.

Dressler, A., 1984. Ap. J., **281**, 512.

Faber, S.M., & Jackson, R.E., 1976. Ap. J., **204**, 668.

Heckman, T.M., Illingworth, G.D., Miley, G.K., & van Breugel, W.J.M., 1985. Ap. J., **299**, 41.

Shaver, P.A., Danziger, I.J., Ekers, R.D., Fosbury, R.A.E., Goss, W.M., Malin, D., Moorwood, A.F.M., Wall, J.V., 1982. In IAU Symp. No. 97, "Extragalactic Radiosources", eds D.S. Heeschen & C.M. Wade (Reidel, Dordrecht), p. 55.

PROPERTIES OF THE X-RAY EMITTING GAS IN EARLY TYPE GALAXIES

C. R. Canizares[1], G. Fabbiano[2] and G. Trinchieri[2,3]

1. Department of Physics & Center for Space Research, M.I.T., Cambridge, USA
2. Harvard/Smithsonian Center for Astrophysics, Cambridge, U.S.A.
3. Osservatorio Astrofisico di Arcetri, Firenze, Italy

We have studied a sample of 81 E and SO galaxies (taken mostly from the literature) that were observed with the *Einstein* Observatory. Fifty-five galaxies are detected in X-rays, most of which come from a hot interstellar medium. It is possible that discrete sources make a significant contribution to the X-ray emission for 21 of the detected galaxies with the lowest X-ray luminosity, L_X, for a given optical luminosity, L_B. We examine the L_X vs. L_B distribution (see Figure 1), and derive approximate values of the central electron density, central cooling time and total mass in gas for the sample. Typical values are $\sim 0.1\mathrm{cm}^{-3}$, $\sim 5 \times 10^6$ yr and $5 \times 10^9 M_\odot$, respectively. The short cooling times suggest the presence of cooling flows, and we consider heating by supernovae and by gravitational processes. Supernovae at the accepted rate would overproduce the observed X-ray luminosity: either the true rate is many times lower or the supernova energy is not well coupled to the hot gas. There are also difficulties in explaining the suppression of a strong galactic wind if supernova heating dominates, unless an external pressure confines the gas. Gravitational heating gives $L_X \propto L_B\sigma^2$, where σ is the line-of-sight velocity dispersion in the galaxy, and we find that this is roughly the case. Gravitational heating exceeds the mean observed L_x by a factor of $\sim 3-4$ if one uses the accepted rate of stellar mass loss and assumes that the gas falls all the way to the center of the galaxy. This could be reconciled if the mass injection rate were lower or if matter drops out of the flow at all radii. The sample properties provide no additional information about the presence or absence of heavy halos in early type galaxies.

A full report of this work is given in *Astroph. J.*, **312**, 503.

T. de Zeeuw (ed.), Structure and Dynamics of Elliptical Galaxies, 431–432.

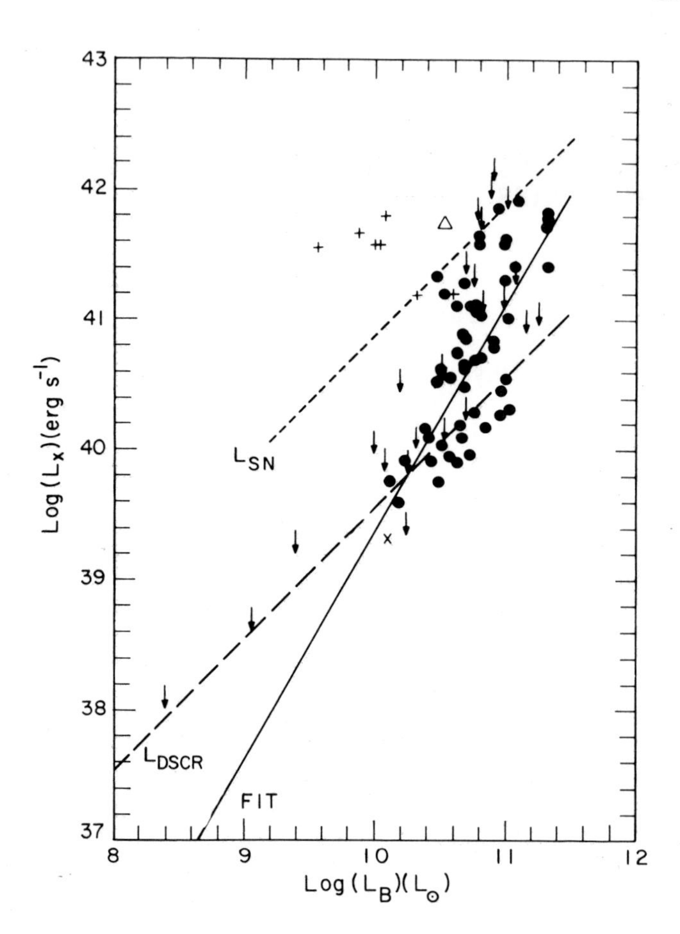

Figure 1. X–ray vs. optical (blue) luminosity of early type galaxies. The filled circles are detections, and the downward arrows are upper limits. The possibly active galaxy NGC 3998 is shown as a triangle, and NGC 5128 (Cen A) is shown as an "X". The plus signs are for galaxies in A1367 identified by Bechtold *et al.* (1983, *Astroph. J.*, **265**, 26); these are not included in our sample because we believe their emission may not be from hot, galactic gas. The dashed line gives L_{dscr}, the estimate of the possible X-ray luminosity from discrete sources. The solid line is the mean L_X vs. L_B fit to the full sample, and the dotted line shows L_{SN}, the luminosity expected from heating by supernovae at the accepted rate, assuming 10^{51} ergs per supernova, and that all the energy emerges as X–rays:

$$L_{SN} = 7.1 \times 10^{41} (R_{SN}/0.22)(L_B/10^{11})\,\text{erg s}^{-1},$$

where R_{SN} is the supernova rate per $(10^{10} L_B 100\ \text{yr})$.

ARE COOLING FLOWS GOVERNING E-GALAXY EVOLUTION?

Edwin A. Valentijn
Kapteyn Astronomical Institute
Groningen, The Netherlands

Gas accretion of intra cluster gas into the potential well of giant elliptical or cD galaxies can provide the material for both nuclear non–thermal activity and continuous, probably low mass, star formation (Fabian *et al.* 1982, Valentijn and Bijleveld 1983). In a few dozen cases it has been observed in the nearby universe that the hydrostatic equilibrium of X-ray emitting gaseous atmospheres around cD galaxies is disturbed by the thermal bremsstrahlung cooling in the central high density area, leading to an inflow of gas into the centrally located gE or cD galaxy ("cooling flows", Jones and Forman 1984). This thermal instability is different from those studied in models of early galaxy formation (Gunn and Gott 1972, Rees and Ostriker 1972) only because of the assumed self–gravitation of the collapsing gas in the galaxy formation scenarios. If however, during early galaxy formation the collapsing gas is accreted by a pre–existing dark potential well, then the processes of early galaxy formation and the observed present epoch cooling flows are likely to be intimately related to each other. The cooling accretion flows could then lead to the original formation of the visible object, and subsequently govern its evolution through the fuelling of star formation and nuclear non–thermal activity.

INPUT PARAMETERS

The velocity dispersions, σ, in the central (<10 kpc) regions of cD's (Tonry 1985) suggest a stellar mass profile $M_*(< R) = \epsilon\, 2\, 10^{10}\, R_{kpc}\, \sigma^2_{300\, km\, sec^{-1}}\, (M_\odot)$, with ϵ a factor $\sim$ 5 representing the softness of the potential well. Dressler's (1979) measurements of the A2029 cD and the analysis of the confinement of X-ray haloes around cD galaxies, however, suggest a total (stellar + dark matter) mass profile $M_{*+d}(< R) = \epsilon\, 17\, 10^{10}\, R_{kpc}\, T_{3\, 10^7 K}\, (M_\odot)$. The dark component thus suggested has $M_{*+d}/M_* = 8$ and a scale size of $\sim 250\, kpc$. We assume that this dark component is pre-existent at $z > 6$ and that it will dominate the potential well throughout the galaxy formation/evolution epochs without evolving significantly itself. The central gas densities in the atmospheres of cD galaxies are observed to be about a factor 100 higher than that of the overall intra cluster gas, which in turn is supposed to accrete its contents from the supercluster pervading medium with an average enhancement of a factor 10 in density (Gunn and Gott, 1972). In a first order approach we can thus couple the central gas density in a cD type potential to that of the supercluster pervading gas by simply taking $\rho_{cen} = 1000 \rho_{sc}$. Radio observations of head tails in

T. de Zeeuw (ed.), Structure and Dynamics of Elliptical Galaxies, 433–434.

the Coma and the Hercules superclusters indicate $\rho_{sc} = 4\,10^{-6}\,cm^{-3}$, which is low enough to follow the Hubble flow. Thus we can parametrize the supercluster gas evolution as $\rho_{sc} = 2\,10^{-6}\,(1+z)^3\,cm^{-3}$ and $T_{sc} = 5\,10^6\,(1+z)^2 K$.

RESULTS

In order to start the accretion (and thus the galaxy formation) process, the temperature of the surrounding gas should be low enough to be bound by the dark component. Using the input parameters the required mass to initiate binding and form a gaseous halo is $M_{req}(< R) = \epsilon\ 1.5\,10^{10}\,R_{kpc}\,(1+z)^2\,(M_\odot)$. When comparing with M_{*+d} this shows that the heaviest systems will start halo formation at z=3.9, while systems with a present halo temperature of $3\,10^7 K$ would start binding a halo at $z = 2.4$. Note, that this turn–on redshift is nowhere pre–programmed in the model since the dark component is assumed to exist from $z = 6$ onwards.

Using the input parameters, the bremsstrahlung cooling time in the centers of the gaseous haloes is $4.6\,10^9\,(1+z)^{-2}$ year. Thus, in all Friedmann cosmologies the cooling time is always short enough to switch on cooling flows immediately after the atmospheres have been formed at $z \leq 4$. However, for $z < 1$ the cooling time approaches or exceeds the object lifetime, especially in $\Omega \geq 1$ cosmologies. This implies that for $z < 1$ the rate of accretion flows becomes critically dependent on other parameters and as $z \rightarrow 0$ both the rate of occurrence of cooling flows and the mass inflow rate in lasting flows decreases. Thus, in an $\Omega = 1$ cosmology, the flows can start at $z < 4$ and will be able to fuel nuclear activity and star formation. At later epochs galaxies are expected to become less active, which seems in agreement with the observed evolution of nuclear activity and optical color indices. For $\Omega = 2$ the turn–on redshift is at $z \sim 3$ which seems not in agreement with the apparent turn–on epoch of quasars. For $\Omega = 0$ cosmologies the predicted effects of cooling flows are not consistent with the observed cosmological evolution, since the cooling times are then always much shorter than the object lifetimes.

The total mass inflow rate can be estimated by assuming total dissipation of energy of the accreted gas, *i.e.* the maximum the galaxy can eat. Over the *total* volume involved this is $\dot{M}_{tot} = 570\,M_d^2\,\rho_{sc}^o\,T_{sc}^{o\ -3/2}\,(M_\odot/year)$, with ρ_{sc}^o and T_{sc}^o the $z = 0$ values of ρ_{sc} and T_{sc}. There is no significant z–dependence in the *total* mass inflow rate and in a $\Omega = 1$ cosmology at most $\sim 6\,10^{12} M_\odot$ of gaseous material can be deposited into the galaxy. The *central* mass inflow rate can be evaluated from the z-evolution of the size of the cooling region and is $\dot{M}_{central} = 2.6\,(1+z)^5\,M_\odot/year$ and is a strong function of z. Both the cosmological enhancement factor of quasars and the evolution of the power of radio sources have been observed to follow a $(1+z)^5$ law, which appears to be consistent with a cooling accretion model as the source of their activity in a $\Omega = 1$, $H_0 = 50$ universe.

REFERENCES

Dressler, A., 1979. *Astrophys. J.*, **231**, 659.
Fabian, A.C., Nulsen, P.E.J., & Canizares, C.R., 1982. *M.N.R.A.S.*, **201**, 933.
Gunn, J.E., & Gott III, J.R., 1972. *Astrophys. J.*, **176**, 1.
Jones, C., & Forman, W., 1984. *Astrophys. J.*, **276**, 38.
Rees, M.J., & Ostriker, J.P., 1972. *M.N.R.A.S.*, **179**, 541.
Tonry, J.L., 1985. *Astron. J.*, **90**, 2431.
Valentijn, E.A., & Bijleveld, W., 1983. *Astron. Astrophys.*, **125**, 223.

METAL-ENHANCED GALACTIC WINDS

J. Patricia Vader
Yale University
Department of Astronomy
Box 6666
New Haven, CT 06511
U.S.A.

ABSTRACT: Constraints on supernova-driven galactic winds from elliptical galaxies at the epoch of star formation are investigated. The occurrence of mass loss is found to depend critically on the supernova rate in the case of dwarf galaxies, while the depth of the potential well is the most important constraint for giant ellipticals. The smallest dwarf ellipticals must have evolved from significantly more massive progenitors in order to have sustained a wind that carried away most of their metal production.

Galactic winds during the star formation epoch of elliptical galaxies offer an attractive explanation for the observed mass-metallicity relation, and in particular for the low metallicities and stellar densities of dwarf ellipticals (dE). These winds, enriched by the metals of all supernovae (SN) that power it, are metal-enhanced with respect to the star-forming gas (Vader 1986). Necessary conditions for the occurrence of a wind and the associated characteristic wind temperatures T are: (i) the hot dilute gas generated by SN explosions occupies most of space (T_p); (ii) a significant fraction of supernova remnants (SNR) overlap (T_w) or are pressure-confined before cooling radiatively (T_c); (iii) the wind energy exceeds the binding energy (T_e); (iv) the flow time of the wind is shorter than the radiative cooling time (T'_c). In terms of T we have

$$\max(T_e, T_c, T'_c,) < T < \min(T_p, T_w) \qquad \text{wind,}$$

$$\max(T_c, T_w) < T < T_p \qquad \text{'puffs'.}$$

A system with steady-state mass loss contains either fewer than 5 overlapping SNR's, each of which escapes directly (puffs), or many non-overlapping pressure confined SNR's (wind). For a given metallicity Z_g of the star-forming gas, each temperature T_x depends at most on the average rate per unit volume η of supernovae that power the wind and the velocity dispersion σ and radius r_e of the galaxy. Any equation $T_x = T_y$ thus defines a characteristic SN rate $\eta_c(\sigma, r_e)$. Star

T. de Zeeuw (ed.), Structure and Dynamics of Elliptical Galaxies, 435–436.

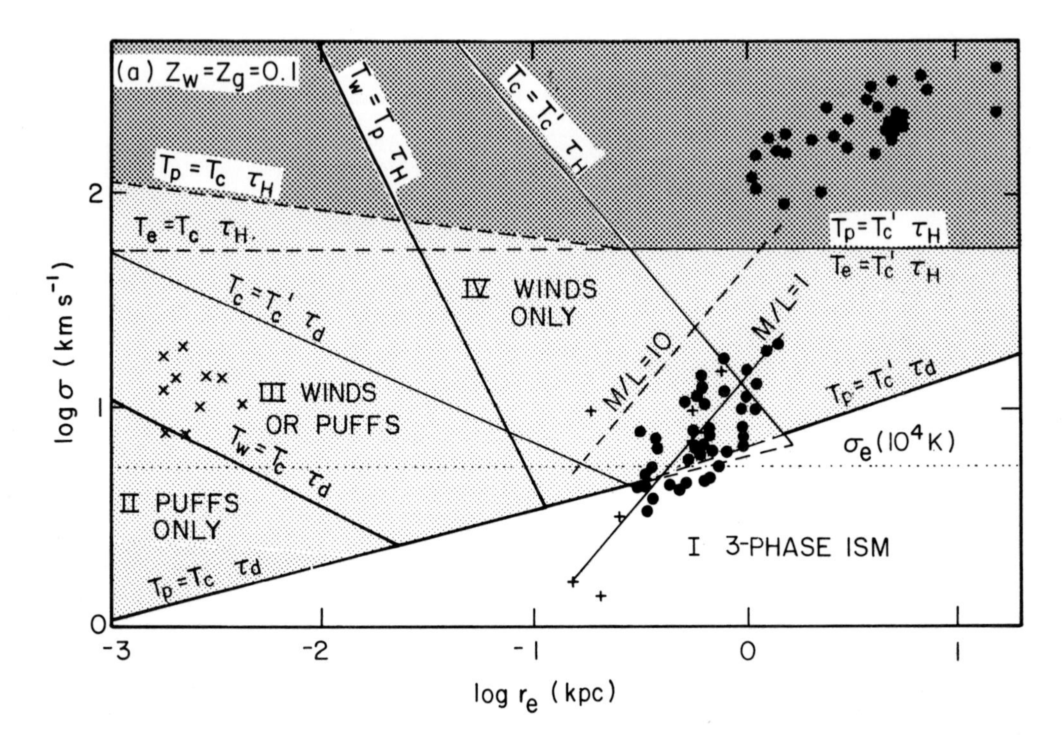

Fig. 1: Mass loss regions in the log σ - log r_e plane for a metallicity of the star-forming gas $Z_g = 0.1Z_\Theta$. Symbols denote: • gE's; • and + dE's; x globular clusters.

formation time scales limited by the dynamical time scale τ_d of the system and the Hubble time $\tau_H = 10^{10}$ yrs yield the limiting values $\eta_L(\sigma, r_e)$, with L = d, H. Equating η_c to η_L yields relations of the form $\sigma = \sigma_L(r_e)$, which are displayed in Fig. 1 as lines labeled $T_x = T_y$ τ_L. Each of these lines marks the transition between different regimes of mass loss. In Fig. 1 we distinguish regions I (mass loss excluded), II (puffs only), III (winds or puffs), and IV (winds only). In the shaded areas mass loss can, but does not necessarily, occur. Giant ellipticals (gE) fall in the heavy-shaded area where a wind occurs independently of η if $T > T_e$. Globular clusters and dE's fall in the light-shaded area where mass loss can only occur given a sufficiently large SN rate. The smallest dE's fall in region I so that, if their low metallicities are due to mass loss, they must indeed have evolved from more massive progenitors.

A full account of this work will be published elsewhere (Vader 1987).

REFERENCES

Vader, J. P. 1986, Ap.J. **305**, 669.
Vader, J. P. 1987, Ap.J.

THE HYDRODYNAMICAL EVOLUTION OF GAS IN YOUNG ELLIPTICAL GALAXIES

R. Kunze, H.H. Loose & H.W. Yorke
University Observatory Göttingen
Geismarlandstraße 11
D-3400 Göttingen
Federal Republic Germany

ABSTRACT. We calculate the partial inflow of gas fuelled by stellar mass loss at an early epoch ($10^9\ yr$ after the birth of the galaxy) during the evolution of an elliptical galaxy assuming a modified King model stellar distribution. The influence of the partial thermalization of stellar mass lost on the amount of gas which can be stored in the nucleus of a typical elliptical during the time of partial inflow is investigated. Masses up to $10^5\,M_\odot$ of cool ($\leq 10^4\,K$) material can be stored in the nucleus of the galaxy before the fast dissipation of the "kinetic bulk energy" of the nuclear gas cloud leads to "thermal" instability and subsequent collapse. A supermassive star can form. A detailed discussion of the model and the results is subject of a forthcoming paper (Kunze et al., 1986).

1. INTRODUCTION

A number of elliptical galaxies show evidence of compact central objects which power their nonthermal activity (Dressel et al., 1982). These objects are believed to have masses ranging from 10^6 to $10^8\,M_\odot$ (massive black holes, supermassive stars, etc.). They have formed at early times in the evolution of the galaxy because active galactic nuclei are observable at high redshifts.

During early epochs of their evolution elliptical galaxies have partial inflow of gas which has its origin in the process of stellar mass loss (Mathews & Loewenstein, 1986 and references therein). The amount of gas which can accumulate before it becomes unstable depends critically on the physics of heat input into the interstellar medium. In earlier works the energy supplied by stellar ejecta and supernova explosions is assumed to be immediately converted into internal energy of the interstellar gas (e.g. Mac Donald & Bailey, 1981; hereafter MB). But the small amounts of ionized gas ($\leq 10^6\,M_\odot$) observed in elliptical galaxies (Phillips et al., 1985) may be a clue to incomplete thermalization of stellar ejecta (White & Chevalier, 1983). We want to check the importance of the assumption of immediate thermalization of stellar ejecta for the resultant amount of gas which can be stored in the core of the galaxy during the time when inflow occurs.

T. de Zeeuw (ed.), Structure and Dynamics of Elliptical Galaxies, 437–438.

2. THE MODEL

We consider a spherical elliptical galaxy (without rotational motions) where the stellar density distribution is taken from King (1972). The mass loss rate from the stars (except from supernovae) is calculated similar to MB. The mass loss rate from supernova explosions (type I) is taken from observations (Tamman, 1977).

We simulate the incomplete thermalization of stellar ejecta by the introduction of a "kinetic bulk energy" to account for the random motion of clumpy mass lost. The efficiencies γ_{sn} and $\gamma_\star$ for creating kinetic bulk energy from SN events and stellar ejecta, respectively, are taken to $\gamma_{sn} = 0.28$ (Spitzer, 1978) and $\gamma_\star = 0.1$ (only 10% of the total kinetic energy of stellar ejecta is used to create random cloud motions, the remainding energy is thermalized instantaneously). We assume that the kinetic bulk energy generated by the processes outlined above is dissipated exponentially and the e–folding time is identified with the local dynamical time scale.

3. RESULTS

The evolution of the partial inflow is followed by an implicit hydrodynamical code including the generation of kinetic bulk energy with the efficiencies specified above. The calculations started from an initially gas–free state. After the gas in the nucleus of the galaxy cooled down to $\leq 10^4\,K$ the nuclear gas cloud is stabilized by the kinetic bulk energy and a shock front at $\sim 10\,pc$ from the center develops. About $10^5\,M_\odot$ gas are gradually accumulated interior the shock. When $M_{gas}(r_{shock}) \geq M_\star(r_{shock})$ collapse sets in due to rapid dissipation of the kinetic bulk energy. The time–scale for the collapse is less than that for accumulation of the gas, therefore the mass of the collapsing gas remains nearly constant. A hydrostatic core forms which is bounded by an accretion shock front.

The further evolution was not followed numerically. Assuming that fragmentation does not occur while the core accretes matter from the surroundings and that all matter will be accreted which is able to fall onto the central object in $t_{ff} \sim 10^4\,yr$ we obtain a mass of $M \sim 10^5\,M_\odot$ for the central object.

REFERENCES

Dressel, L.L., Bania, T.M. and O'Connell, R.W.: 1982, *Astrophys. J.* **259**, 55
King, I.R.: 1972, *Astrophys. J.* **174**, L123
Kunze, R., Loose, H.–H. and Yorke, H.W., 1986: submitted to *Astron. & Astrophys.*
MacDonald, J. and Bailey, M.E.: 1981, *Monthly Notices Roy. Astron. Soc.* **197**, 995 (MB)
Mathews, W.G. and Loewenstein, M.: 1986, *Astrophys. J.* **306**, L7
Phillips, M.M., Jenkins, C.R., Dopita, M.A., Sadler, E.M. and Binette, L.: 1986, *Astrophys. J.* **91**, 1062
Spitzer, L. Jr.: 1978, in *'Physical Processes in the Interstellar Medium'*, John Wiley & Sons, New York, p. 255 ff.
Tamman, G.A.: 1977, *Astrophys. Space Sci.* **66**, 95
White, R.E. III and Chevalier, R.A.: 1983, *Astrophys. J.* **275**, 69

ULTRAVIOLET ENERGY DISTRIBUTIONS OF (32) EARLY–TYPE GALAXIES

F. Bertola, Institute of Astronomy, University of Padova, Italy
D. Burstein, Arizona State University, Tempe, USA
L.M. Buson, Astronomical Observatory, Padova, Italy
S.M. Faber, University of California, Santa Cruz, USA
T.R. Lauer, Princeton University, Princeton, USA

New, self–consistent energy distribution have been generated from 52 short – wavelength and 40 long–wavelength IUE spectra of 31 early–type galaxies, plus the bulge of M31. All galaxies in this sample have measurements of the absorption–line index Mg_2 and central velocity dispersion, and a (1550–V) color is determined by combing the IUE data with photoelectrically–measured V magnitudes.

The galaxies in the sample can be a priori divided into a) star–forming; b) active and c) quiescent galaxies. There exists a well–defined, non–linear relationship between Mg_2 and the (1550–V) color for the 24 quiescent galaxies, with the (1550–V) color becoming bluer with increasing line strength, as originally suggested by Faber. The Mg_2, (1550–V) data for the 4 star–forming galaxies and the 4 active galaxies lie off this relationship.

IUE energy distributions for galaxies with similar Mg_2, (1550–V) values are remarkably similar, and are different in form between quiescent, active and star–forming galaxies. The level of flux at 1550 A is due to essentially the same form of 'contaminating' blue energy distribution for both the quiescent and active galaxies, but not for the star–forming galaxies.

Simple models of continuing star formation and post–asymptotic giant branch (PAGB) evolution show that their energy distributions are similar in the wavelength region covered by IUE. Moreover, both forms of energy distribution are closer matches to that of the 'contaminating' blue energy distribution seen in these galaxies, and both contribute very little at optical wavelengths.

This ambiguity means one cannot yet distinguish between the two main competing models –young stars or old stars– for the source of this blue 'contaminating' stellar population in early–type galaxies.

T. de Zeeuw (ed.), Structure and Dynamics of Elliptical Galaxies, 439.

Lauer, Levison, Bland (seen from behind), Capaccioli and Mamon during a poster session.

LINE-STRENGTH GRADIENTS IN ELLIPTICAL GALAXIES

Roger L. Davies and Elaine M. Sadler
Kitt Peak National Observatory
National Optical Astronomy Observatories
P.O. Box 26732
Tucson, Arizona 85726-6732

ABSTRACT. We have measured line-strength indices as a function of radius in several elliptical galaxies. All of them show strong radial gradients in Mg, but much weaker gradients in Fe and Hβ. The isophotes and contours of constant line-strength have the same flattening. More luminous galaxies have shallower gradients, contrary to the prediction of models of dissipative collapse. Most of the galaxies observed show weak central emission which can partially fill the Balmer absorption lines.

Line-strength gradients in elliptical galaxies have long been proposed as a diagnostic of the processes at work during galaxy formation. Larson (1975) and more recently Carlberg (1984) have shown that an elliptical galaxy which forms by the dissipative collapse of a gas cloud should exhibit a metallicity gradient.

We have measured Hβ, Mg_2, Fe1 (5270 Å) and Fe2 (5335 Å) indices (Burstein *et al.* 1984) from spectra taken by Davies and Birkinshaw (1986) in their study of the kinematics of ellipticals. Each galaxy was observed at four position angles. We find that:

(1) Typical changes in Mg_2 over the range 0.1 to 1.0 r_e are -0.03 to -0.10 which correspond to changes in [Fe/H] of -0.11 to -0.39 using the calibration given by Terlevich *et al.* (1981). Line-strength gradients scale with the galaxy light (i.e. isophotes and contours of constant line-strength have the same flattening).

(2) In this small sample, the brightest galaxies have shallower gradients than fainter ones. This is contrary to the prediction of Carlberg's (1984) dissipative models, and suggests that dissipation was less important in the formation of more luminous galaxies. This is consistent with the formation of giant ellipticals from mergers of low-luminosity ones, as mergers decrease abundance gradients (White 1978).

T de Zeeuw (ed.), Structure and Dynamics of Elliptical Galaxies, 441–442.

(3) Many galaxies have weak LINER emission. The suggestion that the excess blue light in ellipticals comes from young stars rests in part on the anomalously strong Hβ absorption and correcting for the presence of emission increases this anomaly.

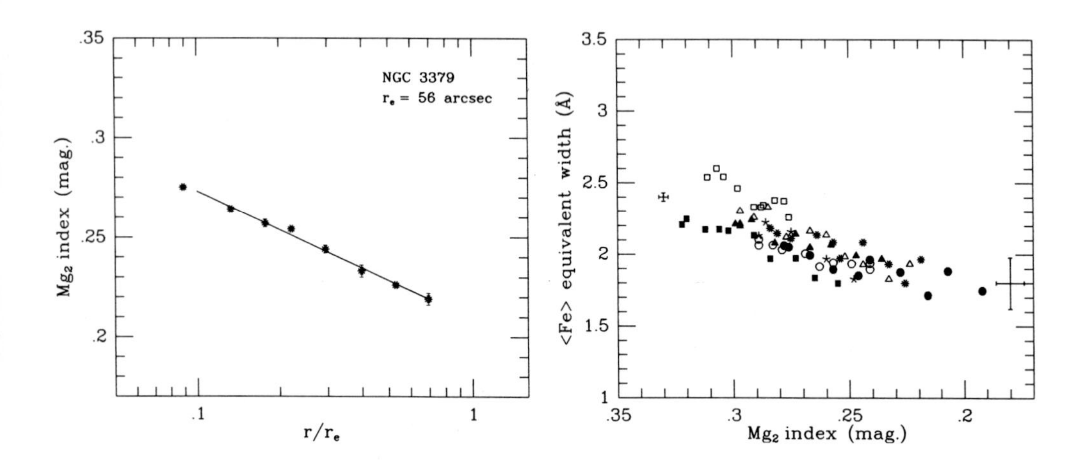

Fig. 1: (a) Mg_2 gradient in NGC 3379 (averaged over four position angles). The line is a least-squares fit from 0.1 to 0.7 r_e. (b) Radial variation of <Fe> with Mg_2 for eight bright ellipticals, each shown by a different symbol. Typical error bars are marked. Note that Mg_2 as defined by Burstein et al. is on a logarithmic scale, while <Fe> is defined as an equivalent width. Our values of Mg_2 and <Fe> may have a small zero-point offset from the Burstein et al. scale.

REFERENCES

Burstein, D., Faber, S.M., Gaskell, C.M. and Krumm, N. 1984, Ap. J., **287**, 586.

Carlberg, R. 1984, Ap. J., **286**, 404.

Davies, R.L. and Birkinshaw, M. 1986, in preparation.

Larson, R.B. 1975, M.N.R.A.S., **173**, 671.

Terlevich, R., Davies, R.L., Faber, S.M. and Burstein, D. 1981, M.N.R.A.S., **196**, 381.

White, S.D.M. 1978, M.N.R.A.S., **184**, 185.

VISUAL–IR COLOR GRADIENTS IN ELLIPTICAL GALAXIES

R. F. Peletier, E. A. Valentijn
Kapteyn Astronomical Institute, Groningen, the Netherlands

R. F. Jameson
Department of Astronomy, University of Leicester, United Kingdom

Dissipational formation theories (e.g. Larson 1974) predict a metallicity change with distance from the center in elliptical galaxies. Several authors have reported small color gradients in visual colors like B–V and B–R. The interpretation of these data is not easy, due to uncertainties caused by the short wavelength–baseline, and by the presence of a hot stellar population suggested by IUE–data (e.g. Burstein *et al.* 1986). Simultaneous measurements of visual and visual–infrared colors provide the means to determine both the average temperature of the giant branch and the turnoff–temperature of the main sequence. This allows to model fractional contributions of different populations, including age– and metallicity–effects. The required continuity of solutions at different radii provides a strong constraint in selecting a more unique overall population model, and relieves the ambiguous interpretation of single measurements.

Our observations have been done with the 1m Kapteyn Telescope at La Palma with the Leicester Photometer, an instrument measuring a visual and an infrared band simultaneously, making it easier to match the beams. The apertures were pointed at different distances from the center, and their diameters were enlarged at larger radii. We observed NGC 3379,4278 and 5813 in B,V,J,H and K. All three galaxies show significant color gradients both in visual and visual-infrared colors, in the sense that the galaxies become bluer going outwards. The B–V profiles are consistent with the CCD–photometry by Peletier *et al.* (1987). Figure 1 shows our data.

Strom *et al.* (1976) and Tinsley (1978) have made some predictions for V–K gradients as a function of B–V gradients in elliptical galaxies. They predict ratios of $\Delta(V-K)/\Delta(B-V)$ of 2.8 (Strom) and 1.6 (Tinsley). Our results of NGC 3379 (2.7) are in agreement with Strom's number. But NGC 4278 (5) and NGC 5813 ($>$ 10) do not agree. They either need an extra hot component in the outer parts or a red component in the center. In the case of NGC 5813 there is good evidence for the presence of dust in the center (Jedrezejewski 1987) and it is well possible that this dust is mainly responsible for the central color gradient. For the three galaxies Mg_2 line–strength gradients have been obtained by Efstathiou & Gorgas (1985, NGC 5813) and Davies & Sadler (1986, NGC 3379 and 4278). We find that our V–J gradients are larger when their Mg_2 gradient is large. A detailed investigation will show how much of the color gradient is due to metallicity differences, to age differences, to dust or to different components which have to be

T. de Zeeuw (ed.), Structure and Dynamics of Elliptical Galaxies, 443–444.

included in parts of the galaxy only. A simple empirical approach as employed by Valentijn & Moorwood (1985) to explain the NGC 3379 data indicates that the main sequence turnoff becomes bluer with increasing radius, while the main giant contribution comes from early M giants throughout the galaxy.

In the three galaxies we find a correlation between visual–infrared color profiles and the change of position angle and ellipticity. This could give some clues about dynamical mixing, or possibly about merging processes in the past.

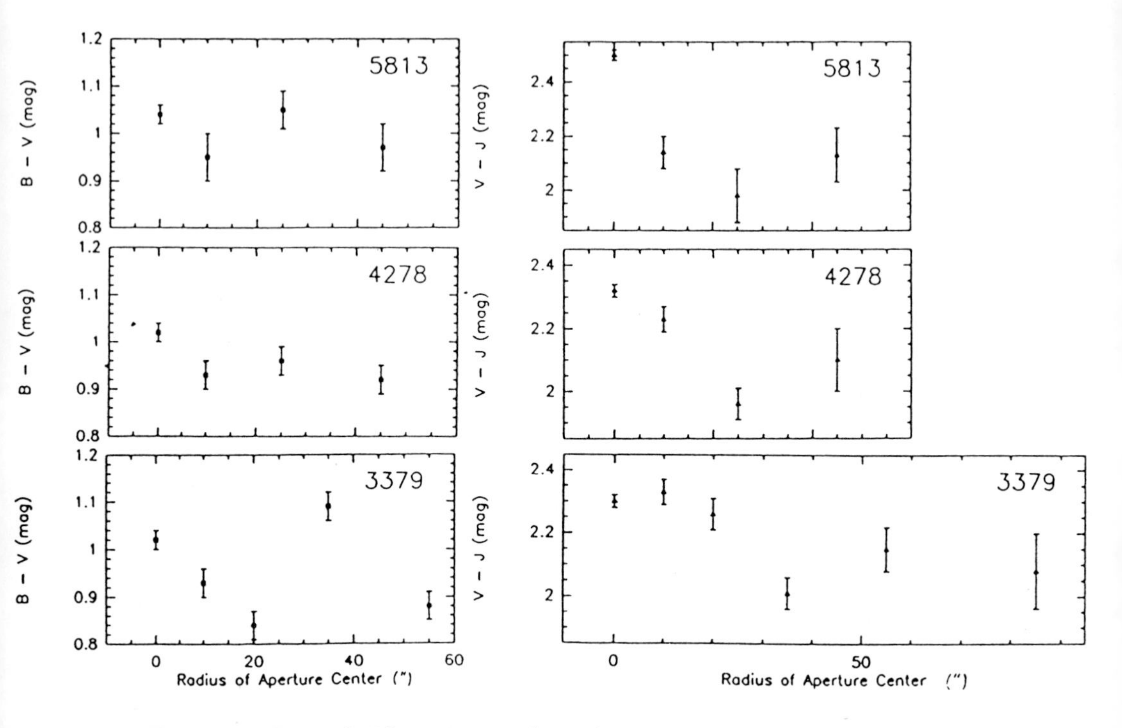

Figure 1. B–V (left) and V–J (right) aperture photometry profiles for NGC 5813, 4278 and 3379.

REFERENCES

Burstein, D., Bertola, F., Buson, F., Faber, S.M., & Lauer, T.R., 1986. in *Stellar Populations*, eds Tosi, M., & Norman, C., (Baltimore).

Davies, R.L., & Sadler, E.M., 1986. in *Stellar Populations*, eds Tosi, M., & Norman, C., (Baltimore).

Efstathiou, G., & Gorgas, J., 1985. *M.N.R.A.S.*, **215**, 37P.

Jedrezejewski, R., 1987. This volume, p. 37.

Larson, R.B., 1974. *M.N.R.A.S.*, **166**, 585.

Peletier, R.F., Davies, R.L., Illingworth, G.D., Davis, L.E., & Cawson, M., 1987, in preparation.

Strom, S.E., Strom, K.M., Goad, J.W., Vrba, F.J., & Rice, W., 1976. *Astroph. J.*, **204**, 684.

Tinsley, B.M., 1978, *Astroph. J.*, **222**, 14.

Valentijn, E.A., & Moorwood, A.F.M., 1985. *Astron. Astroph.*, **143**, 46.

DIRECT IR DETERMINATION OF THE STELLAR LUMINOSITY FUNCTION TO 0.2 $M_\odot$ IN ELLIPTICAL GALAXIES

G. Gilmore, Institute of Astronomy, Cambridge, England

and

K. Arnaud, Center for Astrophysics, Cambridge, MA, USA

SUMMARY. We present a determination of the stellar luminosity function in luminous elliptical galaxies which includes all stars more massive than 0.15 $M_\odot$. This limit corresponds to masses beyond the maximum in the solar neighbourhood stellar mass function, and therefore includes effectively all the luminous mass. Galaxies with X-ray evidence for current massive star formation, also show no evidence for enhanced low mass star formation in their central regions. All elliptical galaxies studied to date have stellar luminosity functions for masses above 0.15 solar masses which do not differ significantly from that in the solar neighbourhood. Elliptical galaxies have stellar bolometric mass-to-light ratios of $2.5<M/L<5.0$.

OBSERVATIONS

The 2.3 micron CO absorption feature is a sensitive gravity indicator, being weak in dwarfs and pronounced in giants cf. Figure 1. We have observed a grid of stars spanning a wide range of gravity and abundance, particularly including super metal rich M giants in Baade's Window. These have been used to synthesise spectral scans of two samples of elliptical galaxies using the technique of Arnaud and Gilmore (1986, MNRAS 220, 759). The first sample (Figure 2a) contains well studied galaxies with no known peculiarities, the second (Figure 2b) galaxies with X-ray evidence for substantial current accretion from a cooling gaseous halo.

RESULTS

There is no evidence for any statistically significant differences in the stellar luminosity function for stars with masses greater than 0.15 solar masses between elliptical galaxies with and without X-ray cooling flows. The best fit model to the stellar CO absorption spectra corresponds to a total bolometric mass-to-light ratio for stars more massive than 0.15 solar masses of: $2.5<M/L<5.0$.

T. de Zeeuw (ed.), Structure and Dynamics of Elliptical Galaxies, 445–446.

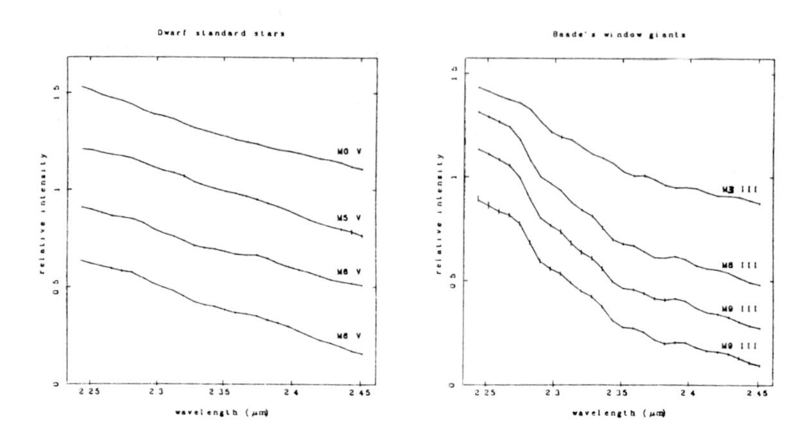

Figure 1. Spectra showing the difference in CO absorption between solar abundance dwarfs (left panel) and metal rich giants (right panel).

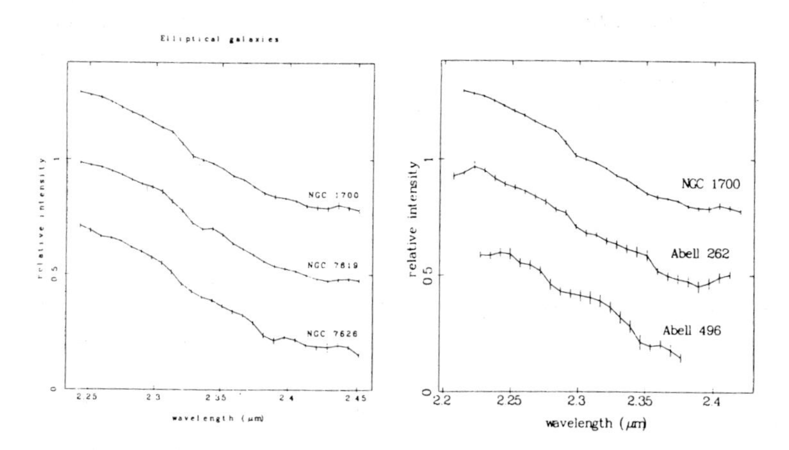

Figure 2. Spectra of elliptical galaxies without (left panel) and with (right panel) X-ray cooling flows. NGC 1700 is shown in both panels for comparison. There is no significant difference in the CO absorption between these galaxies.

DYNAMICS OF THE FORNAX DWARF SPHEROIDAL GALAXY

George Paltoglou and K.C. Freeman
Mount Stromlo and Siding Spring Observatories
The Australian National University

The mass to light ratio of the inner parts of the Fornax system is 3.2 ± 1.1. Its major axis rotation is only 3.4 ± 2.4 km s^{-1} over ± 1 kpc, which argues against its origin as a stripped dwarf irregular.

INTRODUCTION

To estimate the M/L ratio of the Fornax system, we need (i) the surface density profile, which gives the structure and lengthscales, and (ii) the velocity dispersion. New star counts, from UK Schmidt plates, are in excellent agreement with those of Hodge (1971): from a King model fit, the derived core radius is 17.7 arcmin, and the log of the ratio of tidal to core radii R_t/R_c is 0.5.

THE ROTATION OF THE FORNAX SYSTEM

Fornax appears elongated on the sky; its apparent axial ratio is about 0.67. Say it was once a dwarf irregular. These are disklike rotating systems. Fornax itself could then still be a disklike rotating system after the stripping event. Assume it is disklike and has an intrinsic axial ratio of 0.2. Then its inclination to the line of sight is 49° (zero is face on). What would be its rotational velocity ? This depends on the dynamics of the stripping event. We considered several possibilities. As a typical example, say (with hindsight) that the M/L ratio for Fornax is now about 3, and was about 1 as a dwarf irregular. We can then estimate its rotation rate from the Tully Fisher relation. Assume also that it lost half of its total mass and angular momentum in the stripping event, and then readjusted into equilibrium. Then the total rotational velocity difference that we would now observe over ±17 arcmin along the major axis is about 10 km s^{-1}. (The ±17 arcmin is set by the size of the AAT fiber field: 17 arcmin is about 1 core radius).

T. de Zeeuw (ed.), Structure and Dynamics of Elliptical Galaxies, 447–448.

THE OBSERVATIONS

The stars are K giants, chosen from our photographic photometry of the Fornax galaxy. Their mean V magnitude is about 18.5. The stars were chosen in two regions, 17 arcmin SW (17 stars) and 17 arcmin NE (22 stars) from the center and along the major axis. Their velocities were measured with the AAT fiber system. This took an entire night.

THE RESULTS

The velocity difference between the two regions is 3.4 ± 2.4 km s^{-1}, which is significantly less than our estimate of the expected rotation if Fornax was once a dwarf irreegular. The velocity dispersion at a radius of one core radius is 5.0 ± 0.9 km s^{-1} (corrected for the measuring error, which was itself well determined from repeated measurements throughout the night). This dispersion is slightly smaller than previously reported measurements (Seitzer and Frogel 1985, Aaronson and Olszewski, 1986) for carbon stars and globular clusters; however most of the difference results from the different radii at which the measurements were made. Our observed V/σ value is then about 0.34 ± 0.25: for an isotropic oblate system with the apparent axial ratio of Fornax, the V/σ value is about 0.7.

To estimate the mass, we represent Fornax by the King model fitted to the star counts (see above). By projecting the King model on the sky, we can estimate the characteristic velocity dispersion of the King model from our measured velocity dispersion at 17 arcmin radius. The resulting mass to (V) light ratio is 3.2 ± 1.1. This is fairly similar to the M/L value of 2.6 for the globular cluster ω Cen (Freeman and Seitzer, in preparation).

REFERENCES

Aaronson, M. and Olszewski, E, 1986. Astron.J., 92, p 580.
Hodge, P. 1971. Ann.Rev.Astron.Astrophys., 9, p 35.
Seitzer, P. and Frogel, J. 1985. Astron.J., 90, p 1796.

ANISOTROPY OF THE VELOCITY DISPERSION IN ω CENTAURI

G. Meylan
Astronomy Department
University of California
Berkeley CA 94720
U.S.A.

ABSTRACT. By far the brightest and the most massive globular cluster in our Galaxy, ω Cen seems to be, in some of its properties, a kind of transition step between dwarf ellipticals and ordinary globular clusters. For this giant cluster, the comparison between observations and King-Michie multi-mass dynamical models appears possible only using models with strong anisotropy in the velocity dispersion. A more comprehensive description of this work is to be published (Meylan 1986).

1. THE OBSERVATIONS

The present dynamical description of ω Cen uses both surface brightness and velocity dispersion profiles. Precise radial velocities have been obtained with the photoelectric spectrometer CORAVEL at the European Southern Observatory at Cerro La Silla, Chile, in collaboration with astronomers in Geneva, Marseilles, Copenhagen, and ESO (Mayor *et al.* 1986). The number of observations amounts to 540 measurements of 318 member stars, with typical uncertainties of 0.9 km/s.

2. THE MODEL

King-Michie dynamical models, based on an assumed form for the phase-space distribution function, have been constructed in an approach nearly identical to that of Gunn and Griffin (1979). In order to mimic a real cluster, heavy remnants (such as stellar black holes or neutron stars), white dwarfs and main sequence stars have been distributed into ten different subpopulations, each having the energy-angular momentum (E,J) distribution function

$$f_i(E,J) \propto \left[\exp(-A_i E) - 1\right]\exp(-\beta J^2) \qquad (1)$$

In the cluster center, thermal equilibrium is assumed in order to force A_i to be proportional to the mean mass of the stars in the subpopulation considered. A model is specified by a mass function exponent x, and by four parameters: the scale radius r_c , the scale velocity v_s , the central value of the gravitational potential $W_\circ$, and the anisotropy radius r_a . Beyond r_a , the velocity dispersion tensor is mostly radial.

T. de Zeeuw (ed.), Structure and Dynamics of Elliptical Galaxies, 449–450.

3. THE RESULTS

A grid of models has been calculated for a wide range of values of each parameter. **The observations are well fitted only by models with strong anisotropy, i.e. with $r_a \simeq$ 2-3 r_c** (Fig. 1 and 2). This is related to the large value of the half-mass relaxation time $t_{rh} \simeq$ 20-30 $\times 10^9$ yr ($t_r(0) \simeq 1 \times 10^9$ yr).

The mean value of the exponent x of the mass function is $\simeq$ 1.25-1.50, a value close to Salpeter's value of 1.35. Heavy remnants represent from 1 to 9 % of the total mass. In models with no remnants at all ($m_{hr} = 0$), we notice that an increase of the total mass of the white dwarfs is needed to fit the observations. The mean value of the total mass is about $4 \times 10^6 M_\odot$, giving a mean $M/L_v \simeq 3$.

REFERENCES

Gunn, J.E., and Griffin, R.F. 1979, *A. J.* **84**, 752.

Mayor, M., Imbert, M., Anderson, J., Ardeberg, A., Benz, W., Lindgren, H., Martin, N., Maurice, E., Meylan, G., Nordstroem, B., and Prévot, L. 1986 *Astron. Astrophys. Suppl.* to be published .

Meylan, G. 1986 *Astron. Astrophys.* in preparation.

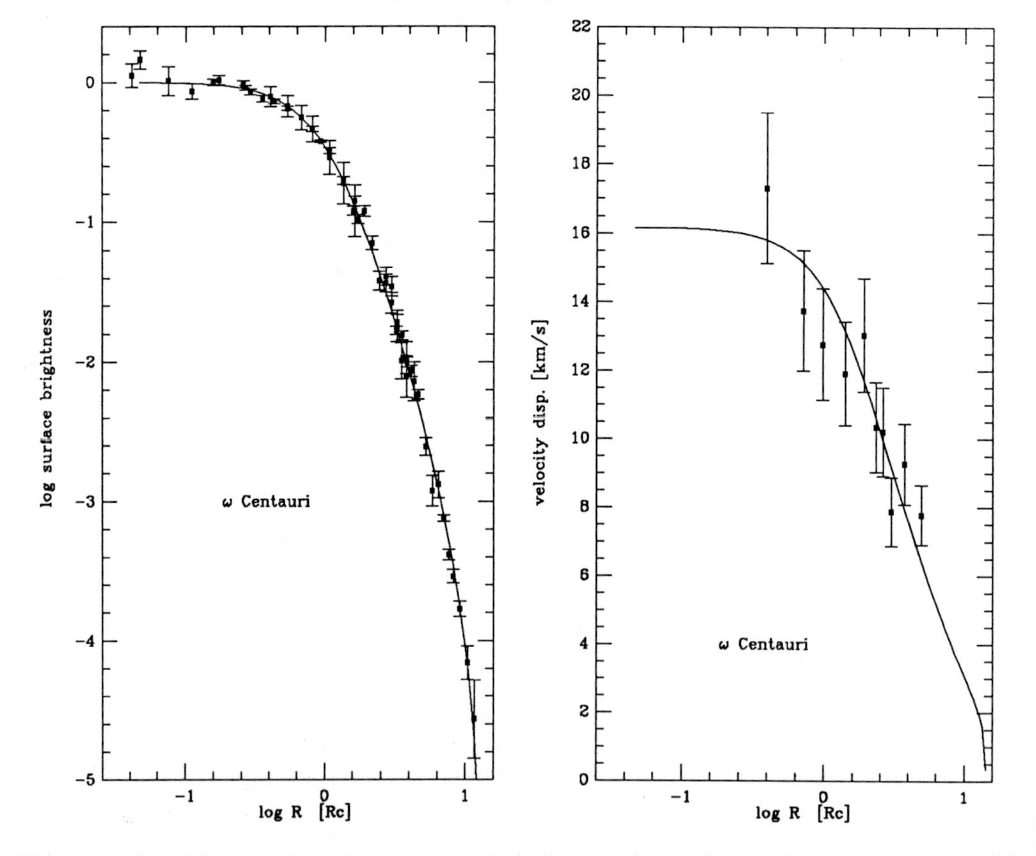

Fig. 1 and 2 Observed and computed surface brightness and velocity dispersion profiles, for the model: $m_{hr} = 0.$, x = 1.25, $W_o = 5.0$, and $r_a = 2.0$.

SPECTROSCOPY OF THE GLOBULAR CLUSTERS IN M87

J.R. Mould, J.B. Oke, and J.M. Nemec
Palomar Observatory, Caltech 105-24
Pasadena CA 91125
U.S.A.

ABSTRACT With a velocity dispersion of 370 ± 50 km/sec the globular cluster system of M87 is kinematically hotter than the stars in the giant elliptical itself. This is consistent with the clusters' shallower density distribution for isotropic orbits. The mean metallicity of the 27 clusters in the sample analyzed here is no more than a factor of 2 more metal rich than the cluster system of the Milky Way, but considerably more metal poor than the integrated starlight in the field at a radius of 1′ from the center of M87. There is no evidence for the existence of young clusters in the system. The mass-radius relation between 1′ and 5′ required to contain the globular clusters joins on to that required to contain the hot gas around M87.

We have obtained multi-slit spectra of a total of 52 objects from Strom *et al* (1981) lying within 7′ of the center of M87. The principal dynamical conclusions are:

1. Twenty-seven of the objects are globular clusters with a mean heliocentric velocity of 1320 km/sec and a dispersion of 370 ± 50 km/sec.

2. The cluster data do not fit an extension of the M/L = constant model of Sargent *et al* (1978) (see Figure 1). They are consistent with $\sigma(r)$ = constant. But the sample size must be doubled in order to make a definite statement on the presence of dark matter in ellipticals.

3. If the velocity dispersion $\sigma(r)$ is constant, $\sigma_{globs} / \sigma_{stars} = 1.4 \pm 0.2$ at 1′, For isotropic orbits this is consistent with the greater radial concentration of the starlight (see Harris 1986).

4. With $\sigma(r)$ = constant, isotropic orbits and a cluster volume density fall off like $r^{-2.1}$,

$$M(r) = 6 \times 10^{10} \text{ r (kpc) } M_{\odot}$$
$$= 3 \times 10^{11} \text{ r (arcmin) } M_{\odot} \text{ for M87 at 15 Mpc.}$$

This mass-radius relation makes the connection between previous optical studies and that inferred from the x-ray gas around M87 by Fabricant and Gorenstein (1983).

We gratefully acknowledge the partial support of this project by NASA grant NGL-05-002-134 and NSF grant AST-8502518.

T. de Zeeuw (ed.), Structure and Dynamics of Elliptical Galaxies, 451–452.

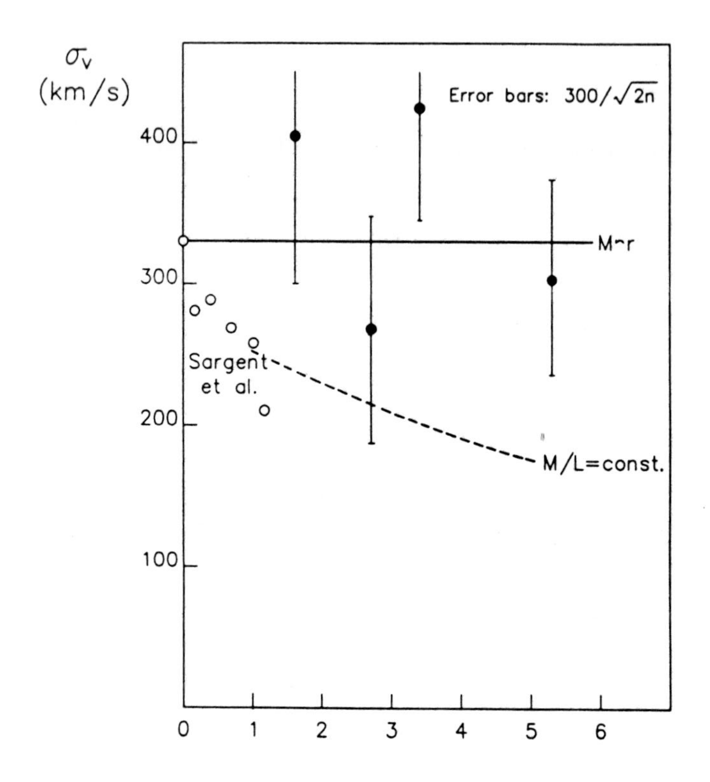

Figure 1. Velocity dispersion profile in 4 radial bins. The abscissa is labelled in minutes of arc.

REFERENCES

Fabricant, D, & Gorenstein, P., 1983. *Astrophys. J.*, **267**, 535.

Harris, W., 1986. *Astron. J.*, **91**, 822.

Sargent, W., Young, P., Boksenberg, A., Shortridge, K., Lynds, C.R., & Hartwick, F.D.A., 1978. *Astrophys. J.*, **221**, 731.

Strom, S., Forte, J.C., Harris, W., Strom, K., Wells, D., & Smith, M., 1981. *Astrophys. J.*, **245**, 416.

THE "JET" OF M89: CCD SURFACE PHOTOMETRY

G. Clark, P. Plucinsky, and G. Ricker
Physics Department and Center for Space Research
Massachusetts Institute of Technology, Cambridge, Mass.

ABSTRACT. We have obtained CCD images in R and V of Malin's "jet" in the weakly radio and X-ray active E0 galaxy M89 (NGC4552). The luminosity of this feature is approximately 1/4% of the total luminosity of the galaxy; its color is bluer than that of the whole galaxy with a V-R value smaller by about 0.15 magnitudes. The likely explanation of the feature, which looks more like a proboscis than a jet, is that it is, in the words of the Toomres (1972), a "tidal relic of a close encounter", seen from a perspective that may hide a drawn out tail.

1. INTRODUCTION

Malin (1979) discovered three faint features of disturbed morphology in the active E0 galaxy M89 (NGC4552) in specially enhanced sky-limited 1.2-m Schmidt photographs:

a) A faint optical "jet" at position angle 115° extending 10 arc min (~36 kpc) from the nucleus;
b) A bulge on the eastern side of the outer envelope
c) Three concentric arcs of the kind now called "ripples" or "shells" at 5', 7' and 10' from the nucleus.

In addition he reported the detection by R. Eckers (unpublished) of variable 6 cm radio emission from the nucleus. Subsequent spectroscopic investigation (Malin 1981) showed the jet free from emission lines and with a color slightly bluer than the rest of the galaxy. Forman, Jones, and Tucker (1985) detected X-rays from an unresolved point source at the nucleus with luminosity of 3×10^{40} ergs s^{-1}.

We have undertaken a detailed study of the faint optical features of M89 by CCD surface photometry. We report here the preliminary results from measurements of the jet.

2. OBSERVATIONS

We used the MASCOT CCD camera attached to the 1.3-m McGraw Hill telescope. Our results are drawn from a 3604-s V and an 1820-s R exposure of the jet region, and 53-s V and R exposures of the nucleus.

The jet is conspicuous in the CCD images of which the V exposure is displayed in contoured form in Figure 1. The surface brightness of the jet is approximately 2% of the sky background.

3. ANALYSIS

The jet emerges radially westward from the outer halo glow at a distance of about 5 arc min

T. de Zeeuw (ed.), Structure and Dynamics of Elliptical Galaxies, 453–454.

from the nucleus, and then bends 30° southwest to a straight uniform section about 5 arc min long and 1/2 arc min wide. The maximum surface brightness of jet is approximately 26.8 V mag arcsec^{-2}.

The ratio of the luminosity of the discernable portion of the jet feature to that of the whole galaxy is $(2.5\pm0.5)\times10^{-3}$. The jet is bluer than the whole galaxy, i.e.

$$(V\text{-}R)_{jet} - (V\text{-}R)_{galaxy} = -0.15\pm0.05 .$$

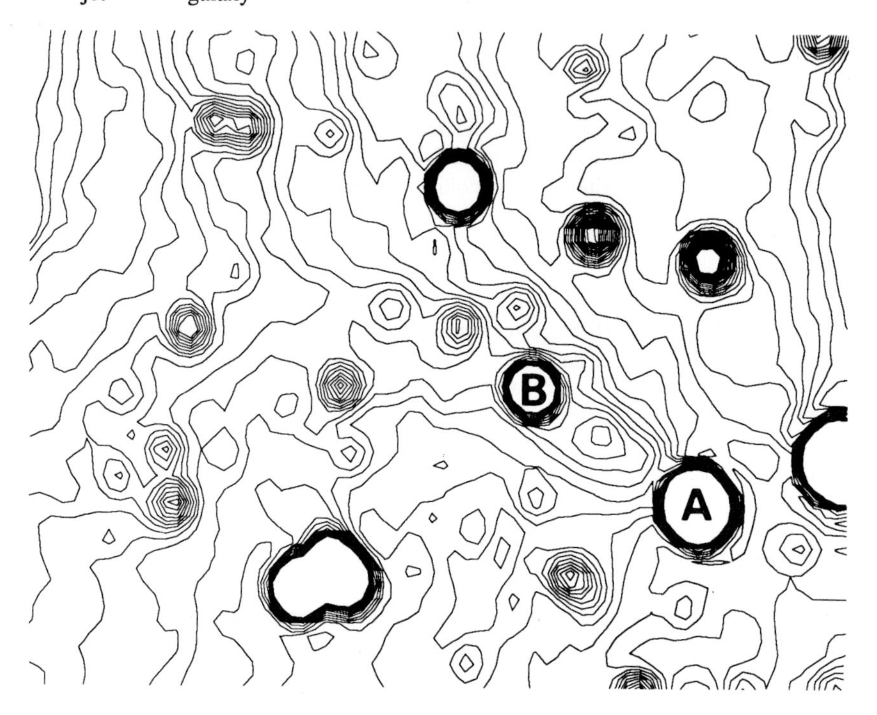

Figure 1. Contour plot of the surface brightness in the vicinity of the "jet" of M89 in the 3604-s V-filter CCD exposure. North is up and east is left. The contours are at intervals of 5 ADU's starting with a maximum of 1500 in the upper left hand corner. The sky level is 1370 ADU's. The separation between foreground stars A and B is 1.2 arc min. The nucleus of M89 is located 8.4 arc min to the east and north of star A. The jet lies along the line defined by the image centers of stars A and B.

4. DISCUSSION

M-89 probably suffered a non-radial encounter with a small galaxy within the last billion years, resulting in a disturbed morphology and a feeding of the nucleus that induced radio and X-ray activity. Modelling of such encounters shows that they can produce tidal streams as well as arc-like features (Quinn 1984). The jet is probably a tidal stream of stars viewed from a perspective that hides a drawn out tail.

REFERENCES

Forman, W., Jones, C., and Tucker, W. 1984, *Astroph. J.*, **293**, 102.
Malin, D. F. 1979, *Nature*, **277**, 279.
Malin, D. F. 1981, *AAS Photo-Bulletin No. 27.*
Quinn, P. J. 1984, *Astroph. J.* , **279**, 596.
Toomre, A., and Toomre, J. 1972, *Astroph. J.*, **178**, 623.

DWARF GALAXIES IN THE FORNAX CLUSTER

Nelson Caldwell
F.L. Whipple Observatory
P.O. Box 97
Amado, AZ 85645

and

Gregory Bothun
Astronomy Department
California Institute of Technology
Pasadena, CA 91125

ABSTRACT. We present the results of an observational study of the dwarf galaxies in the Fornax cluster of galaxies.

We present the results of an observational study of the dwarf galaxies in the Fornax cluster of galaxies. The data consist of optical and infrared photometry, multi-color CCD imaging, and near-infrared spectroscopy. Nearly thirty of the one hundred dwarf galaxies known in this cluster have been observed. The results can be summarized as follows.

(1) Most of the dwarfs are low-surface elliptical systems similar in all respects to those found in the Virgo cluster. There seems to be a slight deficiency of dwarf irregulars in the cluster with respect to the Virgo cluster. The dwarf galaxies that appear to be forming stars are far from the center of the cluster

(2) From UBVJHK photometry and infrared Ca II spectroscopy, we surmise that most of the systems have old, metal-poor stellar populations. There are several elliptical systems which clearly have star formation regions, however, and one very low surface brightness elliptical galaxy is simply too blue to be explained by low metallicity.

(3) A high proportion of the dwarfs are nucleated. The luminosity of the nucleus is well correlated with the color of the galaxy, in the sense that the more nucleated dwarfs tend to be redder. There is rarely any measurable difference in the color of the nucleus and the surrounding halo.

(4) Outside of the nucleus, the light profile falls off exponentially

T. de Zeeuw (ed.), Structure and Dynamics of Elliptical Galaxies, 455–456.

in most cases. Interesting exceptions include a dwarf in which an exceedingly large core radius exists, so large that 25% of the luminous mass of the galaxy resides in the constant density region. General conclusions about the structure of elliptical systems (from the giant ellipticals to the dwarf spheroidals) can now be made. From -23 to -18, a decrease in luminosity is produced by a decrease in radius, since the central surface brightness increases as luminosity decreases. Since the slope of the profile becomes steeper with decreasing luminosity, there is a slight increase in isophotal surface brightness with decreasing luminosity for those galaxies, although the surface brightness at any metric radius outside of the nucleus decreases as the galaxies get less luminous. The light distribution can be represented by a power law in the inner regions and an exponential in the outer regions. The exponential region tends to begin at smaller isophotal radii in the fainter galaxies.

For galaxies with luminosities between -18 and -15, the principal factor in causing a decrease in luminosity is a decrease in the central surface brightness (ignoring the nucleations), thus resulting in a shallower radial light profile in the inner regions. The radius decreases only slowly and hence there is a strong dependence of isophotal surface brightness on luminosity. The light distribution can be well represented by an exponential, although the profiles of the brighter galaxies tend to have more of a power law slope in the inner regions.

Galaxies fainter than -15 show a variety of structures, some with very large core radii, some whose limiting radii are abnormally small, leading to surface brightnesses that are high for their luminosities. The light distribution is approximately exponential for these galaxies.

(5) Dwarf ellipticals in general share the same structure as blue compact dwarfs (BCD's), but are quite distinct from dwarf irregulars, in that the irregulars have significantly larger scale lengths at a given surface brightness. Present day dwarf ellipticals did not evolve from present day dwarf irregulars because the amount of fading due to aging of the stellar population in the irregular would render them much fainter than present day dwarf ellipticals. In fact, they would become undetectable with photographic plates.

(6) Because of the relation between color and brightness of the nucleus, we think the nuclei were probably formed from material that was enriched from stellar mass loss and probably after the main body of the galaxy was formed. Since there is only rarely a color difference between the nucleus and the halo, the formation of these metal-enriched stars probably occurred in the halo as well. Whether a galaxy held on to its interstellar gas for a second round of star formation was probably dependent on how many supernovae exploded in a short amount of time. That is, the success in doing so was random event. One or two of the galaxies observed in the Fornax cluster may be in the process of forming a nucleus.

THE LOCAL DENSITY AND MORPHOLOGY DEPENDENCE OF THE GALAXY LUMINOSITY FUNCTION

Jacek Choloniewski[1], Miroslaw Panek[1,2]

[1] Copernicus Astronomical Center, ul. Bartycka 18,
00–716 Warszawa, Poland
[2] NASA/Fermilab Astrophysics Group, Chicago, USA

We obtained the luminosity function (LF) for samples of galaxies from the CfA North catalogue (Huchra, Davis, Latham and Tonry, 1983). The criteria of selection of samples were the local density (range—more than 2 orders of magnitude) and/or the morphology. No difference in the combined LF for all morphological types is found for subsamples of different density. The LF of elliptical galaxies is found to be less steep at the faint end than the LFs for S and S0 galaxies. E galaxies are on the average brighter than the other morphological types. The LFs measured for early-type galaxies (E+S0) in high and low density regions show marginal difference—the low density LF has a steeper faint end slope. (Such a difference is not found for S galaxies). If this feature is maintained for larger samples it may indicate that the LF determined at the moment of galaxy formation is only weakly influenced by the phenomena present in dense regions. This is because these phenomena would rather leave the opposite imprint on the LF—the tidal stripping in dense regions would populate them with faint remnants of disrupted, bright, low angular momentum galaxies. Mergers could not reverse this trend because they act mainly on the bright galaxies.

REFERENCES

Choloniewski, J., Panek, M. 1985, Fermilab preprint 85/179–A
Huchra, J., Davis, M., Latham, D. W., and Tonry, J. 1983, *Ap. J. Suppl.*, **52**, 89.

T. de Zeeuw (ed.), Structure and Dynamics of Elliptical Galaxies, 457.

George Efstathiou and Ortwin Gerhard.

AN INVESTIGATION OF THE RADIAL DEPENDENCE OF THE GALAXY LUMINOSITY FUNCTION IN ABELL CLUSTERS

Phyllis M. Lugger
Department of Astronomy
Indiana University
Swain West 319
Bloomington, IN 47405
USA

ABSTRACT. The luminosity functions of inner and outer regions of six Abell clusters (A569, A1656, A2147, A2151, A2199, and A2634) were compared. These clusters have a single, reasonably symmetric central concentration of galaxies within the central Mpc. For three other clusters with irregular spatial distributions of galaxies (A779, A1367, and A2197) luminosity functions for high and low density regions were compared. For three of the clusters in the first group (A1656, A2147, and A2199) there is a deficit of bright galaxies, according to the Kolmogorov-Smirnov and Wilcoxon rank-sum nonparametric tests, in a region of radius 0.5 Mpc about the cluster center compared to a concentric annular region with bounds of 0.5 and 1.0 Mpc.

1. DATA SAMPLE

Palomar 48 inch Schmidt R-band plates for nine clusters (A569, A779, A1367, A1656, A2147, A2151, A2197, A2199, and A2634) were digitized on the KPNO PDS and processed using the AUTOPHOT software as described by Lugger (1986). Isophotal magnitudes to the R = 23 mag arcsec^{-2} isophote were measured for all objects in the scanned region. Stars were removed by a classifier based on image size relative to magnitude.

2. LUMINOSITY FUNCTIONS

An analysis of the galaxy surface density distribution in all nine clusters was carried out. For the six clusters (A569, A1656, A2147, A2151, A2199, and A2634) with a single, reasonably symmetric central concentration of galaxies within the central Mpc, luminosity functions of galaxies in inner and outer regions were compared. For the remaining three clusters (A779, A1367, and A2197), luminosity functions of galaxies in high and low density regions were compared. In order to compare galaxy magnitude samples without making any assumption about the form of the luminosity function, two nonparametric tests were used: the Kolmogorov-Smirnov two-sample test and the Wilcoxon rank-sum test.

The figures show the three cases (A1656, A2147, and A2199) where a significant difference was found between the luminosity functions in

T. de Zeeuw (ed.), Structure and Dynamics of Elliptical Galaxies, 459–460.

the two samples. These significant differences were found when comparisons were carried out both including and excluding the brightest galaxy (or first and second brightest galaxies for A1656 and A2147) in the cluster. The (a) figures are for the circular regions 0 - 0.5 Mpc from the cluster centers and the (b) figures are for the annular regions 0.5 - 1.0 Mpc from the centers. For these three cases there is a significant deficit of bright galaxies in the inner (a) region relative to the outer (b) region. The integrated luminosity functions, corrected for background, are indicated by histograms and the differential functions, corrected for background, are indicated by boxes with error bars. The crosses indicate the value of the differential function before background correction.

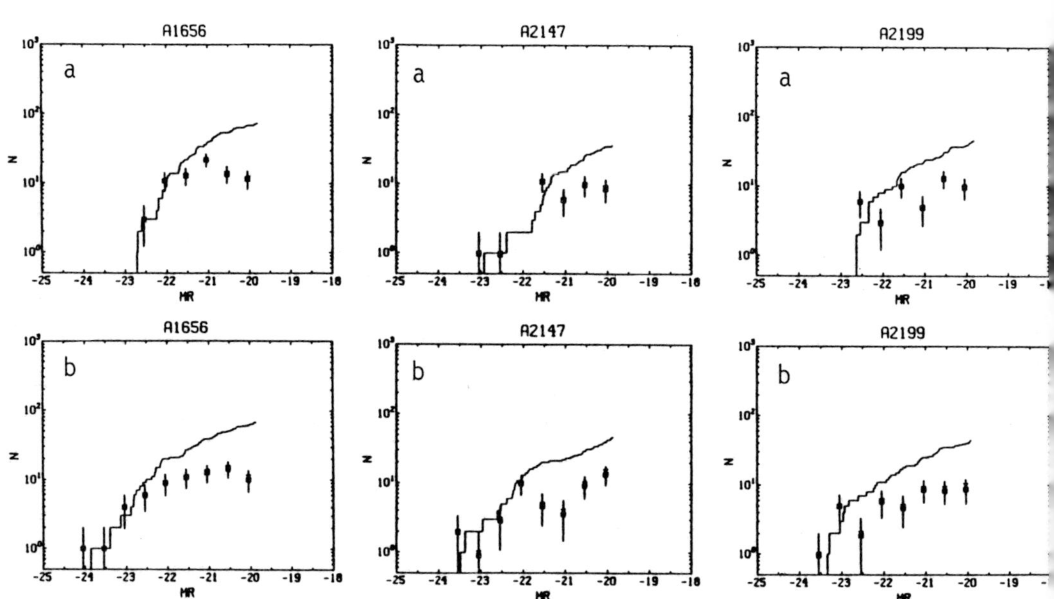

Figure 1. Luminosity functions for (a) inner and (b) outer regions.

3. CONCLUSIONS

The depletion of bright galaxies observed in the cores of A1656, A2147, and A2199 relative to outlying regions is an expected outcome of both galactic cannibalism (Hausman and Ostriker 1978) and tidal stripping. These dynamical processes, which have higher rates in higher density regions, affect the most massive galaxies to the greatest degree. The average surface densities within 0.5 Mpc of the centers of all three clusters place them in the upper half of the range of central densities of the clusters studied here.

REFERENCES

Hausman, M.A. and Ostriker, J.P. 1978, *Ap.J.*, **224**, 320.
Lugger, P.M. 1986, *Ap.J.*, **303**, 535.

ON THE LUMINOSITY FUNCTION OF ELLIPTICAL GALAXIES

V.S. Popov
Central Astronomical Observatory
of the USSR Academy of Sciences
196140 Leningrad, Pulkovo
USSR

Originally the luminosity function of galaxies was derived by E. Hubble (1936). Later studies are by Holmberg (1950), Zwicky (1957, 1964), Kiang (1961), Neyman and Scott (1962), Van den Bergh (1961), Pskovsky (1965), Nezhinsky and Osipkov (1967, 1969), Genkina (1969), Popov (1980), Tammann (1984), and others.

Luminosities of elliptical galaxies are within the range $-10^m.0$ to $-24^m.0$ with a maximum at $M_v = -19^m.0$. They are very close to the integral luminosity function of globular clusters of the Galaxy with the mode $M_v = -7^m.0$, i.e. they converge with weak dwarf galaxies of the Local group of galaxies.

If these observational data are combined with the luminosity function of stars in the Galaxy, with the mode $M_v = +14^m.0$, a new observational fact can be derived.

TABLE

Object	$\mathcal{M}(M_\odot)$	$3\Delta \log \mathcal{M}$	Mode M_v	ΔM_v
Stars	1		14	
		21		21
Globular Clusters	10^7		-7	
		12		12
Elliptical Galaxies	10^{11}		-19	

Each group of objects (stars, globular clusters, elliptical galaxies) has its own maximum on the common curve of the luminosity function. If these maxima are compared with the characteristic mass of the matter in these objects, then we discover $\Delta M_v = 3\Delta \log \mathcal{M}$, where M is the absolute magnitude of the object, and $\mathcal{M}$ is its characteristic mass.

T. de Zeeuw (ed.), Structure and Dynamics of Elliptical Galaxies, 461–462.

REFERENCES

Hubble, E.P., 1936. *The Realm of the Nebulae*, New Haven.
Holmberg, E., 1950. *Lund Medd.*, No. **128**.
Zwicky, F., 1957. *Morphological Astronomy*, Berlin.
Zwicky, F., 1964. *Astroph. J.*, **140**, 1626.
Kiang, T., 1961. *M.N.R.A.S.*, **122**, 262.
Neyman, J., & Scott, E., 1962. *A.J.*, **67**, 119, 582.
van den Bergh, S., 1961. *Z. für Astroph.*, **53**, 219.
Pskovsky, Yu.P., 1965. *Vestnik MGU*, **1**.
Nezhinsky, E.M., & Osipkov, L.P., 1967. *Trudy AO LGU* **24**, 117.
Nezhinsky, E.M., & Osipkov, L.P., 1969. *Trudy AO LGU* **26**, 92.
Genkina, L.M., 1969. *Trudy Astroph. Inst. AN Kaz., SSR*, **12**, 106.
Popov, V.S., 1980. *Izvestia GAO AN SSSR*, **198**, 15.
Tammann, G.A., 1984. The Hubble Diagram, in *Clusters and Groups of Galaxies*, 529 (Dordrecht).

SPECTROPHOTOMETRY OF SHELL GALAXIES

W.D. Pence
The Space Telescope Science Institute
Homewood Campus
Baltimore, MD 21218

Low dispersion spectra of two shell galaxies, NGC 3923 and NGC 3051, have been obtained covering the 5300Å to 10000Å spectral range. These long-slit spectra go through the nucleus of each galaxy and also through 12 shells in NGC 3923 and through 3 shells in NGC 3051. The main results are:

1) In NGC 3051 the surface brightness of the 2 inner shells is nearly constant with radius. These plateaus of luminosity have very abrupt outer cutoffs but have no detectable inner limits.

2) In NGC 3923 all the shells have similar contrast relative to the underlying galaxy which implies that the shells roughly follow the same luminosity distribution as the galaxy itself.

3) Both galaxies contain some gas and dust. In NGC 3923 this is confined to the nucleus and a single large dust spot. The interstellar emission in NGC 3051 is much stronger and spread throughout the galaxy.

4) Both galaxies show a reddening towards the nucleus caused by a metallicity abundance gradient.

5) There is no significant line emission from the shell material; most of the the shell luminosity comes from continuum radiation showing that the shells have a stellar, rather than a gaseous composition.

6) The shells have the same $V-R$ and $R-I$ colors as the underlying galaxy (to within 0.05 mag in NGC 3051 and 0.1 mag in NGC 3923) showing that they must be composed of an old stellar population very similar to that of the underlying galaxy. If the shells were formed by a recent galaxy merger then the infalling material probably came from an elliptical or S0 galaxy. On the other hand, if the merger was with a spiral galaxy, then it must have taken place more than about 10^9 years ago to provide time for the blue population to evolve.

7) The shells only contribute about 0.5% and 5% of the total flux in NGC 3923 and in NGC 3051, respectively. This is in apparent contradiction to the prediction of the merger shell formation theory since a companion galaxy of at least 10% of the primary mass is required to produce the inner shells seen in both galaxies.

The full report on this research has appeared in *Astroph. J.*, **310**, 597.

T. de Zeeuw (ed.), Structure and Dynamics of Elliptical Galaxies, 463.

Maria Petrou and Richard James.

TWO COLOUR CCD PHOTOMETRY OF MALIN-CARTER SHELL GALAXIES

A. Wilkinson
Astronomy Dept., The University, Manchester M13 9PL, England.
W.B. Sparks
RGO, Herstmonceux Castle, East Sussex BN27 1RP, England.
D. Carter
Mount Stromlo Observatory, Canberra ACT 2606, Australia.
D.A. Malin
AAO, P.O. Box 296, Epping NSW 2121, Australia.

1. INTRODUCTION

Shells may be the result of the disruption of a small companion in the potential of a much larger galaxy (e.g., Hernquist & Quinn, 1986; Dupraz & Combes, 1985), the disturbance of a disk system in a tidal encounter (Toomre, 1972), an accumulation of resonant stellar orbits (Binney, pers.comm.) or the result of some shock phenomenon in a hot galactic atmosphere (e.g., Williams & Christiansen, 1985; Wilkinson & Bailey, 1985;). To distinguish between these formation mechanisms, CCD direct images in B and R have been obtained at the Anglo-Australian 3.8m telescope for 66 of the 74 galaxies in the range $01^h\,40' < \alpha < 13^h\,46'$ in the Malin-Carter (1983) catalogue of shell galaxies.

2. IMAGE PROCESSING

Figs. 1a) to d) show different ways of presenting the same image.

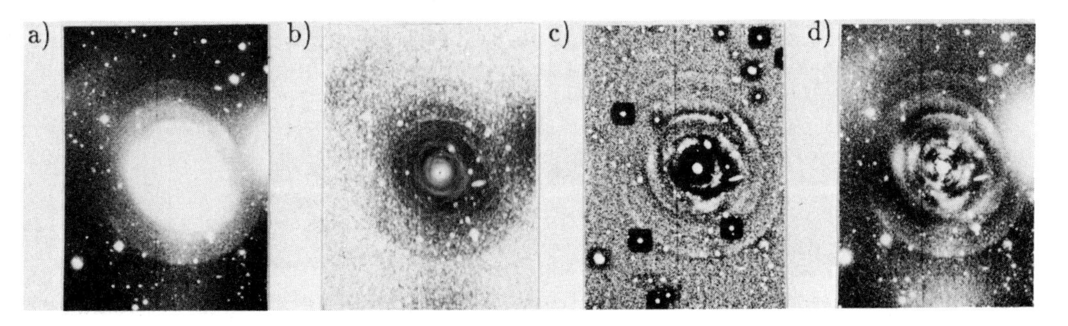

Figure 1. 0422-476 a) Direct CCD image after routine processing. b) Logarithmic gradient. Bright where surface brightness profile is steep, dark areas have shallower profile. c) Unsharp masking. Smoothed image subtracted from normal image. d) Modelling and subtraction of primary galaxy elliptical surface brightness profile.

T. de Zeeuw (ed.), Structure and Dynamics of Elliptical Galaxies, 465–466.

3. RESULTS

Our observations suggest three structure categories. a) *Cone* - Shells aligned with galaxy major axis, interleaved in radius on alternate sides of galaxy. (Fig. 2a). b) *Randomly distributed arcs* - Shells sharper-edged than in (a), irregularly distributed all around rather circular galaxy (Fig. 1a). c) *Irregular* - Shells distributed irregularly, not clearly related to any projected principal axis (Fig. 2b).

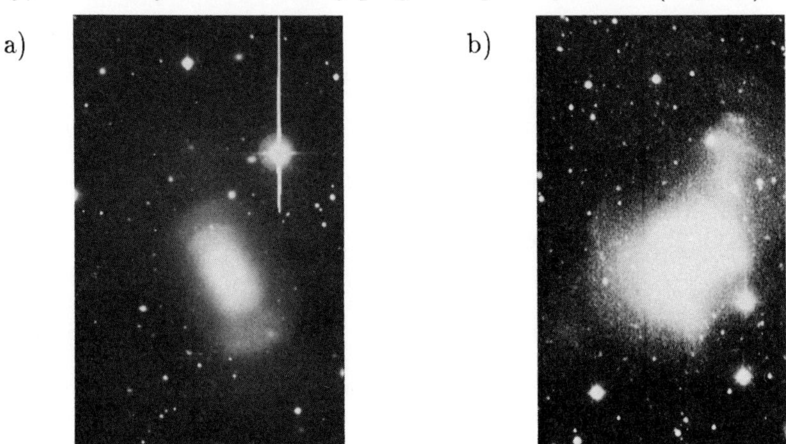

Figure 2. a) 0315-550, direct image, aligned shells b) 0515-541, direct image, irregular shells.

1) The gradient maps show that *shells commonly occur close to the nucleus.* In roughly 20 % of the systems these innermost shells (Fig. 1b) have *spiral morphology.* The galaxy light profiles are quite well represented by an $r^{1/4}$ law, so any inner disk does not dominate the light profile.

2) Despite the very disturbed appearance of the irregular system 1208-337 the *shells still interleave,* with the ratio of shell radii from one side to the other being almost constant at $\simeq 1.2$.

3) 0422-476 is a member of the 'randomly distributed arc' category. *The outer shell here has the same radius all around the galaxy,* although with occasional breaks (Fig. 1a). An ellipse fitted to the upper and lower arcs has a major axis $\simeq$4.3 arcmin but its centre lies within 2 arcsec of the nucleus.

4) The galaxies have an *enormous diversity of central surface brightness.* This suggests that a wide range of potentials allow shells to develop.

REFERENCES

Dupraz, Ch. & Combes, F., 1986. *Astron. Astrophys.*, preprint.
Hernquist, L. & Quinn, P.J., 1986. *Astrophys. J.*, preprint.
Malin, D.F. & Carter, D., 1983. *Astrophys. J.*, **274**, 534.
Toomre,A. & Toomre, J., 1972. *Astrophys. J.*, **178**, 623.
Wilkinson, A. & Bailey, M.E., 1985. *Cosmical Gas Dynamics*, ed. F.D. Kahn., p 63.
Williams, R.E. & Christiansen, W.A.,1985. *Astrophys. J.*, **291**, 80.

SHELLS AND DARK MATTER IN ELLIPTICAL GALAXIES

Lars Hernquist[1] and P.J. Quinn[2]
1) Department of Astronomy, U.C. Berkeley
2) The Space Telescope Science Institute

ABSTRACT. A new method for probing the distribution of matter in elliptical galaxies surrounded by shell systems is described. As an illustration we have applied this technique to the giant elliptical galaxy NGC 3923. If the potential is modeled as the sum of an $r^{1/4}$ law and a non–singular isothermal halo, then the best fit to the shell number and shell distribution gives a halo mass (within the shell system $\sim 100h^{-1}$ kpc) $\sim$ 40 times the mass of the elliptical and a halo core radius $\sim$ 3 times the effective radius of the luminous material.

Shells around elliptical galaxies are formed as the result of interactions with less massive companions (Quinn 1984, Hernquist and Quinn 1986a). Highly ordered shell systems can be generated in nearly radial encounters if the companion is disrupted on the initial pass, through the process of "phase–wrapping" (Quinn 1984). In such cases, the shells are comprised of stars on nearly radial orbits with known turning points, and the shell distribution maps the radial period *vs.* radius which can be used to infer the potential.

At any given time the shell distribution will be comprised of stars with *commensurate* periods. Thus, if stars in the outermost shell have completed τ periods, then stars in the shells consecutively closer to the galaxy will have completed $\tau+1, \tau+2, \tau+3$ ... periods. If shells are labeled according to their distances from the galaxy ($n = 1$ is the outermost shell) then this requirement implies $t = (n+\tau-1-\Theta_n)P(d_n)$, where t is the time since the disruption of the companion, Θ_n is the initial phase of stars in the shell at distance d_n from the galaxy, and $P(d_n)$ is the radial period. It then follows that the expected number of shells between the outermost shell and an interior shell, n, is $N_{shells} = (\tau-.5)P(d_1)/P(d_n)-(\tau-1.5)$ for an initial phase $\Theta = .5$ (Hernquist and Quinn 1986b). For a power law potential $\varphi \propto r^{\nu}$, $P \propto r^{1-\nu/2}$ and the commensurability condition gives the expected shell distribution $\ln d_n/d_1 = (2/(\nu-2))\ln((n+\tau-1.5)/(\tau-.5))$.

A detailed analysis for the sum of an $r^{1/4}$–law plus non–singular isothermal halo yields a good fit to the observed shell system surrounding NGC 3923 if the halo mass (interior to the outermost shell) is $\approx 30-50$ times the luminous mass and the halo core radius is $\approx 2-4$ times the luminous effective radius (see Hernquist and Quinn 1986b).

T. de Zeeuw (ed.), Structure and Dynamics of Elliptical Galaxies, 467–468.

The results for NGC 3923 can be compared with observations of x–ray coronae around early–type galaxies (Forman, Jones, and Tucker 1985). A useful characterization of the halo properties is provided by the quantity $m_{dark}(r)/r$. For $r >> \gamma$, $m_{dark}(r) \sim r$ (for an isothermal distribution) and $m_{dark}(r)/r \sim$ constant $\equiv \Xi$. It is also instructive to compare these data with the Bahcall and Casertano (1985) rotation curve analyses of spiral galaxies. In this case, the halo properties can be characterized by $v_{max}^2/G =$ constant $\equiv \Xi$, where v_{max} is the maximum rotation velocity in the disk, since $v_{max}^2/G \sim m_{dark}(r)/r$. The comparison is shown in Figure 1, for a mass to light ratio $M/L = 6$ for the early type galaxies, and $M/L = 3$ for the spiral galaxies.

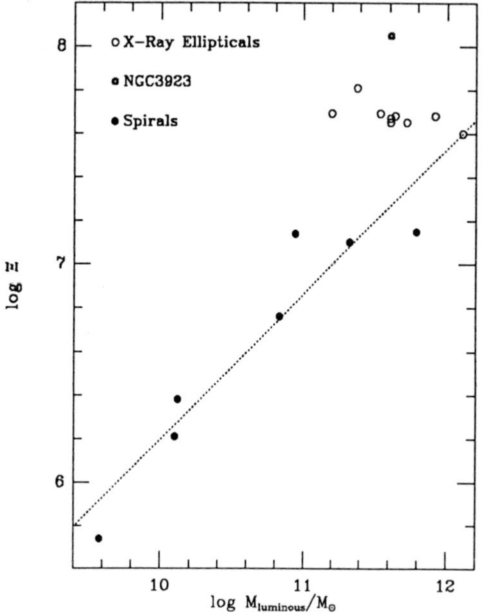

Figure 1. A comparison of the properties of the halo surrounding NGC 3923 with other observations.

The parameters of the halos surronding NGC 3923 and the x–ray galaxies are in good agreement, allowing for the uncertainty in the halo core radii for the x–ray sample (Hernquist and Quinn 1986b). An extrapolation of Ξ for the spirals (dotted line) suggests that the properties of halos surrounding ellipticals may differ from those surrounding spirals of the same luminous mass. However, the uncertainties are not negligible and a larger data–set is required. Future observations may help to determine if ellipticals formed as the result of mergers of spirals or if intrinsic differences in the local distribution of dark matter could have determined the Hubble type of the luminous material.

REFERENCES

Bahcall, J.N. and Casertano, S. 1985. *Astrophy. J. (Letters)*, **293**, L7 .
Quinn, P.J. 1984. *Astroph. J.*, **279**, 596 .
Forman, W., Jones, C., and Tucker, W. 1985. *Astroph. J.*, **293** , 102 .
Hernquist, L. and Quinn, P.J. 1986a. *Astrophy. J.*, submitted .
Hernquist, L. and Quinn, P.J. 1986b. *Astrophy. J.*, in press .

SHELLS AROUND TUMBLING BARS: THE MASS DISTRIBUTION AROUND NGC 3923

Ch. Dupraz[1,2], F. Combes[2], J.-L. Prieur[3]
1. Physikalisches Institut, Universität Köln, West Germany
2. Observatoire de Meudon, F–92190 Meudon, France
Ecole Normale Supérieure, F–75005, Paris, France
3. Mt. Stromlo Observatory, ACT 2606, Canberra, Australia

SHELLS AROUND A TUMBLING PROLATE GALAXY

In previous articles (Dupraz & Combes, 1985, 1986a), we showed that shells form with different geometries around prolate and oblate galaxies. However, theory and observations suggest that some ellipticals could be tumbling bars (Miller & Smith 1980; Möllenhoff & Marenbach 1986). Here we simulate the accretion of a small galaxy by a tumbling bar; the tumble period T_b is kept free. Let T_p be the typical period of motion of a particle in the potential of the elliptical galaxy. Then we find (Dupraz & Combes, 1986*b*):

a) When $T_b > 3T_p$ (Figure 1a), shells form with the geometry of a static *prolate* potential, i.e., aligned with the major axis.

b) When $T_b < 3T_p$ (Figure 1c), the particles feel the time-averaged potential, which is oblate: the shells display the typical *oblate* geometry. But there is no confusion with a static oblate shell galaxy, because the tumbling bars must be seen edge-on for the shells to appear.

c) When $T_b \sim 3T_p$ (Figure 1b), the outer shells form with the oblate geometry, the inner shells with the prolate geometry. In between, no shells form, because particles follow *resonant* (non-radial) motions.

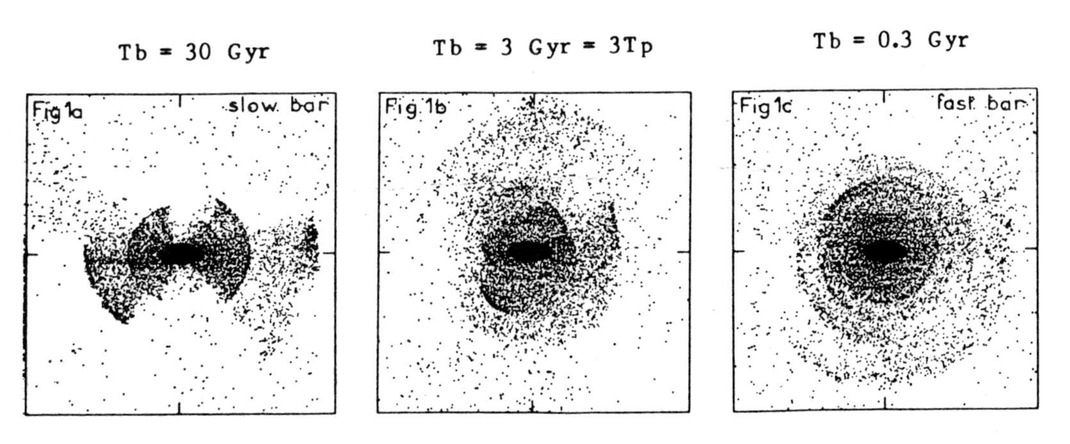

T. de Zeeuw (ed.), Structure and Dynamics of Elliptical Galaxies, 469–470.

MASS DISTRIBUTION (HALO) AROUND NGC 3923

The radial distribution of shells allows a determination of $M(r)$, the mass inside radius r. We apply various methods (Dupraz & Combes, 1986*c*; Hernquist & Quinn 1986) to the best observed shell galaxy NGC 3923, for which 26 shell positions are taken from Prieur et al. (1986).

In Figure 2a, we show the best fitted $M(r)$ functions for the shell system, for 3 values of the free parameter m, the number of shells beyond the outermost one, whether vanished or not. Figure 2b shows curves for the following models:

a) King model alone (representing the luminous component).
b) MOND = MOdified Newtonian Dynamics (Milgrom 1983).
c) FLAG = Finite Length–scale Anti–Gravity (Sanders 1984).
d) REM = REvised MOND (Sanders 1986).

Obviously, *the luminous mass in NGC 3923 is not sufficient:* a halo, or a non–Newtonian theory of gravitation, is needed to account for the shell distribution. As far as inner shells are concerned, any model is discrepant; this is due to the effect of dynamical friction (Dupraz, Combes & Gerhard 1986).

REFERENCES

Dupraz, C. & Combes, F., 1985. In *New Aspects of Galaxy Photometry*, ed. J.-L. Nieto, Lecture Notes in Physics, Springer Verlag, Berlin, p. 151.
Dupraz, C. & Combes, F., 1986*a*, *b*, *c* *Astr. Astrophys.*, preprints.
Dupraz, C., Combes, F. & Gerhard, O.E., 1986. Preprint.
Hernquist, L. & Quinn, P.J., 1986. *Astrophys. J.*, in press.
Milgrom, M., 1983. *Astrophys. J.*, **270**, 365, 371.
Miller, R.H. & Smith, B.F., 1980. *Astrophys. J.*, **235**, 793.
Möllenhoff, C. & Marenbach, G., 1986. *Astr. Astrophys.*, **154**, 219.
Prieur, J.-L., Fort, B., et al., 1986. *Astr. Astrophys.*, preprint.
Sanders, R.H., 1984. *Astr. Astrophys.*, **136**, L21.
Sanders, R.H., 1986. *Mon. Not. R. astr. Soc.*, **223**, 539.

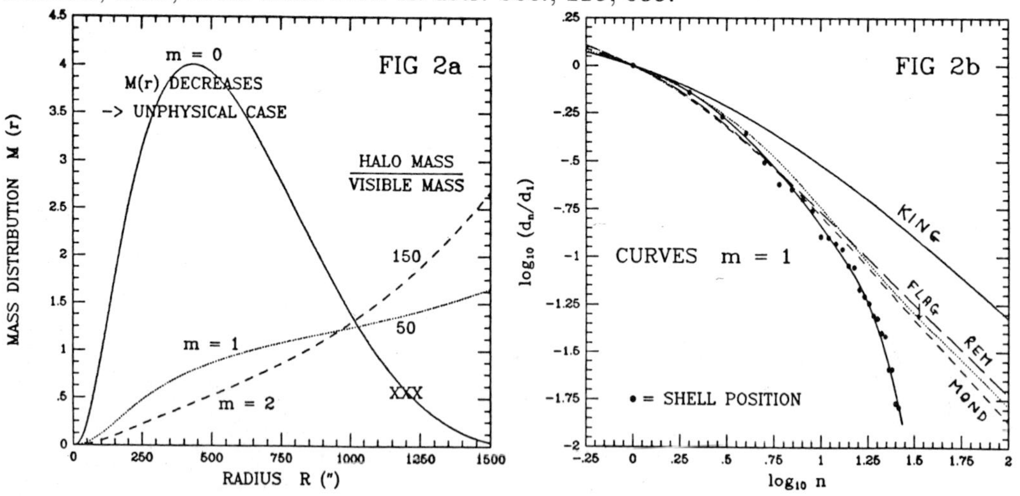

Shell Positions ($''$): 1170, 840, 630, 520, 365, 280, 263, (234), 203, 149, 147, (137), 128, 105, 103, 79, 73, 67, 58, 56, 47, 45, 30, 20, 19.

INITIAL TIDAL EFFECTS ON SHELL FORMATION

R. A. James and Althea Wilkinson,
Department of Astronomy,
University of Manchester, Manchester, M13 9PL, England.

INTRODUCTION

Sharp edged features have been observed by Malin and Carter (1980), Schweizer (1980) and others in the envelopes of many elliptical galaxies. The most promising of current models involves the disruption of a low mass galaxy penetrating close to the centre of a considerably larger ellipsoidal galaxy (Schweizer, 1980, Quinn, 1984, Hernquist and Quinn, 1986, Dupraz and Combes, 1986). The release of stars having a range of energies and with a definite relationship between velocity and position at the point of release generates sharp features by the "phase wrapping" mechanism.

We examine this effect using a self-consistent particle-mesh code to follow the disruption phase in an almost radial encounter. A standard 2-body code continues the integration up to a Hubble time.

NUMERICAL TECHNIQUES

The initial self-consistent model was constructed by the procedure of Wilkinson and James (1982) which allows us to specify the initial ratio of kinetic energy T to potential energy V. Low values of this ratio lead to compact structures following the de Vaucouleurs $r^{1/4}$ law as far as mesh resolution permits (Grimshaw, 1985). The preliminary calculations reported here use 25000 particles and a potential mesh of dimension $33 \times 33 \times 33$ points with a linear scale of 1 kpc per mesh interval and a time step of 2×10^6 years.

We assume that the primary is a Plummer sphere with scale radius 5 kpc and mass 10^{44} grams. The secondary galaxies all start with mass centres at distance 100 kpc from the centre of the primary and with initial systematic velocities directed towards this centre.

RESULTS

A pair of parabolic encounters with a secondary of mass 10^{42} grams shows the effect of the central concentration of the secondary on the break-up. The disruption is a relatively smooth process for an extended object with an initial energy ratio $2T/V$ of 0.5 and a half-mass radius of 4.5 kpc.. It is essentially complete when the secondary enters the 5 kpc sphere defining the primary core. For a compact object with initial energy ratio 0.1 and half-mass radius 1.9 kpc, there is a well-

T. de Zeeuw (ed.), Structure and Dynamics of Elliptical Galaxies, 471–472.

marked disruption phase centred on the transit of the secondary through the centre of the primary. About 60 percent of the secondary is stripped before this stage, and a further 20 percent is released in 2.6×10^6 years. In contrast, a more massive compact model (5×10^{12} grams) loses only 12 percent of its mass in 1.6×10^6 years, emerging with 46 percent of its initial mass still bound. The transition to the 2-body calculation assumes that all stars are unbound, thus introducing a problem in separating stars released during disruption from those which still belong to the secondary. The continuations run to date pick up the model soon after transit and those for the massive secondary are dominated by bound rather than unbound stars. Future code modifications will enable us to separate out the stars released near transit, but at present we restrict ourselves to low mass secondaries. Most of our models generate some shell structure after the disruption phase. The general trend is that shell structure is richer for more compact secondaries and for those with more elliptic orbits. The richest structure is observed for the most elliptic of our orbits with a compact secondary. This has an apocentre distance of 158 kpc and shows 7 visible shells 7.4×10^9 years after disruption.

CONCLUSIONS

The phase wrapping mechanism remains a plausible explanation of shell structure around elliptical galaxies. The efficiency of shell formation is greater with bound companions than in the parabolic case. Compact secondaries give stronger shells than the more extended ones and this is correlated with differences in their disruption history. The disruption process depends on the relative primary and secondary length scales and further work is required to establish if secondary structure is the dominant influence.

The main limitations of our work are the limited resolution employed so far and our inability to deal with more massive and less completely disrupted secondaries correctly. Both these limitations will be removed in future.

ACKNOWLEDGEMENTS

It is a pleasure to thank Lars Hernquist and P. J. Quinn and also Ch. Dupraz and F. Combes for communicating their results to us in advance of publication.

REFERENCES

Dupraz, Ch. & Combes, F., 1986, preprint.
Grimshaw, L., 1985, Ph.D. Thesis, University of Manchester.
Hernquist, L. & Quinn, P. J., 1986, preprint.
Malin, D.F., & Carter, D., 1980, *Nature*, **285**, 643.
Quinn, P. J., 1984, *Astrophys. J.*, **279**, 596.
Schweizer, F., 1980, *Astrophys. J.*, **237**, 303.
Wilkinson, A. & James, R. A., 1982, *M.N.R.a.S.*, **199**, 171.

SHELLS AND ENCOUNTERS OF DISK GALAXIES WITH ELLIPTICALS

Tsvi Piran
Racah Institute for Physics, Hebrew University, Jerusalem, Israel
& The Institute for Advanced Study, Princeton, NJ 08540, USA

Jens Verner Villumsen
California Institute of Technology, Pasadena, CA 91125, USA

INTRODUCTION

Faint shells have been found around a large number of ellipticals (Malin and Carter 1980, 1983; Schweizer, 1980). Quinn (1984) has suggested that shells form around ellipticals when a small disk galaxy is tidally disrupted in a near radial orbit. According to this mechanism the remnants of the nearly two-dimensional disk maintain their dimensionality and form sheets. These sheets will phasewrap around the elliptical, and the shells are seen when the line of sight is tangent to one of these surfaces (Quinn 1984, Hernquist and Quinn 1986a,b,c).

A shortcoming of Quinn's analysis is that the disruption of the disk galaxy is not treated selfconsistently. The disruption of the disk galaxy is done "by hand". The stars of the disk galaxy are treated as test particles in the fixed potential of the elliptical galaxy. After the disk galaxy has been disrupted, self gravity of the disk is unimportant, but it is important during the disruption phase.

Another problem is that the disks have zero thickness. This helps the contrast seen in the shells. Some estimates have been made of the effects of finite thickness, and the conclusion is that the finite thickness is not a problem (Quinn 1984).

NUMERICAL METHODS

We have used a new numerical method, based on expansion of the potential in bispherical coordinates. Solution of the Laplace equation using a Fourier transform is one of the common methods for N-body simulations. This fast method is best suited for a general code, where there is no particular geometry. When the configuration studied has a particular geometry it might be favorable to employ suitable coordinates and solve the Laplace equation using an expansion with the appropriate special functions (Villumsen 1982). Overall, such computations can be much faster than those done with a general purpose code. For the study of interaction between two galaxies the bispherical coordinates (e.g. Morse and Feshbach, 1953) seems the best. In our code the focal points of the bispherical coordinates move along in the computation and follow the centers of the galaxies. We use these coordinates to calculate the gravitational forces. The integration of the particles' orbits is carried out in Cartesian coordinates with a fourth order Adams-Moulton predictor corrector.

T. de Zeeuw (ed.), Structure and Dynamics of Elliptical Galaxies, 473–474.

DISCUSSION

We test Quinn's model with fully self consistent 3-D N-body simulations using this code. The target galaxy is a spherical galaxy with a Hubble profile. The intruder is a *selfconsistent* disk-halo system. The disk is exponential with a finite thickness, and the halo is nearly spherical and isothermal. This system has a Toomre Q value of about 1.5 and is stable. The mass-ratio of disk to elliptical is 1/10 . We have used 5000 particles in the disk, 1000 particles in the halo and 1000 in the elliptical galaxy.

A number of encounters were performed and followed for a few dynamical times after disruption. The present code is not suited for following the evolution long after the disk galaxy is completely disrupted, so we restrict ourselves to the disruption phase. We find that in order to form anything resembling shells, the disruption has to be nearly instantaneous i.e. from a close parabolic or almost radial orbit. If the disruption is slow, the disintegration resembles Roche lobe overflow through the Lagrange point and the particles are scattered widely in phase space. This means that the disk galaxy must be fairly compact so it does not loose its stars too early.

When the disk galaxy has zero thickness, some nearly radial encounters produce sharp edged features. However this happens only in a small number of the encounters that we have simulated. When we perform the same encounters when the disk scale height to scale length is 1/10 as is inferred for our own galaxy, the observed features are much broader and less distinct. James and Wilkinson (1987) find, however, in similar numerical simulations that shells do form even with a finite temperature of the disk. The reason for this discrepancy is not clear. We conclude that Quinn's mechanism will be effective only in a small fraction of galaxy encounters. The shells must have formed by disk stars that have a small velocity dispersion. The sharpness of the shells cannot be explained by phase wrapping of an old, warm disk component. Phase wrapping of a young, cooler component can explain the sharpness of the features.

REFERENCES

Hernquist, L., Quinn, P., 1986a. *Astroph. J.*, in press.
Hernquist, L., Quinn, P., 1986b. *Astroph. J.*, in press.
Hernquist, L., Quinn, P., 1986c. *This volume*, p. 467.
James, R. A. and Wilkinson, A., 1987. *This volume*, p. 471
Malin, D. F., Carter, D., 1980. *Nature*, **285**, 643.
Malin, D. F., Carter, D., 1980. *Astroph. J.*, **274**, 534.
Morse, P. M., and Feshbach, H., 1953. *Methods of Theoretical Physics*, McGraw-Hill, New York.
Quinn, P., 1984. *Astroph. J.*, **279**, 596.
Quinn, P. and Hernquist, L., 1987. *This volume*, p. 249.
Schweizer, F., 1980 *Astroph. J.*, **252**, 303.
Villumsen, J.V., 1982 *M.N.R.A.S.*, **199**, 493.

DYNAMICAL FRICTION AND ORBIT CIRCULARIZATION

Stefano Casertano
Institute for Advanced Study, Princeton

E. Sterl Phinney and Jens V. Villumsen
Caltech, Pasadena

We study the change in shape of the orbit of a satellite sinking (because of dynamical friction) towards the center of a larger galaxy. The galaxy is assumed spherically symmetric with distribution function either *isotropic* or *predominantly radial.* The satellite is a softened point mass.

The orbit evolution is studied analytically using the Chandrasekhar (1943) approximation for the dynamical friction drag and the epicyclic approximation for the orbit of the satellite, and numerically, by direct integration of an N-body system.

Fully self-consistent spherical models have been used for the galaxy. All the models have a King (1966) density profile with concentration parameter 1.255. Anisotropic models are constructed with the prescriptions of Merritt (1985). For these models the ratio of radial to tangential velocity dispersions follows the law $\sigma_r^2/\sigma_t^2 = 1 + r^2/r_a^2$; they are isotropic in the center and strongly radially anisotropic for $r \gg r_a$.

The analytical treatment gives an explicit form for the function $\Upsilon(r) \equiv d\ln(\epsilon)/d\ln(r)$ involving integrals over the distribution function. Although the explicit form of $\Upsilon(r)$ is rather complicated, numerical evaluation is straightforward. Some general properties of $\Upsilon(r)$ are:

- $\Upsilon(r) \geq 0$ (implying circularization) for any realistic *isotropic* system.
- For many realistic *isotropic* systems, $\Upsilon(r) \sim 0.5$ for $r \gg r_c$.
- $\Upsilon(r) \to 0$ as $r \to r_c$.
- Increasing the radial anisotropy makes $\Upsilon(r)$ smaller.

The dashed lines in Fig. 1 give the theoretical $\epsilon(r)$ for an isotropic system (a) and for a marginally unstable system with $r_a = 2r_c$ (b). For even more anisotropic systems ($r_a \leq 1.5r_c$), $\Upsilon(r)$ may be negative, corresponding to orbits becoming *less* circular as they shrink. However, such systems are unstable to bar formation (Barnes 1985).

The validity of the analytic treatment has been tested by a number of N-body simulations with the large galaxy made up of 5000–20000 points, and the satellite being treated as a point mass. The self-gravity of the galaxy is computed by spherical harmonics expansion, whereas the galaxy-satellite interaction is treated by direct summation over the particles. In agreement with the analytical predictions, the eccentricity decreases significantly in the isotropic system, but not for the anisotropic case (see Fig. 1).

T. de Zeeuw (ed.), Structure and Dynamics of Elliptical Galaxies, 475–476.

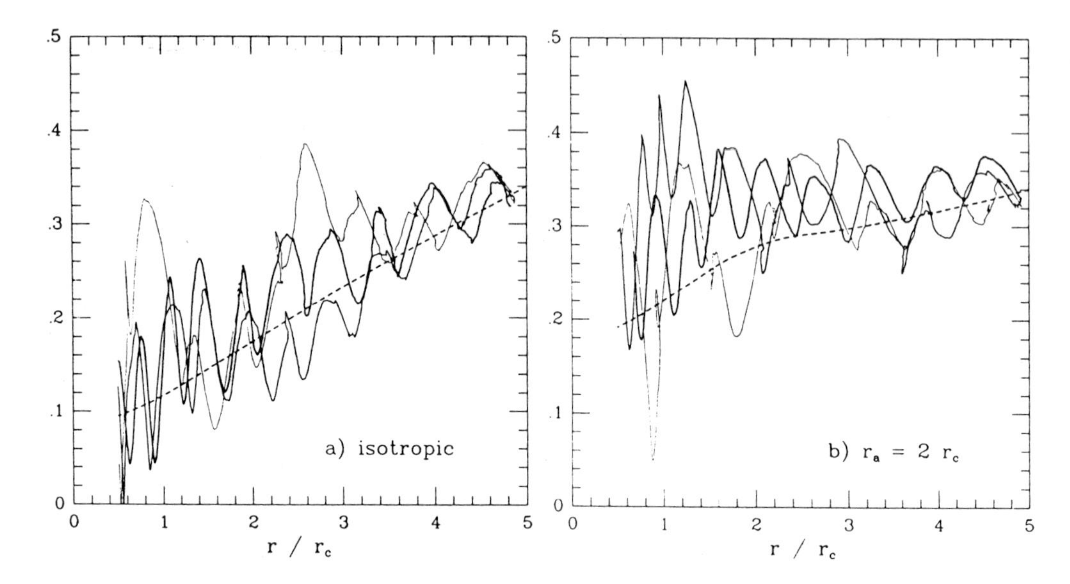

Figure 1. Predicted (dashed) and numerical (solid) variations of eccentricity with radius. Panel a) refers to an isotropic King model, panel b) to an anisotropic model with the same density distribution. The N-body simulations have N=20000 (thick solid), 10000 (medium) and 5000 (thin).

The eccentricity calculated from the numerical experiments is however very noisy, and the agreement is only semi-quantitative. From a detailed comparison of predicted and calculated eccentricity, we find:

- In the isotropic case, ϵ decreases appreciably. The corresponding value of Υ is not well determined, but is within 50% of the analytical prediction.
- There is no measurable decrease in eccentricity in the anisotropic case ($r_a = 2r_c$), consistent with our predictions.

REFERENCES

Barnes, J. 1985, unpublished.
Chandrasekhar, S. 1943, *Ap. J.* **97**, 255.
King, I. R. 1966, *A. J.* **71**, 64.
Merritt, D. 1985, *A. J.* **90**, 1027.

STOCHASTIC STELLAR ORBITS AND THE SHAPES OF ELLIPTICAL GALAXIES

Ortwin E. Gerhard
Max-Planck-Institut für Astrophysik
Karl-Schwarzschild-Str. 1
8046 Garching bei München
West Germany

Abstract Using Melnikov's method to study the appearance of stochastic orbits in perturbed Stäckel potentials, a correlation is found between the observed shapes of elliptical galaxies and the occurence of mainly regular orbits. Some other potential perturbations giving rise to large regions of stochastic orbits, on the other hand, appear to be inconsistent with observations.

Motivation: Most orbits in potentials near those of elliptical galaxies appear to be *regular* box and tube orbits that respect three integrals of motion (Schwarzschild 1979). A simple argument why stochastic orbits cannot occur in large numbers is that they would make the density contours rounder than the potential contours, while Poisson's equation requires the opposite in cases of interest for elliptical galaxies.

There is a family of completely integrable potentials (Stäckel potentials) in which all orbits are regular and belong to the four orbital families that also comprise most orbits in Schwarzschild's model (de Zeeuw 1985). While the proximity of these integrable potentials must clearly play a role for the dynamics of elliptical galaxies (i.e. for the apparent regularity of their phase-spaces, by the KAM-theorem), the fact that the overwhelming majority of potentials near those of elliptical galaxies are non-integrable implies that these galaxies are very unlikely to have an exactly separable potential.

Hence it is important to study perturbations of Stäckel potentials. Some questions of interest are: (i) Is the regular orbital structure of the Stäckel potentials stable to perturbations, i.e. is the region of stochastic orbits thess introduce small? (ii) For which types of perturbations - if not for all - is this true, and what can be learned from this for elliptical galaxies?

Method: In integrable potentials which support several orbit families, small perturbations generally destroy the non-classical integrals of motion because the surfaces on which the different families of orbits touch (the so-called homoclinic surfaces) become infinitely tightly wrapped by the perturbation. In this way, layers of stochastic orbits are introduced in the vicinity of these surfaces. This phenomenon may be shown to occur by Melnikov's method, in which one follows the wrapping and intersection of the perturbed homoclinic surfaces to first order in the perturbation. A new, canonically invariant formulation of Melnikov's method has been derived, which makes application to three-dimensional galaxy potentials

T. de Zeeuw (ed.), Structure and Dynamics of Elliptical Galaxies, 477–478.

possible. In this case, the homoclinic surface is, like any other torus, a three-dimensional phase-space surface, and generally contains a two-parameter family of orbits asymptotic to a one-parameter family of unstable quasi-periodic orbits. Then one has to evaluate a 2-vector of Melnikov integrals along, and as a function of, the unperturbed homoclinic orbits. With the aid of these integrals one may estimate the importance of the resulting stochastic layer, i.e. its width in phase-space. For details see Gerhard (1985, 1986).

Applications to Stäckel potentials: We consider perturbations of Stäckel potentials, which are separable in ellipsoidal coordinates λ, μ, ν. The commonly studied cases correspond to inhomogeneous mass models with homogeneous cores; they then contain the following unstable periodic orbits: z-axial and closed y-loop orbits (for energy $E > E_1$) and y-axial orbits ($E > E_2 > E_1$) (de Zeeuw 1985). Only the z-axial orbits are doubly unstable; for them the theory has to be slightly modified. The y-axial and closed y-loop orbits sire one-dimensional families of unstable quasi-periodic orbits on two-dimensional tori, and the Melnikov integrals must be evaluated as functions of the two-parameter families of orbits asymptotic to them. To this end the equations of motion must be solved and the integrals evaluated alongside. Since this is a difficult numerical problem the computations have sofar been restricted to two-dimensional potentials (describing motion in the equatorial plane of a triaxial galaxy). Then the only unstable periodic orbits are the y-axial orbits, and the Melnikov vector reduces to a single integral. The results of these calculations (Gerhard 1985) are summarized next.

Results: For a particular planar Stäckel potential with density $\rho \propto r^{-2}$ at large radii, several classes of perturbations were studied. The main result is that those perturbations that are consistent with observations of early-type galaxies approximately preserve the regular orbital structure of the integrable potential and lead to only small stochastic layers. Specifically, this was found for (i) $\cos m\phi$ perturbations with $m = 0, 2, 1, 4$ (axisymmetric, elliptic, lopsided, box-shaped) in order of the importance of the resulting stochastic layer, (ii) potential perturbations with moderate ellipticity gradients, and (iii) small figure rotation.

In contrast, other perturbations lead to large stochastic regions and a rapid breakdown of the regular structure of phase-space; e.g. $\cos m\phi$ perturbations with $m = 3$ or ≥ 5. These results suggest that (i) the integrable Stäckel potentials are sufficiently close to galaxy potentials in the studied energy range that the latter may be considered as perturbations of the former, and (ii) that the triaxial-symmetric shapes of ellipticals are determined by the requirement that self-consistent equilibrium models exist rather than by special initial conditions.

References

de Zeeuw, P.T, 1985. *Mon. Not. R. astr. Soc.* **216**, 273.
Gerhard, O.E., 1985. *Astron. Astrophys.* **151**, 279.
Gerhard, O.E., 1986. *Mon. Not. R. astr. Soc.* **222**, 287.
Schwarzschild, M., 1979. *Astrophys. J.* **232**, 236.

COMPLEX INSTABILITY AROUND THE ROTATION AXIS OF TRIAXIAL SYSTEMS

Louis Martinet, Daniel Pfenniger
Geneva Observatory
CH-1290 Sauverny
Switzerland

ABSTRACT. We examine the general instability at large amplitude of the radial periodic orbits along the rotation axis of bulges, spheroids and other rotating triaxial ellipsoidal systems.

1. INTRODUCTION

Complex instability of periodic orbits is a new phenomenon which may only appear in autonomous systems with more than 2 degrees of freedom. It is characterized by the fact that the four eigenvalues associated with the linearized transformation describing the motion close to the periodic orbit are complex and outside of the unit circle. In particular, it is shown that this phenomenon affects the z-axis orbits under the conditions mentioned below.

2. BIFURCATION ON THE Z-AXIS ORBIT FAMILY

Our systematic investigation of periodic orbits in various triaxial potentials led us to examine the stability of the z-axis orbit when the figure rotation and the shape change. In a first approach (Martinet & de Zeeuw, 1987), a quartic potential and a logarithmic potential were considered.

Here we use the 3D gravitational potential corresponding to the triaxial density law (Pfenniger & Udry, this Symposium) $\rho = \rho_c/(1+m^2)^{3/2}$, called $P_{3/2}$, as model of a E galaxy. We consider as well the potential of a previously used SB galaxy model (Pfenniger, 1984), which consists of a $n = 2$ Ferrer's bar imbedded in a Miyamoto disc.

Fig. 1 shows the different zones of stability (s), simple instability (u), double instability (du) and complex instability (cu) for the z-axis orbits in the model $P_{3/2}$ with the Schwarzschild (1979) axial ratios of $8:5:4$ in an energy–rotation (H, Ω_p) diagram. *For not too slowly rotating figures, there exists a critical orbital amplitude above which we have the bifurcation $s \to cu$. This transition prevents the stable anomalous orbits which circle the long axis from joining the z-axis orbits.* We found that these features are *general* for rotating gravitational triaxial systems: completely analogous results are obtained in the SB galaxy, quartic and logarithmic potentials.

T. de Zeeuw (ed.), Structure and Dynamics of Elliptical Galaxies, 479–480.

3. INFLUENCE OF A CONCENTRATED MASS IN THE CORE

A small Plummer sphere was added to the potentials considered above and its influence on our previous results was examined. We represent here the results obtained in the SB galaxy model. Completely similar results were obtained with the $P_{3/2}$ and the logarithmic potentials. The Plummer sphere mass M_p ranges between 0 and 0.0006, while $M_{bar} = 0.1$ and $M_{disc} = 0.9$ $(G = 1)$.

A (GM_p, H) diagram (Fig. 2) summarizes the destabilizing effect of the central mass. When $M_p > 6 \times 10^{-4}$ the z-orbit is practically fully *cu* from $z = 0$. *Therefore this perturbation plays a very important role in the orbital behaviour perpendicular to the equatorial plane.*

In order to illustrate the link between the instability of the z-axis orbits and the possible orbital diffusion of chaotic orbits, orbits starting close to the z-axis and along it and diffusing at larger distances were integrated in the SB galaxy model. The same Plummer sphere as above was put in the core, with an increasing mass. The results are that without perturbation the diffusion time in the spheroid is much longer than the Hubble time t_H. For small M_p $(M_p < 10^{-4})$, the stochastic orbits are still *confined* for a long time. However when $M_p > 10^{-3}$, the diffusion time becomes rapidly much shorter than t_H; then the central mass produces general strongly *cu* z-axis orbits.

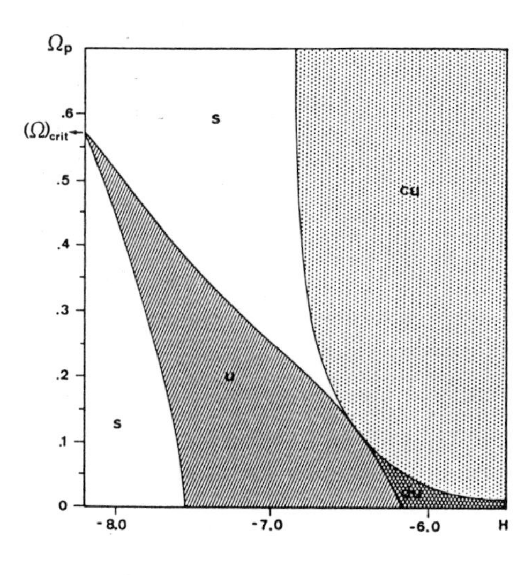

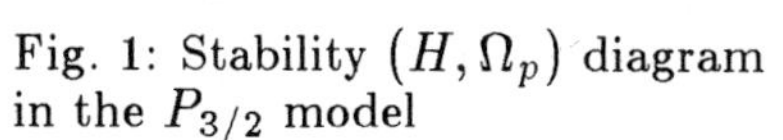
Fig. 1: Stability (H, Ω_p) diagram in the $P_{3/2}$ model

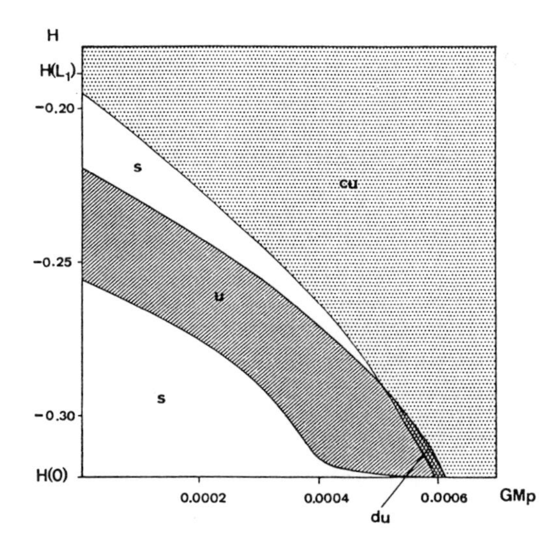

Fig. 2: Stability (M_p, H) diagram in the SB galaxy model

REFERENCES

Martinet, L., de Zeeuw, T.: 1987, in preparation
Pfenniger, D.: 1984, *Astron. Astrophys.* **134**, 373
Schwarzschild, M.: 1979, *Astrophys. J.* **232**, 236

STOCHASTICITY IN MODELS OF ELLIPTICAL GALAXIES

Daniel Pfenniger, Stéphane Udry
Geneva Observatory
CH-1290 Sauverny
Switzerland

ABSTRACT. We measure quantitatively the growth of stochasticity due to various perturbations to a model of elliptical galaxy. This is achieved by computing the Liapunov characteristic exponents of randomly selected orbits.

1. INTRODUCTION

The amount of stochasticity has important implications on the effective orbital relaxation time, which is much shorter for stochastic orbits (Pfenniger, 1986) than classically thought. We examine here the nature of stellar motions by computing orbits in a relatively realistic triaxial model. Stochasticity is measured by computing the Liapunov characteristic exponents (hereafter LCE) of many randomly selected orbits. The effects of changing the axis ratios, rotation speeds or adding various perturbations to the potential can then be estimated quantitatively. The principal differences of our investigation with an analogous work of Gerhard (1985) are first that the orbits here are arbitrary and move in 3D, and secondly that the potentials we have retained are not chosen to be nearly integrable, but to be as much as possible in accordance with the profile of elliptical galaxies.

2. MODELS AND PERTURBATIONS

The retained triaxial density law is $\rho(x, y, z) = \rho_c/(1 + \frac{x^2}{a^2} + \frac{y^2}{b^2} + \frac{z^2}{c^2})^n$, $n = \frac{3}{2}$, which is close to Schwarzschild's model (1979) but is exactly ellipsoidal. The potential Φ of this density stratification, called hereafter $P_{3/2}$, takes a simple form in the confocal ellipsoidal coordinates λ, μ, ν (de Zeeuw, 1984):

$$\Phi(x, y, z) = -2\pi G\rho_c abc \int_0^\infty \frac{du}{\sqrt{(u+\lambda)(u+\mu)(u+\nu)}}. \tag{1}$$

Contrarily to the perfect ellipsoid of de Zeeuw, for which $n = 2$, this potential is non-integrable. It is presently a good compromise between the actual density profile of elliptical galaxies and the need to compute orbits with a not too expensive scheme. This is achieved by using the algorithms R_F and R_D of Carlson (1979).

T. de Zeeuw (ed.), Structure and Dynamics of Elliptical Galaxies, 481–482.

We have considered the following models up to now:

$M_{1.0}$: The $P_{3/2}$ potential with the Schwarzschild axis ratios $(a/b/c) = (8/5/4)$ and a rotation corresponding to a corotation radius sets at 32.

$M_{1.1}$: As $M_{1.0}$ but with an additional central Plummer potential of radius $r_p = 0.01$, and a fractional mass of 2%.

$M_{1.2}$: As $M_{1.0}$ but with an additional $\Phi_3 = \epsilon \cos 3\theta$ azimuthal perturbation of amplitude $\epsilon = 0.0008$,

$M_{1.3}$: As $M_{1.0}$ but with an additional $\Phi_4 = \epsilon \cos 4\theta$ perturbation.

$M_{2.i}$: As $M_{1.i}, i = 0..3$ but with the axis ratios of $(8/4/2.5)$.

The LCE χ_i of a given orbit express the average rate of divergence of the other neighbouring orbits, over a relevant time. The χ_i also express the *sensitivity* to perturbations. In 3D potentials, there are 6 of them which occur in 3 pairs $(\chi_i, -\chi_i)$. To every isolating integral corresponds a pair of vanishing LCE, therefore regular orbits are especially robust to perturbations. The Kolmogorov entropy h, i.e. the rate at which the information about the initial conditions of an orbit is lost, is just equal to the sum of the positive exponents: $h = \sum_{\chi_i > 0} \chi_i$ (Pesin, 1977). This quantity is used below for measuring stochasticity. Its inverse is the typical time-scale for losing the information about the initial conditions.

3. COMPUTATION OF STOCHASTICITY

We have computed the LCE of 100 orbits per model over a time slightly larger than the Hubble time t_H, starting with the same set of random initial conditions. A global estimation of stochasticity is given by the average h of the 100 orbits, $<h>$. Another way to estimate the amount of stochasticity is to count the fraction of orbits with a h larger than some value. $\Gamma_{0.0015}$ is the fraction of orbits for which $h > 0.0015$. Since $t_{<h>} = <h>^{-1}$ is a time-scale, it can be compared to the Hubble time t_H. The computations are summarized in the following table.

Models	$M_{1.0}$	$M_{1.1}$	$M_{1.2}$	$M_{1.3}$	$M_{2.0}$	$M_{2.1}$	$M_{2.2}$	$M_{2.3}$
$<h> \times 10^3$	1.19	2.21	2.53	3.52	1.23	2.91	2.86	3.55
$\Gamma_{0.0015}$	0.06	0.70	0.80	1.00	0.07	0.76	0.90	1.00
$t_{<h>}/t_H$	0.28	0.15	0.13	0.09	0.27	0.11	0.11	0.09

For all the orbits considered, the perturbations systematically *increase* h with respect to the corresponding orbits in the unperturbed models $M_{1.0}$ or $M_{2.0}$. The average time-scale for the loss of information $t_{<h>}$ is always *shorter* than the Hubble time, especially for the perturbed models.

REFERENCES

Carlson, B.C.: 1979, *Numer. Math.* **33**, 1
Gerhard, O.E.: 1985, *Astron. Astrophys.* **151**, 279
Pesin, Ya.B.: 1977, *Russ. Math. Surveys* **32**, 55
Pfenniger, D.: 1986, *Astron. Astrophys.* **165**, 74
Schwarzschild, M.: 1979, *Astrophys. J.* **232**, 236
de Zeeuw, P.T.: 1984, Thesis, Leiden University

NATURAL ACTION–ANGLE VARIABLES

David N. Spergel
Institute for Advanced Study
Princeton, N.J., 08540 USA

Since galaxies are collisionless relaxed systems, actions are an extremely useful tool for understanding their dynamics. There are many potential applications of actions: (1) When orbits in an N-body simulation are characterized by their actions, the six dimensional distribution function, $f(\vec{x}, \vec{p})$, can be reduced to a more tractable three dimensional function, $f(J)$. (2) Actions are adiabatic invariants, and thus are useful for studying slowly evolving systems. Binney, May and Ostriker (1986) have applied this technique to study the response of the spheroid to the disc. (3) The spectral decomposition of an orbit can be used to help generate self–consistent galaxy models (Spergel 1987).

In an integrable potential, a star (or any collisionless particle) moves along a three dimensional torus imbedded in six dimensional phase space. The actions, J_k, are the projected areas of the torus, while the action angles, θ_k, are 2π periodic functions on the surface of the torus. The position and momentum of a particle on the torus can be expanded as a Fourier series:

$$q^j(\vec{\theta}) = \sum_{\vec{m}} C^j_{\vec{m}} \exp(i\vec{m} \cdot \vec{\theta}) = \sum_{\vec{m}} C^j_{\vec{m}} \exp[i(\vec{m} \cdot \vec{\omega})t] \tag{1a}$$

$$p^j(\vec{\theta}) = \sum_{\vec{m}} B^j_{\vec{m}} \exp(i\vec{m} \cdot \vec{\theta}) = \sum_{\vec{m}} B^j_{\vec{m}} \exp[i(\vec{m} \cdot \vec{\omega})t] \tag{1b}$$

where I have utilized the relation $\dot{\theta} = \partial H(J)/\partial J_k = \omega_k$. For further discussion, see Arnold (1977). The actions can now be expressed in terms of the Fourier coefficents of the orbit:

$$J_k = \int_{\gamma_k} \sum_j p_j dq_j = \sum_{\text{lines}} \sum_j m_k B^j_{\vec{m}} C^j_{\vec{m}'} \delta^{(3)}(\vec{m} - \vec{m}') \tag{2}$$

The coefficents $C_{\vec{k}}$ can be found by (1) numerically integrating an orbit, (2) representing its Fourier transform as a series of lines, and (3) then choosing the ω_k's (Binney and Spergel 1984).

While steps (1) and (2) are well defined, the problem remains of how to make the "right" choice of frequencies. Each set of frequency choices corresponds to a set of action–angle coordinates. If different sets of angle coordinates are used on

T. de Zeeuw (ed.), Structure and Dynamics of Elliptical Galaxies, 483–484.

each torus, then action–angle coordinates are no longer useful global descriptions of orbits. We wish to choose angle coordinates so that all but one of the actions of the closed orbits that spawn the major orbital families are zero. This is equivalent to requiring that the frequency choices made in step (3) of the program go smoothly over to the frequencies associated with each of the closed orbits.

In confocal ellipsoidal coordinates, the global structure of the tori is apparent. Two of the actions are always zero for any of the closed orbits and the four classes of orbits can be neatly classified in action space (de Zeeuw 1984). I have explored the implications of de Zeeuw (1984)'s suggestion that the use of confocal coordinates would simplify the calculations of actions. The results have been most gratifying.

A computer program has been developed which calculates an orbit in *any* potential, and then switches from cartesian to confocal ellipsoidal coordinates, where the Fourier spectra are remarkably simple. Most of the power in each of the spectra $[\lambda(\omega), \mu(\omega)$, and $\nu(\omega)]$ lies in one line! (A different line in each coordinate's spectrum.) The frequency of these lines are the fundamental frequencies.

The simple spectra have many computational benefits. In practice, the presence of many lines with similar frequencies requires long integration times so that the two lines can be resolved. Since the choice of coordinates *nearly* decouples the three oscillations, there is not much power in lines that are not harmonics of a single frequency.

Since the momentum in confocal coordinates is often singular near the planes $x = 0$, $y = 0$, and $z = 0$, a coordinate transformation is necessary to remove these singularities. A different change of coordinates is required for each of the orbital families. However, since it is straightforward to classify an orbit, avoiding these numerical singularities is not a problem. The new conjugate momenta and positions can again be expanded in a Fourier series and the actions can be determined.

This technique has been applied successfully in three dimensions to orbits in a Schwarzschild potential. In two dimensions, a primitive form of this technique has been successfully applied to orbits computed in a Bahcall–Soneira galaxy model (Aguilar and Spergel 1987).

REFERENCES

Aguilar, L. and Spergel, D.N., 1987, in preparation.

Arnold, V.I., 1978, *Mathematical Methods of Classical Mechanics*, Chapter 10, Springer, New York.

Binney, J.J., May, A. and Ostriker, J.P., 1986, Oxford preprint.

Binney, J.J. and Spergel, D.N., 1984, *M.N.R.A.S.*, **206**, 159.

de Zeeuw, T., 1984, *Dynamics of Triaxial Stellar Systems*, Ph. D. thesis, Leiden, Netherlands.

Spergel, D.N., 1987, in preparation.

INTEGRALS OF MOTION IN AN ELLIPTICAL GALAXY MODEL

Althea Wilkinson
Astronomy Department, University of Manchester, UK.

Tim de Zeeuw
The Institute for Advanced Study, Princeton, USA.

The structure of a galaxy model is described completely by its phase-space distribution function f. By Jeans' Theorem f can be written as a function of the integrals of motion admitted by the potential of the model. Various independent combinations of the integrals may be used as arguments of f; in many cases the action integrals are to be preferred. For a general N-body model, these can be obtained by numerical integration and subsequent spectral decomposition of each orbit (Binney and Spergel 1984).

The orbital structure in triaxial elliptical galaxies resembles closely that in the general Stäckel potentials, for which the Hamilton-Jacobi equation separates in ellipsoidal coordinates, and all orbits have three exact integrals of motion H, I_2 and I_3, say (de Zeeuw 1985). Thus, for N-body models of elliptical galaxies, approximate integrals of motion can be found by the fitting of Stäckel potentials.

FITTING

In confocal ellipsoidal coordinates (λ, μ, ν) a Stäckel potential V_S is of the form

$$V_S = -\frac{F(\lambda)}{(\lambda-\mu)(\lambda-\nu)} - \frac{F(\mu)}{(\mu-\nu)(\mu-\lambda)} - \frac{F(\nu)}{(\nu-\lambda)(\nu-\mu)},$$

where $F(\tau)$ is an arbitrary function ($\tau = \lambda, \mu, \nu$). Given a potential $V(x^2, y^2, z^2)$, we want to find the Stäckel potential V_S that approximates it most closely. Thus, we have to specify an ellipsoidal coordinate system and the function $F(\tau)$. This can be done by means of the following procedure, described in detail by de Zeeuw and Lynden-Bell (1985):

- Pick coordinates (λ, μ, ν), by specification of two pairs of their foci.
- Transform $V(x^2, y^2, z^2)$ to $V(\lambda, \mu, \nu)$ on a rectangular grid in (λ, μ, ν)-space.
- Construct an auxiliary function $\chi(\lambda, \mu, \nu) = -(\lambda-\mu)(\mu-\nu)(\nu-\lambda)V(\lambda, \mu, \nu)$
- Determine $F(\lambda)$, $F(\mu)$ and $F(\nu)$ by calculation of various weighted averages over $\chi(\lambda, \mu, \nu)$. The weighting functions may be chosen to ensure convergence; most weight can be put where most of the density is.
- Repeat this process for a different choice of the positions of the two pairs of foci, until the best fit is obtained.

T. de Zeeuw (ed.), Structure and Dynamics of Elliptical Galaxies, 485–486.

INTEGRALS

The integrals H, I_2 and I_3 in a Stäckel potential are known explicitly. Their values determine the family (boxes, short–axis tubes, inner long–axis tubes or outer long–axis tubes) to which an orbit belongs. We can obtain approximate integrals for all the particles in the N–body model by substitution of the instantaneous position and velocity coordinates in the expressions for H, I_2 and I_3 in the best–fitting Stäckel potential. Similarly, the action integrals J_λ, J_μ and J_ν can be calculated. This requires a simple quadrature. It follows that all orbits can be classified, and the approximate distribution function $f(H, I_2, I_3)$ or $f(J_\lambda, J_\mu, J_\nu)$ can be calculated by binning.

APPLICATION

We have applied the above procedure to the stationary triaxial galaxy simulated by Wilkinson and James (1982). We have fitted the 20000–body model at two timesteps, the first halfway through its evolution, and the second at the end of the run, i.e., at times equivalent to 0.5 and 0.9 of a Hubble time t_H. In both cases the potential can be fitted accurately with a Stäckel potential out to large radii. Based on $\sim$ 2500 orbits we obtain the orbit classification given in the Table. The uncertainty in the fractions are of the order of at most a few percent. For comparison, the estimates based on the visual classification of Wilkinson and James (1982) for the final timestep are given also.

We conclude that the orbital structure can be found at individual timesteps, and with good accuracy. The distribution function of the model is smooth in action space, and does not change very much during the second half of the run. For N–body models of elliptical galaxies this way of classifying orbits is much faster than spectral stellar dynamics.

Table: Orbit Classification

Orbit Family	$0.5t_H$	$0.9t_H$	By Eye
Boxes	83.8%	91.5 %	75.1 %
Short Axis Tubes	10.8	4.8	7.5
Inner Long Axis Tubes	2.8	1.3	
Outer Long Axis Tubes	1.2	0.4	6.5
Unbound	1.4	2.0	4.3
Unclassified			6.6

REFERENCES

Binney, J.J. & Spergel, D., 1984. *Mon. Not. R. astr. Soc.*, **206**, 159.
de Zeeuw, P.T., 1985. *Mon. Not. R. astr. Soc.*, **216**, 273.
de Zeeuw, P.T. & Lynden–Bell, D., 1985. *Mon. Not. R. astr. Soc.*, **215**, 713.
Wilkinson, A. & James, R.A., 1982. *Mon. Not. R. astr. Soc.*, **199**, 171.

SELF-CONSISTENT MODELS OF PERFECT TRIAXIAL GALAXIES

Thomas S. Statler
Princeton University Observatory
Peyton Hall
Princeton, NJ 08544 USA

We have used Schwarzschild's (1979) method to study the variety of self-consistent solutions available to the family of triaxial "perfect ellipsoids" (de Zeeuw 1985). The time-averaged density at 240 points within the mass model is computed for each of 1065 orbits distributed regularly in phase space and covering all four major orbit families. The underdetermined linear system thereby defined is solved in two ways. First, Lucy's (1974) iterative scheme is used to find a "smooth" solution, lying in the interior of the (mathematically-allowed) solution space. Second, linear programming is used, with linear combinations of the x and z components of the total angular momentum as cost functions, to delineate the boundary of the projection of the solution space in the $L_x - L_z$ plane. The above procedure is applied to 21 figures, with axis ratios, b/a and c/a, chosen at equal intervals of 1/8.

Numerical solutions are found for all axis ratios investigated. In general, the solution spaces are approximately rectangular in the $L_x - L_z$ plane, with $L_{x,min}$ and $L_{z,min}$ nearly zero. The square corners of the solution spaces arise because the short-axis and long-axis tube orbits can each be exchanged for box orbits and the bounding marginal orbits without affecting the other. The simple cases of the sphere, the axisymmetric disk, and the needle can be treated analytically. The solution spaces of the triaxial figures tend to the appropriate limits and the boundaries move monotonically with the axis ratios between the limiting cases.

In the oblate and prolate limits, the projected solution spaces must collapse to zero extent in the L_x and L_z directions, respectively. Our most nearly oblate models show appropriately thin solution spaces, but those closest to the prolate limit do not. The implication is that galaxies that are, in shape, nearly prolate can still support a significant amount of internal streaming around the short axis.

The distribution functions in action space for representative solutions show the following characteristic features:

Minimum-L_x, minimum-L_z solutions have a fairly smooth distribution in the box orbits, which, at low binding energies, cuts off abruptly at the margins dividing the boxes from the tube orbits. At higher binding energies, some tube orbits, particularly the short axis tubes close to the margin are populated at high phase-space densities. Other tube orbits are completely unpopulated.

Minimum-L_x, maximum-L_z solutions have a similar distribution through the box orbits, going to zero more smoothly at the margin between the boxes and the short-axis tubes, and also a disconnected thin region of much higher phase

T. de Zeeuw (ed.), Structure and Dynamics of Elliptical Galaxies, 487–488.

density in the thin-walled short-axis tubes.

Maximum-L_x, minimum-L_z solutions have, again, the same sort of behavior in the box orbits, with the suggestion of a smoother transition to zero at the margin between the boxes and the inner long-axis tubes, and high density regions in the thin-walled outer and inner long-axis tubes.

Solutions by Lucy's method are reasonably smooth over all of phase space, with, at any surface of constant energy, the highest density toward the middle of the box orbits.

Regarding physical plausibility, the Lucy's method solutions tend to have no one orbit or orbit family preferred, and so would not require any special mechanisms. More cannot be said about the plausibility of these solutions without a detailed theory of formation. The minimum-L_x, L_z solutions would require some formation mechanism that would preferentially populate box orbits. This task would seem difficult to accomplish if the bulk of the mass of the galaxy is subject to violent relaxation. The maximum-L solutions require a mechanism to populate the thin-walled tube orbits. Gaseous dissipation will not suffice, since it would populate the elliptic closed orbits, not the orbits with finite extent out of their "equatorial" planes required in the solutions. In principle, these orbits could be populated by the ingestion of a smaller galaxy which is slowed by dynamical friction and tidally stripped as it spirals inward. However, it is not clear whether dynamical friction would damp the "radial" excursions of the victim's orbit faster than the "vertical" ones in a triaxial potential. Furthermore, there are suggestions (Casertano *et al.*, this volume) that a strongly radial velocity ellipsoid in the cannibal galaxy will cause the radial excursions to *increase* rather than decrease.

"Observed" values of v/σ are calculated, under the assumption of maximal streaming around the short axis, for lines of sight along the long and intermediate axes, as upper limits to values observed from other directions. We find that the smooth Lucy's method solutions can account for the positions of most ellipticals in the v/σ *vs.* ϵ diagram, though in some cases maximal streaming and the most favorable viewing geometry are required. A few galaxies, such as NGC 4742, require maximum-L_z models and fortuitous lines of sight, and may be candidates for true oblateness.

Finally, we note that our solutions in which tube orbits do not contribute to the global *pressure* (either because they are unpopulated or because the circulation is all in the same sense and the motion very ordered) show contours of projected velocity dispersion that tend to be elongated contrary to the elongation of the mass distribution. The effect can also be seen in Schwarzschild's Model B (Merritt 1980) and Wilkinson and James' (1982) N-body Model B, both of which are dominated by box orbits. Though it is only a 20% effect at best, it would nevertheless be interesting to look at the dispersions on both axes of some very flattened, slowly-rotating galaxies, such as NGC 4839 and NGC 6909, to determine if the counter-elongation can be useful as an indicator of the shape of the velocity distribution.

REFERENCES

de Zeeuw, T., 1985. *M.N.R.A.S.* **216**, 273.
Lucy, L. B., 1974. *A.J.* **79**, 745.
Merritt, D., 1980. *Ap. J. Supp.* **43**, 435.
Schwarzschild, M., 1979. *Ap. J.* **232**, 236.
Wilkinson, A. and James, R. A., 1982. *M.N.R.A.S* **199**, 171

SELF-CONSISTENT OBLATE-SPHEROID MODELS

J.L. Bishop
Canadian Institute for Theoretical Astrophysics
University of Toronto
Toronto, Ontario M5S 1A1
Canada

ABSTRACT. A method for the construction of models of axisymmetric galaxies is presented. In this formulation we determine the distribution function corresponding to a given gravitational potential and the associated mass density distribution. Although the realization of the model is numerical, the underlying theory is analytic and exact. This method allows us to construct a wide range of models without having to use linear programming and a large amount of computer time. Here we present the results from the application of this method to the "perfect" oblate-spheroid mass model. A large class of valid self-consistent distribution functions which depend on three isolating integrals of the motion is found. The kinematics of many models are consistent with those observed for elliptical galaxies. In particular, models generated by this formulation are in agreement with the observed values of the ratio of the maximum projected rotational velocity to the velocity dispersion along the line of sight versus ellipticity.

FORMULATION OF THE METHOD: We consider a model with a given mass density and gravitational potential which satisfies Poisson's equation and we seek solutions for the distribution function (d.f.) which satisfies Liouville's equation (Bishop 1986a). Self-consistency requires that the matter described by the d.f. be the source of the potential. We choose potentials which admit of three isolating integrals of the motion (I_1, I_2, I_3) and according to Jeans' theorem, the d.f. will depend on them. We make the ansatz: $f(I_1, I_2, I_3) = g(I_1, I_2, I_3)h(I_1, I_2)$. If $g(I_1, I_2, I_3)$ is an assigned function we seek solutions for $h(I_1, I_2)$, the derived function. Thus if B is equal to

$$B(I_1, I_2, \mathbf{x}) = \int g(I_1, I_2, I_3)J(I_1, I_2, I_3, \mathbf{x})dI_3,$$

where $J(I_1, I_2, I_3, \mathbf{x})$ is the Jacobian determinant of the transformation, then the fundamental integral equation is

$$\rho(\mathbf{x}) = \int B(I_1, I_2, \mathbf{x})h(I_1, I_2)dI_1dI_2. \tag{1}$$

Equation (1) relates a function of two variables in configuration space to a function of two variables in integral space. Since $\rho(\mathbf{x})$ and $B(I_1, I_2, \mathbf{x})$ are known the integral

T. de Zeeuw (ed.), Structure and Dynamics of Elliptical Galaxies, 489–490.

equation can be solved for $h(I_1, I_2)$ and the d.f., and its solution, is unique. In order to solve equation (1) numerically, we approximate the continuous stellar system as a discrete system. We divide the configuration and the integral space into a set of finite cells. We choose a set of integrals such that the division of the integral space and the configuration space is accomplished with the same cell structure. All continuous equations are made into discrete equations and equation (1) becomes a set of linear equations which is composed of a lower triangular matrix.

MASS MODEL AND RESULTS: We demonstrate this method with the "perfect" oblate spheroid mass model (de Zeeuw 1985). The corresponding potential admits of three isolating integrals and the motion separates in prolate spheroidal coordinates (λ, μ, ϕ). All orbits are short axis tubes and the motion consists of two librations and one rotation. The double turning points of the orbit are isolating integrals, so $(\lambda^+, \mu^+, \lambda^-) \longleftrightarrow (E, L_z, I_3)$. A natural choice of cell shape has boundaries which are spheroids of constant λ and hyperboloids of constant μ. In each cell there is one double turning point (λ^+, μ^+).

We used $\sim$ 400 cells and $\sim$ 4000 orbits and tested $\sim$ 100 assigned functions. We found a broad range of self-consistent models ($f \geq 0$) (Bishop 1986a,b). We found that rotation curves first rise linearly and then drop off slowly. The maximum values occur near the half-mass radius. The observed values of (V_m/σ_c) versus the ellipticity (ϵ) fall within the allowed region defined by the models (Fig. 1). The lower boundary has approximately equal numbers of stars rotating in each direction. The upper boundary has a range of (N_+/N_-) from 1.9 to 1.002.

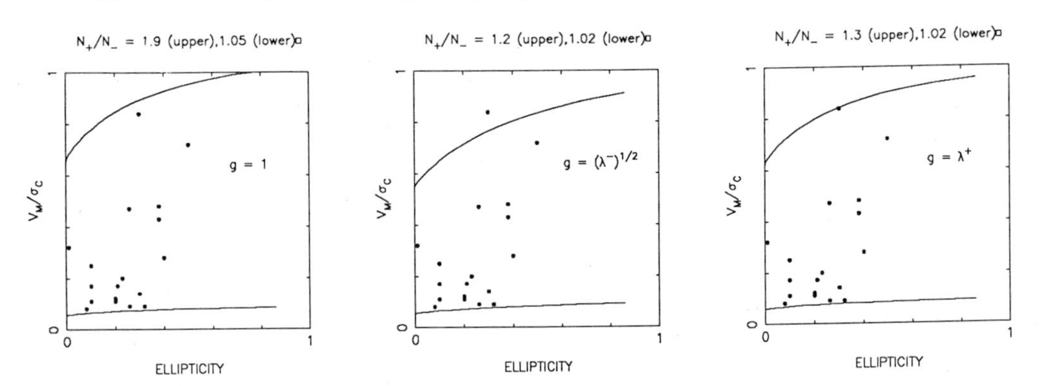

Figure 1. Plot of the ratio of the maximum projected rotational velocity, V_M, to the central line of sight velocity dispersion projected on the sky, σ_c, versus the ellipticity, for the given models. The two lines for each model represent two different rotational states of the same model. N_+/N_- is the ratio of the number of stars moving counter-clockwise to clockwise. Plotted points are values of observed galaxies from Illingworth (1977. *Ap.J.Letters*, **218**, L43.), Schechter and Gunn (1979. *Ap.J*, 229,472.), Davies (1981. *M.N.R.A.S.*, 194, 879.).

REFERENCES

Bishop, J.L. 1986a. *Ap. J.*, **305**, 14.
Bishop, J.L. 1986b. *Kinematics of Perfect Oblate Spheroid Models*, in preparation.
de Zeeuw, P.T. 1985.*M.N.R.A.S.*, **216**, 273.

SELF–CONSISTENT ELLIPTICAL DISKS

Peter Teuben
Kapteyn Astronomical Institute, Groningen, The Netherlands
& The Institute for Advanced Study, Princeton, USA

ABSTRACT. Self-consistent solutions for the perfect elliptic disk have been obtained. Velocity field, dispersion maps and the distribution function in action space have been derived.

Self consistent solutions for perfect elliptic disks (de Zeeuw 1985) have been constructed by means of Schwarzschild's (1979) method. A grid was chosen in elliptic coordinates (λ, μ) with $N_\lambda = 32$ cells in λ and $N_\mu = 8$ cells in μ, so that a complete set of orbits consists of $N_\mu * N_\lambda = 256$ box orbits and $N_\lambda * (N_\lambda + 1)/2 = 528$ tube orbits.

By optimizing the streaming in the disk, two distinct solutions for each model were calculated: a solution with minimum streaming, which *does* however need some tube orbits, and a solution with maximum streaming. This means that such models have non–unique self–consistent solutions.

In the maximum streaming solution, the density along the minor axis outside the focal point is contributed by the thinnest tube orbits only (de Zeeuw, Hunter & Schwarzschild 1987). This makes it possible to calculate the *direct* numerical solution with N_λ tube orbits and $N_\lambda * (N_\mu - 1)$ box orbits (the complete set of box orbits minus the marginal orbits) without linear programming, by solving N_λ simple equations, followed by solving a $N_\lambda * (N_\mu - 1)$ squared triangular matrix equation (cf. Schwarzschild 1986). It has been verified that in the limit $b/a \to 1$, this solution becomes the well known Kuzmin disk with thin tubes only.

Distribution functions in action space have been derived. The maximum streaming solutions (Figure 1) have smooth distribution functions in action space, apart from an allowed discontinuity at the closed loop orbits. In a real system this might plausibly arise through dissipational formation. The minimum angular momentum solutions are all discontinuous over the marginal (unstable) orbits, which indicates that these solutions are probably unphysical.

Associating stars with all orbits and gas with the closed elliptic orbits, mean streaming and velocity dispersion fields for the stars and gas have been derived. Velocity fields for the stars and gas are projected onto the sky and are shown as such hypothetical systems would be observed. There is a significant difference in the kinematics of the stars and gas. This may have some relevance to hot ovally distorted disks with little or no figure pattern speed or to gas in a principal plane of a triaxial elliptical galaxy.

T. de Zeeuw (ed.), Structure and Dynamics of Elliptical Galaxies, 491–492.

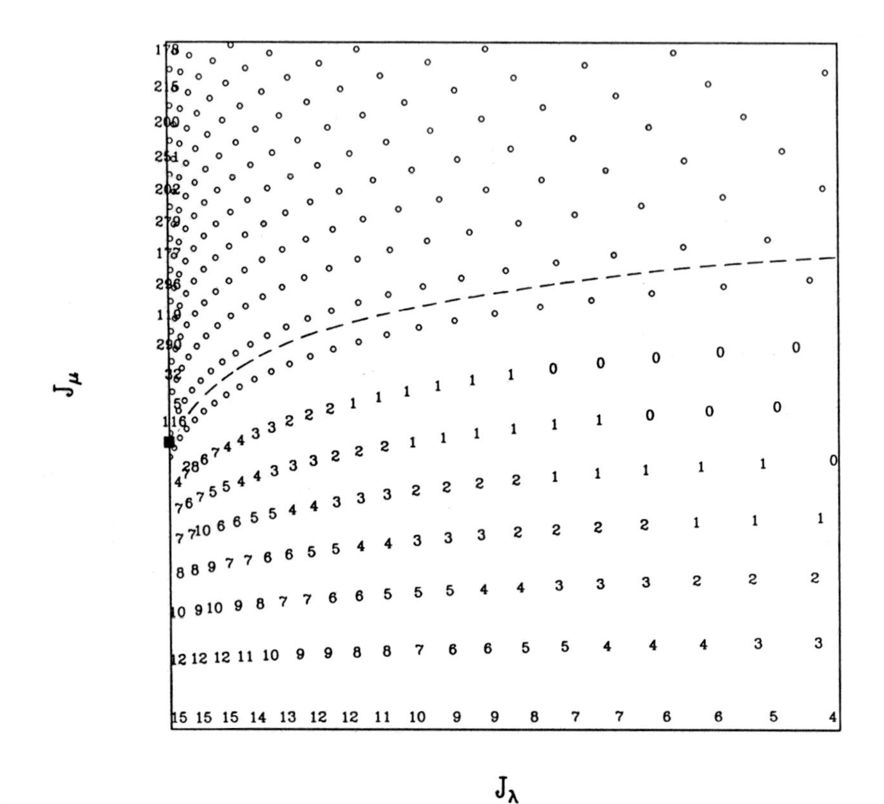

Figure 1. Distribution function in action space for the maximum angular momentum solution for an elliptic disk with $b/a = 0.7$. The dashed line represents the marginal orbits, separating the loop orbits (above) from the box orbits (below), the filled square represents the focal short axial orbit. The open circles are orbits with zero weight from the linear program, the numbers represent the actual value of the distribution function as found by the linear program.

REFERENCES

de Zeeuw, P.T., 1985. *M.N.R.A.S.*, **216**, 273.
de Zeeuw, P.T., Hunter, C., and Schwarzschild, M. 1987. *Astrophys. J.*, in press.
Schwarzschild, M. 1979. *Astroph. J.*, **232**, 236.
Schwarzschild, M. 1986, *Astroph. J.*, **311**, 511.

STELLAR DYNAMICS OF NEEDLES

Scott Tremaine
Canadian Institute for Theoretical Astrophysics
University of Toronto, Toronto M5S 1A1, Canada

Tim de Zeeuw
The Institute for Advanced Study, Princeton, USA

One dimensional "needles" are a limiting case of general triaxial stellar systems. Self-consistent, finite needles can have arbitrary longitudinal density distributions but have a fixed, universal distribution function. All needles are stable to all longitudinal perturbations but neutral to transverse perturbations.

I. POTENTIAL THEORY

Consider an axisymmetric stellar system with density $\rho(R, z)$, where (R, ϕ, z) are the usual cylindrical coordinates. We may always write the density in the form

$$\rho(R, z) = \frac{\lambda(z)}{R_0^2(z)} F\left(\frac{R}{R_0(z)}, z\right), \tag{1}$$

where $\lambda(z) = 2\pi \int R\rho(R, z) dR$ is the linear density, $R_0(z)$ is the boundary of the system, and the normalization condition is $2\pi \int F(s, z) s\, ds = 1$. We assume that the system is very prolate, so that the characteristic length of the needle is $Z \gg R_0$. Thus, there is a region in radius, $R_0 < R \ll Z$, where the z-dependence of the potential $\Phi(R, z)$ resembles that of a wire of linear density $\lambda(z)$:

$$\Phi(R, z) = 2G\lambda(z) \ln R + \Phi_0(z). \tag{2}$$

Now shrink the system to a needle by setting $R_0(z) = \epsilon h(z)/Z$, where $\epsilon \to 0$. The difference between the potential on the axis and at the boundary of the needle is independent of ϵ, and hence with a fractional error of order C^{-1}, with $C \equiv -\ln \epsilon$, we may write the potential anywhere inside the needle as

$$\Phi(z) = -2GC\lambda(z). \tag{3}$$

Note that this approximation fails in a small region near the tip where $\lambda = \mathcal{O}(C^{-1})$.

T. de Zeeuw (ed.), Structure and Dynamics of Elliptical Galaxies, 493–494.

II. EQUILIBRIUM

The distribution of stars in a needle-shaped stellar system is described by the distribution function $f(z, v)$, where v is the velocity in the z-direction and $f(z, v)\, dz\, dv$ is the mass in the phase space interval $(z, v) \to (z + dz, v + dv)$. According to Jeans' theorem, the distribution function must have the form $f(z, v) = f(E)$ where $E = \frac{1}{2}v^2 + \Phi(z)$ is the energy. A self-consistent needle satisfies

$$\lambda(z) = \int_{-\infty}^{\infty} f(E) dv. \tag{4}$$

This integral equation for $f(E)$, together with equation (3) relating the density and potential, can be solved to yield

$$f(E) = \begin{cases} \frac{1}{\pi G C}\left(-\frac{1}{2}E\right)^{1/2}, & E < 0, \\ 0, & E > 0. \end{cases} \tag{5}$$

Thus the distribution function for stellar needles is *unique*, *i.e.*, all stellar needles have the same distribution function, independent of the density distribution $\lambda(z)$.

III. STABILITY

Antonov's (1961) stability criterion for spherical stellar systems with $f = f(E)$ has a straightforward analog for needles: the needle is stable to disturbances along the z-axis if the functional

$$I[\eta] = \int \frac{[\eta(z, v)]^2}{|df/dE|}\, dz\, dv + \int \eta(z, v)\Phi_1(z)\, dz\, dv \geq 0 \tag{6}$$

for all η, where $df/dE < 0$ and $\Phi_1(z)$ is the potential arising from a density $\lambda_1(z) = \int \eta(z, v')\, dv'$. For the needle we find $I[\eta] \geq 0$, which proves stability to longitudinal disturbances. Of course, the needle is neutrally stable to disturbances which alter $\lambda(z)$ but leave the distribution function unaffected, since these simply transform the needle into a different equilibrium state.

The needle is neutrally stable to bending displacements. This result shows that the stability of very prolate systems to bending disturbances cannot be determined in the needle approximation.

REFERENCE

Antonov, V.A., 1961. *Astr. Zh.*, **37**, 918 (Translated in *Sov. Astron.*, **4**, 859).

FORMAL INVERSION OF THE SELF–CONSISTENT PROBLEM FOR TRIAXIAL GALAXIES

Herwig Dejonghe
The Institute for Advanced Study
Princeton, NJ 08540

Triaxial separable potentials $V(\lambda,\mu,\nu) = f_\lambda + f_\mu + f_\nu \geq 0$, with

$$f_\lambda = \frac{f(\lambda)}{(\lambda-\mu)(\lambda-\nu)}, \quad f_\mu = \frac{f(\mu)}{(\mu-\nu)(\mu-\lambda)}, \quad f_\nu = \frac{f(\nu)}{(\nu-\lambda)(\nu-\mu)}, \tag{1}$$

admit three constants of the motion (see de Zeeuw 1985 for more details), for which we can take the following set:

$$E = \sum_{cyc}(f_\lambda - \tfrac{1}{2}v_\lambda^2), \quad J = \sum_{cyc}(\mu+\nu)(f_\lambda - \tfrac{1}{2}v_\lambda^2), \quad K = \sum_{cyc}\mu\nu(f_\lambda - \tfrac{1}{2}v_\lambda^2). \tag{2}$$

Here v_λ, v_μ and v_ν are the velocity components in the (λ,μ,ν) coordinate system, and we recognize in E the binding energy per unit mass.

We define the mass density $\rho(\lambda,\mu,\nu)$ as follows

$$\rho(\lambda,\mu,\nu) = \iiint F(E,J,K)\, dv_\lambda\, dv_\mu\, dv_\nu. \tag{3}$$

The inversion problem has as its goal to determine $F(E,J,K)$ out of $\rho(\lambda,\mu,\nu)$. We have developed a formalism that establishes explicit relations in both directions between $F(E,J,K)$ and a function $\tilde{\rho}$ of six variables. It is always possible to recover the physical $\rho(\lambda,\mu,\nu)$ out of this function, but many different $\tilde{\rho}$'s give the same $\rho(\lambda,\mu,\nu)$.

Assume that the distribution function $F(E,J,K)$ can be written as a combined Laplace–Fourier transform

$$F(E,J,K) = -\frac{1}{2\pi i}\int_{a_0-i\infty}^{a_0+i\infty} e^{aE}\, da \frac{1}{2\pi}\iint_{-i\infty}^{+i\infty} e^{bJ+cK}\, db\, dc\, \mathcal{F}(a,b,c), \tag{4}$$

with $\Re(a_0) > 0$. Performing the integration (3), we find ρ as an integration over a, b and c of $d\rho$, given by

$$d\rho = (2\pi)^{3/2}\mathcal{F}(a,b,c)\left(\prod_{cyc}\frac{1}{\sqrt{a+b(\mu+\nu)+c\mu\nu}}\right) e^{aV + bW + cU}\, da\, db\, dc, \tag{5a}$$

T. de Zeeuw (ed.), Structure and Dynamics of Elliptical Galaxies, 495–496.

with

$$V = \sum_{cyc} f_\lambda, \quad W = \sum_{cyc} (\mu + \nu) f_\lambda, \quad U = \sum_{cyc} \mu\nu f_\lambda. \tag{5b}$$

The kernel in equation (5a) can be seen to permit a simpler form of (5):

$$2^{3/2} \left(\prod_{cyc} \int_{-\infty}^{f_\lambda} \frac{df'_\lambda}{\sqrt{f_\lambda - f'_\lambda}} \right) F(V, W, U) = \tilde{\rho}(V, W, U, \lambda, \mu, \nu), \tag{6}$$

which is a product of three separated Abel integral equations. Hence, we immediately get the inversion of (6):

$$F(V, W, U) = \frac{1}{2^{3/2}\pi^3} \left(\prod_{cyc} \int_{-\infty}^{f_\lambda} \frac{df'_\lambda}{\sqrt{f_\lambda - f'_\lambda}} \frac{\partial}{\partial f'_\lambda} \right) \tilde{\rho}(f'_\lambda, f'_\mu, f'_\nu, \lambda, \mu, \nu). \tag{7}$$

We remark that the mass density is written as an abstract function of 6 variables instead of only the three coordinates. We call it the *augmented mass density* $\tilde{\rho}$. Obviously, when we substitute the values of f_λ, f_μ and f_ν, as given by (1), into $\tilde{\rho}$, we find the usual $\rho(\lambda, \mu, \nu)$. We have assumed that $\rho(f_\lambda, f_\mu, f_\nu, \lambda, \mu, \nu)$ is zero at the boundary to sufficient order. The lower bound of the integrals in (5) and (6) are here formally $-\infty$, but must be assumed to coincide with the boundary of the model in 6–dimensional space.

Not all forms $\rho(f_\lambda, f_\mu, f_\nu, \lambda, \mu, \nu)$ lead, after performing (7), to a distribution function that is a function of three variables V, W and U only, and not of the six f_λ, f_μ, f_ν, λ, μ, ν. It is found that $\rho(f_\lambda, f_\mu, f_\nu, \lambda, \mu, \nu)$ should obey three cyclic partial differential equations of the form

$$\begin{aligned} \frac{\partial^3 \tilde{\rho}}{\partial\nu \partial f_\lambda \partial f_\mu} &+ \frac{1}{\lambda - \nu} \sqrt{f_\lambda} \frac{\partial}{\partial f_\lambda} [\sqrt{f_\lambda} \frac{\partial}{\partial f_\mu} (\frac{\partial \tilde{\rho}}{\partial f_\lambda} - \frac{\partial \tilde{\rho}}{\partial f_\nu})] \\ &+ \frac{1}{\mu - \nu} \sqrt{f_\mu} \frac{\partial}{\partial f_\mu} [\sqrt{f_\mu} \frac{\partial}{\partial f_\lambda} (\frac{\partial \tilde{\rho}}{\partial f_\mu} - \frac{\partial \tilde{\rho}}{\partial f_\nu})] = 0 \quad \text{(cyc.)}. \end{aligned} \tag{8}$$

The construction of explicit dynamical models for triaxial systems hinges critically on our ability to find a solution $\tilde{\rho}$ for (8) that reduces to a given ρ. However, special solutions of (8) can be found for which the inversion (7) takes a particularly simple Eddington form. They are generalizations of the Osipkov–Merritt models. Moreover, (8) is a natural starting point—within this analysis—for a study of the non–uniqueness of the inversion problem.

The augmented mass density carries all the dynamical information, since the distribution function can be calculated from it. Hence, expressions can be given for all the moments as functionals of the augmented mass density. E.g., for the velocity dispersions we find the cyclic relations

$$\tilde{\rho}\tilde{\sigma}_\lambda^2 = \int^{f_\lambda} \tilde{\rho}\, df'_\lambda, \qquad \tilde{\rho} = \frac{\partial}{\partial f_\lambda} (\tilde{\rho}\tilde{\sigma}_\lambda^2). \tag{9}$$

REFERENCE

de Zeeuw, P.T., 1985. *Mon. Not. R. Astr. Soc.*, **216**, 273.

ANALYTIC AXISYMMETRIC DYNAMICAL MODELS WITH THREE INTEGRALS OF MOTION

Herwig Dejonghe & Tim de Zeeuw
The Institute for Advanced Study, Princeton, NJ 08540

We consider a family of inhomogeneous axisymmetric mass models with a simple gravitational potential originally introduced by Kuzmin (1956):

$$V_S = -\frac{GM}{[R^2 + z^2 + a^2 + c^2 + 2\sqrt{a^2c^2 + c^2R^2 + a^2z^2}]^{1/2}}, \tag{1}$$

where $R^2 = x^2 + y^2$, a and c are parameters, G is the gravitational constant, and M is the total mass of the corresponding density distribution.

The potential (1) and the density ρ are most elegantly expressed in terms of prolate spheroidal coordinates (λ, ϕ, ν) (*cf.* de Zeeuw 1985):

$$V_S = -\frac{GM}{\sqrt{\lambda} + \sqrt{\nu}}, \tag{2}$$

and

$$\rho = \frac{Mc^2}{4\pi} \frac{\lambda\nu + a^2(\lambda + 3\sqrt{\lambda\nu} + \nu)}{(\lambda\nu)^{3/2}(\sqrt{\lambda} + \sqrt{\nu})^3}. \tag{3}$$

The surfaces of constant density are smooth and not too far from ellipsoidal for moderate values of the central axis ratio. The axis ratio increases slightly with increasing radius. The spherical limit is Hénon's (1960) isochrone.

As already shown by Kuzmin and Kutuzov (1962), the density distribution $\rho(R, z)$ can be written explicitly as $\rho(R, V_S)$. This fact makes it possible to use a standard inversion technique (Lynden-Bell 1962; Hunter 1975; Kalnajs 1976; Dejonghe 1986) in order to obtain in closed form—as a series of generalized hypergeometric functions—the unique distribution function $f(E, {L_z}^2)$ that depends only on the two classical integrals of motion (the energy E and the component L_z of the angular momentum parallel to the symmetry axis) and that is consistent with the density. This f is nonnegative for all oblate models in the sequence. Numerical tests indicate that $f \geq 0$ for prolate models only in case the central axis ratio is larger than about 0.7. The velocity dispersions can be written in terms of elementary functions.

The potential of all these models is of Stäckel form, so that all orbits in them enjoy an exact third integral of motion I_3, which can be regarded as a generalization

T. de Zeeuw (ed.), Structure and Dynamics of Elliptical Galaxies, 497–498.

of the total angular momentum integral of the spherical limit (de Zeeuw & Lynden–Bell 1985). We have developed a new method for the analytical construction of distribution functions $f(E, L_z{}^2, I_3)$, and apply it to these mass models.

We write f as the sum of two parts: $f(E, L_z^2, I_3) = f_1(E, L_z^2, I_3) + f_2(E, L_z^2)$, where f_1 is of the form

$$f_1(E, L_z^2, I_3) = \sum_{\ell,m,n} c_{\ell mn} E^\ell L_z^{2m} (p + qE + rL_z^2 + sI_3)^n, \tag{4}$$

where $c_{\ell mn}$, p, q, r and s are parameters, and ℓ, m and n are integers. This f_1 can be integrated fairly easily over velocity space, and results in a density ρ_1. The remaining density $\rho_2 = \rho - \rho_1$ is then reproduced by f_2, which is obtained by the standard inversion technique. This produces—for the first time—exact analytic distribution functions for realistic axisymmetric models that depend on all three integrals of motion. The velocity dispersions in these models can again be given explicitly.

Not all values of the parameters in f_1 produce positive distribution functions. However, fine–tuning is possible, and leads to an interesting variety of anisotropic models with positive distribution functions. In particular, we have constructed models with $m = 0$, $p = 0$, $q = 0$ and $r = s/2$. In the spherical limit our distribution functions reduce to the form $f = f_1(E, L^2) + f_2(E)$, where L is the total angular momentum.

REFERENCES

Dejonghe, H., 1986. *Physics Reports*, **133**, 217.
de Zeeuw, P.T., 1985. *Mon. Not. R. astr. Soc.*, **216**, 273.
de Zeeuw, P.T. & Lynden–Bell, D., 1985. *Mon. Not. R. astr. Soc.*, **215**, 713.
Hénon, M., 1960. *Ann. d' Astrophys.*, **23**, 474.
Hunter, C., 1975. *Astron. J.*, **80**, 783.
Kalnajs, A., 1976. *Astrophys. J.*, **205**, 751.
Kuzmin, G.G., 1956. *Astr. Zh.*, **33**, 27.
Kuzmin, G.G., & Kutuzov, S.A., 1962. *Bull. Abastumani Astroph. Obs.*, **27**, 82.
Lynden–Bell, D., 1962. *Mon. Not. R. astr. Soc.*, **123**, 447.

TRIAXIAL SCALE-FREE MODELS OF HIGHLY FLATTENED ELLIPTICAL GALAXIES WITH MASSIVE HALOS

Harold F. Levison & Douglas O. Richstone
Department of Astronomy
University of Michigan

ABSTRACT. Two surveys of dynamical models of highly flattened, triaxial elliptical galaxies with isothermal potentials have been constructed. These models were constructed in order to better understand the range of possible observable dynamical properties of triaxial galaxies. All models have been constructed so that they appear as E6 galaxies when seen from their intermediate axes. However, one set of models is nearly oblate; the other is nearly prolate. The models are constructed with massive halos such that $M/L \propto r$. Triaxial models of either shape can be constructed with their projected axes of rotation at any position angle with respect to the major axes of the galaxies. The most surprising result is that in most models, the position angle of maximum observed rotation is not perpendicular to the position angle of zero rotation.

1. METHODS

The construction of models of elliptical galaxies requires that a phase density be found that is a solution of the collisionless Boltzmann and Poisson equations. We use a modified version of a completely general method developed by Schwarzschild (1979). His technique requires that a complete set of orbits calculated in a potential. Then these orbits are added together to produce the density distribution the creates that potential. However, if the population of stars modeled does not produce the potential, then any emissivity distribution can be substituted in this final step.

We have chosen an isothermal scale-free potential,

$$\Phi = \ln(s)$$

where $s^2 = x^2 + y^2/p^2 + x^2/q^2$. q is set to 0.75. In one survey $p = 0.8$ (almost prolate), in the other $p = 0.9$ (almost oblate). The density distribution that produces this potential looks like an E6 galaxy when 'viewed' from the intermediate (y) axis and falls off as r^{-2} along any radial ray. To better represent real galaxies we let $M/L \propto r$ or

$$\epsilon = \frac{\rho}{r} \propto r^{-3}.$$

T. de Zeeuw (ed.), Structure and Dynamics of Elliptical Galaxies, 499–500.

2. OBSERVABLE PROPERTIES OF TRIAXIAL GALAXIES

These models look like E6 galaxies when viewed from their intermediate axes. They were constructed to compare with observations of real E6 galaxies. E6 galaxies are unique because they are the only ellipticals with known orientation (They are seen along their intermediate axis).

Figure 1 shows the observable properties of a few triaxial galaxies models viewed from their intermediate axis. The model in frame 'a' has a large amount of rotation along its major axis and none along its minor axis. This type of rotation is usually associated with oblate galaxies. Frame 'b' shows a model that is a minor axis rotator. This type of rotation is usually associated with prolate galaxies. However, both these types of models can exist in either potential. The model shown in frame 'c' has an equal amount of angular momentum in the z and x directions. In either triaxial potential it is possible to construct models with their projected axis of rotation at any position angular with respect to the major axis of the galaxy. The model in frame 'd' is more typical of the type of models found. It has rotation along both axes. Note that the axis of maximum rotation and the axis of zero rotation are not perpendicular.

Almost all triaxial models have a significant amount of minor axis rotation. So, if elliptical galaxies are triaxial, then why do so few have significant rotation along their minor axis?

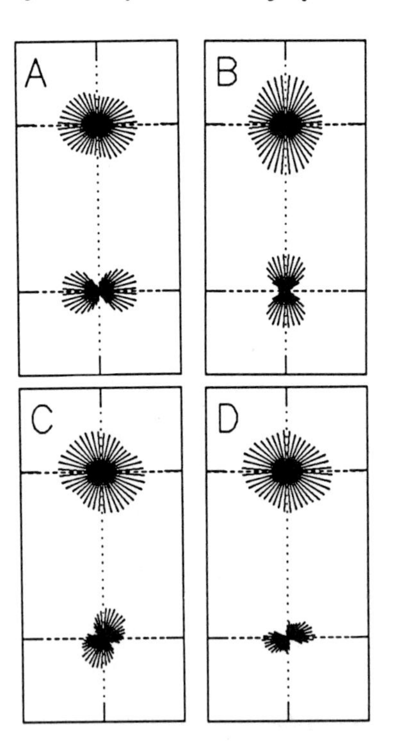

Figure 1: The observable properties of 4 models (see text). The models have flat rotation curves and velocity dispersion profiles. Therefore, the rotation velocity and velocity dispersion are constant along any radial ray and vary only as a function of position angle. On the diagram, the dashed line represents the major axis of the model. The dotted line is the minor axis. In the top part of each frame, the length of the black line at a position angle is proportional to the line of sight velocity dispersion at that position angle. The bottom part of each frame is the rotation velocity.

3. REFERENCE

Schwarzschild, M. 1979, *Ap.J.*, **232**, 236.

BOX MODELS WITHOUT CYLINDRICAL ROTATION

Maria Petrou
Department of Theoretical Physics
1 Keble Road
Oxford, OX1 3NP
United Kingdom

ABSTRACT. We describe here a way of constructing models for box–shaped dynamical systems without cylindrical rotation. We argue that such models may arise as a result of the external heating of a disk.

1. INTRODUCTION

The purpose of this work is to show that it is possible to construct models of box–shaped ellipticals without cylindrical rotation. To show this, we employ the techniques used by Binney and Petrou (1985).

Our method consists in distorting the isochrone sphere into a box, keeping the distribution of particles with respect to their energy unchanged and thus keeping the density profile almost unchanged too. We can then assume that the potential is not very different from that of the isochrone sphere and proceed to compute the response density of the distorted model.

As our purpose is simply to illustrate our point, we do not worry at this stage for the lack of self–consistency.

2. THE DISTRIBUTION FUNCTION

We assume that the distribution function f depends only on the energy E and the angular momentum L_z. We argue that any distribution function which looks like the function shown in Figure 1, when plotted against L_z, for constant E, will give rise to rotationally supported boxes without cylindrical rotation. We choose:

$$f(E, L_z) = f_0(E)(a + (L_z/(J_c + d) - b)^{1/g}), \tag{1}$$

where a, b and d are parameters chosen so that $f \geq 0$ always, g is an odd integer, J_c is the circular angular momentum for given energy and f_0 is:

$$f_0(E) = f_i(E)e^2/G \qquad \text{where} \qquad e = J_c/(J_c + d) \tag{2}$$

$$G = ae^2 + g^2 \frac{(2b^{(2g+1)/g} + (e - b)^{(2g+1)/g} - (e - b)^{(2g+1)/g})}{(2g + 1)(g + 1)} \tag{3}$$

T. de Zeeuw (ed.), Structure and Dynamics of Elliptical Galaxies, 501–502.

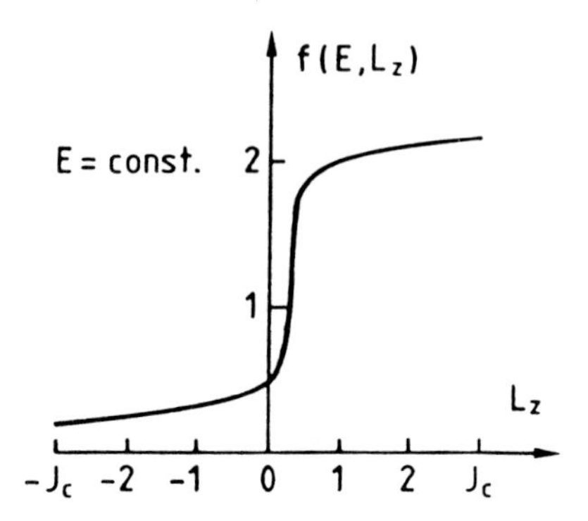

Figure 1. $f(E, L_z)$ versus L_z for $a = 1.2$, $b = 0.1$, $g = 7$, $d = 0.1$ and energy such that $J_c = 3$.

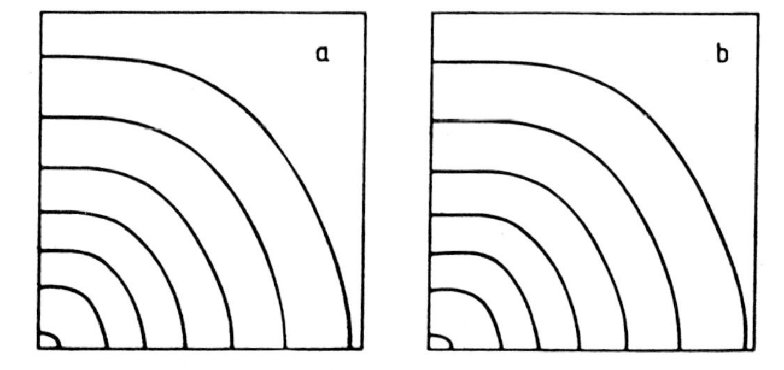

Figure 2. Projected density contours for two models: a) $a = 1.2$, $b = 0.1$, $d = 0.1$, $g = 7$. b) $a = 1.2$, $b = 0.1$, $d = 0.1$, $g = 11$.

and $f_i(E)$ is the distribution function of the isochrone sphere (Hénon 1960). Figure 2 shows the density contours of two models when projected edge on.

3. DISCUSSION

We think that a distribution function such as the one described here, is likely to arise from the external heating (i.e. collisions with other galaxies) of a cold disk. Note that the corresponding profile of Figure 1 for a cold disk is a delta function at $L_z = J_c$. It is plausible, therefore, to assume that external heating of the disk will fatten up the delta function, possibly to the square-like shape of Figure 1.

4. REFERENCES

Binney, J.J. and Petrou, M., 1985, *Mon. Not R. astr. Soc.*, **214**, 449
Hénon, M., 1960, *Ann. d'Astrophys*, **23**, 474

DYNAMICAL APPROACH TO THE $R^{1/4}$ LAW OF ELLIPTICALS

G. Bertin and M. Stiavelli
Scuola Normale Superiore
I-56100 Pisa
Italy

ABSTRACT. The $R^{1/4}$ luminosity law has often been used as an observational constraint in the construction of self-consistent models of ellipticals. In contrast, in this and in the following paper, developing some ideas proposed in the past (BS84, SB85), we investigate certain theoretical arguments that may lead to a dynamical justification of this important empirical law.

Phase space properties of elliptical galaxies suggested by numerical experiments of dissipationless galaxy formation are used as a dynamical constraint (essentially to provide a boundary condition at large radii) in the construction of self-consistent equilibrium models. In particular, from a discussion of the quasi-spherical case, the following simple <u>selection criterion</u> is proposed: the combination of integrals of the motion in the relevant distribution function should guarantee that asympotically at large radii (i) the pressure anisotropy parameter $\alpha = 2 - \langle v_T^2 \rangle / \langle v_r^2 \rangle$ tends to 2 and (ii) the mass density decreases as r^{-4}.

We have already used this working hypothesis to construct a simple family of distribution functions which was found to possess the $R^{1/4}$ law as a built-in property (BS84,SB85). Detailed comparison with photometric and kinematical data has been made by Saglia (1986) for eight ellipticals. Given the interest of these results we now test the generality of our approach in several ways. First we test structural stability by constructing a number of new families of equilibrium distribution functions all following the above defined selection criterion. They are all found to be qualitively similar to our original family and, in particular, to be consistent with the $R^{1/4}$ law in the region of parameter space where the central potential well is sufficiently deep.

Then we address the problem of the dynamical stability of our equilibria by performing N-body experiments using van Albada's (1982) code. No significant instabilities are found in our survey.

All the results described above are restricted to the quasi-spherical case. We finally show that following the prescription

T. de Zeeuw (ed.), Structure and Dynamics of Elliptical Galaxies, 503–504.

of our distribution function we can easily generate N-body equilibria with the qualitative features of our analytic models but with finite flattening, so that (oblate) E2/E3 galaxies are reasonably modelled. Again the agreement with the $R^{1/4}$ law is preserved.

Bertin, G. and Stiavelli, M., 1984. Astron. Astrophys. 137, 26 (BS84)
Saglia, R., 1986. Tesi di laurea, Università di Pisa
Stiavelli, M. and Bertin, G., 1985. Mon.Not.Roy.astron.Soc. 217, 735 (SB85)
van Albada, T.S., 1982. Mon.Not.Roy.astron.Soc. 201, 939.

Table Summary of the results of N-body simulations

run id	N_{par} ($\times 10^3$)	T_{fin} (t_{cr})	$\Delta\mathcal{E}/\mathcal{E}$ % per t_{cr}	$\Delta N/N$	e_f	Δe
P7.7C	40	14.2	0.20	0.00	0.01	0.00
P7.7E	20	40.3	0.02	0.00	0.25	0.01
P7.7G	60	17.1	0.10	0.00	0.18	0.02
P6A	40	28.0	0.07	0.00	0.00	0.01
P6C	40	14.0	0.00	0.00	0.19	0.01
P5.2A	40	51.0	0.00	0.00	0.02	0.01
P5.2B	60	51.1	0.01	0.00	0.00	0.00
P5.2C	20	50.5	0.09	0.00	0.03	0.01

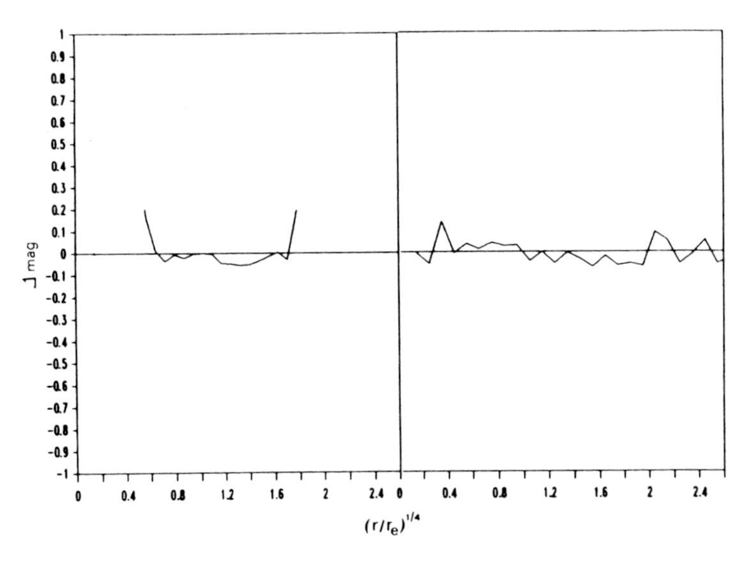

Projected mass distribution for the final equilibrium of the N-body run labeled P7.7G, initialized with the $\Psi = 7.7$ distribution function of BS84. On the left, deviations from the $R^{1/4}$ law are plotted in the range (.1 r_e) to (10 r_e), corresponding to a range of ten magnitudes. On the right, deviations from the initial model are plotted. Note that this case could simulate an E2 galaxy. In the Table e_f denotes final ellipticity of the model, Δe is the ellipticity variation during the run. The rest of the notation is self-explanatory.

STATISTICAL MECHANICS AND EQUILIBRIUM SEQUENCES OF ELLIPTICALS

M. Stiavelli and G. Bertin
Scuola Normale Superiore
I-56100 Pisa
Italy

ABSTRACT. Elliptical galaxies are expected to have undergone incomplete violent relaxation. Here incomplete relaxation is regarded as a process producing a metastable, long-lived state which is dynamically stabilized by the approximate conservation of one global quantity in addition to the total energy and number of particles.

Since the early work of Lynden-Bell (1967) elliptical galaxies have been considered as the final product of incomplete violent relaxation (May and van Albada 1984), nevertheless a satisfactory treatment of incomplete relaxation is still lacking. Recently Tremaine, Henon and Lynden-Bell (1986) have addressed the problem of constraining the relaxation processes of ellipticals using statistical arguments without making use of a maximum entropy principle. Here we study incomplete relaxation from a more conventional point of view, i.e. using the classical (Boltzmann) expression for the entropy, but trying to take into proper account the collisionless nature of the system by specifying the conservation of one additional global quantity Q. Then the distribution function is obtained by extremizing the entropy of the system following the concepts and the procedure outlined by Lynden-Bell (1967) and Shu (1978). If we argue that the additional quantity is of the form $Q = \Sigma_{part}\, q$, i.e. the equal weight sum of one particle contributions, with q being proportional to powers of the particle angular momentum J and energy E, then we must have

$$q = J^{\alpha_1} (-E)^{-3\alpha_1/4} \tag{1}$$

in order to meet the requirements of the dynamical selection criterion formulated by us elsewhere (see Bertin and Stiavelli 1984, and this Symposium). In fact, by extremizing the entropy subject to the conservation of Q we obtain the following family of distribution functions

$$f^{(\alpha_1)} = A \exp \left[-aE - c\, J^{\alpha_1} / (-E)^{3\alpha_1/4}\right] \tag{2}$$

all satisfying our dynamical selection criterion. At this stage the

T. de Zeeuw (ed.), Structure and Dynamics of Elliptical Galaxies, 505–506.

value of α_1 is arbitrary. It is found that in the vicinity of $\alpha_1 = 1$ these distributions are characterized by realistic mass distributions (see Fig. 1). Then the conservation of Q is tested by an extensive study of numerical experiments of dissipationless galaxy formation. Partial conservation during, and especially after the collapse, is found (see Fig. 2).

Bertin, G. and Stiavelli, M., 1984. Astron. Astrophys. 137, 26
Lynden-Bell, D., 1967. Mon.Not.Roy.astron.Soc. 136, 101
May, A. and van Albada, T.S., 1984, Mon.Not.Roy.astron.Soc. 209, 15
Shu, F.H., 1978. Astrophys. J. 225, 83
Tremaine, S., Hénon, M., Lynden-Bell, D., 1986. Mon.Not.Roy.astron. Soc. 219, 285

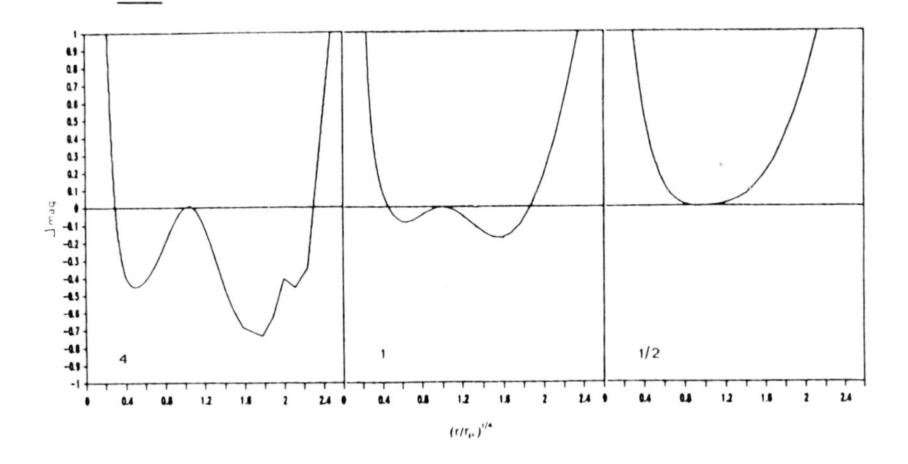

Figure 1. Residuals from the $R^{1/4}$ law in magnitudes for the equilibrium models specified by (2) with $\Psi = 18$ ($\alpha_1 = 4, 1, \frac{1}{2}$).

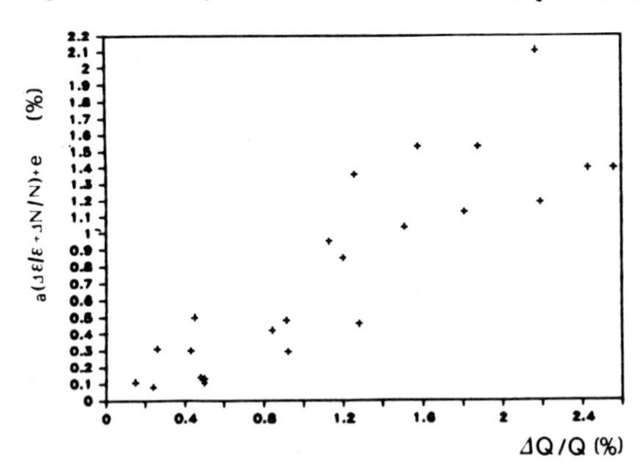

Figure 2. For a set of N-body dissipationless collapses a combination of various factors (total energy and particle number non-conservation and ellipticity) is found to correlate with the non-conservation of Q (see (1) with $\alpha_1 = 1/2$), expressed in percent/crossing time.

ON THE DISTRIBUTION FUNCTION OF ELLIPTICAL GALAXIES

A. Kashlinsky
National Radio Astronomy Observatory and
Department of Astronomy
University of Virginia

Abstract. Distribution functions containing cutoff in energy impose several limitations on systems they describe, e.g., no circular orbits are allowed in the major part of system and spatial boundary is poorly defined. As opposed to these functions, we present here distribution function that describes galaxies of finite extent (i.e., truncated in radius). We discuss properties of systems having this distribution function for spherical and axisymmetric cases and compare their surface brightness, isophotes, and rotation curves with observations of elliptical galaxies. This distribution function can easily be generalized to a triaxial galaxy.

Elliptical galaxies as well as any stellar systems have finite extent and hence they should be described by truncated distribution functions, i.e., the probability to find a star above a certain boundary should be zero. The most commonly used truncated spherical distribution function is the King-Michie one, $f_k \propto (e^{-\varepsilon/\sigma^2} - 1)e^{\alpha J^2}$ which excludes from the system any star with positive energy ε. Prendergast and Tomer (1970) and later Wilson (1975) generalized f_k to a rotating axisymmetric system. However, all the above models do not give adequate fits when applied to observational data. Differently truncated distribution function would produce systems with completely different density profiles (particularly in outer regions) and dynamical properties (e.g., Wilson 1975) and, therefore, it is important to construct consistently truncated distribution function. All distribution functions considered before contained truncation in energy, which means that any star bound to the galaxy must have negative energy with respect to potential at the boundary. However, condition for a star of angular momentum J to be bound to the galaxy of radius R is $\varepsilon < J^2/2R^2$, <u>not</u> $\varepsilon < 0$. Distribution functions with truncation in energy exclude circular orbits from almost half of the system and stars at the boundary would have to move on radial orbits only. This is a rather ad hoc constraint and a more realistic constraint should produce galaxy

T. de Zeeuw (ed.), Structure and Dynamics of Elliptical Galaxies, 507–508.

truncated in radius not energy, since any system that forms by collapse and fragmentation would have finite extent.

As opposed to distribution functions truncated in energy, we construct distribution function that is truncated in radius and *a priori* does not impose constraints on the distribution of stellar orbits (Kashlinsky 1986). Properties of such models are best illustrated by an example of such spherically symmetric system. For a spherical system such a distribution function would be

$f \propto (e^{-\varepsilon/\sigma^2} - e^{-J^2/2R^2\sigma^2})$ and potential can be found by solving Poisson equation $\nabla^2\phi = -4\pi G\int f d^3\underline{v}$ coupled with boundary condition $\phi(r = R) = 0$ that determines R to be used in f which somewhat complicates the problem mathematically. Note that here the distribution of orbits is anisotropic and orbits become predominantly tangential near the boundary. Anisotropy of orbits is a natural product of having stellar system that is truncated in radius and degree of anisotropy does not require any parameters besides specifying the depth of central potential $\phi(o)/\sigma^2$. The resultant velocity dispersion of the system $\sigma^2 = \rho^{-1} \int f v^2 d^3\underline{v}$ is much flatter than that of the King (1966) models which is consistent with observations. Density profile, too, is significantly different from the King one once the surface density drops below $\sim 10^{-4}$ of the central value. We generalize this distribution function to rotating and axisymmetric systems and compare resultant isophotes rotation curves, velocity dispersion profiles, etc., with available observations. The distribution function considered here leads to almost flat rotation curves of elliptical galaxies in agreement with observations. The reason for a much flatter (than in previous models) rotation curve is that outer regions of the galaxy are dominated now by stars on tangential orbits. Because these models are intrinsically anisotropic even in the absence of rotation, it is easy to reproduce an elliptical galaxy of any flattening even for moderate amounts of rotation. Isodensity contours are almost spherical near the center and become more flattened near the boundary. The models can be easily generalized to triaxial galaxies once the relevant integrals of motion are known.

Kashlinsky, A. 1986, in preparation.
King, I. 1966, Astron. J., 71, 64.
Prendergast, K. H. and Tomer, E. 1970, Astron. J., 75, 674.
Wilson, C. P. 1975, Astron. J., 80, 175.

TOWARDS A SELF-CONSISTENT MODEL OF A GALAXY

W.J.L.V. Durodie and F.D. Kahn
Department of Astronomy
University of Manchester
Manchester M13 9PL
United Kingdom

ABSTRACT. Using a theoretical model for the functional distribution of stars in phase space in a spherically symmetric galactic system, it is found upon solving the fundamental equations of stellar dynamics that the rotation curves produced by the model are flat for large distances within the system. The properties of the stellar orbits within such systems are investigated and an N-ring axially and equatorially symmetric model for simulating its dynamics is presented. Poisson's equation is solved by expanding density and potential in Legendre polynomials (c.f. van Albada and van Gorkom, 1977). It is helped to follow the time development of such a system under various forces.

1. BACKGROUND

We are interested in solutions of the self-consistent problem for stellar systems (Richstone and Tremaine, 1984). According to van der Kruit and Allen (1978), it is a fact that not a single galaxy has been found with a Keplerian rotation curve at large radii. This is an important constraint on any theoretical model, together with the fact that the flattening of most ellipticals is not caused by rotation but maintained by an anisotropic velocity distribution (Binney, 1976).

2. MODEL

We define our distribution function to be of the form:

$$f = C\frac{g(J)}{(-E)}$$

where J is the angular momentum, and E is the total energy.

If we choose a "top-hat" distribution for $g(J)$, then the orbits near the centre are more isotropic ($\sigma_r \simeq \sigma_t$), and those in the outer parts more radial ($\sigma_r > \sigma_t$) (c.f. Hénon, 1964; Gott, 1977). The potential can also be separated into two analytic parts and evaluated throughout.

T. de Zeeuw (ed.), Structure and Dynamics of Elliptical Galaxies, 509–510.

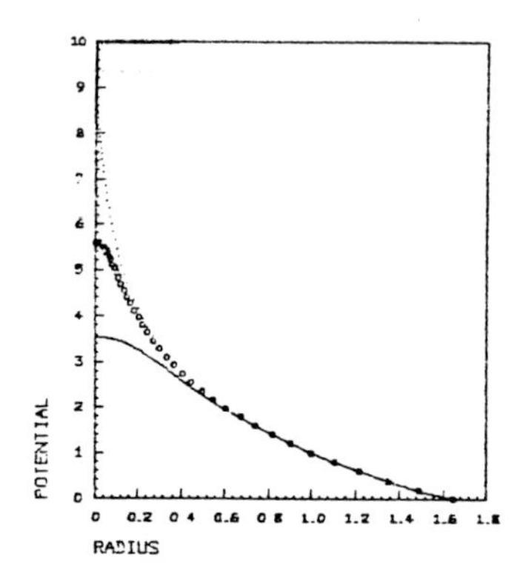

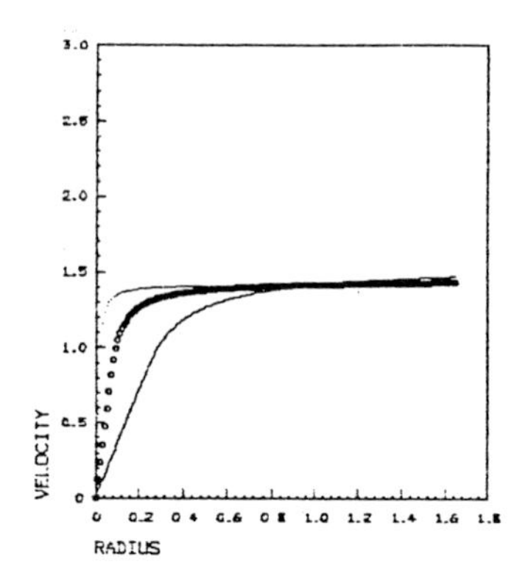

The Figure shows how the potential is gradually softened by the increasing spread in angular momentum, and the corresponding rotation curves.

The orbits of test particles moving in a variety of such systems can be evaluated. These show that the systems with a larger spread in the angular momentum function will be more susceptible to collapse from a passing perturber, as the orbits of their test particles do not precess as rapidly, due to the more softened potential providing less energy for exploring the available phase space.

We produce a model galaxy with N randomly distributed rings. Particles on the same ring and with identical velocities will describe similar orbits around the origin. The potential V and density ρ are expressed as sums of Legendre polynomials.

It is hoped with the aid of our model to show how such a spherical system can be changed into prolate or oblate systems. The source of such distortion could be intrinsic torques or tidal interaction (Peebles, 1969), and can be easily modeled by adding components to the Legendre polynomial coefficients, as the potential outside the system is of the form:

$$V = \sum_n \frac{A_n}{r^{n+1}} P_n(\mu)$$

Our time integration scheme (Hockney and Eastwood, 1981) will allow us to follow the shape of the system and its potential and rotation curve as it develops.

REFERENCES

Binney, J., 1976, *Mon. Not. R. astr. Soc.*, **177**, 19.
Gott, J.R., *Ann. Rev. Astron. Astrophys.*, **15**, 235
Hénon, M., 1964, *Ann. d'Ap.*, **27**, 83.
Hockney, R.W., Eastwood, J.W., 1981, *Computer Simulation Using Particles*, McGraw-Hill Int.
Peebles, P.J.E., 1969, *Astrophys. J.*, **155**, 393.
Richstone, D.O., Tremaine, S., 1984, *Astrophys. J.*, **286**, 27.
van Albada, T.S., van Gorkom, J.H., 1977, *Astron. Astrophys.*, **54**, 121.
van der Kruit, P.C., Allen, R.J., 1978, *Ann. Rev. Astron. Ap.*, **16**, 103.

THE ENVELOPES OF SPHERICAL GALAXIES

Walter Jaffe
Leiden University Observatory
P.O. Box 9513
2300 RA Leiden
Netherlands

ABSTRACT. The light distribution in the envelopes of spherical galaxies seems to be caused by the existence of a break in the energy distribution, N(E) at E = 0. This, in turn is probably caused by the escape of positive energy stars.

Elliptical galaxies, with some exceptions like cDs, show a universal light distribution than can be represented in space as:

$$\varepsilon \propto r^{-2}(r_c + r)^{-2} \quad \text{(Jaffe, 1981).}$$

This suggests a very general mechanism of formation. Here I show that the r^{-4} behavior in the envelope is the result of a sharp break in the energy distribution of particles, N(E), near E = 0, the escape energy. This break, in turn, would be the result of any energy scattering processes whose cross section doesn't vary rapidly near E = 0.

To prove the first point, we write the relation between N(E) and f(E), the phase space density, to be (c.f. Binney, 1982):

$$N(E) = f(E)\ A(E)\ \text{, where,}$$
$$A(E) \propto \int_{r=0}^{\phi(r)=E} (E - \phi(r))^{1/2}\, r^2 dr \quad .$$

In the envelope the potential will be essentially Keplerian, $\phi(r) \sim GM/r$ which, with the above integration, yields $A(E) \propto (-E)^{5/2}$.

Thus if N(E) has a sharp break at E = 0, for example:

$$N(E) = 0 \text{ for } E > 0\text{, and}$$
$$N(E) = 1 \text{ for } E < 0\text{, then}$$

$$f(E) \propto E^{5/2} \text{ near } E=0 \text{ , so, using the standard formula,}$$

$$\rho(r) = \int f(E)\ (E - \phi(r))^{1/2} dE \propto (-\phi)^4 \propto (GM/r)^4$$

in the envelope, which shows the desired behavior.

T. de Zeeuw (ed.), Structure and Dynamics of Elliptical Galaxies, 511–512.

The only requirement is that $N(E) \propto E^0$ as $E \to 0$, i.e. that there is a sharp break in N at the escape energy. Such a break would be the natural result of any scattering process in energy space that doesn't vary radically near E=0; stars scattered to small negative energies stay there while those at small positive energies leave the system.

For example: if energy scattering only occurs near the nucleus, then the typical scattering length, ΔE, will be primarily a function of the velocity, $v = (2(E - \phi))^{1/2}$. For large values of ϕ and E near zero, v is only a slow function of E.

We have numerically calculated N(E) for a case where the scattering rate is:

$S(E \to E') \propto E^{-\alpha} \exp(-(E-E')^2/2\sigma^2)$ and $N(E, t=0) = \delta(E - 1.0)$.

Here are the resultant N(E) curves for $\sigma = 0.4$ and $\alpha = 0.5$ and 1.5:

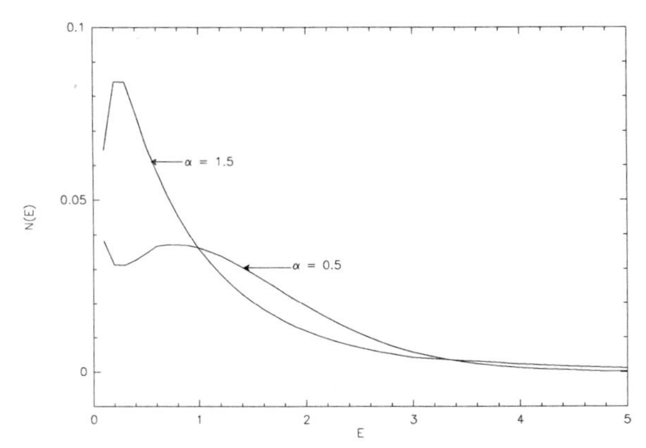

At large radius the residuals of the corresponding surface brightness from the de Vaucouleur's law are, for either α at most a few tenths of a magnitude.

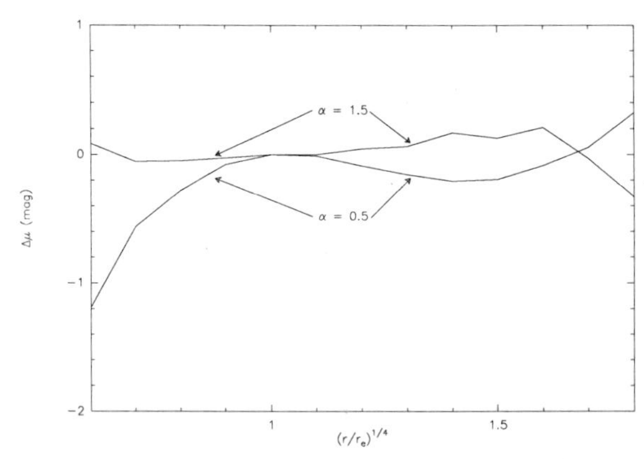

REFERENCES

Binney, J., 1982. Mon. Not. R. Astron. Soc. **200**, 951
Jaffe, W., 1982. Mon. Not. R. Astron. Soc. **202**, 995

SPHERICAL STELLAR SYSTEMS: STRUCTURE AND EVOLUTION

L. M. Ozernoy
Harvard-Smithsonian Center for Astrophysics
Cambridge, MA 02138
and
V. A. Volodin
Lebedev Physical Institute
Moscow 117924, U.S.S.R.

ABSTRACT. We analyze a class of spherical stellar systems with an isotropic, non-isothermal, one-particle distribution function which is truncated in energy ϵ as $(\epsilon_b - \epsilon)^{\nu}$ $(\nu > 0)$ near the energy of truncating ϵ_b and is the Maxwell-Boltzmannian at $|\epsilon| \gg |\epsilon_b|$. The stellar systems which correspond to this distribution function appear to be isothermal in their central parts and polytropic in the external parts. Some qualitative results which follow from such a "polytropic-isothermal" distribution function (PIDF) are briefly discussed. A further extention of PIDF enables us to analyze in a new way the evolution of spherical stellar systems.

Why are the King (1966) models, in spite their simplicity (spherical symmetry, isotropy, one-particle distribution function), capable of reproducing many properties of real stellar systems in a surprisingly wide domain of applications – from globulars to elliptical galaxies and regular clusters of galaxies? Apparently the key element is that the distribution function taken even in the simplest conceivable form – Maxwell-Boltzmannian – is truncated at energies greater than some maximum energy. Physically, this means accounting for the principal fact of rapid evaporation of all those particles which acquire at encounters velocities greater than the local escape velocity.

We propose here a more general distribution function which, being still one-particle, spherically symmetric, isotropic, and Maxwell-Boltzmannian at energies $|\epsilon| \gg |\epsilon_b|$, vanishes as $(\epsilon_b - \epsilon)^{\nu}$ as $\epsilon \to \epsilon_b$, $\nu > 0$ being a free parameter.

We have obtained numerical solutions for the potential and density runs. The shape of each solution is determined by two parameters: the power law index ν and the central dimensional potential W_o.

In the external parts of the stellar system where the local potential is small, it is described approximately by the Lane-Emden equation. The Lane-Emden polytrope of the index $\lambda = \nu + 3/2$ represents the exact solution for the density distribution in the limiting case when $W_o \to 0$.

In the central parts of the system, the density distribution tends, at sufficiently large W_o's, to the isothermal (Emden) solution, independently of ν. The size of the isothermal region, at a given λ and a not too small W_o, does not depend on W_o

T. de Zeeuw (ed.), Structure and Dynamics of Elliptical Galaxies, 513–514.

and for the King models (they are a particular case of our solutions at $\lambda = 5/2$) it comprises about 2% of the overall size of the system. The central, gaussian density peak is typical for the models with great enough $W_\circ$'s (for King models, it appears only when $W_\circ \geq 6$).

The larger (at a fixed λ) is $W_\circ$, the closer is the density distribution in most of the system (except in the narrow [$\sim \exp(-W_\circ/2)$], central core region) to a peculiar, "singular" solution. The latter has the logarithmic singularity at the center and is the limiting, λ-dependent solution which corresponds to $W_\circ \to \infty$. In particular, the King models are transformed at $W_\circ \to \infty$ into the singular solution of the index $\lambda = 5/2$, and not into an isothermal sphere.

All the solutions discussed here, including the limiting ones, are spatially limited and finite. The boundary radius of the system $r_b(\lambda, W_\circ)$ is found to be changing, for any λ and $W_\circ$ satisfying the inequalities $0 < \lambda \leq 3.5$ and $0 \leq W_\circ < \infty$, within some finite limits: $0 < r_{min}(\lambda) < r_b(\lambda, \infty) < r_{max}(\lambda) \equiv r_b(\lambda, 0)$. Note that the finiteness of $r_b(\lambda, W_\circ)$ is valid even for "free" (not tidally truncated) models, *i.e.* even in those cases when the interaction with an external tidal field is negligibly small. (Such systems which are bound due to their intrinsic properties evolve without an appreciable influence of the external tidal field as long as it keeps sufficiently small). As for a tidally truncated system, its boundary radius is merely fixed by the value of the tidal potential and does not depend on $W_\circ$. Thus, at any external conditions of the stellar system (whether in a tidal field or in a free state) its boundary radius remains finite and it is changing gradually during the secular evolution of the system, together with the change of its total energy and number of stars. Such a picture is valid as long as the system does not deviate from the state characterized by the "polytropic-isothermal" distribution function (PIDF) introduced above.

A natural extension of this distribution function, by introducing, as a first step, a scale factor, leads to a two-parameter family whose distribution functions are capable of describing quantitatively the spherical stellar systems at various evolutionary stages, including the core collapse (gravo-thermal catastrophe). This family enables us, in particular, to reproduce the evolutionary sequence for spherical stellar systems which was obtained by Cohn (1980) with the help of the numerical solution of the Fokker-Planck equation. The scale factor introduced into PIDF provides an excellent fit both for dependence of the distribution function on the binding energy and for the star density as a function of the dimensionless potential. However, this results in the fit for the potential run which appears to be insufficiently precise. This defect can be removed if the second step in the generalization of PIDF's is made by taking their linear combination. Already two terms are sufficient to reproduce fairly well all the dependencies mentioned above.

The extensions of PIDF proposed here reveal a new approach to consider both the early and the late stages in the evolution of spherical stellar systems. The King models appear to describe a particular, though a rather long, intermediate evolutionary stage.

A full account of the work will be published elsewhere.

REFERENCES

Cohn, H., 1980. *Astrophys. J.*, **242**, 765.
King, I. R. 1966. *Astron. J.*, **71**, 64.

INSTABILITY THROUGH ANISOTROPY IN SPHERICAL STELLAR SYSTEMS

P.L. Palmer and J. Papaloizou
School of Mathematics
Queen Mary College
Mile End Road, London E1 4NS
United Kingdom

ABSTRACT. We consider the linear stability of spherical stellar systems by solving the Vlasov and Poisson equations which yield a matrix eigenvalue problem to determine the growth rate. We consider this for purely growing modes in the limit of vanishing growth rate. We show that a large class of anisotropic models are unstable and derive growth rates for the particular example of generalized polytropic models. We present a simple method for testing the stability of general anisotropic models. Our anlysis shows that instability occurs even when the degree of anisotropy is very slight.

1. THE EIGENVALUE PROBLEM AND STABILITY TEST

In the limit of low growth rate, σ, the radial orbit resonance, $2\Omega_\theta = \Omega_r$, plays an important role. Here Ω_θ and Ω_r are the angular and radial orbital frequencies. If for small J, $f(E,J)\alpha J^{-s}$, then as $\sigma \to 0$ the eigenvalue problem reduces to

$$\frac{-4\pi G}{\sigma^s} \int K(r,r^1)\psi^1(r^1)dr^1 = \frac{1}{r^2}\frac{d}{dr}\left[r^2\left[\frac{d\psi^1}{dr}\right]\right] - n(n+1)\psi^1/r^2 \qquad (1)$$

where ψ^1 is the perturbed gravitational potential, n the order of the associated spherical harmonic, and the kernel K is positive definite. Further details will be published elsewhere. Equation (1) has a low growth rate spectrum with accumulation point at zero for any s. A similar result occurs for any f such that $f \to \infty$ for $J \to 0$. We have also been able to derive an equation analogous to (1) which can decide stability for more general $f(E,J)$, (more details to be published elsewhere).

We have calculated the low growth rate spectrum for some generalized polytropes, with $f \propto (E_0 - E)^\beta J^{-s}$ for $E \leq E_0$, and $f = 0$ otherwise. These models were studied numerically by Barnes et al. (1986), and Fridman and Polychenko (1984). However, stability boundaries were not well determined in their work.

T. de Zeeuw (ed.), Structure and Dynamics of Elliptical Galaxies, 515–516.

2. NATURE OF THE INSTABILITY

For nearly radial orbits $\Omega_r < 2\Omega_e$, so they precess slowly in the direction of orbital rotation. A bar–like perturbation attracts such an orbit and increases its angular momentum as it does so. The orbit then passes through the bar with an angular momentum greater than it had initially if it started almost radial. If the orbit can reinforce the bar as it passes through, the bar is then able to attract and trap orbits with even lower initial angular momentum, which replace those already trapped but which have moved to higher angular momentum. From this it follows that we expect instability when $df/dJ < 0$.

A similar mechanism has been proposed by Lynden-Bell (1979) for the formation of bars in disc galaxies.

REFERENCES

Lynden-Bell, D., 1979 *Mon. Not. R. astr. Soc.*, **187**, 101.
Barnes, J., Goodman, J., and Hut, P., 1986, *Astrophys. J.*, **300**, 112.
Fridman, A.M., and Polyachenko, V.L., 1984, *Physics of Gravitating Systems I*, Chap. 3, Springer-Verlag, New York.

COLLISIONAL EFFECTS ON THE DENSITY PROFILES OF SPHERICAL GALAXIES

L.A. Aguilar
Harvard-Smithsonian Center for Astrophysics
60 Garden St., Cambridge MA 02138 USA

S.D.M. White
Steward Observatory
University of Arizona, Tucson AZ 85721 USA

ABSTRACT. We use N-body simulations to study the time evolution and the final shape of the density profiles of non-rotating spherical galaxies that have undergone a tidal encounter. We consider models with de Vaucouleurs and King surface density profiles and with isotropic, tangential and radially biased velocity distributions.

King models lose their tidal radius and develop an extended tail very similar to an $r^{1/4}$-law. De Vaucouleurs models, on the other hand, are very robust and are still good fits to the remnants of even very strong encounters resulting in a mass loss of 50%. Although an extended tail of the profile is always regenerated after an encounter, the isophotal radius of the galaxy almost always decreases.

The tidal classes defined by Kormendy (1977) can be interpreted as resulting from a transient radial mixing of stars inside a galaxy after an encounter, and can be used to estimate the time since last encounter.

A galaxy that has suffered tidal encounters does not show any "collision signature" in its luminosity profile or shape of isophotes; the effects of tidal encounters can only be recognized immediately after a close passage, or by a statistical study of galaxy properties of a sample of galaxies.

NUMERICAL EXPERIMENTS. Our initial conditions are 3000-particle random realizations of spherical models with de Vaucouleurs and King surface density profiles. In order to test possible effects due to internal kinematics, three different velocity distributions have been used for the de Vaucouleurs models: isotropic, radially biased, and purely circular orbits. Only isotropic velocity distributions have been used for the King models. All models have the same binding energy.

We have used an N-body code that approximates the forces between the particles in a self-gravitating target galaxy as an expansion in spherical harmonics up to quadrupolar terms. The perturbing galaxy is modelled as a rigid potential that corresponds to a galaxy identical to the target galaxy but with a different mass. For more details see Aguilar and White (1985).

RESULTS. We have calculated the surface density profile projected on the orbital plane of the encounter for all our experiments. Only bound particles have been considered. Figure 1 presents the evolution of the surface density profile for a strong encounter that results in 40% mass loss to the target galaxy.

T. de Zeeuw (ed.), Structure and Dynamics of Elliptical Galaxies, 517–518.

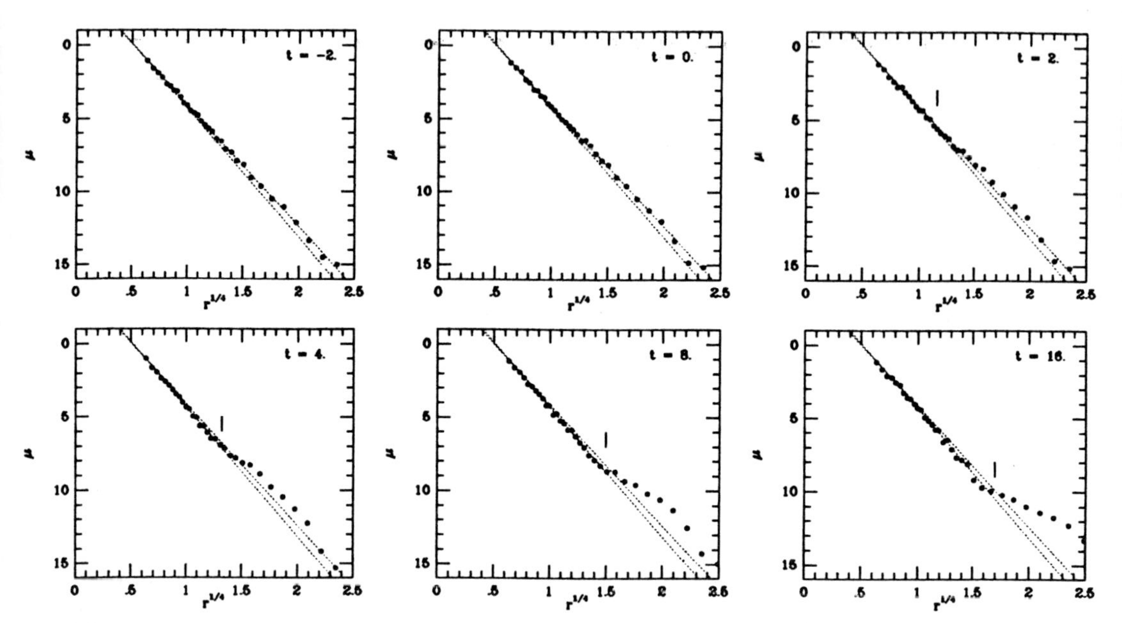

Figure 1. Evolution of the surface density profile of an $r^{1/4}$ model after a strong encounter ($M_p/M_{target} = 1.9$, $p/r_e = 1.0$, and $v_{coll}/v_{rms} = 2.2$). Time is given for each frame in crossing times ($t \equiv 0$ at the instant of closest approach). Upper and lower dotted lines in all frames are $r^{1/4}$-laws fitted to the initial and final (t = 80) frames respectively. The solid tick mark in several frames lies at the radius where the local t_{cross} equals the time elapsed since closest approach.

The evolution of the density profile is quite simple. As time passes, the innermost regions move to the line defined by the fit to the final conditions, while the outer regions produce a "luminous" excess above both, initial and final profiles. The boundary between these two regions evolves on the local dynamical timescale, proving that the bump in the profile consists of particles being radially mixed onto less bound orbits. Notice that the profile at $t = 16t_{cross}$ ($\sim 2.6 \times 10^8$ years for a galaxy with $r_e = 2$ kpc and $v_{rms} = 250$ km/s) is indistinguishable, within the numerical resolution of the simulation, from an $r^{1/4}$-law for $r < 5r_e$. The "tidally distended" objects found by Kormendy have very similar profiles, and if produced by an encounter, indicate a very recent event; we note that Kormendy's T3 tidal class is defined by the presence of a nearby luminous companion.

Collisions using other initial conditions show a similar evolution on the same timescale, the amplitude of the bump being a function of encounter parameters. King models develop an extended tail in the profile and resemble an $r^{1/4}$- law. We conclude that galaxies that have suffered tidal encounters do not show any encounter signature within its observable region, unless observed immediately after the encounter. A statistical study of the relationship between profile shape and nearest neighbor distance would be very useful to see whether "tidal distension" is consistent with the short timescales we infer from our models.

REFERENCES

Aguilar, L.A. and White, S.D.M., 1985. *Astrophys. J.*, **295**, 374.
Kormendy, J. 1977. *Astrophys. J.*, **218**, 333.

COLD COLLAPSE AS A WAY OF MAKING ELLIPTICAL GALAXIES

L.A. Aguilar
Harvard-Smithsonian Center for Astrophysics
60 Garden St., Cambridge MA 02138 USA

D. Merritt and M. Duncan
Canadian Institute for Theoretical Astrophysics
University of Toronto, Toronto ON M5S 1A1 CANADA

SUMMARY

We investigate whether dissipationless collapse starting from very cold, non-rotating initial conditions can produce objects resembling real elliptical galaxies. We also study the effect of various initial geometries on the shape of the final object. Collapses that are initially very cold ($2T/W < 0.1$) are different from warmer collapses, due to the presence of a dynamical instability associated with clumping of nearly-radial orbits (Polyachenko 1981). This instability can produce very elongated bars (1.6 to 2.1 axis ratio) from spherical initial conditions. The instability is also present in models evolved from oblate and triaxial initial conditions. Warm collapses tend to preserve their initial shapes. Cold initial conditions produce objects whose surface density profiles are well fit by a de-Vaucouleurs law; warm collapses, on the other hand, produce a core-halo profile. A large collapse factor seems necessary to produce objects resembling real galaxies; the same collapse factor guarantees the presence of the radial orbit instability. It thus appears that initial flattening is not crucial for producing prolate or nearly prolate galaxies. Oblate galaxies, on the contrary, seem very difficult to form, unless extremely flattened initial conditions are invoked. Preliminary experiments suggest that these results are not changed by realistic amounts of angular momentum.

NUMERICAL EXPERIMENTS

Our initial conditions have a $\rho \propto r^{-1}$ density profile and Gaussian thermal motion independent of position. Some of the spherical models are squeezed along one or two axes to obtain oblate or triaxial initial conditions. All the models have the same binding energy. We used an N-body code that approximates the forces by an expansion in spherical harmonics up to octupolar terms (Aguilar and White 1985). All our runs use 5000 particles and are followed during 10 free-fall times at the initial half mass radius. Several of our experiments have been checked with a direct-summation Aarseth code. The two codes yield results that are in good agreement. We have run collapses with initial virial ratios $2T/W$ ranging from 0.3 to 0.01, and oblate initial conditions ranging from spherical to an axis ratio of 2.5 to 1. One experiment was realized with 1:2:3 triaxial initial conditions.

T. de Zeeuw (ed.), Structure and Dynamics of Elliptical Galaxies, 519–520.

RESULTS

a) Shapes. Table 1 shows the eigenvalues of the inertia tensor for the 80% most bound particles of the final configurations, which we used as estimators of the degree of flattening.

TABLE I

Final axis ratio as a function of initial temperature and axis ratio

$\log 2T/W =$	-0.50	-0.75	-1.00	-1.25	-1.50	-1.75	-2.00
$\epsilon = 1.0$	1.1-1.0-1	1.1-1.0-1	1.6-1.1-1	1.8-1.2-1	1.9-1.2-1	2.0-1.2-1	2.1-1.3-1
1.5	1.2-1.1-1	1.2-1.1-1	1.6-1.1-1	1.9-1.2-1	1.6-1.0-1	1.8-1.0-1	
2.0	1.4-1.4-1	1.7-1.4-1	1.9-1.4-1	1.9-1.3-1	2.2-1.3-1	1.9-1.2-1	
2.5		1.8-1.5-1	2.1-1.5-1	2.3-1.5-1	2.1-1.3-1	2.3-1.4-1	
1:2:3					2.1-1.2-1		

→ Bar Instability

Spherical | Oblate | Prolate | Triaxial

Notice the abrupt onset of non-sphericity for initially spherical models colder than $2T/W \approx 0.1$; the final axis ratio is a slowly varying function of temperature afterwards. The same qualitative behavior is presented by experiments started from non-spherical initial conditions; but a gap of triaxial models now appears at the boundary of the bar instability and widens as we increase the initial flattening.

b) Surface density profiles. The surface density profiles of the final models have been obtained using elliptical annuli with the same axis ratio as the inertia tensor. The results are insensitive to the degree of initial flattening and are a good fit to the de Vaucouleurs law only when $2T/W \leq 0.1$.

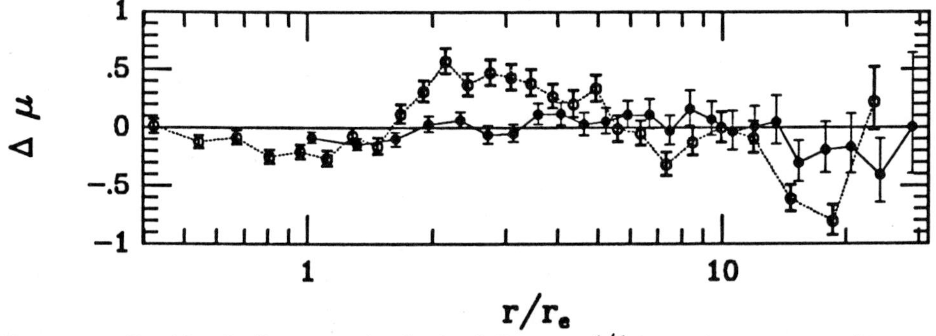

Figure 1. Residuals (in magnitudes) of fits to $r^{1/4}$-laws for a warm (dotted line) and cold (solid line) collapse ($2T/W = 0.32, 0.02$). The error bars are the expected statistical fluctuations due to the number of particles in each bin.

REFERENCES

Aguilar, L.A. and White, S.D.M. 1985. *Astrophys. J.*, **295**, 374.
Polyachenko, V.L. 1981. *Pis'ma Astr. Zh.*, **7**, 142 (translated in *Soviet Astr. Lett.*, **7**, 79).

ANGULAR MOMENTUM IN ELLIPTICAL GALAXIES

Rosemary F.G. Wyse[1,2] and Susana Lizano[2]
1. Space Telescope Science Institute, Baltimore, MD 21218.
2. Astronomy Department, University of California, Berkeley, CA 94720

ABSTRACT. We have used the available published observations of the rotational properties of elliptical galaxies to test theories of galaxy formation which predict an anti-correlation between the angular momentum of a galaxy and its initial overdensity and hence formation epoch. We find that the prediction, at least in its simplest form, is not supported by the data for well studied elliptical galaxies which have a range in rotational support.

DISCUSSION

Major requirements of a theory of galaxy formation are that it offer an explanation of the differences between elliptical and spiral galaxies and an understanding of the correlations among dynamical parameters within a Hubble type. The crucial parameter determining the final morphological type of a proto-galactic density perturbation has recently been proposed to be initial overdensity, relative to the mean overdensity on a galactic scale[1,2]. This proposal, which works best if the universe is dominated by cold dark matter[1], predicts that the nonrotating (elliptical) galaxies form from those density perturbations on a galaxy scale which are very overdense compared to the average – out on the Gaussian tail – and hence collapse at larger redshift than the rms (one-sigma) galaxy-sized perturbation. Initial overdensity is anti-correlated with initial angular momentum, along the Hubble sequence and within a Hubble type, which is the prediction we investigate here.

Davies *et al*[3] have published measurements of the rotation parameter (v/σ) for a sample of elliptical galaxies, and find a large range of rotational support within this Hubble type. We have attempted to derive initial pre-collapse values for the galaxy densities, assuming that the stellar $M/L = 8$, there are dark halos of ten times the stellar masses, and the galaxies underwent dissipationless collapse by a factor 2 in radius. The velocity dispersion is assumed constant, while the rotation velocity increased as expected for non-selfgravitating collapse. An anti-correlation between initial v/σ and density is not compatible with these data, although the possibilities of variable collapse factor and dissipation, together with the spread in masses for the sample are complications. Thus in Figure 1a we show density versus temperature (or velocity dispersion) for these ellipticals. The different symbols reflect different levels of rotational support. The rms 1-sigma relationships for clustering hierarchies of initial power spectrum slope $n = -1, -2$ are as indicated (cold dark matter being somewhere between the two). If rotation were determined mainly by initial overdensity in the way suggested, the low-sigma rapidly rotating

T. de Zeeuw (ed.), Structure and Dynamics of Elliptical Galaxies, 521–522.

ellipticals and the high-sigma slowly rotating ellipticals should be offset from each other, perpendicular to the rms line[1]. This does not appear to be the case. If one were to argue for a variable collapse factor, then one could produce an offset by requiring *ad hoc* that the most highly rotating galaxies collapsed by a factor three more than the slowly rotating galaxies. The likelihood that elliptical galaxies underwent dissipation while forming will cause a spread in the densities, so use of a conserved quantity like mass may be preferred (assuming negligible secondary infall). Mass against temperature is plotted in Figure 1b, the hierarchy lines and symbols as before. Again we see that galaxies of different levels of rotational support are well-mixed. Subsets could be moved relative to each other by adopting a different M/L, but an offset could be achieved only if the slowly rotating ellipticals had an M/L $\gtrsim$ 5 times that of the more rotationally supported ellipticals.

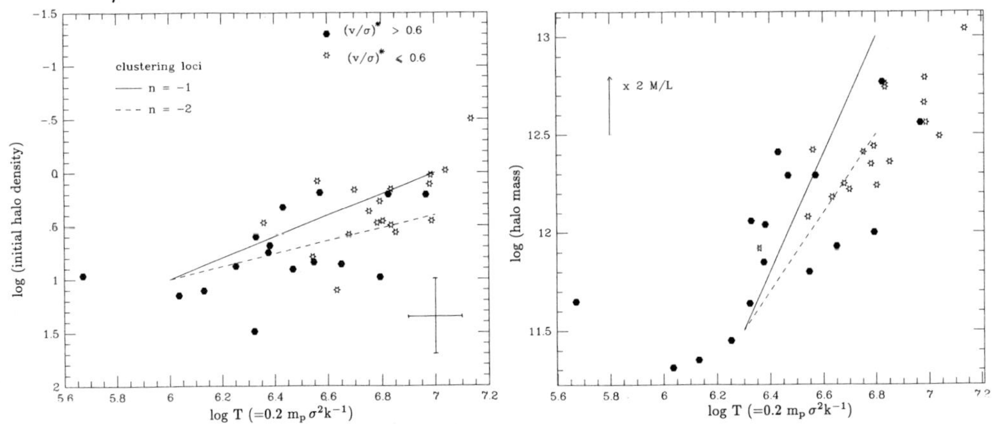

Figure 1. (a) Initial density as a function of temperature, defined as labelled. (b) Total mass against temperature.

CONCLUSIONS

It appears that the simplest theories which predict an anti-correlation between angular momentum and initial density are not strongly supported by the data, at least for well studied elliptical galaxies. As indicated above, theory and data may be reconciled if one introduces a variable amount of dissipation among ellipticals, leading to different collapse factors before star formation. This may be possible if those perturbations which form the slowly rotating galaxies collapsed in the Compton cooling era[4], while those which rotate did not. However, since it is extremely difficult to follow how the angular momentum of a galaxy depends on initial conditions, due to problems of identifying which material will eventually be incorporated, it may be that a more complicated scenario of Gaussian perturbations suffices.

REFERENCES

1. Blumenthal, G.R., Faber, S.M., Primack, J.R., & Rees, M.J., 1985. *Nature*, **311**, 517.
2. Bardeen, J.M., Bond, J.R., Kaiser, N., & Szalay, A.S., 1986. *Ap. J.*, **304**, 15.
3. Davies, R.L., Efstathiou, G., Fall, S.M., Illingworth, G.D., & Schechter, P.L., 1983. *Ap. J.*, **266**, 41.
4. Lizano, S., & Wyse, R.F.G., 1986. In preparation.

VIOLENT RELAXATION AND MIXING IN 1-D GRAVITATIONAL SYSTEMS

Marc Luwel
Vrije Universiteit Brussel
Brussels, Belgium

1. INTRODUCTION

The one dimensional gravitational model consists of N mass sheets with surface density m_i, parallel to the (y, z)-plane and constrained to move along the x-axis under influence of their mutual gravitational force $F_{ij} = -2\pi G m_i m_j \,\mathrm{sgn}(x_i - x_j)$. In order to study the evolution of this one-dimensional system, the N Newtonian equations of motion are integrated numerically, using an "exact" double precision algorithm.

2. STATIONARY WATERBAG CONFIGURATION

A one dimensional gravitational system in a stationary waterbag configuration occupies a region in the two dimensional phase-space with a constant phase density, enclosed by a single energy contour. Such a system, with a bimodal mass distribution, containing N_ℓ light particles with mass m_ℓ and N_h heavy particles with mass m_h, is studied. In order to follow its macroscopic evolution, the equipartition ratio $k(t)$, defined as the ratio of the average kinetic energies of the light and the heavy species, is measured during the integration (Luwel and Severne 1985).

For a system in a stationary waterbag configuration with both mass groups initially well mixed, no violent relaxation is observed. Moreover, the mixing ratio, which initially has its well mixed value $k(t = 0) = m_\ell/m_h$, shows no evolution towards equipartition ($k_{eq} = 1$) on a timescale of order Nt_d, where t_d is the crossing or dynamical timescale (Figure 1).

In systems initially in a stationary waterbag configuration, two particle effects turn out to be *inoperative*, at least on a timescale of order Nt_d (Figure 2). Consequently, in the absence of any macroscopic evolution in such a system, one would expect that each particle remains in its motion close to a stationary phase space trajectory of nearly constant energy. But plots of the particle trajectories and of the evolution of the individual particle energies show that under influence of mean field fluctuations the particles migrate slowly through large energy intervals, on a timescale t_{mix} of order $10t_d$ (Figure 2).

T. de Zeeuw (ed.), Structure and Dynamics of Elliptical Galaxies, 523–524.

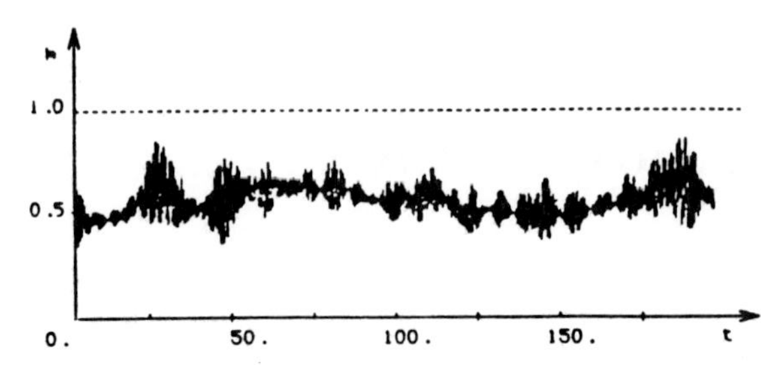

Figure 1. Evolution of the equipartition ratio $k(t)$ for system I containing $N_\ell = 100$ light particles with mass $m_\ell = 1/200$ and $N_h = 50$ heavy particles with mass $m_h = 2m_\ell$, and initially in a stationary waterbag configuration with both mass groups distributed uniformly.

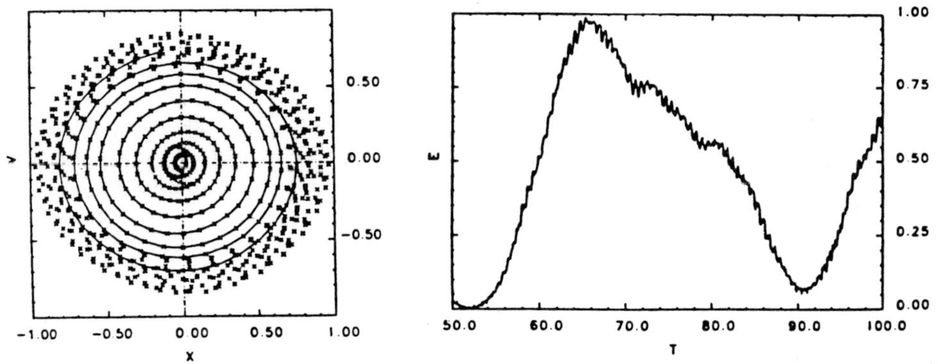

Figure 2. a) Phase orbit of a typical particle of system I. For $50t_d < t < 80t_d$ the crosses give the phase positions at successive time intervals of $0.04t_d$; they are connected by a continuous line for $54t_d < t < 65t_d$. b) Evolution of the energy of the same particle.

3. CONCLUSIONS

This hitherto unobserved mixing process, which becomes clearly evident due to the inefficiency of two particle effects in systems initially in a stationary waterbag configuration, is also observed as an evolution phase intermediate between the violent and the collisional phase in systems with less peculiar initial conditions. Moreover, this mixing process turns out to be the missing dynamical element of Shu's theory of violent relaxation (Shu 1978). Numerical experiments on systems characterized by two phase densities show that, after an initial evolution towards the Lynden-Bell distribution, these systems start evolving towards the Shu distribution until discrete particle effects become dominant (Severne and Luwel 1986).

REFERENCES

Luwel, M. and Severne, G., 1986 to be published.
Severne, G. and Luwel, M., 1985 *Astron. Astroph.*, **152**, 305.
Shu, F., 1978 *Astroph. J.*, **225**, 83.

THE DYNAMICAL EVOLUTION OF A STAR CLUSTER

G. Som Sunder and R.K. Kochhar
Indian Institute of Astrophysics
Bangalore 560034
India

ABSTRACT. Using the second order tensor virial equations and the equation for the rate of change of the kinetic energy tensor, we follow the dynamical evolution of a star cluster with anisotropic velocity distribution. We show that the cluster executes finite amplitude oscillations both in size and eccentricity. However, unlike in the isotropic case, the amplitude now depends on the initial eccentricity.

1. THE BASIC EQUATIONS

$$\frac{1}{2}\frac{d^2 I_{ii}}{dt^2} = 2K_{ii} + W_{ii}, \tag{1}$$

$$\frac{d}{dt}K_{ii} = \frac{1}{2}\frac{W_{ii}}{I_{ii}}\frac{dI_{ii}}{dt}. \tag{2}$$

Here various symbols have their usual meaning, and summation convention is not used. It can be seen that

$$M_{ii} I_{ii} = const. \tag{3}$$

See Chandrasekhar & Elbert (1972) and Som Sunder & Kochhar (1985, 1986=Paper 1, Paper 2) for details. The total energy $E = K + W$ is conserved.

We consider homogeneous spheroidal systems with axes $a_1 = a_2$, and a_3. Then

$$I_{ii} = \frac{1}{5}Ma_i^2; \qquad W_{ii} = -\frac{3}{10}\frac{GM}{a_1^2 a_3}a_i^2 A_i. \tag{4}$$

Here A_i are the index symbols (Chandrasekhar & Elbert 1972). The gravitational energy of a spheroid is

$$W = -\frac{3}{5}\frac{GM^2}{a_1}S(y), \tag{5}$$

where $S(y)$ is a function of the eccentricity: $y = e^2$ for $y > 0$ (oblate spheroids). For prolate spheroids $S(y) = 1$. Equations (1) and (2) now reduce to

$$\frac{d^2 a_1}{dt^2} = \frac{2[Q + S(y_0)]}{a_1^3} - \frac{3}{2}\frac{A_1}{a_1^2} \tag{6}$$

T. de Zeeuw (ed.), Structure and Dynamics of Elliptical Galaxies, 525–526.

$$\frac{d^2a_3}{dt^2} = \frac{2\,[Q + S(y_0)]}{a_3^2}(1 - y) - \frac{3}{2}\frac{A_1}{a_1^2} \tag{7}$$

where $Q = E/|W_0|$. Here subscript 0 refers to values at $t = 0$; a_1, a_3 are in the units of $a_0 (= a_1$ at time $t)$; t is in the units of $\sqrt{(a_0^3/GM)}$. Special cases of equations (6)–(7) have been considered by various authors (see Paper 2).

2. RESULTS AND DISCUSSION

We have numerically solved equations (6) and (7) subject to the conditions $da_1/dt = da_3 dt = 0$ at $t = 0$, and for various values of the initial eccentricity y_0, and Q. Note that equilibrium corresponds to $K = 0.5|W|$, or, for a sphere, to $Q = -0.5$.

Figure 1 shows the behavior of the a_1-axis and y for $Q = 0.75$. $Q = -0.75$ corresponds to $y_0 = 0$ to $K = 0.25|W|$ so that the kinetic energy is less than (or gravitational energy is more than) the equilibrium value. Consequently the a_1 axis contracts. This contraction is accompanied by an increase in the pressure, which causes the system to expand once again. These oscillations have a period $P \approx \pi/[2(-Q)^{\frac{3}{2}}] \approx 3.5$. The amplitude of the oscillations increases with decreasing y_0. The eccentricity also decreases initially. At about 0.5 P however there is a steep rise and a fall, after which the system slowly returns to $y \approx y_0$. The system however does not in general return to the exact initial conditions after a cycle, except for the spherical case. Other cases are discussed in Paper 2.

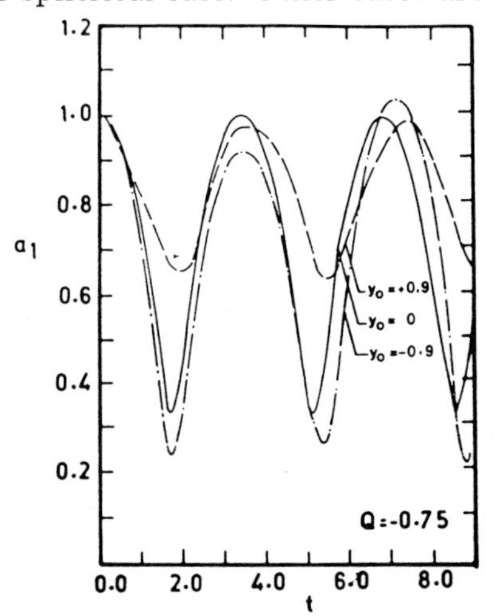

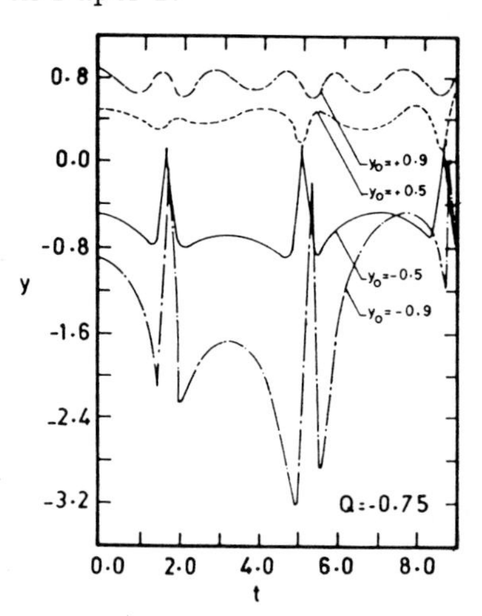

Figure 1. a) Behavior of the a_1-axis with time for $Q = -0.75$ and for various values of initial eccentricities y_0. b) Behavior of the eccentricity with time for $Q = -0.75$. The ordinate y measures the eccentricity; it is positive (negative) for oblate (prolate) spheroids.

REFERENCES

Chandrasekhar, S. and Elbert, D.D., 1972. *M.N.R.A.S.*, **155**, 435.
Som Sunder, G. and Kochhar, R.K., 1985. *M.N.R.A.S.*, **213**, 381 (Paper 1).
Som Sunder, G. and Kochhar, R.K., 1986. *M.N.R.A.S.*, **221**, 553 (Paper 2).

NUMERICAL INVESTIGATION OF THE DENSITY DISTRIBUTION OF STARS AND THE DISPERSION OF VELOCITIES IN SPIRAL GALAXIES

Hong-Nan Zhou
Astronomy Department, Nanjing University, Nanjing
People's Republic of China

We use a three–component model of a spiral galaxy, given by (Huang et al., 1979):

1. An outer halo with a gravitational potential of the form

$$\Phi_h(R) = \frac{-GM_1}{\sqrt{R^2 + b_1^2}}. \tag{1}$$

where $R^2 = x^2 + y^2 + z^2$, $M_1 = 1.2 \times 10^{11} M_\odot$ and $b_1 = 1.1$ Kpc.

2. A nucleus, with a gravitational potential given by

$$\Phi_n(R) = \frac{-GM_0}{\sqrt{R^2 + b_0^2}}. \tag{2}$$

where $M_0 = 1.1 \times 10^7 M_\odot$, and $b_0 = 0.61 \times 10^{-2}$ Kpc.

3. A self–gravitating disk containing N stars. The density distribution in the z–direction is assumed to be:

$$\rho(r, z) = \frac{\alpha}{2}\sigma(r) \exp(-\alpha|z|). \tag{3}$$

where $\alpha = 2.1$ Kpc^{-1} is the equivalent semi–thickness of the galaxy, and $\sigma(r)$ is the surface density. We consider three different initial expressions for $\sigma(r)$:

   i. A Toomre (1963) disk.
   ii. A uniform distribution.
   iii. An exponential distribution.

The model is covered with a cubic mesh which is divided in 32 × 32 × 8 cells. We use 20,000 particles to simulate the stars in the disk. The evolution of the disk is followed by means of the Particle-Mesh method.

REFERENCES

Huang, K.L., Huang, J.H., & Peng, Q.H., 1979. *Acta Astronomica Sinica*, **20**, 232.
Toomre, A., 1963. *Astroph. J.*, **18**, 385.

T. de Zeeuw (ed.), Structure and Dynamics of Elliptical Galaxies, 527.

Illingworth and Freeman, after the last session.

APPENDICES

SOLUTION OF THE PROBLEM OF STABILITY OF A STELLAR SYSTEM WITH EMDEN'S DENSITY LAW AND A SPHERICAL DISTRIBUTION OF VELOCITIES

V. A. Antonov
Zhdanov State University
Leningrad, U.S.S.R.

ABSTRACT. Applying a criterion previously derived by the author, the stability of stellar systems with an isotropic distribution velocity distribution and Emden's polytropic density law is demonstrated for the exponent $n = 3/2$.

§I.

Emden's law connecting density and potential has the following form [1]:

$$\nu = c(U - H)^n. \tag{1}$$

In this case, $\nu = \nu(r)$ is the stellar density of a spherical system; $U = U(r)$ is the potential; H is its value at the boundary of the system; c and n are constants. By varying the exponent n, it is possible to obtain models which are more or less concentrated toward the center. In the present paper we shall assume that the phase density is a function of the energy integral only, *i.e.*, the velocity diagram is spherically symmetric at any distance from the center. In this case, in order to obtain Emden's law, it is necessary to give the phase density

$$\Psi = c_1(U - \frac{v^2}{2} - H)^{n-\frac{3}{2}}, \tag{2}$$

where c_1 is a certain new constant. This can be verified easily by integration with respect to the velocities. Furthermore, expression (2) is the only one corresponding to (1) under our assumptions, as can be seen from the theory of integral equations [2].

In what follows, we shall assume that $n \geq 3/2$. Otherwise, equation (2) gives a very artificial form of the velocity diagram; moreover, the usual definition of Lyapunov stability is not suitable here.

We shall prove the following theorems concerning stability with respect to the regular forces in stellar systems in which the phase density is a decreasing function of the energy integral:

1. For the stability of a stellar system, generally only perturbations preserving the spherical symmetry of the system can be dangerous (see Section IV).

T. de Zeeuw (ed.), Structure and Dynamics of Elliptical Galaxies, 531–548.

2. A sufficient condition for the stability of a stellar system is that $d^3\nu/du^3 \geq 0$ for all values of r (see Sections II and III). In particular, this condition encompasses Emden's law with $n > 2$.
3. We shall give below a proof that is valid for Emden's law with $n < 3$ only. Thus, the points 2 and 3 overlap.
   Now, let $n < 3$. Consider the stability criterion given in [3]:

$$-\int\int \frac{q^2}{(dF/dE)} d\vec{r}d\vec{v} - mG\int\int\int\int \frac{q(\vec{r},\vec{v})\cdot q(\vec{r}_1,\vec{v}_1)}{|\vec{r}-\vec{r}_1|} d\vec{r}d\vec{r}_1 d\vec{v}d\vec{v}_1, \tag{3}$$

where $\Psi = f(E)$ is the phase density. We have to prove that this expression is positive for any function q which can be represented in the form

$$q = \vec{v}\, grad_r f + grad_r U\, grad_v f, \tag{4}$$

where f is a differentiable function of the phase coordinates. As in the following formulae, if no limits are indicated, the integrals in equation (3) extend over the whole physical and phase volume occupied by the system.

The positivity of expression (3) can be proved with a very weak limitation imposed on q, namely,

$$\int\int q\, d\vec{r}d\vec{v} = 0. \tag{5}$$

By use of the Cauchy inequality, we have

$$\int \frac{q^2}{|dF/dE|} d\vec{v} \geq \Big(\int q d\vec{v}\Big)^2 \Big/ \int |dF/dE| d\vec{v}. \tag{6}$$

The absolute value sign is necessary here since $F(E)$ is a decreasing function of E.

Let us prove an auxiliary equality. Write the relation between the stellar density and the phase density as

$$\nu = \int \Psi(\tfrac{1}{2}\vec{v}^2 - U)\, d\vec{v}. \tag{7}$$

Differentiate equation (7) with respect to U

$$\frac{d\nu}{dU} = -\int \Psi'(\tfrac{1}{2}\vec{v}^2 - U)\, d\vec{v},$$

or

$$\int |df/dE| d\vec{v} = \frac{d\nu}{dU}.$$

Taking into consideration condition (6), we see that the stability will be proved if we succeed in demonstrating the positivity of the expression which is easily obtained from (3) and that is not greater than the latter:

$$\int \frac{p^2}{(d\nu/dU)} d\vec{r} - mG\int\int \frac{p(\vec{r})\cdot p(\vec{r}_1)}{|\vec{r}-\vec{r}_1|} d\vec{r}d\vec{r}_1, \tag{8}$$

where

$$p(\vec{r}) = \int q \, d\vec{v}. \tag{9}$$

Let us limit our proof, for the time being, to the case of perturbations of the kind of radial pulsations which do not disturb the spherical symmetry of the system. The stability for perturbations of a more general form will be demonstrated later on.

Thus, let $p = p(r)$. If we disregard the coefficient $-mG$, the second term in expression (8) represents twice the energy of the field created by the "charge" distributed in space with a density $p(r)$. However, it is known [4] that the energy of such a field can be represented in a different form, namely,

$$\frac{1}{8\pi} \int |grad\, W_p|^2 d\vec{r} = \frac{1}{2} \int_0^\infty \left(\frac{dW_p}{dr}\right)^2 dr, \tag{10}$$

where W_p is the potential that corresponds, mathematically, to the charge p. It is easy to find that $dW_p/dr = 4\pi/r^2\tau$, where

$$\tau(r) = \int_0^r r^2 p \, dr. \tag{11}$$

The first term of (8) can be expressed in terms of the same function of τ. Indeed, if we differentiate equation (11), we obtain

$$p = \frac{1}{r^2} \frac{d\tau}{dr}. \tag{12}$$

Substituting these results in (8), we obtain

$$4\pi \int \left[\left(\frac{d\tau}{dr}\right)^2 / r^2 \frac{d\nu}{dU}\right] dr - 16\pi^2 mG \int \frac{\tau^2}{r^2} dr. \tag{13}$$

From (5) and (11) we have the boundary condition that τ becomes zero on the boundary of the system and outside it. From (11) it follows also that $\tau(0) = 0$, this zero being of the third order.

We have reached a variation problem typical for oscillatory phenomena. We are looking for the minimum of

$$\int \left[\left(\frac{d\tau}{dr}\right)^2 / r^2 \frac{d\nu}{dU}\right] dr \Big/ 4\pi mG \int \frac{\tau^2}{r^2} dr = \lambda,$$

and we have to prove that $\lambda > 1$. By the ordinary rules, the solution of this problem is given by the differential equation

$$\frac{d}{dr}\left(\frac{d\tau}{dr} / r^2 \frac{d\nu}{dU}\right) + 4\pi mG\lambda \frac{\tau}{r^2} = 0, \tag{14}$$

where λ turns out to be the first eigenvalue for the above mentioned boundary conditions $\tau(0) = \tau(b) = 0$, with b equal to the radius of the system.

Let us take the comparison function $\tau_0(r)$, which satisfies the initial condition $\tau_0(0) = 0$ and the differential equation

$$\frac{d}{dr}\left(\frac{d\tau_0}{dr}/r^2\frac{d\nu}{dU}\right) + 4\pi mG\frac{\tau_0}{r^2} = 0. \tag{15}$$

In the case of Emden's law, τ_0 can be found in an explicit form; it is equal to

$$\tau_0 = (n-3)r^2\frac{dU}{dr} + 4\pi mG(n-1)r^3\nu, \tag{16}$$

which can be verified easily by direct substitution in (15), taking into account the Poisson equation.

Let us now multiply (15) by τ, and (14) by τ_0; let us subtract one from the other and integrate from 0 to b. We obtain

$$\int_0^b\left[\tau_0\frac{d}{dr}\left(\frac{d\tau}{dr}/r^2\frac{d\nu}{dU}\right) - \tau\frac{d}{dr}\left(\frac{d\tau_0}{dr}/r^2\frac{d\nu}{dU}\right)\right]dr + 4\pi mG(\lambda-1)\int_0^b\frac{\tau\tau_0}{r^2}\,dr = 0.$$

We apply partial integration, and take into account that τ_0 has a zero of the third order for $r = 0$. Then it follows from (16):

$$\left(\tau_0\frac{d\tau}{dr}/r^2\frac{d\nu}{dU}\right)_{r=b} + 4\pi mG(\lambda-1)\int_0^b\frac{\tau\tau_0}{r^2}\,dr = 0. \tag{17}$$

However, inside the range $(0,b)$ τ is positive as the first eigenfunction, so that $(d\tau/dr)_{r=b} < 0$. As for τ_0, it is positive because of (16) (this is where the assumption $n < 3$ is required). On account of (2), we have $d\nu/dU > 0$. From (17) we then obtain $\lambda > 1$, which was required to prove stability.

§II.

The derivation of the stability criterion changes little if, instead of a system with a spherically symmetrical distribution of velocities, we consider a system of spherical form in which the phase density is a function of the energy and of the area integral in the orbital plane, and which is subject only to perturbations which would preserve spherical symmetry, (pulsations). Instead of dF/dE one must now write $\partial F/\partial E$.

Now, let us assign a certain scalar quantity V_s to each star. We shall discuss the distribution of this variable in detail below. Let us assume that V_s has the dimension of velocity although it does not vary with the motion of the star and, in turn, it does not affect its motion. Let us construct a phase density $\check{\Psi}(\vec{r},\vec{v},V_S)$, which depends on V_s as on a parameter. The choice of this fictitious phase density is, to a large degree, arbitrary, but in the integration with respect to V_s it must give the true phase density,

$$\int_{-\infty}^{+\infty}\check{\Psi}(\vec{r},\vec{v},V_S)\,dV_S = \Psi(\vec{r},\vec{v}). \tag{18}$$

Since we are considering a stationary case, we can express $\check{\Psi}$ in terms of the integrals of motion, and write $\check{\Psi}(E, I, V_S)$ instead of $\check{\Psi}(\vec{r}, \vec{v}, V_S)$. Here $E = \Phi + \frac{1}{2}\vec{v}^2$ is the energy integral,* and

$$I = (\vec{r} \times \vec{v})^2 = (rV_t)^2$$

is the square of the angular momentum, where V_t is the tangential velocity.

Let us further assume that $\check{\Psi}$ has the form

$$\check{\Psi} = \check{F}(\check{E}, I), \tag{19}$$

where $\check{E} = E + \frac{1}{2}V_S^2$. Substituting in (18), we obtain

$$\Psi = \int_{-\infty}^{+\infty} \check{F}(E + \tfrac{1}{2}V_S^2), I)\, dV_S.$$

If Ψ is a sufficiently smooth function of E, this integral equation will determine a function $\check{F}(\check{E}, I)$, which is decreasing with respect to $\check{E}$.

With respect to the fictitious phase density $\check{\Psi}$, the stability problem is formulated in the same way as with respect to the true Ψ. It is obvious that the stability of $\check{\psi}$ implies the stability of Ψ. We thus arrive at the functional

$$\begin{aligned} &-\iiint \frac{q^2}{(d\check{F}/d\check{E})}\, d\vec{r} d\vec{v} dV_S \\ &- mG \iiiiiint \frac{q(\vec{r}, \vec{v}, V_s) \cdot q(\vec{r}_1, \vec{v}_1, V_s)}{|\vec{r} - \vec{r}_1|} d\vec{r} d\vec{r}_1 d\vec{v} d\vec{v}_1 dV_s dV_{s1}, \end{aligned} \tag{20}$$

with

$$(q = \vec{v}\, grad_r f - grad_r \phi\, grad_v f).$$

Let us divide the whole phase space into "layers" with almost identical values of I within the limits of each layer. Essentially, this means a classification of stars with respect to I, which is an integral of motion even in a pulsating gravitational field; this cannot be said of E. We shall indicate the number of the layer by the subscript index $q = \sum_i q_i$, where q_i differs from zero only for $I_i < (rV_t)^2 < I_i + \Delta I_i$. We shall represent $\check{F}$ in exactly the same way:

$$\check{F} = \sum_i F_i(\check{E}),$$

where F_i differs from zero for $I_i < (rV_t)^2 < I_i + \Delta I_i$.

It is understood that I_i and ΔI_i depend on i only. The functional (20) takes the form

$$\begin{aligned} &-\sum_i \iiint \frac{q_i^2}{(dF_i d\check{E})} d\vec{r} d\vec{v} dV_s \\ &- mG \iiiiiint \frac{\sum_i q_i(\vec{r}, \vec{v}, V_s) \cdot \sum_i q_i(\vec{r}_1, \vec{v}_1, V_{s1})}{|\vec{r} - \vec{r}_1|} d\vec{r} d\vec{v} dV_s d\vec{r}_1 d\vec{v}_1 dV_{s1}. \end{aligned} \tag{21}$$

* The potential energy Φ equal to minus the potential is introduced in place of the potential U since in this way the formulae become more symmetrical.

Let us substitute, for each q_i, the condition of orthogonality to an arbitrary function of $\check{E}$. This condition is weaker than is required. Indeed,

$$\sum_i \int\int\int q_i\sigma_i(\check{E})\, d\vec{r}d\vec{v}dV_s = \int\int\int q\sigma(\check{E}, I)\, d\vec{r}d\vec{v}dV_s$$

$$= \int\int\int (\vec{v}\, grad_r f - grad_r \Phi\, grad_v f)\, \sigma(\check{E}, I)\, d\vec{r}d\vec{v}dV_s$$

$$= \int\int\int [\vec{v} grad_r (f\sigma) - grad_r \Phi\, grad_v (f\sigma)] d\vec{r}d\vec{v}dV_s,$$

since the terms with $grad_r\sigma$ and $grad_v\sigma$ cancel. If one transforms to area integrals, it becomes clear that the last integral equals zero.

In turn, each q_i is now expanded into two terms in such a manner that both of them conserve orthogonality to the arbitrary functions of $\check{E}$, and, moreover, have the following properties:

$$q_i = Q_i + P_i,$$
$$Q_i = \frac{dF_i}{d\check{E}} (A_i(\check{E}) + B_i(r)), \tag{22}$$

$$\int\int P_i d\vec{v}dV_s = 0. \tag{23}$$

First, we must prove that such an expansion is always possible; secondly, we have to find the functions $A_i(\check{E})$ and $B_i(r)$. In order that Q_i is orthogonal to an arbitrary function of $\check{E}$, it suffices to require orthogonality of Q_i to the functions of the form 1 for $\check{E} < h$ and 0 for $\check{E} > h$; h is an arbitrary parameter that may have values between the limiting values of $\check{E}$.

We integrate first with respect to v and V_s:

$$\int\int Q_i\, dv dV_S = 2\pi \int\int Q_i V_t\, dV_t dV_r dV_S,$$

where V_r is the radial and V_t is the tangential velocity. The range of variation of V_t in which Q_i differs from zero can be easily found from the equalities

$$V_t^2 = \frac{I}{r^2}, \qquad 2V_t \Delta V_t = \frac{\Delta I}{r^2}.$$

Consequently,

$$\int\int_{\check{E}<h} Q_i\, d\vec{v}dV_S = \frac{\pi \Delta I}{r^2} \int\int Q_i\, dV_i dV_S.$$

We introduce "polar" coordinates, with $V^2 = V_r^2 + V_S^2$ and $\tan\theta = V_S/V_r$; also, $\check{E} = \frac{1}{2}V^2 + \Phi + I_t/2r^2$. Then

$$\int\int_{\check{E}<h} Q_i\, d\vec{v}dV_s = \frac{2\pi^2 \Delta I_i}{r^2} \int_{\Phi_1}^{h} \frac{dF_i}{d\check{E}} (A_i(\check{E}) + B_i(r)) d\check{E}, \tag{24}$$

where

$$\Phi_1 = \Phi + \frac{I_i}{2r^2}. \tag{25}$$

Physically, Φ_1 denotes the energy of a star that possesses the angular momentum $\sqrt{I}$ and that has its apocenter or pericenter at a distance r. Therefore, in particular, the regions of phase space where $\Phi_1 > 0$ are not occupied by stars in a stationary system. Φ_1 has the following properties:

$$\text{for} \quad r \to 0 \qquad \Phi_1 \to +\infty,$$
$$\text{for} \quad r \to \infty \qquad \Phi_1 \to 0,$$
$$r^2 \frac{d\Phi_1}{dr} = r^2 \frac{d\Phi}{dr} - \frac{I_i}{r},$$

which is obtained by differentiation of (25). Making use of the Poisson equation, we see that $r^2 d\Phi_1/dr$ is an increasing function. Consequently, Φ_1 has only one minimum at a certain point $r = r_0$.

Equation (24) may be written in the following form:

$$\int\int_{\check{E}<h} Q_i \, d\vec{v} dV_S = \frac{2\pi^2 \Delta I_i}{r^2} [M_i(h) - M_i(\Phi_1) + B_i(r)(F_i(h) - F_i(\Phi_1))], \tag{26}$$

where

$$M_i(\Phi_1) = \int_0^{\Phi_1} \frac{dF_1}{d\check{E}} A_i(\check{E}) \, d\check{E}. \tag{27}$$

Finally, we integrate (26) with respect to r, and equate the result to zero:

$$\int_{r_p}^{r_a} [M_i(h) - M_i(\Phi_1) + B_i(r)(F_i(h) - F_i(\Phi_1))] \, dr = 0. \tag{28}$$

Here r_a is the large root, and r_p is the small root of the equation $\Phi_1(r) = h$. We shall call these quantities reciprocal. We differentiate (28) with respect to h, taking into account that a substitution of r_a or r_p for r in the integrand changes the quantity Φ_1 into h, and the whole integrand into zero. Therefore, in (28), only the expression under the integral sign is differentiated with respect to h. We obtain

$$\int_{r_p}^{r_a} (F_i'(h) A_i(h) + B_i(r) F_i'(h)) \, dr = 0,$$

or

$$(r_a - r_p) A_i(h) + \int_{r_p}^{r_a} B_i(r) \, dr = 0. \tag{29}$$

As for condition (23), it can be written in the form

$$\int\int Q_i\, dv dV_S = p_i, \tag{30}$$

where $p_i = \int\int q_i\, d\vec{v} dV_S$.

The integration in (30) is carried out as above (just as in (24) and (26)), and we obtain

$$p_i = -\frac{2\pi^2 \Delta I_i}{r^2}(M_i(\Phi_i) + B_i(r)F_i(\Phi_i)). \tag{31}$$

The feasibility of the required expansion will be proved if we succeed in solving the equations (29) and (31) for the functions A and B.

For this purpose, let us write (29) in a somewhat different form, taking into consideration that r_a can be chosen arbitrarily in the region $r > r_0$. Let us denote r_a by r, and the reciprocity relation by the sign $\sim$, so that $\Phi_1(\tilde{r})$ is always equal to $\Phi_1(r)$. Equation (29) then takes the form

$$(r - \tilde{r})A_i(\Phi_1) + \int_{\tilde{r}}^{r} B_i(r)\, dr = 0,$$

and the same result is obtained for $r < r_0$. Thus,

$$(r - r_0)A_i(\Phi_1) + \int_{r_0}^{r} B_i(r)\, dr = (\tilde{r} - r_0)A_i(\Phi_1) + \int_{r_0}^{r} B_i(r)\, dr. \tag{32}$$

Since (32) means that the expression on the left side does not change when r is changed to $\tilde{r}$, this expression represents a unique function of Φ_1:

$$(r - r_0)A_i(\Phi_1) + \int_{r_0}^{r} B_i(r)\, dr = \xi(\Phi_1).$$

We differentiate ξ with respect to r, and we isolate $B_i(r)$:

$$B_i(r) = -\frac{d}{dr}[(r - r_0)A_i(\Phi_1)] + \xi'(\Phi_1)\frac{d\Phi_1}{dr}. \tag{33}$$

Substituting (33) in (31), we obtain

$$M_i(\Phi_1) - F_i(\Phi_1)\frac{d}{dr}[(r - r_0)A_i(\Phi_1)] = -\frac{r^2 p_i}{2\pi^2 \Delta I_i} - F_i(\Phi_1)\xi'(\Phi_1)\frac{d\Phi_1}{dr}. \tag{34}$$

The lefthand side of (34) can be written in the form of a total derivative. Indeed, if one recalls the definition (27) of $M_i(\Phi_1)$, then

$$\frac{d}{dr}[(r - r_0)M_i(\Phi_1) - (r - r_0)F_i(\Phi_1)A_i(\Phi_1)]$$
$$= M_i(\Phi_1) - F_i(\Phi_1)\frac{d}{dr}[(r - r_0)A_i(\Phi_1)].$$

We integrate (34) from r_0 to r, introducing the function

$$\tau_i(r) = \int_0^r r^2 p_i \, dr, \tag{35}$$

so that

$$(r - r_0)(M_i(\Phi_1) - F_i(\Phi_1)A_i(\Phi_1)) = \\ = -\frac{\tau_i(r) - \tau_i(r_0)}{2\pi^2 \Delta I_i} - \int_{r_0}^{r} F_i(\Phi_1)\xi'(\Phi_1)\frac{d\Phi_1}{dr}\, dr. \tag{36}$$

Substituting, in (36), $\tilde{r}$ for r, we obtain

$$(\tilde{r} - r_0)(M_i(\Phi_1) - F_i(\Phi_1)A_i(\Phi_1)) \\ = -\frac{\tau_i(\tilde{r}) - \tau_i(r_0)}{2\pi^2 \Delta I_i} - - \int_{r_0}^{\tilde{r}} F_i(\Phi_1)\xi'(\Phi_1)\frac{d\Phi_1}{dr}\, dr. \tag{37}$$

The integral at the end of (36) permits replacement of the integration variable

$$\int_{r_0}^{r} F_i(\Phi_1)xi'(\Phi_1)\frac{d\Phi_1}{dr} dr = \int_{\Phi_1(r_0)}^{\Phi_1(r)} F_i(t)\xi'(t)dt = \int_{r_0}^{\tilde{r}} F_i(\Phi_1)\xi'(\Phi_1)\frac{d\Phi_1}{dr}\, dr.$$

We subtract (37) from (36), thus eliminating the function $\xi(\Phi_1)$:

$$M_i(\Phi_1) - F_i(\Phi_1)A_i(\Phi_1) = -\frac{\tau_i(r) - \tau_i(\tilde{r})}{2\pi^2 \Delta I_i (r - \tilde{r})}. \tag{38}$$

From (27) we find

$$A_i(\Phi_1) = \frac{M_i'(\Phi_1)}{F_i'(\Phi_1)}.$$

Substituting the righthand side of this equation in (38), we multiply by

$$\frac{F_i'(\Phi_1)}{[F_i(\Phi_1)]^2} \cdot \frac{d\Phi_1}{dr}$$

and we integrate from r_0 to r. We obtain

$$\frac{M_i(\Phi_1)}{F_i(\Phi_1)} = \frac{1}{2\pi^2 \Delta I_i} \int_{r_0}^{r} \frac{\tau_i(r) - \tau_i(\tilde{r})}{r - \tilde{r}} \, \frac{F_i(\Phi_1)}{[F_i(\Phi_1)]^2} \, \frac{d\phi_1}{dr}\, dr + c \tag{39}$$

where c is a constant.

Eliminating $M_i(\Phi_1)$ from (31) and (39), we obtain

$$B_i(r) = -\frac{r^2 p_i}{2\pi^2 \Delta I_i F_i(\Phi_1)} - \frac{1}{2\pi^2 \Delta I_i} \int_{r_0}^{r} \frac{\tau_i(r) - \tau_i(\tilde{r})}{r - \tilde{r}} \frac{F'(\Phi_1)}{(F_i(\Phi_1))^2} \frac{d\Phi_1}{dr} dr - c. \tag{40}$$

From (38) we also obtain

$$A_i(\Phi_1) = \frac{1}{2\pi^2 \Delta I_i} \int_{r_0}^{r} \frac{\tau_i(r) - \tau_i(\tilde{r})}{r - \tilde{r}} \frac{F_i'(\Phi_1)}{(F_i(\Phi_1))^2} \frac{d\Phi_1}{dr} + c + \frac{\tau_i(r) - \tau_i(\tilde{r})}{2\pi^2 \Delta I_i (r - \tilde{r}) F_i(\Phi_1)}. \tag{41}$$

Thus, the possibility of the required expansion is proved.

§III.

Let us return to the functional (21). Since we have written $q_i = P_i + Q_i$, the terms of the first summation in (21) may be given in the form

$$\int\int\int \frac{q_i^2}{(dF_i/d\check{E})} d\vec{r} d\vec{v} dV_s = \int\int\int \frac{P_i^2 + 2P_i Q_i + Q_i^2}{(dF_i/d\check{E})} d\vec{r} d\vec{v} dV_s$$
$$= \int\int\int \frac{P_i^2 + Q_i^2}{(dF_i/d\check{E})} d\vec{r} d\vec{v} dV_s + 2 \int\int\int (A_i(\check{E}) + B_i(r)) P_i \, d\vec{r} d\vec{v} dV_s.$$

The integral $\int\int\int A_i(\check{E}) P_i \, d\vec{r} d\vec{v} dV_S$ becomes zero due to the orthogonality property of P_i, and the integral $\int\int\int B_i(r) P_i \, d\vec{r} d\vec{v} dV_S$ becomes zero on account of (23). Thus the expression

$$-\sum_i \int\int\int \frac{Q_i^2}{(dF_i/d\check{E})} d\vec{r} d\vec{v} dV_s - mG \int\int\int\int\int\int \frac{\sum Q_i(\vec{r}, \vec{v}, V_S) \cdot \sum Q_i(\vec{r}_1, \vec{v}_1, V_{s1})}{|\vec{r} - \vec{r}_1|} d\vec{r} d\vec{r}_1 d\vec{v} d\vec{v}_1 dV_s dV_{s1}, \tag{42}$$

is not greater than (21).

Let us substitute the expression (22) for Q_i in equation (42). We obtain

$$\int\int\int \frac{Q_i^2}{(dF_i/d\check{E})} d\vec{r} d\vec{v} dV_s = \int\int\int Q_i (A_i(\check{E}) + B_i(r)) \, d\vec{r} d\vec{v} dV_s$$
$$= \int\int\int Q_i B_i(r) \, d\vec{r} d\vec{v} dV_s = \int\int\int q_i B_i(r) \, d\vec{r} d\vec{v} dV_s$$
$$= \int p_i B_i(r) d\vec{r} = 4\pi \int_{a_1}^{b_1} r^2 p_i B_i(r) \, dr,$$

again making use of (23) and of the property of orthogonality. Substituting (40), we obtain

$$\int\int\int \frac{Q_1^2}{(dF_i/d\check{E})} d\vec{r}d\vec{v}dV_s = -\frac{2}{\pi\Delta I_i}\int_{a_i}^{b_i} \frac{r^1 p_i^2}{F_i(\Phi_1)} dr + 4\pi \int_{a_1}^{b_i} r^2 p_i H_i(r) dr. \tag{43}$$

Here, $H_i(r)$ is the remaining integral

$$H_i(r) = \frac{1}{2\pi^2\Delta I_i}\int_{r_0}^{r} \frac{\tau_i(r)-\tau_i(\tilde{r})}{r-\tilde{r}} \frac{d}{dr}\left(\frac{1}{F_i(\Phi_1)}\right) dr + c. \tag{44}$$

The interval (a_i, b_i) marks the region of the physical space in which stars with angular momentum $\sqrt{I_i}$ can move (with $b_i = \tilde{a}_i$). It is clear from the definition (35) of $\tau_i(r)$ that it must satisfy the boundary conditions $\tau_i = 0$ for $r \geq b_i$ and $r \leq a_i$.

Let us carry out the partial integration of the second integral on the right side of (43):

$$\int_{a_i}^{b_i} r^2 p_i H_i(r)\, dr = \int_{a_i}^{b_i} \tau_i'(r) H_i(r)\, dr = -\int_{a_i}^{b_i} \tau_i(r) H_i'(r)\, dr.$$

We substitute the last result in (43), taking into consideration (44). As for the last integral in (42), it is transformed as the field energy, as explained above. We finally obtain from (42)

$$\begin{aligned}\sum_i \frac{2}{\pi\Delta I_i}\int_{a_i}^{b_i} \frac{[\tau_i'(r)]^2}{F_i(\Phi_1)} dr - 16\pi^2 mG\int_0^\infty \frac{[\sum_i \tau_i(r)]^2}{r^2} dr \\ + \sum_i \frac{2}{\pi\Delta I_i}\int_{a_i}^{b_i} \tau_i(r)\frac{\tau_i(r)-\tau_i(\tilde{r})}{r-\tilde{r}} \frac{d}{dr}\left(\left(\frac{1}{F_i(\Phi_1)}\right) dr.\end{aligned} \tag{45}$$

Let us prove the auxiliary inequality

$$\int_{a_i}^{b_i} \tau_i(r)\frac{\tau_i(r)-\tau_i(\tilde{r})}{r-\tilde{r}} \frac{d}{dr}\left(\frac{1}{F_i(\Phi_1)}\right) dr - \int_{a_i}^{b_i} \frac{[\tau_i(r)]^2}{r} \frac{d}{dr}\left(\frac{1}{F_i(\Phi_1)}\right) dr > 0. \tag{46}$$

For this purpose, let us inspect the integral

$$\int_{a_i}^{b_i} \frac{\tau_i(r)-\tau_i(\tilde{r})}{r-\tilde{r}} \frac{r\tau_i(\tilde{r})-\tilde{r}\tau_i(r)}{r-\tilde{r}} \frac{d}{dr}\left(\frac{1}{F_i(\Phi_1)}\right) dr. \tag{47}$$

The fractions $[\tau_i(r) - \tau_i(\tilde{r})]/(r - \tilde{r})$ and $[r\tau_i(\tilde{r}) - \tilde{r}\tau_i(r)]/(r - \tilde{r})$ remain unchanged when $\tilde{r}$ replaces r. Consequently, they are single-valued functions of Φ_1. The integral (47) is equal to zero, because we have the derivative of a function of Φ_1 under the integral sign, while $\tau_i(a_i) = \tau_i(b_i) = 0$. Let us add (47) to the difference considered in (46), and let us take into account that

$$\tau_i(r)\frac{\tau_i(r) - \tau_i(\tilde{r})}{r - \tilde{r}} - \frac{[\tau_i(r)]^2}{r} + \frac{\tau_i(r) - \tau_i(\tilde{r})}{r - \tilde{r}}\,\frac{r\tau_i(\tilde{r}) - \tilde{r}\tau_i(r)}{r - \tilde{r}} = -\frac{[r\tau_i(\tilde{r}) - \tilde{r}\tau_i(r)]^2}{r(r - \tilde{r})^2}.$$

Then we obtain

$$\int_{a_i}^{b_i}\left[\tau_i(r)\frac{\tau_i(r) - \tau_i(\tilde{r})}{r - \tilde{r}} - \frac{[\tau_i(r)]^2}{r}\right]\frac{d}{dr}\left(\frac{1}{F_i(\Phi_1)}\right)dr = \int_{a_i}^{b_i}\frac{[r\tau_i(\tilde{r}) - \tilde{r}\tau_i(r)]^2}{r(r - \tilde{r})^2}\,\frac{F_i'(\Phi_1)}{[F_i(\Phi_1)]^3}\,\frac{d\Phi_1}{dr}\,dr. \tag{48}$$

We write the integral $\int_{a_i}^{b_i}$ in the form of a sum, $\int_{r_0}^{b_i} + \int_{a_i}^{r_0}$, and we replace the integration variable r by $\tilde{r}$ in the first one, taking into consideration that $d\Phi_1 = (d\Phi_1/dr)dr = (d\Phi_1/d\tilde{r})d\tilde{r}$. Instead of (48) we now have

$$\int_{a_i}^{r_0}\left(\frac{r\tau_i(\tilde{r}) - \tilde{r}\tau_i(r)}{r - \tilde{r}}\right)^2\left(\frac{1}{r} - \frac{1}{\tilde{r}}\right)\frac{F_i'(\Phi_1)}{[F_i(\Phi_1)]^2}\,\frac{d\Phi_1}{dr}\,dr \geq 0$$

taking into account that $F_i'(\Phi_1) < 0$, and that for $a_i < r < r_0$, we have $\tilde{r} > r$ and $d\Phi_1/dr < 0$. Thus, the inequality (46) has been proved.

According to a well-known algebraic inequality,

$$\left(\sum_i \tau_i\right)^2 < \sum_i \nu_i \sum_i \frac{\tau_i^2}{\nu_i}, \qquad \text{or} \qquad \left(\sum_i \tau_i\right)^2 < \nu \sum_i \frac{\tau_i^2}{\nu_i}, \tag{49}$$

which we shall use for the transformation of the last integral in formula (45). The inequalities (46) and (49) show that it suffices, for the stability proof, to prove the positiveness of the functional

$$\sum_i\left[\frac{2}{\pi\Delta I_i}\int_{a_i}^{b_i}\frac{[\tau_i'(r)]^2}{F_i(\Phi_1)}\,dr + \frac{2}{\pi\Delta I_i}\int_{a_i}^{b_i}\frac{[\tau_i(r)]^2}{r}\,\frac{d}{dr}\left(\frac{1}{F_i(\Phi_1)}\right)dr - 16\pi^2 mG\int_{a_i}^{b_i}\frac{\nu(\tau_i[r])^3}{r^2\nu_i}\,dr\right]. \tag{50}$$

Let us prove the positiveness of each square bracket in (50). For this purpose, we shall estimate the minimum

$$\min \frac{\frac{2}{\pi \Delta I_i} \int_{a_i}^{b_i} \frac{[\tau_i(r)]^2}{F_i(\Phi_1)}\,dr + \frac{2}{\pi \Delta I_i} \int_{a_i}^{b_i} \frac{[\tau_i(r)]^2}{r} \frac{d}{dr}\left(\frac{1}{F_i(\Phi_1)}\right) dr}{16\pi^2 mG \int_{a_i}^{b_i} \frac{\nu(\tau_i[r])^3}{r^2 \nu_i}\,dr} = \lambda.$$

We find again the classical variation problem from the theory of oscillations.

The corresponding differential equation is

$$\frac{2}{\pi \Delta I_i} \frac{d}{dr}\left(\frac{\tau_i'(r)}{F_i(\Phi_1)}\right) - \frac{2}{\pi \Delta I_i} \frac{\tau_i(r)}{r} \frac{d}{dr}\left(\frac{1}{F_i(\Phi_1)}\right) + 16\pi^2 mG\lambda \frac{\nu}{r^2 \nu_i} \tau_i(r) = 0, \quad (51)$$

where λ is the first eigenvalue. Let us form the comparison function $N_i = r^3 \nu_i$. It is easy to express ν_i in terms of F_i through integration with respect to the velocities and with respect to V_S, just as in the derivation of (24):

$$\nu_i = \frac{2\pi^2 \Delta I_i}{r^2} \int_{\Phi_1}^{0} F_i(E)\,dE,$$
$$N_i = 2\pi^2 \Delta I_i r \int_{\Phi_1}^{0} F_i(E)\,dE. \quad (52)$$

The ν_i and, consequently, the N_i, become zero at the spatial boundaries of the ith layer. It is easy to find the differential equation satisfied by N_i. We have

$$\frac{dN_i}{dr} = 2\pi^2 \Delta I_i \int_{\Phi_1}^{0} F_i(E)\,dE - 2\pi^2 \Delta I_i r \frac{d\Phi_1}{dr} F_i(\Phi_1),$$

$$\frac{d}{dr}\left(\frac{N_i'(r)}{F_i(\Phi_1)}\right) = r^2 \nu_i \frac{d}{dr}\left(\frac{1}{F_i(\Phi_1)}\right) + \frac{(d(r^2\nu_i)/dr}{F_i(\Phi_1)} - 2\pi^2 \Delta I_i \left(r \frac{d^2\Phi_1}{dr^2} + \frac{d\Phi_1}{dr}\right),$$

but from (52) it follows that

$$\frac{1}{F_i(\Phi_1)} \frac{d}{dr}(r^2 \nu_i) = -2\pi^2 \Delta I_i \frac{d\Phi_1}{dr}.$$

Thus, we have

$$\frac{d}{dr}\left(\frac{N_i'(r)}{F_i(\Phi_1)}\right) - \frac{N_i(r)}{r} \frac{d}{dr}\left(\frac{1}{F_i(\Phi_1)}\right)$$
$$= -2\pi^2 \Delta I_i \left(r \frac{d\Phi_1}{dr^2} + 2 \frac{d\Phi_1}{dr}\right) = -2\pi^2 \Delta I_i \left(4\pi mGr\nu + \frac{I_i}{r^4}\right),$$

and from this we obtain the desired equation

$$\frac{2}{\pi\Delta I_i}\frac{d}{dr}\left(\frac{N_i'(r)}{F_i(\Phi_1)}\right) - \frac{2}{\pi\Delta I_i}\frac{N_i(r)}{r}\frac{d}{dr}\left(\frac{1}{F_i(\Phi_1)}\right) + 16\pi^2 mG\frac{\nu}{r^2\nu_i}N_i(r) = -\frac{4\pi I_i}{r^4}. \tag{53}$$

Let us multiply (51) by N_i and (53) by τ_i, then subtract one from the other and integrate from a_i to b_i. After partial integration the boundary terms disappear on account of $\tau_i(b_i) = \tau_i(a_i) = 0$ and $N_i(b_i) = N_i(a_i) = 0$. We obtain

$$16\pi^2 mG(\lambda - 1)\int_{a_i}^{b_i}\frac{\nu}{r^2\nu_i}\tau_i N_i\,dr = 4\pi I_i\int_{a_i}^{b_i}\frac{\tau_i}{r^4}\,dr,$$

hence $\lambda > 1$, and this concludes the stability proof under the assumptions specified above.

Let us note that the character of the dependence of the phase density on I is unessential, because it is always possible to introduce a sufficiently detailed classification of stars by the kinetic momentum, in order to make it possible to neglect the change of I inside each separate layer.

As an example of an application of the preceding discussions, let us consider the simplest case, when the phase density is a function of the energy integral only. Then the fictitious phase density must be a function of $\check{E}$. Let us consider what limitations on the stellar density are imposed by our condition that $\check{\Psi} = \check{F}(\check{E})$ must be a decreasing function. Let us integrate it with respect to the velocities and with respect to V_S:

$$\begin{aligned}\nu &= \int\int \check{F}(\check{E})d\vec{v}dV_s \\ &= \int\int\int\int \check{F}\left(\Phi + \tfrac{1}{2}(u^2 + v^2 + w^2 + V_s^2)\right) du dv dw dV_s \\ &= \int_0^\infty \check{F}(\Phi + \tfrac{1}{2}W^2)\sigma(W)\,dW.\end{aligned}$$

Here $\sigma(W)$ is the surface of the four–dimensional hypersphere $u^2 + v^2 + w^2 + V_S^2 = W^2$. It is well known that $\sigma(W) = 2\pi^2 W^3$. Therefore,

$$\nu = 2\pi^2\int_0^\infty \check{F}(\Phi + \tfrac{1}{2}W^2)W^3\,dW.$$

We differentiate this equality twice with respect to Φ:

$$\frac{d\nu}{d\Phi} = -4\pi^2\int_\Phi^\infty \check{F}(t)\,dt, \qquad \frac{d^2\nu}{d\Phi^2} = 4\pi^2\check{F}(\Phi).$$

Thus, the condition of decreasing $\check{F}$ is equivalent to the condition $d^3\nu/d\Phi^3 \leq 0$, which is satisfied for the Emden laws with $n > 2$. Actually,

$$\nu = c(\Phi_0 - \Phi)^n, \qquad \frac{d^3\nu}{d\Phi^3} = -cn(n-1)(n-2)(\Phi_0 - \Phi)^{n-3} < 0.$$

This proves the stability of the Emden laws with $n > 2$ for a spherically symmetrical distribution of velocities with respect to spherically symmetrical pulsations.

§IV.

Let us now consider perturbations that violate the spherical symmetry of the system. We assume that $dF/dE < 0$. Any perturbation may always be made into a spherically symmetrical one by averaging over all orientations. We subtract this average and we obtain the supplementary density p, with the property

$$\int p\, d\omega = 0. \tag{54}$$

Here, $d\omega$ is an element of the unit sphere.

Let us expand p into a series of spherical functions

$$p = \sum p_n^{(m)} Y_n^{(m)}(\theta, \phi). \tag{55}$$

The spherical function with the index $n = 0$ (it is equal to a constant) is not included in the series (55) on account of (54). As is known from theory [5], the potential corresponding to the density distribution $p_n^{(m)}(r) Y_n^{(m)}(\theta, \phi)$ is proportional to the same spherical function $Y_n^{(m)}(\theta, \phi)$.

On account of the known property of orthogonality of spherical functions in the functional characterizing the stability [we take it in the form (8)], all the terms containing products of two different spherical functions disappear. It is therefore possible to substitute, in (8), each term of the series (47) separately, and that is what we are going to do.

As has been explained already, the functional (8) may be presented in the form

$$\begin{aligned} &\int\int \frac{r^2p^2}{(d\nu/dU)}\, dr d\omega - \frac{mG}{4\pi}\int\int |grad W_p|^2\, dr d\omega \\ &= \int \frac{r^2p^2}{(d\nu/dU)}\, dr d\omega - \frac{mG}{4\pi}\int\int [(\frac{\partial W_p}{\partial r})^2 + |grad_1 W_p|^2] r^2\, dr d\omega. \end{aligned} \tag{56}$$

Here, $grad_1 W_p$ is the projection of $grad\, W_p$ on a plane perpendicular to the radius. But, as has been stated above, $W_p = w_n^{(m)}(r) Y_n^{(m)}(\theta, \phi)$, so that

$$\int\int [(\frac{\partial W_p}{dr})^2 + |grad_1 W_p|^2] r^2\, d\omega dr = \int_0^\infty [r^2(\frac{dw_n^{(m)}}{dr})^2 + n(n+1)[w_n^{(m)}]^2]\, dr.$$

We assume here that the spherical functions are normalized such that

$$\int\int [Y_n^{(m)}(\theta,\phi)]^2\, d\omega = 1. \tag{57}$$

Then the theory of spherical functions gives [6]

$$\int\int |grad Y_n^{(m)}(\theta,\phi)|^2\, d\omega = n(n+1). \tag{58}$$

By partial integration it is easy to obtain

$$\int_0^\infty \left[r^2\left(\frac{dm_n^{(m)}}{dr}\right)^2 + n(n+1)[\omega_n^{(m)}]^2\right] dr = \int_0^\infty \left[r\frac{d\omega_n^{(m)}}{dr} + (n+1)\omega_n^{(m)}\right]^2 dr.$$

Let us now use Poisson's equation to find the connection between $w_n^{(m)}$ and $p_n^{(m)}$.

$$4\pi\rho = \Delta W_p,$$

$$4\pi p_n^{(m)} Y_n^{(m)}(\theta,\phi) = \frac{1}{r^2}\frac{d}{dr}\left(r^2\frac{dw_n^{(m)}}{dr}\right)Y_n^{(m)}(\theta,\phi) - \frac{n(n+1)}{r^2}w_n^{(m)}(r)Y_n^{(m)}(\theta,\phi). \tag{59}$$

Let us introduce the auxiliary function

$$\tau_n^{(m)}(r) = \frac{d\omega_n^{(m)}}{dr} + \frac{n+1}{r}\omega_n^{(m)}. \tag{60}$$

Then

$$\int\int \left[\left(\frac{\partial\omega_p}{\partial r}\right)^3 + |grad_1\omega_p|^3\right] dr d\omega = \int_0^\infty r^2[\tau_n^{(m)}]^2\, dr. \tag{61}$$

On the basis of (59) and (60) we have

$$4\pi p_n^{(m)} = \frac{dw_n^{(m)}}{dr^2} + \frac{2}{r}\frac{dw_n^{(m)}}{dr} - \frac{n(n+1)}{r^2}w_n^{(m)} = \frac{d\tau_n^{(m)}}{dr} - \frac{n-1}{r}\tau_n^{(m)}. \tag{62}$$

We express (56) through τ with the aid of (61), (62) and of the normalization condition (57); we obtain*

$$\frac{1}{(4\pi)^2}\int_0^\infty \frac{[r d\tau/dr - (n-1)\tau]^2}{(d\nu/dU)}\, dr - \frac{mG}{4\pi}\int_0^\infty r^2\tau^2\, dr. \tag{63}$$

---

* To shorten the notation we shall omit the subscript on τ in the remainder of this article

Let us note that τ vanishes outside the system and on its boundary. Indeed, in the outer space W_p coincides, accurately, except for a constant factor, with the harmonic function $r^{-n-1}Y_n^{(m)}(\theta,\phi)$. It then follows from (60) that $\tau = 0$. It is possible, therefore, in (63), to put b, the radius of the system, as the upper boundary. For $r = 0$ continuity is required of $\tau(r)$, because at the center of the system W_p becomes zero.

We are back at the evaluation problem of a certain minimum

$$min \int_0^b \frac{[r d\tau/dr - (n-1)\tau]^2}{(d\nu/dU} dr \Big/ 4\pi mG \int_0^b r^2\tau^2\, dr = \lambda. \tag{64}$$

Let us write the differential equation corresponding to (64)

$$\frac{d}{dr}\Big(\frac{r^2 d\tau/dr}{(d\nu/dU)}\Big) - (n-1)\frac{d}{dr}\Big(\frac{r}{(d\nu/dU)}\Big) = -\frac{(n-1)^2}{(d\nu/dU)}\tau + 4\pi mG\lambda r^2\tau = 0, \tag{65}$$

where λ is the first eigenvalue.

Let us take $N = r^{n-1}\nu$ as the comparison function. Let us construct a differential equation for it analogous to (65). We have

$$\frac{dN}{dr} = (n-1)r^{n-2}\nu + r^{n-1}\frac{d\nu}{dU}\frac{dU}{dr},$$

$$\frac{d}{dr}\Big(\frac{r^2(dN/dr)}{(d\nu/dU)}\Big) = (n-1)r^{n-1}\nu\frac{d}{dr}\Big(\frac{r}{(d\nu/dU)} + (n-1)^2\frac{r^{n-1}\nu}{(d\nu/dU)}$$

$$+ (n-1)r^n\frac{dU}{dr} + \frac{d}{dr}\Big(r^{n+1}\frac{dU}{dr}\Big),$$

$$\frac{d}{dr}\Big(\frac{r^2(dN/dr)}{(d\nu/dU)}\Big) - (n-1)\frac{d}{dr}\Big(\frac{r}{(d\nu/dU)}\Big)N + 4\pi mGr^2N - \frac{(n-1)^2}{(d\nu/dU)}N$$

$$= 2nr^n\frac{dU}{dr} + r^{n+1}\frac{d^2U}{dr^2} + 4\pi mGr^{n+1}\nu.$$

Eliminating d^2U/dr^2 with the aid of Poisson's equation, we obtain the desired equation

$$\frac{d}{dr}\Big(\frac{r^2(dN/dr)}{(d\nu/dU)}\Big) - (n-1)\frac{d}{dr}\Big(\frac{r}{(d\nu/dU)}\Big)N + 4\pi mGr^2N - \frac{(n-1)^2N}{(d\nu/dU)} = 2(n-1)r^n\frac{dU}{dr}. \tag{66}$$

We multiply (65) by N, and (66) by τ, we subtract one from the other, and we integrate from 0 to b, taking into consideration that $N(b) = 0$. Then we obtain

$$4\pi mG(\lambda - 1)\int_0^b r^2\tau N\, dr = -2(n-1)\int_0^b r^n\frac{dU}{dr}\tau\, dr.$$

The right side is not negative, because $dU/dr < 0$ and n is a positive integer. Therefore, for $n > 1$, we have $\lambda > 1$, and for $n = 1$, we have $\lambda = 1$. In the latter case it would be possible to show that instability occurs, because (66) vanishes for the function τ, accurate to a constant factor. However, then

$$\tau = N = \nu, \qquad p_1^{(m)} = \frac{1}{4\pi}\frac{d\nu}{dr}.$$

Multiplying $\tau_i^{(m)}$ by the spherical function which, for $n = 1$, equals x/r, y/r, or z/r, we obtain

$$p = c\frac{x}{r}\frac{d\nu}{dr} = c\frac{\partial\nu}{\partial x},$$

etc. It is clear that we are dealing simply with a displacement of the system as a whole. Thus, perturbations of a "higher order" are not dangerous for the stability of the system, when $F(E)$ is a decreasing function. In particular, this conclusion is applicable to a system with Emden's density law, if $n \geq \frac{3}{2}$.

REFERENCES

1. Emden, E., 1907. *Gaskugeln*, Leipzig.
2. Mikhlin, S.G., 1959. *Lectures on Linear Integral Equations*, Moscow, Fizmatgitz (State Publ. House for Phys. and Math. Literature).
3. Antonov, V.A., 1960. *Astr. Zh.*, **37**, 918.
4. Landau, L.D., & Lifshitz, E.M., 1960. *Field Theory*, Moscow, Fizmatgitz (State Publ. House for Phys. and Math. Literature).
5. Idelson, N.K., 1936. *Theory of the Potential*, Moscow, Gostekhizdat (State Publ. House for Scient. and Techn. Literature).
6. Hobson, E., 1952. *Theory of Spherical and Ellipsoidal Functions*, Moscow, Foreign Liter. Publ. House.

ON THE INSTABILITY OF STATIONARY SPHERICAL MODELS WITH PURELY RADIAL MOTION

V.A. Antonov
Zhdanov State University
Leningrad, U.S.S.R.

Models of spherical, self-gravitating stellar systems in which all the trajectories pass initially near the center are interesting to examine from different perspectives. Such models are compatible not only with the idea of explosive cosmogony, but also with the concept of formation of stars from a gaseous, spherical cloud, if they are not affected during formation by significant peculiar motions and thus "fall" into the center. Another question is whether such a system is capable of remaining stable for a long period of time. We will show that such a system is unstable with respect to regular forces and must be reconstructed during the course of one period of revolution of an individual star.

The system may be represented as being composed of separate radial streams, each with its own energy integral E. For the sake of convenience, let us assume that there exists a finite value N of such streams, although this is not important. It is easy to see that the absolute value of the velocity in each stream is given by the formula

$$|V_\nu| = \sqrt{2\left(E_\nu - \Phi(r)\right)},$$

whereas the partial density of the matter is

$$\rho_\nu = \frac{h_\nu}{r^2 |V_\nu(r)|},$$

where ν is the number of the stream, E_ν and h_ν are certain constants characteristic of it, and $\Phi(r)$ is the gravitational potential of the system as a whole. $\rho_\nu = 0$ should be considered to be the turning point. Taking into account Poisson's equation, we derive the basic equation for determining the model structure:

$$\frac{d}{dr}\left(r^2 \frac{d\Phi}{dr}\right) = 4\pi G \sum_{\nu=1}^{N} \frac{h_\nu}{\sqrt{2\left(E_\nu - \Phi(r)\right)}}, \tag{1}$$

From: *The Dynamics of Galaxies and Star Clusters*, Proceedings of a National Conference in Alma-Ata, October 23-26, 1972, p. 139–143, ed. Omarov, T.B. ("Nauka" of the Kazakh S.S.R., Alma Ata, 1973). Translated from the Russian by Alex Shprintsen, CITA, Toronto.

T. de Zeeuw (ed.), Structure and Dynamics of Elliptical Galaxies, 549–552.

which already appeared in the work of Agekyan [1] for the case of a single stream. There, the asymptotic behavior in the center was observed to be

$$\Phi \sim |\ln r|^{2/3},$$

which is also correct in the general case. This singularity is rather weak and presents no obstacles to the following reasoning. We cannot solve equation (1) in its general form, but we do not require a specific kind of solution. In order to examine the instability, we will make use of the quasi-classical approximation. Only the different orientation of the constant-phase surfaces distinguishes the present problem from the problem of the development of annular perturbations in a cylinder [2]. Namely, let us take the variation of the potential proportional to the sectorial harmonic $Y_n(\theta, \phi)$, i.e.

$$\delta\Phi = \sin^n \theta \cos n\phi \cdot \chi(r), \tag{2}$$

where $\chi(r)$ is a certain slowly-varying function. Then, Poisson's equation may be written in the following form:

$$n(n+1)\delta\Phi = \sin^n \theta \cos n\phi \frac{d}{dr}\left(r^2 \frac{d\chi}{dr}\right) - 4\pi G r^2 \delta\rho. \tag{3}$$

In the approximation under consideration, $n >> 1$, and on the right hand side of (3), we must leave the terms which are proportional to n^2. The additional radial displacement of particles, like the force $\partial\delta\Phi/\partial r$ which acts in this direction, does not increase at all with n. On the basis of general properties of symmetry [3], it may be stated immediately that the transverse displacement must depend on the angular coordinate like grad $Y_n(\theta, \phi)$, and in the zone $|\theta - \frac{\pi}{2}| \sim n^{-1/2}$ where expression (2) is significantly different from zero, $\partial Y_n/\partial\theta$ is of a smaller order than $\partial Y_n/\partial\phi$. In calculating the divergence, which is necessary for deriving the continuity equation, this difference will increase even further. Thus, it is sufficient to limit oneself to displacements along θ near the equator. In relating the linear displacement x_ν to the particles of each stream, we derive the simplified form of the continuity equation:

$$\delta\rho_\nu = -\frac{\rho_\nu}{2r}\left(\frac{\partial x_\nu^+}{\partial\phi} + \frac{\partial x_\nu^-}{\partial\phi}\right). \tag{4}$$

The superscripts + and − in (4), just as below, are brought in to distinguish between the two halves of the stream, as their velocities V_ν are of opposite sign. Meanwhile, Poisson's equation, after the truncation of low-order terms, has the following form:

$$n^2\delta\Phi = -\frac{\partial^2\delta\Phi}{\partial\phi^2} = -4\pi G r^2 \sum_{\nu=1}^{N} \delta\rho_\nu. \tag{5}$$

Using (4) and (5), it is now an easy task to derive an equation of motion along θ in terms of the transverse deflection $x_\nu(r)$ and the angular momentum per unit mass $J_\nu(r)$:

$$\frac{dx_\nu^+}{dt} = \frac{J_\nu^+ + V_\nu x_\nu^+}{r}, \qquad \frac{dx_\nu^-}{dt} = \frac{J_\nu^- - V_\nu x_\nu^-}{r},$$

$$\frac{dJ_\nu^+}{dt} = \frac{dJ_\nu^-}{dt} = 2\pi G r \sum_{\mu=1}^{N} \rho_\mu (x_\mu^+ + x_\mu^-). \tag{6}$$

Instability is proved by constructing the Lyapunov functional

$$H(t) = \sum_{\nu=1}^{N} \int \frac{x_\nu^+ J_\nu^+ + x_\nu^- J_\nu^-}{r} dm_\nu, \tag{7}$$

where $dm_\nu = r^2 \rho_\nu dr$ stands for the mass element along a certain diameter (more precisely, along the elementary cone), whereas the integrals are taken between the turning point of the corresponding stream. Differentiating (7) and keeping in mind the invariance of dm_ν, we get after some calculations

$$\frac{dH}{dt} = \sum_{\nu=1}^{N} \int \frac{(J_\nu^+)^2 + (J_\nu^-)^2}{r^2} dm_\nu + 2\pi G r^2 \int \left[\sum_{\nu=1}^{N} \rho_\nu (x_\nu^+ + x_\nu^-) \right]^2 dr > 0.$$

This means that even in our scheme there exist no stable modes at all.

Earlier [4] we had proven the stability of a wide range of models with a spherically-symmetric velocity distribution. It would be interesting to trace the loss of stability in response to a gradual increase in the radial prolateness of the velocity ellipsoid. At present, the difficulty lies in the relative complexity of the known models.

REFERENCES

1. Agekyan, T. A., 1961. "Spherical Stellar and Galactic Systems in Early Stages of Evolution", *Leningrad University Bulletin*, **1**, 152-161.
2. Mikhailovskii, A. B. & Fridman, A. M., 1971. "Stream Instability in Gravitating Media", *Journal of Experimental and Theoretical Physics*, **61**, #2, 457-468.
3. Lyubarskii, G. Ia., 1957. *Group Theory and its Application to Physics*, Moscow, Gostekhizdat (State Publ. House for Scient. and Techn. Literature).
4. Antonov, V. A., 1962. "Solution to the problem concerning the stability of a stellar system with Emden's density law and with a spherical velocity distribution", *Leningrad University Bulletin*, **19**, p. 96-111, (Translation in this volume, p. 531).

DISCUSSION

A. M. Fridman: These results appear obvious to me. If during the course of motion along circular orbits, there exists an analog to elastic forces, then during the course of radial motion, there is nothing to prevent an adhesion to neighboring orbits.

V. A. Antonov: From time to time, one should not forego the opportunity to support intuition with solid evidence. In matters of this kind, certainty is quite relative, and only becomes established after careful consideration.

A. M. Fridman: The only thing that bothers me is that you are allowing for perturbed motion in a single direction only.

V. A. Antonov: This is not an arbitrary assumption. It follows with certainty from the comparison of components of grad $\delta\Phi$ with varying directions when $n >> 1$.

QUADRATIC INTEGRALS OF MOTION AND STELLAR ORBITS IN THE ABSENCE OF AXIAL SYMMETRY OF THE POTENTIAL

G.G. Kuzmin
Tartu Astrophysical Observatory
202444 Toeravere, Estonia, USSR

As is well known, in the case of an axially symmetric and time-invariant gravitational potential, if the potential satisfies one particular additional constraint, there exist three isolating integrals of motion: the energy integral, the area integral, and the third integral which is quadratic in the velocities. This work discusses the case in which there exist quadratic integrals in the absence of axial symmetry of the potential. Such a case has already been examined by Eddington [1], but in their explicit form, the integrals were introduced by Clark [2].

Let us use, following Eddington and Clark, an ellipsoidal coordinate system q_1, q_2, q_3. The coordinate surfaces (which are orthogonal) represent a family of second-order confocal surfaces. Let this coordinate system be co-axial with Cartesian coordinates x, y, z. Moreover, let z be the axis with which all the coordinate surfaces (hyperboloids as well as ellipsoids) have a real intersection. Then, on the z-axis, either $z = q_1$ (hyperboloids of two sheets), or $|z| = q_2$ (hyperboloids of one sheet), or $|z| = q_3$ (ellipsoids). The domains of the q_i are $-\alpha \leq q_1 \leq \alpha$, $\alpha \leq q_2 \leq \beta$, $\beta \leq q_3$, where α and β are parameters of the coordinate system (determined by the intersection of the focal curves with the z-axis).

Let v_i be the velocities and p_i the generalized (specific) momenta along the coordinate line q_i. In the ellipsoidal coordinate system:

$$\frac{dq_i}{dt} = \sqrt{g_i} v_i = g_i p_i, \qquad (i = 1, 2, 3) \tag{1}$$

(t is the time), while

$$g_i = \frac{(q_i^2 - \alpha^2)(q_i^2 - \beta^2)}{(q_i^2 - q_k^2)(q_i^2 - q_l^2)}, \qquad (k, l \neq i). \tag{2}$$

Obviously, $0 \leq g_i \leq 1$, and as can easily be shown, $\sum g_i = 1$. On the z-axis, $g_i = 1$ in its "own" domain, i.e. where z (or $|z|$) equals q_i, and $g_i = 0$ outside this domain.

From: *The Dynamics of Galaxies and Star Clusters*, Proceedings of a National Conference in Alma-Ata, October 23-26, 1972, p. 71–75, ed. Omarov, T.B. ('Nauka" of the Kazakh S.S.R., Alma-Ata, 1973). Translated from the Russian by Alex Shprintsen at CITA, Toronto.

T. de Zeeuw (ed.), Structure and Dynamics of Elliptical Galaxies, 553–556.

Clark [2], using Eddington's work, presented differential equations which the potential Φ must satisfy if the three isolating quadratic integrals are to exist. However, he did not find solutions to these equations. The solution has the following simple form:

$$\Phi = \sum_i g_i \phi_i, \qquad \phi_i = \phi(q_i). \tag{3}$$

Formally, the solution contains three arbitrary functions ϕ_i. But they should be viewed as parts of one single function $\phi(q)$ where $q \geq -\alpha$. On the z-axis, it is obvious that $\Phi = \phi(z)$, or $\phi(|z|)$ when $|z| \geq \alpha$.

The three independent quadratic isolating integrals of motion have the following form:

$$I_j = \sum_i h_{ij} \left(\frac{1}{2} p_i^2 - \phi_i \right), \qquad (j = 1, 2, 3), \tag{4}$$

where

$$\begin{gathered} h_{ij} = a_{ij} g_i, \\ a_{i1} = 1, \qquad a_{i2} = \frac{q_k^2 + q_l^2}{\alpha^2 + \beta^2}, \qquad a_{i3} = \frac{q_k^2 q_l^2}{\alpha^2 \beta^2}. \end{gathered} \tag{5}$$

The properties of the coefficients h_{ij} are analogous to the properties of the coefficients g_i as presented above. Obviously, I_1 is the energy integral.

Using Poisson's equation for the density ρ, we derive the expression:

$$\rho = \sum_{i,k} g_i g_k \rho_{ik}, \tag{6}$$

where

$$4\pi G \rho_{ii} = \Psi_i = \Psi(q_i), \quad 4\pi G \rho_{ik} = \frac{2 \int_{q_i}^{q_k} \Psi(q) q dq}{q_k^2 - q_i^2}. \tag{7}$$

Here G is the gravitational constant, and $\Psi(q)$ is an arbitrary function linked to the arbitrary function $\phi(q)$ by the equation:

$$\phi'' + \left(\frac{1}{q^2 - \alpha^2} + \ldots \right) q\phi' - \left[\frac{2\alpha^2}{(q^2 - \alpha^2)^2} (\phi - \phi_\alpha) + \ldots \right] + \Psi = 0, \tag{8}$$

where the ellipses stand for a similar term for β, and $\phi_\alpha = \phi(\alpha)$. Obviously, on the z-axis, $4\pi G\rho = \Psi(z)$ (or $\Psi(|z|)$ when $|z| \geq \alpha$).

The expression for the density (6) (as well as the expression for the potential (3)) is a generalization of the expression derived for axial symmetry [3]. Theorems for the non-negativity of the density and the finiteness of the mass remain valid: the density is everywhere non-negative if it is non-negative on the z-axis; the mass is finite if $\Psi q^3 \to 0$ when $q \to \infty$ (both conditions are necessary and sufficient). The theorem concerning the equatorial plane of symmetry also remains valid [4]. The potential and the density are obviously symmetric relative to the planes $x = 0$ and $y = 0$. But if the stellar system is self-gravitating (so that $\Psi = \Psi[\phi(q), q^2]$), then it is also symmetric to the plane $z = 0$ ($\phi(-q) = \phi(q)$ when $-\alpha \leq q \leq \alpha$).

Having set up the function $\Psi(q)$ (or $\phi(q)$), we derive a model for the mass distribution, allowing for the integral of motion (4). If $\Psi(q)$ decreases rapidly with $|q|$, then the models are greatly flattened along z. Such models are treated similarly to the greatly flattened, axisymmetric models [3]. Ellipsoidal models are also possible. They may be derived by taking:

$$\Psi(q) = 4\pi G\rho_0 \left(1 + \frac{q^2}{\gamma^2}\right)^{-2}. \tag{9}$$

Then

$$\rho = \rho_0 \left(1 + \frac{x^2}{\beta^2 + \gamma^2} + \frac{y^2}{\alpha^2 + \gamma^2} + \frac{z^2}{\gamma^2}\right)^{-2} \tag{10}$$

(ρ_0 is the central density, γ is a parameter). This is the generalization of the axisymmetric ellipsoidal models examined earlier [3].

Let us now turn to the problem of stellar orbits. From expression (4) for the integrals of motion I_j, we find:

$$\frac{p_i^2}{2} = \frac{Q_i}{(q_i^2 - \alpha^2)(q_i^2 - \beta^2)} + \phi_i, \quad Q_i = Q(q_i), \tag{11}$$

where

$$Q(q) = I_1 q^4 - (\alpha^2 + \beta^2) I_2 q^2 + \alpha^2\beta^2 I_3. \tag{12}$$

Using p_i, it is possible to express the remaining three independent integrals of motion:

$$I_{j+3} = \sum_i \int \frac{\partial p_i}{\partial I_j} dq_i + \begin{cases} -t, & j=1, \\ 0, & j=2,\ 3. \end{cases} \tag{13}$$

The first of these integrals is non-conservative, the other two are conservative but non-isolating. With fixed I_1, I_2, I_3, the expressions for the integrals (13) become the equations of the orbit.

The orbits are bounded by the surfaces $q = \text{const}$. We derive the values of q by solving the equation

$$Q(q) + (q^2 - \alpha^2)(q^2 - \beta^2)\phi(q) = 0. \tag{14}$$

Three solutions for q^2 correspond to a finite orbit (assuming there is equatorial symmetry). The negativity of the third derivative of the second term with respect to q^2 is the condition for there to be no more than 3 solutions. There exist four classes of orbits: 1) box orbits, 2) horizontal annular, 3) external vertical annular, and 4) internal vertical annular orbits. The three solutions for q^2 are to be found respectively in the domains: q_1, q_2, q_3; $q_1, q_3(2)$; $q_2, q_3(2)$; $q_2(2), q_3$.

Box orbits are reversible: after entering the corner of the box, the direction of stellar motion is reversed. Annular orbits are irreversible: motion is possible in two opposite directions, but a reversal of direction is not possible. The phase density of stars, which in a stationary state is a function of the isolating integrals I_j, may depend on them in the domain of irreversible orbits in a double-valued fashion. This makes possible centroid motion, despite the quadratic nature of the integrals.

On the boundaries between the four indicated classes, orbits are unstable, and there exist asymptotic orbits.

REFERENCES

1. Eddington, A.S., 1916. *Mon. Not. R. astr. Soc.*, **76**, 37.
2. Clark, G.L., 1936. *Mon. Not. R. astr. Soc.*, **97**, 182.
3. Kuzmin, G.G., 1956. *Astr. Zh.*, **33**, 27. (*Tartu Astr. Obs. Teated*, **2**).
4. Kuzmin, G.G., 1963. *Publ. Tartu Astroph. Obs.*, **34**, 18.

INDICES

SUBJECT INDEX

OBJECT INDEX

NAME INDEX